Microscopic and Macroscopic Simulation: Towards Predictive Modelling of the Earthquake Process

Edited by
Peter Mora
Mitsuhiro Matsu'ura
Raul Madariaga
Jean-Bernard Minster

Springer Basel AG

Reprint from Pure and Applied Geophysics
(PAGEOPH), Volume 157 (2000), No. 11/12

Editors:

Peter Mora
Department of Earth Sciences
The University of Queensland
Brisbane
Australia
e-mail: mora@earth.uq.edu.au

Mitsuhiro Matsu'ura
Department of Earth and Planetary Physics
The University of Tokyo
Tokyo
Japan
e-mail: matsuura@eps.s.u-tokyo.ac.jp

Raul Madariaga
Laboratoire de Géologie
École Normale Supérieure
Paris
France
e-mail: madariag@geologie.ens.fr

Jean-Bernard Minster
Scripps Institution of Oceanography
University of California at San Diego
La Jolla
USA
e-mail: jbminster@ucsd.edu

A CIP catalogue record for this book is available from the Library of Congress,
Washington D.C., USA

Deutsche Bibliothek Cataloging-in-Publication Data

**Microscopic and macroscopic simulation: towards predictive modelling of the
earthquake process** / ed. by Peter Mora ...- Basel ; Boston ; Berlin : Springer Basel AG 2001
   (Pageoph topical volumes)
   ISBN 978-3-7643-6503-5   ISBN 978-3-0348-7695-7 (eBook)
   DOI 10.1007/978-3-0348-7695-7

© 2001 Springer Basel AG
Ursprünglich erschienen bei Birkhäuser Verlag, Basel - Boston - Berlin 2001

Printed on acid-free paper produced from chlorine-free pulp

9 8 7 6 5 4 3 2 1

Contents

Part III: Macroscopic Simulation: Long Time Scale Phenomena (Earthquake Cycle)

Pure appl. geophys. 157 (2000) 1817–1819
0033–4553/00/121817–03 $ 1.50 + 0.20/0

Pure and Applied Geophysics

Introduction

The holy grail of earthquake research has always been earthquake prediction. Numerical simulation models for the complete earthquake process represent a new thrust in earthquake science that offers an outstanding opportunity for gaining the knowledge needed to advance towards this grand goal. At the same time, such numerical models provide a means to develop the next generation of earthquake hazard forecasting and analysis system based on predictive understanding of the phenomena.

Simulation of the entire earthquake process is an ambitious challenge. The crust is a complex system with many interacting physical processes occurring over a wide range of space and time scales. Advances in high performance computer technology and numerical simulation methodology are bringing this vision within reach.

Simulations at the micro-scale are a means to extend knowledge of fault zone behaviour beyond the domain accessible to laboratory observations. Since fault behaviour controls earthquake nucleation, rupture and arrest, advancement of this knowledge provides the key to model the entire earthquake process and earthquake cycle.

Simulations at a macro-scale allow dynamic rupture to be probed which, through comparison with observations, can be used to place constraints on the fault zone constitutive relation. Simulations of rupture and the ensuing seismic wave radiation also provide a basis to predict strong ground motion patterns of earthquakes. Ultimately, macroscopic models could be used to simulate the entire earthquake cycle from stress buildup, to quasi-static nucleation, dynamic rupture, arrest and healing. This provides a new opportunity to probe the physics of earthquakes and develop the scientific underpinning for a new generation of earthquake hazard forecasting systems.

On January 31 through February 5, 1999, some 70 scientists from around the world met in Brisbane and Noosa Australia for the Inaugural Workshop of the APEC Cooperation for Earthquake Simulation (ACES) to discuss this integrated simulation-based approach for understanding earthquakes. The workshop initiated the international cooperative program to construct the numerical simulation models and assimilate data into the models. This collection of articles is an outcome of the workshop and presents a cross-section of cutting-edge research in the field.

Part I contains articles covering microscopic simulation. These span simulation and laboratory studies of the behaviour of fault gouge layers, through to studies of fracture and the underlying mechanisms for nucleation and catastrophic failure with an aim towards prediction. Two kinds of microscopic numerical models are used. The particle-based models which simulate rock as a large system of interacting particles, and the mesoscopic damage evolution and nonlinear models which simulate a damage field equation or simplified rules. Both approaches yield interesting discoveries and provide a foundation for subsequent research into fault behaviour, nucleation and catastrophic failure.

Part II collects articles on short time scale macroscopic simulation. These range from simulation studies of rupture and strong motion through to development of finite-element methods. A variety of numerical simulation studies investigate the sensitivity of rupture propagation, arrest and slip distribution to parameters of the frictional constitutive relation, stress field and geometry of faults. These results provide important constraints on the frictional constitutive relation and yield insights into the complexity of rupture. Simulations of strong ground motion explain observed damage patterns and strong motion data, and demonstrate the important influence of 3-D structure on patterns of strong ground shaking. Finite-element methods and a parallel computational approach that aims to allow for large-scale simulation of complex 3-D fault structures are presented.

Part III contains articles on macroscopic simulation of long time scale phenomena and the earthquake cycle. These range from an approach for 3-D modelling of stress accumulation and release to a theoretical study of critical point phenomena and numerical experiments with a variety of methods to study the earthquake cycle. While the simulations of the earthquake cycle are presently highly simplified, they yield several interesting results that progress towards an improved understanding of spatial and temporal evolution in seismicity, and hence, towards laying a scientific foundation for intermediate term earthquake prediction.

Part IV collects articles on data assimilation, scaling observations or relations and earthquake hazard forecasting. Scaling studies motivate future directions in numerical simulation and lead to interesting possibilities including that of deterministic modelling of earthquake occurrence for earthquake prediction. An article presents detailed crustal deformation observations and describes the assimilation of this new data into simulation models and the implications to forecasting of future seismic activity. Several articles present approaches to earthquake hazard assessment, forecasting and prediction based on underlying physical or combined statistical and physical models for seismicity or crustal response. The results are encouraging that an improved understanding of the physics of earthquakes gained through numerical simulation can advance towards the grand goal of earthquake forecasting and prediction.

The unified numerical simulation approach to progress earthquake science is in its infancy. The scientific challenges are great but the cooperative approach spanning across disciplines and national boundaries gives cause for optimism. New participation in ACES to extend its existing synergies is welcomed.

We wish to thank the scientific participants of The APEC Cooperation for Earthquake Simulation (ACES) and the contributors to this book. We express appreciation to the Australian, Chinese, Japanese and USA governments for supporting the establishment of ACES. We gratefully acknowledge funding support by the Australian government's Department of Industry, Science and Resources, The University of Queensland, Japan's Science and Technology Agency through its Research Organisation for Information Science and Technology, the Chinese Ministry of Science and Technology, and the National Science Foundation of China. We acknowledge with appreciation additional workshop sponsorship provided by SGI (Silicon Graphics). Special thanks to QUAKES team members (Tracy Paroz, David Place, Steffen Abe, Dion Weatherley and Steven Jaumé) and Kim Olsen who provided assistance to the Editors. Peter Mora would also like to thank Evelyne Meier.

REFERENCES

1-st ACES Workshop Proceedings (1999), ed. Mora, P. (ACES, Brisbane, Australia, ISBN 1 86499 121 6), 554 pp.
APEC Cooperation for Earthquake Simulation: http://quakes.earth.uq.edu.au/ACES
ACES Inaugural Workshop: http://quakes.earth.uq.edu.au/ACES_WS

Peter Mora
QUAKES
Department of Earth Sciences
The University of Queensland
4072 Brisbane, Qld
Australia
mora@earth.up.edu.au

Mitsuhiro Matsu'ura
Department of Earth and
  Planetary Physics
The University of Tokyo
7-3-1, Hongo, Bunkyo-Ku
Tokyo 113-0033
Japan
matsuura@eps.s.u-tokyo.ac.jp

Raul Madariaga
Laboratoire de Géologie
Ecole Normale Superieur
24 Rue Lhomond
F-75231 Paris, Cedex 05
France
madariag@geologie.ens.fr

Jean-Bernard Minster
Institute of Geophysics and
  Planetary Physics
Scripps Institution of Oceanography
University of California at San Diego
La Jolla, CA 92093-0225
U.S.A.
jbminster@ucsd.edu

Part I
Microscopic Simulation

Pure appl. geophys. 157 (2000) 1821–1845
0033–4553/00/121821–25 $ 1.50 + 0.20/0

Pure and Applied Geophysics

Numerical Simulation of Localisation Phenomena in a Fault Zone

DAVID PLACE[1] and PETER MORA[2]

Abstract—A lattice solid model was developed to study the physics of rocks and the nonlinear dynamics of earthquakes and is applied here to the study of fault zone evolution. The numerical experiments involve shearing a transform fault model initialised with a weak and heterogeneous fault zone. During the experiments, a fault gouge layer forms with features that are similar to those observed in recent laboratory experiments involving simulated fault gouge (BEELER *et al.*, 1996). This includes formation of Reidel (R_1) shears and localisation of shear into bands. During the numerical experiments, as in the laboratory experiments, decreases of the gouge layer strength correlate with decreases in gouge layer thickness. After a large displacement, a re-organisation of the model fault gouge is observed with slip becoming highly localised in a very narrow basal shear zone. This zone is such that it enhances the rolling-type micro-physical mechanism that was responsible for the low heat and fault strength observed in previous numerical experiments (MORA and PLACE, 1998, 1999) and proposed as an explanation of the heat flow paradox (HFP). The long time required for the self-organisation process is a possible reason why the weak gouge layers predicted by the numerical experiments, and which could explain the HFP, have not yet been observed in the laboratory. The energy balance of a typical rupture event is studied. The seismic efficiency of ruptures of the gouge layer is found to be low (approximately 4%), substantially lower than previous estimates and compatible with typical field-based estimates.

Key words: Localisation, earthquake dynamics, lattice solid model, particle-based model, friction.

Introduction

The "lattice solid model" consists of a lattice of interacting particles and was motivated by the molecular dynamics method. The model is similar to the Discrete Element Model proposed by CUNDALL and STRACK (1979) but was motivated by a desire to simulate earthquake processes (cf., PLACE and MORA, 1999). The model was developed in order to study fracturing, wave propagation in complex discontinuous media, faulting (DONZÉ *et al.*, 1993) and the stick-slip instability which is responsible for earthquakes (MORA and PLACE, 1994). Here, the model is applied to the study of fault zone evolution. Numerical studies are performed in two

---
[1,2]Queensland University Advanced Centre for Earthquake Studies (QUAKES), Department of Earth Sciences, The University of Queensland, St Lucia, Brisbane, 4072, Qld, Australia. www. http://quakes.earth.uq.edu.au
[1] E-mail: place@earth.uq.edu.au
[2] E-mail: mora@earth.uq.edu.au

dimensions, using a refined version of the lattice solid model (PLACE and MORA, 1999).

A long-standing paradox in earthquake studies has been the low heat flow observed around the San Andrews fault, compared with the theoretical value computed using the value of rock friction measured in laboratory experiments. In recent numerical studies using the lattice solid model (MORA and PLACE, 1998, 1999), a dynamic solution to the heat flow paradox was proposed: When an artificial fault gouge was specified, a rolling-type mechanism was observed which significantly reduced both the amount of heat generated during simulated earthquakes and the strength of the model faults. The experiments consisted of a narrow fault gouge placed between two elastic books, and they were conducted for two different gouge setups ("Gouge 2" being made of more angular grains than "Gouge 1").

Without the presence of a fault gouge the actual heat generated was in accordance with the theoretical prediction. When a fault gouge was present, a significant reduction in heat was observed (when more angular grains are used or when the intrinsic coefficient of friction is increased, a greater reduction in heat is observed, cf., Fig. 1). Grains of model rock in the gouge zone were rolling and jostling in preference to slipping against each other when this heat reduction was observed. In the experiments, a pre-existing narrow fault gouge consisting of unbreakable grains was initialised, hence the rolling type mechanism could have been due to the setup of the numerical experiment. Consequently, questions were raised such as: Is a thin shear band with limited particle sizes and shapes a necessary condition for the rolling-type mechanism to occur? Can a natural fault

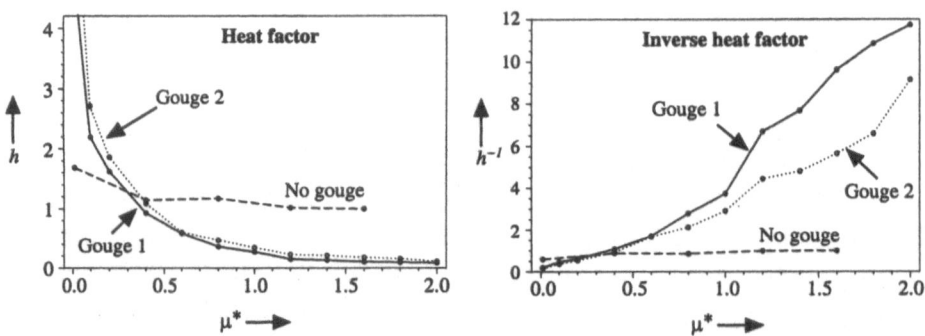

Figure 1

Heat factor h and (right) inverse heat factor h^{-1} as a function of the intrinsic coefficient of friction μ^* for the two experiments with a gouge layer (Gouge 1 and Gouge 2) and the experiment without a gouge layer (i.e., single fault). The heat factor h represents the factor between the actual heat generated and the heat that should theoretically be generated, assuming the fault friction equals the intrinsic friction. A value of h less than 1.0 means that the generation of heat is lower than the theoretical expectation, based on the assumption that the model fault friction is equal to the friction specified between surface particles in the model.

gouge zone organise itself such that slip is localised in such a thin weak region where slip occurs mainly through the rolling-type mechanism?

Localisation of slip in a thin shear band has been observed in laboratory experiments although no significant reduction in the fault strength was observed (BEELER et al., 1996). In the following, numerical experiments are performed to study the evolution of a model fault zone and to study localisation phenomena.

The results presented in this paper expand from those in MORA and PLACE (1999), in which the Heat Flow Paradox (HFP) was studied and in a final plot of model fault strength, a weakening was observed which correlated with highly localised slip. The paper mentioned showed that the weakening occurred at a time when slip had become highly localised, but a presentation or analysis of this phenomena was beyond the scope of this article. These results were put into the context of the HFP and represent a subset of the work presented here. In this paper, a detailed analysis of the localisation phenomena observed in the model fault gouge is presented, and the seismic efficiency and fracture orientation is studied.

Results are compared with laboratory experiments to obtain insights on the scalability of the numerical and laboratory experiments and the validity of the numerical results. The lattice solid model that allows frictional behaviour (cf., PLACE and MORA, 1999) to be accurately simulated, is used here to perform the numerical experiments. Experiments consist of a weak and heterogeneous fault zone placed between two elastic blocks being subjected to shear and normal stress. The gouge region constitutes a zone of weakly bonded unbreakable grains of various shapes and sizes. Grains are composed of several particles linked to one another by strong bonds, and bonds between surface particles of different grains were allowed to break after a prescribed separation was exceeded. During the numerical experiments, fractures with orientations similar to Reidel (R_1) shears are observed as the weak gouge zone breaks down and evolves. After a large displacement, slip becomes localised in a band. After localisation takes place, the strength of the fault decreases due to the rolling-type mechanism. Results show that the strength can again decrease significantly after a much larger displacement when the slip re-localises on a narrower and considerably weaker basal shear zone.

The Lattice Solid Model

The particles in the lattice solid model interact with one another as do particles in molecular dynamics. Particles specify the smallest indivisible units of the system and in the following, represent building blocks of rock that are approximately grain size in diameter. This approach enables the nonlinear behaviour of discontinuous solids such as rocks to be simulated with relative simplicity, while capturing within a time step the frictional discontinuities such as the transition between static and dynamic frictional states (which also ensures the correct calculation of the heat generated during slip (cf., PLACE and MORA, 1999)).

Particles in consolidated regions are arranged in a two-dimensional triangular lattice and linked by elastic bonds. Bonds can break if the separation exceeds a given threshold r_{cut}. A bond, acting as a spring, exerts a normal force on the two bonded particles given by

$$\mathbf{F}_{ij} = k(r_0 - r_{ij})\mathbf{e}_{ij}, \qquad (1)$$

where $\mathbf{F}_{ij}$ is the force on particle i due to bonded particle j, k is the spring constant, r_0 is the equilibrium separation, r_{ij} is the separation between particle i and particle j, and $\mathbf{e}_{ij}$ is the unit vector pointing from particle i to j. From Equation (1), when the separation r_{ij} is greater than r_0, the two bonded particles attract each other, otherwise the two particles repel one another. When unbonded particles come into contact, only the repulsive part of the normal force given by Equation (1) is applied (i.e., the normal force is zero if the separation r_{ij} is greater than r_0, that is, when they are not in contact; see also MORA and PLACE, 1994).

A viscosity is introduced in order to damp reflected waves from the edges of the lattice. The viscous forces are proportional to the particle velocities and are given by

$$\mathbf{F}_i^v = -\nu\dot{\mathbf{x}}_i, \qquad (2)$$

where ν is the viscosity coefficient and $\dot{\mathbf{x}}_i$ the velocity of particle i. The viscosity is frequency-independent and does not fundamentally alter the dynamics of the system if carefully chosen (MORA and PLACE, 1994).

In addition, a frictional force is added to unbonded particles that come into contact. The frictional force opposes the direction of slip and its magnitude is no greater than the dynamical frictional force given by

$$F_{ij}^F = \mu|\mathbf{F}_{ij}|, \qquad (3)$$

where μ is the coefficient of friction and $\mathbf{F}_{ij}$ is the normal force (Eq. (1)) between particle i and j. The dynamical frictional forces and the static frictional forces (e.g., the force required to stop slip between particles) are computed using an approach involving the resolution of a nonlinear system (PLACE and MORA, 1999).

The numerical integration is based on a modified velocity Verlat scheme which uses a half time-step integration approach (PLACE and MORA, 1999). The total energy of the system is computed at each time step as the sum of the kinetic energy, the energy lost to the artificial viscosity (which we normally consider as lost energy of the radiated waves), the energy lost due to bond breaking (fracture energy), the work done by frictional forces (heat generated) and the energy added into the system (e.g., to maintain a constant pressure and speed of the driving plates). The total energy is verified to remain constant within an error of 1% after, 1,000,000 time steps.

Localisation Experiment

Numerical experiments are performed to study fault gouge evolution in time. The experiments are conducted over a period of time that corresponds to a total displacement of 1400 units of length. The experiments are scale-less and one unit of length corresponds to a particle diameter. The width of the model fault gouge is approximately 50 times the particle diameter which would correspond to 1 mm in comparison with the BEELER *et al.* laboratory experiment (cf., BEELER *et al.*, 1996).

Initialisation and Setup

In the following numerical experiments, a gouge is placed between two elastic blocks with rough surfaces. The gouge is made up of unbreakable grains of different sizes and shapes. Because rotation is not modelled at the particle scale and frictional forces are applied at the particle centres, single particle dynamics are unrealistic. Grouping of particles into unbreakable grains allows more realistic rock behaviour to be simulated. Namely, grains in the model can rotate and frictional forces are applied to their surface particles. While single particles have an infinite effective shear stiffness (PLACE and MORA, 1999), grains can be deformed when subjected to shear stresses (as a result of the lattice geometry). Consequently, grains in the model offer a better representation of rock grains than do particles themselves. Grains are bonded to one another by relatively weak bonds with $r_{cut} = 1.075$ r_0 and single particles are removed from the gouge zone (Fig. 2). Thus, the gouge region is weak and will break down during the simulation. Large grains, composed of groupings of smaller grains which are bonded together, will be naturally formed early in the simulation.

The solid is composed of 256×64 particles in which the gouge region has an initial thickness of 50 particle diameters or about 12 times the longest grain length (the gouge thickness is defined as the separation between the fault surfaces, see Fig. 2). Four different shapes of grain of model rock are used: elongated hexagon (composed of 10 particles), hexagon (7 particles), diamond (4 particles) and triangle (3 particles). A homogeneous distribution of grains (inversely proportional to the grain size[3]) is first used to randomly fill the gouge region. Diamond and triangle grain shapes are then used to fill as much of the remaining porosity as possible. Consequently, the final distribution of grain shapes differs from the initial distribution. This initialisation procedure results in the final distribution of grain shapes shown in Table 1.

Roughness of each model fault surface is initialised using a power-law spatial spectrum of surface heights with exponent 0.5 (MORA and PLACE, 1994) using a maximum height variation of 5 particle diameters. Each block is pushed from its

---

[3] Size is measured here in terms of number of particles.

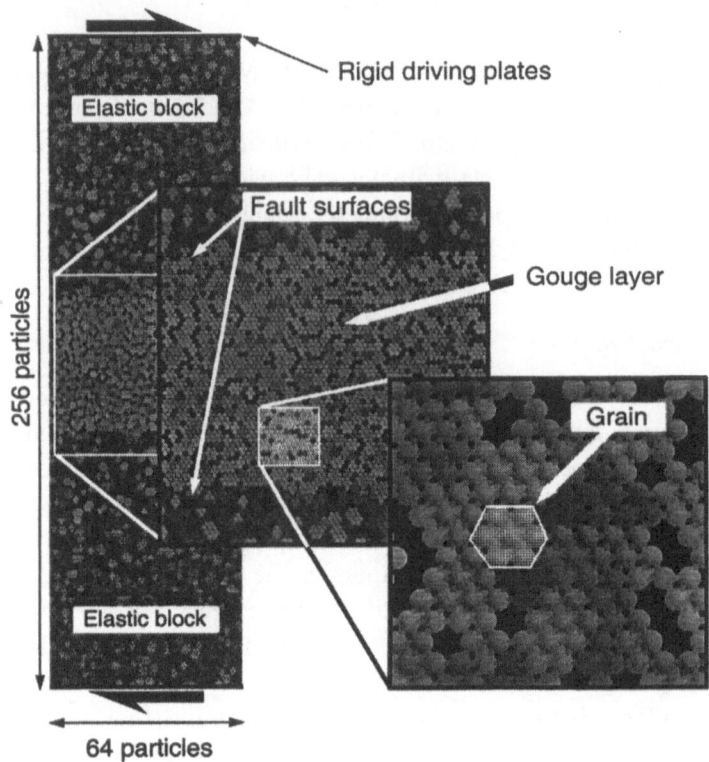

Figure 2

Experiment setup: The simulated gouge is initialised by specifying a weak region composed of unbreakable grains of rock which are weakly bonded together. Boundary conditions are circular in the x-direction.

Table 1

Grain distributions used in the localisation numerical experiment inside and outside the gouge region

Grain shape	Gouge region	Elastic region
Elongated hexagons	6.20%	4.20%
Hexagons	24.4%	14.3%
Diamonds	29.5%	16.0%
Triangles	39.9%	16.5%
Single particles	0.00%	49.0%

rigid driving plate at a constant velocity $V_\ell = 0.00025 \ V_p$ where V_p represents the P-wave velocity (approximately equal to 1.0 for a spring constant $k = 1$ and a particle mass $M = 1$, see MORA and PLACE, 1994). Hence, each driving plate is displaced by an amount r_0 after 4000 units of time ($d = V_\ell t$). Localisation phenomena are studied over a displacement of approximately 1400 particle diameters (21 full rotations of the circular model along the x axis)[4]. A constant normal stress is maintained in the solid[5] and is equivalent to 1500 MPa using $V_p = 3\sqrt{3}$ km/s and a density $\rho = 3000$ kg.m^{-3} (see PLACE and MORA, 1999, for an explanation of scaling of units in the model).

The observed coefficient of friction of the fault (μ_f) is given by

$$\mu_f = \frac{\tau_f}{\sigma_n} = \frac{(\bar{\mathbf{F}}_{z_1} - \bar{\mathbf{F}}_{z_2}) \cdot \mathbf{e}_x}{(\bar{\mathbf{F}}_{z_1} - \bar{\mathbf{F}}_{z_2}) \cdot \mathbf{e}_z}. \tag{4}$$

where τ_f and σ_n are respectively the shear stress and the normal stress measured on the driving plates, $\bar{\mathbf{F}}_{z_i}$ is the average value of the force for a row of particles at $z = z_i$ (z_1 for the upper edge and z_2 for the lower edge of the lattice). The average coefficient of friction of the fault at a given time t, denoted $\mu_f^*(t)$, is computed using a window average with triangular weights of the observed coefficient of friction, namely

$$\mu_f^*(t) = \frac{1}{\Delta t''} \sum_{t' = t - \Delta t''}^{t + \Delta t''} \left(1 - \frac{|t - t'|}{\Delta t''} \right) \mu_f(t'), \tag{5}$$

where $2\Delta t''$ is the window length (or period of time) on which the observed coefficient of friction is averaged. The window length $\Delta t''$ is given by

$$\Delta t'' = \frac{1}{2 V_\ell} \Delta d. \tag{6}$$

Typically, Δd is set to 30 r_0 in order to filter variations in the observed coefficient of friction due to stick-slip cycles.

The observed coefficient of friction represents the macroscopic friction of the fault due to the combined effects of intrinsic friction between grains, surface roughness, and gouge layer dynamics and microstructure.

Results

Figure 3 shows the fault friction as a function of the displacement for numerical experiments using a coefficient of friction $\mu = 1.0$ and $\mu = 0.6$. Stick-slip cycles can be seen as the characteristic sawtooth shapes observed in laboratory experiments. During the stick phase, the stress builds up in the solid. When enough energy is stored in the deformation of blocks to overcome the static fault friction, a synthetic earthquake occurs and stress is released (slip phase). In Figure 3, six periods can be

---

[4] The displacement is measured in terms of particle diameters of the driving plate on the upper (or lower) block.

[5] The normal stress is applied and measured on the rigid driving plates.

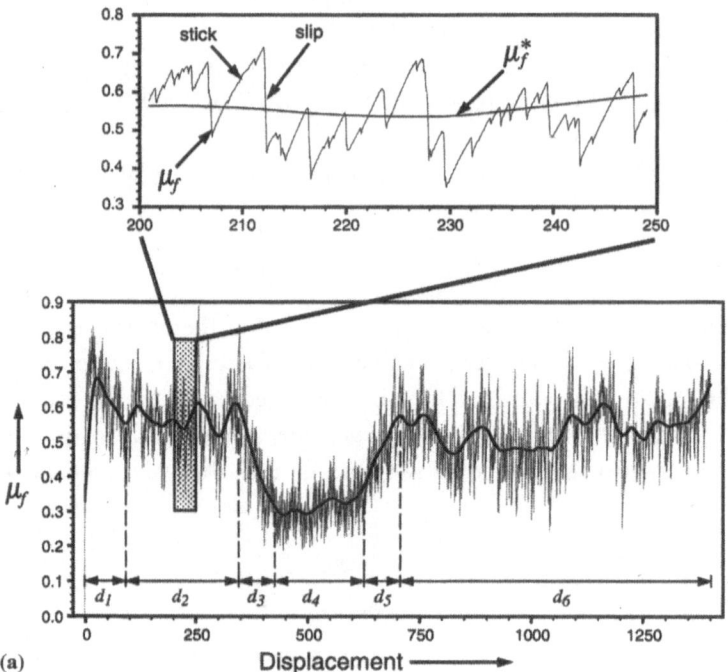

(a)

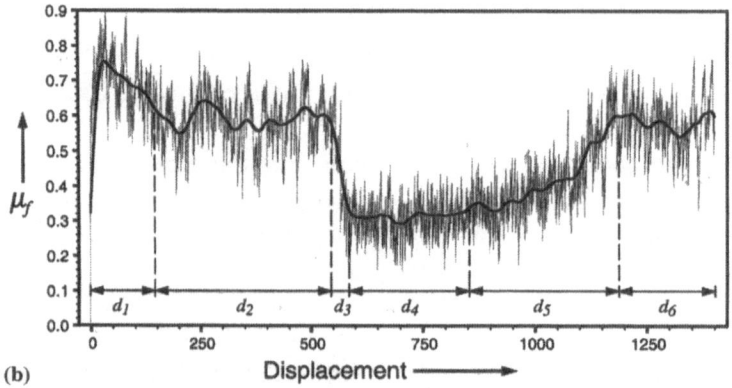

(b)

Figure 3
Plot of the instantaneous fault friction μ_f (thin light line) and average fault friction μ_f^* (solid bold line).
The top (a) and bottom (b) plots show the fault friction respectively using $\mu = 1.0$ and $\mu = 0.6$ for the
intrinsic coefficient of friction. The displacement represents the displacement of the upper (or lower)
block shown in particle diameters (r_0). The magnification box on Figure 3a shows the fault friction from
$d = 200\ r_0$ to $d = 250\ r_0$ and allows the characteristic saw-tooth shapes of the stick-slip cycles to be seen.

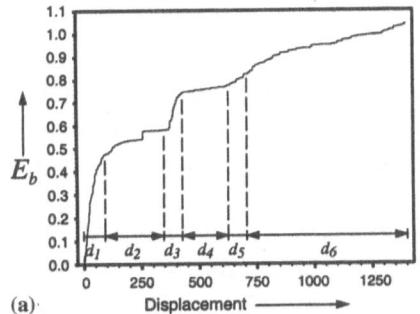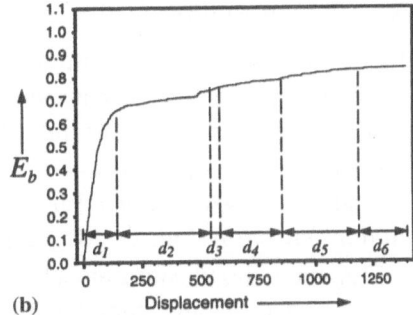

(a)   (b)

Figure 4
Plot of the averaged energy lost due to fracture (fracture energy) for a coefficient of friction $\mu = 1.0$ (a) and $\mu = 0.6$ (b).

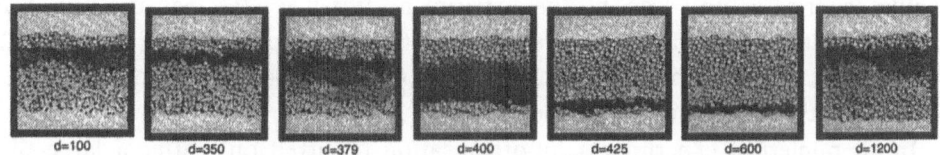

Figure 5
Snapshots of the localisation experiment using $\mu = 1.0$. The shade of grey represents the time averaged horizontal particle velocity. Dark grey represents small velocities and highlights the active shear zone. The lightest grey corresponds to the large velocities of the upper and lower blocks and any gouge material that is moving in unison with the blocks (i.e., locked zones of the gouge layer that are not shearing).

distinguished: d_1 is the period during which the fault friction initially decreases; d_2, d_4 and d_6 are the periods during which the fault friction is almost constant; d_3 is the period during which slip is re-localising and friction is rapidly dropping; and finally d_5 is a period during which the fault friction increases.

The fault strength is at its maximum near the beginning of the simulation and decreases progressively (period d_1 in Fig. 3). This decrease in the fault strength correlates with a large amount of new fracturing in the active shear zone. Large grains are breaking down to form smaller grains and accommodate slip in that region. During this period, the rate of gouge generation, shown using the energy lost due to fracture (period d_1 in Figs. 3 and 4), is the highest.

During period d_2 (Figs. 3 and 4), the rate of gouge generation and the observed coefficient of friction is approximately constant. Snapshots $d = 100$ and $d = 350$ in Figure 5 show that slip is localised within a relatively wide shear band during periods d_1 and d_2 (approximately 25% the thickness of the gouge layer).

Gouge Self-organisation

Figure 6 shows the distribution of slip in the gouge region, with each trace representing the average displacement rate of each row of particles during the simulation as a function of distance from the lower fault surface and displacement of the fault. Initially (periods d_1 and d_2) slip is localised into a relatively wide band located at $z = 25 \rightarrow 45$ for $\mu = 1.0$ and $z = 17 \rightarrow 33$ for $\mu = 0.6$ (see also snapshots $d = 100$ to $d = 379$ in Fig. 5). After a large displacement ($d = 300 \, r_0$ for $\mu = 1.0$ and $d = 550 \, r_0$ for $\mu = 0.6$), slip re-localises in a narrower shear zone (period d_3) located at the base of the gouge zone (snapshot $d = 379$ to $d = 425$ in Fig. 5). When this occurs, fault friction μ_f^* drops dramatically from $\mu_f^* \sim 0.6$ to $\mu_f^* \sim 0.3$.

For the experiment using $\mu = 1.0$, the process of re-localisation correlates with an increase of the rate of gouge generation (period d_3 in Fig. 4a) and is due to fractures occurring throughout the gouge layer as the shear band re-locates, and bonds breaking between grains inside the new active shear band. This increase in fracture energy is not seen in the experiment using $\mu = 0.6$ because slip was distributed at the beginning of the simulation (beginning of period d_1 in Fig. 6b). Hence, most of the bonds between grains were already broken when slip re-localised into the narrower shear zone.

In all numerical experiments, re-organisation occurred only after a large displacement (at least 5 full rotations of the blocks). Such a re-organisation into a weak microstructure could be occurring on large displacement (mature) faults. In field studies of the San Andreas fault system, principle faults seem to be weak, with extreme slip localisation apparently being a common feature (e.g., CHESTER *et al.*, 1993). A re-arrangement of grains such that slip is re-localised into a narrower and substantially weaker shear band has not yet been observed in laboratory experiments. A possible reason is the long time required for self-organisation of gouge layers such that slip is highly localised and the rolling-type mechanism is enhanced. Alternatively, the self-organisation may occur only with specific model faults. Grains of model rock have a high intrinsic roughness and, unlike real rocks, they are not allowed to fracture until they reach a grinding limit, which can be at the micron scale or below, where plastic deformations take place. These limitations can play an important role in slip localisation and gouge self-organisation. Highly localised slip was observed only during numerical experiments with high normal stress (1500 MPa). However, due to the long and unpredictable time required for the fault gouge to self-organise, the numerical experiments with lower normal stresses may not have been run for a sufficiently long time.

Gouge Thickness

Figures 7 and 8 show that variations in the gouge friction correlate with variations in the gouge thickness, with the gouge compacting when the friction of

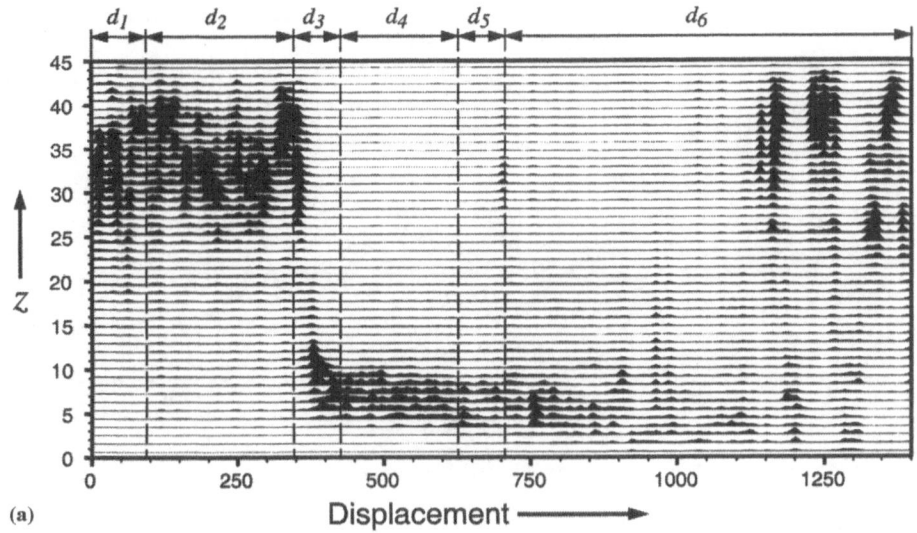

(a)

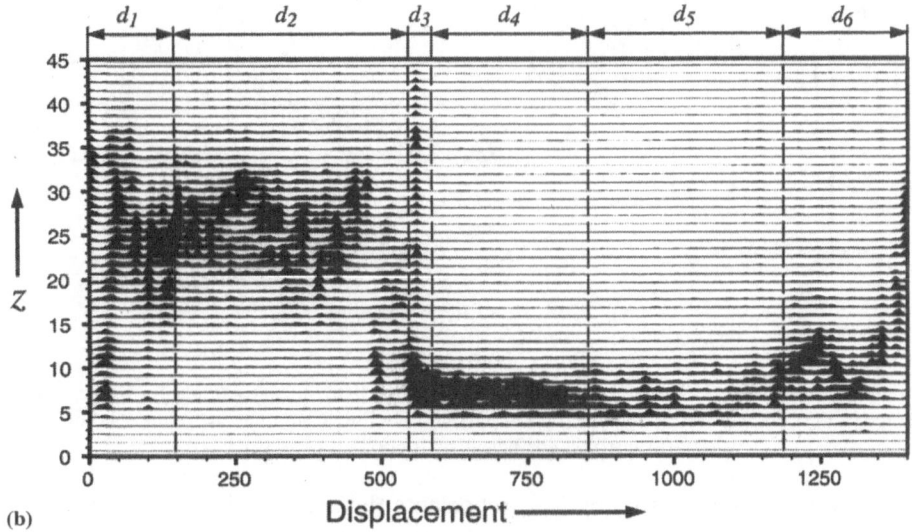

(b)

Figure 6

Plots showing the distribution of slip in the gouge region (a: $\mu = 1.0$, b: $\mu = 0.6$). Forty-five traces are displayed where each trace is the average instantaneous displacement rate of particles located at the corresponding z-coordinate. Black regions depict low displacement rates and highlight the active shear band whereas white regions with no offset of the trace depict the plate driving rate (cf., Fig. 5).

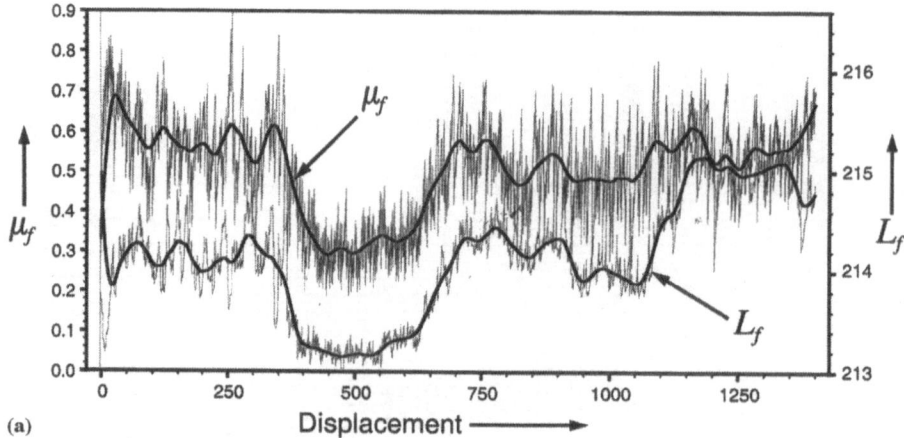

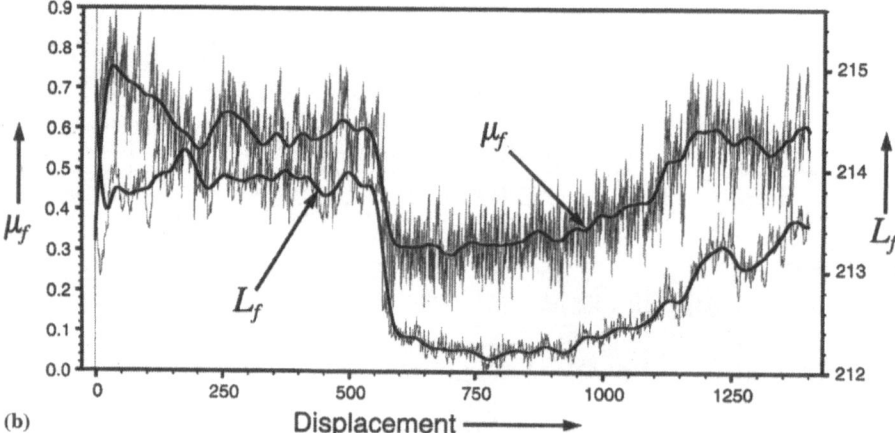

Figure 7

Plot of the separation L_f between the driving plates (thin light line: instantaneous value, solid line: average value) compared with the fault friction μ_f (thin light line: instantaneous value, solid line: average value). (a) intrinsic coefficient of friction $\mu = 1.0$; (b) $\mu = 0.6$.

the gouge decreases. In Figure 7, L_f represents the actual height of the model, that is, the separation between the upper and lower driving plates. Since the normal stress on the driving plates is constant, the variation of the gouge thickness can be measured at the driving plates by assuming that the elastic blocks will remain at a constant height L_b.[6] Hence, the actual gouge thickness is equal to $L_g = L_f - L_b$ and the gouge thickness variations equal the model height variations (i.e., $\Delta L_g = \Delta L_f$).

---

[6] Note that during synthetic earthquake events, the height of the elastic blocks changes due to the passage of seismic waves, however the average value of L_b remains constant since a constant normal stress is applied on the driving plates.

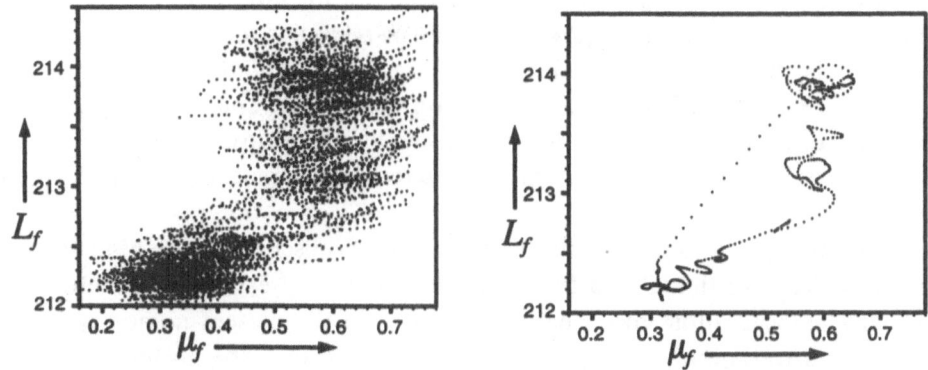

Figure 8

Plot of the separation L_f between the driving plates as a function of the fault friction μ_f during the numerical simulation using an intrinsic coefficient of friction $\mu = 0.6$. Left plot: instantaneous value (i.e., correlation between μ_f and L_f; right plot: average value (i.e., correlation between μ_f^* and L_f^*).

Figure 7 also shows that when the fault is weak, variations in gouge thickness are smaller. When a slip event occurs, grains of model rock are sliding and rotating past one another, causing the gouge thickness to change. When slip is distributed over a relatively broad region, the variations in the gouge layer thickness are relatively large due to the involvement of many grains in the slip process. When slip re-localises into a narrower shear band, fewer grains are involved in these re-arrangements, and the variations in gouge thickness become smaller.

It must be noted that dilatation or compaction can occur without a significant change in the fault friction: for example the dilation occurring at $d = 1100 \ r_0$ (Fig. 7a) does not correlate with a large increase of the fault friction.

Comparison with Laboratory Experiments

In the laboratory experiments performed by BEELER et al. (1996), a simulated gouge is placed between two blocks of rock and subjected to a shear and normal stress. Shear stress is induced by the displacement of the two blocks of rock at a constant velocity. The simulated gouge was composed of grains of rock of different sizes which break as the system evolves. The gouge zone had a thickness of 1 mm, a width of 400 mm and was subjected to a displacement of 400 mm (one full rotation of the disc-shaped sample). The ratio between the gouge layer width and the largest grain size in the numerical experiments is similar to the ratio for the laboratory experiments. Although the laboratory experiments were performed with considerably lower normal stress (15 MPa in the BEELER et al. experiments), similar features in the apparent fault friction are observed[7].

---

[7] Note that in numerical experiments, it was found that the applied normal stress has little effect on the general aspect of the apparent fault friction (cf., PLACE and MORA, 1999).

During periods d_1 and d_2, the observed friction of the model gouge layer as well as its general form as a function of displacement is similar to that of the observed friction in the BEELER et al. laboratory experiments (i.e., both the numerical and laboratory results manifest an initial rise to a maximum value of $\mu_f^* \sim 0.7$ and a subsequent drop to a lower value ~ 0.6 followed by smaller oscillations around this value). Also the variation in gouge thickness with the apparent friction is a common feature.

Slip localisation into a basal shear zone has been observed in the BEELER et al. experiment but was thought to be due to the laboratory experimental setup. The systematic re-localisation into a basal shear zone in the present numerical experiments demonstrates that this phenomena is not due to the experimental setup, and is more likely due to the specific interactions between the fault surface and gouge grains.

Although the general aspect of the laboratory experiment results correlates well with the numerical experiment results, the microscopic properties of the lattice solid differ in some points due to the model limitations. The scale of the grain of the rocks is different due to the limited number of particles in the numerical simulation: the gouge thickness is only 50 particles whereas in the laboratory experiment grains of rock can reach microns for a gouge thickness of 1 mm (a model size of 5000×1000 particles would be required). However, numerical simulations performed on larger models exhibit similar initial features with the localisation of slip in a broad central region (although it was not possible to observe a re-localisation of slip due to the high computational time required) suggesting that the experiments are scalable. The porosity of the medium in the numerical experiment is too high compared to the BEELER et al. experiments (which have a porosity of 1% compared to 16–20% in the numerical experiments) and is more comparable to a high porosity sandstone. In the following the porosity of the simulated gouge is calculated and compared with the minimum possible porosity, which is not zero in the model rock.

Porosity

Figure 9 shows the porosity of the gouge region compared with the thickness of the gouge zone. The gouge porosity varies between 22% and 26%, and has an average value of approximately 24% which is comparable to a high porosity sandstone. Variations in porosity during the simulation correlate with gouge layer thickness as expected (i.e., the more compact the gouge layer, the lower the porosity).

The relatively high porosity is to be expected in view of computer limitations which restrict the range of grain sizes and shapes that can be modelled. This limitation implies that the minimum possible porosity of the gouge layer is 5.5% rather than 0% as explained in the following.

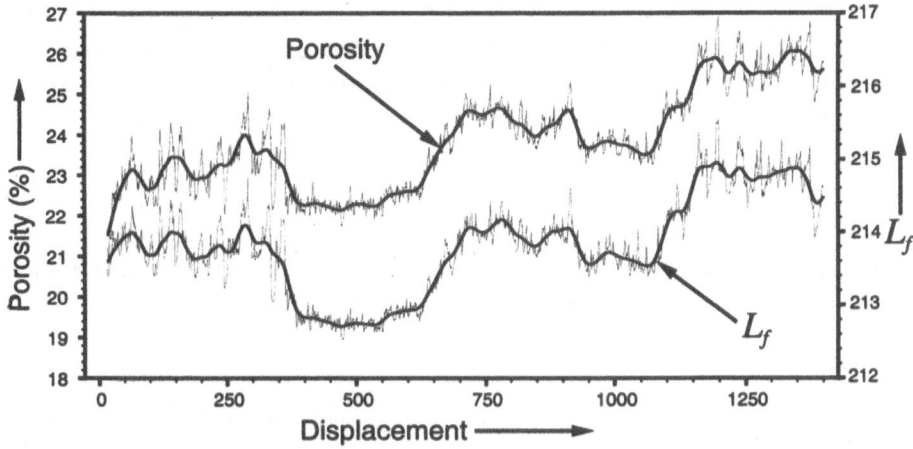

Figure 9
Plot of the porosity of the gouge region (thin light line: instantaneous value, solid line: average value) compared with the separation L_f between the driving plates (thin light line: instantaneous value, solid line: average value).

Grains that are composed of particles bonded together are considered to have a zero porosity (i.e., the area between bonded particles is not considered as vacant space). Hence, a fully bonded triangular lattice (i.e., without gouge layer) would have a 0% porosity, and due to the circular shape of particles, a lattice consisting of single unbonded particles has a theoretical porosity p of 9.3%. This value is computed as the difference between the area of a hexagon (space filling unit of the lattice) and that of the largest circle it encloses (i.e., a particle). Namely,

$$p = \frac{\text{hexagon area} - \text{particle area}}{\text{hexagon area}} = \frac{6r_0/\sqrt{3} - \pi r_0^2}{6r_0/\sqrt{3}} \sim 0.093. \qquad (7)$$

This value can be used to calculate the porosity of close-packed arrangements of the different shaped grains in the gouge layer. This yields a theoretical minimum porosity of 5.5% for the gouge layer (i.e., porosity of a close-packed arrangement with the given distribution of grain shapes specified in Table 1, assuming that grains themselves have a zero porosity). This theoretical minimal porosity is not representative of what could be expected for the porosity in more random arrangements. For example, the initialisation procedure for the weak heterogeneous fault zone attempted to fill space randomly using the specified distribution. The initialisation procedure is estimated to yield a porosity (if all weak bonds between grains were broken), of approximately 23.9%. Hence, porosity during the simulation ranges from 16.5% to 20.5% above that of the least porous state, and fluctuates around that of a random but densely packed arrangement of the grain shapes being modelled.

Future work is required to determine whether or not the high porosity of the fault gouge plays an important role in the rolling-type mechanism observed. Using particles of various sizes and granular friction (work in progress) a more realistic porosity of the medium can be specified.

Heat

Assuming that the energy goes mainly into heat and seismic waves, the average frictional stress $\bar{\tau}_f$ can be written as

$$\bar{\tau}_f = \bar{\tau}_s + \bar{\tau}_h, \tag{8}$$

where $\bar{\tau}_h$ and $\bar{\tau}_s$, respectively represent the components of frictional stress that go into heat generation and seismic waves. The actual heat generated (E_h) is computed as the work done by the intrinsic friction and is given by

$$E_h(t) = \int_0^t \sum_{ij \in F} F_{ij}^F \dot{x}_{ij}^T \, dt, \tag{9}$$

where $\dot{x}_{ij}^T$ is the slip velocity between the two particles i and j. The ratio of the heat being generated relative to a fault with friction μ defines a measure of the fault gouge layer heat generation relative to a simple reference. The effective coefficient of friction (producing heat) is defined as

$$\bar{\mu}_h = \mu \frac{\overline{E'_h}}{\overline{E'_{th}}}, \tag{10}$$

where $\overline{E'_h}$ and $\overline{E'_{th}}$ are respectively the average rate of heat actually generated and heat theoretically generated as defined in the following. The ratio between these two values should be close to unity if the heat generated remains close to the theoretical value. In this case, the effective macroscopic coefficient of friction $\bar{\mu}_h$ should be equal to the microscopic coefficient of friction μ. The theoretical heat generated E_{th} is computed as the heat that would be generated by rubbing two smooth surfaces against one another with a coefficient of friction μ. Thus, $E_{th} = \mu \sigma_n 2LV_e t$ where L is the length of the driving plates.

The actual heat generated during the numerical experiments (Fig. 10) is computed using Equation (9) and is compared with the theoretical heat. Three periods are distinguished: before the re-localisation (period p_1 which corresponds to the period $d_1 + d_2 + (d_3/2)$ in Fig. 3), after the re-localisation and while slip continues to occur within the narrow basal shear band (period p_2 which corresponds to the period $(d_3/2) + d_4 + (d_5/2)$ in Fig. 3), and a final period when slip again occurs on a less localised shear band (period p_3 which corresponds to the period $(d_5/2) + d_6$ in Fig. 3). Table 2 displays values of the fault friction $\bar{\mu}_f$ and the effective coefficient of friction $\bar{\mu}_h$ (generating heat).

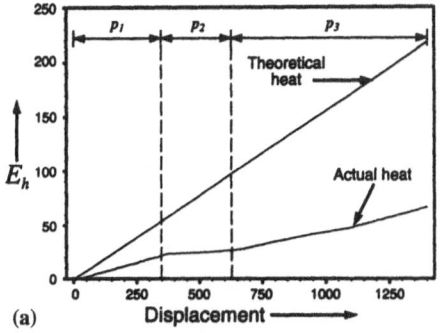
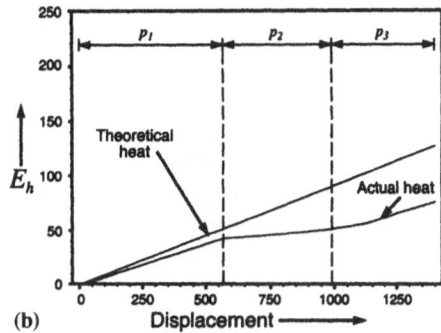

(a)   (b)

Figure 10

Plot of the theoretical and actual heat generated during the simulations for an intrinsic coefficient of friction $\mu = 1.0$ (a) and $\mu = 0.6$ (b).

The fault friction decreases from $\bar{\mu}_f \sim 0.6$ (period p_1) to $\bar{\mu}_f \sim 0.3$ (period p_2) when slip re-localises. The heat reduction factor given by $\mu/\bar{\mu}_h$ is 11 for the experiment using $\mu = 1.0$ and 4 for the experiment using $\mu = 0.6$ during period p_2. When the fault restrengthens (period p_3), the fault friction increases to $\bar{\mu}_f \sim 0.5$ which is somewhat lower than the initial fault friction of $\bar{\mu}_f \sim 0.6$. Figure 6a and the two snapshots $d = 600$ and $d = 1200$ in Figure 5 show that slip eventually alternates between two shear bands located at $z = 20 \rightarrow 45$ and from $z = 1 \rightarrow 12$ during period p_3.

The dominant mechanism capable of explaining the reduction in fault friction and heat observed during period p_2 is a rolling-type mechanism which minimises the amount of slip between grain surfaces during slip of the fault (MORA and PLACE, 1998, 1999).

Another mechanism observed is a "bouncing mechanism" (Fig. 11) that allows particles to slip past one another when the normal stress is reduced. In order to estimate the reduction in heat due to the bouncing mechanism, the heat generated assuming particles have a linear trajectory is computed. For instance, in Figure 11

Table 2

Values of the fault friction and the effective coefficient of friction $\bar{\mu}_h$ for each of the three periods distinguished in two numerical localisation experiments

μ	Period	$\bar{\mu}_f$	$\bar{\mu}_h$
1.0	p_1	0.58	0.40
	p_2	0.35	0.09
	p_3	0.53	0.33
0.6	p_1	0.62	0.49
	p_2	0.34	0.16
	p_3	0.53	0.40

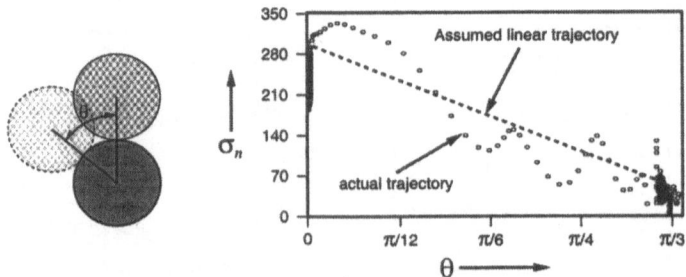

Figure 11

Dynamics of slip between two particles. The normal stress (σ_n) is plotted as a function of θ (right) which represents displacement along the particle surface (left).

the assumed particle trajectory follows a line extending from ($\theta = 0$, $\sigma_n = 290$) to ($\theta = \mathrm{II}/3$, $\sigma_n = 35$). The heat generated if the particle trajectory follows this path is compared with the heat actually generated. In the simulation with $\mu = 1.0$, a typical simulated earthquake event was selected during period p_1 (i.e., when the heat reduction factor was 2.5). All particle trajectories were computed for this event and the trajectory displaying maximal bouncing-type behaviour was selected for analysis (i.e., the trajectory where most slip occurred at a lower inter-particle normal stress relative to that before and after the event). For this trajectory, bouncing was responsible for a reduction in heat of 10%. This result suggests that at most, a small fraction of the heat reduction observed in the simulation is due to the bouncing mechanism (see also MORA and PLACE, 1998).

The numerical experiments indicate that shear band localisation is accompanied by improved efficiency of the rolling-type mechanism. The fractures that occur in the shear zone cause large grains to break down to form grains which can more easily rotate. The re-localisation occurring during period p_2 suggests that a natural fault gouge could self-organise such that slip is localised in a thin and weak region. In the numerical experiments, this re-localisation is stable for a long period of time (during a displacement of 250 particle diameters using $\mu = 1.0$ and 400 particle diameters using $\mu = 0.6$), but eventually the active shear band broadens and re-strengthens or shear re-localises into a broader and stronger band elsewhere in the gouge layer (cf., Fig. 6).

Results show that, after the re-localisation, slip can eventually de-localise into a broader and stronger shear band (period d_6 in Fig. 6). This suggests that the re-localisation into a weak and thin shear band could be an unstable phenomena. On the other hand, at the end of period d_6 in Figure 6a slip is switching between the initial shear band and the weak and thin shear band where slip re-localised. This suggests that slip could eventually re-localise once again into a weak and thin shear band later in the numerical simulation. Future work is required to study the switching between states and their stability.

Seismic Efficiency

The seismic efficiency η represents the amount of energy transformed into seismic waves. Assuming that all the kinetic energy is generated by waves, the seismic efficiency is defined by

$$\eta = \frac{E_k}{E_k + E_h + E_b},\tag{11}$$

where E_k, E_h and E_b represent the kinetic, heat and fracture energy, respectively. The heat energy E_h is defined in the model by the work done by the intrinsic friction. The fracture energy is the potential energy lost when a link between two particles breaks. The sum of these energies represents the loss in potential energy during a synthetic event. Hence, Equation (11) represents the percentage of the energy transformed into seismic waves during events. The total kinetic energy in the model is the sum of the instantaneous kinetic energy and the cumulative loss in kinetic energy due to the artificial viscosity that damps radiated seismic waves.

The potential problem with definition (11) is that some energy lost to the artificial viscosity would potentially have transferred into heat rather than radiated waves in the absence of artificial damping. If E_k^w represents the amount of kinetic energy of radiated seismic waves, the seismic efficiency is equal to

$$\eta = \frac{E_k^w}{E_k + E_h + E_b}.\tag{12}$$

We assume that part of the energy is consumed by internal re-organisation of the gouge layer (e.g., rotational kinetic energy). In the absence of the artificial viscosity, this internal kinetic energy would eventually be radiated as waves and/or transform into heat and/or be stored in elastic potential energy once the dynamic rupture event had ceased. In the following, we will analyse the energy partitioning of the system during one event in order to estimate the component of internal kinetic energy transformed into seismic waves (i.e., E_k^w).

In the absence of artificial damping, radiated waves will remain trapped in the closed system after dynamic rupture has ceased. Hence, the value of the kinetic energy in the model after the rupture has ceased represents the kinetic energy of the radiated waves. Figure 12 shows the kinetic energy inside and outside the gouge region, the fracture energy, heat energy and the sum of the external work done and the potential energy (the sum of these energies is verified to be constant). The plot of the kinetic energy generated inside the gouge region $(E_k)_{\text{inside}}$ shows a sharp increase that corresponds to the nucleation of the synthetic event. After the nucleation, part of this energy is radiated outside the gouge region (i.e., when $(E_k)_{\text{outside}}$ increases). $(E_k)_{\text{outside}}$ has a sinusoidal shape due to the fundamental mode of oscillation of the closed system. The values of $(E_k)_{\text{inside}}$ and $(E_k)_{\text{outside}}$ decrease due to the passage of the seismic waves through the gouge region, which has a physical attenuation effect due to minor grain slippage being induced by the passage of waves, resulting in some heat production (i.e., the heat energy continues to slightly increase after the rupture has ceased due to this effect). The time when the generation of heat and the kinetic energy inside the gouge region ($(E_k)_{\text{inside}}$) becomes relatively flat marks the moment when the rupture has ceased (heat is only generated by the passage in the gouge region of the waves trapped in the system). At this time, the total kinetic energy ($(E_k)_{\text{outside}} + (E_k)_{\text{inside}}$) gives an estimate of the energy transformed into seismic waves during this single event. From Figure 12, the total kinetic energy after the rupture has ceased is approximately equal to 0.01, $E_b \sim 0$ and $E_h \sim 0.3$. Hence the seismic efficiency for this event is $\eta \sim 0.01/0.31 = 3.2\%$.

An observationally based estimate of the lower bound of the seismic efficiency η is 0.15 (KANAMORI, 1996). One recent theoretical study (MELOSH, 1996) assumed a value of $\eta = 0.5$ and can probably be considered as an upper bound. Note that the lower bound above should be considered as approximate in view of uncertainties (e.g., probably one would be safer in assuming that η is greater than a few percent but less than 0.5). The estimated seismic efficiency in the rupture is compatible with the assumed lower and upper bounds for seismic efficiency of 0.15 and 0.5, within the uncertainties (i.e., likely range of order $\eta \in$[few percent, 0.5]).

During this numerical experiment (with artificial viscosity) the average fault friction is $\bar{\mu}_f \sim 0.58$ and the average effective fault friction (generating heat) is $\bar{\mu}_h \sim 0.40$. The average seismic efficiency $\bar{\eta}$, assuming that all energy goes into heat and seismic waves (cf., MORA and PLACE, 1998), is equal to

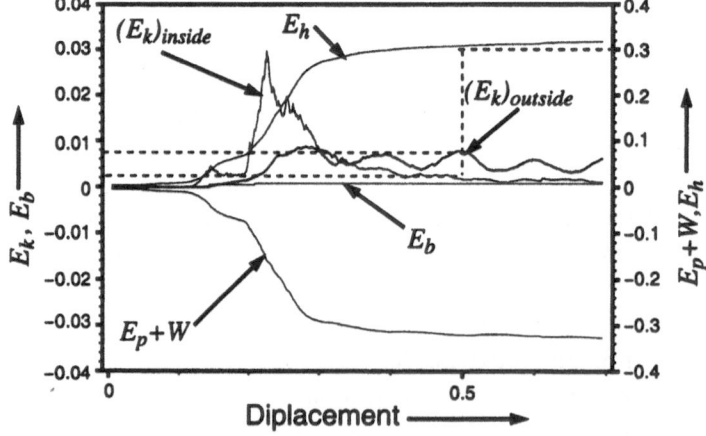

Figure 12

Plot of the kinetic energies outside ($(E_k)_{\text{outside}}$) and inside ($(E_k)_{\text{outside}}$) the fault gouge. Superimposed is the sum of the fracture energy and heat energy ($E_b + E_h$) generated during the event.

$$\bar{\eta} = \frac{\bar{\mu}_f - \bar{\mu}_h}{\bar{\mu}_f}. \tag{13}$$

Using this approach, an average seismic efficiency of $\bar{\eta} = 31\%$ is obtained for the numerical experiment, which is greater than the seismic efficiency for the single event studied above by one order of magnitude. This difference between the seismic efficiency of a single event and the average seismic efficiency can be due to various factors: (1) Equation (13) assumes that the fracture energy lost is zero; (2) Equation (13) assumes that all kinetic energy and energy lost by viscosity corresponds to the energy of seismic waves radiated whereas some energy may go into internal re-organisation and produce heat as explained above, and (3) aseismic events or some seismic events may have a greater seismic efficiency. Fracture energy is found to be small in comparison to the heat and kinetic energy, consequently the first factor is not relevant to the experiment. The second factor implies that Equation (13) represents an upper bound of the average seismic efficiency. But it alone cannot explain the difference between the seismic efficiency of a single event and the average seismic efficiency (from Fig. 12 approximately a third of the kinetic energy generated during the event remains present in the form of seismic waves, hence in this event at least 30% of the kinetic energy is radiated as seismic waves). Hence aseismic events and some seismic events may have a higher seismic efficiency by up to a factor of 3.

Reidel Shears

The direction of the two principal axes of stress, having an orientation θ from the x axis (Fig. 13), can be calculated from the normal and shear stress (TURCOTTE and SCHUBERT, 1982, p. 82) applied on the boundary of the lattice. Namely

$$\tan 2\theta = \frac{2\sigma_{xy}}{\sigma_{xx} - \sigma_{yy}}, \tag{14}$$

where σ_{xy} is the shear stress ($\tau_f = \mu_f \sigma_n$), σ_{yy} is the normal stress, and σ_{xx} horizontal stress which is given by (cf., TURCOTTE and SCHUBERT, 1982, p. 108)

$$\sigma_{xx} = \frac{v}{1-v} \sigma_{yy}, \tag{15}$$

where v is the Poisson's ratio and is equal to 1/3 (cf., MORA and PLACE, 1994). From the above, we obtain

$$\tan 2\theta = \frac{4\mu_f \sigma_n}{-\sigma_n} = -4\mu_f. \tag{16}$$

Hence, the principal stress angle is oriented at

$$\theta = -\frac{\tan^{-1}(4\mu_f)}{2} \sim -33.7°, \tag{17}$$

from the x axis for the case of $\mu_f \sim 0.6$ (and $-27.2°$ for $\mu_f \sim 0.35$).

The gouge actually deforms on shear planes which are oriented at $\pi/4 - \phi/2$ from the angle of the maximal compressional stress σ_1 (BEELER and TULLIS, 1995; BYERLEE and SAVAGE, 1992; MARONE et al., 1992), where ϕ is the friction angle. The friction angle can be computed from the apparent friction μ_f (which is different than the true friction on the inclined planes) using

$$\sin \phi = \mu_f. \tag{18}$$

From the above, a friction angle of 36.9° is obtained for $\mu_f = 0.6$ (and 20.5° for $\mu_f = 0.35$). Hence Reidel R_1 shears when $\mu_f = 0.6$ are oriented at $\theta_{R_1} = \theta + (\pi/4 - \phi/2) \sim -7.15°$ and R_2 shears are oriented at $\theta_{R_2} = \theta - (\pi/4 - \phi/2) \sim -60.3°$ from x axis (cf., Fig. 13) and when $\mu_f = 0.35$, $\theta_{R_1} \sim -7.55°$ and $\theta_{R_2} \sim -62°$. During the localisation experiments, Reidel R_1 shears are observed (Fig. 14) although at somewhat higher angles than expected (from $-20°$ to $-10°$, cf., the theoretical value of $\sim -7°$).

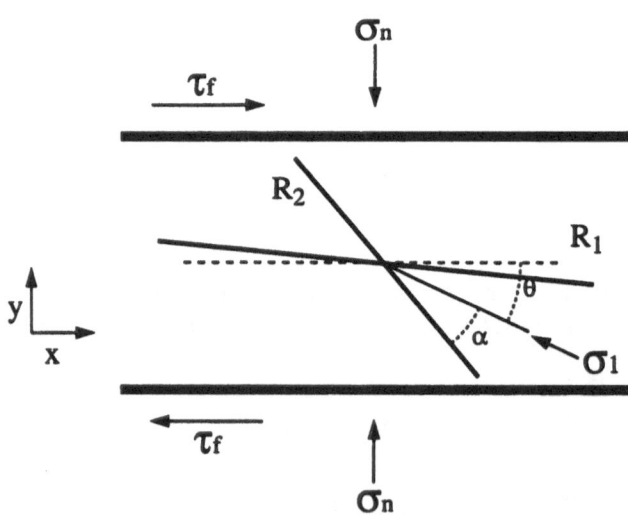

Figure 13

Orientation of Reidel R_1 and R_2 shears. The angle θ represents the orientation of the maximal compressional stress σ_1, and α represents the orientation of the two conjugate Reidel shears from the maximal compressional stress orientation (i.e., $\alpha = \pm (\pi/4 - \phi/2)$ where ϕ is the friction angle).

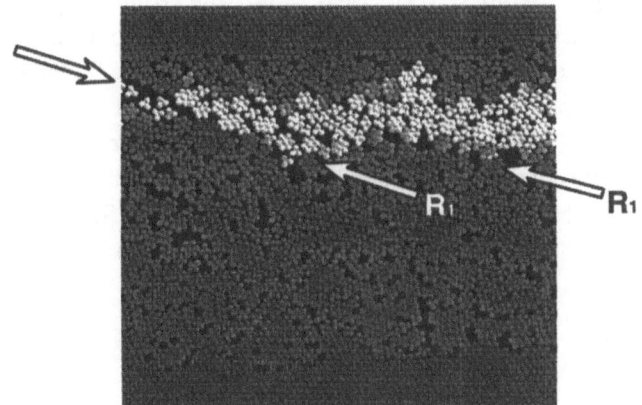

Figure 14
Snapshot of a localisation experiment in which the shade of grey represents the particle displacement
during the last 10,000 time steps. Hence, lines separating areas with contrasted colours highlight shear
zones. Two R_1 shears zones can be distinguished in the snapshot.

Discussion

It must be noted that many microphysical processes that occur in the fault
gouge can strongly affect the macroscopic behaviour. Only a few of these processes
(such as friction) are simulated in these numerical experiments. For instance,
healing, chemical processes, thermal expansion or fluid-solid interaction may effect
the macroscopic behaviour of the fault gouges. The model is being refined to allow
these different microphysics to be easily incorporated in the model. Preliminary
results suggest that thermal expansion and pore fluid pressure affect the macroscopic behaviour (ABE *et al.*, 2000), however further work is required to quantify
the changes in macroscopic behaviour.

Conclusions

Numerical studies of fault gouge layers indicate localisation phenomena and
yield similar results to those obtained in laboratory experiments, at least in the
earlier stages of the numerical experiments. This includes fracturing with orientations similar to Reidel (R_1) shears and decreases in gouge layer thickness during
periods of lower fault strength, which correlate with those times when slip has
become more localised within the gouge layer. After a large displacement, a
re-organisation is observed at which time the slip is highly localised along a very
narrow shear zone located near the base of the gouge layer. When this occurs, the

model fault gouge becomes very weak and the rolling type micro-physical mechanism for shearing is highly efficient. This is consistent with the suggestion of MORA and PLACE, 1999 that a fault gouge can self-organise to form a narrow and weak band capable of explaining the heat flow paradox (HFP). The prolonged time required for the self-organisation process offers a possible explanation of why the weak gouge layers predicted in the numerical experiments have not yet been observed in the laboratory. Seismic efficiency of a typical rupture event is 3.2% which is consistent with observationally based estimates.

Acknowledgements

This research was funded by the Australian Research Council. Supplementary funding was provided by The University of Queensland (UQ) and by the sponsors of QUAKES. Principle computations were made using QUAKES' 8 processor Silicon Graphics Origin 2000. Supplementary computations were made using UQ's Origin 2000.

REFERENCES

ABE, S., MORA, P., and PLACE, D. (2000), *Extension of the Lattice Solid Model to Incorporate Temperature-related Effects*, Pure appl. geophys. *157*, 1867–1887.

BEELER, N. M., and TULLIS, T. (1995), *Implications of Coulomb Plasticity for the Velocity Dependence of Experimental Faults*, Pure appl. geophys. *144* (2), 251–276.

BEELER, N. M., TULLIS, T., BLANPIED, M., and WEEKS, P. J. (1996), *Frictional Behavior of Large Displacement Experimental Faults*, J. Geophys. Res. *101*, 8697–8715.

BYERLEE, J., and SAVAGE, J. C. (1992), *Coulomb Plasticity within the Fault Zone*, Geophys. Res. Lett. *19* (23), 2341–2344.

CHESTER, F. M., EVANS, J. P., and BIEGEL, R. L. (1993), *Internal Structure and Weakening Mechanisms of the San Andreas Fault*, J. Geophys. Res. *98*, 771–786.

CUNDALL, P. A., and STRACK, O. D. L. (1979), *A Discrete Numerical Model for Granular Assemblies*, Géotechnique *29*, 47–65.

DONZÉ, F., MORA, P., and MAGNIER, S. A. (1993), *Numerical Simulation of Faults and Shear Zones*, Geophys. J. Int. *116*, 46–52.

KANAMORI, H., (1996) personal communication.

MARONE, C., HOBBS, B., and ORD, A. (1992), *Coulomb Constitutive Laws for Friction: Contrasts in Frictional Behavior for Distributed and Localized Shear*, Pure appl. geophys. *139* (2), 195–214.

MELOSH, H. J. (1996), *Dynamical Weakening of Faults by Acoustic Fluidization*, Nature *379*, 601–606.

MORA, P., and PLACE, D. (1994), *Simulation of the Frictional Stick-slip Instability*, Pure appl. geophys. *143*, 61–87.

MORA, P., and PLACE, D. (1998), *Numerical Simulation of Earthquake Faults with Gouge: Towards a Comprehensive Explanation for the Heat Flow Paradox*, J. Geophys. Res. *103*, 21,067–21,089.

MORA, P., and PLACE, D. (1999), *The Weakness of Earthquake Faults*, Geophys. Res. Lett. *26*, 123–126.

PLACE, D., and MORA, P. (1999), *The Lattice Solid Model: Incorporation of Intrinsic Friction*, J. Int. Comp. Phys. *150*, 332–372.

TURCOTTE, D., and SCHUBERT, G., *Geodynamics*, *Applications of Continuum Physics to Geological Problems* (John Wiley & Sons, New York 1982).

(Received October 3, 1999, revised March 31, 2000, accepted April 15, 2000)

 To access this journal online:
http://www.birkhauser.ch

Pure appl. geophys. 157 (2000) 1847–1866
0033–4553/00/121847–20 $ 1.50 + 0.20/0

⌐Pure and Applied Geophysics

Shear Heating in Granular Layers

KAREN MAIR[1,2] and CHRIS MARONE[1]

Abstract—Heat-flow measurements imply that the San Andreas Fault operates at lower shear stresses than generally predicted from laboratory friction data. This suggests that a dramatic weakening effect or reduced heat production occur during dynamic slip. Numerical studies intimate that grain rolling or localization may cause weakening or reduced heating, however laboratory evidence for these effects are sparse. We directly measure frictional resistance (μ), shear heating and microstructural evolution with accumulated strain in layers of quartz powder sheared at a range of effective stresses ($\sigma_n = 5$–70 MPa) and sliding velocities ($V = 0.01$–10 mm/s). Tests conducted at $\sigma_n \geq 25$ MPa show strong evidence for shear localization due to intense grain fracture. In contrast, tests conducted at low effective stress ($\sigma_n = 5$ MPa) show no preferential fabric development and minimal grain fracture hence we conclude that non-destructive processes such as grain rolling/sliding, distributed throughout the layer, dominate deformation. Temperature measured close to the fault increases systematically with σ_n and V, consistent with a one-dimensional heat-flow solution for frictional heating in a finite width layer. Mechanical results indicate stable sliding ($\mu \sim 0.6$) for all tests, irrespective of deformation regime, and show no evidence for reduced frictional resistance at rapid slip or high effective stresses. Our measurements verify that the heat production equation ($q = \mu\sigma_n V$) holds regardless of localization state or fracture regime. Thus, for quasistatic velocities ($V \leq 10$ mm/s) and effective stresses relevant to earthquake rupture, neither grain rolling/sliding or shear localization appear to be a viable mechanism for the dramatic weakening or reduced heating required to explain the heat flow paradox.

Key words: Friction, shear heating, fault strength, shear localization, fault gouge.

1. Introduction

An enduring paradox in earthquake studies is that the heat flow measured near the San Andreas Fault is much lower than that predicted using stress field, plate velocity and standard laboratory derived friction values (e.g., LACHENBRUCH and SASS, 1980). The heat production equation (1) relates shear heating to the work done due to frictional shear:

$$q = \tau V = \mu\sigma_n V \qquad (1)$$

[1] Department of Earth, Atmospheric and Planetary Sciences, Massachusetts Institute of Technology, Cambridge, MA 02139, USA. E-mail: karen@barre.mit.edu cjm@westerly.mit.edu
[2] Now at: Department of Earth Sciences, University of Liverpool, Liverpool, L69 3GP, UK.

where q is rate of heat production per unit area; τ is shear stress, which during steady sliding equals friction coefficient (μ) multiplied by effective normal stress (σ_n); and V is sliding velocity.

Solutions to this heat-flow debate (summarized in SCHOLZ, 1996, 2000) require a reduction in q and generally invoke mechanisms to reduce fault friction. For example the addition of clays may lubricate the fault zone reducing friction, however, it is unclear whether clays are pervasive in natural fault zones (CHESTER et al., 1993). Alternatively, frictional melting due to high temperatures caused by rapid sliding on the fault may reduce shear stress (McKENZIE and BRUNE, 1972). SIBSON (1975) argued that pseudotachylytes observed in the field were evidence of this type of frictional melting during seismic slip events.

There is little experimental data to test these hypotheses since few laboratory studies have systematically measured both friction and shear heating over a wide range of conditions. However, melts formed in the laboratory during high velocity frictional sliding were found to: (1) have very low viscosity, which may lubricate fault surfaces (SPRAY, 1993); and (2) be associated with rapid changes in friction (TSUTSUMI and SHIMAMOTO, 1997). Even below the melting point, temperature may affect the stability and velocity dependence of friction (e.g., STESKY, 1978; LOCKNER et al., 1986; BLANPIED et al., 1995). YOSHIOKA (1985, 1986) and BROWN (1998) showed that unstable stick-slip sliding on simulated faults generates less heat than stable sliding, suggesting that different mechanisms may operate in each regime.

Mature fault zones generally contain significant quantities of wear material or gouge. However, previous experimental work (e.g., BLANPIED et al., 1998; BROWN, 1998) has mainly concentrated on the shearing of bare surfaces with no gouge. There are very few direct observations of shear heating in granular layers (LOCKNER and OKUBO, 1983), hence this topic clearly warrants further investigation. Numerical experiments of faults containing granular material show low-heat generation compared to faults with no gouge (MORA and PLACE, 1998). This may suggest a dramatic weakening in faults with gouge. MORA and PLACE (1999) show evidence of significant weakening associated with the onset of localized slip in faults with gouge which is mainly attributed to grain rolling. Hence they (MORA and PLACE, 1998) propose that the mechanism for low-heat generation in faults with gouge is rolling and bouncing of grains leading to minimal slip between individual grains. This raises questions regarding the role of localization fabrics in heating and how grain fracture (absent in these numerical simulations) would affect shear heating compared to a non-destructive mechanism such as grain rolling/sliding.

The purpose of this paper is to present new laboratory results investigating shear heating in granular layers as a function of accumulated strain, localization, normal stress and sliding velocity. Results indicate that the heat production equation holds for distributed and localized shear and that for a given value of q, heat production is comparable for grain rolling/sliding and grain fracture regimes.

2. Experimental Technique

We shear layers of quartz powder between rough steel surfaces in a servo-controlled direct shear apparatus (Fig. 1). The quartz powder simulates the fault gouge present in mature fault zones. The granular material has an initial grain size of 50–150 μm and sub-rounded—angular grains. The rough steel surfaces approximate the boundary conditions of rough rock and inhibit boundary slip.

We apply a uniform normal force to the side blocks, then apply a constant or stepped load point velocity to the middle block. The resulting shear stress, normal stress, shear displacement and layer thickness changes are continuously measured by transducers during deformation. Friction is nominally defined as shear stress

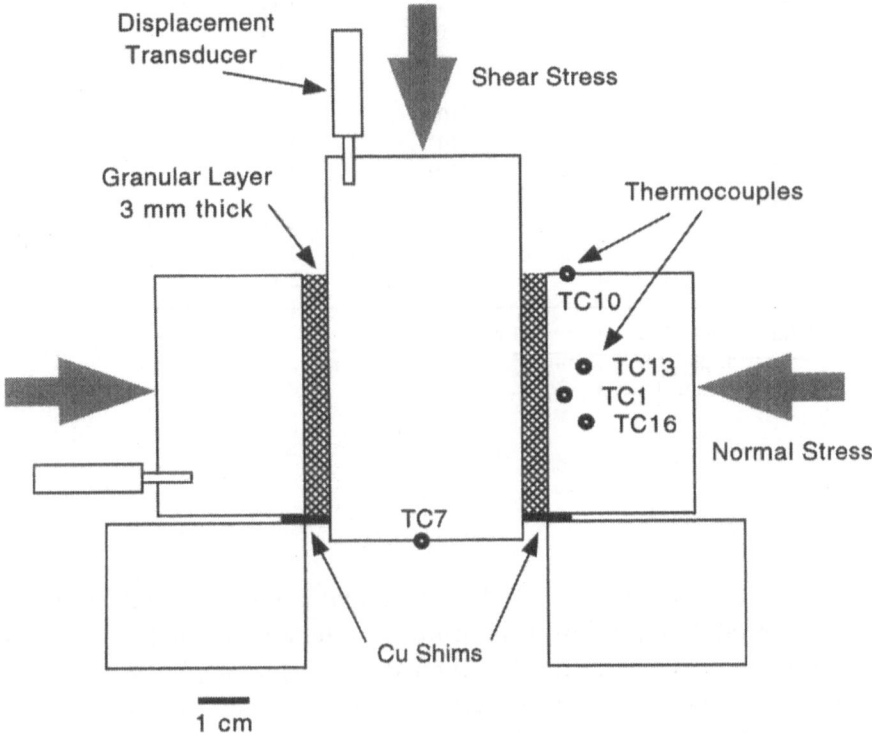

Figure 1

Schematic diagram of the sample geometry illustrating the 3-mm thick granular quartz layers sandwiched between rough steel blocks. Shear and normal stress orientations are indicated and the locations of thermocouples are marked. Three K-type thermocouples (TC1, TC13, TC16) located in blind access holes in one steel side block measure temperature centered on the 50×50 mm² granular layer at distances 2 mm, 6 mm, 6 mm from the center of the layer respectively. Two additional surface thermocouples (TC7, TC10) were attached directly to the center and side blocks, 19.5 mm and 3 mm from the center of the granular layer, respectively. Thermocouples are calibrated giving absolute temperatures correct to within 1°C and data are generally presented as temperature rise with respect to the initial conditions.

Table 1

Experiments

Experiment	σ_n, MPa	V, mm/s	$V\sigma_n$, $\times 10^5$ W/m^2
Velocity Stepping			
m156	25	0.3–3	–
m181	30	0.3–3	–
m147	40	0.3–3	–
m150	40	0.3–3	–
m177	50	0.3–3	–
m178	60	0.3–3	–
m179	70	0.3–3	–
m180	70	0.3–3	–
Constant Velocity			
m288	5	5.0	0.25
m290	5	5.0	0.25
m276	25	1.0	0.25
m277	25	1.0	0.25
m269	25	10.0	2.5
m270	25	10.0	2.5
m278	25	10.0	2.5
m291	50	0.5	0.25
m292	50	0.5	0.25
m287	50	5.0	2.5
m289	50	5.0	2.5

Initial gouge layer thickness 3 mm.
Shear displacement 20 mm.
σ_n = Normal stress, V = Velocity.

divided by normal stress. Additional experimental details are given in MAIR and MARONE (1999a). Temperature is measured by multiple sensor locations near the gouge layer to investigate heat conduction through the sample assembly (Fig. 1).

Velocity stepping tests were carried out to investigate the thermal and mechanical response at differing normal stresses. Temperature changes give important information on shear heating, whereas the friction rate parameter indicates fault zone structure and stability. Constant velocity tests were carried out to study the influence of normal stress and sliding velocity on shear heating for a given $V\sigma_n$ product. Shear heating was studied for different deformation regimes (e.g., localized versus distributed shear, and fracture versus non-fracture conditions). Experimental conditions covered the following range: $0.001 \leq V \leq 10$ mm/s; $5 \leq \sigma_n \leq 70$ MPa; slip 0–20 mm (Table 1). All tests were carried out at room temperature and humidity.

3. *Experimental Results*

3.1 *Deformation of Granular Layers*

The coefficient of friction (shear stress/normal stress) and layer thickness are plotted as a function of shear displacement for a typical high velocity test in Figure 2. Friction increases on initial loading, with some macroscopic strain hardening and then reaches approximately steady state (neutral behavior) after a few millimeters of slip. Second-order perturbations associated with decade step changes in loading velocity are superimposed. At a step increase in velocity, friction shows an abrupt increase and subsequent exponential decay to a new steady state. The converse is observed for a step decrease in velocity. Detailed analyses over a wide range of conditions ($0.001 \leq V \leq 10$ mm/s, $25 \leq \sigma_n \leq 70$ MPa) indicate that steady-state friction is ~ 0.6 after a few mm slip for all tests and yield no evidence for reduced friction at rapid slip or enhanced normal stress (MAIR and MARONE, 1999a). This result applies for the thermal conditions and displacements of our experiments, which overlap with those for which numerical simulations (MORA and PLACE, 1998) have found dramatic weakening. For larger net slip and greater shear heating, GOLDSBY and TULLIS (1998) have reported dramatic weakening.

The friction rate parameter $(a - b)$ is determined directly from our velocity stepping experiments (see inset to Fig. 3). $(a - b)$ indicates the localization state of the gouge as well as fault zone stability (e.g., MARONE and KILGORE, 1993). All of

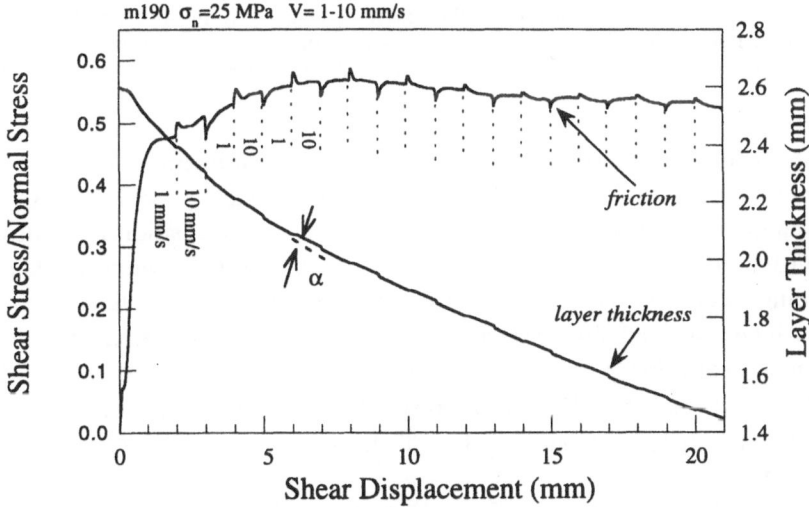

Figure 2

Friction and layer thickness response to step changes in sliding velocity plotted as a function of shear displacement. Velocity steps (1–10 mm/s) are indicated as dotted lines and $\sigma_n = 25$ MPa. Layer thickness decreases systematically with shear displacement but individual dilation (α) and compaction events associated with velocity steps are seen.

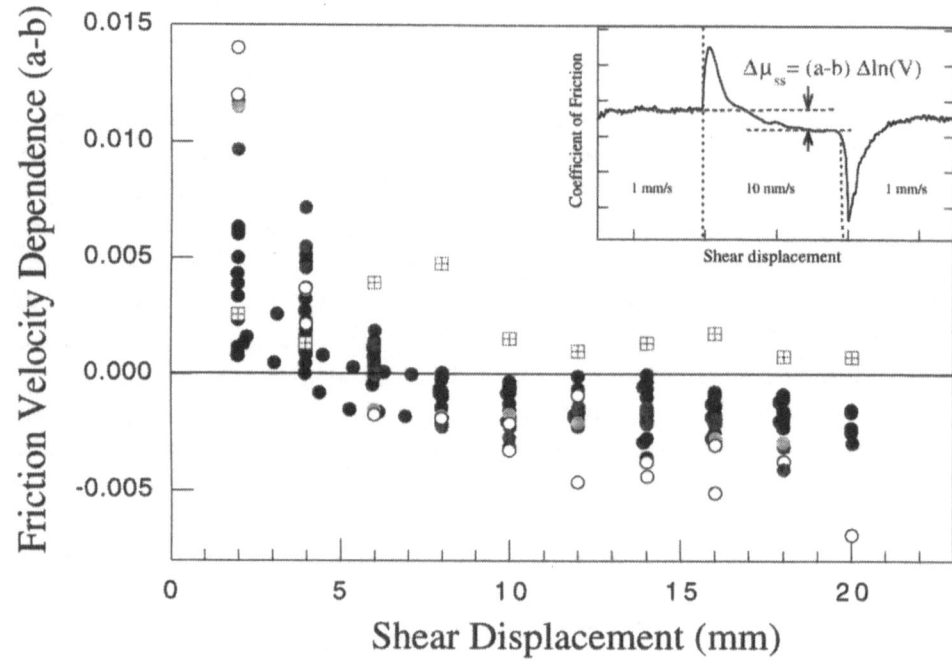

Figure 3

Friction velocity dependence as a function of shear displacement for a range of conditions. Inset shows how $(a-b)$ is measured from a velocity stepping experiment. Black circular markers indicate $\sigma_n = 25$–30 MPa, $0.001 \leq V \leq 10$ mm/s; dark grey circles indicate $\sigma_n = 40$ MPa; light grey circular markers indicate $\sigma_n = 50$–60 MPa; open circles represent $\sigma_n = 70$ MPa (data replotted from MAIR and MARONE, 1999a). Open square symbols represent new data for $\sigma_n = 5$ MPa. Note the transition from velocity strengthening to velocity weakening in all tests except those carried out at 5 MPa where velocity strengthening persists throughout the test.

our tests (with the exception of the $\sigma_n = 5$ MPa experiment) undergo a gradual transition from velocity strengthening $(a - b > 0)$ to velocity weakening $(a - b < 0)$ with accumulated slip (Fig. 3). This is interpreted as a transition from distributed to localized shear at ~ 5–10 mm slip (MARONE, 1998). Previous work indicates that sliding velocity $(0.001 \leq V \leq 10$ mm/s) has little systematic influence on $(a - b)$, however higher normal stress systematically increases both initial strengthening and subsequent weakening by a small amount (MAIR and MARONE, 1999a). The test carried out at $\sigma_n = 5$ MPa shows distinctly different behavior, with velocity strengthening, interpreted as distributed shear, persisting throughout the entire experiment.

Direct (SEM) observations of granular layers deformed at low normal stress $(\sigma_n = 5$ MPa; $V = 5$ mm/s) for 20 mm slip show little or no grain size reduction with respect to the starting powder and no fabric can be identified (Fig. 4a). In contrast, initially identical layers subjected to a higher normal stress $(\sigma_n = 50$ MPa; $V = 0.5$ mm/s) and 20 mm slip show distinct fabric development highlighted by oblique

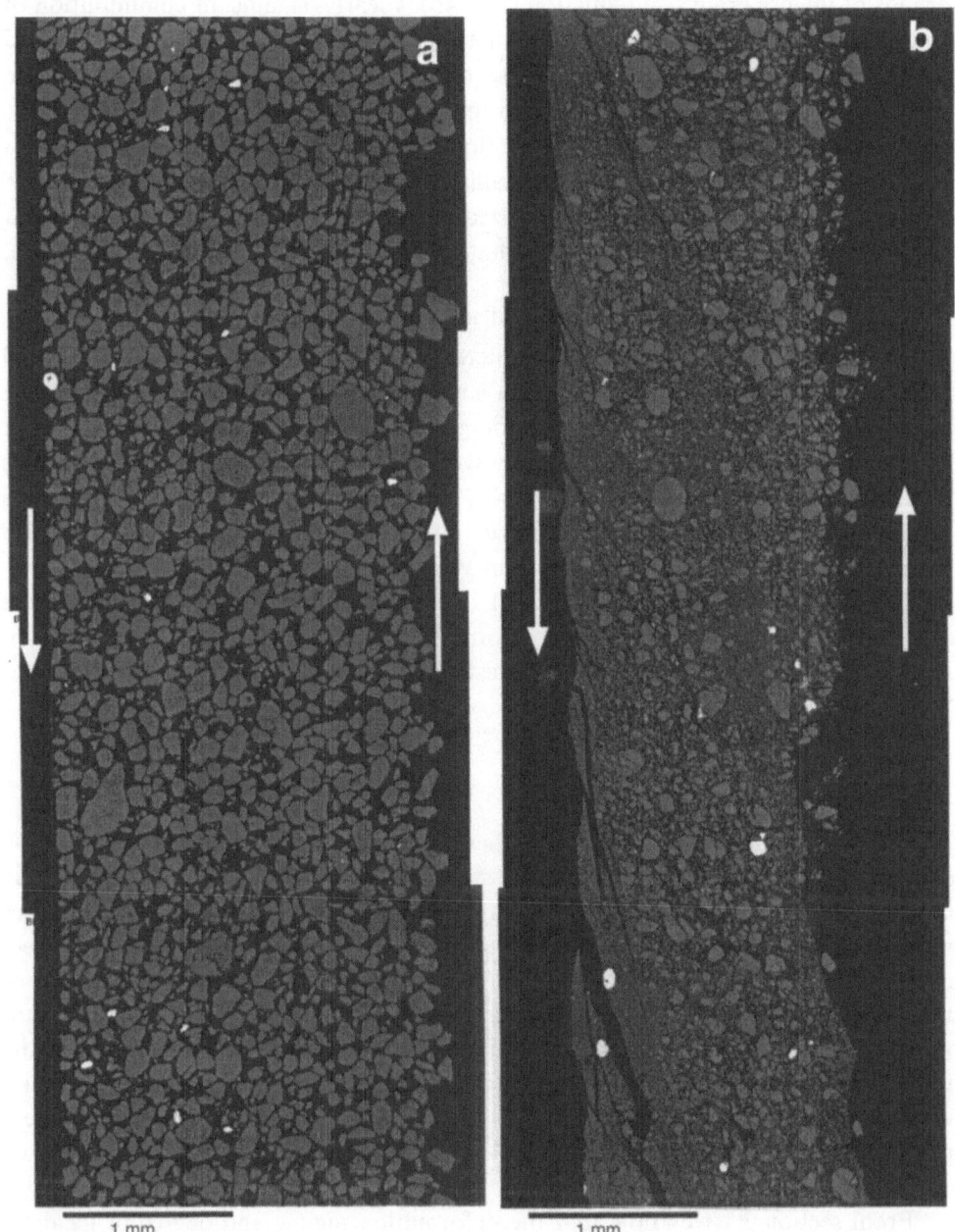

Figure 4

SEM photomicrographs of initially identical granular quartz layers deformed under different conditions: a) $\sigma_n = 5$ MPa, $V = 5$ mm/s, slip $= 20$ mm/s, b) $\sigma_n = 50$ MPa, $V = 0.5$ mm/s, slip $= 20$ mm. The sense of shear is indicated by arrows. Note the intense grain size reduction and development of heterogeneous fabric in b) which is not apparent in a). Both layers were originally 3 mm thick hence layer thinning is clearly enhanced at high σ_n.

strands of intense grain size reduction (Fig. 4b). Clearly, significant comminution of grains has occurred at $\sigma_n = 50$ MPa, whereas almost no grain fracture has ensued at $\sigma_n = 5$ MPa. We suggest from these observations that deformation at 'high' normal stress ($\sigma_n \geq 25$ MPa) is achieved mainly through grain fracture whereas the dominant deformation mechanism at 'low' normal stresses ($\sigma_n = 5$ MPa) is a non-destructive process such as grain rolling or sliding. Moreover 'high' and 'low' normal stresses result in highly localized and homogeneously distributed shear respectively, consistent with the interpretations of friction velocity dependence (Fig. 3).

MAIR and MARONE (1999b) studied progressive deformation as a function of accumulated slip for identical tests conducted at 'high' normal stresses ($25 \leq \sigma_n \leq 70$ MPa) revealing the following. At 5 mm slip (i.e., within the velocity strengthening regime, Fig. 3), shear is distributed homogeneously throughout the granular layer whereas at 20 mm slip (i.e., the velocity weakening regime, Fig. 3) distinct fabrics highlighted by grain size reduction have developed indicating the transition to localized shear. This independent observation is consistent with the interpretation given above for the transition in friction velocity dependence.

The microstructural and mechanical observations correlate well, therefore we can study the influence of different deformation regimes and localization state on shear heating by comparing data from tests conducted at 'high' ($\sigma_n \geq 25$ MPa) and 'low' ($\sigma_n = 5$ MPa) normal stresses.

3.2 Heat Production

From equation (1) it is clear that heat production (q) should be affected by changes in applied normal stress or sliding velocity. We test this theory by independently varying sliding velocity and normal stress, focussing on two cases where the product $V\sigma_n$ is equal to 2.5×10^5 W/m^2 and 0.25×10^5 W/m^2, respectively. Figure 5 shows heat production (q) due to frictional heating (calculated as the product of shear stress and sliding velocity (τV) measured during constant velocity tests) as a function of shear displacement. After the first few mm of slip, the frictional work done (q) is comparable for tests with different sliding velocities and normal stresses but the same product ($V\sigma_n$). Two levels of 'steady-state' heat production, equal to 1.4×10^5 and 0.14×10^5 W/m^2, are associated with the two cases examined.

From section 3.1 we know that the deformation regime and degree of localization depend strongly on normal stress. By altering the normal stress we can study the non-fracture (5 MPa) and fracture (50 MPa) regimes. We can isolate the influence of deformation regime on shear heating (Table 1; Fig. 5) by varying normal stress (5, 50 MPa) and velocity (5, 0.5 mm/s) in a complementary way, such that the product remains constant (i.e., $\sigma_n V = 0.25 \times 10^5$ W/m^2), hence the work done is the same for both regimes. An important observation (Fig. 5) is that heat

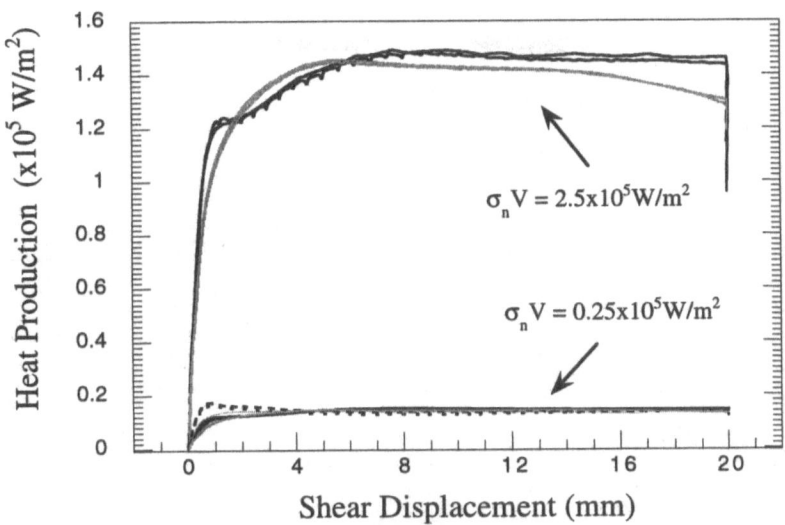

Figure 5

Heat production q ($= \tau V$) versus shear displacement for a range of tests. Dotted lines indicate $\sigma_n = 5$ MPa; black solid lines $\sigma_n = 25$ MPa; grey lines $\sigma_n = 50$ MPa. Sliding velocity is varied in a complementary way to give $V\sigma_n = 2.5 \times 10^5$ W/m² or 0.25×10^5 W/m² (see Table 1). After a few mm slip, heat production q is comparable for a given product irrespective of the individual values of V or σ_n.

production is similar for the fracture and non-fracture regimes. In the next section we will directly measure the temperature rise produced and thereby test if the heat production equation holds for all our tests.

3.3 Temperature Measurements: Velocity Stepping Tests

Temperature change is monitored throughout the tests to reveal temperature signals associated with differing heat production. In Figure 6, temperature change measured at thermocouple TC1 is plotted as a function of shear displacement where velocity is stepped between 0.3–3 mm/s and $\sigma_n = 25$–70 MPa. The overall trend shows temperature increasing with accumulated slip, consistent with continual shear heating throughout the test. There is a positive relation between the overall temperature rise and increasing normal stress.

Superimposed onto these are smaller temperature excursions associated with individual step changes in loading velocity (0.3–3 mm/s). The increase in temperature is greater and more rapid at higher velocity, consistent with the enhanced heat production at higher velocity described by equation (1). The temperature changes lag behind the velocity perturbations due to thermal inertia, finite diffusivity, and the distance between the fault zone and the thermocouple. This explains why the rapid increase in temperature associated with the faster (shaded) velocity reaches the sensor during the subsequent low velocity (unshaded) part. The rate of

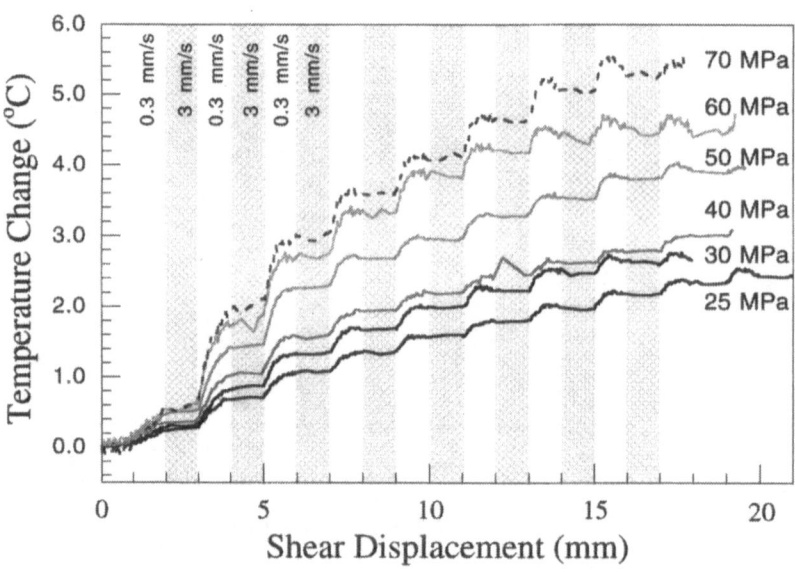

Figure 6

Temperature change (measured at TC1) as a function of shear displacement for velocity stepping tests where σ_n ranges between 25–70 MPa and velocity is stepped between 0.3–3 mm/s. High velocity periods are indicated by the shaded regions. Temperature perturbations lag the associated velocity steps. Note the effects of both normal stress and velocity perturbations on temperature change.

temperature rise (the slope of the perturbations) systematically increases with normal stress, consistent with results from bare surface experiments (BROWN, 1998). At any specific shear displacement, the temperature rise observed is systematically larger at higher normal stress. This is anticipated from equation (1) due to the enhanced heat production at high σ_n. For a given test, the temperature rise associated with an individual velocity step is largest at the beginning of the test and gradually decreases in size with increasing slip. A related observation is made for the cooling associated with a step down in velocity. Initially, heating is larger than cooling even during the slow velocity periods, however with additional slip, the heating decreases until net cooling during the slow steps is observed. This effect is enhanced at high normal stress.

3.4 Temperature Measurements: Constant Velocity Tests

Recognizing that the influences of individual velocity steps may be difficult to decompose we evaluate a simpler case, where V is constant throughout the test. Temperature is monitored simultaneously at several locations in order to determine heat conduction in the sample setup and to constrain modelling parameters for the heat-flow solution presented later. In these tests, we vary normal stress and velocity in a complementary manner such that the product is constant, focussing on two cases $V\sigma_n = 2.5 \times 10^5$ W/m^2 and $V\sigma_n = 0.25 \times 10^5$ W/m^2. We see from Figure 5, that heat production (q) for a constant $V\sigma_n$ product is comparable. Therefore, we can independently study the influence of normal stress and sliding velocity on temperature change for a given value of q (Table 1).

In Figure 7, temperature change is plotted as a function of time where conditions are as follows: a) $V = 5$ mm/s, $\sigma_n = 50$ MPa; b) $V = 10$ mm/s, $\sigma_n = 25$ MPa. Figure 8 shows two additional cases: a) $V = 1$ mm/s, $\sigma_n = 25$ MPa; and b) $V = 5$ mm/s, $\sigma_n = 5$ MPa.

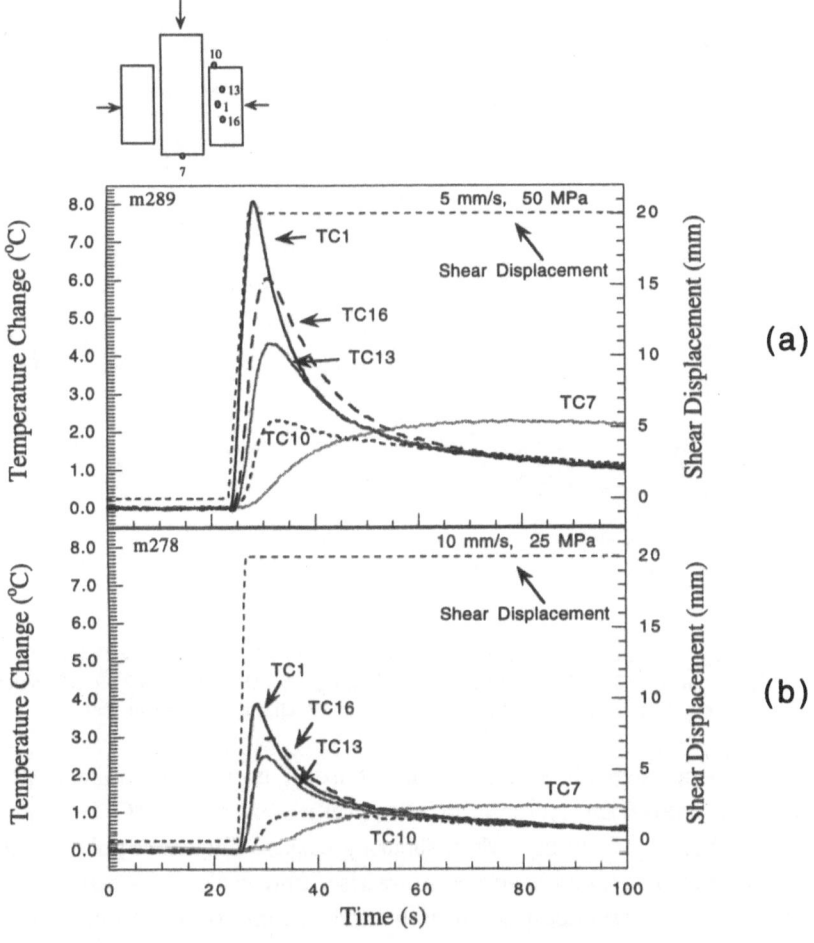

Figure 7

Temperature change measured at different sensor locations as a function of time for constant velocity tests where velocity and normal stress are: a) $V = 5$ mm/s, $\sigma_n = 50$ MPa; b) $V = 10$ mm/s, $\sigma_n = 25$ MPa. $V\sigma_n = 2.5 \times 10^5$ W/m^2 in both cases. Shear displacement is also plotted (dotted line) as a function of time. Thermocouple locations, shown in subplot, clearly show the significance of heat conduction through the sample. Sliding velocity strongly influences the rate of heating.

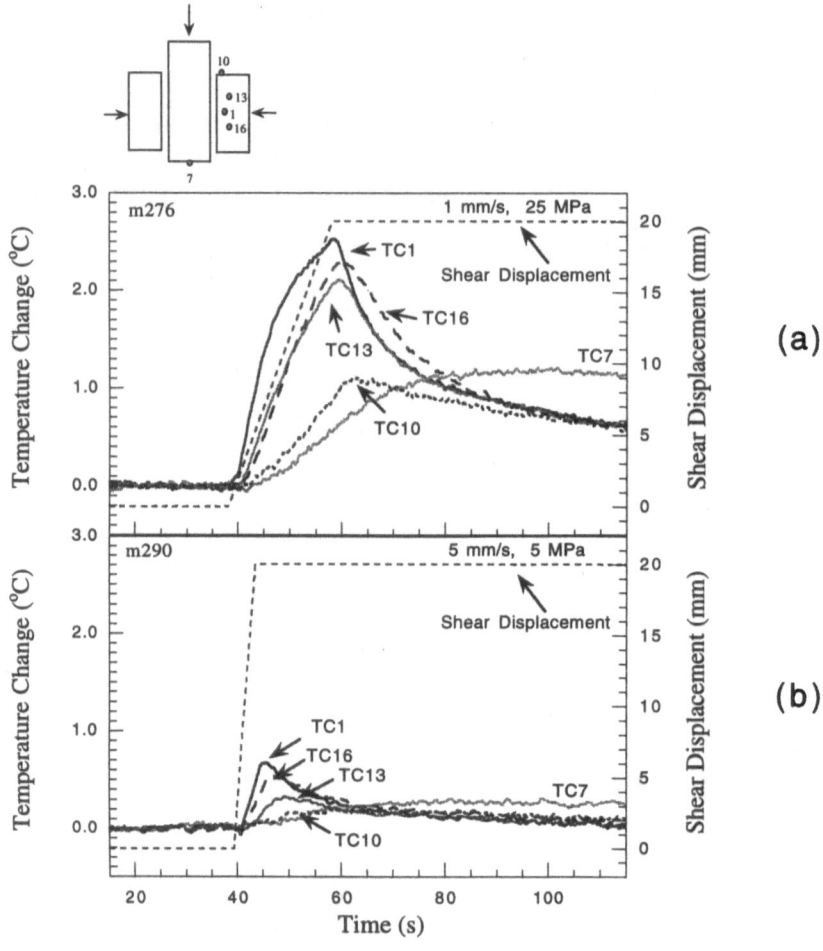

Figure 8
Temperature change as a function of time for constant velocity tests where velocity and normal stress are: a) $V = 1$ mm/s at $\sigma_n = 25$ MPa; and b) $V = 5$ mm/s, $\sigma_n = 5$ MPa. $V\sigma_n = 0.25 \times 10^5$ W/m^2 in both cases. Details are as in Figure 7 but note the change in vertical scale.

The main aspects of all four cases in Figures 7 and 8 are similar. All show a rapid (though nonlinear) increase in temperature of a few °C throughout the heating stage then rapid cooling when shearing ceases. Thermocouples farther from the granular layer have smaller temperature rises and peak temperature is achieved after a longer time interval consistent with expectations from thermal diffusion. An interesting observation is that the temperature rise at TC16 always exceeds that measured at TC13 although the thermocouples are the same distance from the fault zone. This is because the bottom of the middle block that has already participated in shearing is hotter than the top section that has not. As the middle block moves down it carries heat with it hence TC16 is hotter than TC13. Another point is that

TC7 eventually reaches higher temperatures than TC10 despite being farther from the source. Here we must invoke the contribution from the second layer. TC7, located equidistant from both fault zones, receives heat of the same amplitude at the same time from both layers. In contrast, the other sensors receive heat of different amplitudes from both layers at different times and hence the contribution of the second layer is generally delayed and significantly smaller.

The different values of heat productions in Figures 7 and 8 influence the temperature rise (note the change in scale). For a given heat production (i.e., Fig. 7 or 8, respectively), the rate of temperature rise is influenced by sliding velocity (i.e., plot a or b). Although the heat production is essentially comparable in a) and b), in each figure, the peak temperature change is larger in the lower velocity tests in each case. This is due to the finite time required for thermal diffusion through the sample. In the faster tests the duration of heating and hence the length of a test is shorter than the diffusion time required for the heat pulse to reach the sensor. Hence the peak temperature is apparently low. This effect is discussed below.

In summary, the total temperature rise for 20 mm slip (measured at TC1) is plotted versus normal stress in Figure 9. This plot highlights the influence of both normal stress and sliding velocity. Note in particular that an increase in normal

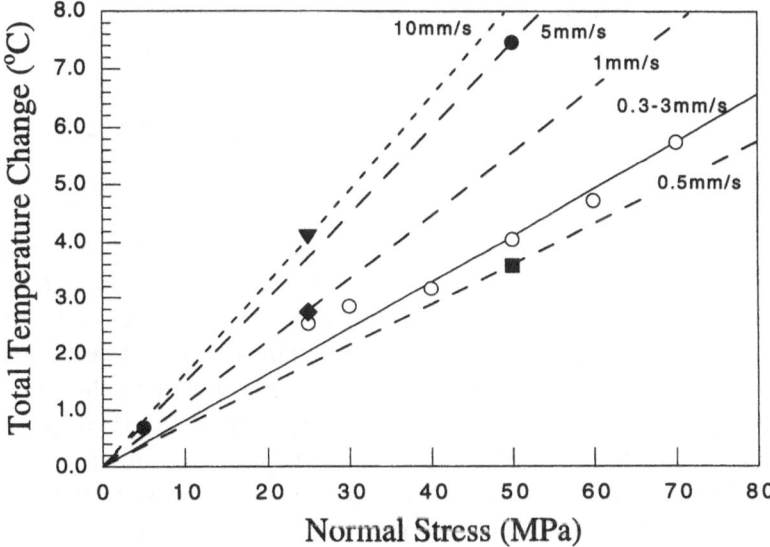

Figure 9
Total temperature change after 20 mm slip versus applied normal stress for both velocity stepping and constant velocity tests. All the data are measured at TC1. The solid line is best-fit regression to data (indicated by open circles) from test at range of σ_n between 25–70 MPa and $V = 0.3$–3 mm/s. Solid line is forced to cut origin since temperature rise will be zero for zero heat production. Other markers show total temperature change after 20 mm slip for a range of constant velocity tests. Dotted lines are visual guides linking data for a given velocity and the origin. Velocity conditions are marked. A systematic temperature increase is observed as a function of both increasing normal stress and velocity.

stress shown by open circles gives a linear increase in temperature. An increase in velocity yields a systematic although not linear increase in temperature. This is due to the finite time required for thermal diffusion.

4. Heat-flow Solution

4.1 Method

We use a 1-D heat-flow solution for frictional heating in a layer of finite thickness to investigate the variation in shear heating predicted for the conditions of our tests. The equation governing 1-D unsteady transport of heat can be written:

$$\frac{\partial T}{\partial t} = \frac{Q(x, t)}{\rho C} + \kappa \frac{\partial^2 T}{\partial x^2} \tag{2}$$

where T is temperature; t is time; ρ is density; C is specific heat; Q is the rate of heat generation per unit volume (i.e., $Q = q/w$); $\kappa = k/\rho C$ is thermal diffusivity where k is thermal conductivity; and x is distance. Equation (2) indicates that the rate of temperature increase at a distance x away from the source is a balance between a *heat production* term and a *heat conduction* term. We solve for temperature using the approach of CARDWELL et al. (1978) who consider that heating is uniform over a fault of finite width w:

$$Q(x_0, t_0) = \frac{\sigma_f D}{ws} \left[H\left(x_0 + \frac{w}{2}\right) - H\left(x_0 - \frac{w}{2}\right) \right], \qquad 0 < t_0 < s$$

$$= 0, \qquad\qquad\qquad\qquad\qquad\qquad t_0 < 0, t_0 > s. \tag{3}$$

In equation (3), σ_f is frictional stress on the fault; D is fault displacement; s is slip duration; and H is the Heavyside step function. We assume that granular layer thickness ($w/2 = 1$ mm) is constant throughout the test and that temperature change is averaged over this entire layer. We choose material constants ($k = 9.3$ W/(m · C); $\kappa = 7.8 \times 10^{-7}$ mm^2/s) which lie between tabulated values for steel and sandstone and give the best fit to our experimental data. Heat production is calculated from equation (3) using our measured values of friction ($\mu = 0.6$), normal stress (σ_n), distance from the center of the fault zone (x) and velocity (V) (Table 2). Simulations are designed to reproduce the conditions of our experiments (Figs. 6, 7, 8) and yield temperature change as a function of time (or slip).

4.2 Simulation Results

Simulations were run duplicating the range of stress conditions and velocity stepping histories of the data presented in Figure 6. The temperature rise 2 mm from the center of the layer (i.e., the position of TC1) is plotted as a function of

Table 2

1-D shear heating simulations

V, mm/s	σ_n, MPa	x, mm	t, s
0.3–3	25	2	29
0.3–3	50	2	29
0.3–3	70	2	29
0.5	50	2	40
1	25	2	20
5	5	2	4
5	50	2	4
10	25	0	2
10	25	2	2
10	25	3	2
10	25	4	2
10	25	6	2

x is the distance from center of the fault zone (i.e., the heat source) to the sensor.

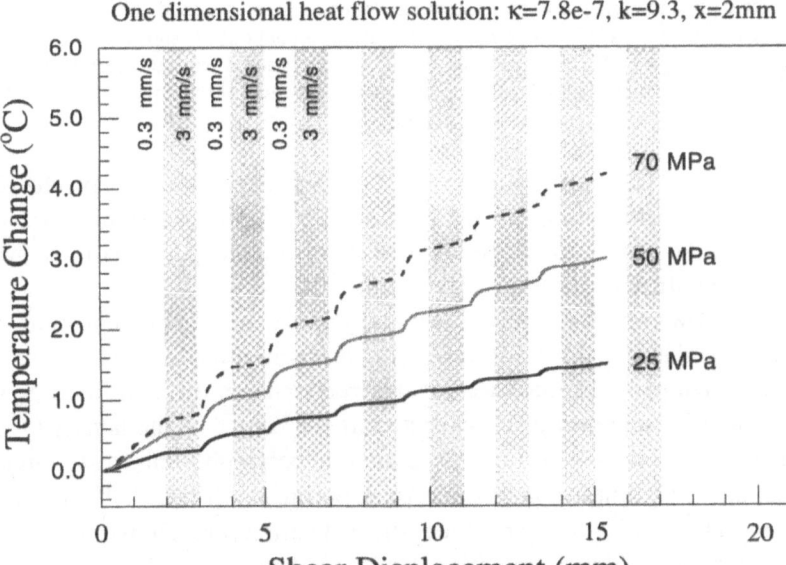

Figure 10

One-dimensional heat-flow solution showing temperature (associated with step changes in heat production) versus shear displacement. Conditions are $V = 0.3–3$ mm/s, $\sigma_n = 25, 50, 70$ MPa. Temperature change is calculated 2 mm from the center of the fault zone. The high velocity periods are indicated by the shading. Note the overall influence of normal stress and also how the shape of the temperature perturbations changes with increasing slip. The thermal properties used in the simulation are indicated in the text.

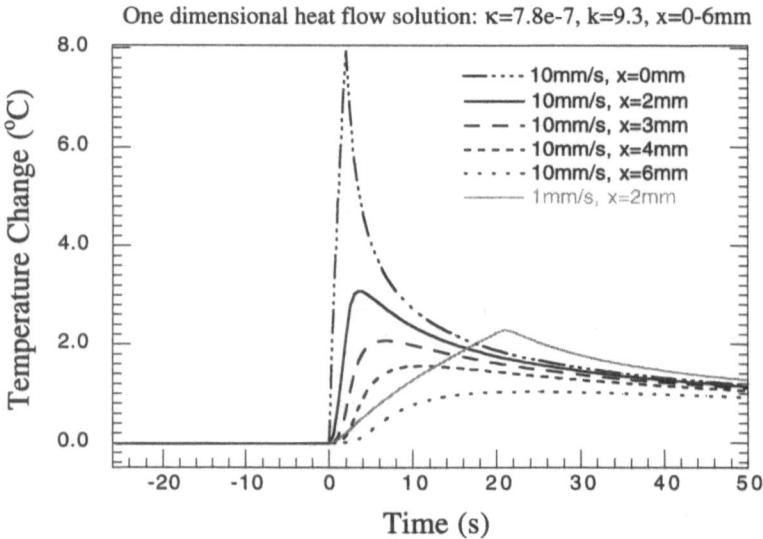

One dimensional heat flow solution: κ=7.8e-7, k=9.3, x=0-6mm

Figure 11

Temperature change versus time for shear heating within a layer of finite thickness (2 mm) using parameters documented in Table 2. These 1-D shear heating simulations are shown for the two velocities $V = 1$, 10 mm/s, and $\sigma_n = 25$ MPa for varying distances (x) measured from the center of the granular layer. One simulation calculates the temperature change at the center of the 2 mm wide layer (i.e., $x = 0$ mm).

time in Figure 10. The main features of the data in Figure 6 are reproduced. Also apparent are the perturbations in temperature associated with individual velocity steps. Importantly, for a given test, these temperature perturbations gradually decrease in size with increasing displacement.

Figure 11 shows temperature rise versus time for a series of constant velocity simulations. The distance from the center of the fault zone (x) is varied (2–6 mm) to simulate the temperature measured at different sensors. In addition, one simulation is run for the temperature rise expected at the center of the layer ($x = 0$ mm). Heat production is constant during a time interval corresponding to 20 mm slip and is zero for all other times, comparable to experiments (Figs. 7, 8). Heating and cooling associated with the period of shear heating are shown in all curves. Maximum temperature decreases with the distance from the fault zone and the time required to reach maximum temperature increases systematically with distance. Results are qualitatively comparable to data presented in Figures 7, 8. The temperature change expected at the center of the layer is ∼ 8°C. If the layer, where heat is produced, was thinner e.g., a shear band ∼ 0.1 mm thick as suggested from microstructural observations (Fig. 4b), temperature rise would instead be ∼ 12°C.

The total temperature changes obtained from simulations of temperature change (at TC1) following 20 mm slip are plotted in Figure 12 (solid symbols) as a function of steady-state heat production. Experimental data (from TC1) for the relevant

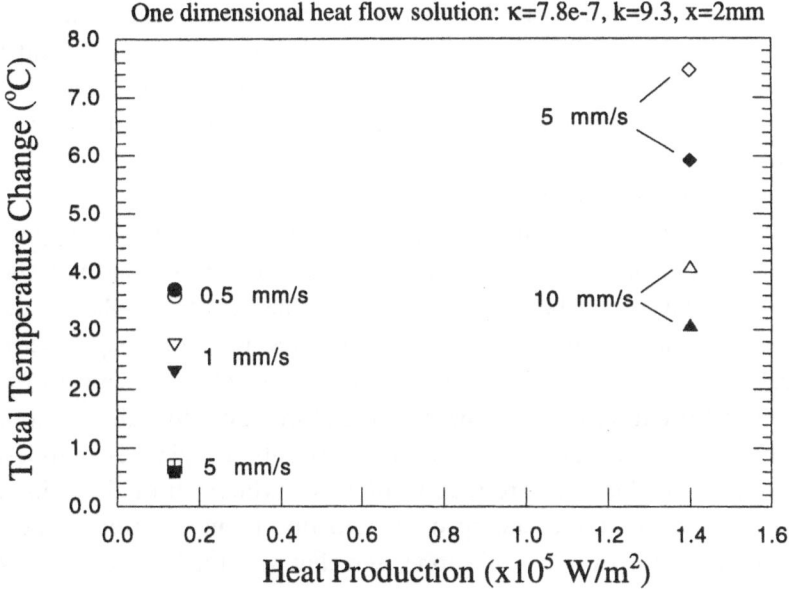

Figure 12

Total temperature change during 20 mm slip versus heat production (q). Experimental data shown as open symbols for the conditions annotated. One-dimensional heat-flow solutions for the same conditions are indicated by the corresponding solid markers. Thermal properties used in the simulations are shown. In general, the model approximates the data well.

conditions (open symbols) are superimposed. For a given heat production the temperature change is smaller for high velocity tests due to the finite time required for thermal diffusion through the sample. In general the one-dimensional heat flow results are in good qualitative agreement with our experimental data (Fig. 12). The agreement between the data and models is best for the low q case, however the model underpredicts data for higher q. This likely due to the 1-D nature of the model. A more complex model would be required to fit other aspects of the data, however, given the focus here on the experimental data, additional modelling was considered beyond the scope of the current work.

5. Discussion

We have carried out experiments on granular layers over a wide range of conditions where we measure the quantities important in the heat production equation. Our study shows that temperature increases systematically as a function of both normal stress and sliding velocity, as observed previously in granular layers at lower normal stress conditions (LOCKNER and OKUBO, 1983) and on bare granite surfaces (BLANPIED et al., 1998; BROWN, 1998). We verify that the heat production equation holds for all tests. The conditions studied span regimes where deformation is interpreted to be dominated by very different mechanisms of grain fracture ($\sigma_n \geq 25$ MPa) and grain rolling ($\sigma_n = 5$ MPa). Interestingly, heat production and temperature rise are comparable in both regimes. In addition, we investigate the influence of distributed versus localized slip. The degree of localization has no dramatic effect on heating. Moreover, we find no evidence for dramatic velocity weakening at high velocity or normal stress.

The main aspects of the temperature data for all our tests are well approximated by a simple one-dimensional heat-flow solution (Fig. 12). Importantly, tests at high and low normal stress can be described by the same heat-flow solution that lacks any complexities in microstructure. This indicates that the bulk thermal and mechanical properties of a granular layer are comparable irrespective of localization state or deformation regime. The data therefore imply that the frictional work done in shearing is equivalent whether the deformation regime is non-destructive (e.g., rolling) or involves the production of new surfaces through grain fracture. Hence we conclude that the energy dissipated due to new surface generation is minor compared to the overall work done, agreeing with the work of LOCKNER and OKUBO (1983).

A possible alternative explanation is that energy is dissipated differently at low normal stress. In previous studies on bare surfaces (BLANPIED et al., 1998; BEELER et al., 1996), the work done due to fault dilation was essentially ignored since dilation was indeed minimal, however faults containing granular material show significantly larger dilation. If fault zone dilation is enhanced more at low normal stress as preliminary observations of granular layers suggest (MAIR and MARONE, manuscript in preparation), more work may be required to achieve this dilation. This energy could in theory compensate for the energy involved in grain fracture at high normal stress and alleviate the necessity that fracture is a minor energy sink. This model is speculative and clearly requires quantitative testing.

Our friction and microstructural data are consistent with previous observations (e.g., MARONE et al., 1990; BEELER et al., 1996) indicating that velocity strengthening is associated with distributed shear and velocity weakening is associated with localized deformation, respectively. An additional result is an association between a broad particle size distribution and localization (fracture regime, Fig. 4b) compared to a narrow range of grain sizes associated with distributed deformation (rolling regime, Fig. 4a). These observations correspond to a conceptual model based on numerical studies by MORGAN and BOETTCHER (1999) who propose that a broad grain size distribution (or larger number of grains) may assist shear localization due to a 'lubrication' effect of small grains acting as rollers. For a distribution of similarly sized grains (or a small number of grains) they anticipate the entire granular layer deforming by distributed shear.

Extrapolation of our laboratory results to natural fault systems is speculative due to differences in loading conditions, e.g., net slip. We can, however, directly compare our work to recent numerical studies investigating shear in granular layers for a similar range of conditions and deformation regimes (e.g., MORA and PLACE, 1998, 1999). MORA and PLACE (1998) suggest significant weakening and reduced shear heating is associated with a switch from sliding to rolling mechanisms. In our experiments, we interpret grain rolling to be an important, perhaps the dominant deformation mechanism operating at low normal stress. However, we see no evidence for weakening or low heat production as proposed by MORA and PLACE (1998). This may indicate that grain rolling is not the sole deformation mechanism operating in our tests at low normal stress and that deformation is accommodated by a combination of rolling and sliding. MORA and PLACE (1999) state that the weakening observed in their numerical simulations is due to a self-organization process which appears to be localized slip. We show that dramatic weakening does not occur in our experiments for either localized or distributed slip. Thus, the particular self-organization processes thought to be responsible for weakening in the numerical experiments (MORA and PLACE, 1999) may not be occurring in our laboratory tests. In summary, based on our study of shear heating in granular layers, we conclude that fault strength and heat flow are consistent with theoretical predictions for a wide range of slip velocities and effective stresses.

6. Conclusions

The heat production equation holds for all conditions studied for both localized and distributed shear. For a given q, the heat generated is similar in regimes where fracture and non-fracture are thought to dominate. The presence or absence of localized shear structures has little influence on heat production or temperature rise. If our results can be extrapolated to the San Andreas Fault, they would predict a strong fault with significant heating and hence additional mechanisms must operate to explain the heat-flow paradox.

Acknowledgements

We are grateful to U. Mok, S. Karner, K. Frye, N. Beeler and M. Blanpied for scientific discussions and technical expertise that benefited this study. N. Chatterjee provided help with the SEM work. We thank reviewers and editors for comments that improved the manuscript. This work was funded by National Science Foundation Grant EAR–9805327 and Petroleum Research Foundation Grant 33306–AC2.

REFERENCES

BEELER, N. M., TULLIS, T. E., BLANPIED, M. L., and WEEKS, J. D. (1996), *Frictional Behavior of Large Displacement Faults*, J. Geophys. Res. *101*, 8697–8715.

BLANPIED, M. L., LOCKNER, D. A., and BYERLEE, J. D. (1995), *Frictional Slip of Granite at Hydrothermal Conditions*, J. Geophys. Res. *100*, 13,045–13,064.

BLANPIED, M. L., TULLIS, T. E., and WEEKS, J. D. (1998), *Effects of Slip, Slip Rate and Shear Heating on the Friction of Granite*, J. Geophys. Res. *103*, 489–511.

BROWN, S. R. (1998), *Frictional Heating on Faults: Stable Sliding versus Stick Slip*, J. Geophys. Res. *103*, 7413–7420.

CARDWELL, R. K., CHINN, D. S., MOORE, G. F., and TURCOTTE, D. L. (1978), *Frictional Heating on a Fault Zone with Finite Thickness*, Geophys. J. R. Astron. Soc. *52*, 525–530.

CHESTER, F. M., EVANS, J. P., and BIEGEL, R. L. (1993), *Internal Structure and Weakening Mechanisms of the San Andreas Fault*, J. Geophys. Res. *98*, 771–786.

GOLDSBY, D. L., and TULLIS, T. E. (1998), *Experimental Observations of Frictional Weakening During Large and Rapid Slip* (Abstract), EOS Trans. AGU *79*, Fall Meet. Suppl., F610.

LACHENBRUCH, A. T., and SASS, J. H. (1980), *Heat Flow and Energetics of the San Andreas Fault Zone*, J. Geophys. Res. *85*, 6185–6222.

LOCKNER, D. A., and OKUBO, P. G. (1983), *Measurements of Frictional Heating in Granite*, J. Geophys. Res. *88*, 4313–4320.

LOCKNER, D. A., SUMMERS, R., and BYERLEE, J. D. (1986), *Effects of Temperature and Sliding Rate on Frictional Strength of Granite*, Pure appl. geophys. *124*, 445–469.

MAIR, K., and MARONE, C. (1999a), *Friction of Simulated Fault Gouge for a Wide Range of Velocities and Normal Stresses*, J. Geophys. Res. *104*, 28,899–28,914.

MAIR, K., and MARONE, C. (1999b), *Frictional and Microstructural Observations of Fault Gouge at High Velocity* (Abstract), EOS Trans. AGU *80*, Spring Meet. Suppl., S329.

MARONE, C. (1998), *Laboratory-derived Friction Laws and their Application to Seismic Faulting*, Ann. Rev. Earth Planet. Sci. *26*, 643–696.

MARONE, C., and KILGORE, B. (1993), *Scaling of Critical Slip Distance for Seismic Faulting with Shear Strain in Fault Zones*, Nature *362*, 618–621.

MARONE, C., RALEIGH, C. B., and SCHOLZ, C. H. (1990), *Frictional Behavior and Constitutive Modelling of Simulated Fault Gouge*, J. Geophys. Res. *95*, 7007–7025.

MCKENZIE, D. P., and BRUNE, J. N. (1972), *Melting of Fault Planes during Large Earthquakes*, Geophys. J. R. Astron. Soc. *29*, 65–78.

MORA, P., and PLACE, D. (1998), *Numerical Simulation of Earthquake Faults with Gouge: Toward a Comprehensive Explanation for the Heat-flow Paradox*, J. Geophys. Res *103*, 21,067–21,089.

MORA, P., and PLACE, D. (1999), *The Weakness of Earthquake Faults*, Geophys. Res. Lett. *26*, 123–126.

MORGAN, J. K., and BOETTCHER, M. S. (1999), *Numerical Simulations of Granular Shear Zones Using the Distinct Element Method: I Shear Zone Kinematics and the Micromechanics of Localization*, J. Geophys. Res. *104*, 2703–2719.

SCHOLZ, C. H. (1996), *Faults without Friction*, Nature *381*, 556–557.

SCHOLZ, C. H. (2000), *Evidence for a Strong San Andreas Fault*, Geology *28*, 163–166.

SIBSON, R. H. (1975), *Generation of Pseudotachylytes by Ancient Seismic Faulting*, Geophys. J. R. Astron. Soc. *43*, 775–794.

SPRAY, J. G. (1993), *Viscosity Determinations of Some Frictionally Generated Silicate Melts: Implications for Fault Zone Rheology at High Strain Rates*, J. Geophys. Res. *98*, 8053–8068.

STESKY, R. M. (1978), *Rock Friction—Effect of Confining Pressure, Temperature and Pore Pressure*, Pure appl. geophys. *116*, 690–704.

TSUTSUMI, A., and SHIMAMOTO, T. (1997), *High-velocity Frictional Properties of Gabbro*, Geophys. Res. Lett. *24*, 699–702.

YOSHIOKA, N. (1985), *Temperature Measurements during Frictional Sliding of Rocks*, J. Phys. Earth *33*, 295–322.

YOSHIOKA, N. (1986), *Fracture Energy and the Variation of Gouge and Surface Roughness during Frictional Sliding of Rocks*, J. Phys. Earth *34*, 335–355.

(Received November 8, 1999, revised March 3, 2000, accepted April 15, 2000)

Pure appl. geophys. 157 (2000) 1867–1887
0033–4553/00/121867–21 $ 1.50 + 0.20/0

❘Pure and Applied Geophysics

Extension of the Lattice Solid Model to Incorporate Temperature Related Effects

STEFFEN ABE,[1] PETER MORA[1] and DAVID PLACE[1]

Abstract—The elastic and frictional properties of solids are temperature-dependent. Thus, heat has without doubt a major influence on the dynamics of earthquakes, particularly considering the high temperatures generated during large slip events. In order to provide a foundation for the study of these heat related effects, the Lattice Solid Model for the study of earthquake dynamics is extended to incorporate the generation and transfer of heat. The thermal and elastic properties of 2- and 3-D lattice solids in the macroscopic limit are derived. To verify the numerical implementation of heat transfer, a simulation has been performed in a simple case and the results compared to a known analytical solution for the same problem. Thermal expansion and a simple approximation of a temperature-dependent pore fluid pressure are implemented in the 2-D Lattice Solid Model. Simulations confirm that these effects influence the dynamics of the slip of a fault with fault gouge. Whereas thermal expansion has only minor influence on the dynamics of fault rupture, the influence of the increase in the pore fluid pressure generated by slip heating is more significant. The simulations show that the temperatures generated during slip events accord with those expected for real earthquakes as inferred from geologic evidence.

Key words: Lattice Solid Model, earthquake dynamics, simulation, thermal effects.

1. Introduction

The importance of thermal effects on the dynamics of earthquakes has been pointed out by various authors (SIBSON, 1975; LACHENBRUCH, 1980; SHAW, 1995). The Lattice Solid Model (MORA and PLACE, 1994, 1998; PLACE and MORA, 1999), a particle-based numerical model for the simulation of earthquake dynamics, has been used successfully to investigate various dynamic features of earthquakes, such as the weakness of earthquake faults and the anomalously low heat flow on the San Andreas fault (MORA and PLACE, 1998; PLACE and MORA, 1999). Here the Lattice Solid Model has been extended to include the generation and transport of heat and two of these temperature related effects: thermal expansion and the increase of pore fluid pressure caused by heat.

---
[1] QUAKES, The University of Queensland, Qld 4072, Brisbane, Australia. E-mail: steffen@ quakes.earth.uq.edu.au, mora@quakes.earth.uq.edu.au, place@quakes.earth.uq.edu.au

2. Overview of the Lattice Solid Model

The Lattice Solid Model (MORA and PLACE, 1994; PLACE and MORA, 1999) consists of spherical particles interacting with their nearest neighbors by elastic and frictional forces. The particles can be linked together, in which case the elastic forces are attractive or repulsive, depending on whether the particles are closer or more distant than the equilibrium distance r_0 as given by Equation (1)

$$\mathbf{F} = \begin{cases} k(r - r_0)\mathbf{e} & r \leq r_{\text{cut}} \\ 0 & r > r'_{\text{cut}} \end{cases}, \tag{1}$$

where k is the spring constant for the elastic interaction between the particles, r is the distance between the particles, r_0 the equilibrium distance, r_{cut} the breaking separation for bonds and $\mathbf{e}$ is a unit vector in the direction of the interaction. The basic structure of the lattice and the directions of the vectors $\mathbf{e}^\alpha$ of links for a 2-D and a 3-D triangular lattice are shown in Figures 1 and 2. Those links are broken if the distance between the particles exceeds a threshold breaking distance r_{cut}. If two particles are not linked together the elastic force between the particles is purely repulsive

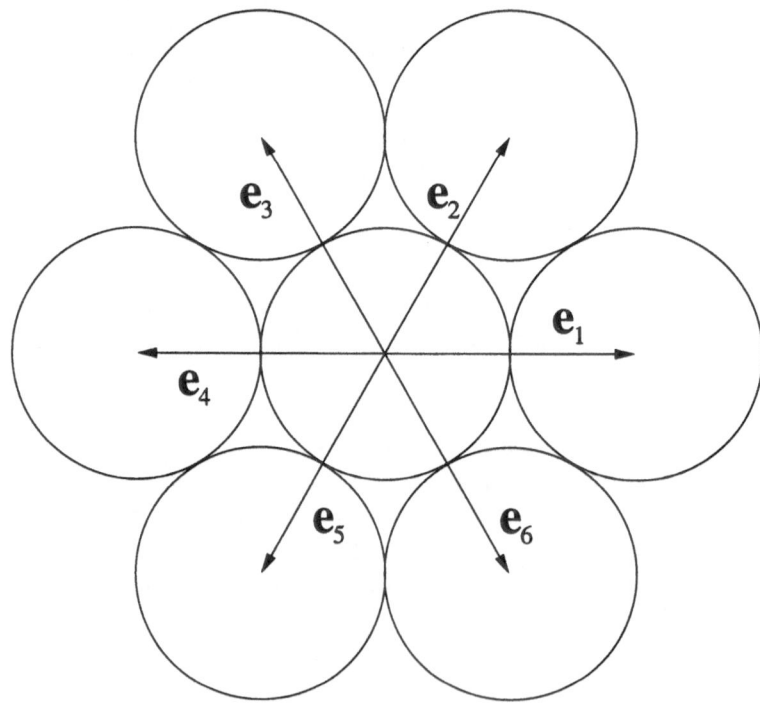

Figure 1
Particles and direction vectors $\mathbf{e}^\alpha$ for a 2-D triangular lattice ($\alpha = 1, 6$).

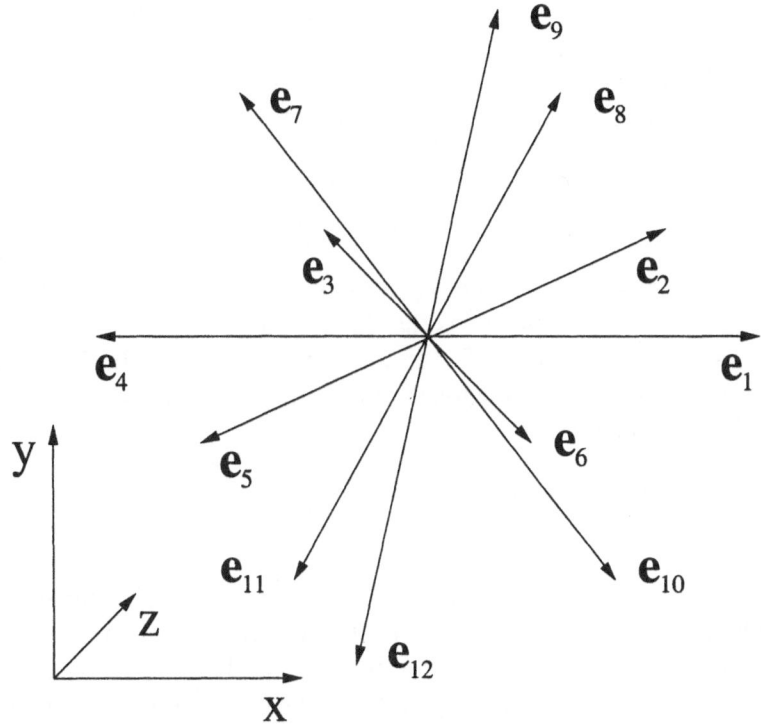

Figure 2
Direction vectors e^α for a 3-D close-packed lattice ($\alpha = 1, 12$).

$$\mathbf{F} = \begin{cases} k(r - r_0)\mathbf{e} & r \le r_0 \\ 0 & r > r_0 \end{cases}.$$

(2)

To avoid a buildup of kinetic energy in the closed model originating from waves reflected at the boundaries of the model, or circulating in the model in the case of periodic boundary conditions, an artificial viscosity is introduced (MORA and PLACE, 1994).

An intrinsic friction between particles is incorporated in the model (PLACE and MORA, 1999). Two unbonded interacting particles can be in static or dynamic frictional contact. If they are in static contact, no movement between the particles takes place until the shear force overcomes the static frictional force and the interaction between the particles changes to a dynamic frictional contact. In this state the particles are moving past each other and a dynamic frictional force opposes this movement. The action of this force is released as heat (PLACE and MORA, 1999).

3. Implementation of Heat Transport in the Lattice Solid Model

Using a Cartesian coordinate system, the behavior of temperature in a solid with heat sources can be described by the heat flow equation

$$\frac{\partial T}{\partial t} = \kappa_{ij} \frac{\partial^2 T}{\partial x_i \, \partial x_j} + \frac{1}{\rho c_p} H(\mathbf{x}, t), \tag{3}$$

where ρ is the density of the solid, c_p is the heat capacity, $H(\mathbf{x}, t)$ the local heat production or source term and κ_{ij} is the thermal diffusivity tensor. The x_i refer to the x- and y-dimensions in the 2-D case and the x, y and z in the 3-D case. In case of an isotropic medium, κ_{ij} is diagonal, consequently a scalar value κ can be used and Equation (3) simplifies to

$$\frac{\partial T}{\partial t} = \kappa \frac{\partial^2 T}{\partial \mathbf{x}^2} + \frac{1}{\rho c_p} H(\mathbf{x}, t), \tag{4}$$

where the source term in the Lattice Solid Model is the frictional heat being generated by particles rubbing together. In the numerical simulation, the solution of Equation (4) is split into two steps. First, the heat produced by interparticle friction is added to the particles and subsequently the heat is extrapolated in time by an explicit finite difference method.

$$T'^I(t) = T^I(t) + \frac{1}{\rho c_p} H^I(t) \tag{5}$$

$$T^I(t + \Delta t) = T'^I(t) + \xi \kappa \Delta t \sum_j \frac{T^J(t) - T^I(t)}{(r^{IJ})^2}, \tag{6}$$

where $T^I(t)$ is the temperature of particle I at time t, $T^I(t + \Delta t)$ is the temperature at time $t + \Delta t$, and r^{IJ} is the distance between touching particles I and J. This method, which is equivalent to the well known FTCS Finite Difference scheme (forward time, centered space), is numerically stable under the condition $\kappa \Delta t / (r^{IJ})^2 < \frac{1}{2}$. The constraint $\Delta t \leq \varepsilon (r_0/v_p)$ (e.g., MORA and PLACE, 1994; PLACE and MORA, 1999) is imposed on the time step by the computation of the elastic interactions in the model. Typically $\varepsilon < 0.2$ must be chosen to ensure numerical stability and sufficient accuracy of the dynamic computations. Considering the thermal diffusivity of typical rocks is approximately $\kappa = 10^{-6}$ m^{2}s^{-1}, and assuming a P-wave velocity $v_p = 5000$ ms^{-1}, this guarantees that the stability condition is satisfied for $r_0 \gtrsim 4 \cdot 10^{-11}$ m which is always the case (ABE *et al.*, 1998).

The only heat source implemented in the model is frictional heating, i.e., there is no heat generated by the dissipation of elastic wave energy in the present version of the model.

3.1 Scaling Relations

In order to compare the results obtained by simulations in arbitrary, numerically convenient model units with real earthquake processes, a conversion between model units and real world units is necessary. The variables with the tilde ($\sim$) will refer to the values in model units and those without to the values in real world (e.g., MKS) units.

The model contains three thermal parameters: the heat capacity c_p, the thermal diffusivity κ and the temperature T. From the definition of kinetic and thermal energy

$$E_{th} = mc_p T \tag{7}$$

$$E_{kin} = \frac{m}{2} v^2, \tag{8}$$

a scaling relation can be obtained for T and c_p

$$\frac{\tilde{T}}{T} = \left(\frac{\tilde{v}}{v}\right)^2 \frac{c_p}{\tilde{c}_p}. \tag{9}$$

From Fourier's Law for the 1-D heat flow q through an area A in an isotropic solid (e.g., HOLMAN, 1990)

$$q = \kappa \rho c_p A \frac{\partial T}{\partial x}, \tag{10}$$

and the heat flow through an area in the lattice, a second scaling relation can be obtained. The heat flow through an area in the lattice can be written as the sum of the heat flows through the particles in this area

$$\tilde{q} = \sum_i \tilde{k} \tilde{\varrho} \tilde{c}_p \frac{d\tilde{T}_i}{d\tilde{x}_i}, \tag{11}$$

where $d\tilde{T}_i$ is the temperature difference between the particles and $d\tilde{x}_i$ is the separation of the particles which, combined with Equation (10), yields the second scaling relation

$$\frac{\tilde{\kappa}}{\kappa} = \zeta \frac{\tilde{v}}{v} \frac{\tilde{r}_0}{r_0} \frac{\tilde{\rho}\tilde{c}_p}{\rho c_p}, \tag{12}$$

where ζ is a factor which takes into account the geometric properties of the lattice. In the current simulations we set $\tilde{\rho}\tilde{c}_p = 1$ for numerical convenience so Equation (12) simplifies to

$$\frac{\tilde{\kappa}}{\kappa} = \zeta \frac{\tilde{v}}{v} \frac{\tilde{r}_0}{r_0} \frac{1}{\rho c_p}. \tag{13}$$

The two equations given by (13) and (9) are sufficient to determine the scaling for the two thermal parameters T and k from the scaling of the mechanical parameters v and r_0.

4. Thermal and Elastic Properties of the Lattice Solid in the Macroscopic Limit

In this section the following summation convention is used: If an index occurs multiple times in a term, summation over this index is implicitly assumed, except if the other side of the equation contains terms where this index occurs without summation over it.

4.1 Thermal Properties.

Heat flow in a general solid without heat sources is given by

$$\frac{\partial T}{\partial t} = q = \kappa_{ij} \frac{\partial^2 T}{\partial x_i \, \partial x_j}, \tag{14}$$

where q is the heat flow and κ_{ij} is the tensor of thermal diffusivity. In a lattice, the heat flow is given by

$$q = \sum_\alpha q^\alpha = \sum_\alpha \kappa^\alpha \frac{(T - T^\alpha)}{r_\alpha^2}, \tag{15}$$

where q^α is the heat flow in direction α, κ^α the heat diffusivity in that direction and T^α the temperature of the particle in that direction.

The temperature of the particles can be expanded relative to continuous differentiable temperature field T by

$$T^\alpha = T + r e_k^\alpha \frac{\partial T}{\partial x_k} + \frac{r^2}{2!} e_k^\alpha e_\ell^\alpha \frac{\partial^2 T}{\partial x_k \, \partial x_\ell} \cdots . \tag{16}$$

Using Equation (15) and (16) and ignoring higher order terms, the heat flow at a particle can be written as

$$q = \kappa^\alpha r e_k^\alpha \frac{\partial T}{\partial x_k} + \kappa^\alpha \frac{r^2}{2!} e_k^\alpha e_\ell^\alpha \frac{\partial^2 T}{\partial x_k \, \partial x_\ell}. \tag{17}$$

Assuming that all κ^α are equal, this expression simplifies to

$$q = \kappa r \left[\sum_\alpha e_k^\alpha \frac{\partial T}{\partial x_k} + \frac{r}{2} e_k^\alpha e_\ell^\alpha \frac{\partial^2 T}{\partial x_k \, \partial x_\ell} \right]. \tag{18}$$

Using the geometric relations for a triangular lattice in 2-D (FRISCH *et al.*, 1986a) and 3-D

$$\sum_\alpha e_i^\alpha = 0 \tag{19}$$

$$e_i^\alpha e_j^\alpha = \begin{cases} 3\delta_{ij} & \text{2-D lattice} \\ 4\delta_{ij} & \text{3-D lattice} \end{cases}, \tag{20}$$

we obtain the relation

$$q = \xi\tilde{\kappa}r\delta_{ij}\frac{\partial^2 T}{\partial x_i\,\partial x_j}, \tag{21}$$

where the $\xi = 1.5$ in the 2-D case and $\xi = 2$ in 3-D. Thus, the heat flow is isotropic in both lattices. The factor ξ is introduced by the lattice geometry and imposes that $\tilde{\kappa} = \kappa/3$.

4.1.1 Numerical verification test

To verify that the implementation of the 3-D-triangular lattice solid converges to the relation derived above (Equation 21), the analytic solution for a known heat flow problem was compared to the numerical solution for the same problem using different lattice sizes.

For a temperature distribution $T(x, t)$ in a homogeneous medium the extension l in x-direction given by the initial condition

$$T(x, 0) = \sin\left(\frac{\pi x}{l}\right), \tag{22}$$

and boundary conditions

$$T(0, t) = 0 \tag{23}$$

$$T(l, t) = 0, \tag{24}$$

the temperature $T(x, t)$ can be obtained analytically (CARSLAW and JAEGER, 1959) and is given by

$$T(x, t) = \sin\left(\frac{\pi x}{l}\right)e^{-\pi^2\kappa t/l^2}. \tag{25}$$

The tests with lattice sizes in the x-dimension of 64 to 512 particles indicate that the numerical solution converges towards the analytical solution with increasing lattice size (Fig. 3).

4.2 Elastic Properties of the Lattice Solid

Assuming it is possible to expand the displacement of particles in terms of a continuous differentiable displacement field u_i, we can write

$$u_i^\alpha = u_i + r e_k^\alpha \frac{\partial u_i}{\partial x_k} + \frac{r^2}{2!} e_k^\alpha e_\ell^\alpha \frac{\partial^2 u_i}{\partial x_k \, \partial x_\ell} + \cdots, \tag{26}$$

where r is the equilibrium distance between particles. It should be noted that in Equation (26) the third term on the right-hand side is not summed over α because the left-hand side u_i^α contains α as an index. Within the limit of small displacements relative to r, the total force on a given particle is therefore

$$F_i = \sum_\alpha F_i^\alpha = k^\alpha (u_j^\alpha - u_j) e_j^\alpha e_i^\alpha, \tag{27}$$

where u_j^α is the displacement of the particle located along the α-direction relative to the particle being considered, and k_α is the microscopic spring constant in that direction. Hence, using Equation (26) we obtain

$$F_i = k^\alpha e_i^\alpha e_j^\alpha e_\ell^\alpha \frac{\partial u_j}{\partial x_\ell} + k^\alpha \frac{r^2}{2!} e_i^\alpha e_j^\alpha e_k^\alpha e_\ell^\alpha \frac{\partial^2 u_j}{\partial x_k \, \partial x_\ell}. \tag{28}$$

For all lattices considered here, the geometric relation (FRISCH *et al.*, 1986a)

$$e_i^\alpha e_j^\alpha e_k^\alpha = 0, \tag{29}$$

is satisfied. Thus, Equation (28) simplifies to

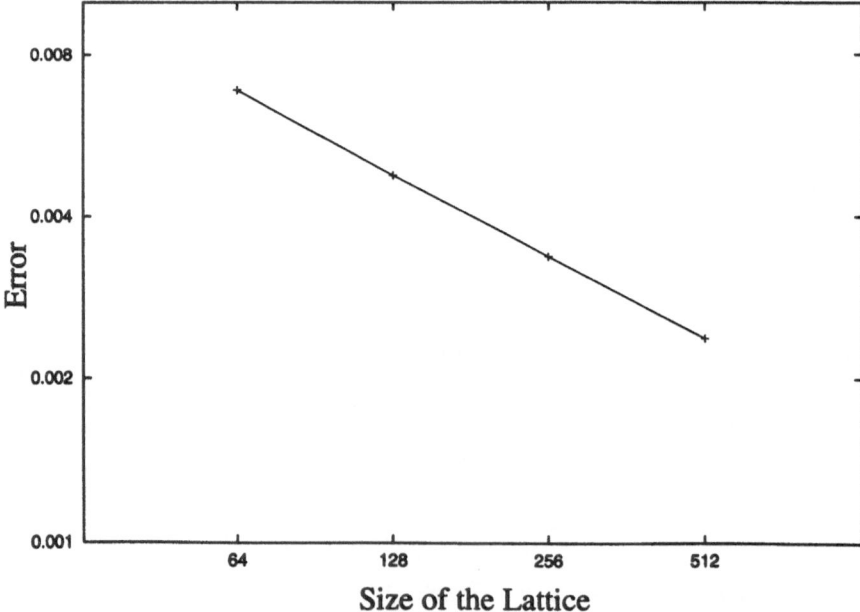

Figure 3
Log-log plot of the cumulative error after 1000 time steps depending on the lattice size.

$$F_i = \tilde{k}^\alpha \frac{r^2}{2!} e_i^\alpha e_j^\alpha e_k^\alpha e_\ell^\alpha \frac{\partial^2 u_j}{\partial x_k \, \partial x_\ell}. \tag{30}$$

Since

$$a_i = \frac{F_i}{m} = \frac{c_{ijk\ell}}{\rho} \frac{\partial^2 u_j}{\partial x_k \, \partial x_\ell}, \tag{31}$$

this expression coupled with the relation between mass and density allows us to determine the relation between the microscopic spring constant k and the macroscopic Hooke's tensor

$$c_{ijk\ell} = \frac{1}{\rho} \tilde{k}^\alpha \frac{r^2}{2!} e_i^\alpha e_j^\alpha e_k^\alpha e_\ell^\alpha, \tag{32}$$

where $\tilde{\rho}$ is the lattice density.

Given the Hooke tensor, the velocities of wave propagation in the medium can be computed (see JANUZEMIS, 1967, p. 374ff). From the equation of motion

$$\rho \frac{\partial^2 u_i}{\partial t^2} = c_{ijk\ell} \frac{\partial^2 u_k}{\partial x_j \, \partial x_\ell}, \tag{33}$$

and the description of a plane wave propagating in direction $\mathbf{n}$

$$u_k = A_k \, e^{i\omega((\mathbf{n}\mathbf{x}/c) - t)}, \tag{34}$$

the following relation can be derived:

$$c^2(\mathbf{n}) A_i = c_{ijk\ell} n_j n_\ell A_k. \tag{35}$$

Defining the *acoustic tensor* Q_{ik} for wave propagation in direction $\mathbf{n}$ as

$$Q_{ik}(\mathbf{n}) = c_{ijk\ell} n_j n_\ell, \tag{36}$$

and substituting (36) into (35) we obtain

$$Q_{ik}(\mathbf{n}) A_k = c^2(n) A_i. \tag{37}$$

Thus the wave speeds in this direction can be computed as the square roots of the eigenvalues of Q_{ik}.

4.2.1 2-D triangular lattice

In the 2-D case (triangular lattice) the lattice density is given by

$$\tilde{\rho} = \frac{\sqrt{3}}{2} \frac{m_i}{r_0^2}, \tag{38}$$

where m_i is the mass of a particle. Combined with the geometric relation

$$e_i^\alpha e_j^\alpha e_k^\alpha e_\ell^\alpha = \frac{b}{D(D+2)} (\delta_{ij}\delta_{k\ell} + \delta_{ik}\delta_{j\ell} + \delta_{i\ell}\delta_{jk}) \tag{39}$$

which is applicable for triangular lattices in 2-D and the FCHC lattice in 3-D in which b is the number of lattice directions and D is the number of dimensions (i.e., $b = 6$, $D = 2$, for the 2-D lattice and $b = 12$, $D = 3$ for the FCHC lattice), the macroscopic Hooke tensor of the medium by Equation (32) becomes

$$c_{ijk\ell} = \tilde{k}(\delta_{ij}\delta_{k\ell} + \delta_{ik}\delta_{j\ell} + \delta_{i\ell}\delta_{jk}), \tag{40}$$

where $\tilde{k} = k(\sqrt{3}/4)$.

4.2.2 The 3-D case

In order to achieve isotropic behavior in the 3-D case it would be necessary to implement a lattice which contains second-order interactions in a face centered cubic (FCC) structure (face centered hypercube, FCHC, FRISCH *et al.*, 1986b). First-order, or nearest neighbor, interactions involve particles separated by a distance d_{ij} smaller than a given maximum distance $d_{ij} \leq r^1_{\max}$ whereas second-order interactions involve particles separated by a distance between a minimum distance and a maximum distance $r^2_{\min} \leq d_{ij} \leq r^2_{\max}$ where the minimum distance for the second-order interactions is greater than the maximum distance for first-order interactions (DONZÉ, 1994). For the 3-D triangular lattice without second-order interactions used here, the wave velocities v_p, v_{s_1} and v_{s_2} have been computed numerically from Equation (37) by inserting the appropriate values for the geometric relations into Equation (32), computing the acoustic tensor Q_{ik} for different directions **n**, and solving for the eigenvalues. The resulting velocity surfaces which are shown in Figure 4 demonstrate the strong anisotropy of the shear wave velocity. This contrasts to the isotropy (spherical velocity surfaces, Fig. 5) in the case of the 3-D lattice with the second-order interactions (FCHC) and the 2-D triangular lattice.

5. Frictional Heating and Temperature at the Fault

The heat generated during a time step δt by friction between two particles with an intrinsic coefficient of friction μ as they move past each other with a tangential velocity v, can be computed as the work done

$$H = \mu F_n v \delta t, \tag{41}$$

where F_n is the normal force between the particles. This heat is then distributed evenly to both particles.

In order to compare the temperatures generated along the fault in the model with those inferred from geological evidence, simulations have been performed with an intrinsic friction, $\mu = 0.8$, a normal pressure of 300 MPa and the thermal parameters derived from those used by Lachenbruch and Sass to compute the heat flow in the San Andreas fault zone (e.g., LACHENBRUCH and SASS, 1980), i.e.,

Figure 4
Velocity surfaces for 3-D FCC lattice without 2nd-order interactions (the part above the x-y plane is cut off).

$$K = 2.5 \text{ Wm}^{-1}\text{K}^{-1}$$

$$\kappa = 10^{-6}\text{m}^2\text{s}^{-1},$$

This heat conductivity K, together with an assumed density of 3000 kgm^{-3}, leads to a heat capacity $c_p = 833$ Jkg^{-1}K^{-1}. These parameters are similar to those given for average rock ($\kappa = 1.18 \cdot 10^{-6}$m^{2}s^{-1}) (e.g., CARSLAW and JAEGER, 1959) and to those measured for granite close to the San Andreas fault (e.g., SASS et al. 1992) which showed heat conductivities $K \approx 2$–3 Wm^{-1}K^{-1}. The model used for the following simulations consisted of 64 × 64 particles in a 2-D triangular lattice with a gouge layer averaging approximately 6 particle diameters thick, i.e., $6r_0$ (Fig. 6). Inside the gouge layer, the particles are bonded together in groups of 3, 4, 7 and 9 particles, i.e., triangles, diamonds, hexagons and elongated hexagons.

Because of limits in computation time, a driving speed of $v_{\text{plate}} \approx 0.024\% \, v_p$ was used, equivalent to approximately 1.3 m s^{-3} which is around 10^8 times higher than realistic tectonic speeds (cm/year). With these parameters the temperature increase in a single slip event ranged 300–400 K (Fig. 7), however because of the high driving speed, the time between subsequent events was too short to allow the region to cool down sufficiently. Thus the temperature accumulates over multiple slips and rises to unrealistically high values.

In a second simulation, the diffusivity was increased by a factor of 10^3 to compensate for the too high driving speed. Thus there was sufficient time between events for the fault to cool down and the maximum temperatures to remain in a range of about 400 K above the initial temperature (Fig. 8). Considering that a normal geotherm of ≈ 30 K/km results in an ambient temperature at seismogenic depths of about 5 km to 15 km in the absence of abnormal heat flows from below of 400 K to 700 K, the maximum temperature predicted by the model along a fault would be between 800 K to 1100 K after an earthquake. This is below the temperatures of at least 1500 K which are inferred from the observation of partial melting on certain exhumed ancient earthquake faults and which are also expected from theoretical computations (e.g., McKenzie and Brune, 1972). There are two possible explanations. The theoretical computations for the estimation of the maximum temperature show a very high temperature gradient close to the fault,

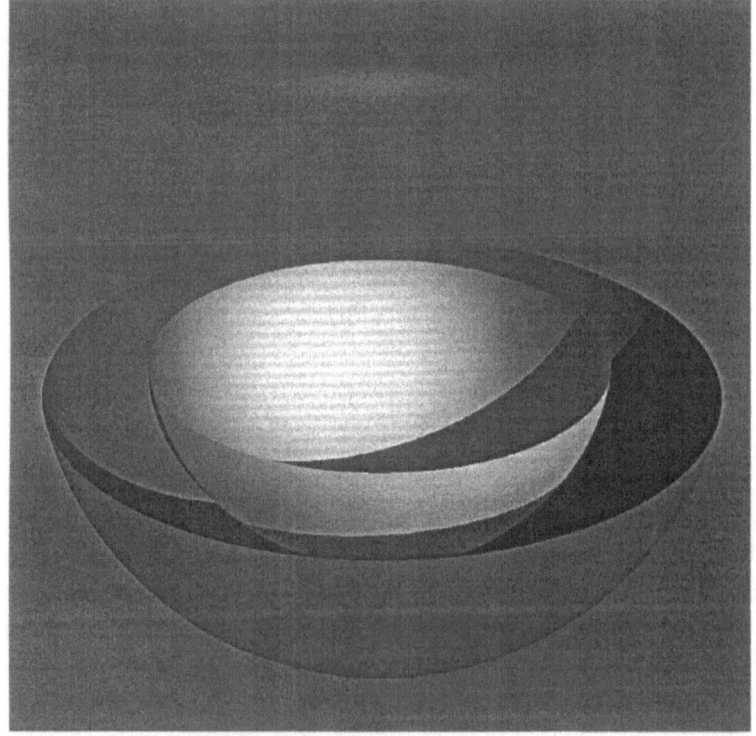

Figure 5
Velocity surfaces for the isotropic 3-D FCHC lattice with 2nd-order interactions (the part above the x-y plane is cut off). Only two surfaces are visible because the 2 slower velocities are identical ($v_{s_1} = v_{s_2}$).

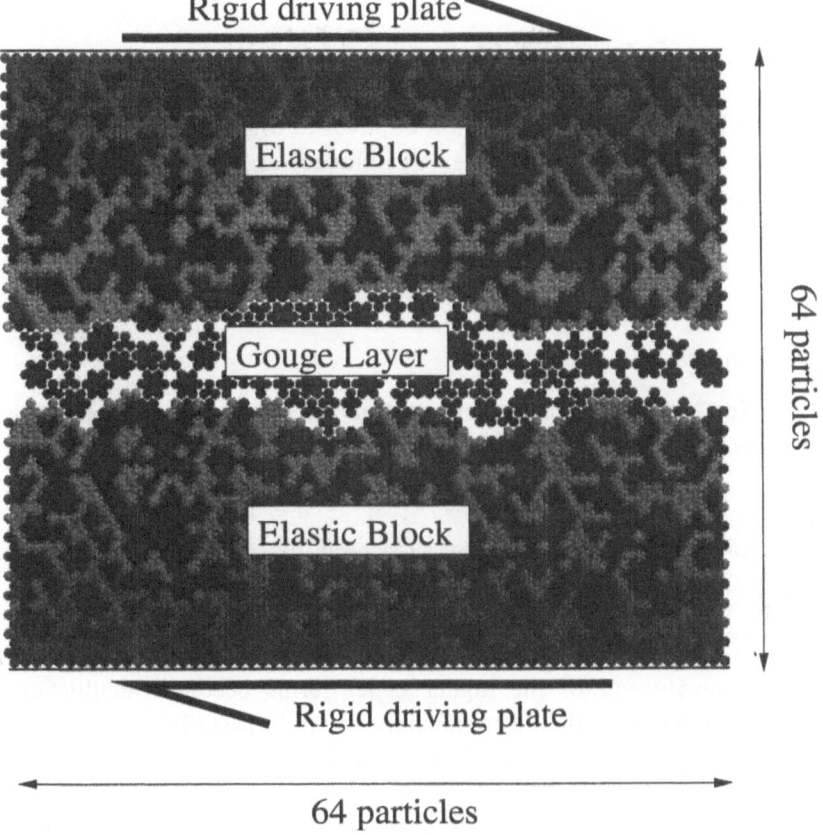

Figure 6
Model setup used for the simulations.

which leads to an underestimation of the maximum temperature in the model because of the inherent discretization error (i.e., finite r_0). The maximum temperature in the model is the temperature of a whole particle, i.e., an average over a region the size of a particle diameter which was set to 1 m in this simulation. The other reason is that in the model, the actual heat generated during the slip is generally reduced by a factor of 2 due to rolling and jostling movements of the grains (e.g., MORA and PLACE, 1998) whereas the theoretical computations have been performed without taking into account this heat reduction. While this should not influence the heat generated by the slip between a single particle pair, it will influence the amount of heat produced along the fault gouge during the entire slip event.

6. Thermal Expansion

If a fault or a fault system is heated so that an inhomogeneous temperature distribution occurs, the geometry will be changed by thermal expansion of the material. This influences the distribution of stress in the system and thus potentially the dynamics of fault slip and rupture. An influence can be expected particularly in cases when the system is close to a critical state where even small changes, such as geometric changes caused by thermal expansion which will not exceed 1%, can produce an influence.

6.1 Implementation

The simplest approach for implementation of thermal expansion would be to use a linear relation between the temperature and the radius of a particle

$$r(T) = r_0(1 + \alpha(T - T_0)), \tag{42}$$

however the high temperature increases of several hundred degrees which can occur during large slip events necessitate use of a nonlinear relationship

$$r(T) = r_0(1 + \alpha_0 + \alpha_1(T - T_0) + \alpha_2(T - T_0)^2 + \cdots), \tag{43}$$

because the contribution of the higher order terms becomes significant. In the model, a cubic relation with the parameters for polycrystalline quartz (e.g., TOU-LOUKIAN *et al.*, 1977)

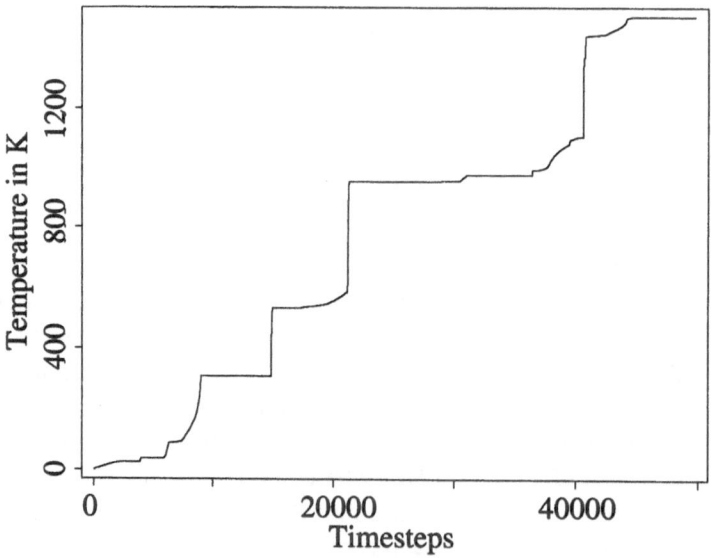

Figure 7
Maximum temperature in a model with realistic thermal diffusivity and excessively large driving speed.

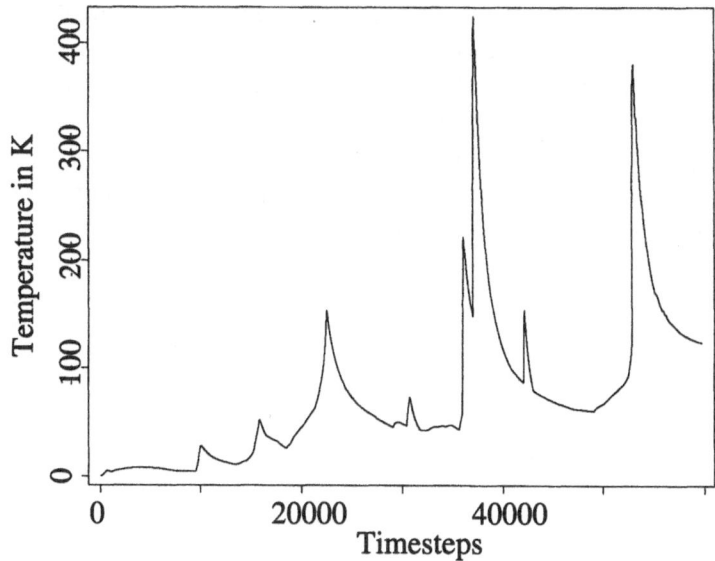

Figure 8
Maximum temperature in a model with thermal diffusivity scaled to fit driving speed.

$$\alpha_0 = -0.00236$$
$$\alpha_1 = 6.912 \cdot 10^{-6}$$
$$\alpha_2 = 5.559 \cdot 10^{-11}$$
$$\alpha_3 = 1.312 \cdot 10^{-11},$$

is used. Thus at 800 K, a temperature which can be reached during slip events (e.g., MCKENZIE and BRUNE, 1972), the contribution of the higher order terms is approximately 70% of the expansion.

6.2 Results

A number of simulations have been performed to determine the influence of thermal expansion on the dynamics of slip events. For these simulations we used a 2-D model consisting of 64×64 identical particles with a well developed fault gouge similar to the model used previously (Fig. 6). To restrict the influence of the thermal expansion to a single major slip event, a long simulation was run without thermal expansion. Subsequently the model was restarted immediately before a major slip event. Thus an identical initial state of the model was obtained for each of the comparisons between an event with and without the influence of thermal expansion. Figure 9 shows that there is no significant difference in the size of the main events although the time of the event is influenced by the thermal expansion. One of the simulations with thermal expansion also revealed a small precursory event which did not occur when the simulation was started from the same initial conditions but without thermal expansion.

To investigate the influence of thermal expansion on the statistical distribution of event sizes, longer simulations have been performed with and without thermal expansion. A statistical analysis shows that there is no change in the distribution of small to medium events and the change for large events is probably within the bounds of the statistical error. On a log-log plot of frequency vs. magnitude, the negative slope is $B \approx 0.65$ for both cases, changing for large events to $B \approx 1.2$ with thermal expansion and $B \approx 1.35$ without it[1] (Fig. 10).

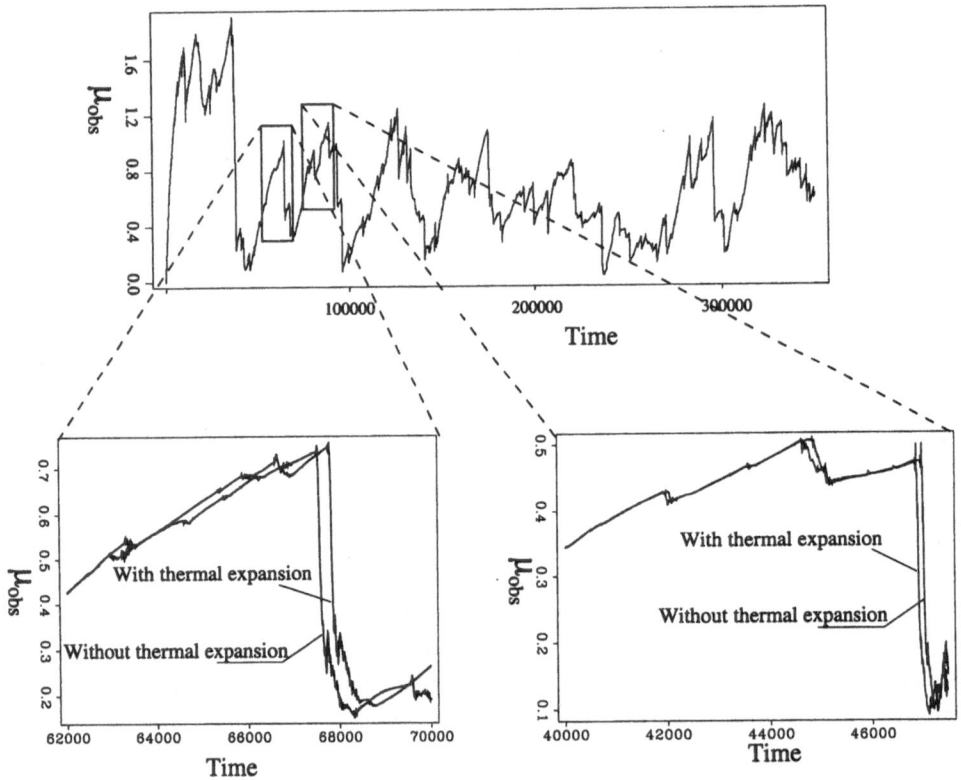

Figure 9
Observed coefficient of friction μ_{obs} during slip events with and without thermal expansion. It is measured as $\mu_{obs} = \tau/\sigma_n$ where σ_n is the normal stress and τ the shear stress necessary to maintain a constant driving velocity.

---

[1] The B-value is the negative slope of $\log(N)$ vs. energy so $B = \frac{2}{3}b$ where b is the negative slop of $\log(N)$ vs. magnitude.

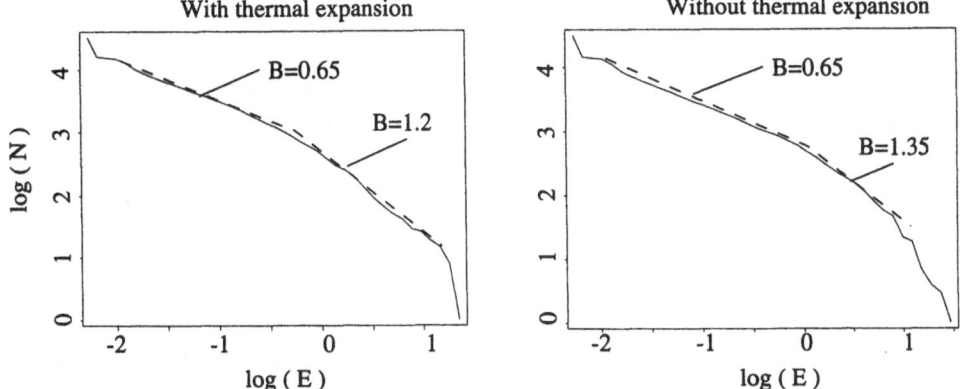

Figure 10
Log-log plot of the event size distributions for a model with and without thermal expansion.

7. Temperature-dependent Pore Fluid Pressure

If a pore fluid is present in a fault zone, the heat generated during a slip event can cause thermal expansion of the fluid, leading to a local dynamic increase in pore fluid pressure along the fault. This can reduce the effective normal stress and the resulting frictional force. If the reduction in frictional force is significant, then this process will influence the dynamics of fault slip.

7.1 Implementation

In the presence of a pore fluid, the heat flow equation given by (4) becomes

$$\frac{\partial T}{\partial t} = \kappa \frac{\partial^2 T}{\partial \mathbf{x}^2} + \frac{1}{\rho c_p} H(\mathbf{x}, t) + \frac{1}{\rho c_p} \frac{\partial}{\partial \mathbf{x}} (\rho_{fl} c_{fl} \mathbf{w} T), \qquad (44)$$

where ρ and c_p are the density and heat capacity of the aggregate, i.e., rock and fluid together, ρ_{fl} and c_{fl} are the parameters for the fluid only, and $\mathbf{w}$ is the convective velocity of the pore fluid. As shown by Lachenbruch (i.e., LACHENBRUCH, 1980), the convective term can be ignored safely so Equation (44) simplifies to

$$\frac{\partial T}{\partial t} = \kappa \frac{\partial^2 T}{\partial \mathbf{x}^2} + \frac{1}{\rho c_p} H(\mathbf{x}, t), \qquad (45)$$

which is identical to Equation (4). Assuming that the porosity of the material is small enough for the contribution of the fluid to the heat capacity of the aggregate to be ignored, the same parameters as in the case without a pore fluid can be used in the numerical solution of Equation (45).

The fluid pressure can be computed from

$$\frac{\partial P}{\partial t} = \frac{1}{\beta}\left[\gamma\frac{\partial T}{\partial t} - \dot{D}\right] + \alpha_P\frac{\partial^2 P}{\partial \mathbf{x}^2} \tag{46}$$

where β represents the compressibility, γ the thermal expansion of the fluid and α_P the Darcian diffusivity (LACHENBRUCH, 1980). If the dilatational strain rate of the pore space $\dot{D}$ is assumed to be sufficiently small and is therefore ignored, Equation (46) simplifies to

$$\frac{\partial P}{\partial t} = \frac{\gamma}{\beta}\frac{\partial T}{\partial t} + \alpha_P\frac{\partial^2 P}{\partial \mathbf{x}^2}. \tag{47}$$

The first term on the right-hand side represents the source term, i.e., the pressure increase caused by the increase in temperature, and the second term is the transport term. Thus Equation (47) has the same structure as the heat flow equation given by (3) and is solved numerically using the same method. The same considerations about numerical stability also apply, i.e., for realistic values of α_P, the constraints placed on the time step for accurate dynamics ensure that the numerical solution of the pressure equation is stable.

Assuming that the pore fluid is water, the value of the pressure increase for conditions at seismogenic depths in the crust, i.e., 150–300 MPa and 600–900 K, is $\gamma/\beta = 1.5$–2 MPa K^{-1}. Thus an immediate temperature increase of a few hundred degrees would generate a pore fluid pressure which exceeds the lithostatic pressure (Fig. 11).

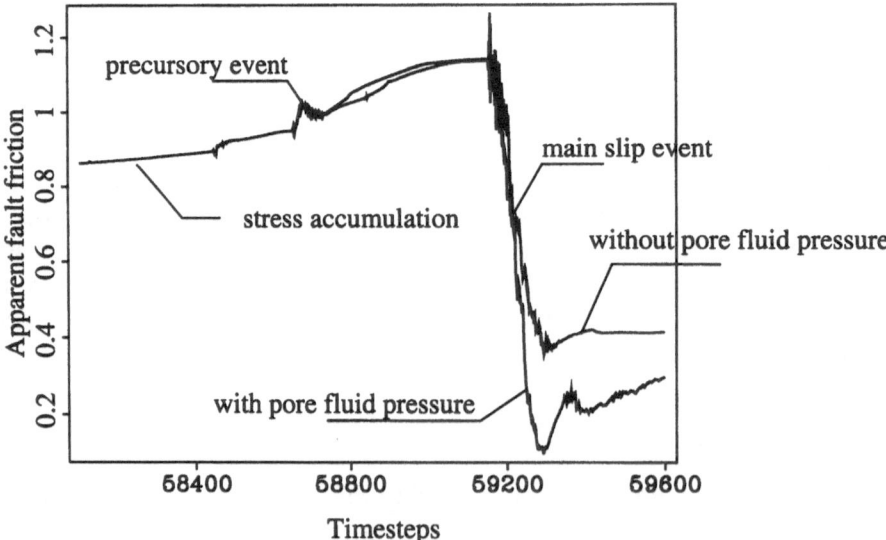

Figure 11
Difference of the development of slip events with and without reduced friction due to an increased pore fluid pressure.

The relation between frictional stress τ, normal stress σ_n, coefficient of dynamic friction μ and the pore fluid pressure P can be expressed as (NUR and BYERLEE, 1971)

$$\tau = \mu\left(\sigma_n - \left(1 - \frac{\beta_g}{\beta_r}\right)P\right), \tag{48}$$

where β_g is the compressibility of the solid grains and β_r is the compressibility of the aggregate including the fluid. Assuming that the compressibility of the solid grains is far smaller than the compressibility of the aggregate, i.e., the compression takes place at the expense of pore volume (LACHENBRUCH, 1980) we can assume

$$\frac{\beta_g}{\beta_r} \to 0. \tag{49}$$

Thus, equation (48) simplifies to

$$\tau = \mu(\sigma_n - P). \tag{50}$$

If it is further assumed that the frictional interaction vanishes if the fluid pressure exceeds the normal stress and that any effects occurring at high fluid pressures such as hydraulic fracturing can be ignored, the effective coefficient of friction between two particles in the numerical model can be computed as

$$\mu_{\text{eff}} = \begin{cases} \dfrac{\sigma_n - P}{\sigma_n} & \text{if} \quad P < \sigma_n \\ 0 & \text{if} \quad P \geq \sigma_n. \end{cases} \tag{51}$$

7.2 Results

To investigate the influence of the pore fluid pressure on the dynamics of a fault, two sets of simulations have been performed. First, single slip events have been investigated by restarting the model immediately before a slip event with and without pore fluid pressure included in the computations. The comparison of those slip events starting from identical initial conditions shows that the inclusion of the pore fluid pressure significantly influences the dynamics of large events, whereas small events which do not produce sufficient heat are not significantly influenced. The reduction of friction during large events generally leads to an increase in size for those events.

To investigate the influence of the increase in size of the largest events generated by the reduced friction, a long simulation has been performed and the statistical distribution of event sizes has been computed. Figure 12 shows that the magnitude-frequency relation remains unchanged for small events, however that there is an increase in the number of large events and the maximum event size. The tendency

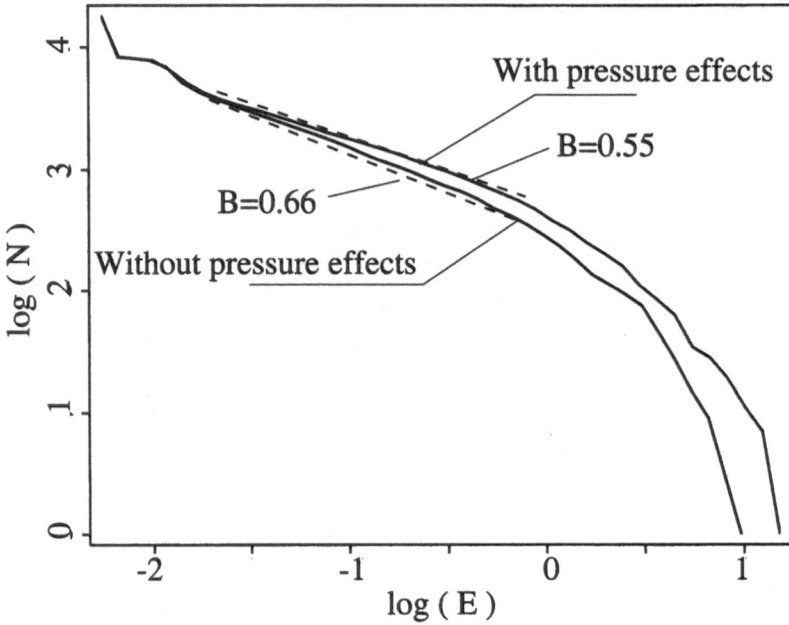

Figure 12
Change in the event size distribution caused by a reduction in friction due to an increase in pore fluid pressure.

towards an increase in the size of large events can be explained by the reduction in friction during large events caused by the frictional heating and resultant increase in pore fluid pressure. Small events do not generate enough heat to produce a significant increase in pore fluid pressure.

8. Conclusions

It has been shown that the Lattice Solid Model is capable modeling temperature related effects in simulations of the dynamics of earthquakes. The results suggest that temperature related effects can have a major influence on the dynamics of single slip events and can affect the statistical distribution of events sizes. The influence of the reduction of friction generated by an increase of the pore fluid pressure is particularly important. The reduction in friction generated by the pore fluid pressure during slip heating leads to an increase in the size of large events, whereas small events are not influenced because they produce insufficient heat to generate a significant increase in pore fluid pressure.

The dynamics of slip events is not strongly influenced by thermal expansion in the current model nonetheless a comprehensive model of earthquake dynamics should include this effect as its role may become important under certain conditions.

Acknowledgements

This work was supported by the Australian Research Council, The University of Queensland and the sponsors of QUAKES.

REFERENCES

ABE, S., MORA, P., and PLACE, D. (1998), *Introduction of thermal effects into the Lattice Solid Model: preliminary results.* In QUAKES Tech. Report #3 (The University of Queensland, Brisbane) pp. 314–321.

CARSLAW, H. S., and JAEGER, J. C., *Conduction of Heat in Solids* (Clarendon Press, Oxford 1959).

DONZÉ, F. V. (1994), *A Discrete Numerical Model for Brittle Rock Deformation.* Ph.D. Thesis.

FRISCH, U., HASSLACHER, B., and POMEAU, Y. (1986a), *Lattice Gas Automata for the Navier-Stokes Equation*, Phys. Rev. Lett. *56*, 14, 1505–1508.

FRISCH, U., D'HUMIÉRE, D., HASSLACHER, B., LALLEMAND, P., POMEAU, Y., and RIVET, J. (1986b), *Lattice Gas Hydrodynamics in Two and Three Dimensions*, Complex Systems *1*, 649–707.

HOLMAN, J. P., *Heat Transfer* (McGraw-Hill 1990).

JANUZEMIS, W., *Continuum Mechanics* (The Macmillan Company, New York 1967).

LACHENBRUCH, A. H., and SASS, J. H. (1980), *Heat Flow and Energetics of the San Andreas Fault Zone*, J. Geophys. Res. *85*, 6185–6222.

LACHENBRUCH, A. H. (1980), *Frictional Heating, Fluid Pressure, and Resistance to Fault Motion*, J. Geophys. Res. *85*, 6097–6112.

MCKENZIE, D., and BRUNE, J. N. (1972), *Melting on Fault Planes during Large Earthquakes*, Geophys. J. R. astr. Soc. *29*, 65–78.

MORA, P., and PLACE, D. (1994), *Simulation of the Stick-slip Instability*, Pure appl. geophys. *143*, 61–87.

MORA, P., and PLACE, D. (1998), *Numerical Simulation of Earthquake Faults with Gauge: Towards a Comprehensive Explanation for the Low Heat Flow*, J. Geophys. Res. *103*, B9, 21,067–21,089.

NUR, A., and BYERLEE, J. D. (1971), *An Exact Effective Stress Law for Elastic Deformation of Rock with Fluids*, J. Geophys. Res. *76*, 6414–6419.

PLACE, D., and MORA, P. (1999), *The Lattice Solid Model to Simulate the Physics of Rocks and Earthquakes: Incorporation of Friction*, J. Comp. Physics *150*, 332–372.

SASS, J. H., LACHENBRUCH, A. H., and MOSES JR., T. H. (1992), *Heat Flow from a Scientific Research Well at Cajon Pass, California*, J. Geophys. Res. *97*, 5017–5030.

TOULOUKIAN, Y. S., KIRBY, R. K., TAYLOR, R. E., and LEE, T. Y. R., *Thermophysical Properties of Matter*, vol. 13 (IFI/Plenum 1977).

SIBSON, R. H. (1975), *Generation of Pseudotachylite by Ancient Seismic Faulting*, Geophys. J. R. astr. Soc. *43*, 775–794.

SHAW, B. E. (1995), *Frictional Weakening and Slip Complexity in Earthquake Faults*, J. Geophys. Res. *100*, B9, 18,239.

(Received October 3, 1999, revised March 22, 2000, accepted April 6, 2000)

Pure appl. geophys. 157 (2000) 1889–1904
0033–4553/00/121889–16 $ 1.50 + 0.20/0

Pure and Applied Geophysics

Hybrid Modelling of Coupled Pore Fluid-solid Deformation Problems

HIDE SAKAGUCHI[1] and HANS-BERND MÜHLHAUS[1]

Abstract—A hybrid formulation for coupled pore fluid-solid deformation problems is proposed. The scheme is a hybrid in the sense that we use a vertex centered finite volume formulation for the analysis of the pore fluid and a particle method for the solid in our model. The pore fluid formally occupies the same space as the solid particles. The size of the particles is not necessarily equal to the physical size of materials. A finite volume mesh for the pore fluid flow is generated by Delaunay triangulation. Each triangle possesses an initial porosity. Changes of the porosity are specified by the translations of the mass centers of particles. Net pore pressure gradients are applied to the particle centers and are considered in the particle momentum balance. The potential of our model is illustrated by means of a simulation of coupled fracture and fluid flow developed in porous rock under biaxial compression condition.

Key words: Discrete element method, pore fluid flow, fracture, irregular shape of particles.

1. Introduction

Many important geophysical processes as well as a large number of geotechnical problems embody coupling between solid and fluid momentum transfer; at times heat and reactive mass transfer must be considered as well. For example, high pore fluid pressure induces fracture development in the rock mass and expansion of pore fluid by heat results in the opening of fractures or joints to reduce frictional resistance enhancing the mobility of fault zones. On the other hand, fractures within rocks may act as conduits for fluid flow, and the development of fractures and their deformation control the permeability and the fluid migration in the rock mass.

Although the influence of fluids on fault mechanics has been under discussion for a long time (LOGAN, 1992), the definitive model for predicting the outcome of natural or industrial scenarios or even to interpret field observations does not exist. The main obstruction to modelling has been the lack of an appropriate constitutive relationship and numerical tools to handle the difficulties arising from strong

[1] CSIRO, Division of Exploration and Mining, 39 Fairway (PO Box 437), Nedlands WA, Australia. E-mail: hide@ned.dem.csiro.au

material inhomogeneity, brittle fracture and the complexities in flow and deformation interaction at various scales. In the following we outline a simple, particle based model for coupled pore fluid-solid deformation problems.

The scheme is hybrid in the sense that we use a vertex centered finite volume formulation for the discretization of the pore fluid and a particle method for the solid. The coupling between pore fluid flow and solid deformation occurs in three ways: 1) The volume changes of the solid are considered in the mass balance of the pore fluid; 2) each particle includes in its momentum balance the influence of the net gradient of the hydraulic head applied as an external force at the mass center of each particle; 3) the deformation of the solid determines the change of the porosity which directly relates to the permeability of the matrix material. Fractures modelled by bond breakage in the particle method thus appear as zones of high permeability.

In the following section we give details of the particle based model coupling with fluid flow. In section 3 the validation of the calculation is given, and in section 4 the potential of the new method is illustrated by means of example solutions related to failure and localization in brittle materials under undrained biaxial compression.

2. Formulation

In this study rocks are basically described as assemblies of bonded particles. The procedure for the establishment of the domain of each boundary value problem and the solutions of the equations of motion are similar to the ones employed in the discrete element method (CUNDALL and STRACK, 1979). However in the contact search algorithm, advantage is taken of the fact that in solid mechanics problems most of the contacts are permanent (see e.g., SAKAGUCHI, 1995 for details). In the following we first give a brief description of the assumed interaction relationships between particles and outline of the fluid flow coupling scheme. Finally, we turn to an outline of an algorithm for the generation of randomly shaped macro-particles. Assemblies of bonded macro-particles may represent, for instance, particular rock or ceramics fabrics, while assemblies of loose macro-particles are considered in comminution and flow problems involving angular granules.

2.1 Contact Mechanics and Yield Criterion

We consider two granules i and j which have translational velocities v^i and v^j and angular velocities s^i and s^j. Continuum versions of the following formulations have been derived by MÜHLHAUS and OKA (1996) and MÜHLHAUS and HORNBY (1997). The relative velocity (Fig. 1) at the contact is

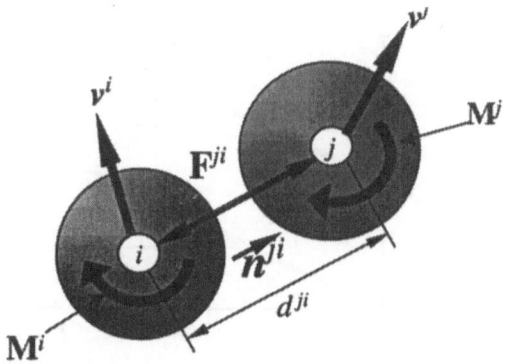

Figure 1
Contact geometry for spherical granules.

$$g^{ji} = v^{ji} + d^{ji}n^{ji} \times s^{ji}, \tag{1}$$

where $v^{ji} = v^j - v^i$, $s^{ji} = s^j + s^i$, n^{ji} is the unit vector along the connecting line between the mass centers of the granules and d^{ji} is the distance between the mass centers. To facilitate the treatment of nonlinear particle interactions we formulate the constitutive relationships of the contacts in rate form (KLEIBER, 1989). The relevant co-rotational rates of the contact force (F^{ji}) and moment (M^{ji}) vectors read:

$$\overset{\triangledown}{F}{}^{ji} = \dot{F}^{ji} - W^{ji}F^{ji}, \tag{2}$$

$$\overset{\triangledown}{M}{}^{ji} = \dot{M}^{ji} - W^{ji}M^{ji}, \tag{3}$$

$$W^{ji}_{\alpha\beta} = \frac{1}{d^{ji}}(v^{ji}_\alpha n^{ji}_\beta - v^{ji}_\beta n^{ji}_\alpha), \tag{4}$$

where indices α and β denote components of a vector or tensor in the global coordinate system. As is expressed in (4), W^{ij} in (2) or (3) is the spin of the connecting line between the mass centers of the particles i and j. The purpose for using co-rotational rates is to separate the effect of rigid rotation (where two or more particles in contact rotate as a single, rigid body) from the kinematics relevant for the calculation of the relative velocity at the particle contacts.

Elastic contacts are described in the simplest possible way by

$$\overset{\triangledown}{F}{}^{ji}_{el} = K_s(g^{ji} - (g^{ji} \cdot n^{ji})n^{ji}) + K_n(g^{ji} \cdot n^{ji})n^{ji}, \tag{5}$$

$$\overset{\triangledown}{M}{}^{ji}_{el} = d^{ji^2}K_r(s^j - s^i), \tag{6}$$

where K_s, K_n and K_r are constant tangent, normal and rotational stiffnesses. Note that the above formulation is physically meaningful only if the magnitude of the relative displacements is infinitesimal.

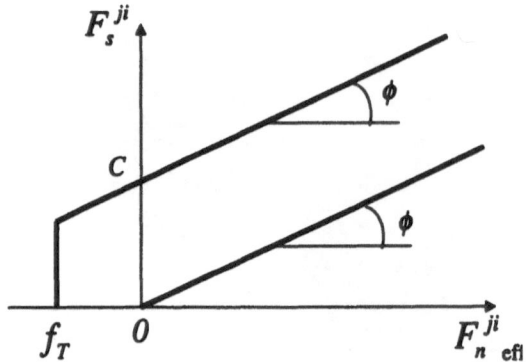

Figure 2
Contact yield criterion.

The total contact forces and moments are in general made up of an elastic and a viscous part, so that

$$F^{ji} = F_{el}^{ji} + F_{v}^{ji}, \tag{7}$$

$$M^{ji} = M_{el}^{ji} + M_{v}^{ji}. \tag{8}$$

In analogy to (5) and (6) we assume

$$F_{v}^{ji} = \eta_s(g^{ji} - (g^{ji} \cdot n^{ji})n^{ji}) + \eta_n(g^{ji} \cdot n^{ji})n^{ji}, \tag{9}$$

$$M_{v}^{ji} = d^{ji^2}\eta_r(s^j - s^i), \tag{10}$$

where η_s, η_n and η_r are constant tangent, normal and rotational viscosities.

The contact viscosities may have a real physical significance e.g., for viscoelastic contacts or in fast granular flows. In the latter case, the viscosities are defined by the normal and tangent coefficients of restitution (MÜHLHAUS and HORNBY, 1997). In connection with the solution of quasi-static problems by dynamic relaxation, the significance of the viscosities is purely numerical and the values of the viscosities are chosen exclusively in view of numerical efficiency.

In the 2-D analyses intended here, F^{ji}, g^{ji}, s^j and M^{ji} have shear, normal (denoted by s and n) and rotational components ($F_s^{ji}, F_n^{ji}, 0$), ($g_s^{ji}, g_n^{ji}, 0$), $(0, 0, s^j)$ and $(0, 0, M^{ji})$, respectively.

The contact yield criterion in the present model is illustrated in Figure 2, where C and f_T designate the cohesive force and tensile strength of the contact, ϕ is Coulomb's angle of friction and F_{neff}^{ji} is an effective normal force defined as

$$F_{\text{neff}}^{ji} = F_n^{ji} + \frac{2\alpha^{ji}}{d^{ji}}|M^{ji}|, \tag{11}$$

where $\alpha^{ji} \geq 1$ is a dimensionless material parameter. Through the inclusion of the moment term in equation (11) we consider the influence of possible load eccentricities with respect to the mathematical center of the contact (a similar criterion was used by VAN MIER et al. (1995), within the context of a lattice model for concrete fracture). Upon initial yield, C and f_T are put equal to zero and

$$F_s^{ji} = \tan \phi F_{\text{neff}}^{ji}. \tag{12}$$

Separation of the particles i and j occurs if the effective normal force $F_{\text{neff}}^{ji} = 0$.

The configurations of the particle system are determined as is standard by integration of the equations of motion

$$\sum_{\text{contacts}} F^{ji} + \gamma^i - m^i v^i = 0 \tag{13}$$

and

$$\sum_{\text{contacts}} \left(\frac{D^i}{2} n^{ji} \times F^{ji} + M^{ji} \right) + \mu^i - \frac{1}{8} m^i D^{i^2} s^i = 0, \tag{14}$$

where m^i and D^i are the mass and diameter of the i-th particles; γ^i and μ^i are external particle forces and particle moments, respectively.

2.2 Vertex Centered Finite Volume Scheme for Fluid Flow

2.2.1 Spatial discretization and finite difference scheme

The formulation embodies full coupling between solid and fluid momentum transfer. A flow calculation is based on local porosities and permeabilities on a triangular mesh. As is depicted in Figure 3, the triangular mesh is generated by Delaunay triangular connecting the mass centers of each particle and its nearest neighbors. For each triangular domain, the flow rate is calculated, as follows.

The two-dimensional form of Gauss' divergence theorem applied to an arbitrary function $F(x_i)$ reads,

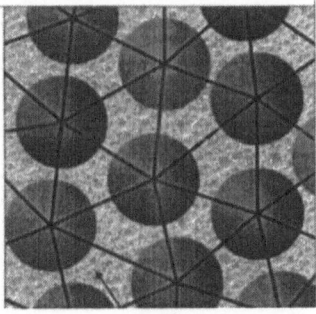

Figure 3
Discretized space: Triangular mesh and particles.

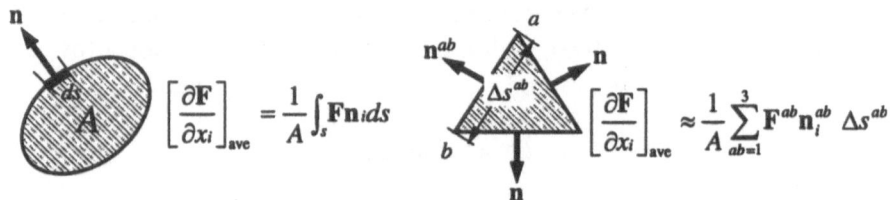

(a) general region   (b) triangular region

Figure 4
Line integral over a small region.

$$\int_S F n_i \, ds = \int_A \frac{\partial F}{\partial x_i} \, da, \tag{15}$$

where S is the length of a closed loop, n_i is the outward unit normal of S, A is the area of a region surrounded by S, the average value of the gradient of the function $\partial F / \partial x_i$ over a small region A can be expressed as

$$\left[\frac{\partial F}{\partial x_i}\right]_{ave} = \frac{1}{A} \int_S F n \, ds. \tag{16}$$

For a triangular region for A, the equation above may be approximated by

$$\left[\frac{\partial F}{\partial x_i}\right]_{ave} \approx \frac{1}{A} \sum_{ab=1}^{3} F^{ab} n_i^{ab} \Delta_s^{ab}, \tag{17}$$

where Δs is the length of a side of the triangular region (Fig. 4) and $(\cdot)^{ab}$ is the value of $(\cdot)$ at the center of the side ab.

Applying this finite volume formulation in (17) to Darcy's law for a porous medium, i.e.,

$$v_i = K \frac{\partial P}{\partial x_i}, \tag{18}$$

(note that $P < 0$ in compression) where v_i is the specific discharge vector, P is the pressure and K is the permeability, we obtain for the average specific discharge vector over a small triangular region

$$v_i \approx \frac{1}{A} \sum_{ab=1}^{3} K^{ab} P^{ab} n_i^{ab} \Delta s^{ab}, \tag{19}$$

where

$$P^{ab} = \frac{P^a + P^b}{2}. \tag{20}$$

In our model, the pressure P is treated as a nodal quantity and the permeability K may have a different value on each side of the triangular zone or element. If the contact force of a bond reaches the contact yield strength, then bond breakage takes place and the permeability of the respective side is increased correspondingly.

The flow rate through the side ab of the triangle abc in Figure 5 is the projection of the specific discharge v_i in the normal direction of the side ab multiplied by the length of ab, Δs^{ab} as,

$$q^{ab} = v_i \cdot n_i^{ab} \Delta s^{ab}, \tag{21}$$

where n_i^{ab} is the unit normal vector on the side ab. Then, as is illustrated in Figure 6, the nodal flow rate is given as

$$q^a = \frac{q^{ab}}{2} + \frac{q^{ca}}{2}. \tag{22}$$

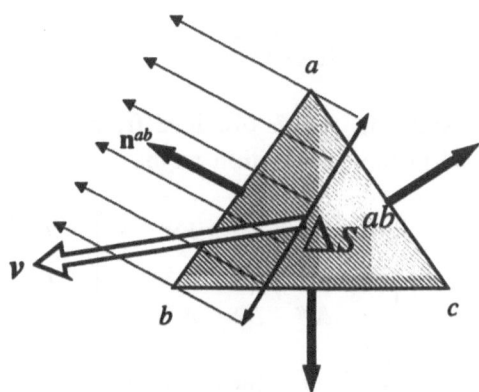

Figure 5
Flow rate through the side ab.

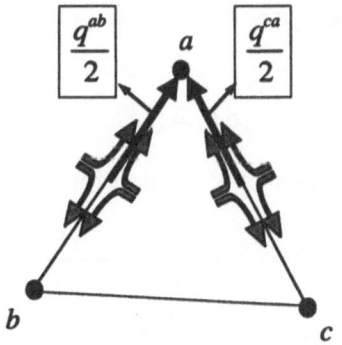

Figure 6
Nodal flow rate.

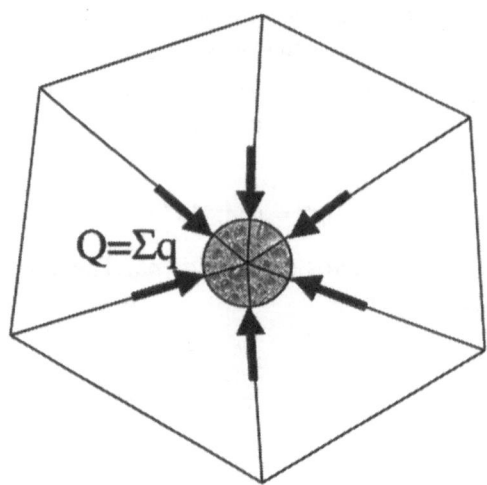

Figure 7
Nodal flow rate Q.

Consequently, the nodal flow rate is obtained from the nodal pressure for the triangular abc. Expressed in matrix notation we have

$$
\begin{bmatrix} q^a \\ q^b \\ q^c \end{bmatrix} = \begin{bmatrix} \cdots & K^{ab} & K^{bc} & K^{ca} & \cdots \\ \cdots & n_i^{ab} & n_i^{bc} & n_i^{ca} & \cdots \\ \cdots & \Delta^{ab} & \Delta^{bc} & \Delta^{ca} & \cdots \end{bmatrix} \begin{bmatrix} P^a \\ P^b \\ P^c \end{bmatrix}. \tag{23}
$$

As is depicted in Figure 7, one nodal point is generally shared by several triangles. Therefore, the net nodal flow rate Q is the sum of q on the node of all the triangles.

$$
Q = \sum q. \tag{24}
$$

Note that $Q = 0$ for incompressible flow. The fluid volume change ΔV_f in small time step Δt is related to Q as

$$
\Delta V_f = Q \Delta t. \tag{25}
$$

In the present notation the volumetric relation for a Biot coefficient of $\alpha = 1$ reads:

$$
\Delta P = -B_w \left[\frac{Q}{V_f} \Delta t + \frac{\Delta V_p}{V_p} \right], \tag{26}
$$

where B_w is the bulk modulus of pore fluid, V_f is the fluid volume contained in a cell and V_p is the pore volume (Fig. 8). We assume that a $t = 0$, $V_f = V_p$. Note that in our notation, ΔV_v and ΔV_p are positive in extension.

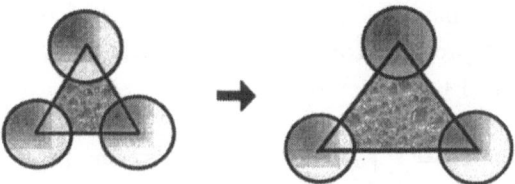

Figure 8
Pore volume change.

In general, the order of magnitude of the coefficient B_w is not larger than that of the solid, so that fully explicit integration of (26) is possible for the same time Δt in mechanical part. In all other cases, the initial explicit step is used to update the right-hand side of (26). This is repeated in each time step until convergence occurs. The resulting fully implicit formulation is robust even for B_w in the order of solid. However, fully implicit or Crank Nicholson schemes are also necessary when Q or ΔV_p changes abruptly as in the case of bond breakage.

2.2.2 Mechanical coupling

The coupling between pore fluid flow and solid deformation occurs in three ways:

1. The deformation and bond fracture of the solid influences the permeability of the matrix material. Hydraulically, fractures are modelled as zones of high permeability.
2. The volume changes of the solid material are considered in the mass balance of the pore fluid.
3. Each particle includes in its momentum balance the resultant of the pore fluid surface tractions.

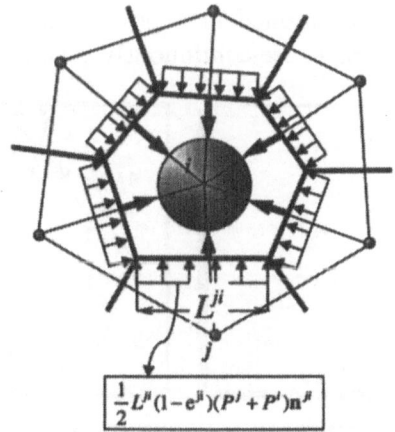

Figure 9
Pore fluid surface tractions.

The first two items have been discussed above. To consider the problem of applying the net fluid pressure force to i-th particle center F_P^i, we consider the polygon as depicted in Figure 9. We have

$$F_P^i = \sum_{ji} \left(\frac{1}{2} L^{ji}(1 - \phi^{ji})(P^j + P^i)n^{ji} \right), \tag{27}$$

where P^i, P^j are nodal point pressures, L^{ji} is the length of the side of the Voronoi polygons intersecting the line which connects the mass centers of particles i and j, ϕ^{ji} is the local porosity which is determined by the ratio of the solids part and on the voids part on the line connecting the mass centers of particles i and j, and n^{ji} is the unit vector into the direction of the line. If the pressure around a particle i is uniform, then $F_P^i = 0$.

The formulation above is based on compressible Darcy flow in porous media. Note that no restrictions other than $\alpha = 1$ (for a Biot coefficient see DETOURNAY and CHENG, 1985) are imposed regarding the compressibility of the pore fluid or gas. The theory therefore also holds for problems involving gas-solid interactions, in connection with the simulation of blasting operations. Since, as is seen in Equation (26), the fluid dealt in this method is compressible and the vertex centered finite volume scheme itself is suitable for rapid flow.

2.3 Macro-particle Generation

On the scale of grains and micro-cracks the structure of rock appears highly disordered and complex in general. A major advantage of the present model is that such complexities can be easily dealt with. In the following we describe a simple algorithm for the generation of complex shaped macro-particles (SAKAGUCHI and MÜHLHAUS, 1997).

Seed particles (SP) are positioned in a given domain by a uniform random process in which coordinate pairs are continuously generated until the total number

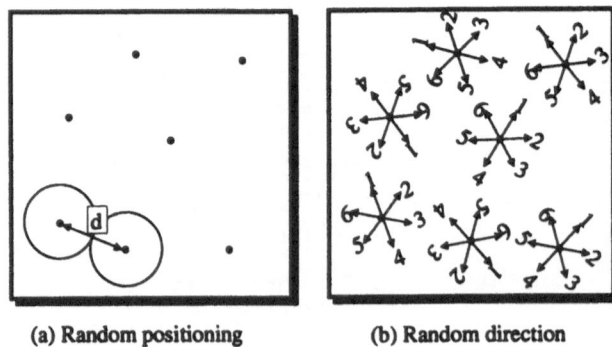

(a) Random positioning   (b) Random direction

Figure 10
Seed particles and directions of particle growth.

of seed particles reaches a preset value N. For mono-sized elemental-particles (EP) the minimum distance between neighboring SPs is d (= the distance between the mass centers of adjacent grains in the reference configuration) as depicted in Figure 10.

Once all the SPs are positioned, macro-particles (MP) are grown according to the following random process (Fig. 11):

1. $MP \# i$, $1 \leq i \leq N$ is selected at random. Each of the particles constituting the MP is endowed with six directions (Fig. 11(b)) into which new particles may be placed. A specific direction is ruled out if the corresponding position is already occupied by a particle. Let $N_g(i)$ be the number of particles with at least one free direction.

2. A particle j on $MP \# i$, $0 \leq j \leq N_g(i)$ is selected at random. Available directions are stored in $N_0(i, j, k)$, $1 \leq k \leq 6$ from which a particular direction is selected and $N_0(i)$ is updated accordingly (Figs. 11(a)–(c)).

3. Update $N_g(i)$, i.e., $N_g(i) = N_g(i) - 1$. $MP \# i$ is complete if $N_g(i) = 0$. The process is repeated until $N_g(i) = 0$ for $i = 1$ to N.

The resulting particle shapes are usually irregular (Fig. 12) with a significant proportion of non-convex particle shapes. The proportion of convex particle shapes would increase if, in step 2 above, more than one occupation direction were to be tested and populated.

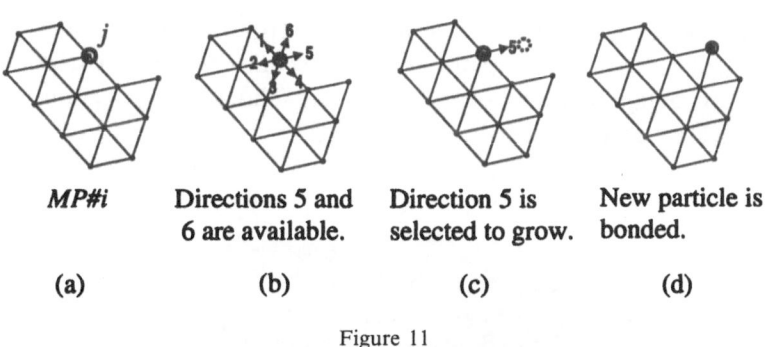

$MP\#i$	Directions 5 and 6 are available.	Direction 5 is selected to grow.	New particle is bonded.
(a)	(b)	(c)	(d)

Figure 11
Growth of macro-particles (MP).

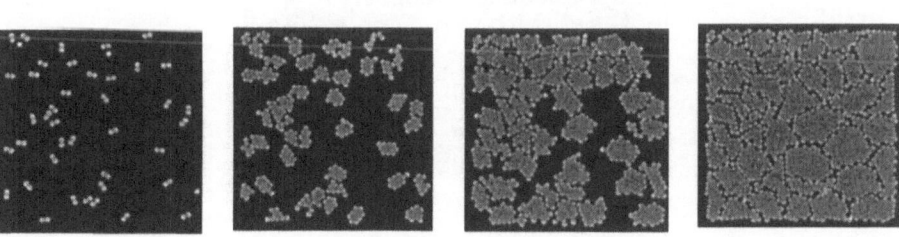

Figure 12
Irregular shaped macro-particles.

Finally the *MP*s are connected using a Delaunay triangular tessellation technique (WATSON, 1992).

3. Verification

The following simulation verifies the flow scheme coupling with the solid deformation in our model.

A square block of porous rock saturated with water is considered. The geometry and boundary conditions are depicted in Figure 13. The rock material is simulated as a random assembly of 960 adhesive particles with purely elastic bonds. Bond strength f_T is set to a high value in order to expect no bond breakage but only consolidation. The initial porosity of the particle system is 0.5. In the reference configuration the pore pressure is zero everywhere. The specimen is confined by the vertical walls at $X = 0.0$ and $X = 1.0$ both mechanically and hydraulically. At the top ($Y = 1.0$) and the bottom ($Y = 0.0$) boundaries, dimensionless constant pressures of 1.0 and 0.0 are applied respectively, to set the hydraulic gradient $i = 1.0$. At the bottom boundary, particles are mechanically fixed in Y direction. No gravitational force is considered here.

Figures 14(a) and (b) show the evolution of the pore-pressure profile of the region. Initially the high pressure gradient is concentrated in the vicinity of the boundary at $Y = 0.0$ in (a), and the body forces induced by the local pore pressures result in the local consolidation in Y direction. However, in the later part of the simulation (1000 steps), the pressure gradient becomes linear and steady-state flow conditions are obtained with homogeneous consolidation throughout the region. Thus the linearity of Darcy flow is reproduced.

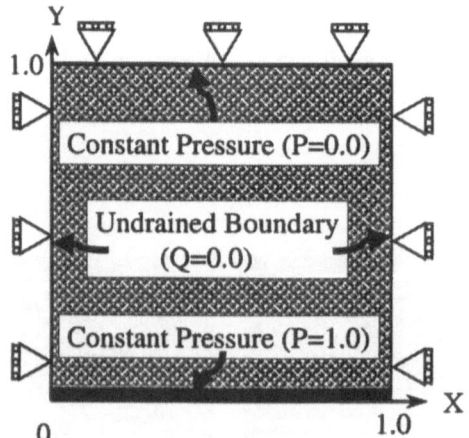

Figure 13
Boundary condition and initial condition.

Pore Pressure Distribution

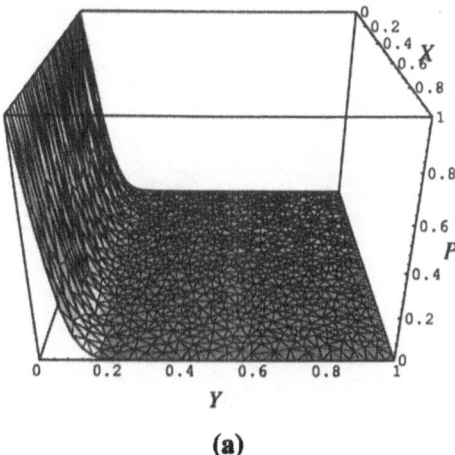

(a)

Pore Pressure Distribution

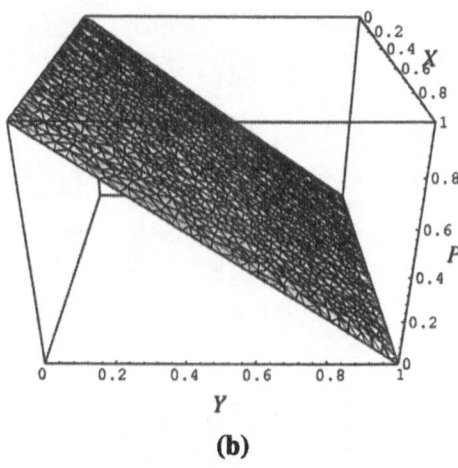

(b)

Figure 14
Profiles of pore pressure distribution: (a) 100 steps, (b) 1000 steps.

4. Biaxial Compression Test

The procedure described below simulates an undrained biaxial compression test of porous rock by confining a sample within a rectangular box with moving boundaries (Fig. 15).

A random assembly of 500 macro-particles (each macro-particle consists of 10 to 20 micro-particles) is packed in a rectangular box. Macro-particles are also bonded together to simulate a homogeneous porous rock. In this case the initial length of inter macro-particles bonds can be different to each other. However,

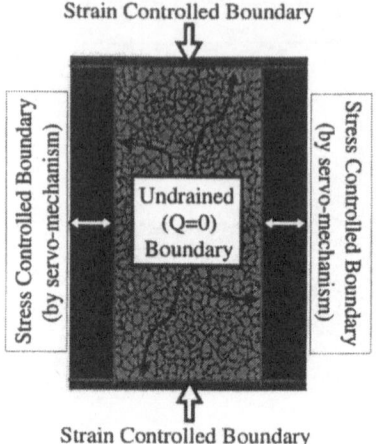

Figure 15
Boundary conditions for mechanical deformation and fluid flow.

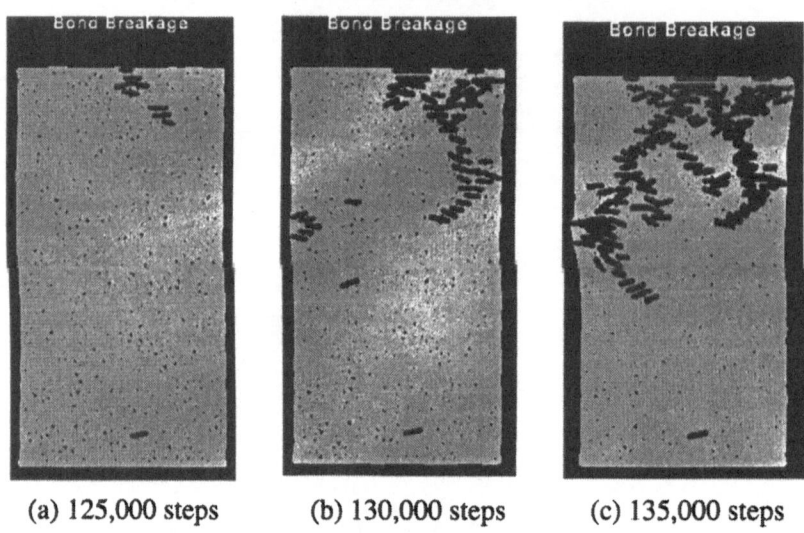

(a) 125,000 steps   (b) 130,000 steps   (c) 135,000 steps

Figure 16
Bond breakage distribution near peak load.

pre-stress on each bond is always set to be zero. The top and bottom boundaries simulate loading platens in a strain-controlled condition by specifying the vertical velocities. On the other hand, the left and right boundaries are controlled by a numerical servo-mechanism to maintain a constant confining stress horizontally at the boundaries. For hydraulic boundary conditions, all walls are impermeable ($Q = 0.0$), that is undrained.

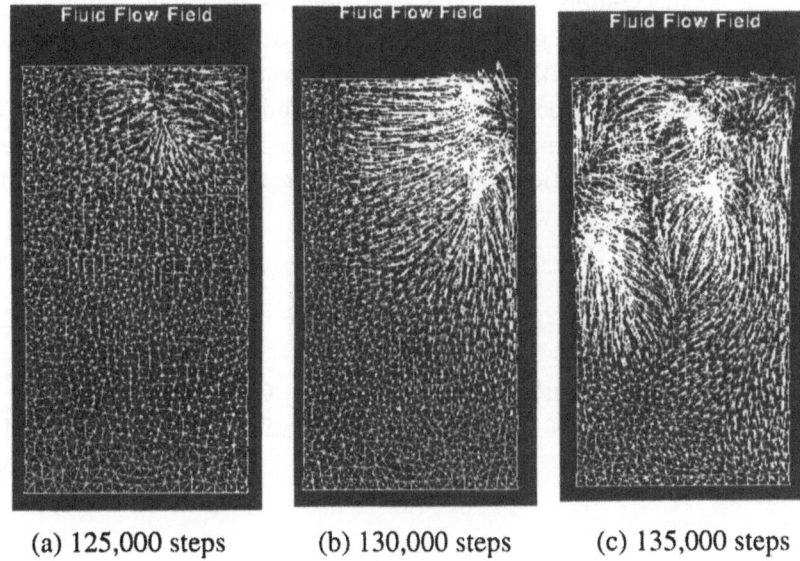

(a) 125,000 steps   (b) 130,000 steps   (c) 135,000 steps

Figure 17
Velocity field of fluid flow near peak load.

Figure 16 shows the propagation of the bond breakage near peak load at 130,000 timesteps. Clearly bond breakages are localized and are lined up in the approximate directions of the principal stresses, which are conjugate to each other, to form shear fault cracks.

As bond breakages initiate, local pore fluid migration occurs. Figure 17 is the velocity field of fluid flow. Arrows indicate the instantaneous velocity of fluid flow which averaged over each triangular region.

Careful observation of Figures 16 and 17 tells us that fluid migrates into a local crack containing the lower pore pressure and the higher permeability due to the volume expansion and the bond breakage. Thus, this fluid concentration into the local crack takes part in further propagation and branching of the local crack.

Note that biaxial compression tests without fluid flow tend to show only one pair of conjugate shear fault crack which is thought to be more or less the effect of mechanical/geometric boundary condition. Generally, one starts from a connor to another diagonal connor.

5. Conclusion

The formulation of the modelling of couple pore fluid-solid deformation problems has been described. The results obtained above are in full agreement with what is known from corresponding experiments with geo-materials. Especially the simulation of the biaxial compression test which suggests that fluid migration plays a crucial role on the fracture propagation in water-saturated rocks and cannot be ignored in the prediction of the fault zone formation.

The main advantage of coupled solid particle-fluid/gas systems is that they do not suffer from the same unrealistic restrictions regarding fracture and evolving fluid pathways as do conventional finite element based models for poro-elasticity or poro-plasticity.

Acknowledgements

The support of the Australian Geodynamics Cooperative Research Centre (AGCRC) and the Centre of Offshore and Foundation Systems (COFS), UWA is gratefully acknowledged.

REFERENCES

CUNDALL, P. A., and STRACK, O. D. L. (1979), *A Discrete Numerical Model for Granular Assemblies*, Géotechnique *29*, 47–65.

DETOURNAY, E., and CHENG, A. H. D., *Fundamentals of poroelasticity*. In *Comprehensive Rock Engineering, Vol 2: Analysis and Design Methods* (eds. Brown, Fairhurst, and Hoek) (Pergamon 1985) pp. 113–172.

KLEIBER, M., *Incremental Finite Element Modelling in Nonlinear Solid Mechanics* (Polish Scientific Publishers 1989).

LOGAN, J., *The influence of fluid flow on the mechanical behaviour of faults*. In *Rock Mechanics* (eds. Tillerson, and Wawersik) (Balkema 1992) pp. 141–149.

MÜHLHAUS, H.-B., and OKA, F. (1996), *Dispersion and Wave Propagation in Discrete and Continuous Models for Granular Materials*, Int. J. Solids Structures *33* (19), 2841–2858.

MÜHLHAUS, H.-B., and HORNBY, P., *On the reality of antisymmetric stresses in fast granular flow*. In *Proc. IUTAM Conf. On Granular and Porous Media* (eds. Fleck, and Cooks) (Kluwer 1997) pp. 299–311.

SAKAGUCHI, H., *Pattern formation in granular media* (Doctoral dissertation, Kyoto University, Japan 1995).

SAKAGUCHI, H., and MÜHLHAUS, H.-B., *Mesh free modelling of failure and localization in brittle materials*. In *Deformation and Progressive Failure in Geomechanics* (eds. Asaoka, Adachi, and Oka) (Pergamon 1997) pp. 15–21.

VAN MIER, J. G. M., SCHLAGEN, E., and VERVURT, A., *Lattice type fracture models for concrete*. In *Continuum Models for Materials with Microstructure* (ed. Mühlhaus) (Wiley 1995) pp. 341–377.

WATSON, D. F., *CONTOURING: A Guide to the Analysis and Display of Spatial Data* (Pergamon 1992).

(Received August 20, 1999, revised March 23, 2000, accepted April 6, 2000)

Pure appl. geophys. 157 (2000) 1905–1928
0033–4553/00/121905–24 $ 1.50 + 0.20/0

Pure and Applied Geophysics

Numerical Simulation of Rock Failure and Earthquake Process on Mesoscopic Scale

Yu-Cang Wang,[1] Xiang-Chu Yin,[1,2] Fu-Jiu Ke,[1] Meng-Fen Xia[1] and
Ke-Yin Peng[2]

Abstract—On the basis of the lattice model of Mora and Place, Discrete Element Method, and Molecular Dynamics approach, another kind of numerical model is developed. The model consists of a 2-D set of particles linked by three kinds of interactions and arranged into triangular lattice. After the fracture criterion and rules of changes between linking states are given, the particle positions, velocities and accelerations at every time step are calculated using a finite-difference scheme, and the configuration of particles can be gained step by step. Using this model, realistic fracture simulations of brittle solid (especially under pressure) and simulation of earthquake dynamics are made.

Key words: Numerical simulation, discrete element method, lattice model, rock failure, earthquake process.

Introduction

Earthquake prediction still remains one of the most difficult problems, due to many reasons of which the poor understanding of the physical process of earthquakes is a major one. Essentially, an earthquake is a rapidly occurring fracture of the rocks in the interior of the earth (Knopoff, 1993) (intraplate earthquake), or is a slip instability controlled by friction (Brace and Byerlee, 1966) (an earthquake in plate boundaries). Either a damage theory to predict precisely the failure of such brittle solids is still in a very rudimentary stage, or full understanding of friction phenomenon has not come about, which may be the scientific reason for the difficulty of earthquake prediction.

Similarly, in seismology, it is still very difficult to study analytically the whole earthquake occurrence (spatial and temporal distribution of seismicity) with a large number of coupled fault systems in a geologically complex area.

---

[1] LNM, Institute of Mechanics, Chinese Academy of Sciences, Beijing, China, 100080.
[2] Center for Analysis and Prediction, China Seismological Bureau, Beijing, China, 100036. E-mail: WYC: yin@lnm.imech.ac.cn; YXC: yinxc@btamail.net.cn; KFJ: kefj@lnm.imech.ac.cn; XMF: xiam@lnm.imech.ac.cn; PKY: xcyin@public.bta.net.cn

In recent years, numerical simulation began to play a more important role in studying such complex phenomena. There are already many successful numerical models proposed to simulate earthquakes, such as the spring-block model (BUR-RIDGE and KNOPOFF, 1967; CARLSON and LANGER, 1989; CARLSON *et al.*, 1989), cellular automata (BARRIERE and TURCOTTE, 1991; LOMNITZ-ADLER, 1993; SAMMIS and SMITH, 1999), and SOC model (BÅK and TANG, 1989). These models only simulate the general aspects and statistical behaviors of the earthquake process.

Some simulations, such as finite element method (TANG, 1997), finite difference (DAY, 1982; MADARIAGA *et al.*, 1997), boundary integral equation method (FUKUYAMA and MADARIAGA, 1995) are based on the macroscopic and continuous media. But the earth's crust is far from being continuous, it contains many discontinuities of different sizes.

Molecular Dynamic simulation (MD, HERRMANN, 1989) provides a natural way to study the discrete behavior on a molecular/atomic scale, but when modeling earthquake phenomena, the scale must be many magnitudes coarser. It is difficult and unnecessary to model the earthquake process on a molecular/atomic scale. Discrete Element Method (DEM, CUNDALL and STRACK, 1979) is suitable to model the dynamic behavior of an assemblage of blocks and granular materials with discontinuities, and was widely used in engineering and geology.

Several kinds of discrete models have been used in mechanical and engineering fields to model failure and damage evolution of materials. Lattice models are commonly used. Central-force lattice or truss (ASHURT and HOOVER, 1976; CURTIN and SCHER, 1990a,b) is the simplest one, in which only axial forces can be transmitted. These models are simple and easy to handle. However, if the simulation is not on a molecular/atomic scale, negligence of transverse forces and moments may cause some problems, especially under compressive loads. On one hand, it may yield an unrealistic failure mode. On the other hand, if all link rigidity parameters are identical, Poisson's ratio is equal to $1/3$, so central-force lattice has only one parameter, while an isotropic elastic solid is defined by two parameters. In the beam lattice (frame) (MONETTE and ANDERSON, 1994), bending rigidity is included to model forces due to the relative rotations of the links at nodes. In some lattice models (BORN and HUANG, 1954; ASKAR and CAKMAK, 1968; BATHURST and ROTHENBURG, 1988), axial and transverse forces can be transmitted.

Theoretically, there should be three kinds of relative displacements and therefore three kinds of interactions in order to make a perfect description of relative displacement between two adjacent particles. Therefore three kinds of rigidity parameters should be introduced. However, in order to make a realistic simulation, the three lattice parameters must match the two macroscopic elastic parameters and the failure criterion must be chosen carefully. Although a few researchers (JAGOTA and DAWSON, 1988a,b; JAGOTA and SCHERR, 1993; TOI and CHE, 1994; TOI and KIYOSUE, 1995) used models with three kinds of interactions transmitted, they did not describe how to determine the lattice parameters.

Lattice models have been applied to model earthquake phenomena. For example, MORA and PLACE (1993, 1994, 1998) developed a lattice called "the Lattice Solid Model (LSM)" to study earthquake dynamics whose initial version involved central forces, and used it to simulate the effect of fault gouge and surface roughness on friction and tried to explain the heat flow paradox. SCOTT (1996) modeled seismicity and stress rotation using the discrete lattice with central and shear forces transmitted.

On the basis of MORA and PLACE's LSM, DEM and Molecular Dynamic simulation, we studied another kind of discrete model. In this paper we describe our model first, then give the lattice parameters and some preliminary simulations on the fracture of brittle solid (especially under compressive loads) and the earthquake process.

About our Model

In our study, the material is also discretized into a number of round particles linked by bonds and arranged into a 2-D triangular lattice (Fig. 1). The particles are the smallest mesoscopic units which cannot be broken. The sizes of particles range from millimeter (grains) to kilometer (geological blocks). Radial forces F_r, tangential forces F_s and bending moment M are transmitted between the adjacent particles, and every particle is described by three variables: positions x, y and spin θ, so there are three kinds of relative displacements between every adjacent particle pair: radial displacement Δr, shear displacement Δs, and angular displacement $\Delta \theta$. If we use the linear relation of force and displacement, we have

$$
\begin{aligned}
F_r &= K_r\, \Delta r \\
F_s &= K_s\, \Delta s \\
M &= K_m\, \Delta \theta
\end{aligned}
\tag{1}
$$

where K_r, K_s, K_m are radial, tangential and bending rigidity.

Elastic Properties and Chosen of Mesoscopic Parameters

We demonstrated (see appendix) that if K_r, K_s, K_m are identical, our model has isotropic elastic properties, and under the condition of small strain, the mesoscopic parameters K_r, K_s, K_m should be chosen according to the macroscopic Young modulus of elasticity E, and Poisson's ratio v,

$$
K_r = \frac{\sqrt{3}E}{3(1 - v)}
\tag{2}
$$

$$
K_s = \alpha K_r
\tag{3}
$$

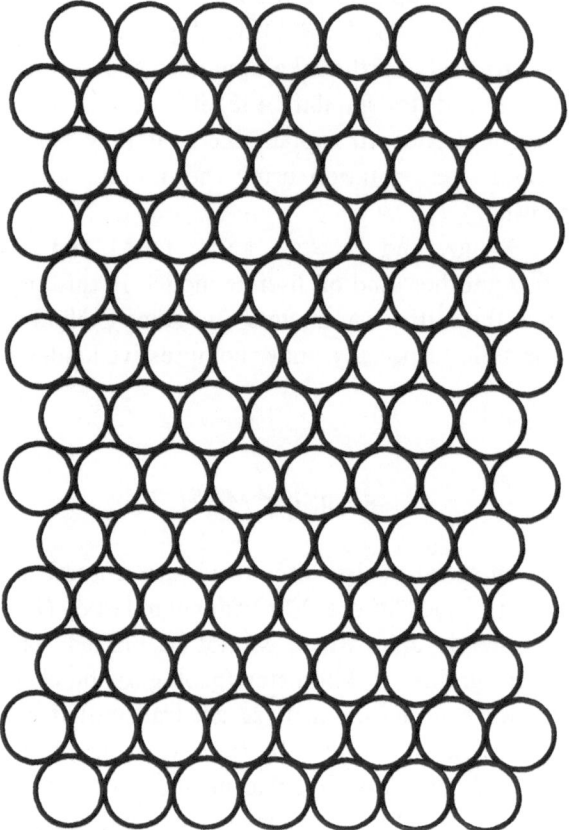

Figure 1
2-D Discrete lattice model.

$$K_m = \frac{\sqrt{3} r_0^2 E}{18} \tag{4}$$

where

$$\alpha = \frac{1 - 3\nu}{1 + \nu} \tag{5}$$

$$\nu = \frac{1 - \alpha}{3 + \alpha} \tag{6}$$

and r_0 is the diameter of the particles. The equations above reflect the connection between the mesoscopic and macroscopic parameters. From the equations above we also know that the materials of ν from 0 to 1/3 can be modeled, and if the tangential rigidity is neglected ($K_s = 0$), $\nu = 1/3$.

Fracture Criterion and States after Fracture

The link between two adjacent particles can break independently in the following three different ways:
When radial extension force exceeds the maximum value F_{r0},
or, when tangential force exceeds the maximum value F_{s0},
or, when moment exceeds the maximum value M_0.

Generally three kinds of interactions always exist at the same time. We use the following empirical criterion similarly to judge if the link will break:

$$\frac{F_r}{F_{r0}} + \frac{|F_s|}{F_{s0}} + \frac{|M|}{M_0} \geq 1. \tag{7}$$

In the simulation, we choose $F_{r0} = K_r(\Delta r)_m = K_r r_0 \varepsilon_m$, $\varepsilon_m = 0.005-0.010$, $F_{s0} = CF_{r0}$ and $M_0 = Dr_0 F_{r0}$, where $C = 0.3-0.7$ and $D = 1/6$ are chosen empirically. The effect of radial force on the tangential and bending fracture is considered.

There are four kinds of linking states between neighboring particles:
(A) Intact, (B) Sliding, (C) Locked by static friction, (D) Departing.
Table 1 shows whether or not the forces and moment can be transmitted in each case.

The states can change according to:
(a) A → B, when (7) is met, and $r \leq r_0$ and $V_t \neq 0$
(b) A → D, when (7) is met, and $r \geq r_0$
(c) B → C, when $V_t = 0$
(d) C → B, when $f_s \geq \mu_s F_r$
(e) B → D, when $r \geq r_0$
(f) C → D, when $r \geq r_0$
(g) D → B, when $r \leq r_0$ and $V_t \neq 0$
(h) D → C, when $r \leq r_0$ and $V_t = 0$
where f_s stands for static frictional force, μ_s refers to static frictional coefficient, and V_t, is relative tangential velocity.

Table 1

f_d and μ_d are sliding frictional force and sliding frictional coefficient, respectively.

	F_r	F_s	M
A	can	can	can
B	can, only when $r \leq r_0$	sliding frictional force $f_d = \mu_d F_r$	cannot
C	can, only when $r \leq r_0$	static frictional force	cannot
D	cannot	cannot	cannot

Calculation of Static Friction

When two particles are locked by static friction, if the tangential rigidity is chosen to be infinite (PLACE and MORA, 1999), the frictional forces are calculated according to all forces acting on the particles. This is a difficult step considering that frictional forces are all inter-dependent when infinite shear rigidity is chosen. In our model, the tangential rigidity is not infinite, but a finite one. The static frictional force can be easily decided by the relative tangential displacement according to the principle of DEM. Figure 2 illustrates how the shear forces between bonded particles are calculated. Similarly, static frictional forces are calculated from the relative displacements at the contact surfaces (cf., Fig. 3).

Solution of Equation of Motion

The differential scheme of MD is used as in MORA's model, with the only difference of adding the equation of θ. Due to the changes of linking states between two adjacent particles (e.g., breaking and sliding), the forces and accelerations may be discontinuous. These discontinuities are dealt with from $t^- + \Delta t$ to $t^+ + \Delta t$, and no breaking and sliding occurs from t^+ to $t^- + \Delta t$. However, the positions and velocities are continuous. These quantities are calculated in the following way.

First, calculate the positions at $t + \Delta t$

$$\begin{cases} \vec{X}_i(t + \Delta t) = \vec{X}_i(t) + \Delta t \, \vec{X}'_i(t) + \frac{(\Delta t)^2}{2} \, \vec{X}''_i(t^+) \\ \theta_i(t + \Delta t) = \theta_i(t) + \Delta t \, \theta'_i(t) + \frac{(\Delta t)^2}{2} \, \theta''_i(t^+) \end{cases} \qquad (8)$$

Second, calculate the forces (and moments) and accelerations at $t^- + \Delta t$,

$$\begin{cases} \vec{X}''_i(t^- + \Delta t) = \vec{F}_i(t^- + \Delta t)/m_i \\ \theta''_i(t^- + \Delta t) = M_i(t^- + \Delta t)/I_i \end{cases} \qquad (9)$$

where m_i and I_i are the mass and rotational inertia of the i-th particle. Then we calculate the velocities at $t + \Delta t$

$$\begin{cases} \vec{X}'_i(t + \Delta t) = \vec{X}'_i(t) + \frac{\Delta t}{2} [\vec{X}''_i(t^+) + \vec{X}''_i(t^- + \Delta t)] \\ \theta'_i(t + \Delta t) = \theta'_i(t) + \frac{\Delta t}{2} [\theta''_i(t^+) + \theta''_i(t^- + \Delta t)] \end{cases} \qquad (10)$$

Finally, judge whether breaking or sliding occurs between any particles pair, if so, update the forces and accelerations from $t^- + \Delta t$ to $t^+ + \Delta t$ according to the rules mentioned above.

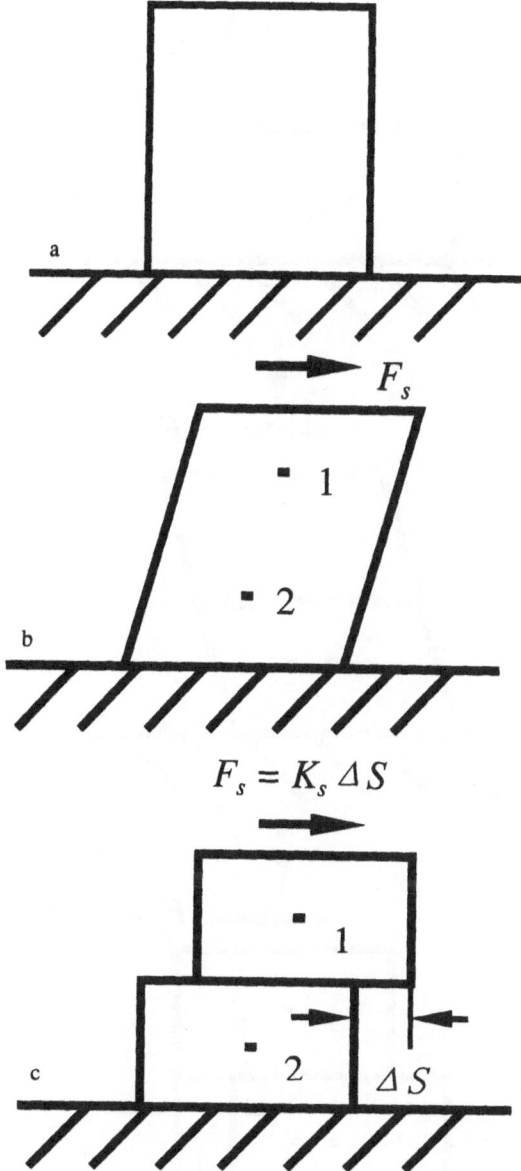

Figure 2
Sketch of calculation of shear forces using the Discrete Element Method in which a sample is subjected to shear load. The initial sample (inset a), composed of two bonded blocks, is subjected to a shear force (inset b). By assuming that the blocks are rigid, the resulting shear force between block 1 and block 2 can be calculated from the relative displacement of the blocks (inset c).

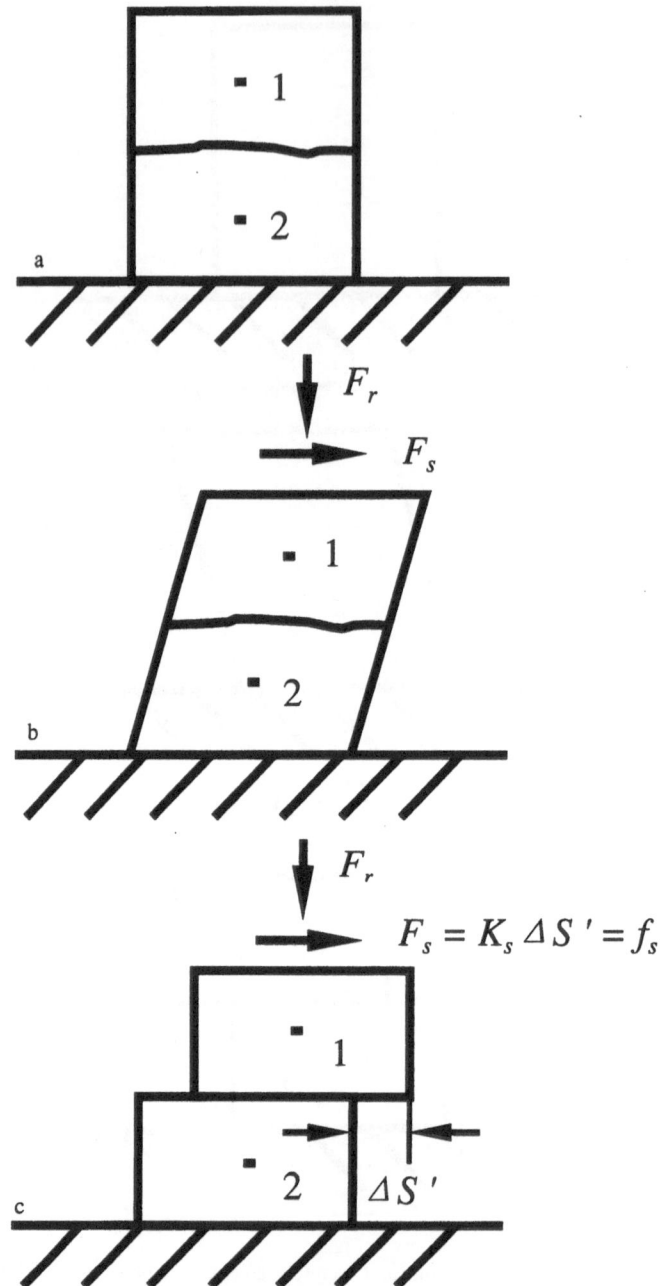

Figure 3

Sketch of calculation of static frictional force. The initial sample (inset a) is composed of two unbonded blocks locked by static friction. When subjecting the sample to a shear force, shear deformations occur (inset b). By assuming that the blocks are rigid, the frictional forces locking the two blocks can be calculated from the relative displacement of the two blocks (inset c).

Viscosity is also introduced, so $\vec{F}_i = \vec{F}_i^1 + \vec{F}_i^\eta$, $\vec{F}_i^1$ includes elastic forces and frictional forces, $\vec{F}_i^\eta$ is viscous force. We use

$$\vec{F}_{ij}^\eta = -\eta(\vec{V}_i - \vec{V}_j) \tag{11}$$

where η is coefficient of viscosity. Equations (9), (10) and (11) indicate that due to the viscosity, computation of the accelerations at $t^- + \Delta t$ requires the velocities at $t + \Delta t$, however the velocities update also requires accelerations at $t^- + \Delta t$, so iteration must be used to calculate $\vec{X}_i''(t + \Delta t)$.

First, we choose a group of approximate trial solutions $\tilde{\vec{X}}_i'(t + \Delta t)$

$$\tilde{\vec{X}}_i'(t + \Delta t) = \vec{X}_i'(t) + \Delta t \vec{X}_i''(t) \quad (i = 1, N)$$

then use the following iteration scheme. (take $\tilde{x}_i'(t + \Delta t)$ as an example),

Calculate $\vec{F}_i^\eta$ (Eq. 11)

Calculate $\vec{F}_i(t^- + \Delta t)$ and $x_i''(t^- + \Delta t)$ (Eq. 8) (Eq. 9)

Calculate $x_i'(t + \Delta t)$ (Eq. 10)

Compare $\tilde{x}_i'(t + \Delta t)$ with $x_i'(t + \Delta t)$:

If $\left| \dfrac{\tilde{x}_i'(t + \Delta t) - x_i'(t + \Delta t)}{x_i'(t + \Delta t)} \right| \geq \varepsilon$ then

$\tilde{x}_i'(t + \Delta t) = \tilde{x}_i'(t + \Delta t)$ and repeat iteration

If $\left| \dfrac{\tilde{x}_i'(t + \Delta t) - x_i'(t + \Delta t)}{x_i'(t + \Delta t)} \right| \leq \varepsilon$ then

$\tilde{x}_i'(t + \Delta t) = \tilde{x}_i'(t + \Delta t)$ and stop iteration

where ε stands for an iteration precision. The convergence rate is rapid and three to four iterations were adequate.

Some Preliminary Results

Uniaxial Compressive Test

The failure process of brittle rocks under compressive stress was modeled. In our numerical test, the number of particle is 25×51, $m = 1.0$, $r_0 = 4.0$, $K_r = 1000$, $dt = 0.01$, $v = 0.2$, $\eta = 0.6$. K_s, K_m and E are decided according to equations (2)–(6). Figure 4 shows the results of a homogeneous, intact sample subjected to increased uniaxial compressive stress on the top and bottom ($\sigma = 0.2\ t$). The black lines mean that the links between two particles are intact, the grey ones stand for broken but

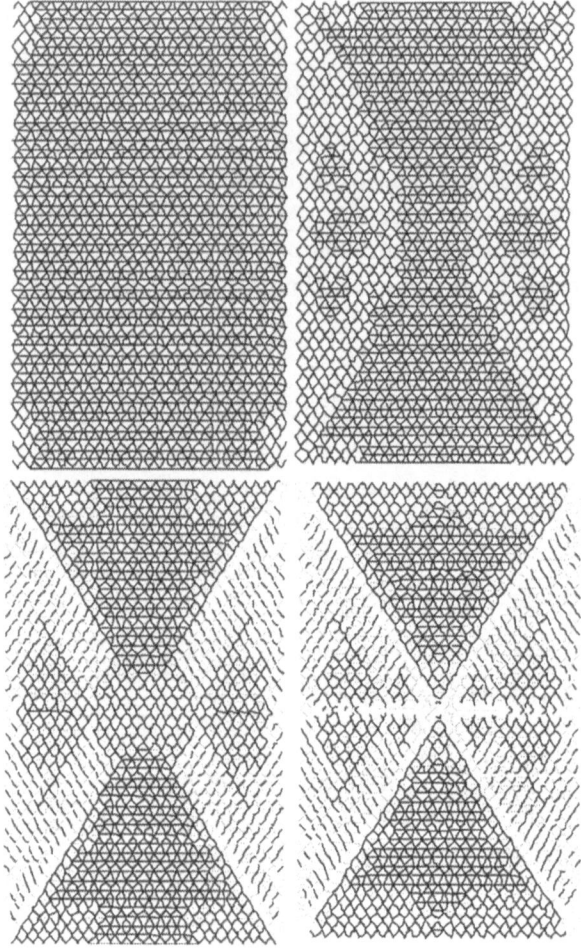

Figure 4

The fracture process of a homogeneous sample (all of the rigidity parameters and fracture parameters are identical) subjected to uniaxial compressive load.

contacting links, and disappearances of lines represent broken links. First, extensive cracks ($r \geq r_0$ when broken) appear on the corners, then spread inside. Later, shear cracks ($r < r_0$ when broken) spread along the diagonals, forming macroscopic conjugate X-shaped failure patterns with two relative intact pyramidal parts and two relative fractured parts.

If disorder is introduced, the results may be different. For example, Figure 5 shows the sample with 2% randomly distributed initial defects. By defects, we mean the bonds are pre-broken but the adjacent particles are still in contact with one

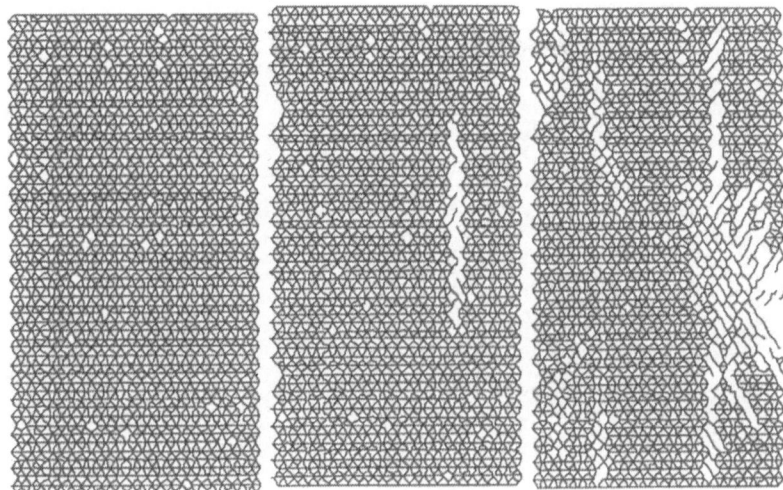

Figure 5
Brittle cleavage of material sample subjected to uniaxial compressive load, 2% bonds are randomly broken before loading.

another. In this case, vertical cleavage appears abruptly, then follows collapse of the sample, which is often observed in brittle rock experiments.

In addition, the dynamic expansion of closed oblique crack is also simulated (Fig. 6). It is seen that tensile cracks expand from the tips of the oblique crack. The results are also similar with rock experiments (BRACE, 1960; BRACE and BOMBAL-AKIS, 1963).

Figure 6
Extension of closed oblique crack under uniaxial pressure, tensile cracks ($r \geq r_0$ when broken) expand from the tips of the oblique crack.

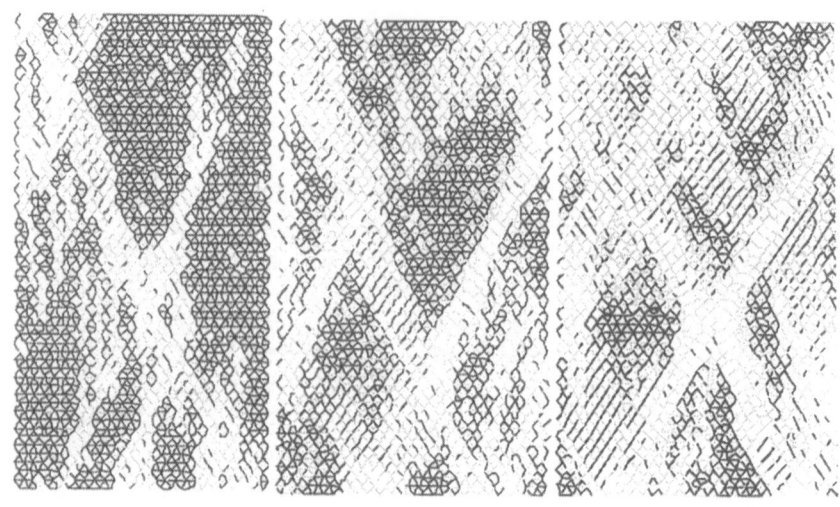

Figure 7
Effects of confining pressure on macroscopic failure modes. Axial stress $\sigma_1 = 0.2t$, confining pressure $\sigma_2 = \lambda\sigma_1$ for $t < 10$ and $\sigma_2 = $ const for $t > 10$. From left to right, $\lambda = 0.1, 0.2, 0.4$, confining pressure is increasing, the angles between two groups of shear bands are increasing, and more shear bands can be seen.

Effects of Confining Pressure

If there is confining pressure, the failure pattern is represented by two groups of shear bands nearly along two diagonal directions. The effect of confining pressure on failure patterns is obvious in Figures 7 and 8, with an increase of the confining pressure, the failure process tends to be more ductile; the strength becomes larger; there are more shear bands in each group, and the angle between two groups of main shear bands also increases.

Simulation of Seismicity

Considering that earthquakes in China are mainly intraplate earthquakes, we used a model different from the earthquake model for transform faults (such as the San Andreas Fault). In this model (Fig. 9) faults of different sizes are distributed in random directions and positions; the model is subjected to an increased compressive load and confining pressure. An earthquake is defined as an event releasing potential energy, such as the breaking of a single bond or sliding between two particles, but if the adjacent events occur in successive time steps, these are considered to be part of a single larger event. The magnitudes are decided according to the potential energy E released by the broken bonds

$$M = c_1 + c_2 \log E \qquad (12)$$

where c_1 and c_2 are constants. In seismology, $\log E = 11.8 + 1.5 \ M$, so $c_1 = -7.8667$, $c_2 = 0.6667$. Here we still choose $c_2 = 0.6667$, c_1 is chosen to make the magnitude of the smallest event be 0. In this way, seismic activity, such as frequency (Fig. 10), energy release (Fig. 11) and M-t charts (Fig. 12) are gained using this model.

Some factors that influence b values are discussed. The random precracks are distributed according to $P(l) \propto cl^{-\gamma}$, where c is a constant, representing the density of cracks, l is the length of faults (SCHOLZ and COWIE, 1990), and $\gamma > 0$. b values are calculated according to the well-known G-R relation. In Figure 13 it is easy to see that the more cracks there are (the bigger c is), the bigger the b values are,

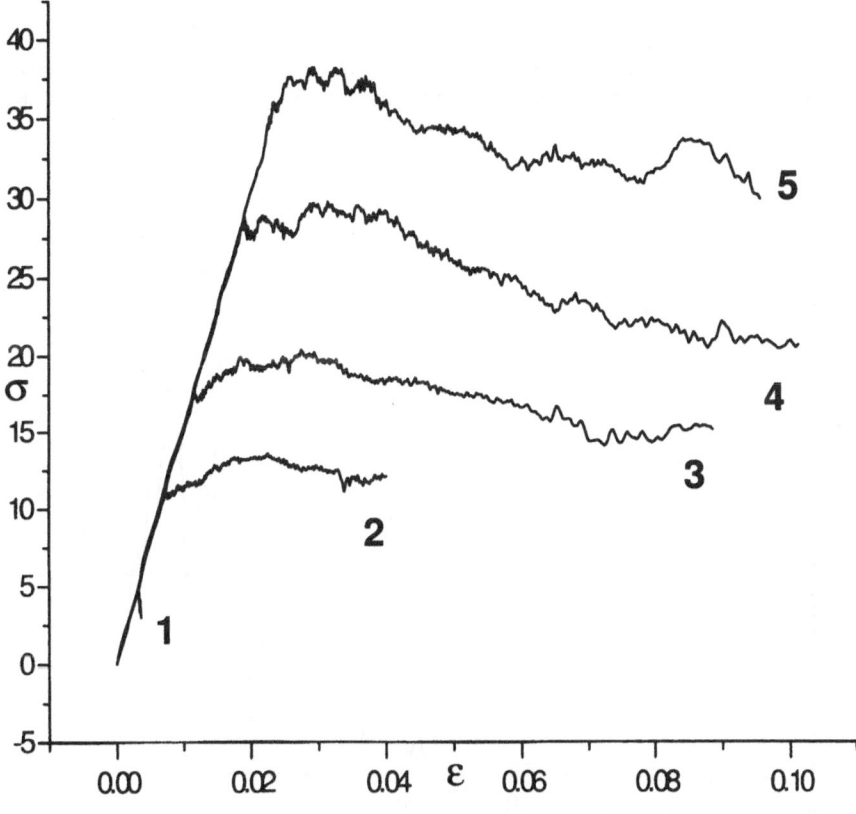

Figure 8
The simulated axial stress-strain curves with different confining pressure. Curve 1 has no confining pressure ($\lambda = 0$), brittle fracture is seen, from 2 to 5, $\lambda = 0.1$, 0.2, 0.4, 0.6. With the increase of confining pressures, the strengths also increase. The fracture processes tend to be more ductile.

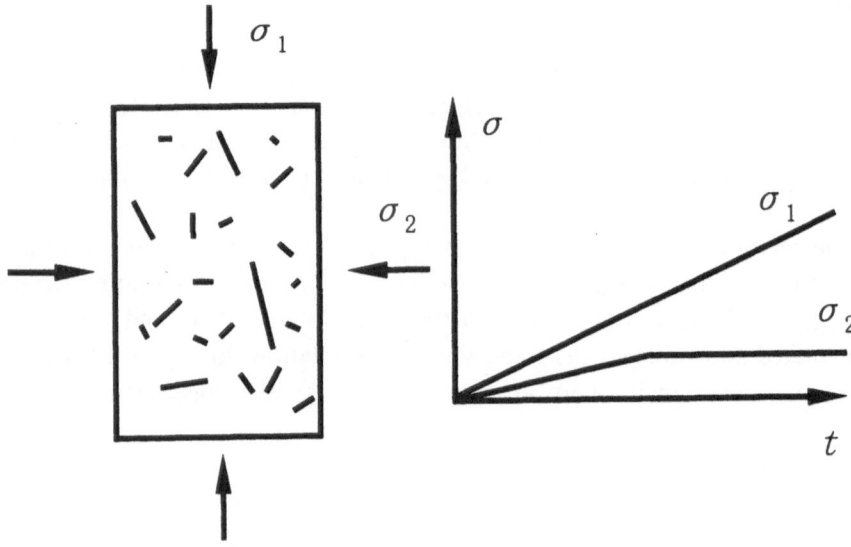

Figure 9
An earthquake model with random cracks. The model is subjected to increased compressive load
$\sigma_1 = 0.2\, t$ and confining pressure which increases first and then remains constant ($\sigma_2 = 0.2\sigma_1$ for $t < 10$
and $\sigma_2 = \text{const}$ for $t > 10$).

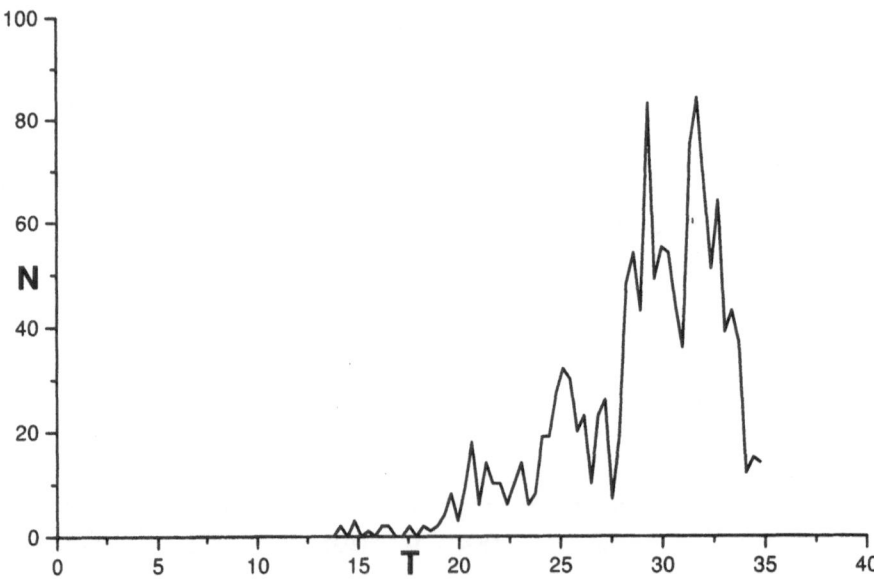

Figure 10
The variations of frequency (numbers of earthquakes during per unit time) of modeled earthquakes with
time.

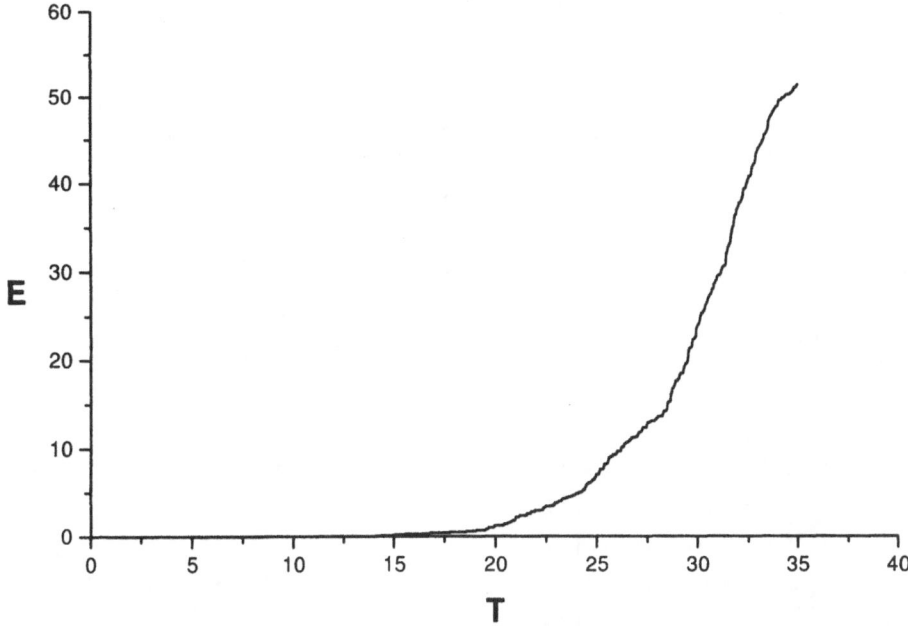

Figure 11
Plot of accumulated energy released by breaking of bonds or sliding between particle pairs, accelerated
releases of energy correspond to bigger earthquakes.

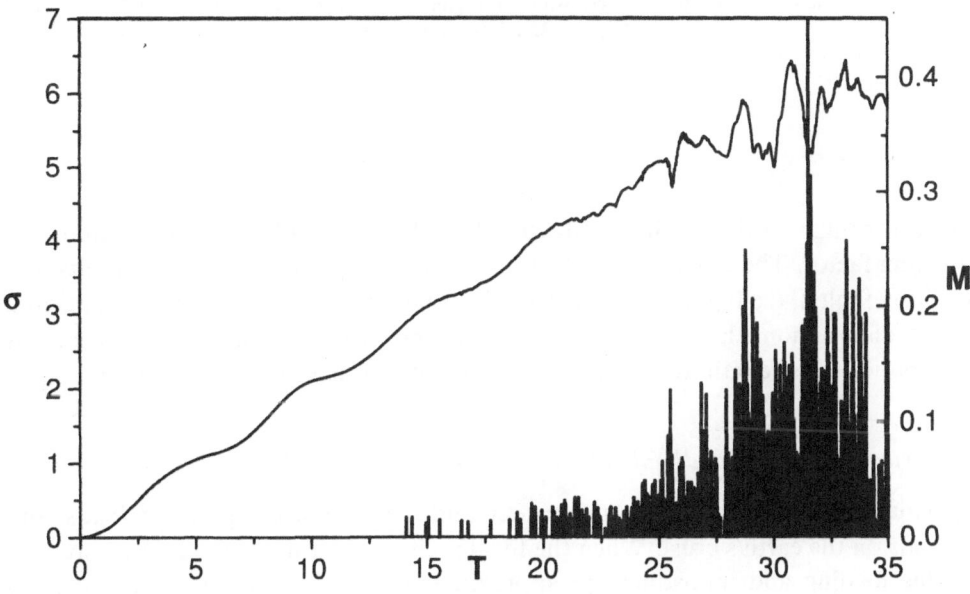

Figure 12
Modeled magnitudes of events and main compressive stress curves. The stress drops correspond to
earthquakes.

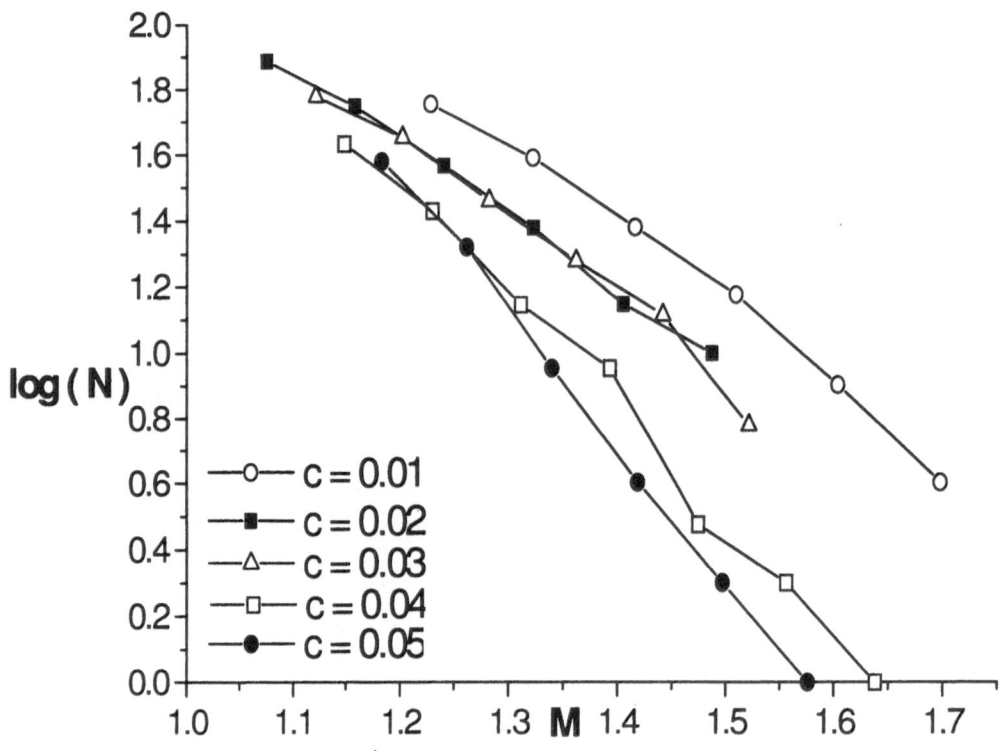

Figure 13

The effect of crack density on *b* values, with increasing of crack density *c*, *b* values (the slope of log $N - M$ relation) increase.

which is similar to the results of MOGI (1976). In Figure 14 γ is also an important effecting factor. The bigger γ is (which means that there are more short faults and less long faults), the bigger the *b* values are. In fact, besides the density of faults and length distribution, there must be other effecting factors, such as inhomogeneous distribution of strength and elastic constants, which deserves further study.

Numerical Study of Load/Unload Response Ratio Theory

Tidal forces generated by the moon and the sun exert periodic loads and unloads on the earth's crust. When the focal region is in a stable state, the responses to this loading and unloading are approximately equal, however when the focal region is unstable and near fracture, the responses are quite different. According to this principal, YIN proposed Load/Unload Response Ratio Theory (LURR, YIN *et*

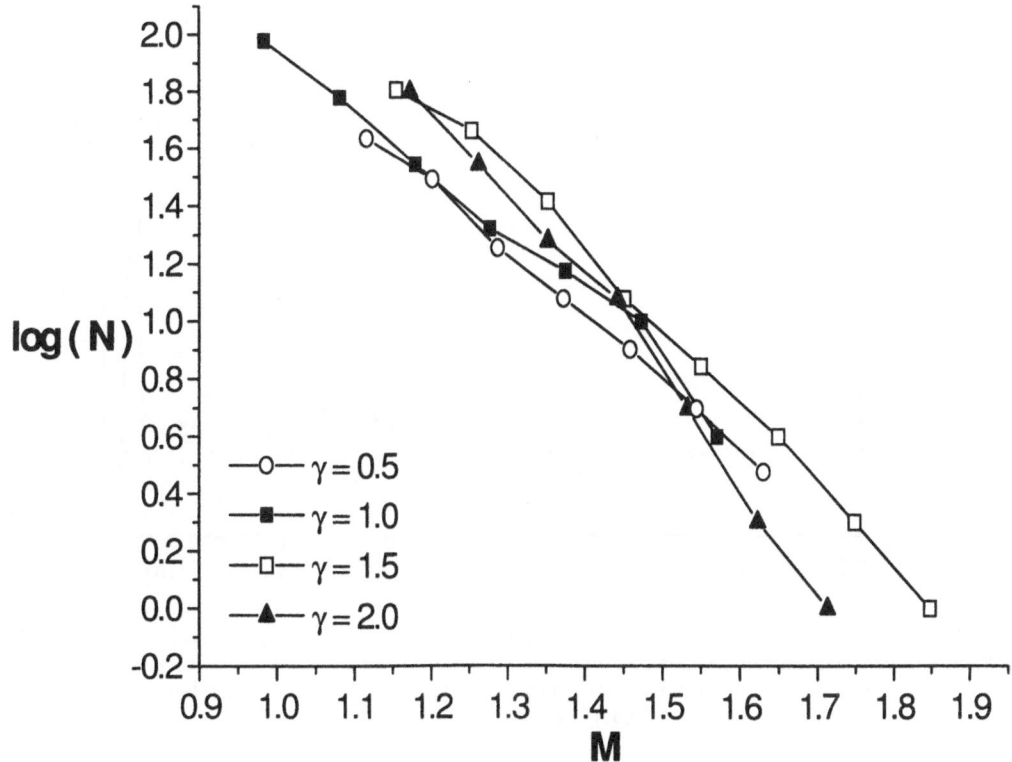

Figure 14

The effect of distribution of length of cracks on b values, with increasing γ (there are more short cracks and less long ones), b values increase.

al., 1995) to judge the imminence of earthquake. This method is successfully applied for medium-term earthquake prediction (YIN *et al.*, 2000, this issue).

We simulated LURR theory using the model above. The main compressive stress is increased with a small additional sine-shaped disturbance, which modeled loading and unloading. LURR values are decided by

$$LURR = \frac{\sum \sqrt{E_+}}{\sum \sqrt{E_-}} \tag{13}$$

where E_+ and E_- are earthquake energy released during loading and unloading periods, respectively. Figure 15 presents the results. Figure 16 is an $M\text{-}t$ chart corresponding to Figure 15. It is clearly seen that LURR values fluctuate about

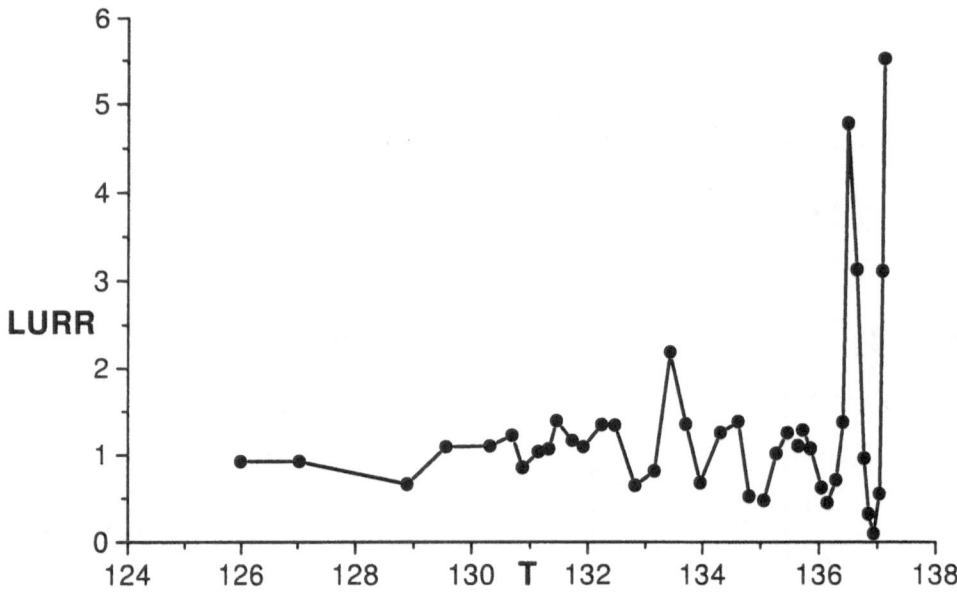

Figure 15
Simulated variations of Load/Unload Response Ratio (LURR) with time, LURR values fluctuate about 1 when there is no large earthquake or when seismicity is weak, before the largest earthquake ($T = 137.05$), LURR values rise markedly.

1 when there is no large earthquake or when the seismicity is low, but before the largest earthquake in the sequence, LURR values rise markedly, which is similar to real earthquake cases (YIN *et al.*, 2000, this issue). This indicates that LURR is a quantitative parameter to judge the closeness of an earthquake.

Conclusion

The preliminary simulations indicate that the model is capable of simulating rock failure, especially under compressive loads. However when modeling the earthquake process, our simulations are still coarse, since the number of particles is very small. Simulations have a narrow range of grain sizes, whereas grain sizes in the crust have a broad distribution, therefore large scales of simulations are needed in future work.

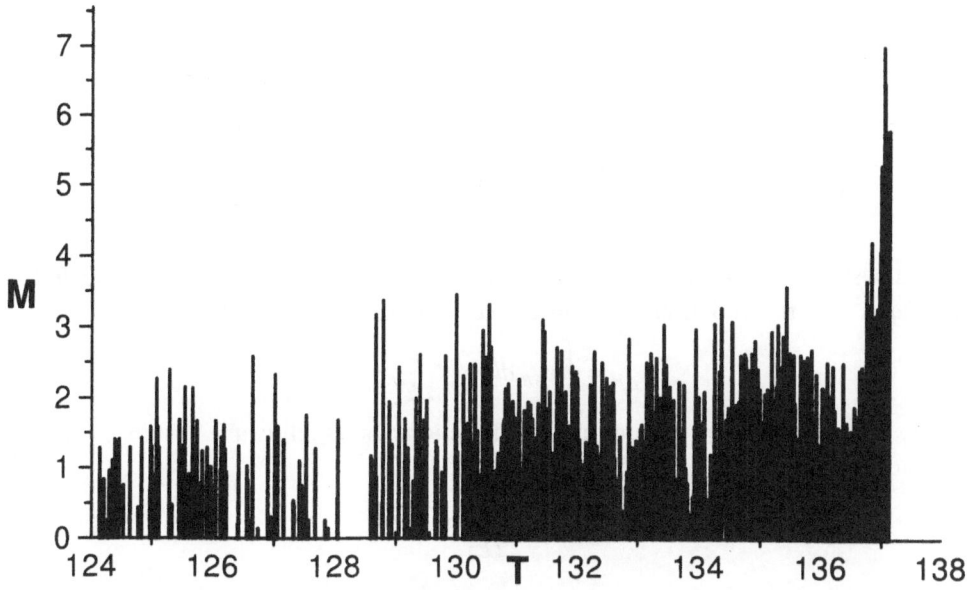

Figure 16
M-t chart corresponding to Figure 15.

Acknowledgements

This research was funded by the National Scientific Foundation of China (Grant No. 19732060), the Earthquake Association Foundation (Grant No. 95-07-435), and the Laboratory of Nonlinear Mechanics, Institute of Mechanics, CAS. The authors are grateful to Prof. MORA and Dr. PLACE for their valued suggestions.

Appendix

In order to make a realistic simulation, the lattice rigidity parameters K_r, K_s, K_m cannot be chosen arbitrarily, but must be chosen according to the elastic parameters of the continuum E and v. We determine these parameters by letting the nodal displacements of the lattice equal the displacements of corresponding points in the continuum subjected to specifically chosen strain conditions (so-called patch test, BATHE and WILSON, 1976).

First, consider the case of uniaxial elongation (Fig. 17). The stresses, strains and displacements (u, v) are

$$\begin{cases} \sigma_x = 2q \\ \sigma_y = 0 \\ \tau = 0 \end{cases}, \quad \begin{cases} \varepsilon_x = \dfrac{2q}{E} \\ \varepsilon_y = -\dfrac{2qv}{E} \\ \gamma = 0 \end{cases} \quad \text{and} \quad \begin{cases} u = \dfrac{2q}{E}x \\ v = -\dfrac{2qv}{E}y \end{cases}$$

and the strain energy density is $\phi = 2q^2/E$, where q is a small value such that $\varepsilon_x = 2q/E \ll 1$, the radial elongation between particles A and B is

$$\Delta r_{AB} = \int \varepsilon_{AB}\, dl_{AB} = \frac{q}{2E}(3-v)r_0$$

where ε_{AB} is extension strain in AB direction, r_0 equilibrium distance between two adjacent particles.

$$\Delta r_{AC} = \frac{q}{2E}(3-v)r_0$$

$$\Delta s_{AB} = \Delta s_{AC} = \frac{\sqrt{3}q}{2E}(1+v)r_0$$

$$\Delta r_{BC} = \frac{2q}{E}vr_0$$

$$\Delta s_{BC} = 0$$

where Δs_{AB} represents transverse displacement between particles A and B. The potential energy stored in triangle ABC

$$E_p = \tfrac{1}{2}[\tfrac{1}{2}K_r(\Delta r_{AB}^2 + \Delta r_{AC}^2 + \Delta r_{BC}^2) + \tfrac{1}{2}K_s(\Delta s_{AB}^2 + \Delta s_{AC}^2 + \Delta s_{BC}^2)] = \tfrac{1}{2}r_0 \frac{\sqrt{3}r_0}{2}\phi,$$

$$K_r(3v^2 - 2v + 3) + K_s(v^2 + 2v + 1) = \frac{4\sqrt{3}}{3}E. \tag{14}$$

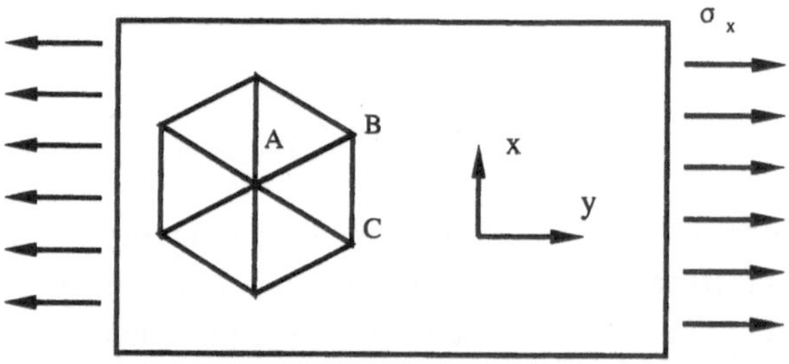

Figure 17
The lattice and continuum model subjected to uniaxial tensile stress. The nodal displacements of the lattice equal the displacements of corresponding points in the continuum, subjected to some specifically chosen strain conditions. Lattice parameters can be decided such that the lattice has realistic elastic properties.

In the second case (Fig. 18), the stresses become

$$\begin{cases} \sigma_x = 0 \\ \sigma_y = 0 \\ \tau = -p \end{cases}$$

where p is a small value such that $p/E \ll 1$. Similarly, we have

$$K_r + K_s = \frac{2\sqrt{3}}{3}\frac{E}{1+v}. \tag{15}$$

From (13) and (14), K_r and K_s can be solved

$$K_r = \frac{\sqrt{3}E}{3(1-v)} \tag{16}$$

$$K_s = \frac{1-3v}{1+v}K_r. \tag{17}$$

In both cases above, the rigid rotation $\omega_z = \frac{1}{2}(\partial v/\partial x - \partial u/\partial y) = 0$, $\Delta\theta = 0$, so K_m is not involved.

In the third case (Fig. 19), the object is subjected to a bending moment:

$$\begin{cases} \sigma_x = 6hy \\ \sigma_y = 0 \\ \tau = 0 \end{cases} \quad \text{and} \quad \begin{cases} \varepsilon_x = \dfrac{6h}{E}y \\ \varepsilon_y = -\dfrac{6hv}{E}y \\ \gamma = 0 \end{cases}$$

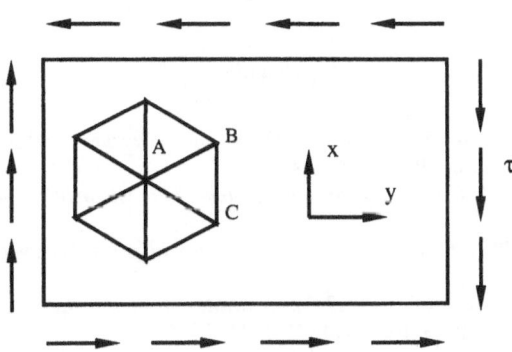

Figure 18
The lattice and continuum model subjected to shear stress.

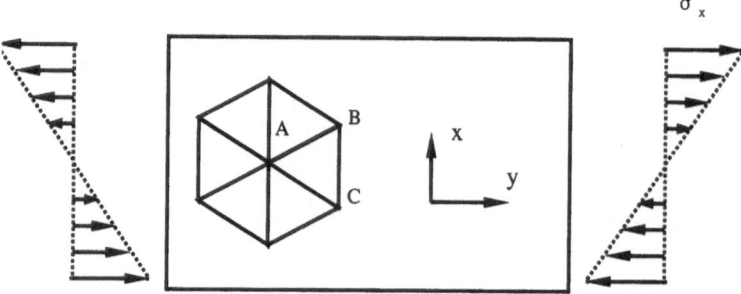

Figure 19
The lattice and continuum model subjected to bending moment.

where h is a small constant such that $h/E \ll 1$. In this case, the strain energy density is

$$\phi = \frac{18h^2}{E} y^2$$

and the rigid rotation is $\omega_z = -6hx/E$, K_m is involved in, and the energy caused by the bending between two particles must be taken into account. Similarly, we can obtain

$$K_m = \frac{\sqrt{3} r_0^2 E}{18}. \tag{18}$$

Lastly, if the lattices in Figures 17–19 are turned over an arbitrary angle a, the parameter a does not appear in equations (14)–(18). Hence, such a lattice is isotropic if K_r, K_s, K_m are identical for all particle pairs.

References

ASHURT, W. T., and HOOVER, W. G. (1976), *Microscopic Fracture Studies in the Two-dimensional Triangular Lattice*, Phys. Rev. B *14*, 1465–1473.

ASKAR, A., and CAKMAK, A. S. (1968), *A Structural Model of a Micropolar Continuum*, Int. J. Eng. Sci. *6*, 583–589.

BÅK, P., and TANG, C. (1989), *Earthquakes as a Self-organized Critical Phenomena*, J. Geophys. Rev. *94*, 15,635–15,637.

BARRIERE, B., and TURCOTTE, D. L. (1991), *A Scale—Invariant Cellular—Automata Model for Distributed Seismicity*, Geophys. Res. Lett. *18*, 2011–2014.

BATHE, K. J., and WILSON, E. L., *Numerical Methods in Finite Element Analysis* (Prentice-Hall, Inc. Englewood Cliffs, NJ 1976).

BATHURST, R. J., and ROTHENBURG, L. (1988), *Micromechanical Aspects of Isotropic Granular Assemblies with Linear Contact Interactions*, J. Appl. Mech. *55*, 17–23.

BORN, M., and HUANG, K., *Dynamical Theory of Crystal Lattice* (Oxford Univ. Press, New York, NY 1954).

BRACE, W. F. (1960), *An Extension of the Griffith Theory of Fracture to Rocks*, J. Geophys. Rev. *65*, 3477–3480.

BRACE, W. F., and BOMBALAKIS, E. G. (1963), *A Note on Brittle Crack Growth in Compression*, J. Geophys. Rev. *68*, 3709–3713.

BRACE, W. F., and BYERLEE, J. D. (1966), *Stick-slip as a Mechanism for Earthquakes*, Science *153*, 990–992.

BURRIDGE, R., and KNOPOFF, L. (1967), *Model and Theoretical Seismicity*, B.S.S.A. *57*, 341–371.

CARLSON, J. M., and LANGER, J. S. (1989), *Mechanical Model of an Earthquake Faults*, Phys. Rev. A *40*, 6470–6484.

CARLSON, J. M., LANGER, J. S., and SHAW, B. E. (1989), *Dynamics of Earthquake Faults*, Rev. Mod. Phys. *66*, 657–670.

CUNDALL, P. A., and STRACK, O. D. L. (1979), *A Discrete Element Model for Granular Assemblies*, Geotechnique *29*, 47–65.

CURTIN, W. A., and SCHER, H. (1990a), *Brittle Fracture in Disordered Materials: A Spring Network Model*, J. Mater. Res. *5*, 535–553.

CURTIN, W. A., and SCHER, H. (1990b), *Mechanics Modeling Using a Spring Network*, J. Mater. Res. *5*, 554–562.

DAY, S. M. (1982), *Three-dimensional Finite Difference Simulation of Faults Dynamics: Rectangular Faults with Fixed Rupture Velocity*, B.S.S.A. *72*, 705–727.

FUKUYAMA, E., and MADARIAGA, R. (1995), *Integral Equation Method for Plane Crack with Arbitrary Shape in 3-D Elastic Medium*, B.S.S.A. *85*, 614–628.

HERRMANN, D. W., *Computer Simulation Methods in Theoretical Physics*, second edition (Springer-Verlag, Berlin, New York 1989).

JAGOTA, A., and DAWSON, P. R. (1988a), *Micromechanical Modeling of Powder Compacts—I. Unit Problems for Sintering and Traction Induced Deformation*, Acta Metall. *36*, N,2551–2561.

JAGOTA, A., and DAWSON, P. R. (1988b), *Micromechanical Modeling of Powder Compacts—II. Truss Formulation of Discrete Packing*, Acta Metall. *36*, 2563–2573.

JAGOTA, A., and SCHERR, G. W. (1993), *Viscosities and Sintering Rates of a Two-dimensional Granular Composite*, J. Am. Ceram. Soc. *76*, 3123–3135.

KNOPOFF, L. (1993), *Self-organization and the Development of Pattern: Implications for Earthquake Prediction*, Proceedings of the American Philosophical Society *137*, 3.

LOMNITZ-ADLER, J. (1993), *Automation Models of Seismic Fracture: Constraints Imposed by Magnitude-frequency Relation*, J. Geophys. Rev. *98*, 17,745–17,756.

MADARIAGA, R., OLSEN, K. B., and ARCHULETA, R. J. (1997), *3-D Finite-difference Simulation of a Spontaneous Rupture*, Seismol. Res. Lett. *68*, 312.

MOGI, K. (1976), *Regional Variations in Magnitude Frequency Relation of Earthquakes*, Bull. Earthquake Res. Inst. *45*, 313–325.

MONETTE, L., and ANDERSON, M. P. (1994), *Elastic and Fracture Properties of the Two-dimensional Triangular and Square Lattice*, Modelling Simul. Mater. Sci. Eng. *2*, 52–66.

MORA, P., and PLACE, D. (1993), *A Lattice Solid Model for the Nonlinear Dynamics of Earthquakes*, Int. J. Mod. Phys. C *4*, 1059–1074.

MORA, P., and PLACE, D. (1994), *Simulation of the Frictional Stick-slip Instability*, Pure appl. geophys. *143*, 61–87.

MORA, P., and PLACE, D. (1998), *Numerical Simulation of Earthquake Faults with Gouge: Towards a Comprehensive Explanation for the Heat Flow Paradox*, J. Geophys. Res. *103*, 21,067–21,089.

PLACE, D., and MORA, P. (1999), *The Lattice Solid Model to Simulate the Physics of Rocks and Earthquake: Incorporation of Friction*, J. Comp. Phy. *150*, 332–372.

SAMMIS, C. G., and SMITH, S. W. (1999), *Seismic Cycles and the Evolution of Stress Correlation in Cellular Automation Models of Finite Faults Networks*, Pure appl. geophys. *155*, 307–334.

SCOTT, D. R. (1996), *Seismicity and Stress Rotation in a Granular Model of the Brittle Crust*, Nature *381*, 592–595.

SCHOLZ, C. H., and COWIE, P. A. (1990), *Determination of Geologic Strain from Fault Slip Data*, Nature *346*, 837–839.

TANG, C. A. (1997), *Numerical Simulation of Rock Failure and Associated Seismicity*, Int. J. Rock Mech. Min. Sci. *34*, 249–262.

TOI, Y., and CHE, J. S. (1994), *Computational Damage Mechanics Models for Brittle Microcracking Solids Based on Mesoscopic Simulation*, Eng. Fracture Mech. *48*, 483–498.

TOI, Y., and KIYOSUE, T. (1995), *Damage Mechanics Models for Brittle Micro-cracking Solids Based on Three-dimensional Mesoscopic Simulation*, Eng. Fracture Mech. *50*, 11–27.

YIN, X. C., CHEN, X. Z., SONG, Z. P., and YIN, C. (1995), *A New Approach to Earthquake Prediction: The Load/Unload Response Ration (LURR) theory*, Pure appl. geophys. *145*, 701–715.

YIN, X. C., WANG, Y. C., PENG, K. Y., BAI, Y. L., WANG, H. T., YIN, X. F. (2000), *Development of a New Approach to Earthquake Prediction: Load/Unload Response Ratio (LURR) Theory*, Pure appl. geophys. *157*, 2365–2383.

(Received July 31, 1999, revised February 20, 2000, accepted April 28, 2000)

 To access this journal online:
http://www.birkhauser.ch

Pure appl. geophys. 157 (2000) 1929–1943
0033–4553/00/121929–15 $ 1.50 + 0.20/0

Pure and Applied Geophysics

Damage Localization as a Possible Precursor of Earthquake Rupture

H. L. Li,[1] Y. L. Bai,[1] M. F. Xia,[1,2] F. J. Ke[1,3] and X. C. Yin[1,4]

Abstract—Based on the concepts of statistical mesoscopic damage mechanics, the rupture of a heterogeneous medium is investigated in terms of numerical simulations of a network model, subjected to simple shear loading. The heterogeneities are simulated by varying the sizes and fracture strains of the elements of the network. Progressive damage is governed by a damage field equation and a dynamic function of damage (DFD). From the damage field equation, a criterion for damage localization can be derived, and the DFD can be extracted from the simulations of the network. Importantly, the DFD intrinsically governs the damage localization. Both stress-free and periodic boundary conditions for the network are examined. It is found that damage localization may be the underlying mechanism of eventual rupture and thus could be used as a possible precursor of earthquake rupture.

Key words: Heterogeneous medium, rupture, statistical mesoscopic damage mechanics, dynamic function of damage, damage localization.

1. Introduction

Many efforts have been made in an attempt to extract a realistic picture of the mechanical nature of earthquake rupture from seismological, geological and other geophysical data. Investigations have been carried out from the viewpoint of fracture mechanics (RICE, 1980; EVANS, 1978), stick-slip rate-dependent fault models (DIETERICH, 1978; RICE and RUINA, 1983), and the dynamic of fault rupture (DAS and AKI, 1977). Especially recently, earthquake nucleation and the effects of heterogeneity under simple (unidirectional) shear loading were examined numerically (DIETERICH, 1992; KEMENY and HAGAMAN, 1992). These data indicate that earthquake rupture may result from the progressive damage of microstructures with varying sizes and strengths within fault sections. Thus the

[1] State Key Laboratory of Non-linear Mechanics, Institute of Mechanics, Chinese Academy of Sciences, Beijing 100080, China. E-mail: lihl@lnm.imech.ac.cn; baiyl@lnm.imech.ac.cn
[2] Department of Physics, Peking University, Beijing 100871, China. E-mail: xiam@lnm.imech.ac.cn
[3] Department of Applied Physics, Beijing University of Aeronautics and Astronautics, Beijing 100083, China. E-mail: kefj@lnm.imech.ac.cn
[4] Center for Analysis and Prediction, China Seismological Bureau, Beijing 100036, China. E-mail: yinxc@btamail.net.cn

statistical nature of the progressive damage of a fault, owing to the heterogeneous distribution of microdefects, offers us a real challenge.

Recently, statistical mesoscopic damage mechanics has been developed to describe the collective behavior of distributed microcracks and microvoids nucleated at the sites of the original defects in solids (BAI *et al.*, 1991). Also a transscale (from meso- to macroscopic) analysis has been proposed to correlate the important microstructural effects and the macroscopic mechanical behavior of solids, especially rupture. It has been found that, as a precursor of brittle rupture, the population of microdamage tends to form macroscopically localized damage (BAI *et al.*, 1998). This may provide an informative way to understand the abrupt rupture of a fault containing numerous defects.

In the present study, the basic concepts and theoretical formulation of the statistical mesoscopic damage mechanics are combined with the numerical simulation of a network model under simple shear loading to reveal the intrinsic mechanism governing the rupture of a heterogeneous medium. The network model contains elastic bar-elements (chains) with varying sizes and fracture strains resembling the heterogeneity. The eventual rupture in such a model consists of many breakings of chains nearly at the same time, after which the network cannot sustain further loading. The data of the numerical simulations are analyzed from the viewpoint of statistical mesoscopic damage mechanics. The important microstructural effects are included in a function referred to as the dynamic function of damage (DFD). It is found that the DFD may be an inherent factor which governs damage localization and the concentration of microfractures prior to the eventual rupture. Moreover, a criterion for the emergence of damage localization can be expressed by means of the DFD. This may provide a way to give an alarm before the rupture.

2. Basic Formulation of Statistical Mesoscopic Damage Mechanics

In order to deal with the statistical nature of the progressive failure of a fault, we need a quantitative model which can reveal the collective effects of nucleation and the growth of microdamages created at the sites of original defects. Subsequently a transscale analysis should be carried out to estimate to what extent this collective behavior will influence the macroscopic rupture.

Evolution Equation of Microdamage

The collective behavior of microdamages can be described by the following fundamental equation in the light of statistical mesoscopic damage mechanics (BAI *et al.*, 1991)

$$\frac{\partial n}{\partial T} + \sum_{i=1}^{l} \frac{\partial (n \cdot P_i)}{\partial p_i} = n_N - n_A, \tag{1}$$

where n is the number density of microdamage and T is time. p_i are the independent variables describing the state of microdamages. For example, p_i can be the current size of microdamage c, and the macroscopic spatial coordinates x of the element where the microdamages locate, etc. $P_i = \dot{p}_i$ are the rates of variables p_i. n_N and n_A are nucleation and healing rate densities of microdamage, respectively.

Macroscopic Damage

The number density of microdamages can be related to a macroscopic measurement of damage by the following definition.

$$D = \int_0^\infty \psi n \, dc, \tag{2}$$

where ψ is the average failure volume of a microdamage. When we assume that there is a spherical failure volume surrounding each microdamage, $\psi = (\pi/6) \, c^3$.

Dynamic Function of Damage (DFD)

With the definition of equation (2) in mind, one can turn the equation of number density of microdamages n, equation (1), into the governing equation for the macroscopic damage field

$$\frac{\partial D}{\partial T} + \nabla \cdot (D\tilde{v}) = f \tag{3}$$

with the dynamic function of damage f defined by

$$f = \int_0^\infty n_N \psi \, dc + \int_0^\infty (nA\psi' - n_A\psi) \, dc \tag{4}$$

where $\psi' = \partial \psi / \partial c$, and $\tilde{v}$ is the macroscopic velocity of the element where the microdamages locate. One key point of the present theoretical formulation is that all mesoscopic dynamics of microdamage, such as nucleation and healing rate densities of microdamage n_N and n_A, as well as the average growth rate of microdamage A are all included in the dynamic function of damage (DFD). The DFD serves to bridge the gap between the collective effects of mesoscopic damage mechanisms and the evolution of the macroscopic continuum damage field.

From the left-hand side of equation (3), the dynamic function of damage may be expressed as a function of the macroscopic mechanical variables and damage D. Without loss of generality, the DFD is assumed to be a function of damage D and far-field shear stress τ in the simple shear case $f = f(\tau, D)$ as we discuss hereafter in this paper.

Prediction of Damage Localization

According to the idea that damage localization may occur beyond a threshold and lead to eventual rupture (Fig. 1), it is of critical interest to propose a criterion which corresponds to this threshold. With the derivation given in (BAI *et al.*, 1998), here we only cite the expression for a simple lower bound of the criterion for damage localization. It is assumed that the damage localization would occur once the relative gradient of damage begins to increase with time. Under unidirectional shear loading, this is expressed as

$$\frac{\partial}{\partial T}\left[\frac{\partial D}{\partial Y}\Big/D\right] > 0 \tag{5}$$

where T and Y are temporal and spatial coordinates, respectively. (Y-direction is perpendicular to the section of damage localization.) Actually, it is

$$\left(\frac{\partial\left(\frac{\partial D}{\partial Y}\right)}{\partial T}\right)\Big/\left(\frac{\partial D}{\partial Y}\right) > \left(\frac{\partial D}{\partial T}\right)\Big/D. \tag{6}$$

Under quasi-static simple shear conditions, with the help of equation (3), the above inequality gives a simple lower bound to damage localization as

$$f_D > \frac{f}{D}. \tag{7}$$

where $f_D = (\partial f/\partial D)$. It is noteworthy that the dynamic function of damage f not only seems to characterize the evolution of collective microdamages, but also to intrinsically govern the damage localization process. This is why the DFD is emphasized in the analysis. In the next sections, we will show how these concepts and the criterion derived from the statistical mesoscopic damage mechanics can be

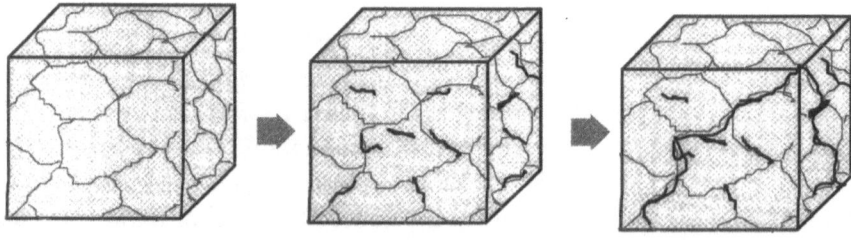

Figure 1
An illustration: damage localization leads to eventual rupture of solids.

extracted from the numerical results when we use a network model under unidirectional shear stress to simulate the progressive failure of a fault section. We show how to predict the occurrence of damage localization prior to rupture, based on the data from numerical simulations.

3. A Network Model

In this section, we use a two-dimensional network model (XIA et al., 1996; LIANG et al., 1997) shown in Figure 2 to simulate the rupture process of a heterogeneous medium under simple shear loading. The network consists of bar-elements (chains). With increasing shear loading, the chains are stretched or contracted linear elastically until breaking, when their stretching strains exceed their fracture strains. Surmise that all chains have the same elastic modulus. Once some chains break, their stresses will be redistributed to their neighbors. This may induce further breaking or the network may attain a new quasi-equilibrium state. We presume that this would resemble the progressive degradation to eventual rupture of the heterogeneous medium. In order to simulate the heterogeneous medium, the chains are assumed to be randomly distributed in the network (see Fig. 2) with their lengths and fracture strains constrained by distribution functions (see Fig. 3).

The parameters adopted in the numerical simulations are as follows:
- The total number of bar-elements in each network sample is 4120, and the total number of nodes is 1423;
- The fracture strains of chains are uniformly distributed between 0 and 2.0×10^{-3}, with the mean value being 10^{-3} (Fig. 3(a));

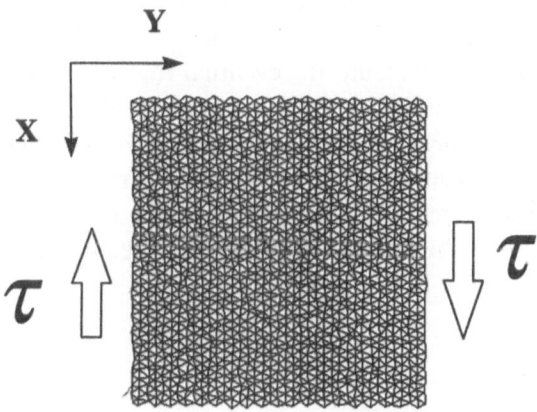

Figure 2
A sample of the network model under simple shear loading.

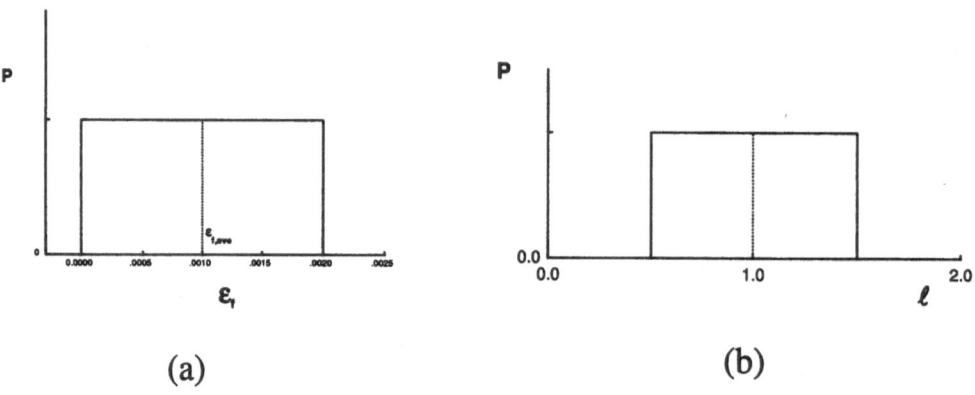

Figure 3
An uniform distribution of (a) fracture strains ε_f and (b) lengths ℓ of chains (P = numbers of chains with ε_f (or ℓ)/the total of chains).

– The lengths of chains are uniformly distributed between 0.5 and 1.5, with a mean length of 1.0, the length scale (Fig. 3(b)).
At each step, we calculate the following quantities:
– Far-field stress τ;
– The fraction of broken chains, as the damage D;
– The far-field displacement and the corresponding strain γ.
Here stress τ is dimensionless by dividing by the elastic modulus of chains.

4. Numerical Results of the Network Model

As described in the previous section, the parameters which represent the statistical features of chains in samples of network are specified. However, the samples of network are different from each other due to fluctuations in length and fracture strain of chains. As a result, the eventual rupture patterns and the value of peak stress for different samples may differ distinctly. We wonder whether general intrinsic characteristics, which govern the concentration of microfractures and the eventual rupture, can be captured for all these random samples.

From the obtained numerical data, the dynamic function of damage can be deduced. In fact, in one-dimensional approximation, equation (3) can be rewritten as follows

$$f = \frac{\partial D}{\partial T} + \frac{D}{1+\theta} \dot{\theta}, \qquad (8)$$

where θ is the volumetric strain. The above equation (8) suggests the feasibility of calculating the DFD f inversely, if the data of macroscopic variables such as τ, D

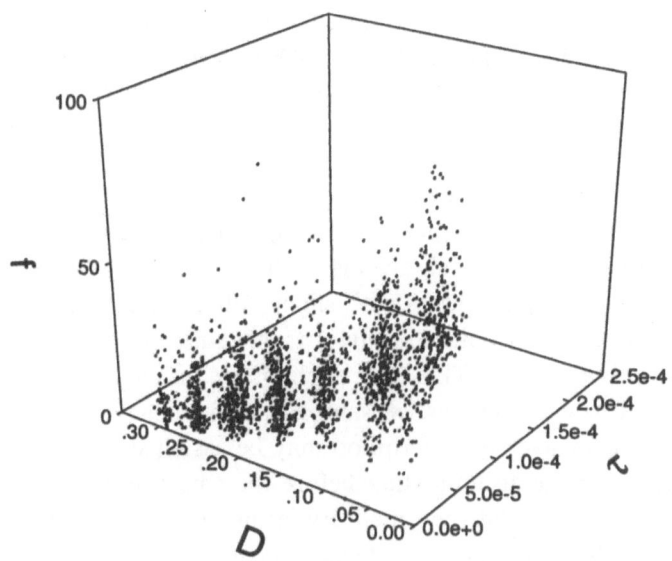

Figure 4
Dynamic function of damage (DFD) $f = f(D, \tau)$ under stress-free boundary conditions.

and θ can be obtained at different times. For example, these data can be obtained either from elaborately controlled experiments, seismological or geophysical observations, or as we do in this paper, from the numerical simulation. Then, we can deduce the DFD effectively according to equation (8). This provides a substitute for directly calculating the DFD from equation (4). Actually, the calculation of the DFD from equation (4) would involve the *in situ* experimental observation of the microdamage evolution (i.e., direct measurement of the nucleation rate n_N, growth rate A and healing rate n_A). The complex mechanisms of microdamage evolution and their interaction may elude the direct estimate of the effects of nucleation, growth and coalescence, and thus, make the calculation of the DFD from equation (4) rather difficult, especially for heterogeneous media such as rocks.

Free Boundary Conditions

Thirty-five samples of the network model were run, and the DFD was calculated from equation (8) based on the data of τ, D and θ for all samples. All obtained data of DFD are plotted in the space (f, D, τ) in Figure 4. Now, we fit the data into the following expression

$$f = a(1 + c \cdot D + b \cdot e^{m \cdot D})\sigma \qquad (9)$$

and obtain the fitting parameters $a = 1.229 \times 10^4$, $b = 0.2831$, $c = -15.31$, $m = 10.90$, and the relative deviation between the original and fitting values of f is estimated to be 0.6563. (The relative deviation $s = \sqrt{\Sigma_{i=1}^{N} d_i^2/(N-1)}$, where $d_i = (f_{\text{original}} - f_{\text{fitting}})/f_{\text{fitting}}$, and N is the total number of data.) This rather large relative deviation is reminiscent of the relative deviation of the fracture strains of network chains, $1/\sqrt{3}$, a value in the same magnitude. Actually, the deviation of the calculated DFD indicates the existence of the sample-specific behavior of samples with heterogeneous microstructures. The dynamic evolution amplifies the differences between the samples as progressive damage of microstructure proceeds, and the deviation from the statistical average behavior becomes increasingly more significant. As a result, even samples with slight initial mesoscopic differences may exhibit significantly different failure behavior from sample to sample as the samples approach the eventual failure. This may explain the rather large deviation of the calculated DFD when the effects of sample-specific behavior are all included. However, a statistically average description may be assumed to properly characterize the process of damage accumulation before the emergence of damage localization when the deviation is relatively small. By fitting an expression to the calculated DFD data, we search for an approximate expression of the statistical average DFD which works well before the sample-specific behavior manifests itself.

Then, equation (9) is substituted into equation (7) to examine the occurrence of damage localization. This leads to the threshold of damage localization

$$D_c = 0.1530 \pm 0.0045. \tag{10}$$

(Even for the calculated DFD data including the effect of the most significant sample-specific behavior, we obtained a critical value $D_c = 0.1575$.) What will really happen beyond this critical value?

We will make this point more clear by the run of other samples, and examine the validity of the criteria (7) and (10). Three additional samples (other than the 35 samples mentioned above) were run and their results are shown in Figures 5 (a)–(c). The critical "time" predicted by the criteria (7) and (10) is also illustrated by a cross in these figures. It can be seen that the failure patterns of the three samples are clearly different from each other due to fluctuations in microstructure. The stress-strain curves also differ in a distinct way. For one of the samples (Fig. 5(b)), a sharp stress drop such as a foreshock can be observed much earlier than the final rupture. However, for one of the other samples (Fig. 5(c)), the process between the damage localization predicted by criterion (10) and the final unstable rupture is extremely fast. However, the damage localization given by criterion (7) does provide an alarm ahead of eventual rupture for all three samples. In Figures 6(a)–(c), the evolution of damage patterns is shown. Thusly, we can identify minor concentration of microdamages in the network just after the predicted damage localization. More importantly, it is convincing that the criterion for damage localization based on the DFD data can serve as a precursor of abrupt rupture in random samples sufficiently well. This may provide evidence that the DFD intrinsi-

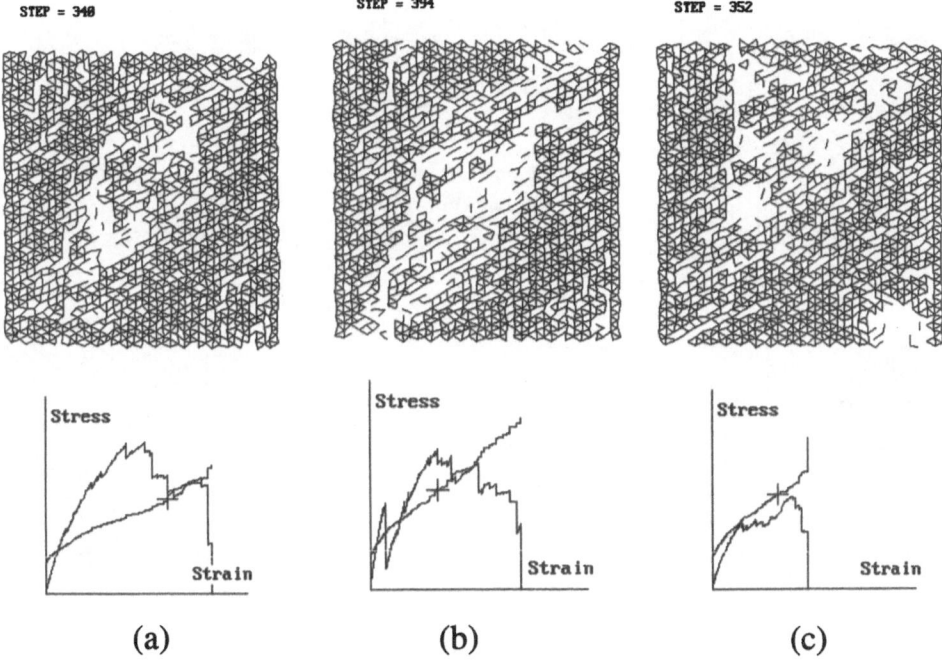

Figure 5

Rupture patterns and stress-strain curves of three samples, and the prediction given by the criterion for damage localization (marked by +) under stress-free boundary conditions. The curve with + indicates the variation of damage D with strain.

cally governs damage localization prior to rupture for all the samples, though their macroscopic mechanical behaviors appear to be decidedly different from each other.

Periodic Boundary Conditions

In order to approximately simulate the behavior of faults parallel to shear, periodic boundary conditions are also considered. In this case, the position and displacement of each node on the upper boundary in Figure 2 are assumed to be the same as that of its counterpart on the lower one.

Analogous to what we have done for the stress-free boundary case, the DFD data were obtained from simulations involving 5 random samples according to equation (8), and were shown in Figure 7. These data in the (f, D, τ) space are also fitted in formula (9) with $a = 3558$, $b = 0.9563$, $c = -22.18$, $m = 6.449$, and the relative deviation between the original and fitting values of f is 0.3184.

The threshold value for damage localization is given by criterion (7) as

$$D_c = 0.1998 \pm 0.0043. \tag{11}$$

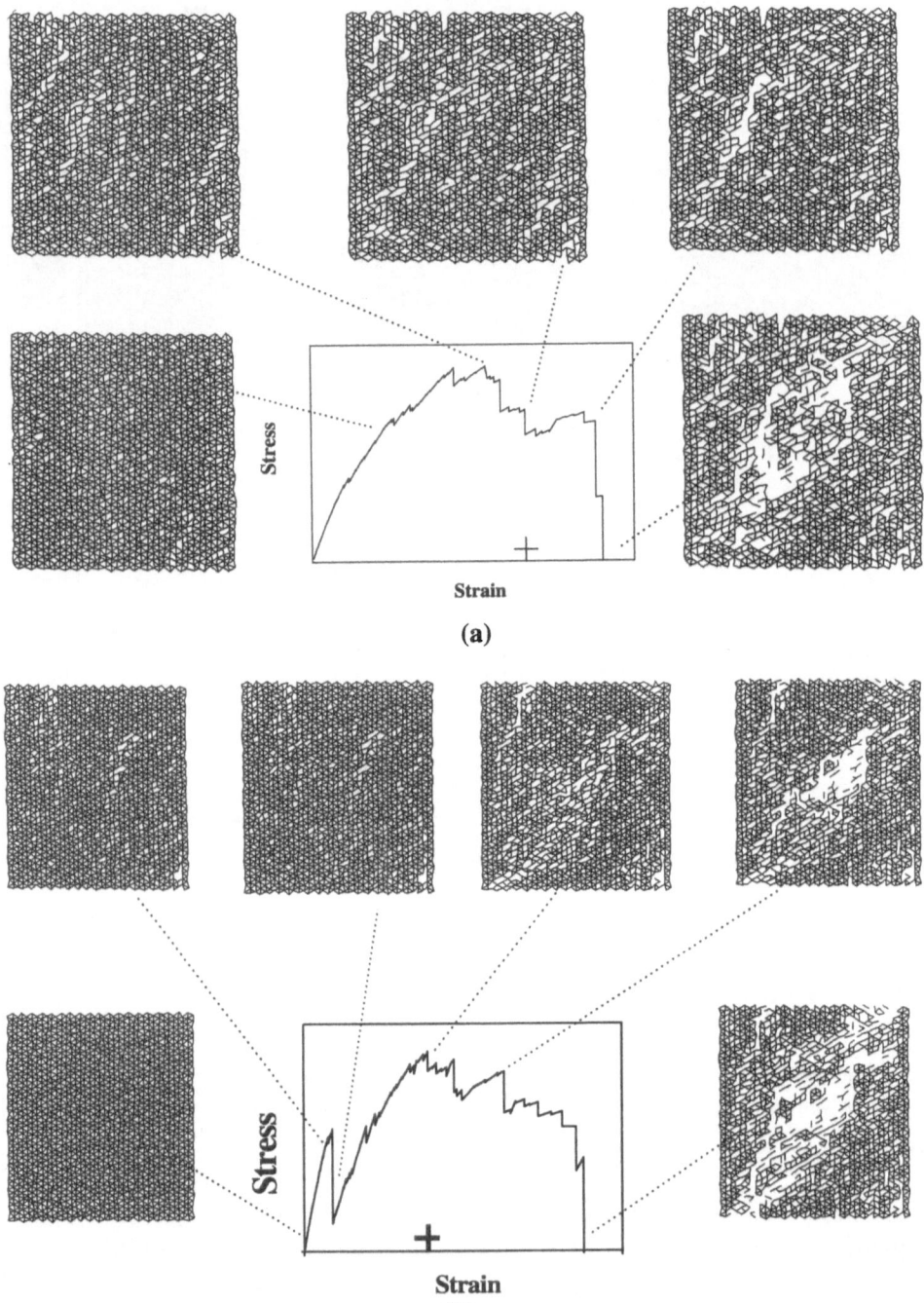

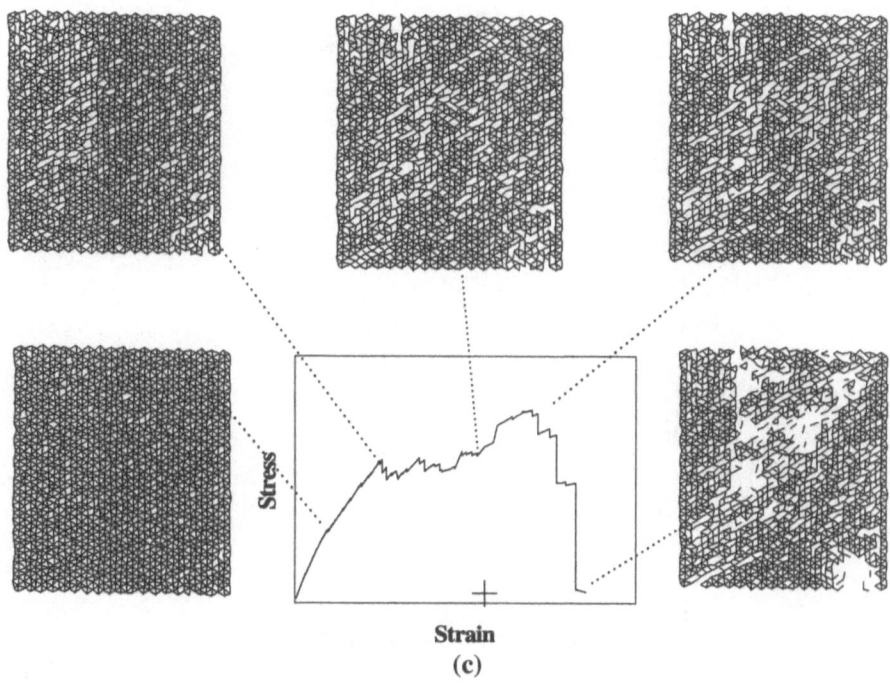

Figure 6
Damage patterns of the three samples at several times before and after predicted damage localization (marked by +). (a) Sample No. 1; (b) No. 2; (c) No. 3 (under stress-free boundary condition).

The rupture patterns of four other random samples are shown in Figures 8(a)–(d) with the stress-strain curves as well as the alarm given by equation (11). Figure 9 delineates the evolution of damage patterns of two samples with periodical boundary conditions. Compared to the stress-free boundary cases, the failure patterns of these samples are more likely to be parallel to the direction of shear stress. The threshold value of damage is larger than that of the stress-free boundary case. This indicates that the periodic constraint may retard the occurrence of damage localization. Nonetheless, with both stress-free and periodic boundary conditions, criterion (7) for damage localization always gives a proper alarm prior to the rupture.

The DFD, linking mesoscopic dynamics (nucleation, growth and coalescence) of microdamage and mesoscopic damage evolution, intrinsically characterizes the dynamic evolution of the heterogeneous damage (roughly speaking, the time rate of damage D). It is the dynamic evolution which amplifies the slight initial differences in microstructure, and results in the emergence of damage localization. Moreover, theoretical investigations (BAI et al., 1998) revealed that, for the same initial damage distribution, either homogeneous damage field or damage localization phenomenon may be observed, entirely depending on the expression of the DFD.

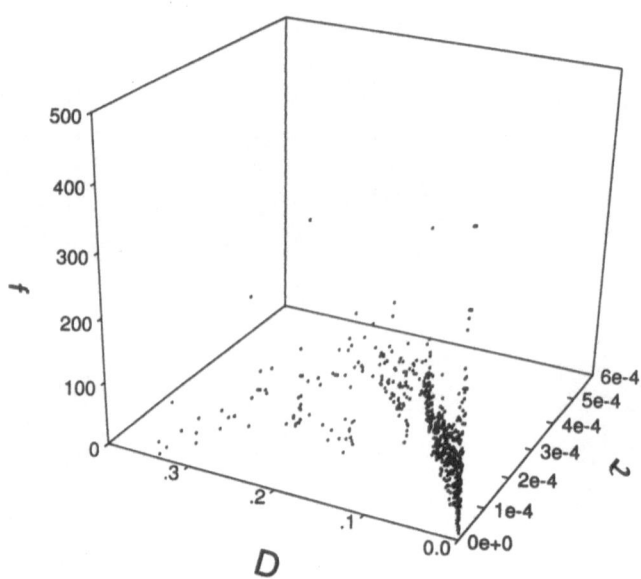

Figure 7
Dynamic function of damage (DFD) under periodic boundary conditions.

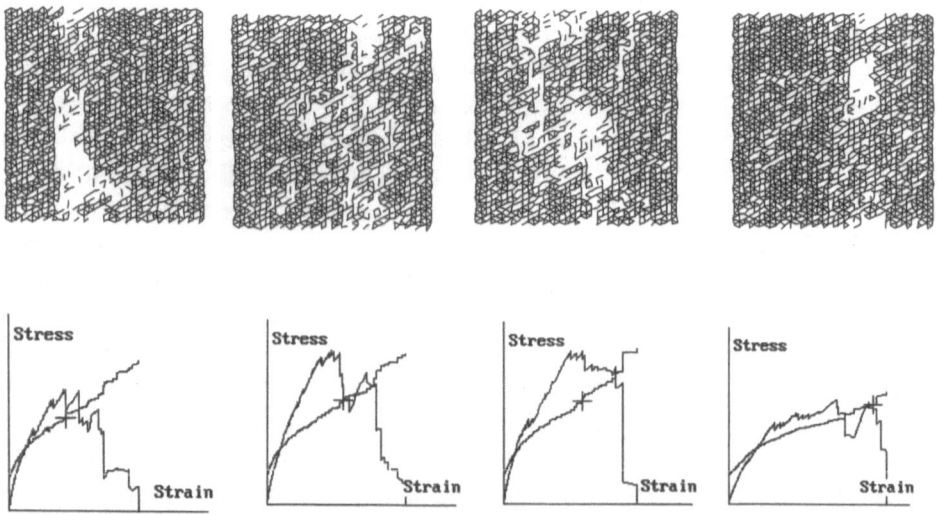

Figure 8
Rupture patterns and stress-strain curves of four samples, and the prediction given by the criterion for damage localization (marked by +) under periodic boundary condition. The curve with + indicates the variation of damage D with strain.

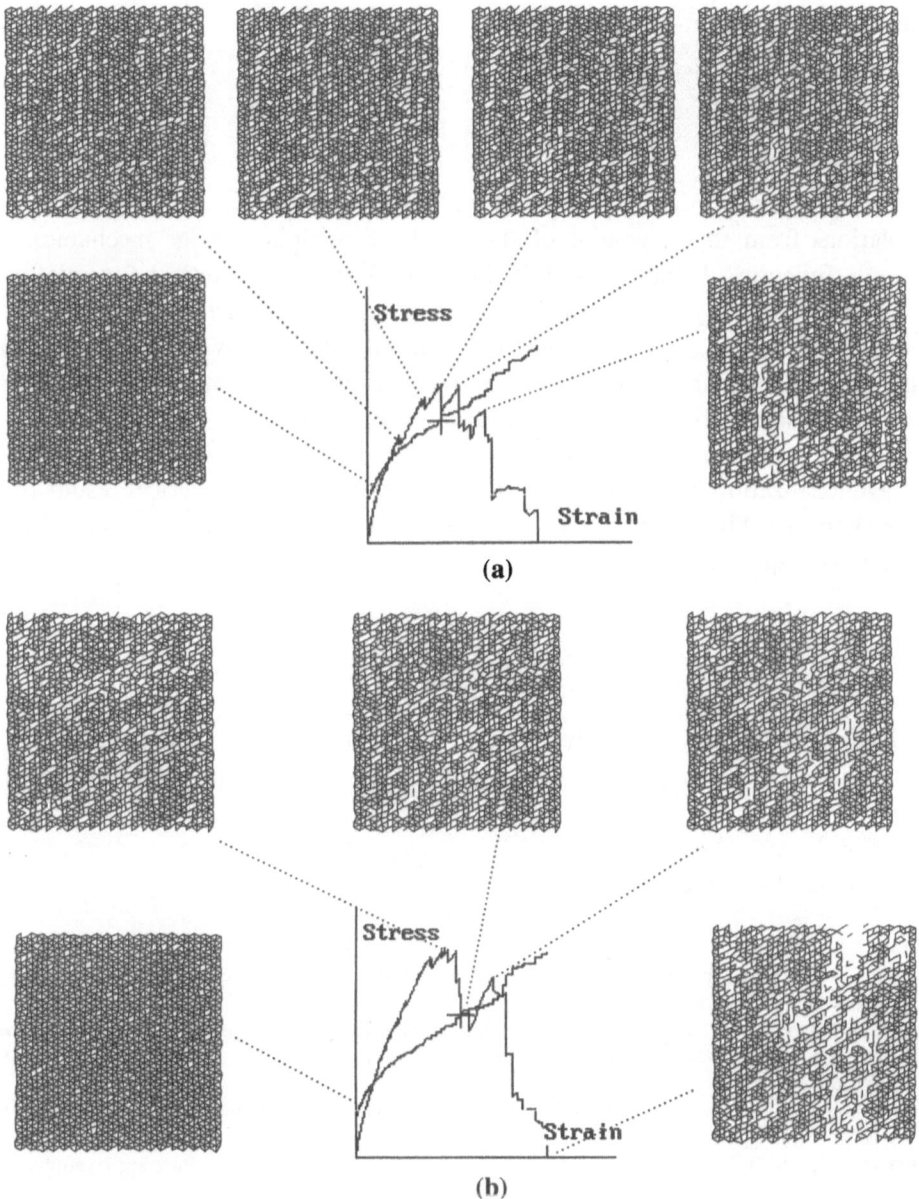

Figure 9
Damage patterns of the two samples at several times before and after predicted damage localization
(marked by +). (a) Sample No. 1; (b) No. 2 (under periodic boundary condition) The curve with +
indicates the variation of damage D with strain.

Thus, the DFD plays a key role in governing the occurrence of damage localization prior to eventual rupture.

5. Conclusions

The progressive damage of fault sections was investigated with numerical simulations from the viewpoint of statistical mesoscopic damage mechanics. The dynamic function of damage (DFD) introduced in the theoretical formulation is found to intrinsically govern the evolution of the heterogeneous damage field as well as damage localization prior to eventual rupture. With the information provided by the DFD, we may effectively predict the emergence of damage localization before the spatial concentration of microfractures can be observed. As a possible precursor for eventual rupture, the effectiveness of the criterion for damage localization based on the DFD is confirmed by the numerical results of the network model. Therefore damage localization may serve as a possible precursor of earthquake rupture.

Acknowledgements

This work is supported by National Natural Science Foundation of China (19891180, 19732060 and 19972004) and Chinese Academy of Sciences (KJ951-1-201).

REFERENCES

BAI, Y. L., KE, F. J., and XIA, M. F. (1991), *Formulation of Statistical Evolution of Microcracks in Solids*, Acta Mechanica Sinica 7, 59–66.

BAI, Y. L., XIA, M. F., KE, F. J., and LI, H. L., *Damage field equation and criterion for damage localization.* In *Rheology of Bodies with Defects* (ed. Wang, R.) (Kluwer Academic Publishers 1998) pp. 55–66.

DAS, S., and AKI, K. (1977), *Fault Planes with Barriers: A Versatile Earthquake Model*, J. Geophys. Res. 82, 5658–5670.

DIETERICH, J. (1978), *Time Dependent Fraction and the Mechanics of Stick Slip*, Pure appl. geophys. 116, 790–806.

DIETERICH, J. (1992), *Earthquake Nucleation on Defects with Rate- and State-dependent Strength*, Tectonophysics 211, 115–134.

EVANS, A. G., *Acoustic emission sources in brittle solids.* In *Fundamentals of Acoustic Emission* (ed. Ono, K.) (Materials Dept., UCLA, Los Angeles 1978) pp. 209–227.

KEMENY, J. M., and HAGAMAN, R. M. (1992), *An Asperity Model to Simulate Rupture along Heterogeneous Fault Surfaces*, Pure appl. geophys. 138, 549–567.

LIANG, N. G., LIU, H. Q., and XU, H. Q., *A Multi-scale network model and parameter optimization of discontinuous fiber reinforced composites.* In *Macro/Micro/Meso Mechanical Properties of Materials* (eds. Tokuda, M., and Bingye, XU) (Mie Academic Press 1997) pp. 269–275.

RICE, J. R., *The mechanics of earthquake rupture*. In *Physics of the Earth's Interior* (eds. Dziewonski, A., and Boschi, E.) (North Holland, Amsterdam 1980) pp. 555–649.

RICE, J. R., and RUINA, A. L. (1983), *Stability of Steady Fractional Slopping*, J. Appl. Mech. *105*, 343–349.

XIA, M. F., SONG, Z. Q., XU, J. B., ZHAO, K. H., and BAI, Y. L. (1996), *Sample-specific Behavior in Failure Models of Disordered Media*, Commun. Theor. Phys. *25*, 49–54.

(Received July 22, 1999, revised February 1, 2000, accepted April 7, 2000)

 To access this journal online:
http://www.birkhauser.ch

Pure appl. geophys. 157 (2000) 1945–1957
0033–4553/00/121945–13 $ 1.50 + 0.20/0

Pure and Applied Geophysics

Evolution-induced Catastrophe and its Predictability

Yu-Jie Wei,[1] Meng-Fen Xia,[1,2] Fu-Jiu Ke,[1,3] Xiang-Chu Yin[1,4] and
Yi-Long Bai[1]

Abstract—Both earthquake prediction and failure prediction of disordered brittle media are difficult and complicated problems and they might have something in common. In order to search for clues for earthquake prediction, the common features of failure in a simple nonlinear dynamical model resembling disordered brittle media are examined. It is found that the failure manifests evolution-induced catastrophe (EIC), i.e., the abrupt transition from globally stable (GS) accumulation of damage to catastrophic failure. A distinct feature is the significant uncertainty of catastrophe, called sample-specificity. Consequently, it is impossible to make a deterministic prediction macroscopically. This is similar to the question of predictability of earthquakes. However, our model shows that strong stress fluctuations may be an immediate precursor of catastrophic failure statistically. This might provide clues for earthquake forecasting.

Key words: Evolution-induced catastrophe, sample-specificity, stress fluctuations, earthquake prediction, predictability.

1. Introduction

Earthquake prediction, the contemporary scientific challenge, is drawing more and more attention. This is mainly due to two factors: the importance of earthquake prediction and the difficulty of the problem.

Currently, it seems that efforts to identify deterministic precursors of earthquakes have been mostly unsuccessful. This arouses the discussion of whether an earthquake is predictable or not (Geller *et al.*, 1997). We still believe that earthquakes do have some incubating phase, although it seems still to be quite far beyond our knowledge. With many intriguing but fragmentary observations of possible precursory phenomena, people are frustrated again and again. Therefore,

---

[1] State Key Laboratory of Nonlinear Mechanics, Institute of Mechanics, Chinese Academy of Sciences, Beijing 100080, China. E-mail: weiyj@lnm.imech.ac.cn; E-mail: baiyl@lnm.imech.ac.cn
[2] Department of Physics, Peking University, Beijing 100871, China. E-mail: xiam@lnm.imech.ac.cn
[3] Department of Applied Physics, Beijing University of Aeronautics and Astronautics, Beijing 100083, China. E-mail: kefi@lnm.imech.ac.cn
[4] Center for Analysis and Prediction, China Seismological Bureau, Beijing 100036, China. E-mail: yinxc@btamail.net.cn

perhaps we should seek alternative approaches to earthquake forecasting and seek sound bases for the approaches. If we assert that an earthquake is unpredictable now, we may throw the baby out with the bath water! (BOWMAN and SAMMIS, 1999).

Actually, recent investigations have highlighted the key role of the heterogeneity of the crust and the nonlinear evolution of aggregations in the nucleation of earthquakes. The lithosphere of the solid earth can be regarded as a hierarchy of blocks with different length scales, from major tectonic plates to grains of rocks, which are interlinked by weak boundary zones or interfaces (TURCOTTE, 1992). *"This implies that exact predictions are not possible. But it does not imply that earthquakes cannot be predicted (forecast) with considerable accuracy"* (TURCOTTE, 1999). In fact, precursory clustering of earthquakes was investigated in terms of an aggregation model of crack growth and ultimate fusion on a multiplanar earthquake fault system with nonlinear rheology (YAMMASHITA and KNOPOFF, 1992). Other interesting attempts to understand earthquake forecasting in a nonlinear heterogeneous system include the stress release model (VERE-JONES, 1978), cellular automaton model (SORNETTE and SAMMIS, 1995; SAMMIS and SMITH, 1999), particle-based lattice solid model (MORA and PLACE, 1999), load/unload response ratio (YIN *et al.*, 1995), and so on. Therefore, the complexity of earthquake prediction seems to be rooted in the nonlinear evolution of systems with disordered heterogeneity on multiple scales.

Aimed at such a class of heterogeneous materials whose failure is a nonlinear evolution, we construct a model based on brittle fracture. In comparison with earthquakes, there are two similar features (BAI *et al.*, 1994a,b; XIA *et al.*, 1996a,b, 1997; KE *et al.*, 1998):

1. Evolution-induced catastrophe (EIC). The failure usually exhibits an abrupt transition from globally stable accumulation of damage (GS) to catastrophic failure.
2. Sample-specificity. There may be a great diversity in catastrophe thresholds, i.e., catastrophic failure shows different behaviors sample-to-sample under identical macroscopic conditions. This leads to uncertainty of macroscopic failure.

In this paper a statistical analysis is performed for a simple nonlinear dynamical model of damage and failure, which demonstrates evolution-induced catastrophe and sample-specificity. Next an approach to distinguish the two phases of nonlinear evolution, namely, from GS to EIC is explored. The results show that the magnitude of fluctuations in the governing stress field appears to be a possible precursor of EIC. This may indicate that statistical warning signs in fluctuations before catastrophic failure may shed a light on earthquake forecasting.

The next section will introduce a nonlinear dynamical model of damage and failure called the coupled-pattern model. Thereafter, its evolution-induced catastrophe and sample-specificity will be discussed. Finally, we try to explore precursors of the eventual failure.

2. Coupled-pattern Model

Consider a macroscopic system consisting of mesoscopic units. The system should be described by patterns mesoscopically. They are the pattern of material properties, the pattern of damage and the pattern of the governing stress field. The behavior of the system can be simulated by coupled-evolution of these patterns.

As an example, we consider a ring model (one-dimensional chains with periodic boundary condition) with N units (XIA et al., 1994).

(1) Patterns

After denoting the strength of unit i by σ_{ci}, the pattern of material properties is expressed by $\Sigma_c = \{\sigma_{ci}, i = 1, 2, \ldots, N\}$. In order to take heterogeneity into account, the values of σ_{ci} are assumed to be random in the chain, but follow a distribution function $f(\sigma_c)$. We take the average strength $\bar{\sigma}_c = \int_0^\infty \sigma_c f(\sigma_c)\, d\sigma_c = 1$. A sample is specified by its pattern Σ_c mesoscopically. Then, samples with identical distribution functions $f(\sigma_c)$ are identical macroscopically.

The damage pattern is denoted by $X = \{x_i, i = 1, 2, \ldots, N\}$, where $x_i = 0$ for an intact unit and $x_i = 1$ for a broken unit. Macroscopically, the system is described by the damage fraction.

$$p = \frac{1}{N}\sum_{i=1}^{N} x_i. \tag{1}$$

It is interesting to examine the damage process with increasing stress. The stress pattern $\Sigma = \{\sigma_i, i = 1, 2, \ldots, N\}$, where σ_i is the ratio of the current stress on unit i to $\bar{\sigma}_c$. Macroscopically, the governing parameter is the nominal stress σ_0 given by

$$\sigma_0 = \frac{1}{N}\sum_{i=1}^{N} \sigma_i. \tag{2}$$

There are two levels which describe damage evolution as shown in Table 1.

Table 1

Description for meso-macro levels

	Material property	Damage	Stress
Meso-patterns	Strength pattern Σ_c	Damage pattern X	Stress pattern Σ
Macro-parameters	Distribution of strength $f(\sigma_c)$	Damage fraction p	Nominal stress σ_0

(2) Load-sharing Rules

The stress pattern Σ is determined from the damage pattern X according to load-sharing rules, which describe how the nominal stress of broken units is to be transferred to nearby intact units. Formally, the load-sharing rules can be expressed by "interactions" between intact units and broken units. For simplicity, we adopt a model in which the interaction is equal to the summation of intact-broken pair interactions. In this way, the stress on unit i can be given by

$$\sigma_i = (1 - x_i)\left[1 + \sum_{j \neq i} x_i a(|i - j|)\right]\sigma_0 \tag{3}$$

where $a(|i-j|)$ is the contribution of unit j to unit i and depends on the "distance" $|i-j|$. From equations (2) and (3), $a(|i-j|)$ satisfies

$$p = \frac{1}{N}\sum_i\sum_j(1 - x_i)x_j a(|i-j|) \tag{4}$$

where only the contributions of terms with $i \neq j$ are included, because $(1 - x_i)x_j = 0$ as $i = j$.

In the following, we examine three types of the contribution function a, which roughly cover all possible stress redistributions.

Rule I: Global mean field

This is the simplest load-sharing rule,

$$a = \frac{1}{N(1 - p)}. \tag{5}$$

Then,

$$\sigma_i = \frac{\sigma_0}{(1 - p)}. \tag{6}$$

The stress is uniformly shared by all intact units. This is a rule without stress fluctuations.

Rule II: Local interactions

$$a(|i - j|) = \begin{cases} \text{const} & \text{for} \quad |i - j| \leq \Delta \\ 0 & \text{otherwise} \end{cases} \tag{7}$$

This rule depends on a fixed local scale Δ. As $\Delta \to N$, the model approaches the global mean field model.

Rule III: Cluster load-sharing

This is a type of local mean field model with a cutoff due to broken clusters. A broken cluster consists of l connecting broken units with intact units at its two sides. The nominal stress of a broken cluster is shared by its two neighboring intact clusters equally. A uniform stress distribution is assumed for each intact cluster. Thus, a unit in an s-intact cluster separating an l- and an r-broken cluster supports a stress

$$\sigma = \left(1 + \frac{l+r}{2s}\right)\sigma_0. \tag{8}$$

In this rule, local scale is determined by the local pattern of damage. Therefore, there may be multiple scales during the course of damage evolution. Then, the evolution shows a nonlocal feature essentially (without a characteristic scale).

Generally, for rules II and III, the stress pattern Σ depends on the details of the damage pattern X, and there are stress fluctuations in the system. The stress fluctuations play an essential role in damage and failure.

(3) Evolution Dynamics

The evolution of damage in the system with disordered heterogeneity is very complicated. In order to investigate the universal behavior of such kind of nonlinear system, we introduce simplified dynamics as follows:

For a sample specified by a strength pattern $\Sigma_c = \{\sigma_{ci}, i = 1, 2, \ldots, N\}$, the evolution of the damage pattern $X = \{x_i, i = 1, 2, \ldots, N\}$ is governed by the stress pattern $\Sigma = \{\sigma_i, i = 1, 2, \ldots, N\}$ according to the dynamics:

$$x_i(t+1) = x_i(t) + \Delta x_i(t) \tag{9}$$

where

$$\Delta x_i(t) = S(\sigma_i(t) - \sigma_{ci}), \qquad i = 1, 2, \ldots, N \tag{10}$$

and

$$S(y) = \begin{cases} 1, & \text{for} \quad y \geq 0 \\ 0, & \text{for} \quad y < 0 \end{cases} \tag{11}$$

where t is the ordinal number. At each step, as pattern X changes, pattern Σ must be redetermined from the new pattern X according to the specified load-sharing rule. This is to say, the stress pattern Σ and the damage pattern X evolve in a coupled way. From equations (1) and (9), we can obtain a "time series" of the damage fraction $p(t)$.

For a sample with an initially intact pattern $X(0) = \{x_i(0) = 0, i = 1, 2, \ldots, N\}$, as the nominal stress σ_0 increases from $\sigma_0 = 0$ slowly, the first damage event will occur at a value of $\sigma_0 = \sigma_0^{(1)}$, depending on distribution function $f(\sigma_c)$. Keeping $\sigma_0 = \sigma_0^{(1)}$, the system will evolve and approach an equilibrium pattern $X^{(1)}$ (with damage fraction $p^{(1)}$). As σ_0 increases, the above procedure repeats. Subsequently,

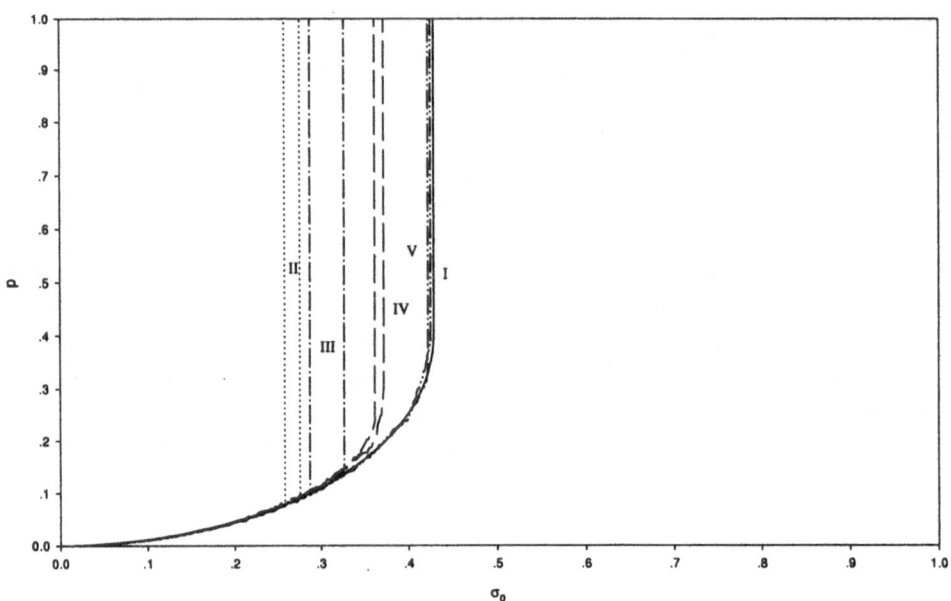

Figure 1

Equilibrium damage fraction p as a function of imposed nominal stress σ_0 (with identical distribution functions $f(\sigma_c)$). I. Globally mean field; II. Cluster load-sharing rule, $N = 20,000$; III. Cluster load-sharing rule, $N = 2000$; IV. Local interaction, $\Delta = 2$, $N = 2000$; V. Local interaction, $\Delta = 100$, $N = 2000$.

we can obtain a series of stresses $\sigma_0^{(k)}$, equilibrium damage fractions $p^{(k)}$ and equilibrium damage patterns $X^{(k)}$. Eventually, failure will occur at $\sigma_0 = \sigma_0^{(k_f)} = \sigma_{0f}$.

3. *Evolution-induced Catastrophe and the Difficulty of Prediction*

Figure 1 shows the equilibrium damage fraction p as a function of the imposed nominal stress σ_0, based on various load-sharing rules, as σ_0 increases quasi-statically. The strength distribution function $f(\sigma_0)$ adopted is a Weibull distribution

$$f(\sigma_c) = mb\sigma_c^{m-1} \exp(-b\sigma_c^m) \tag{12}$$

with $m = 2$ and $b = 1$ (then $\bar{\sigma}_c = 1$). We also examined cases for various $f(\sigma_c)$. It is found that the system exhibits the following common behaviors.

(1) *Evolution-induced catastrophe*

The evolution of the system manifests two different modes: globally stable (GS) and evolution-induced catastrophic (EIC). There is a catastrophic threshold σ_{0f} for each sample. When $\sigma_0 < \sigma_{0f}$, the sample shows a stable accumulation of damage, called the globally stable mode (GS). This means that there are equilibrium patterns X with damage fraction $p < 1$ according to prescribed stresses σ_0. However, at $\sigma_0 = \sigma_{0f}$, the sample turns to catastrophic failure abruptly. This is to say, there are no equilibrium patterns with $p < 1$ and the sample will go to failure ($p = 1$) unavoidably. This is called evolution-induced catastrophe (EIC). EIC is one of the fundamental modes of failure. σ_{0f} (and the relative damage fraction p_f) indicates the point of transition from the GS to the EIC mode.

(2) Sample-specificity

In the simulation, the samples with identical strength distribution function $f(\sigma_c)$ are regarded as identical samples macroscopically, however their mesoscopically disordered heterogeneity is specified by the strength pattern Σ_c. The failure behavior of a system following the global mean field rule is uniquely determined by function $f(\sigma_c)$. From equation (6), the equilibrium condition for specified σ_0 is given by equation $p = \int_0^{\sigma_0/(1-p)} f(\sigma_c) \, d\sigma_c$ and the failure threshold σ_{0f} can be deduced. That is to say, the macroscopically identical samples display identical behavior regardless of their mesoscopic details.

However, for samples with mesoscopic disorder and following load-sharing rules other than the global mean field rule, the macroscopic behavior, shows a clear difference sample-to-sample. Especially, there is a great diversity in catastrophic threshold σ_{0f} for the macroscopically identical samples. This distinct feature is called the sample-specificity of EIC. Sample-specificity results in uncertainty of material failure macroscopically.

(3) Strength distribution

The distribution function of the catastrophic threshold $W(\sigma_{0f})$ (shown in Fig. 2) can provide a probabilistic prediction of the catastrophe. For samples following the global mean field rule, there is a distinct failure strength σ_{0f} (Fig. 2). The macroscopic behaviors (p vs. σ_0 curves in Fig. 1) under other load-sharing rules deviate from that of the global mean field model. Catastrophe thresholds σ_{0f} for both global interaction and cluster load-sharing rules are usually considerably lower than that given by the global mean field model, and show a greater diversity (see Fig. 2).

Due to the sample-specificity of catastrophic failures, the macroscopic prediction of failure cannot be deterministic as in the global mean field model. Consequently, it is clear that the strengths of samples are statistically meaningful. This is helpful for material sciences, however not sufficient for rupture prediction.

4. *Warning Signs of Catastrophe*

From the distinct differences in strength distributions, for the global mean field rule and the two other rules we are aware that the stress fluctuations play an essential role in evolution-induced catastrophe and sample-specificity. The main effects of stress fluctuations are as follows:

1. A significant reduction of catastrophe threshold from that derived from the global mean field model (Figs. 1 and 2).
2. The sample-specificity, especially the great diversity of catastrophe thresholds (Figs. 1 and 2).

The stress fluctuations can be measured by $\delta\sigma/\bar{\sigma}$, where

$$\delta\sigma = \left[\frac{1}{N(1-p)} \sum_{i=1}^{N} (\sigma_i - \bar{\sigma})^2 (1 - x_i) \right]^{1/2} \tag{13}$$

is standard deviation of stress on intact units in the system, and

$$\bar{\sigma} = \frac{\sigma_0}{1-p} \tag{14}$$

is the mean stress of intact units.

Figure 3 delineates the stress fluctuations $\delta\sigma/\bar{\sigma}$ as a function of p (the data are collected from the processes shown in Fig. 1). From Figure 3(a), we can see that, in the GS regime, the system remains in a state with a low level of stress fluctuations (< 1), nonetheless the fluctuations strengthen with increasing p. As the

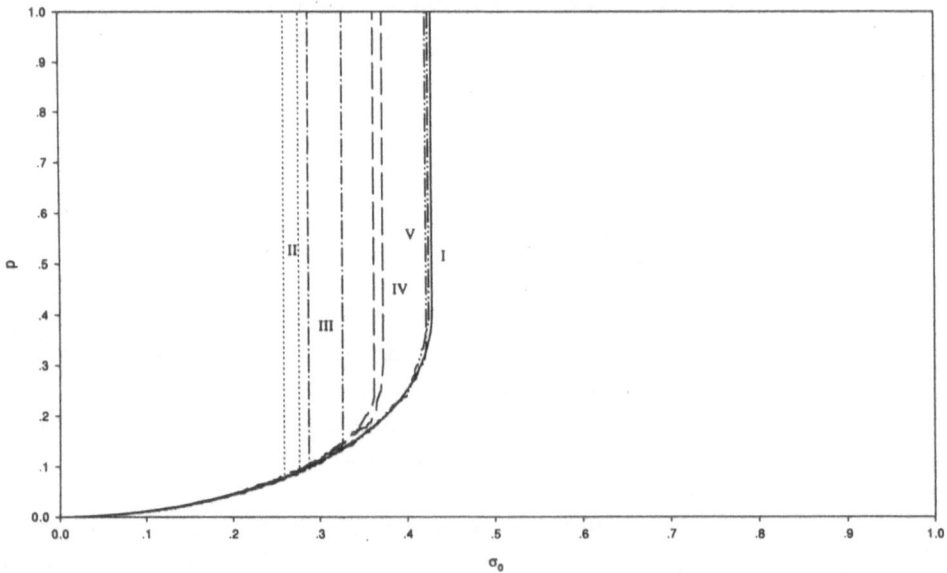

Figure 2

Statistical distribution function of the catastrophic threshold $W(\sigma_f)$. — Globally mean field; ---- Cluster load-sharing rule, $N = 20,000$; ···· Cluster load-sharing rule, $N = 2000$; ---- Local interaction, $\Delta = 2$, $N = 2000$.

system turns to the EIC regime at $\sigma_0 = \sigma_{0f}(p = p_f)$, the stress fluctuations become considerably stronger than that in the GS regime. Figure 3(b) shows similar behavior. For the local interaction model with sufficiently large Δ (see Fig. 3(c)), the behavior should approach that given by the global mean field model, and the stress fluctuations will remain at a low level.

Although it is impossible to find a deterministic criterion for catastrophic failure macroscopically due to sample-specificity, the above-mentioned results provide

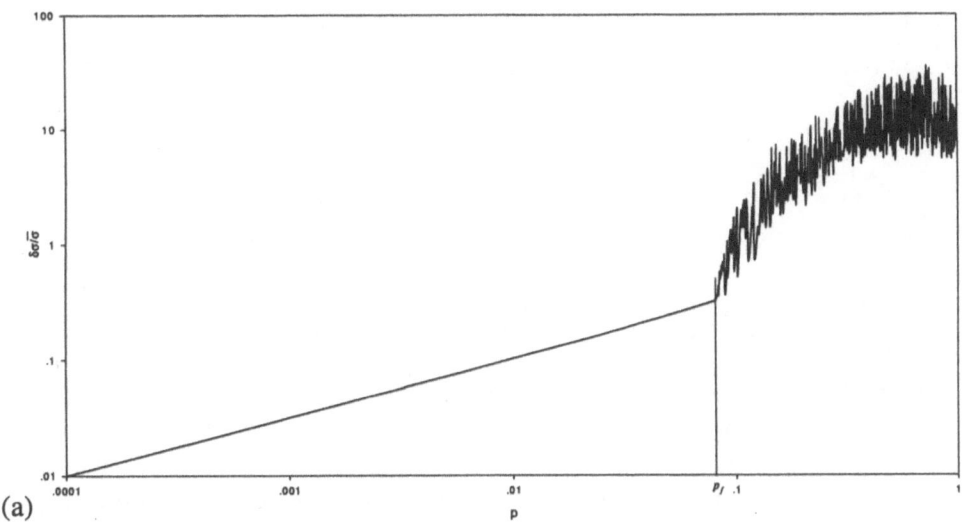

(a)

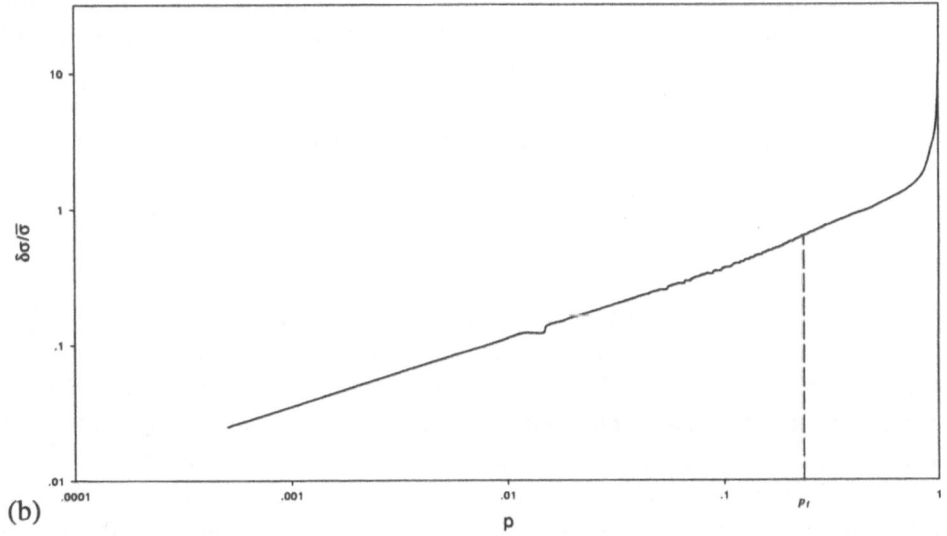

(b)

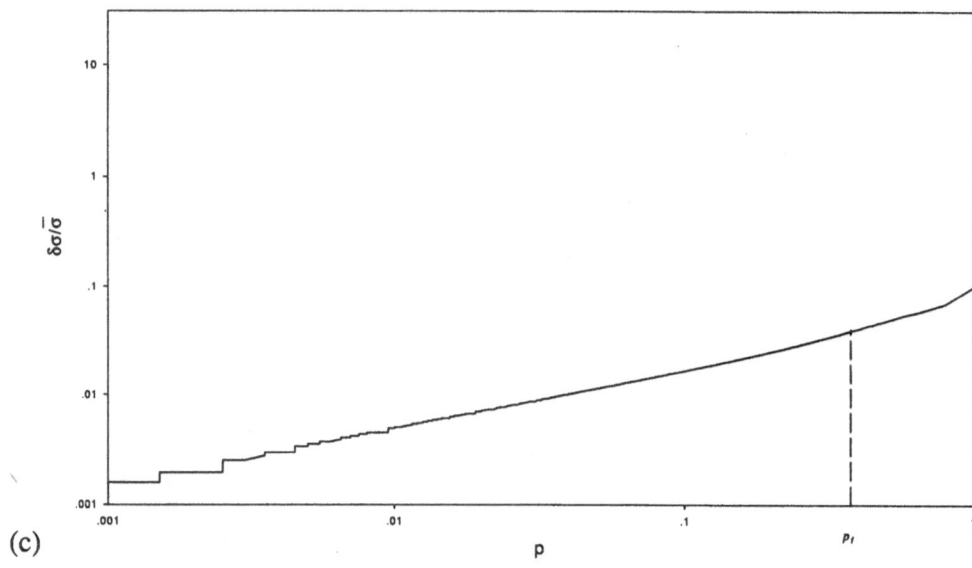

(c)

Figure 3

Stress fluctuation $\delta\sigma/\bar{\sigma}$ as a function of the damage fraction p. (a) Cluster load-sharing rule, $N = 20,000$; (b) Local interaction, $\Delta = 2$, $N = 2000$; (c) Local interaction, $\Delta = 100$, $N = 2000$.

clues to failure forecasting statistically. In this way we could possibly provide statistical warning signs based on an ensemble analysis. Let θ_{GS} be the maximum of stress fluctuations $\delta\sigma/\bar{\sigma}$ in the GS regime and θ_{EIC} be that in the EIC regime. The ensemble distributions $\Pi_{GS}(\theta)$ and $\Pi_{EIC}(\theta)$ are shown in Figure 4. From Figures 4(a), 4(b) and 4(c), we can see that the distribution functions $\Pi_{GS}(\theta)$ and $\Pi_{EIC}(\theta)$ are well separated from each other. Thus, statistically, we can set up a warning level of stress fluctuations to forecast the occurrence of catastrophic failure. For the local interaction model with sufficiently large Δ (say $\Delta = 100$ for $N = 2000$, see Fig. 4(d)), the distribution functions $\Pi_{GS}(\theta)$ and $\Pi_{EIC}(\theta)$ overlap each other. However, in this case, the behavior of the system is close to that of the global mean field model and failure prediction can be made from a deterministic criterion (see Fig. 1).

5. Discussion

The prediction of catastrophe is a problem of the first importance for engineering and natural hazards. However prediction is very difficult, especially for heterogeneous media like the crust. This is mainly due to the nonlinear evolution far from equilibrium and the catastrophic failure exhibited as a cascade from smaller scales to larger ones. Even in the EIC regime, there may be no significant signs in the initial stage of the catastrophic mode. The characteristics of catastrophe become visible only after it is enhanced strongly during nonlinear evolution. Secondly, due to sample-specificity, there is no deterministic, macroscopic criterion for catastrophe. However, for a system with disordered heterogeneity on multiple scales, it is usually impossible to secure a detailed description mesoscopically. Furthermore, sample-specificity implies that there is a sensitive link between macroscopic and mesoscopic scales. In other words, the macroscopic behaviors of a system may be sensitive to subtle details on mesoscopic scales. Consequently, an approach to failure prediction based upon statistical averages may be insufficient and may even be inappropriate.

In order to overcome the difficulty resulting from sample-specificity, we must find some delicate features which can be adopted to distinguish the GS and EIC

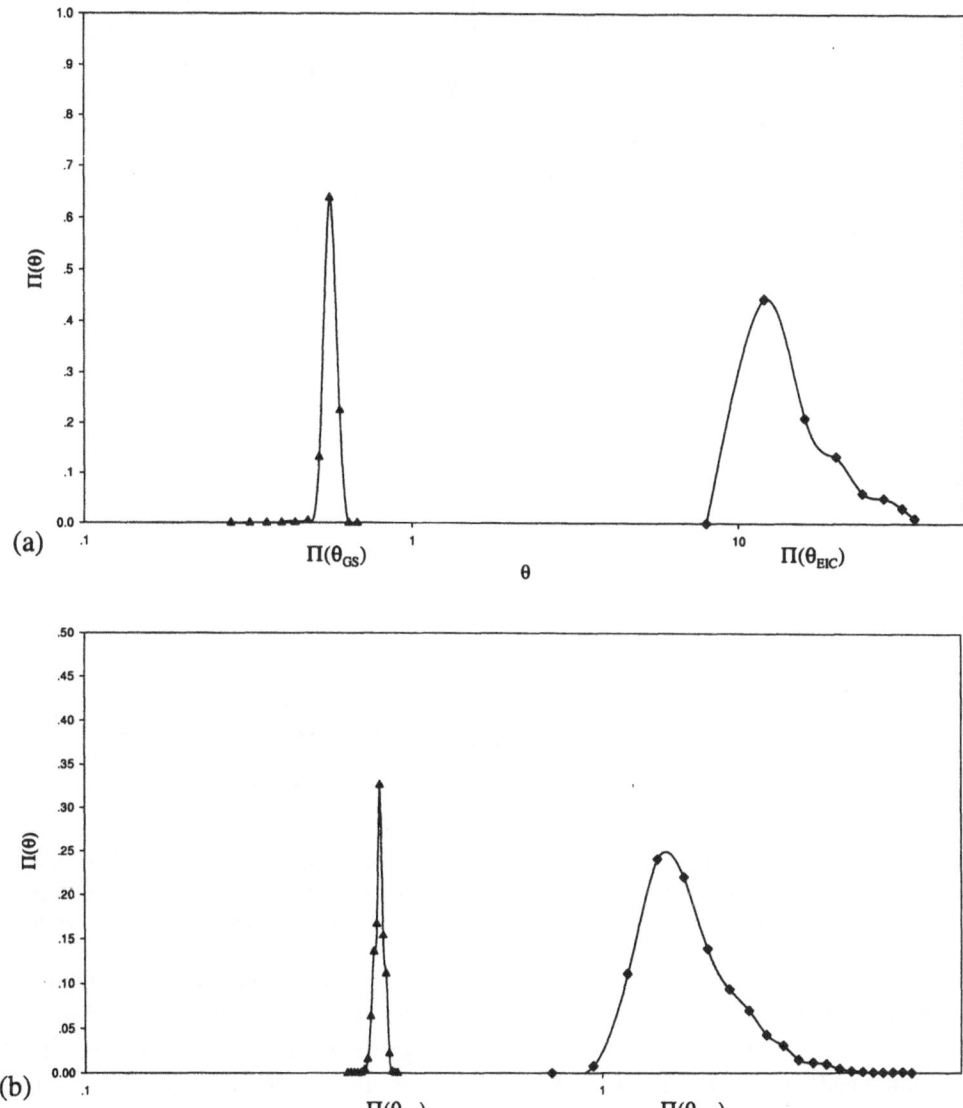

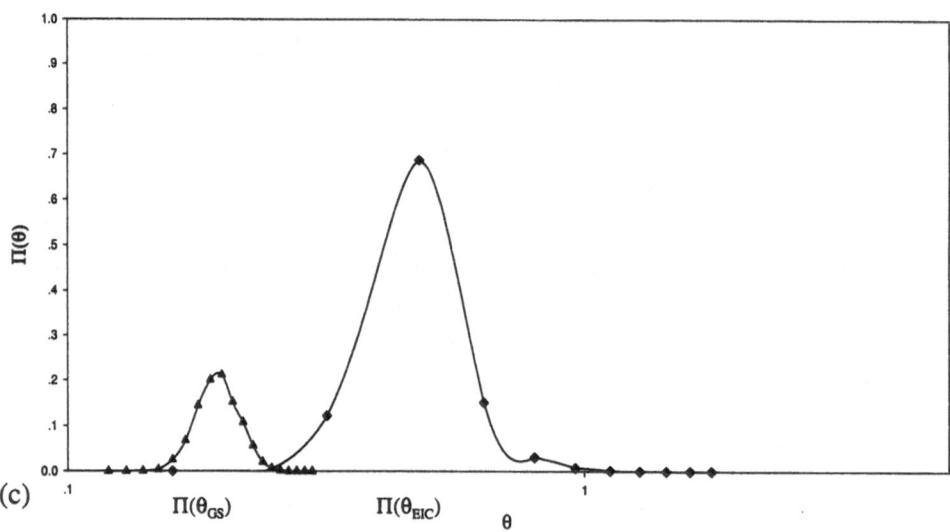

(c)

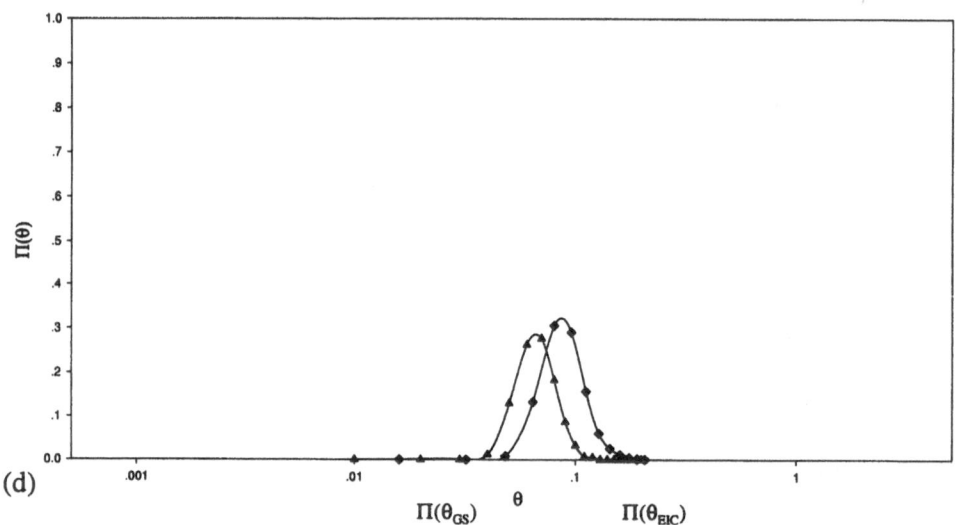

(d)

Figure 4

Statistical distribution function of maximum stress fluctuation θ_{GS} and θ_{EIC} in the GS and EIC regimes, respectively. Δ: $\Pi(\theta_{GS})$ $\Diamond$: $\Pi(\theta_{EIC})$. (a) Cluster load-sharing rule, $N = 20,000$; (b) Local interaction, $\Delta = 2$, $N = 2000$; (c) Local interaction, $\Delta = 10$, $N = 2000$; (d) Local interaction, $\Delta = 100$, $N = 2000$.

modes. Statistically, monitoring the fluctuations of the governing stress field in a sample may be a possible approach as discussed in the previous section. The important feature is that the fluctuations of the governing stress field will statistically be enhanced to a higher level as the evolution mode transits from the GS to EIC. Therefore, as the fluctuations of the governing field go beyond a warning level, we could note that the evolution of the system is entering EIC mode and catastrophic failure may occur. This approach might provide clues for earthquake prediction.

Acknowledgements

This work is supported by the National Fundamental Research Project "Nonlinear Science" and National Natural Science Foundation of China (No. 19732060, 19972004, 19572072 and 19577102). Computation was supported by the State Key Laboratory of Scientific and Engineering Computing.

REFERENCES

BAI, Y. L., KE, F. J., and XIA, M. F. (1994a), *Deterministically Stochastic Behavior and Sensitivity to Initial Configuration in Damage Fracture*, Science Bulletin (in Chinese) *39*, 892–895.

BAI, Y. L., LU, C. S., KE, F. J., and XIA, M. F. (1994b), *Evolution-induced Catastrophe*, Phys. Lett. A *185*, 196–200.

BOWMAN, D., and SAMMIS, CH. (1999), An E-mail contribution for the debate on web site www.nature.com/debates/earthquake/, *A case for intermediate-term earthquake prediction: Don't throw the baby out with the bath water!*

GELLER, R. J., JACKSON, D. D., KAGAN, Y. Y., and MULARGIA, F. (1997), *Earthquakes Cannot be Predicted*, Science *275*, 1616–1617.

KE, F. J., FANG, X., XIA, M. F., ZHAO, D. W., and BAI, Y. L. (1998), *Contingent Sensitivity to Configuration in a Nonlinear Evolution System*, Prog. in Nat. Sci. *8*, 170–173.

MORA, P., and PLACE, D., *Accelerating energy release prior to large events in simulated earthquake cycles: Implications for earthquake forecasting*. In *Proc. ACES Inaugural Workshop*, 31 Jan.–5 Feb., 1999, Brisbane and Noosa, QLD, Australia, pp. 195–200.

SORNETTE, D., and SAMMIS, C. G. (1995), *Complex Critical Exponents from Renormalization Group Theory of Earthquakes: Implications for Earthquake Predictions*, J. Phys. *15*, 607–619.

SAMMIS, C. G., and SMITH, S. (1999), *Seismic Cycles and the Evolution of Stress Correlation in Cellular Automaton Models of Finite Fault Networks*, Pure appl. geophys. *155*, 307–334.

TURCOTTE, D. L., *Fractals and Chaos in Geology and Geophysics* (Cambridge University Press, Cambridge 1992).

TURCOTTE, D. L., *The physics of earthquakes: Is it a statistical problem?* In *Proc. ACES Inaugural Workshop*, 31 Jan.–5 Feb., 1999. Brisbane and Noosa, QLD, Australia, pp. 95–98.

YAMMASHITA, T., and KNOPOFF, L. (1992), *Model for Intermediate-term Precursory Clustering of Earthquakes*, J. Geophys. Res. *97* (B13), 19,873–19,879.

VERE-JONES, D. (1978), *Earthquake Prediction: A Statistician's View*, J. Phys. Earth *26*, 129–146.

XIANG-CHU YIN, XUE-ZHONG CHEN, ZHI-PING SONG, and CAN YIU (1995), *A New Approach to Earthquake Prediction—The Load/Unload Response Ratio (LURR) Theory*, Pure appl. geophys. *145* (3/4), 701–715.

XIA, M. F., BAI, Y. L., and KE, F. J. (1994), *Statistical Description of Pattern Evolution in Damage-fracture*, Science in China (A) *37*, 331–340.

XIA, M. F., KE, F. J., BAI, J., and BAI, Y. L. (1997), *Threshold Diversity and Trans-scales Sensitivity in a Nonlinear Evolution Model*, Phys. Lett. A *236*, 60–64.

XIA, M. F., SONG, Z. Q., XU, J. B., ZHAO, K. H., and BAI, Y. L. (1996a), *Sample-specific Behavior in Failure Models of Disordered Media*, Communication in Theoret. Phys. *25*, 49–54.

XIA, M. F., BAI, Y. L., and KE, F. J. (1996b), *A Stochastic Jump and Deterministic Dynamics Model of Impact Failure Evolution with Rate Effect*, Theoret. Appl. Fract. Mech. *24*, 188–196.

(Received July 22, 1999, revised February 1, 2000, accepted April 15, 2000)

Part II
Macroscopic Simulation: Short Time Scale Phenomena (Rupture and Strong Motion)

Pure appl. geophys. 157 (2000) 1959–1979
0033–4553/00/121959–21 $ 1.50 + 0.20/0

Pure and Applied Geophysics

Dynamic Propagation and Interaction of a Rupture Front on a Planar Fault

EIICHI FUKUYAMA[1] and RAÚL MADARIAGA[2]

Abstract—We investigate the propagation and interaction of a rupture front that propagates on a planar fault using a boundary integral equation method. We show first that the rupture velocity is controlled by a delicate balance between consumed fracture energy and supplied elastic strain energy. A very sharp boundary in parameter space separates models in which ruptures stop spontaneously from those in which rupture propagates at super-shear speeds. The transition zone (or bifurcation) is shown to be stable with reference to small-scale heterogeneities of the stress field. Using the relations derived from this analysis we examined the mechanism to generate high slip rate when two rupture fronts collide. We found that collision at slow rupture velocities causes abrupt stress drop and generates high slip rates. However, these features tend to be moderated by large slip-weakening distances. Finally, we simulated rupture front focusing at the initial stages of an earthquake, a phenomenon that may cause high slip rate pulses and therefore generate high frequency seismic waves. We assume a pre-slip region, in which stress has decreased quasi-statically to the dynamic friction level. Due to this pre-slip, strong stress concentration has developed around the pre-slip area and a dynamic rupture starts at a certain point on the rim of the pre-slip region. We observe rupture front focusing that generates high slip rate pulses. We also studied a double pre-slip model, in which two pre-slip regions exist close to each other before the earthquake and found that multiple pre-slips enhance the focusing effects.

Key words: Dynamic earthquake rupture, rupture front interaction, boundary integral equation method.

Introduction

In the classical dynamic faulting models, rupture was treated as propagating unilaterally or bilaterally (HASKELL, 1964) along a rectangular fault. Later, in the 1980s, several waveform inversion methods were developed (e.g., HARTZELL and HEATON, 1983; FUKUYAMA and IRIKURA, 1986), assuming simple circular or elliptical rupture propagation due to insufficient resolution of waveform data. ARCHULETA (1984) found that the Imperial Valley earthquake propagated at super-shear speeds with a complex stress drop pattern. Numerous recent inversions

[1] National Research Institute for Earth Science and Disaster Prevention, Tsukuba, Japan. E-mail: fuku@bosai.go.jp
[2] Laboratoire de Géologie, URA CNRS 1316, Ecole Normale Supérieure, Paris, France. E-mail: madariag@geologie.ens.fr

of strong motion have revealed very complex slip functions, stress drops and kinematics of rupture propagation (see, e.g., IDE and TAKEO, 1997; WALD and HEATON, 1994; COTTON and CAMPILLO, 1995). OLSEN et al. (1997) modeled the 1992 Landers earthquake and found strong rupture focusing effects around high stress concentration regions in the WALD and HEATON (1994) kinematic model of the Landers earthquake.

The initial phase of earthquake ruptures has been frequently discussed based on the observation of very small events (IIO, 1992), medium size shocks (ELLSWORTH and BEROZA, 1995) and large earthquakes (UMEDA, 1992). A large variety of initial phases has been recognized and it seems difficult to explain all of them with a single initial rupture model. At least two kinds of initial phases have been recognized: A slow emerging phase observed mainly for small earthquakes (IIO, 1992) and sharp onset phases (UMEDA, 1992; ELLSWORTH and BEROZA, 1995). These phases may represent a transition from slow rupture initiation to fast dynamic rupture in the source region.

These kinds of rupture initiation and interaction phenomena have not been studied in detail because they require precise large-scale numerical computation of spontaneous rupture propagation based on fracture dynamics. DAS and KOSTROV (1983) were the first to study in some detail the break of a single asperity and seismic wave radiation from this rupture. They showed that once rupture initiates it propagates around the rim of the asperity and the two fronts meet on its opposite side. In their computations, they could not thoroughly analyze the interaction of rupture fronts when they meet because their computations had limited spatial resolution. Close scrutiny of the far-field displacement function that they computed reveals that rupture front interaction produced large slip velocity pulses.

FUKUYAMA and MADARIAGA (1995, 1998) developed a boundary integral equation method for the computation of spontaneous crack propagation on a planar fault in three-dimensional elastic medium which is very appropriate to study this kind of phenomena. In the study of rupture interaction phenomena, the rupture velocity is a key parameter. Thus we examine the relation between the rupture velocity and the initial stress as well as the critical slip weakening distance following DAY (1982) who investigated it by using the so-called S value, the ratio of the strength excess to stress drop. We demonstrate that rupture propagation in a homogeneous stress field is controlled by a delicate balance between strain energy release and fracture energy. Perfect balance between these two values defines a bifurcation in parameter space that we will explore in some detail.

Recently, dynamic rupture propagation on non-planar (curved and branched) fault becomes possible (TADA et al., 2000; AOCHI et al., 2000). As pointed out by RAVI-CHANDAR and YANG (1997), non-planar fault configuration plays an important role during the dynamic rupture propagation in microscopic point of view. However, we limit our discussion on a planar fault problem in order to obtain the macroscopic feature of the rupture dynamics in 3-D.

In this article we assume that the asperity is close to rupture due to stress concentration on its rim. We compute the stress distribution after quasi-static pre-slip and then, at some arbitrary point on the rim, we assume a dynamic rupture starts to propagate. We investigate in detail the rupture process around the asperity employing a friction model that includes a finite slip weakening zone in order to regularize the singularity in the rupture front dynamics. The observed radiation complexity is entirely due to the heterogeneity of the initial stress distribution.

We also consider a multi-asperity model, in which dynamic rupture interaction occurs between the asperities. This model simulates some of the basic features of the cascade model proposed by ELLSWORTH and BEROZA (1995). Our numerical simulations produce high slip velocities and strong radiation when rupture fronts interact strongly. Our computations include three-dimensional effects which have not been completely considered in previous studies (HARRIS and DAY, 1993; YAMASHITA and UMEDA, 1994; UMEDA et al., 1996b). This mechanism can produce high frequency waves with low stress drop or with small moment release.

Dynamic Rupture Simulation Method

Let us consider a fault plane in a homogeneous unbounded elastic medium. The fault plane is located on the $x_3 = 0$ plane in the Cartesian coordinate system (x_1, x_2, x_3). We solve it using the boundary integral equation method (BIEM) proposed by FUKUYAMA and MADARIAGA (1998). In this formulation, the stress distribution is linearly related to the slip velocity on the fault surface. Inside the slipping part of the fault we assume the simple triangular slip-weakening friction law proposed by IDA (1972) and depicted in Figure 1. In our BIEM the α-component of stress $T_\alpha^{\ell m n}$ and the β-component of slip velocity $V_\beta^{\ell m n}$ at $(\ell \Delta x, m \Delta x, n \Delta t)$ are related by the following discrete boundary element relations:

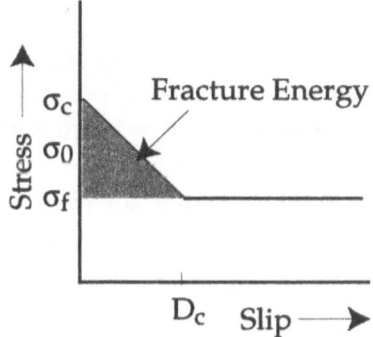

Figure 1
Slip-weakening frictional law used in the present study.

$$T_\alpha^{\ell mn} = -\frac{\mu}{2c_T} V_\alpha^{\ell mn} + \sum_{ij} \sum_{k<n} B_{\alpha\beta}^{ijk\ell mn} V_\beta^{ijk} + \sigma_0^{\ell m} \qquad (1)$$

$$T_\alpha^{\ell mn} = \begin{cases} \dfrac{\sigma_f^{\ell m} - \sigma_c^{\ell m}}{D_c} (\Delta t V_\alpha^{\ell mn} + D_\alpha^{\ell mn}) + \sigma_c^{\ell m} & (\text{if } 0 \le \Delta t V_\alpha^{\ell mn} + D_\alpha^{\ell mn} < D_c) \\ \sigma_f^{\ell m} & (\text{otherwise}) \end{cases} \qquad (2)$$

where

$$D_\alpha^{\ell mn} = \Delta t \sum_{k=0}^{n-1} V_\alpha^{\ell mk} \qquad (3)$$

and $B_{\alpha\beta}^{ijk\ell mn}$ is a kernel which was derived by FUKUYAMA and MADARIAGA (1998). Δx and Δt are the spatial and temporal sampling intervals, respectively. $\sigma_c^{\ell m}$ and $\sigma_f^{\ell m}$ are the peak frictional stress (static friction) and the final stress, respectively. $\sigma_0^{\ell m}$ is the initial stress at the point $(\ell\Delta x, m\Delta x, 0)$. The summation convention rule is applied. Greek and Roman suffices take the values of 1 or 2; and 1, 2, or 3, respectively.

Equation (1) represents the boundary condition on the crack, while Equation (2) shows the constitutive relation (or friction law) acting on the crack. Note that $D_\alpha^{\ell mn}$ is known at time $n\Delta t$. We study spontaneous rupture propagation by solving Equations (1) and (2) simultaneously at each time step to obtain stress and slip velocity distribution. For simplicity we assumed that slip can only occur in the x_1 direction so that only $T_1 (= \sigma_{31})$ is relevant in the friction law.

In the following computations all the parameters are normalized in the same way as done by FUKUYAMA and MADARIAGA (1998). We use 0.25 as the Courant-Friedrichs-Lewy (CFL) ratio $w (= c_T \Delta t / \Delta x)$, where c_T represents the S-wave velocity, and the P-wave velocity is assumed to be 1.73 c_T.

What Controls the Rupture Velocity?

Before studying complex rupture models, we must verify that we have proper resolution and control of the rupture velocity because this is the key parameter in the discussion of focusing and defocusing of rupture fronts. It is crucial to understand what controls the rupture velocity.

Since we use the slip-weakening friction law (2), fracture energy (G) is defined as

$$G = \frac{1}{2}(\sigma_c - \sigma_f)D_c. \qquad (4)$$

Fracture energy is consumed during rupture. This energy is supplied by the decrease of the elastic strain energy (ΔE) caused by stress drop on the fault. Thus the

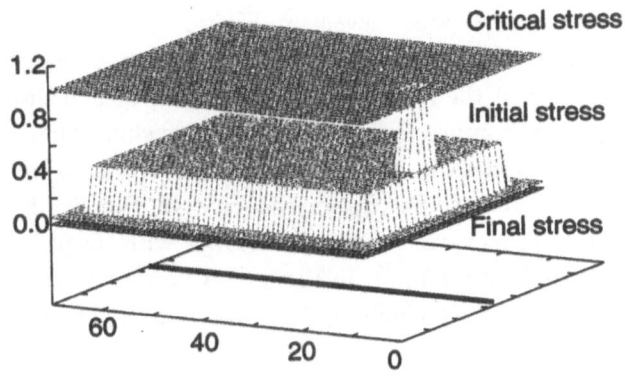

Figure 2
Stress fields for the rupture velocity test. Critical stress (σ_c) is set at 1.0 everywhere. Initial stress (σ_0) is set at 0.05 at the edge of the fault and at 1.0 at the initiation point. In other regions the initial stress is uniform and set to a value that we change in the simulations.

rupture velocity is controlled by a delicate balance between fracture energy (G) and strain energy release (ΔE). If fracture energy is much greater than strain energy release, rupture cannot propagate while, on the other hand, if strain energy release is much greater than fracture energy, the rupture velocity accelerates and reaches a terminal velocity. When fracture energy and strain energy release are balanced, rupture propagates with constant velocity.

For the mode II (in-plane) shear crack in two-dimensional medium, BURRIDGE (1973) theoretically predicted that rupture velocity is less than Rayleigh wave velocity, and intersonic rupture velocity is possible when static friction is low. ANDREWS (1976b, 1985) confirmed Burridge's result numerically by using a finite difference method as well as a boundary integral equation method. BURRIDGE et al. (1979) examined the intersonic rupture velocity in detail taking into account the slip-weakening friction, and suggested $\sqrt{2}c_T$ becomes a critical velocity in the in-plane rupture.

In order to confirm that our numerical method resolves energy balance accurately, we conducted several numerical tests. We considered the simple rectangular fault shown in Figure 2. Rupture initiates on one end of the fault and propagates unilaterally to the opposite end. Fault size is $64\Delta x \times 32\Delta x$ and the radius of initial patch is $2\Delta x$. Since the slip is in the x_1 direction, the dominating rupture mode is in-plane. In this case, the terminal velocity is the P-wave velocity. Under these conditions, we studied 1) the dependence of the rupture velocity on the initial stress level and 2) on the slip-weakening distance. The former experiment is based on that done by DAY (1982), while the second one was produced for circular faults by MADARIAGA et al. (1998) using the finite difference method.

In Figure 3A, we present the rupture front propagation at various initial stress levels with constant slip-weakening distance $D_c = 0.1$ in non-dimensional units. The critical (σ_c) and final (σ_f) stress levels in Figure 1 are set at 1.0 and 0.0 in non-dimensional units, respectively. The horizontal axis shows the rupture front location along the axis of the rectangular fault (the thick line shown in Figure 2). The ordinate shows the rupture time or the non-dimensional time at which rupture arrives at a given position along the fault. Thus the average slopes of the broken curves in Figure 3A indicate the rupture velocity. In Figure 3B rupture velocity normalized by the shear-wave velocity is shown with respect to the distance normalized by the width of the fault. In Figure 3 we observe the following features: If the initial stress (σ_0) is smaller than 0.412, rupture does not propagate; this is the

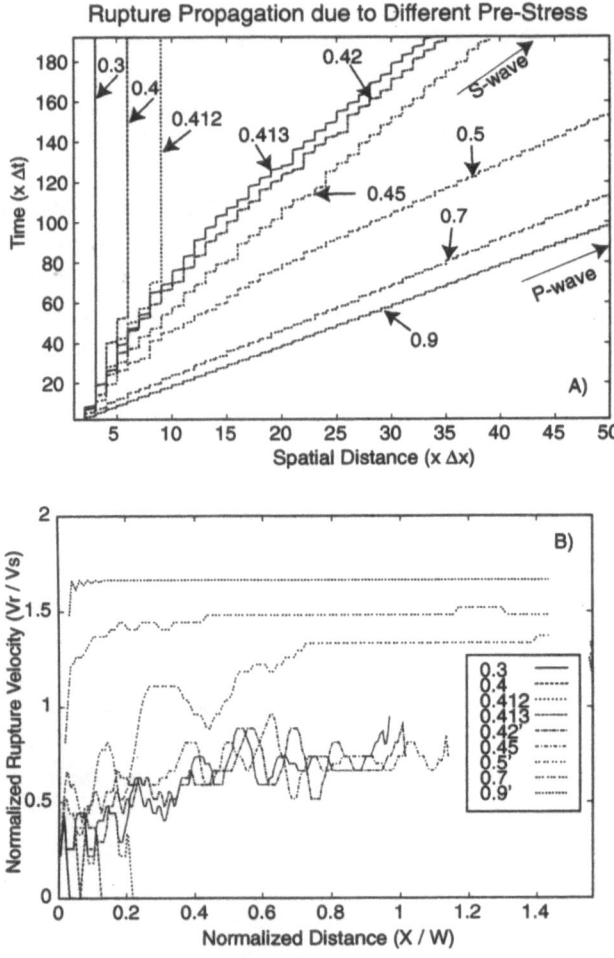

Figure 3
A) Time of arrival of the rupture front as a function of position along the axis of the fault shown with a thick line in Figure 2. B) Rupture velocity normalized by the shear-wave velocity as a function of distance normalized by the fault width. In this computation, D_c was fixed at 0.1 non-dimensional units and σ_0 was varied from 0.3 to 0.9.

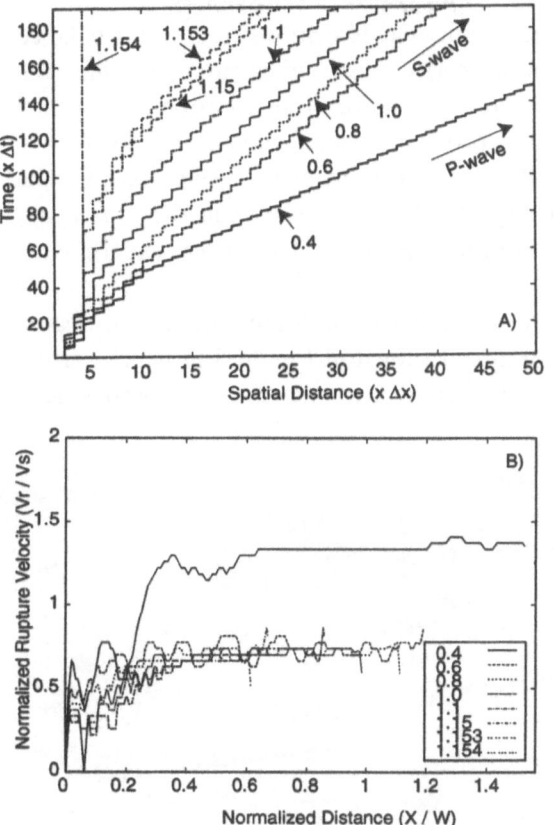

Figure 4
A) Time of arrival of the rupture front as a function of position along the thick line in Figure 2. B) Rupture velocity normalized by the shear-wave velocity as a function of distance normalized by the fault width. In this computation, σ_0 was fixed to be 0.5 non-dimensional units and D_c varied from 0.4 to 1.154.

case when fracture energy is much larger than strain energy release. On the other hand, if the initial stress is larger than 0.5, rupture accelerates and the velocity reaches intersonic speed (between P- and S-wave velocities). This corresponds to the large strain energy release case. It should be noted that there is a narrow zone between $0.412 < \sigma_0 < 0.5$ in which rupture propagates with speeds lower than the Rayleigh wave velocity, and in this range fracture energy and strain energy are balanced so that rupture propagates stably.

In Figure 4 we show the rupture front propagation for various slip-weakening distances with constant initial stress ($\sigma_0 = 0.5$). We also find a sharp bifurcation boundary: With a large slip-weakening distance ($D_c > 1.154$), rupture stops rapidly; while the rupture velocity becomes intersonic when the slip-weakening distance is small ($D_c < 0.4$). Within a narrow band defined between these two values of D_c rupture propagates with sub-Rayleigh wave velocity.

The above numerical experiments demonstrate that a balance between fracture energy and elastic strain energy release is essential, keeping the initial crack size constant. This balance defines a sharp bifurcation boundary so that the parameter region in which rupture propagates with sub-shear velocity is very narrow. It is important to remark that waveform inversion results for large earthquakes systematically report rupture speeds that are mostly sub-shear. It is also very difficult to observe super-shear rupture velocity in laboratory experiments (ROSAKIS *et al.*, 1999).

In order to explore the stability conditions for the bifurcation, following DAY (1982) we consider the model shown in Figure 5 in which we introduce a series of rectangular asperities and troughs in the direction perpendicular to the axis of

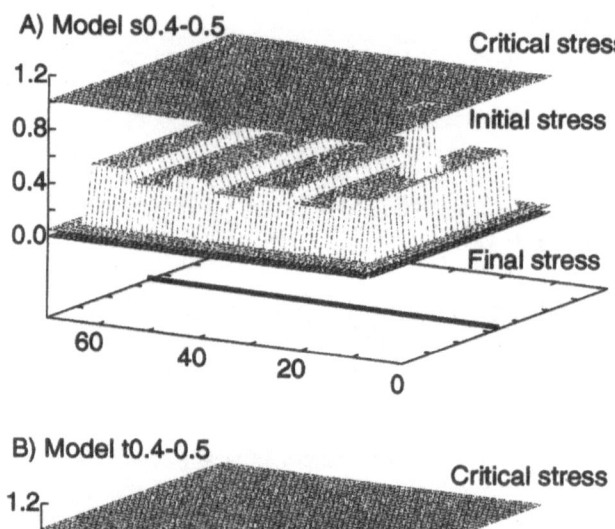

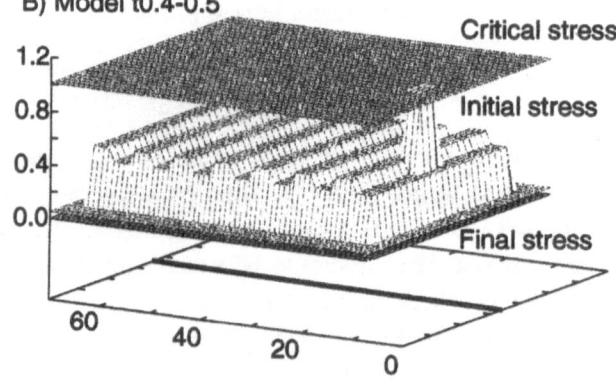

Figure 5
Stress distributions for the rupture velocity test. Critical stress (σ_c) was set at 1.0 everywhere. Initial stress (σ_0) was set at 0.05 at the edge of the fault and at 1.0 at the initiation point. On the rest of the fault the initial stress was set between 0.4 and 0.5 in regular steps. A) Model s0.4–0.5 has wider stress fluctuations and B) Model t0.4–0.5 has narrower fluctuations.

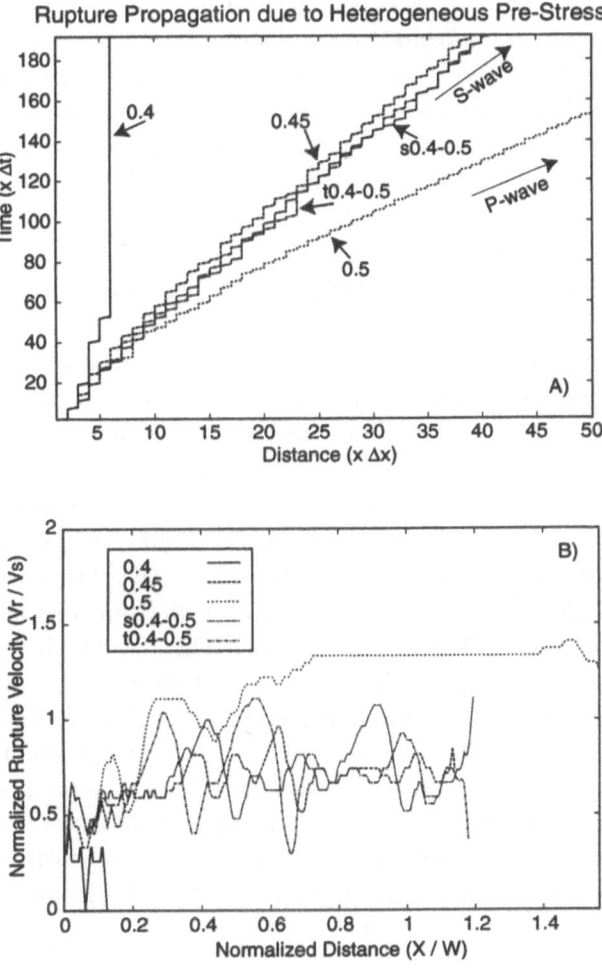

Figure 6
A) Time of arrival of the rupture front as a function of position along the thick line in Figure 5. B) Rupture velocity normalized by the shear-wave velocity as a function of distance normalized by the fault width. In this computation, D_c was fixed to be 0.1 non-dimensional units and σ_0 is shown in Figure 5. Results for models s0.4–0.5 and t0.4–0.5 are shown as well as results for the uniform initial stress of 0.4, 0.45, and 0.5.

rupture propagation. The initial stress profile is now rough with a variation between 0.4 and 0.5 non-dimensional stress units. We studied two models: model s0.4–0.5 (Fig. 5A) that has relatively broad troughs; and model t0.4–0.5 (Fig. 5B) that has narrower troughs. Figure 6 shows the result of rupture propagation for both models. The models with homogeneous stress of 0.4, 0.45, and 0.5 (Fig. 3) are also shown as a reference. Stable sub-shear rupture propagation is observed in the models with a rough initial stress profile as long as the average initial stress is 0.45, the bifurcation value. In the simulations by DAY (1982) the rupture velocity varied rapidly locally although the average velocity was sub-Rayleigh. In the present computation, on the other hand, the local rupture velocity is still sub-shear. This can be interpreted as follows: In our simulations the time delay between strain energy supply and fracture energy consumption allows them to compensate with each other and to be balanced at each time step. In DAY (1982) the representative size of stress heterogeneity was too large to compensate both energies.

Rupture Collision and Slip Rate

In the previous section we found a relation between the rupture velocity and the initial stress level as well as the critical slip-weakening distance. In this section we consider rupture collision: The initial stress field is similar to that in the previous section except that now we introduce two initiation points as shown in Figure 7. Two ruptures start simultaneously at both ends of the fault, so that the two rupture fronts propagate towards the center of the fault and "collide" with each other. Using the relations obtained in the previous section, we control the rupture velocity just before the collision and observe its effects in the slip rate field.

Figure 8 shows slip (left) and slip velocity (right) plotted along the thick line in Figure 7. In each panel, slip or slip velocity distributions at 21 successive moments are shown, with 10 of these before the collision, 1 during the collision and 10 following the collision. To the upper left of each panel, the collision time is indicated. We compute 3 cases with different initial stress levels close to the bifurcation value determined in the previous section: (A) 0.413, (B) 0.45 and (C) 0.5. From Figure 3, we estimate that the rupture velocity is sub-Rayleigh for (A), near the Rayleigh-wave velocity for (B) and close to the P-wave velocity for (C).

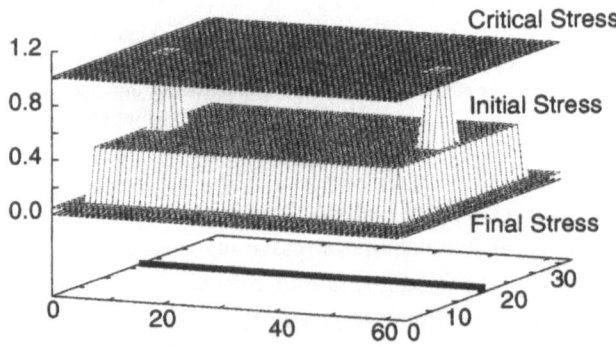

Figure 7

Stress distribution for the rupture velocity test. Critical stress (σ_c) is set at 1.0 everywhere. Initial stress (σ_0) is set at 0.05 at the edge of the fault and at 1.0 at the two nucleation points. On the rest of the fault initial stress is set to be a constant.

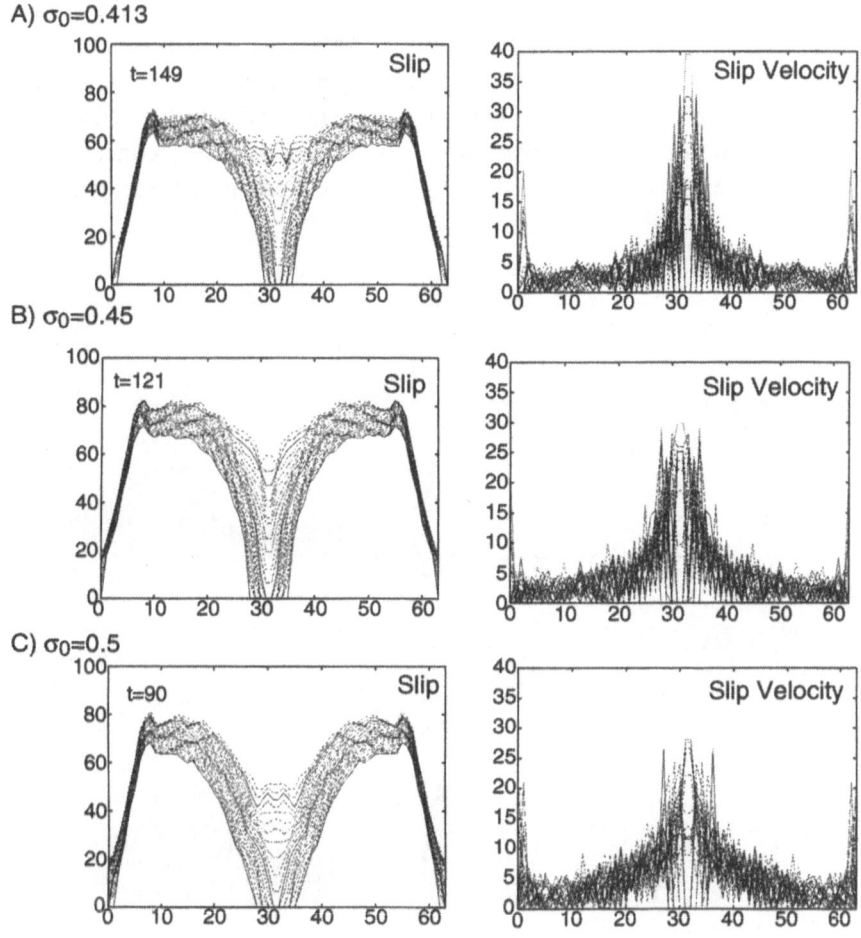

Figure 8

Slip and slip velocity plotted along the thick line in Figure 7. In each panel, slip/slip velocity distributions for 21 successive time steps are shown. The sections are centered about the time step when the two rupture fronts collide. We picked up 10 time steps before and after the collision. A) Initial stress is 0.413, so that rupture propagates with sub-Rayleigh wave velocity. B) Initial stress is 0.45, and rupture propagates at the Rayleigh-wave velocity. C) Initial stress is 0.5 and rupture propagates at the P-wave velocity. To the upper left of each panel we show the time when collision occurs. In each case, D_c is set at 0.1 non-dimensional unit.

We observe that when the rupture velocity is high (Fig. 8C), the slip velocity during the collision is rather low. This happens because stress concentrations ahead of the running ruptures are not large enough. On the other hand, when the rupture velocity is low (Fig. 8A), stress is large enough ahead of the rupture front so that during rupture interaction, very large and rapid stress drop occurs producing high slip velocities on the fault.

Figure 9 shows the slip and slip velocity distribution at 21 successive time steps for three different slip-weakening distances (A) $D_c = 0.4$, (B) 1.0 and (C) 1.15. Following Figure 4, rupture propagates with super-shear velocity for the case (A), Rayleigh wave velocity for (B) and sub-Rayleigh velocity for (C). High slip velocities at collision are observed when D_c is small. Stress concentration was relatively high for $D_c = 1.15$, but the large slip-weakening distance prevents rapid

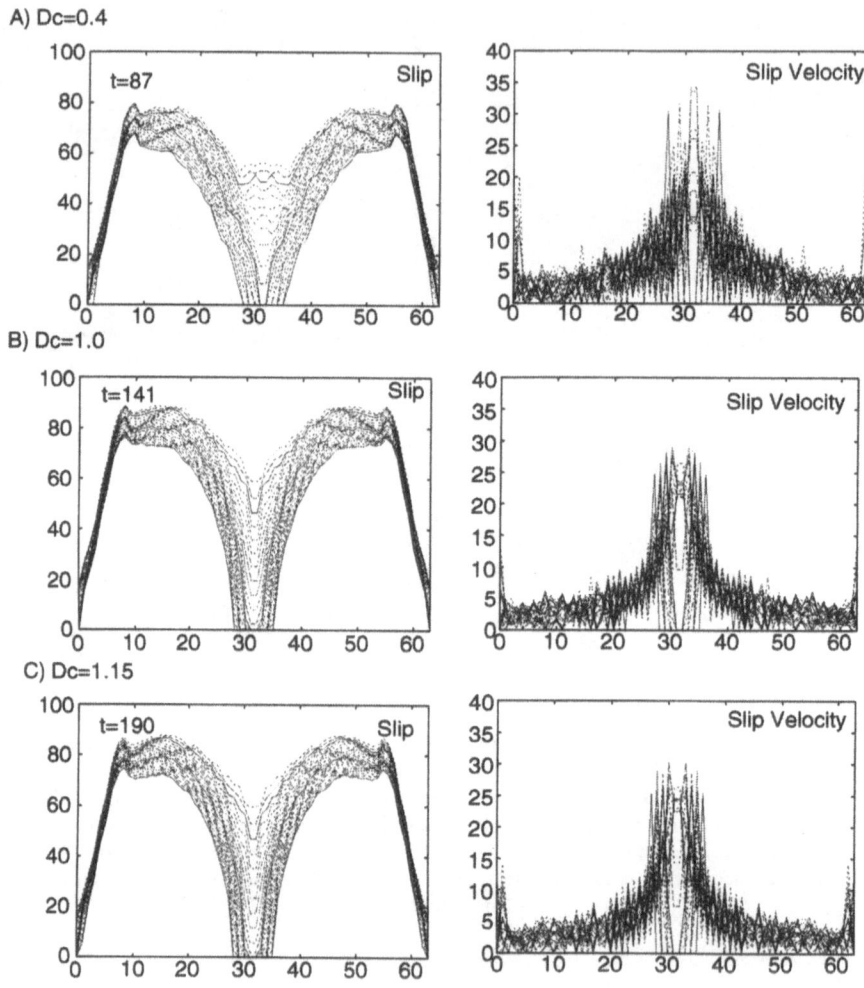

Figure 9

Slip and slip velocity plotted along the thick line in Figure 7. In each panel, slip/slip velocity distributions for 21 successive time steps are shown. The sections are centered about the time step when the two rupture fronts collide. We picked up 10 time steps before and after the collision. A) Slip-weakening distance D_c is 0.4, so that rupture propagates with super-shear wave velocity. B) D_c is 1.0 and rupture propagates with the Rayleigh-wave velocity. C) D_c is 1.15 and rupture propagates with sub-Rayleigh wave velocity. To the upper left of each panel we show the time when collision occurs. In each case, σ_0 is set at 0.5 non-dimensional unit.

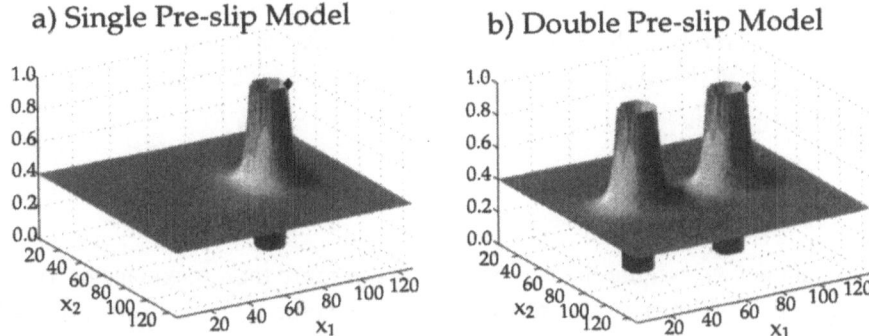

a) Single Pre-slip Model   b) Double Pre-slip Model

Figure 10

Initial stress (σ_i) distribution for (a) the single pre-slip model and (b) the double pre-slip model. In both models, the critical stress (σ_c) and frictional stress (σ_f) are assumed to be 1.0 and 0, respectively. Diamonds indicate the rupture starting point where the critical stress is assumed to be slightly below the initial stress.

slip, so that the slip velocity remains low. This indicates that even if stress concentrations are on the same order of magnitude, higher slip rates are observed when the slip-weakening distance (or fracture energy) is smaller.

Description of the Localized Asperity Model

We investigate three-dimensional effects on the propagation of rupture due to a heterogeneous distribution of initial stress. The stress heterogeneity is due to a zone that has undergone slip due to previous earthquakes or to quasi-static pre-slip. Because of pre-slip, stress is strongly localized along the perimeter of the region. We consider two models. These models are schematically illustrated in Figure 10.

Single Pre-slip Model

In this model the initial shear stress at the center of the asperity has been lowered by pre-slip. Stress has been raised by the stress concentration outside the pre-slip area and it decays outward as shown in Figure 10 (see ANDREWS, 1976a for a two-dimensional version of this model). A simplified 3-D version of this model was carefully studied by DAS and KOSTROV (1983). We define the asperity as the zone of high stress concentration that surrounds the area in which the pre-slip occurred. In this model and hereafter, we assume that the critical stress (σ_c) and the frictional stress (σ_f) in Equation (2) are equal to 1.0 and 0.0, respectively. This model has been designed in order to study the three-dimensional effects of rupture propagation.

We simulate a dynamic fracture episode that starts some time after the pre-slip process. We assume that the dynamic fracture initiates at one point along the rim of the asperity and that it propagates spontaneously under control of the slip-weakening friction.

Double Pre-slip Model

In this model we assume that pre-slip occurs in more than one place on the fault and that these pre-slip patches are surrounded by static stress concentrations. The stress concentration around each asperity has the same shape as in the single asperity model. Other, more complex stress distributions, can of course be considered but we want to study focusing effects due to inhomogeneous stress distributions deduced from simple pre-slip fields.

At one point on the rim of one of the asperities, a rupture process starts dynamically. It breaks the rim of the first asperity, then rupture progresses outward and approaches the stress concentrations around other asperities. Since elastic waves propagate faster than the rupture front, the other asperities start to break due to stress perturbations produced by the elastic waves emitted by the rupture front of the first asperity.

In order to emphasize the qualitative features of our model, we show results only for two asperities, although it is not difficult to extend our computations to more general multi-asperity models.

Results of Computation

In the single pre-slip case, as shown in Figure 11a, rupture propagates closely following the area of stress concentration around the rim of the initial slip zone. After moving around the asperity as proposed by DAS and KOSTROV (1983), the two rupture fronts meet with each other at the opposite side of the asperity with respect to the initiation point (A in Fig. 11a). At that time a very high slip velocity concentration occurs. After this focusing phenomenon, rupture propagates beyond the asperity and also towards its center. Inside the asperity the pre-slip zone starts to move until the rupture front concentrates near the center of the asperity (B in Fig. 11a) where very high slip velocities are observed. This effect is due to rupture front focusing. If the rupture front focuses, stress concentration in front of the rupture increases and then rapid stress drop occurs. This high stress drop produces in turn very high slip velocities. In two dimensions this kind of rupture focusing can only occur when two isolated ruptures connect with each other (DAS and AKI, 1977). However, in three dimensions we emphasize that rupture focusing and high slip rates can occur even if rupture starts from a single point.

In more realistic situations several asperities may co-exist and one of them starts to break. As shown in Figure 11b, the rupture process is initially very similar to the single asperity case until the second asperity starts to break. This second break starts before the rupture front reaches the asperity triggered by the elastic waves

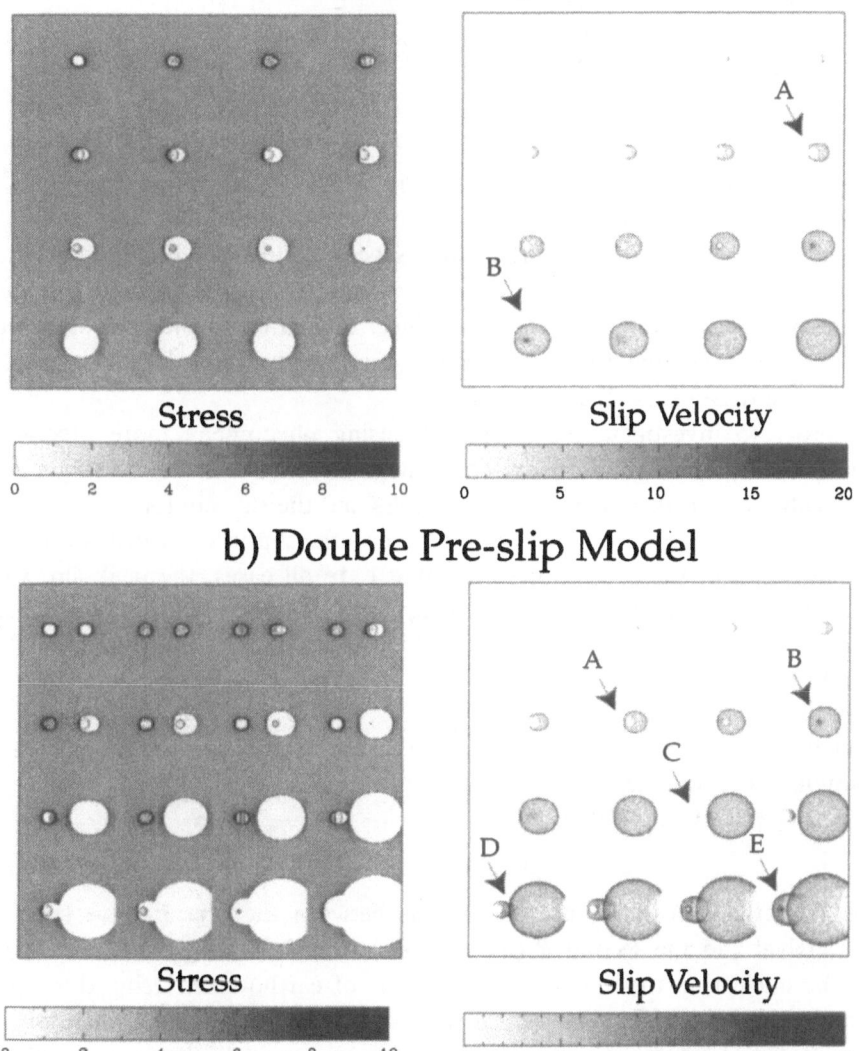

Figure 11

a) Result of computation for the single pre-slip model. Left panel shows the stress distribution snapshots and on the right is the slip velocity snapshots. Time passes from top left to right and then downward with an equal time step ($10\Delta t$). b) Result of computation for the double pre-slip model. Left panel shows stress distribution snapshots and on the right is slip velocity snapshots. Time passes from top left to right and then downward with an equal time step ($16\Delta t$).

Table 1

Parameters changed for the single pre-slip model. Rupture focusing time (t_r) is also indicated.

Model	σ_0	D_c	t_r
#1	0.3	1.2	137
#2	0.4	1.2	118
#3	0.5	1.2	110
#4	0.4	0.8	101
#5	0.4	1.0	109
#6	0.4	1.4	136

emitted from the initial rupture (C in Fig. 11b). After that, the two ruptures grow separately until they connect with each other. At this point the slip velocity becomes very large (D in Fig. 11b). This is because in three dimensions, the two ruptures interfere along an elongated zone with a certain time delay; consequently the zone of high slip velocities becomes bigger compared to the previous case and emits the strongest high frequency waves. Finally, the unbroken area inside the pre-slip zone disappears (E in Fig. 11b).

In order to investigate the rupture focusing phenomena more precisely, we conducted several computations by changing the parameters for the single pre-slip model. Table 1 indicates the model parameters and the rupture focusing time which corresponds to B in Figure 11a. Figure 12 shows the slip velocity distribution along the x_1 direction at the time shown in Table 1. In all cases, the peak slip velocity does not change significantly. This suggests that this kind of pre-slip model produces rupture focusing and generates high slip velocities even if initial stresses and slip-weakening distances vary slightly. The rupture focusing can be recognized as a stable phenomenon.

Finally, we estimated the physical scale of this phenomenon. Let us consider a magnitude 7 earthquake. We first assumed the following empirical relation proposed by UMEDA et al. (1996a):

$$\log(T_p) = 0.55M - 3.4 \tag{5}$$

where T_p is the time difference in seconds between the first P-wave (nucleation phase) arrival and the second P (main phase), and M is magnitude. Equation (5) scales the nucleation time (T_p) and magnitude of earthquake. Using this equation we obtain the nucleation time of 2.8 seconds for M7 earthquake. Assuming that this nucleation time corresponds to the collision time (t_r) shown in Table 1 for model #2, we ascertain that the diameter of the pre-slip region is 4 km by assuming S-wave velocity (c_T) is 3.5 km/s. In this case, critical slip-weakening distance (D_c) becomes 1.1 m if we assume that rigidity (μ) and stress drop ($\Delta\sigma = \sigma_i - \sigma_f$) are 35 GPa and 10 MPa, respectively. The nucleation size of 4 km for magnitude 7 earthquake looks reasonable (e.g., SHIBAZAKI and MATSU'URA, 1998).

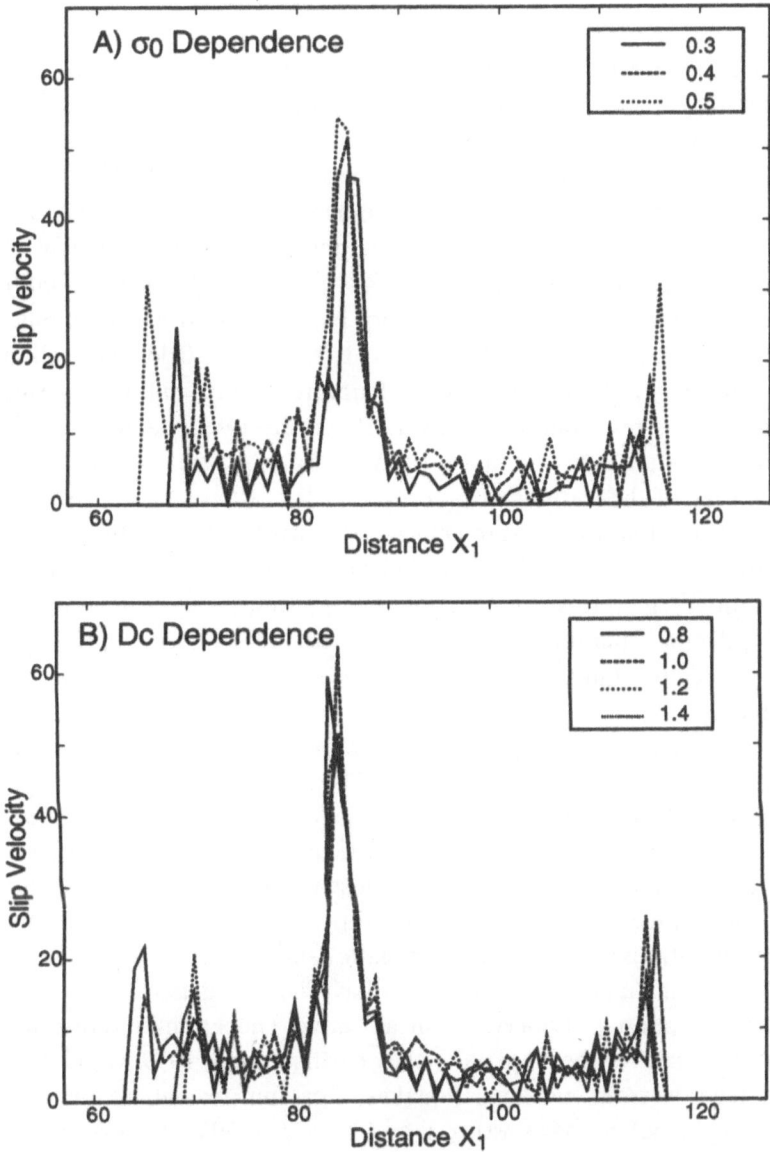

Figure 12

Slip velocity distribution along the x_1 direction across the rupture focusing point (B in Fig. 11). A) Initial stresses of 0.3, 0.4 and 0.5 with a constant slip-weakening distance of 1.2. B) Slip-weakening distances of 0.8, 1.0, 1.2 and 1.4 with a constant initial stress of 0.4.

Discussion and Conclusion

In this paper we have examined with an improved 3-D BIEM several basic phenomena that occur in seismic ruptures propagating along a flat finite fault embedded in a three-dimensional elastic medium. Our work is a continuation of the pioneering work by DAY (1982) and DAS and KOSTROV (1983). When the initial stress is heterogeneous the rupture phenomenology is considerably richer than could be expected from line faults in two-dimensional elastic medium. The most important finding, discussed also by MADARIAGA and OLSEN (2000), is that earthquakes probably propagate under very particular critical conditions. If this is the case, ruptures very closely follow the initial stress field and will propagate following strong stress concentrations. Rupture fronts will then easily collide, producing strong stress concentrations as shown by DAS and KOSTROV (1983).

Rather than studying a complex rupture situation, we studied simple models that isolate each of the features of 3-D ruptures that we want to illustrate. Thus, we first demonstrated that the rupture velocity is controlled by the initial stress and the slip-weakening distance. In our non-dimensional variables, these two parameters control strain energy release and fracture energy, respectively. We showed that the boundary integral equation method we use can resolve the rupture fronts adequately in order to study the delicate balance between strain energy release and fracture energy.

We showed that a critical phenomenon occurs at a certain level of stress for a given value of the slip-weakening distance. The transition from non-propagation to fast propagation is not only very sharp but also very narrow, because for values of the initial stress only 10% higher than the critical ruptures jump to speeds exceeding the S-wave speed and terminate at speeds close to the P-wave speed of the medium. This particular range of initial stresses and slip-weakening distances defines the only acceptable conditions for rupture propagation in the earth. Inversions of strong motion data from numerous earthquakes have indicated that super-shear rupture speeds are exceptional. Otherwise shear wave shocks like those inferred by ARCHULETA (1984) for the Imperial Valley earthquake of 1979 would dominate the near field records from most earthquakes. For this reason, as discussed in the accompanying paper by MADARIAGA and OLSEN (2000), we expect that critical rupture conditions dominate within the earth.

Next we showed that the criticality conditions are not local but they concern the stress distribution over a large region. For that purpose, following DAY (1982), we studied a rough initial stress distribution that oscillates 10% about the critical initial stress value. We observe again that rupture propagates critically. This suggests that for realistic stress distributions it is not the local stress but a certain averaged value that determines the propagation of rupture. The slip-weakening distance represents a certain averaging length scale over which stresses are measured. This distance may be commensurate with the initial stress patch derived from modern friction laws

(DIETERICH, 1978) but we have not yet determined the averaging length with accuracy.

Using the criticality conditions derived for a simple rectangular fault we examined the collision of two propagating ruptures. This is a phenomenon that occurs very frequently in three-dimensional ruptures whenever the stress distribution is heterogeneous. Rupture collisions or coalescence generates high slip velocities that may be at the origin of acceleration peaks observed in the near field. We showed that slip rates are enhanced when the stress concentration ahead of the crack is sufficient and the slip-weakening distance is small.

In the last part of our work, we demonstrated that the rupture process is strongly guided by the pre-stress field on a fault. Heterogeneous pre-stress produces segmented rupture fronts that propagate very closely following the zones of stress concentration. In the case of a single asperity, rupture surrounds the asperity as proposed by DAS and KOSTROV (1983) and produces very strong radiation when the two rupture zones that follow the stress concentration meet on the opposite side of the asperity edge. This mechanism could be the origin of the seismically observed initial phase in large earthquakes. Moreover, in large earthquakes, asperities are likely to be more densely distributed than for smaller earthquakes. Therefore in large earthquakes high frequency waves tend to be emitted at the initial stage.

Acknowledgements

We thank Dr. K. B. Olsen for very useful discussions concerning the origin of rupture complexity. Dr. Taku Tada and an anonymous reviewer's comments were beneficial. This work was supported by the Crustal Activity Modeling Project (CAMP) (EF), by the Environment Program of the European Community under project SGME (RM) and by the Programme National des Risques Naturels of INSU-CNRS (RM).

REFERENCES

ANDREWS, D. J. (1976a), *Rupture Propagation with Finite Stress in Antiplane Strain*, J. Geophys. Res. *81*, 3575–3582.

ANDREWS, D. J. (1976b), *Rupture Velocity of Plane-strain Shear Cracks*, J. Geophys. Res. *81*, 5679–5687.

ANDREWS, D. J. (1985), *Dynamic Plane-strain Shear Rupture with a Slip-weakening Friction Law Calculated by a Boundary Integral Method*, Bull. Seismol. Soc. Am. *75*, 1–21.

AOCHI, H., FUKUYAMA, E., and MATSU'URA, M. (2000), *Spontaneous Rupture Propagation on a Non-planar Fault in 3-D Elastic Medium*, Pure appl. geophys. *157*, 2003–2027.

ARCHULETA, R. J. (1984), *A Faulting Model for the 1979 Imperial Valley Earthquake*, J. Geophys. Res. *89*, 4559–4585.

BURRIDGE, R. (1973), *Admissible Speeds for Plane-strain Self-similar Shear Cracks with Friction but Lacking Cohesion*, Geophys. J. R. astr. Soc. *35*, 439–455.

BURRIDGE, R., CONN, G., and FREUND, L. B. (1979), *The Stability of a Rapid Mode II Shear Crack with Finite Cohesive Traction*, J. Geophys. Res. *85*, 2210–2222.

COTTON, F., and CAMPILLO, M. (1995), *Frequency Domain Inversion of Strong Motions: Application to the 1992 Landers Earthquake*, J. Geophys. Res. *100*, 3961–3975.

DAS, S., and AKI, K. (1977), *Fault Plane with Barriers: A Versatile Earthquake Model*, J. Geophys. Res. *82*, 5658–5670.

DAS, S., and KOSTROV, B. V. (1983), *Breaking of a Single Asperity: Rupture Process and Seismic Radiation*, J. Geophys. Res. *88*, 4277–4288.

DAY, S. M. (1982), *Three-dimensional Simulation of Spontaneous Rupture: The Effect of Nonuniform Prestress*, Bull. Seismol. Soc. Am. *72*, 1881–1902.

DIETERICH, J. H. (1978), *Time-dependent Friction and the Mechanics of Stick-slip*, Pure appl. geophys. *116*, 790–806.

ELLSWORTH, W. L., and BEROZA, G. C. (1995), *Seismic Evidence for an Earthquake Nucleation Phase*, Science *268*, 851–855.

FUKUYAMA, E., and IRIKURA, K. (1986), *Rupture Process of the 1983 Japan Sea (Akita-Oki) Earthquake Using a Waveform Inversion Method*, Bull. Seismol. Soc. Am. *76*, 1623–1640.

FUKUYAMA, E., and MADARIAGA, R. (1995), *Integral Equation Method for Plane Crack with Arbitrary Shape in 3-D Elastic Medium*, Bull. Seismol. Soc. Am. *85*, 614–628.

FUKUYAMA, E., and MADARIAGA, R. (1998), *Rupture Dynamics of a Planar Fault in a 3-D Elastic Medium: Rate- and Slip-weakening Friction*, Bull. Seismol. Soc. Am. *88*, 1–17.

HARRIS, R. A., and DAY, S. M. (1993), *Dynamics of Fault Interaction: Parallel Strike-slip Faults*, J. Geophys. Res. *98*, 4461–4472.

HARTZELL, S. H., and HEATON, T. H. (1983), *Inversion of Strong Ground Motion and Teleseismic Waveform Data for the Fault Rupture History of the 1979 Imperial Valley, California, Earthquake*, Bull. Seismol. Soc. Am. *73*, 1553–1583.

HASKELL, N. A. (1964), *Total Energy and Energy Spectral Density of Elastic Wave Radiation from Propagating Faults*, Bull. Seismol. Soc. Am. *54*, 1811–1841.

IDA, Y. (1972), *Cohesive Force across the Tip of a Longitudinal-shear Crack and Griffith's Specific Surface Energy*, J. Geophys. Res. *77*, 3796–3805.

IDE, S., and TAKEO, M. (1997), *Determination of Constitutive Relations of Fault Slip Based on Seismic Wave Analysis*, J. Geophys. Res. *102*, 27,379–27,391.

IIO, Y. (1992), *Slow Initial Phase of the P-wave Velocity Pulse Generated by Micro-earthquakes*, Geophys. Res. Lett. *19*, 477–480.

MADARIAGA, R., OLSEN, K. B., and ARCHULETA, R. J. (1998), *Modeling Dynamic Rupture in a 3-D Earthquake Fault Model*, Bull. Seismol. Soc. Am. *88*, 1182–1197.

MADARIAGA, R., and OLSEN, K. B. (2000), *Criticality of Rupture Dynamics in Three Dimensions*, Pure appl. geophys. *157*, 1981–2001.

OLSEN, K. B., MADARIAGA, R., and ARCHULETA, R. J. (1997), *Three-dimensional Dynamic Simulation of the 1992 Landers Earthquake*, Science *278*, 834–838.

RAVI-CHANDAR, K., and YANG, B. (1997), *On the Role of Microcracks in the Dynamic Fracture of Brittle Materials*, J. Mech. Phys. Solids *45*, 535–563.

ROSAKIS, A. J., SAMUDRALA, O., and COKER, D. (1999), *Cracks Faster than the Shear Wave Speed*, Science *284*, 1337–1340.

SHIBAZAKI, B., and MATSU'URA, M. (1998), *Transition Process from Nucleation to High-speed Rupture Propagation: Scaling from Stick-slip Experiments to Natural Earthquakes*, Geophys. J. Int. *132*, 14–30.

TADA, T., FUKUYAMA, E., and MADARIAGA, R. (2000), *Non-hypersingular Boundary Integral Equations for 3-D Non-planar Crack Dynamics*, Computational Mechanics (in press).

UMEDA, Y. (1992), *The Bright Spot of an Earthquake*, Tectonophys. *211*, 13–22.

UMEDA, Y., YAMASHITA, T., ITO, K., and HORIKAWA, H. (1996a), *The Bright Spot and Growth Process of the 1995 Hyogo-ken Nanbu Earthquake*, J. Phys. Earth *44*, 519–527.

UMEDA, Y., YAMASHITA, T., TADA, T., and KAME, N. (1996b), *Possible Mechanisms of Dynamic Nucleation and Arresting of Shallow Earthquake Faulting*, Tectonophys. *261*, 179–192.

WALD, D. J., and HEATON, T. H. (1994), *Spatial and Temporal Distribution of Slip for the 1992 Landers, California, Earthquake*, Bull. Seismol. Soc. Am. *84*, 668–691.

YAMASHITA, T., and UMEDA, Y. (1994), *Earthquake Rupture Complexity due to Dynamic Nucleation and Interaction of Subsidiary Faults*, Pure appl. geophys. *143*, 89–116.

(Received August 16, 1999, revised March 14, 2000, accepted April 6, 2000)

To access this journal online:
http://www.birkhauser.ch

Pure appl. geophys. 157 (2000) 1981–2001
0033–4553/00/121981–21 $ 1.50 + 0.20/0

Criticality of Rupture Dynamics in 3-D

RAUL MADARIAGA[1] and KIM B. OLSEN[2]

Abstract—We study the propagation of seismic ruptures along a fault surface using a fourth-order finite difference program. When prestress is uniform, rupture propagation is simple but presents essential differences with the circular self-similar shear crack models of Kostrov. The best known is that rupture can only start from a finite initial patch (or asperity). The other is that the rupture front becomes elongated in the *in-plane* direction. Finally, if the initial stress is sufficiently high, the rupture front in the in-plane direction becomes super-shear and the rupture front develops a couple of "ears" in the in-plane direction. We show that we can understand these features in terms of single nondimensional parameter κ that is roughly the ratio of available strain energy to energy release rate. For low values of κ rupture does not occur because Griffith's criterion is not satisfied. A bifurcation occurs when κ is larger than a certain critical value, κ_c. For even larger values of κ rupture jumps to super-shear speeds. We then carefully study spontaneous rupture propagation along a long strike-slip fault and along a rectangular asperity. As for the simple uniform fault, we observe three regimes: no rupture for subcritical values of κ, sub-shear speeds for a narrow range of supercritical values of κ, and super-shear speeds for $\kappa > 1.3\kappa_c$. Thus, there seems to be a certain universality in the behavior of seismic ruptures.

Key words: Rupture dynamics, finite-difference modeling, spontaneous rupture propagation.

Introduction

Earthquakes are complex at all scales. Recent studies of a number of earthquakes that were well-recorded in the near field reveal that both slip and stress drop in these earthquakes are very complex. The first evidences of these complexities were discussed by DAS and AKI (1977a), AKI (1979), DAY (1982) and many others. BEROZA and MIKUMO (1996), BOUCHON (1997), IDE and TAKEO (1997), among others, showed that stress drop for all well studied earthquakes is highly variable in space and can be both positive (where fault slips) or negative (stress increases) where there is no slip. Some of these properties were indeed introduced in the original "asperity" model of KANAMORI and STEWART (1976) and in the "barrier" model of DAS and AKI (1977b) who realized that the simple uniform stress drop

---
[1] Laboratoire de Géologie, Ecole Normale Supérieure, 24 rue Lhomond, 75231 Paris Cedex 05, France. E-mail: madariag@geologie.ens.fr

[2] Institute for Crustal Studies, University of California, Santa Barbara, CA 93106, U.S.A. E-mail: kbolsen@crustal.ucsb.edu

models were exceedingly simple representations of the earthquake source. MADARIAGA (1979) showed that seismic radiation from these models would be very complex and that high frequency radiation would be quite different from that predicted by the simple circular crack models of KOSTROV (1964) and MADARIAGA (1976).

The study of complex fault models has gained new attention in recent years in the wake of the suggestion by CARLSON and LANGER (1989) that earthquakes may become spontaneously complex due to nonlinear effects in friction. Other alternatives to the origin of complexity were discussed by RICE (1993) and RICE and BEN-ZION (1996), who suggested that at least some of the complexity found by Carlson and Langer was due to the lack of a continuum limit in the velocity weakening friction law they used. Applying a simple regularized friction law with velocity weakening, COCHARD and MADARIAGA (1996) found that heterogeneity appears spontaneously only in a limited parameter range of rate-weakening. More recently, SHAW and RICE (2000) have studied the conditions for the development of complexity in well-posed numerical simulations.

In order to study complexity we must model earthquakes accurately, taking into account all the relevant length scales in the problem, both (intrinsic) length scales associated with friction as well the (extrinsic) scale associated with fault size and asperity distribution. This is not a simple task because accurate numerical modeling of ruptures requires vast amounts of computer resources. Although DAS and KOSTROV (1983) and DAY (1982) made calculations of complex faults with heterogeneous distributions of stress and rupture resistance, these models contained insufficient resolution to study the more difficult problem of the interaction between intrinsic and extrinsic length scales. Recent development of finite difference methods in 3-D (HARRIS and DAY, 1993; MIKUMO and MIYATAKE, 1995; OLSEN et al., 1997) or Boundary Integral Equation methods (FUKUYAMA and MADARIAGA, 1998), together with the availability of fast parallel computers, has opened the way to the study of accurate rupture propagation models.

In this paper we will report on our study of the nature of three-dimensional rupture propagation for a couple of very simple classical models of rupture: the rectangular fault and rectangular asperity. With our new computational capabilities we study the propagation of rupture in these models for a broad parameter range and demonstrate that rupture propagation is controlled by a simple nondimensional number. The nondimensional number has a critical value below which ruptures die very quickly. For nondimensional values slightly above critical, ruptures grow at speeds close to the shear wave speed as most earthquakes do. For larger values, ruptures grow faster than the speed of shear waves, which is rarely observed unambiguously in earthquakes. By examining the stress field of recent events we suggest that earthquake ruptures occur at nondimensional numbers that are most often critical. Although we cannot prove it for the moment, we believe that this may be the actual self-organized criticality of earthquake ruptures.

Modeling Complex Earthquakes in 3-D

An essential requirement to study dynamic faulting is an accurate and robust method for the numerical modeling of seismic sources. In our recent work we have used a fourth-order formulation of the velocity-stress method (MADARIAGA, 1976; OLSEN et al., 1997; MADARIAGA et al., 1998) in order to study dynamic rupture propagation on a planar shear fault embedded in a heterogeneous elastic half-space.

Rupture propagation on a major earthquake fault is controlled by the properties of the friction law on the fault. Friction controls the initiation, development of rupture and the healing of faults. Laboratory experiments at low slip rates were analyzed by DIETERICH (1978), who proposed models of rate- and state-dependent friction, and by OHNAKA and SHEN (1999) who concluded that their experiments could be explained with a simpler slip-weakening friction law. Actually, from the point of view of earthquake observations, these two models of friction are essentially indistinguishable as shown by OKUBO (1989). Basically, both slip-weakening and rate- and state-dependent friction contain finite length scales that control the behavior of the rupture front. Although, these intrinsic length scales are of very different origin, we can hypothesize that their effects on rupture is similar. The length scale, that we will generically refer to as D_c, is the most important ingredient in our study. Without this length scale earthquake faulting makes little sense due to lack of energy conservation in the fault system.

Because of the equivalence of friction laws at high slip rates, we used a simple slip-weakening friction law in which slip is zero until the total stress reaches a peak value (yield stress) that we denote with T_u. Once this stress has been reached, slip D increases and $T(D)$ decreases:

$$
\begin{aligned}
T(D) &= T_u\left(1 - \frac{D}{D_c}\right) \quad \text{for} \quad D < D_c \\
T(D) &= 0 \qquad\qquad\qquad \text{for} \quad D > D_c
\end{aligned}
\tag{1}
$$

where D_c is a characteristic slip distance. This friction law has been applied in numerical simulations of rupture by ANDREWS (1976), DAY (1982) and many others. We note that in (1) and in the following we refer all our stresses to a reference value equal to the residual friction at high slip. This is why $T(D) = 0$ in the second equation of (1).

Figure 1 presents the geometry of the fault model we study. The most important feature of the friction law is slip-weakening that occurs near the rupture zone on a so-called breakdown or slip-weakening zone just behind the rupture front. The propagation of the rupture front is completely controlled by the friction law and the distribution of initial stress on the fault.

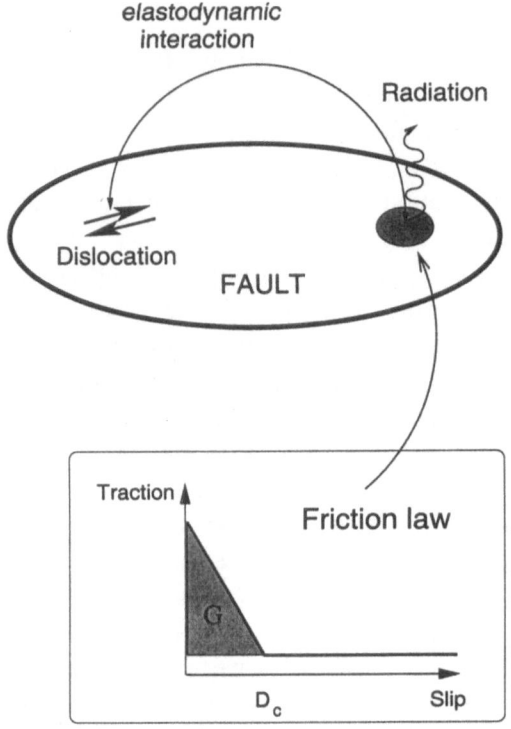

Figure 1

Geometry of a simple rupture propagating along a flat fault embedded in a 3-D elastic medium. Rupture propagation is controlled by elastodynamic interactions and is damped by seismic radiation. The boundary conditions on the fault are dominated by friction. We use a simple slip-weakening friction of peak stress T_u, slip-weakening D_c and energy release rate G.

Spontaneous Rupture of a Uniform Fault

MADARIAGA *et al.* (1998) studied spontaneous rupture, starting from a circular asperity of radius R that is ready to break (with stress T_u), and is surrounded by a fault surface at a constant effective stress level ($T_e < T_u$). Rupture is initiated at time $t = 0$ by instantaneously triggering stress relaxation inside the asperity of radius R according to the friction law of equation (1). This instantaneous rupture of the circular patch is less physical than starting from a critical fault and letting the rupture grow spontaneously. Unfortunately, rupture of a critical fault is an expensive numerical problem because rupture must start from one point on the edge, encircle the asperity and then grow outside the initial patch. Such a problem would take excessive time to solve in our currently available computer.

As ANDREWS (1976) and many other authors have noted, in order for rupture to expand stress must be high over a finite zone, sometimes referred to as the minimum rupture patch. Once rupture has broken the asperity, it will grow or stop depending on the values of the stress field inside (T_u) and outside the asperity (T_e), the shear modulus μ, the half-width of the asperity $L = R$ and the slip-weakening constant D_c of the friction law in (1).

ANDREWS (1976) computed the minimum patch for a 2-D in-plane fault from energy considerations and found that the critical half-width of the fault was:

$$L_c = \frac{1}{\pi(1 - v)} \frac{T_u \mu}{T_e^2} D_c,$$ (2)

where v is Poisson's modulus. He derived this relationship by equating the available strain energy in the vicinity of the fault to the energy needed to propagate rupture (i.e., $G = 1/2T_u D_c$) at one edge of the fault. A similar expression for a circular fault was computed by DAY (1982) who found that the minimum radius of the initial patch was:

$$R_c = \frac{7\pi}{24} \frac{T_u \mu}{T_e^2} D_c.$$ (3)

It is well-known in fluid mechanics (see, e.g., LANDAU and LIFSHITZ, 1959) that expressions like (2) and (3) can also be derived from dimensional analysis. In doing that we introduce a nondimensional number κ defined as follows:

$$\kappa = \frac{T_e^2}{\mu T_u} \frac{L}{D_c},$$ (4)

where L is a characteristic length scale, for instance the half-width of the fault, and T_e, T_u, μ and D_c were defined earlier. These are the only parameters in this problem, thus there is no need to introduce additional nondimensional numbers to study the bifurcation.

When considered as a function of L, expression (4) together with (2) or (3) defines a bifurcation at a certain critical value L_c, so that ruptures grow only for values of $L \geq L_c$. For in-plane faults κ_c can be derived from ANDREWS (1976) expression reproduced in equation (2). We obtain $\kappa_c = 1/[\pi(1 - v)] \simeq 0.424$ for $v = 1/4$, the usual value of Poisson's ratio for crustal materials. For anti-plane rupture the minimum patch can be computed from IDA's (1972) expressions, we find $\kappa_c = 1/\pi = 0.318$. Thus, as expected, it is easier to initiate shear fracture in anti-plane mode than in in-plane mode. For a circular fault the critical number can be estimated from (3) as $\kappa_c = 7\pi/24 = 0.916$. From numerical simulations MADARIAGA et al. (1998) found a $\kappa_c \simeq 0.6$. The difference probably comes from the loading conditions assumed by DAY (1982) who studied the rupture of a quasi-statically loaded crack, while MADARIAGA et al. (1998) triggered rupture by suddenly loading a circular asperity.

We realize that κ_c defines a typical Hopf bifurcation. For $\kappa < \kappa_c$ rupture does not grow beyond the initial asperity; while for $\kappa > \kappa_c$ it grows indefinitely at increasing speed without ever stopping. This is a typical bifurcation first noted by

GRIFFITH (1929). However there is a further complication: if the parameter κ is considerably larger than κ_c the rupture front in the in-plane direction jumps to speeds higher than the shear-wave speed. The jump to super-shear speeds was first reported for plane faults by ANDREWS (1976), who attributed it to the interaction between an S wave running ahead of the crack and the rupture front itself. In MADARIAGA et al. (2000) we show that in 3-D the jump from sub-Rayleigh to super-shear speeds is due to an instability of the rupture front that spreads along the rupture front like a wave. For the initially circular shear fault, the rupture front acquires a very nice elliptical shape with two "ears" elongated along the in-plane direction (see the bottom snapshots in Figure 2).

Rupture in the simple uniform model studied here will never stop once it starts. Stopping ruptures in 2-D was discussed by HUSSEINI et al. (1975) who showed that ruptures would stop either because they met unbreakable barriers, or because they ran out of steam when they entered regions with low stress (a sort of anti-asperity). In 3-D there are many other possibilities for rupture to stop because stress distributions can be very complex as mentioned above.

Very Long Rectangular Barrier Model on a Flat Fault

We study two simple models that will put in evidence a simple nondimensional number that controls rupture growth in 3-D. In order to understand rupture we must introduce external length scales. These length scales are naturally associated either with the distribution of rupture resistance in the "barrier" model or with the distribution of initial stress in the "asperity" model. We will demonstrate that although the same nondimensional number controls rupture processes in both models, the critical value of the bifurcation differs.

We study first a model of faulting with only one external length scale. We assume that rupture is guided by the presence of two strong barriers defining a long rectangular fault zone (see the upper part of Fig. 3). Rupture resistance is constant ($T_u = 1$) inside a zone of width W. Outside this zone rupture resistance is sufficiently high (effectively ∞), so that ruptures never break outside the fault zone. For numerical reasons and to avoid end effects we also put barriers along the direction of faulting. We assume that the entire fault plane is subject to constant "effective" shear stress T_e. The only additional ingredient in our model is the friction law defined above (1). This problem has two stress parameters T_u and T_e although only their ratio is relevant to rupture. Two length scales are involved: W, the half-width of the fault and D_c, the slip-weakening distance of the friction law. Using the same dimensional arguments that led to (4), we define the nondimensional number

$$\kappa = \frac{T_e^2}{\mu T_u} \frac{W}{D_c}, \tag{5}$$

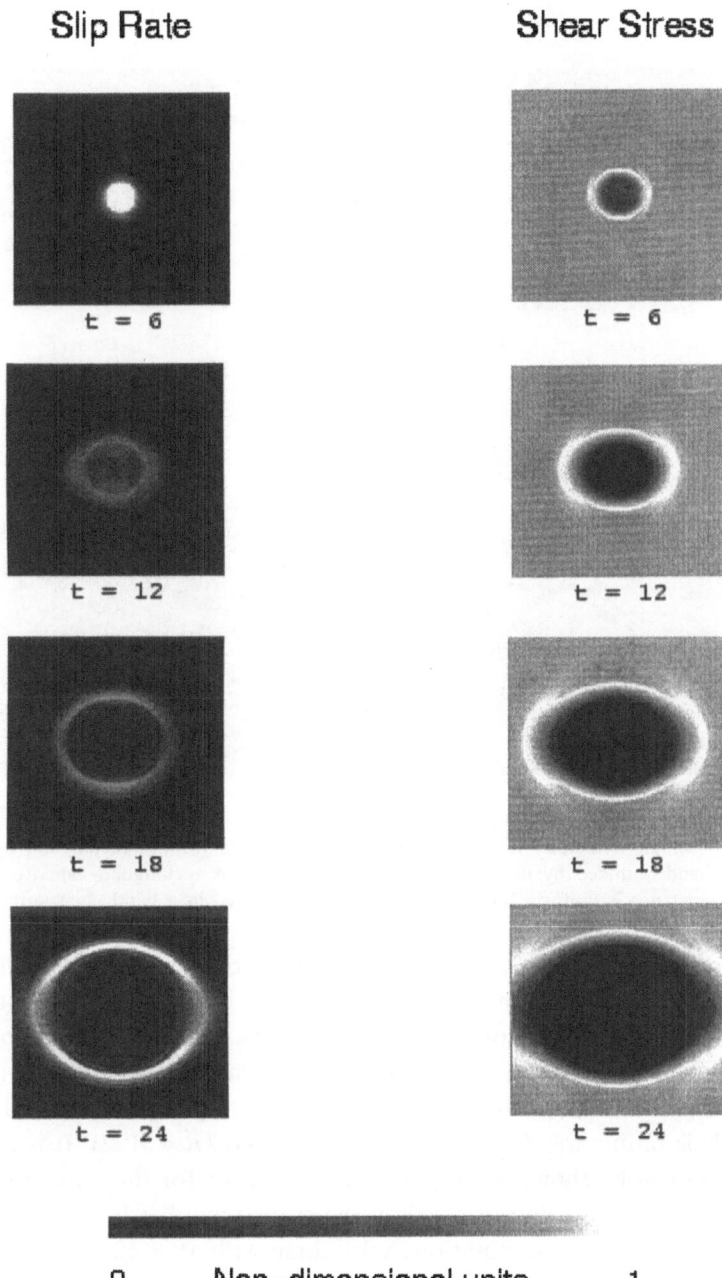

Figure 2
Rupture growth on a flat perfectly uniform fault embedded in a homogeneous elastic medium. Rupture starts from a finite initial asperity and then grows at subsonic speed in all spatial directions. After an interim rupture along the in-plane direction jumps at a speed that is higher than that of shear waves. The snapshots on the left show slip rate on the fault; and those on the right show the corresponding shear stress field. Snapshots are shown at four successive instants of time. Time is measured in units of time for a shear wave to propagate across the asperity. Around time 18, rupture jumps to super-shear speed along the in-plane direction (horizontal axis).

Rectangular Barrier Model

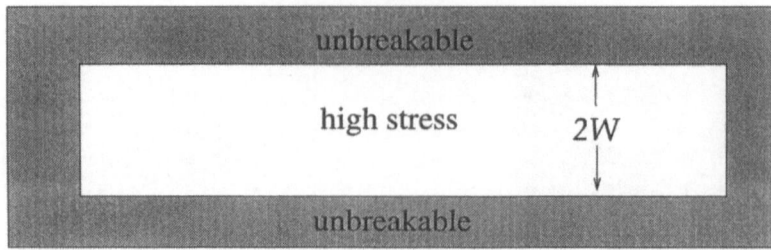

Rectangular Asperity Model

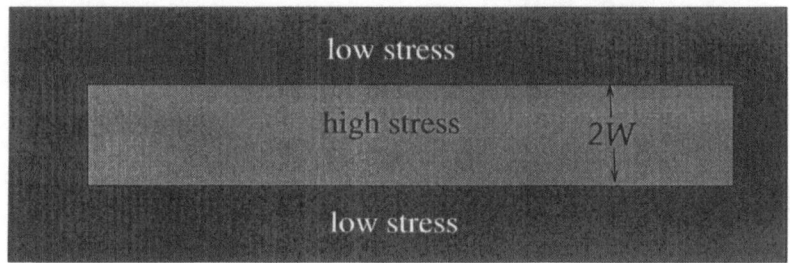

Figure 3
Two models of simple dynamic shear faults studied in this paper. (top) A rectangular fault loaded by uniform stress and bounded by unbreakable barriers. (bottom) A rectangular asperity where rupture resistance is uniform but initial stress is concentrated in a band or asperity.

in which we have replaced L by the half-width W of the fault. We will now show that this nondimensional number completely controls rupture propagation.

In the numerical simulations we surrounded the fault by a finite homogeneous elastic medium with absorbing boundary conditions. As in the unbounded fault plane, we triggered rupture by instantaneously starting slip on a circular patch of radius R. This radius was fixed in our computations ($R = W/2$). It is smaller than W because as will be shown below the critical number for the initiation of rupture from a circular patch is always less than the critical number for propagation along a rectangular fault. All computations were done with $W = 48$ grid points.

In Figure 4 we present snapshots of the slip rate and shear stress fields on the fault at the same instant of time $t = 5W/\beta$. All these snapshots were calculated with a value of $D_c = 0.416 \ WT_u/\mu$. The control parameter for these calculations was the ratio T_e/T_u that varied from 0.55 on the top panel of Figure 4 up to 0.7 at the bottom. In the top panel ($\kappa = 0.726$) rupture has already stopped at time $t = 5W/\beta$. This shows clearly that seismic ruptures can stop spontaneously. The stress field shown on the right hand of the top panel is very close to the final static stress change around the ruptured part of the fault. In the middle panels ($\kappa = 0.864$) we show the results for a fault that is very close to critical. Here, rupture advances at subsonic speed along the longitudinal direction of the fault. The rupture front in this model is almost steady state and will simply continue to propagate forever at this speed unless it encounters stress or frictional heterogeneity. The slip rate indicates that the width of the rupture front is probably controlled by the fault width, as suggested by DAY (1982). The stress field on the right-hand side manifests a peak in stress that propagates slightly ahead of the rupture front. This is the shear-wave front that establishes that rupture is sub-shear in this case. Finally in the bottom two panels ($\kappa = 1.176$) we show slip rate and stress change for a super-critical rupture. Here the rupture front has jumped ahead of the shear-wave front and is close to a steady state. Rupture will continue forever at super-shear speeds unless it encounters stress or strength heterogeneities.

The results shown in Figure 4 are clearly reminiscent of a bifurcation in the behavior of the system for a particular value of the effective stress T_e/T_u. These results were found for a particular value of the slip-weakening distance $D_c/W =$

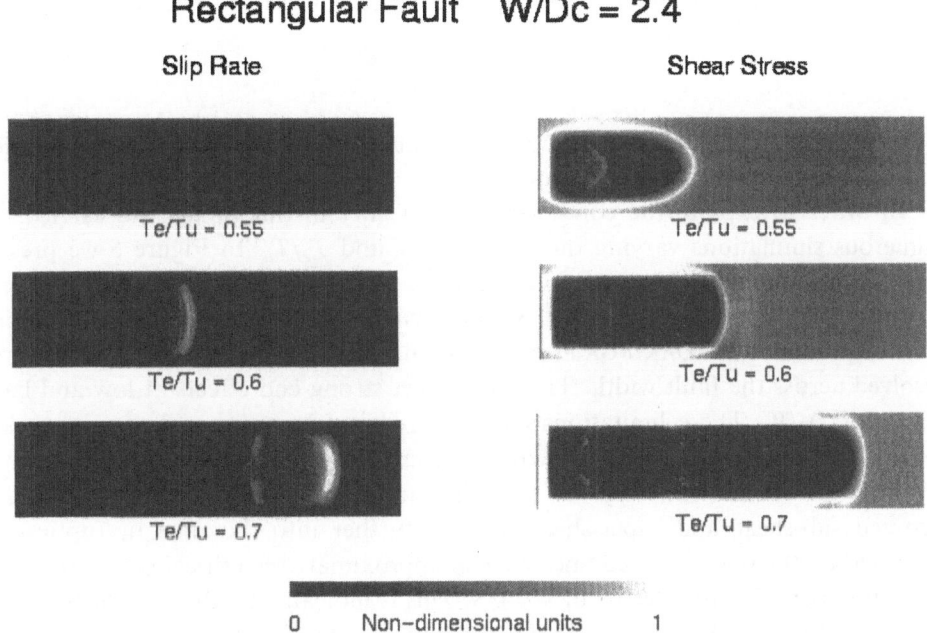

Rectangular Fault   W/Dc = 2.4

Slip Rate   Shear Stress

Te/Tu = 0.55   Te/Tu = 0.55

Te/Tu = 0.6   Te/Tu = 0.6

Te/Tu = 0.7   Te/Tu = 0.7

0   Non−dimensional units   1

Figure 4

Snapshots of the shear stress (right) and slip rate (left) fields on a fault bounded by two unbreakable barriers as shown in Figure 3. The critical value for this configuration is $\kappa_c = 0.8$ as shown in Figure 5. The top panels show ruptures at subcritical ($\kappa = 0.726$), slightly super critical ($\kappa = 0.864$) and very super critical ($\kappa = 1.176$) values of the nondimensional parameter κ.

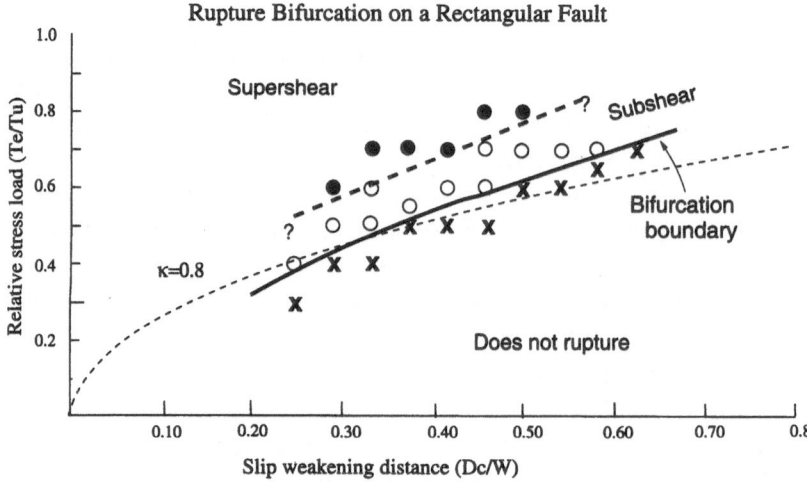

Figure 5

Bifurcation conditions for the growth of a rectangular fault. The crosses represent parameter values for which rupture stops spontaneously. The empty circles are parameter values for which rupture propagates along the fault with a rupture velocity below the shear-wave speed. Filled circles are parameter combinations that produce super-shear fracture. The thick continuous and discontinuous lines separate these parameter fields. The thin dashed line is the approximate critical value $\kappa_c = 0.8$ and corresponds very well with the thick line, the critical boundary found experimentally.

$0.416\ T_u/\mu$; changing this ratio affects the value of T_e/T_u at which the bifurcation between non-growth and growth occurs.

In order to explore the conditions of criticality in this model, we carried out numerous simulations varying the ratios D_c/W and T_e/T_u. In Figure 5 we present the results of our study. Numerical computations are reproducible and feasible only in the range shown. For $D_c/W < 0.30$ the accuracy of the rupture front resolution is inadequate (see MADARIAGA *et al.*, 1998) and for $D_c > 0.7\ W$, faulting is not well resolved across the fault width. Thus we expect strong end effects at low and large values of D_c/W. These limitations could be reduced by using a denser numerical grid. However, there is a clear systematic separation between the region of growth and non-growth indicated by the thick line in Figure 5, and a similar boundary between sub-shear and super-shear fracture further into the zone of rupture, as indicated by the thick, dashed line. We can approximate the critical boundary using the expression (5) with a value of $\kappa = 0.8$. This is indicated by the thin dashed lined marked with the value $\kappa = 0.8$. This is the critical value for this particular model of rupture, although as remarked earlier its exact value can only be approximately determined because of end effects. We note that it is larger than the critical value we inferred for the circular fault ($\kappa_c = 0.6$). This explains why it is sufficient to start the rupture from a circular patch of radius $R = W/2$.

Very Long Rectangular Asperity on a Flat Fault

In order to study the condition for rupture propagation in a heterogeneous initial stress field, we turn to a very simple rectangular asperity model described in the bottom panel of Figure 3. In this model the initial stress field contains a long asperity of width W loaded with a longitudinal "effective" shear stress T_e. The asperity is surrounded by a fault plane in which stress is very low (only $0.1 * T_u$). At time $t = 0$ rupture is initiated by forcing rupture over a circular patch of radius $R > R_c$ where R_c is the critical size for the circular asperity discussed earlier in the paper. Depending on the value of T_e, and the slip-weakening distance D_c, rupture either grows along the asperity or stops very rapidly. As for the rectangular fault we again have a bifurcation in rupture behavior, controlled by a critical value.

In Figure 6 we show the slip rate and stress fields on the fault at the same instant of time $t = 5W/\beta$, for three values of the load T_e/T_u. In all these simulations the slip-weakening distance was kept fixed at $D_c/W = 0.833\ T_u/\mu$. On the top right we show the results for $T_e/T_u = 0.7$. Rupture started in this case near the left end of the asperity and stopped growing very rapidly. At time $t = 5\ W/\beta$ rupture has already stopped and what we see on the right-hand panel is the static shear stress field left over by the rupture. The second row shows the results obtained for

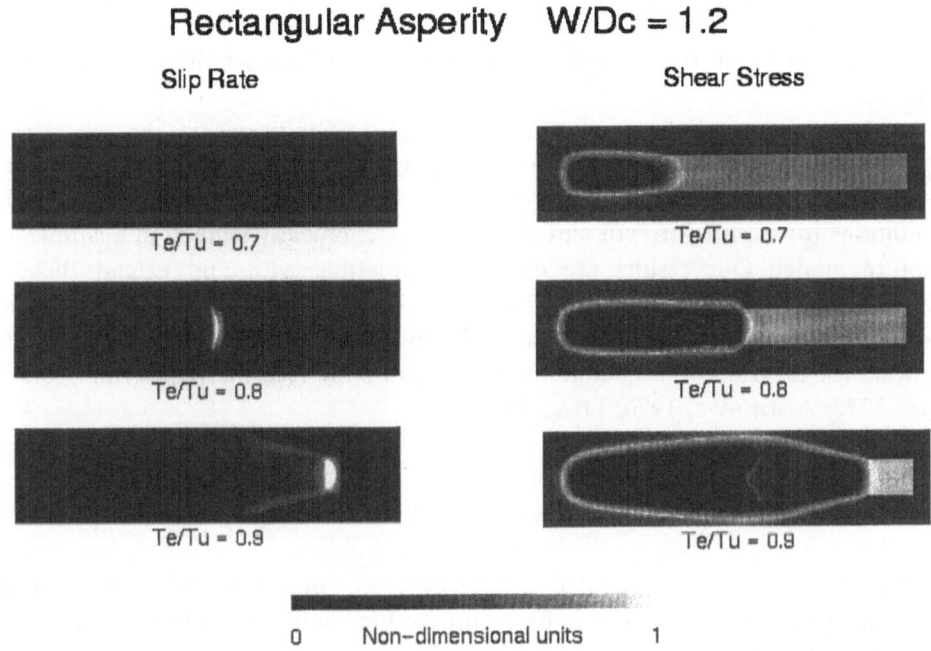

Rectangular Asperity W/Dc = 1.2

Slip Rate   Shear Stress

Te/Tu = 0.7   Te/Tu = 0.7

Te/Tu = 0.8   Te/Tu = 0.8

Te/Tu = 0.9   Te/Tu = 0.9

0   Non–dimensional units   1

Figure 6

Snapshots of the shear stress and slip rate fields on a fault containing a very long and narrow asperity (clear in the stress snapshots). The top panels show ruptures at subcritical ($\kappa = 0.588$), slightly super critical ($\kappa = 0.768$) and very super critical ($\kappa = 0.972$) values of the parameter κ.

$T_e/T_u = 0.8$, where rupture propagates along the asperity at sub-shear speeds. We see that the rupture zone extends beyond the asperity, leaving an elongated final fault shape. Finally, at the bottom of Figure 6 we show snapshots of slip rate and stress for a super-critically loaded asperity with $T_e/T_u = 0.9$. In this simulation the rupture front is running faster than the shear wave, producing a wake that spreads deeply into the lower prestress zone. Thus, as for the rectangular fault model, we are again in the presence of a bifurcation: when κ is smaller than critical as in the case in the top panels of Figure 6, rupture stops. When rupture is lightly super-critical as in the central panel, rupture grows slowly at speeds close to the shear-wave velocity. Finally, for very super-critical conditions, as in the bottom panels, rupture becomes faster than the shear wave speed, producing strong super-shear head waves in the low stress areas that surround the asperity.

For the rectangular asperity model we omitted the systematic exploration of the parameter space that we carried out for the rectangular fault (Fig. 5). Such parameter study is computationally expensive and we expect the effects to be similar. The sole difference between the asperity is lower for the former than for the latter. We find $\kappa_c = 0.7$ for the asperity model which may be compared to 0.8 for the barrier model. This is expected since an asperity compared to a barrier of the same dimensions is easier to break because rupture will extend into the low stress areas surrounding the asperity. This is of course impossible in the barrier model. Finally, as shown in Figure 6, we find again that for large enough κ ruptures grow initially at very high speeds and then jump to a speed higher than the shear-wave velocity.

Although many authors have studied rupture conditions in the presence of asperities and barriers, most notably DAS (1981), DAS and KOSTROV (1983), KOSTROV and DAS (1989) and DAY (1982), we believe that this is the first time that conditions for rupture growth and arrest are systematically studied in a simple 3-D rupture model. Our results are entirely compatible with and extend those of HUSSEINI et al. (1975) who studied conditions for rupture of 2-D faults. As we will show the nondimensional number κ is closely related to the minimum patch that appears for all friction laws containing a length scale (see the pioneering work by IDA, 1972; ANDREWS, 1976; DAY, 1982).

Spontaneous Rupture on a Realistic Fault: The Landers 1992 Earthquake

OLSEN et al. (1997) generated a fully dynamic rupture model of the Landers earthquake of July 1992 that was based on the kinematic model inverted by WALD and HEATON (1994) using a combination of seismic and geodetic data with minimum wavelengths of 3–4 km. From the kinematic slip model they computed the stress drop on the fault and found that it was very heterogeneous, containing both large positive stress drops where the fault slipped and very negative values where it did not slip. They generated from this stress drop field an initial stress field that is shown in the top panel of Figure 7, superimposed with the patch in the upper right part of the fault where rupture was initiated in the simulation. The range of variation of initial stress in this and the following panels of Figure 7 is roughly from -12 MPa to 12 MPa. In addition to the initial stress field, the only other assumption in the dynamic simulation is that the friction law (1) was the same everywhere on the fault plane. After trial and error modeling, OLSEN et al. (1997) fixed T_u at 12 MPa, which approaches the maximum value of the initial stress field shown in the first panel of Figure 7. The only free parameter that was left to be determined was the slip-weakening distance D_c. OLSEN et al. found that rupture would be similar to the kinematic one only if D_c was close to 0.8 m. For smaller values the rupture would break the fault at super-shear speed, leaving a uniform final stress field, and a slip distribution that was completely different from that of WALD and HEATON (1994). For larger D_c, on the other hand, rupture would simply not propagate along the fault. Thus again, we are dealing with a critical phenomenon, as in the two previous examples discussed above.

We can make a quick calculation of the value of κ for the OLSEN et al. model. As we mentioned above, $T_u = 12$ MPa and $D_c = 0.8$ m so that the energy release rate $G = 4.8$ MJm^{-2}. This is a large value but on the order of magnitude of those proposed by AKI (1979) and OHNAKA and SHEN (1999). In order to compute κ we need an average value for T_e which is estimated from the initial stress field to be 4 MPa (see MADARIAGA, 1979 for a discussion of different ways to compute the average stress drop). Then, using the average value $\mu = 3.45 \times 10^{10}$ Pa and a fault half width of 15 km, we find $\kappa = 0.72$ for the OLSEN et al. (1997) Landers earthquake model. We note that in order to estimate κ for Landers we used a fault half-width of 15 km instead of 7.5 km in order to take into account approximately the effect of the free surface. κ for the Landers earthquake is well below the critical values we obtained above for the rectangular fault models ($\kappa = 0.8$). This is not surprising since the rupture in Landers propagated mainly on those parts of the fault where stresses were close to the maximum of $T_e = 10$ MPa. All this seems to indicate that rupture in the Landers earthquake reproduces the general behavior of the kinematic model of WALD and HEATON only if κ is very close to critical. There are many corrections applicable to this calculation, however we believe that the main result will stand further scrutiny: rupture in the Landers earthquake was very close to critical, enough to make the rupture grow but not enough for the rupture to become super-shear.

The result that the Landers earthquake rupture was close to critical may be obtained in another way. Careful study of rupture propagation in the initial stress field shown at the top of Figure 7 reveals that rupture only propagates in regions of high stress and sufficient width. It takes a very minor modification of the initial stress field to either stop it or guide it into a different area of the fault.

Stress change

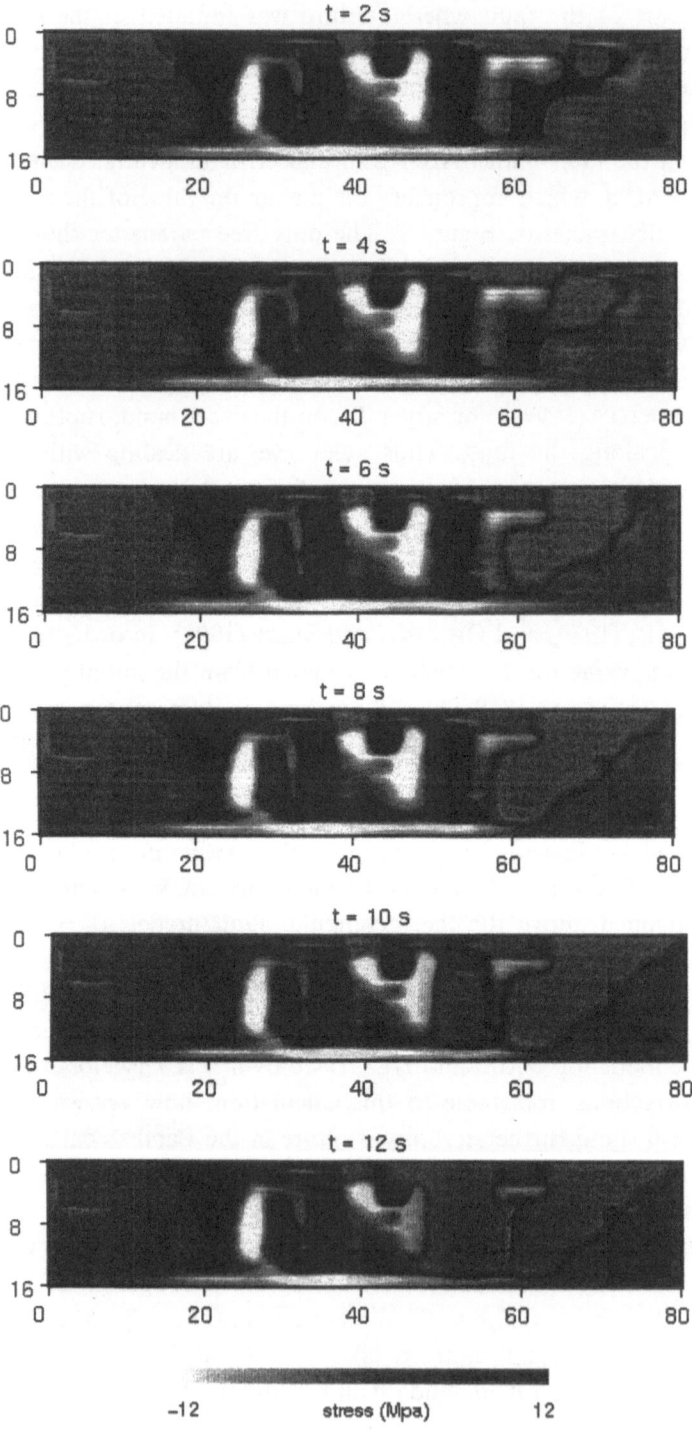

stress (Mpa)

In Figures 7 and 8 we show the longitudinal shear stress and slip rate fields at several successive instants of time at 2 s intervals. Only the initial instants of rupture simulation are shown here. Note that slip rate is faint and weak. As proposed by HEATON (1990) the instantaneous rupture width seems not to be controlled by the overall width of the fault but by the size of the local asperities or concentrations of the initial stress distribution on the fault. This is in agreement with the observations of BEROZA and MIKUMO (1996) and with theoretical results of COCHARD and MADARIAGA (1994). Rupture in Figure 8 is never super-shear, it barely makes it through the fault just as one would expect from a critical phenomenon.

Next we analyzed the details of rupture. The stress field is not very favorable to rupture initiation because the stresses in the vicinity of the initial patch are relatively weak. Therefore as shown in Figures 7 and 8, the initial rupture takes a long time before reaching the vertical high stress zone to the northwest (left) of the initial patch. Upon entering this zone, rupture propagates downwards at high speeds because the loaded zone is relatively wide. Upon hitting the bottom of the fault it stops because there is no lateral communication in this area.

As a final remark concerning criticality, we observe that rupture can initiate from many points on the fault for the initial stress field shown in Figure 7. We have generated models of rupture that start from any high stress spot on the fault, all indicating that the final stress and slip distributions were almost insensitive to the actual kinematics of rupture. As long as we stay close to critical conditions rupture seems to be controlled mainly by the initial stress field. We have yet to complete the study of rupture in Landers although we expect that we can control the way rupture grows by simple manipulation of the initial stress field (the kinematics).

In conclusion, the initial stress field controls rupture propagation very closely. As long as rupture conditions are very close to critical, rupture extends following a relatively clear pattern of infiltration. It penetrates places where stress is high over large patches and completely avoids the zones where stresses are low. The overall kinematics are then similar to those determined by WALD and HEATON (1994), COHEE and BEROZA (1994) or COTTON and CAMPILLO (1995).

Discussion: Where Does the Nondimensional Number Come from?

Determining the origin of the nondimensional number that controls the bifurcation is a complex problem that will require the study of many models with other

---

Figure 7

Snapshots of the shear stress field on the fault of the Landers 1992 earthquake. Snapshots were calculated at regular intervals of 2 s. The initial shear stress is shown in the upper snapshot, modified by the stress drop shortly after rupture initiation.

Slip rate

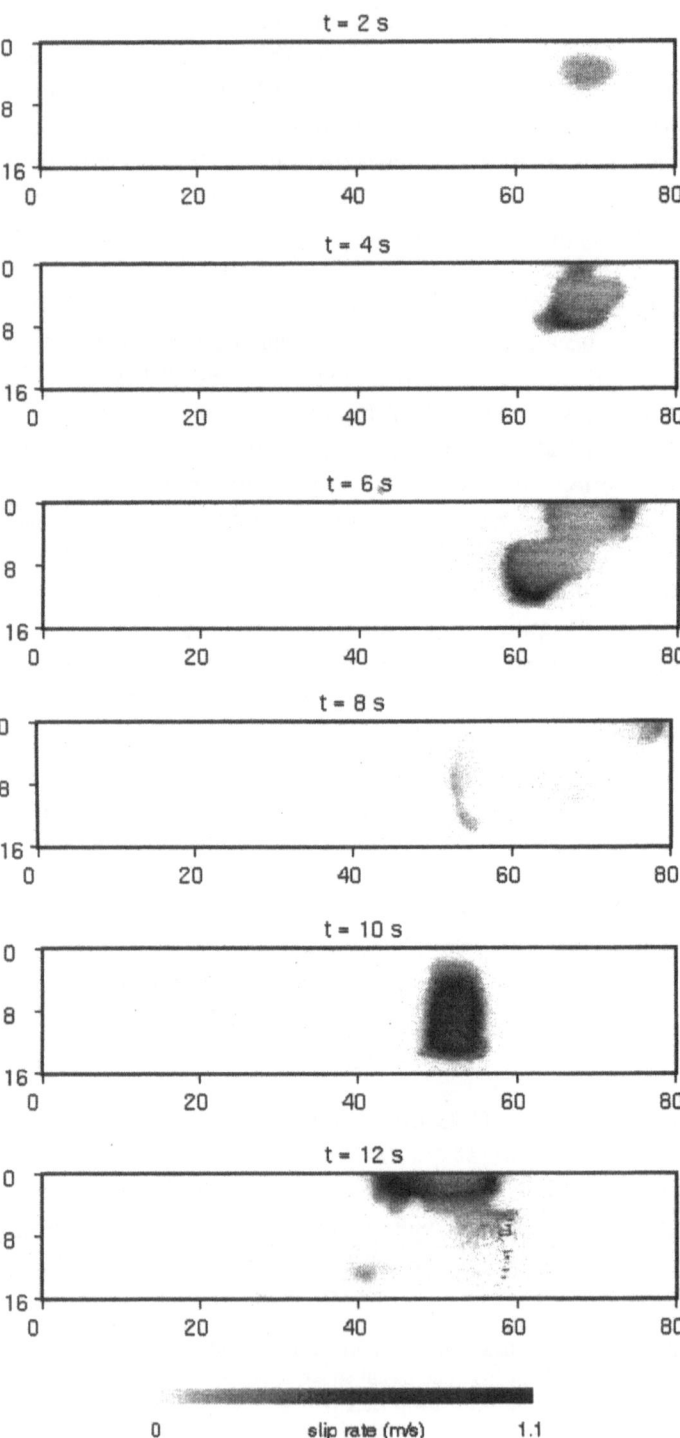

geometries and more complex stress distributions. The fact that we have a bifurcation in the behavior of the system means that we must have a single nondimensional control parameter that determines whether a rupture will grow or stop. The determination of such a number can be very complex as shown by SCHMITTBUHL *et al.* (1996) who carried out an exhaustive study of the parameters' space of the BURRIDGE and KNOPOFF (1967) model. In the rectangular barrier and asperity models the stress field is uniform such that we expect no fractional powers; for this reason we successfully applied the simple dimensional analysis approach to find nondimensional numbers (see, e.g., LANDAU and LIFSHITZ, 1959, page 63).

Following IDA (1972) and ANDREWS (1976) we can easily show that the nondimensional number κ derives from the competition between a measure of the strain energy released during rupture and the energy dissipated in fracture. Rupture resistance, or energy release rate for our slip-weakening friction model (1) is

$$G = \frac{1}{2} T_u D_c. \tag{6}$$

Strain energy change ΔU per unit fault surface in a rupture zone of characteristic length W is on the order of:

$$\Delta U = \frac{1}{2} \langle D \rangle T_e \simeq A T_e^2 / \mu R \tag{7}$$

where $\langle D \rangle \simeq T_e / \mu W$ is the average slip on a fault of length W, T_e is the effective stress drop and A a numerical coefficient on the order of 1. Thus, our definition (5) implies that

$$\kappa \simeq \Delta U / G. \tag{8}$$

For simple uniformly-stressed models, ANDREWS' (1976) critical rupture patch implies that $\kappa_c \simeq A$. The critical value is simply the numerical coefficient A that relates these two measures of energy. As we mentioned earlier, it is more difficult to estimate the critical value for faults in 3-D, because the shape of the fault and the rupture mode will produce variations in A that are not easy to calculate. For the long rectangular fault we prefer the determination of Figure 5 which yields $\kappa = 0.8$. For the asperity we found $\kappa = 0.7$.

In a more realistic situation, as for the Landers earthquake, stress is complex and we expect the numerical value of κ_c to be different because somehow the measure of energy must integrate spatial variations of stress. We believe, however, that this number should be of order 1 as we obtained in our analysis. For more general models with fractal distributions of stress and fracture resistance G, we expect that the simple expression for κ may include fractional powers.

Figure 8

Snapshots of the slip rate field on the fault of the Landers 1992 earthquake computed at regular intervals of 2 s, starting with the initial stress field shown in the upper snapshot of Figure 7.

We can now speculate on the behavior of active faults in the earth. We have seen from these simple examples that the essential requirement for rupture growth is that $\kappa > \kappa_c$ where the critical value of κ depends on the geometry of the problem but is of order 1. Once κ becomes substantially larger than critical, ruptures tend to become super-shear very rapidly although this is rarely observed in nature. Why is that? We believe that ruptures are in general sub-shear because faults always stay close to the critical condition. The reason may be very simple: as soon as the conditions for rupture in a fault area are fulfilled, a rupture will occur unloading the fault. When the system is reloaded to criticality another earthquake will occur and so on. The size of these earthquakes is not determined by the load but by the connectivity of the stress or rupture resistance field. In other words, the earthquake will propagate to the extent the stress field permits! Under special circumstances stress can be high and the fault remains in supercritical condition. Only in those exceptional circumstances will super-shear speeds be observed (e.g., ARCHULETA, 1984). The normal behavior for earthquakes is to break under critical conditions at speeds below the shear wave speed. This eliminates the longstanding dilemma of why earthquakes stop, and why there are earthquakes of vastly different sizes on the same fault segment.

Conclusions

We have carefully studied rupture growth on a simple flat fault embedded in a homogeneous elastic medium of rigidity μ. It emerges from our studies that rupture is controlled by a nondimensional number $\kappa = T_e^2/\mu T_u \times L/D_c$ where L is a characteristic size of the stress field, for instance the patch (asperity) radius or the width of the fault or asperity, T_e is a characteristic stress load on this patch, and $T_u \times D_c$ is a measure of energy release rate on the fault.

We then estimated κ for the dynamic simulation of the 1992 Landers earthquake by OLSEN et al. (1997) that was based on the WALD and HEATON (1994) kinematic model. From a rough estimate we found that κ was slightly less than 1, implying that the Landers earthquake rupture occurred under conditions that were almost critical. Considerable additional evidence in favor of criticality was examined, leading to the hypothetical suggestion that earthquake ruptures occur as soon as the stress distribution on the fault becomes critical and that faults rarely stay in a meta-stable state that would allow large super-critical states on the fault.

In our analysis to date we have assumed uniform rupture resistance (constants T_u and D_c). All the complexity in our models arises from the heterogeneity of initial stress. As important as stress heterogeneity is probably the small-scale geometry of faulting. Its integration in fault models is difficult because most of the observations of fault rupture are still limited to the range of frequencies less than 1 Hz or about 3 km wavelength. This is too coarse a resolution to observe effects of complex geometry other than major fault segmentation as in Landers or Kobe. High-frequency seismic radiation is probably the only source of information regarding small-scale geometry.

The implications of this simple number are wide ranging and require extensive tests and analysis of modeling. We are currently carrying out such tests for the case of the Landers earthquake.

Acknowledgements

We thank J. R. Rice for very useful discussions concerning the origin of the nondimensional number κ. The computations in this study were carried out in the parallel computer of Département de Modélisation Physique et Mathématique of Institut de Physique du Globe de Paris. R. Madariaga's work was supported by the Environment Program of the European Community under project SGME and by CNRS (Centre National de la Recherche Scientifique) under contract 99PNRN13AS of the PNRN program. K. Olsen's work was supported by NSF Grant EAR 9628682 and by the Southern California Earthquake Center (SCEC). SCEC is funded by NSF Cooperative Agreement EAR-8920136 and USGS Cooperative Agreements 14-08-0001-A0899 and 1434-HQ-97AG01718. This is ICS contribution 0345-93EQ and SCEC contribution number 529.

References

AKI, K. (1979), *Characterization of Barriers of an Earthquake Fault*, J. Geophys. Res. *84*, 6140–6148.
ANDREWS, J. (1976), *Rupture Velocity of Plane Strain Shear Cracks*, J. Geophys. Res. *81*, 5679–5687.
ARCHULETA, R. (1984), *A Faulting Model for the 1979 Imperial Valley Earthquake*, J. Geophys. Res. *89*, 4559–4585.
BEROZA, G., and MIKUMO, T. (1996), *Short Slip Duration in Dynamic Rupture in the Presence of Heterogeneous Fault Properties*, J. Geophys. Res. *101*, 22,449–22,460.
BOUCHON, M. (1997), *The State of Stress on Some Faults of the San Andreas System as Inferred from Near-field Strong Motion Data*, J. Geophys. Res. *102*, 11,731–11,744.
BURRIDGE, R., and KNOPOFF, L. (1967), *Model and Theoretical Seismicity*, Bull. Seismol. Soc. Am. *67*, 341–371.
CARLSON, J., and LANGER, J. (1989), *Mechanical Model of an Earthquake Fault*, Phys. Rev. *A40*, 6470–6484.
COCHARD, A., and MADARIAGA, R. (1994), *Dynamic Faulting under Rate-dependent Friction*, Pure appl. geophys. *142*, 419–445.
COCHARD, A., and MADARIAGA, R. (1996), *Complexity of Seismicity due to Highly Rate-dependent Friction*, J. Geophys. Res. *101*, 17,581–17,596.
COHEE, B., and BEROZA, G. (1994), *Slip Distribution of the 1992 Landers Earthquake and its Implications for Earthquake Source Mechanics*, Bull. Seismol. Soc. Am. *84*, 692–712.
COTTON, F., and CAMPILLO, M. (1995), *Frequency Domain Inversion of Strong Motions: Application to the 1992 Landers Earthquake*, J. Geophys. Res. *100*, 3961–3975.
DAS, S. (1981), *Three-dimensional Spontaneous Rupture Propagation and Implications for the Earthquake Source Mechanism*, Geophys. J. Roy. astr. Soc. *67*, 375–393.

DAS, S., and AKI, K. (1977a), *A Numerical Study of Two-dimensional Spontaneous Rupture Propagation*, Geophys. J. Roy. astr. Soc. *50*, 643–668.

DAS, S., and AKI, K. (1977b), *Fault Plane with Barriers: A Versatile Earthquake Model*, J. Geophys. Res. *82*, 5658–5670.

DAS, S., and KOSTROV, D. (1983), *Breaking of a Single Asperity: Rupture Process and Seismic Radiation*, J. Geophys. Res. *88*, 4277–4288.

DAY, S. (1982), *Three-dimensional Simulation of Spontaneous Rupture: The Effect of Non-uniform Prestress*, Bull. Seismol. Soc. Am. *72*, 1881–1902.

DIETERICH, J. (1978), *Time-dependent Friction and the Mechanics of Stick-slip*, Pure appl. geophys. *116*, 792–806.

FUKUYAMA, E., and MADARIAGA, R. (1998), *Rupture Dynamics of a Planar Fault in a 3-D Elastic Medium: Rate- and Slip-weakening Friction*, Bull. Seismol. Soc. Am. *88*, 1–17.

GRIFFITH, A. A. (1929), *The Phenomenon of Rupture and Flow in Solids*, Phil. Trans. Roy. Soc. London *A221*, 163–198.

HARRIS, R., and DAY, S. (1993), *Dynamics of Fault Interaction: Parallel Strike-slip Faults*, J. Geophys. Res. *98*, 4461–4472.

HEATON, T. (1990), *Evidence for and Implications of Self-healing Pulses of Slip in Earthquake Rupture*, Phys. Earth. Planet. Int. *64*, 1–20.

HUSSEINI, M. I., JOVANOVICH, D. B., RANDALL, M. J., and FREUND, L. B. (1975), *The Fracture Energy of Earthquakes*, Geophys. J. Roy. astr. Soc. *43*, 367–385.

IDA, Y. (1972), *Cohesive Force across the Tip of a Longitudinal-shear Crack and Griffith's Specific Surface Energy*, J. Geophys. Res. *77*, 3796–3805.

IDE, S., and TAKEO, M. (1997), *Determination of the Constitutive Relation of Fault Slip Based on Wave Analysis*, J. Geophys. Res. *102*, 27,379–27,391.

KANAMORI, H., and STEWART, G. S (1976), *Seismological Aspects of the Guatemala Earthquake of February 4, 1976*, J. Geophys. Res. *83*, 3427–3434.

KOSTROV, B. (1964), *Self-similar Problems of Propagation of Shear Cracks*, J. Appl. Math. Mech. *28*, 1077–1087.

KOSTROV, B., and DAS, S., *Principles of Earthquake Source Mechanics* (Cambridge University Press, Cambridge, UK 1989).

LANDAU, L., and LIFSHITZ, E., *Fluid Mechanics* (Pergamon Press, London 1959).

MADARIAGA, R. (1976), *Dynamics of an Expanding Circular Fault*, Bull. Seismol. Soc. Am. *66*, 639–667.

MADARIAGA, R. (1979), *On the Relation between Seismic Moment and Stress Drop in the Presence of Stress and Strength Heterogeneity*, J. Geophys. Res. *84*, 2243–2250.

MADARIAGA, R., OLSEN, K. B., and ARCHULETA, R. J. (1998), *Modeling Dynamic Rupture in a 3-D Earthquake Fault Model*, Bull. Seismol. Soc. Am. *88*, 1182–1197.

MADARIAGA, R., OLSEN, K. B., and PEYRAT, S. (2000), *Rupture Dynamics in 3-D: A Review*, Annali di Geofisica, accepted for publication.

MIKUMO, T., and MIYATAKE, T. (1995), *Heterogeneous Distribution of Dynamic Stress Drop and Relative Fault Strength Recovered from the Results of Waveform Inversion*, Bull. Seismol. Soc. Am. *85*, 178–193.

OKUBO, P. (1989), *Dynamic Rupture Modeling with Laboratory-derived Constitutive Relations*, J. Geophys. Res. *94*, 12,321–12,335.

OHNAKA, M., and SHEN, L.-F. (1999), *Scaling of Rupture Process from Nucleation to Dynamic Propagation: Implications of Geometric Irregularity of the Rupturing Surfaces*, J. Geophys. Res. *104*, 817–844.

OLSEN, K., MADARIAGA, R., and ARCHULETA, R. (1997), *Three-dimensional Dynamic Simulation of the 1992 Landers Earthquake*, Science *278*, 834–838.

RICE, J. R. (1993), *Spatio-temporal Complexity of Slip on a Fault*, J. Geophys. Res. *88*, 9885–9907.

RICE, J., and BEN-ZION, Y. (1996), *Slip Complexity in Earthquake Fault Models*, Proc. Natl. Acad. Sci. USA *93*, 3811–3818.

SHAW, B., and RICE, J. R. (2000), *Existence of Continuum Complexity in the Elastodynamics of Repeated Fault Ruptures*, J. Geophys. Res., in press.

SCHMITTBUHL, J., VILOTTE, J. P., and ROUX, S. (1996), *A Dissipation-based Analysis of an Earthquake Fault Model*, J. Geophys. Res. *101*, 27,741–27,764.

WALD, D., and HEATON, T. (1994), *Spatial and Temporal Distribution of Slip for the 1992 Landers, California Earthquake*, Bull. Seismol. Soc. Am. *84*, 668–691.

(Received August 4, 1999, revised April 4, 2000, accepted April 5, 2000)

 To access this journal online:
http://www.birkhauser.ch

Pure appl. geophys. 157 (2000) 2003–2027
0033–4553/00/122003–25 $ 1.50 + 0.20/0

❙Pure and Applied Geophysics

Spontaneous Rupture Propagation on a Non-planar Fault in 3-D Elastic Medium

HIDEO AOCHI,[1,*] EIICHI FUKUYAMA[2] and MITSUHIRO MATSU'URA[1]

Abstract—We constructed a new calculation scheme of spontaneous rupture propagation on non-planar faults in a 3-D elastic medium using a boundary integral equation method (BIEM) in time domain. We removed all singularities in boundary integral equations (BIEs) following the method proposed by FUKUYAMA and MADARIAGA (1995, 1998) for a planar fault in a 3-D elastic medium, and analytically evaluated all BIEs for a basic box-like discrete source. As an application of the new calculation scheme, we simulated rupture propagation on a bending fault subjected to uniform triaxial compression and examined the effect of fault bend upon the dynamic rupture propagation. From the numerical results, we found that rupture propagation is decelerated or arrested for some combination of inclined angle of the bending fault and absolute value of the fault strength. The most significant effect of bending is the nonuniform distribution of pre-loaded shear stress due to different orientation of the fault plane under a uniform tectonic stress regime. Our results also indicate that low absolute shear stress level is required to progress the rupture propagation ahead of the inclined fault.

Key words: Boundary integral equation, spontaneous rupture propagation, non-planar fault.

Introduction

Large earthquakes generally occur on a complex interacting fault system rather than a simple planar fault, as inferred from the observations of hypocenter distribution of aftershocks and surface fault traces. In order to explain observed strong ground motion and local crustal deformation near the source, one planar fault model cannot often be a good approximation. For example, the 1992 Landers earthquake, California, has been investigated very well from various points of view. AYDIN and DU (1995) observed the surface fault traces of this earthquake and examined the relation between static stress field and local crustal deformation caused by many fault bends. WALD and HEATON (1995) also supplied a fault model

---

[1] Department of Earth and Planetary Physics, University of Tokyo, 7-3-1 Hongo, Bunkyo-ku, Tokyo 113-0033, Japan. E-mail: matsuura@eps.s.u-tokyo.ac.jp

[2] National Research Institute for Earth Science and Disaster Prevention, 3-1 Tennodai, Tsukuba, Ibaraki 305-0811, Japan. E-mail: fuku@bosai.go.jp

* H. A. is now at Laboratoire de Géologie, École Normale Supérieure, 24 rue Lhomond 75231 Paris Cedex 05, France. E-mail: aochi@geologie.ens.fr

of three planar faults which rotated gradually and intersect with each other, and analyzed the dynamic process of rupture propagation from one fault to another by inverting seismic waveform data. BOUCHON *et al.* (1998a) estimated a dynamic stress change on the bending fault from the slip history determined by seismic waveform inversion. They reported that stress concentration due to dynamic rupture propagation reached the order of 20 to 30 MPa on the 30-km-long northernmost segment of the fault before rupture started there, and concluded that unfavorable orientation of the fault might have caused the high stress concentration and led to the arrest of rupture.

In the case of the 1995 Kobe earthquake, Japan, rupture seems to start at the discontinuous point of fault segments and propagates bilaterally. YOSHIDA *et al.* (1996) gave a source model consisting of 2 or several subfaults to explain geodetic and seismic data and SEKIGUCHI *et al.* (1996) and IRIKURA *et al.* (1996) also gave a source model of 3 to 5 subfaults for inversion of strong motion data. Based upon the result from seismic inversion analysis, BOUCHON *et al.* (1998b) have investigated dynamic stress change during the earthquake. Their results show that distribution of shear stress is highly heterogeneous and that its average value is about 3.3 MPa.

The importance of the complex geometrical structure of faults has been long recognized, and several seismologists have tried modeling spontaneous dynamic rupture propagations on non-planar faults. Using finite difference methods (FDM), for example, HARRIS and DAY (1993) and KASE and KUGE (1998) simulated parallel or perpendicular strike-slip faults in 2-D medium, and subsequently HARRIS and DAY (1999) and MAGISTRALE and DAY (1999) investigated them in 3-D medium. However these approaches are limited in fault geometry due to the spatial grid. On the other hand, the boundary integral equation method (BIEM) enables us to handle complicated fault geometries such as bending and branching. However, since this kind of computation requires complex and advanced techniques with high performances to estimate stress field accurately, as well as to treat numerous freedoms, modeling of dynamic rupture process on complex non-planar faults in 3-D medium is not well established. KOLLER *et al.* (1992), TADA and YAMASHITA (1996, 1997), KAME and YAMASHITA (1997), BOUCHON and STREIFF (1997) and SEELIG and GROSS (1997) have simulated spontaneous dynamic rupture propagation on preexisting non-planar faults with different boundary integral equations (BIEs) in a 2-D framework. KAME and YAMASHITA (1999a,b) and SEELIG and GROSS (1999a,b) have applied their methods to the problem that rupture proceeds along unknown paths in 2-D. Further, TADA *et al.* (2000) have formulated a BIE for arbitrary 3-D non-planar faults, removing any singularities in time domain.

One of our aims in this paper is to construct a numerical simulation scheme of spontaneous dynamic rupture propagation for non-planar faults in 3-D elastic medium. The mathematical method developed here is based on a stress BIEM in time domain which rises from COCHARD and MADARIAGA (1994) and FUKUYAMA and MADARIAGA (1995). Since they have formulated BIEs directly for stress components instead of displacement field removing any strong singularities, their methods enable one to estimate stress field accurately on the fault to simulate spontaneous rupture propagation successfully. First, we extend the on-plane BIEs for a 3-D planar fault, derived by FUKUYAMA and MADARIAGA (1995, 1998), to estimate stress field at an arbitrary point in 3-D medium. Next, we discretize the BIEs and analytically evaluate them for a box-like source element. Finally, using the numerical computation scheme constructed here, we simulate spontaneous rupture propagation on a bending fault in 3-D medium, and discuss the dynamic effects of the geometry such as a fault bend.

Mathematical Formulation Based on 3-D BIEM

Regularization of Boundary Integral Equation

In this section, we give a theoretical framework of numerical computation using BIEM. In order to describe a non-planar fault in 3-D medium, we may take two methodologies. The first one, which is shown in Figure 1(a), is to describe the fault shape exactly by introducing a curved local coordinate system (TADA et al., 2000). This, however, will lead to a very complicated formulation in discretized form, which looks very difficult to compute numerically. Another way that we use here is to approximate a curved fault surface with a combination of small planar patches as shown in Figure 1(b). In this case, the stress on a planar patch (thick line) can be attributed to the slip on the same patch (thick line) through the on-plane kernel, which is already derived by FUKUYAMA and MADARIAGA (1995, 1998), and to those on the other patches (thin lines) through off-plane kernels. TADA and YAMASHITA (1996) pointed out that in the case of a 2-D in-plane shear fault, a smoothly curved crack (Fig. 1(a)) and its approximate expression with a series of small line patches (Fig. 1(b)) produce different normal traction distributions on the

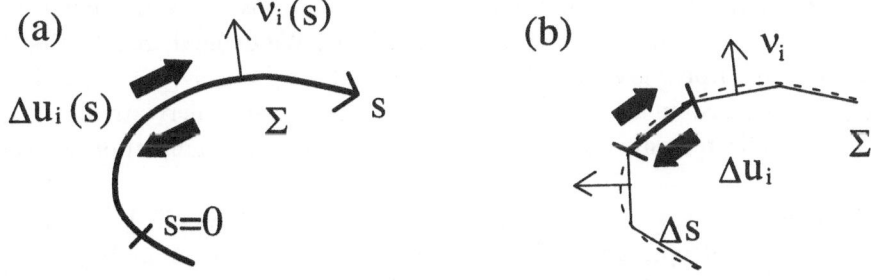

Figure 1

Two representations of a curved fault Σ. (a) Description of a fault by introducing curved local axis (s) along the fault. (b) Approximation of a curved fault with a series of small planar fault elements (Δs). In this case, off-plane stress expression for a flat source is required with the geometry in Figure 2.

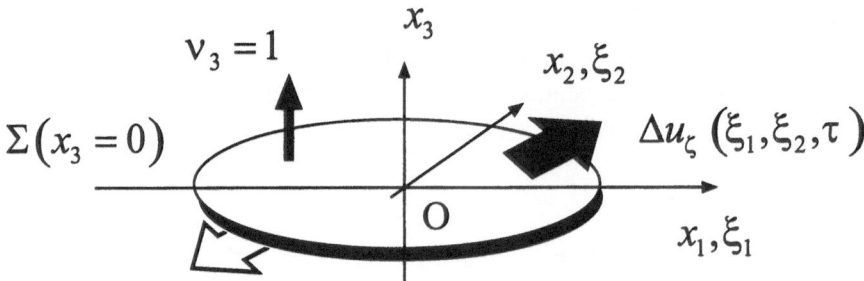

Figure 2
Geometry of the coordinate system with a fault plane $x_3 = 0$ and its slip discontinuity $\Delta u_\zeta(\vec{\xi}, \tau)$.

fault. Thus we should treat the paradox very carefully in 3-D BIEM, too. In this study, we use a simple slip-weakening law as a fracture criterion, which has no explicit dependence on normal traction.

Now, we derive the expressions of off-plane stress components for a planar fault and then obtain the analytical discrete off-plane kernel suitable for numerical simulation. It usually begins with a representation theorem such as equation (3.2) in AKI and RICHARDS (1980). The displacement field $u_i(\vec{x}, t)$ in a (x_1, x_2, x_3)-coordinate system is written by a spatio-temporal convolution of slip discontinuity $\Delta u_j(\vec{\xi}, \tau)$ over the fault Σ

$$u_i(\vec{x}, t) = \int_{-\infty}^{t} d\tau \int_{\Sigma} \Delta u_j(\vec{\xi}, \tau) c_{jkpq} v_k(\vec{\xi}) \frac{\partial G_{ip}(\vec{x} - \vec{\xi}, t - \tau)}{\partial \xi_q} d\Sigma \qquad (1)$$

where $\vec{x}$ and t denote position and time at a point in the medium, and $\vec{\xi}$ and τ are position and time at a point on the fault Σ. $G_{ip}(\vec{r}, t - \tau)$ is a Green's function, v_k is normal vector of Σ, and c_{ijkl} are elastic coefficients. We define hereafter that Latin and Greek subscripts vary from 1 to 3 and 1 to 2, respectively, $r = \|\vec{r}\| \equiv \|\vec{x} - \vec{\xi}\|$, $\gamma_i \equiv (x_i - \xi_i)/r$, a comma between subscripts denotes spatial derivatives $\partial/\partial \xi_i$, and an overdot indicates time derivative $\partial/\partial \tau$. We also use the summation convention rule.

We assume that a planar fault is embedded on the $x_3 = 0$ plane in a 3-D homogeneous, isotropic and infinite elastic medium as shown in Figure 2. Slip discontinuity occurs only on the fault without tensile element ($\Delta u_3 \equiv 0$) after $t = 0$. As long as we consider shear rupture on the fault-like earthquakes, the mathematical formulation under these assumptions is enough to model any fault with arbitrary structure and shape in 3-D. The Green's function in equation (1) is called

"Stokes tensor," the concrete expression of which is given by equation (A-1) in Appendix A. For the stress representation obtained by spatial derivatives of equation (1)

$$\tau_{pq}(\vec{x}, t) = - \int_0^t \int_\Sigma \Delta u_\zeta(\vec{\xi}, \tau) c_{\zeta 3 st} c_{pqin} \frac{\partial^2}{\partial x_n \, \partial x_t} G_{is}(\vec{r}, t - \tau) \, d\Sigma \, d\tau, \tag{2}$$

we use a regularization method similar to that used by COCHARD and MADARIAGA (1994) and FUKUYAMA and MADARIAGA (1995, 1998) for removing very strong singularities due to the second derivatives of Green's function in equation (2). Applying the techniques of Laplace transforms in time and the integration by parts, we obtain the regularized expression of stress in the form of convolutions of some derivatives of slip with weak singularities. In Appendix A, we briefly explain the derivation for the 3η element and list the final expressions for other components. All the expressions derived here are consistent with the special case of the general formulation for arbitrary shaped fault in 3-D medium derived by TADA et al. (2000), although they were independently formulated. Our equations are suitable for further numerical computations.

Discretization

For numerical simulations we discretize the slip velocity field on the fault with the box-like constant functions. We assume constant slip velocity within a spatio-temporal grid (spatial grid size $= \Delta s \times \Delta s$ and temporal interval $= \Delta t$)

$$\Delta \dot{u}_\zeta(\vec{\xi}, \tau) = \sum_{l,m,n} V_\zeta^{mn} d(\xi_1, \xi_2, \tau; \xi_1^l, \xi_2^m, \tau^n), \tag{3}$$

where function $d(\cdot)$ consists of 8 Heaviside functions $H(\cdot)$

$$d(\xi_1, \xi_2, \tau; \xi_1^l, \xi_2^m, \tau^n) = \begin{cases} 1 \text{ for } \xi_1^l - \Delta s/2 \leq \xi_1 < \xi_1^l + \Delta s/2 \\ \quad \text{and } \xi_2^m - \Delta s/2 \leq \xi_2 < \xi_2^m + \Delta s/2 \\ \quad \text{and } \tau^n - e_t \Delta t \leq \tau < \tau^n + (1 - e_t) \Delta t \\ 0 \text{ otherwise} \end{cases} \tag{4}$$

where e_t $(0 \leq e_t \leq 1)$ is the time collocation coefficient appeared in COCHARD and MADARIAGA (1994) and discussed in detail by TADA and MADARIAGA (2000). Therefore we have to obtain discrete kernels of regularized BIEs, equations (A-4), (A-5), (A-6) and (A-7), for a basic Heaviside function in slip velocity field;

$$\Delta \dot{u}_\zeta(\vec{\xi}, \tau) = V_\zeta H(\xi_1, \xi_2, \tau). \tag{5}$$

We analytically evaluate all the integrals, and the details of expressions are summarized in Appendix B.

Finally, we can write the discrete boundary integral equations in the following form

$$\tau_{pq}(\vec{x}, t) = \frac{\mu}{4\pi\beta} \sum_{\zeta=l}^{2} \sum_{l,m,n} P_{pq/\zeta}(\vec{r}, t - \tau^n) V_{\zeta}^{lmn} \tag{6}$$

where (l, m, n) represents the discrete position (l, m) in (ξ_1, ξ_2)-coordinate and time n of a source grid, and $P_{pq/\zeta}$ is a discrete kernel depending on the position vector $(\vec{r} = (x_1 - \xi_1^l, x_2 - \xi_2^m, x_3)^T)$, the time lag $(t - \tau^n)$, the slip velocity direction (ζ), and the stress component (pq). Stress components on the left-hand side of equation (6) are usually estimated at the center of the spatial grid on the fault, and can be freely estimated for time t as we will discuss below.

Discretization Parameters

As for this discretization method in equation (4), it is well known that there exist two parameters for discretization: the Courant-Friedrichs-Lewy (CFL) ratio w_c and the time collocation coefficient e_t. The CFL ratio w_c is defined by $w_c \equiv c\Delta t/\Delta x$, where c is the velocity of P-wave (α) or S-wave (β). It controls the length of time step Δt relative to a given spatial grid size Δs. The time collocation coefficient e_t defines the relative location of the time $(t = t^k)$ within a time step Δt. For example, the time to evaluate stress corresponds to the beginning of the current time step $(t^k, t^k + \Delta t)$ for $e_t = 0$, and is collocated at the end between $(t^k - \Delta t, t^k)$ for $e_t = 1$. COCHARD and MADARIAGA (1994) used $w_\beta = 0.5$ and $e_t = 1$ for 2-D anti-plane problems, and KAME and YAMASHITA (1997) used $w_\alpha = 0.5$ and $e_t = 1$ for 2-D in-plane problems. In those situations, the current slip velocity on a grid (V_ζ^{ijk}) effects the stress value on its own grid within its own time step (τ_{pq}^{ijk}), by expressing the point and time $(\vec{x}; t)$ as a discrete form $(x_1^i, x_2^j, x_3; t^k)$ in the similar expression for the discretized source $(\xi_1^l, \xi_2^m, 0; \tau^n)$. Thus we can extract the instantaneous term explicitly out of equation (6) as

$$\tau_{pq}^{ijk} = \frac{\mu}{4\pi\beta} \sum_{\zeta=1}^{2} \left[P_{pq/\zeta}^{000} V_\zeta^{ijk} + \sum_{l,m,n} P_{pq/\zeta}^{(i-l)(j-m)(k-n)} V_\zeta^{lmn} \right], \tag{7}$$

where the instantaneous kernel $P_{pq/\zeta}^{000}$ exists only in the 3ζ component for shear source V_ζ, and $P_{pq/\zeta}^{(i-l)(j-m)(k-n)}$ is the discrete kernel of $P_{pq/\zeta}(\vec{r}, t - \tau^n)$. Consequently we can solve the above equation on each spatial grid independently, i.e., using an explicit scheme. FUKUYAMA and MADARIAGA (1998) also examined the stability of numerical computations depending upon the CFL ratio, and reported that the value of w_α should be less than 0.5 for the numerical computations of 3-D planar faults using the explicit schemes. TADA and MADARIAGA (2000) investigated the stability of numerical computations which depend on both the CFL ratio w_α and the time collocation coefficient e_t for all modes in 2-D problems, and reported that the best combination of the two parameters is localized. With the explicit discrete BIE (7), we will evaluate the stress at the end of one time step, assuming $e_t = 1$ and $w_\alpha = 0.5$.

Fault Constitutive Relation and Normalization

Since the stress is a function of the current slip velocity in the explicit BIE (7), we can mathematically give various types of the fault constitutive relation, e.g., slip-dependent friction law or velocity-dependent friction law, as a fracture criterion on the fault. In this paper we introduce a slip-weakening law defined by

$$\sigma(w) = \tau_r + \Delta\tau_b(1 - w/D_c)H(D_c - w), \tag{8}$$

where σ is shear strength, w is net slip on the fault, τ_r is residual stress level, and $\Delta\tau_b$ and D_c are the constitutive parameters called the breakdown strength drop and the critical weakening displacement, respectively. When shear rupture occurs on the fault, some fraction of the total stress stored there is released with the progress of fault obeying equation (8). This type of the constitutive relation during shear fracture was originally proposed by IDA (1972) and PALMER and RICE (1973), observed in laboratory experiments by OKUBO and DIETERICH (1984) and OHNAKA *et al.* (1987), theoretically modeled by MATSU'URA *et al.* (1992) and also inferred in the field by IDE and TAKEO (1997) and Bouchon *et al.* (1998b) through seismological data analysis.

Based on the slip-weakening law, we normalize the equations by the critical slip displacement D_c, the breakdown strength drop $\Delta\tau_b$, and initial crack size L, and then introduce a normalization constant Υ

$$\Upsilon = \frac{\alpha}{4\pi\beta} \frac{\mu}{\Delta\tau_b} \frac{D_c}{L}. \tag{9}$$

Furthermore we normalize time with travel time of P wave over an initial crack, $T_\alpha = L/\alpha$, and then slip velocity V_ζ is normalized with $V_\alpha = D_c/T_\alpha$. For example, taking $D_c = 10$ cm, $\Delta\tau_b = 10$ MPa, $L = 400$ m and shear modulus $\mu = 20$ GPa, then $\Upsilon \approx 0.07$ which has almost the same value as that proposed by MATSU'URA *et al.* (1992), ELLSWORTH and BEROZA (1995), and OHNAKA (1996). When we approximate the initial grid size L with 5 grids, Δs becomes 80 m and Δt is 0.01 s ($T_\alpha = 0.1$ s) for an assumed P-wave velocity $\alpha = 4$ km/s. As a result, the normalized slip velocity $V_\alpha = 1$ m/s.

Explicit Time Stepping Scheme

We explain the method of numerical computation based on the BIE in equations (7) and (8). For simplicity, we restrict the direction of slip vector to one direction on the fault. If slip occurs in the ξ_1 direction, we have only to consider $\Delta u'_1$ and τ'_{31}. Here we use a prime (') to denote the quantities in the local coordinate (ξ_1, ξ_2, ξ_3) fixed on the non-planar fault $(\xi_3 \equiv 0)$.

Hereafter we use the normalized form of BIE and apply the same symbol V_ζ to denote the normalized slip velocity instead of equation (7). At a time step k, we have known all past slip velocities V'^{lmn}_1 $(n \leq k - 1)$. On some fault patch (i, j) in the local coordinate (ξ_1, ξ_2), the BIE and the general fracture criterion are written in the normalized form

$$T'^{ijk}_{31} = T^{0}_{31}{}' + \Upsilon \left[P^{000}_{31/1} V'^{ijk}_1 + \sum_{n=0}^{k-1} \sum_{l,m} P'^{ijk:lmn}_{31/1} V'^{lmn}_1 \right] \tag{10}$$

$$S^{ij} = T_r + (1 - W'^{ijk}_1)H(1 - W'^{ijk}_1). \tag{11}$$

Here we express all the normalized stresses, strength and slip as T, S and W instead of τ, σ and w, respectively, and T^{0}_{31} is a pre-loaded initial shear stress on the fault. Note that $P'^{ijk:lmn}_{31/1}$ is no longer the same as $P^{(i-l)(j-m)(k-n)}_{31/1}$ in equation (7) since we consider it on the non-planar faults. For the above equations, in general, there are two different states which depend on current slip velocity V'^{ijk}_1. When the fault is stuck $(V'^{ijk}_1 \equiv 0)$, shear strength is required to be larger than applied shear stress $(S_{ij} > T'^{ijk}_{31})$. On the other hand, during fracturing $(V'^{ijk}_1 \neq 0)$, shear strength and applied shear stress must be equal $(S^{ij} = T'^{ijk}_{31})$. Since the unknown parameter is only the current slip velocity V'^{ijk}_1 at the time k, we can solve the equation one by one for every fault patch (i, j). Then we continue this procedure at the next time step $(k + 1)$.

Later we provide 61×61 grids on the fault and continue calculating until the rupture front reaches the end of the computation region. We used Origin 2000 (R 10000, 250 MHz) at Earthquake Information Center of Earthquake Research Institute, University of Tokyo, Japan, and it takes more than 4 hours CPU time to calculate 120 time steps directly. It takes a much longer time than that for a 3-D planar fault, because of the anti-symmetry of the kernels. We have to calculate the kernels between each two points on the non-planar fault, so that more memory is required at the same time. We are able to shorten the computation time to about 30 minutes for the same problem shown in later sections by using FFT technique for convolution, regarding one spatial axis along which the fault structure is univariable (i.e., ξ_2-direction), though it requires about 2 GB memory which is generally twice as many as that without the FFT technique.

Model Setting

Fault Geometry and Initial Stress Field

We consider a simple physical situation in which a bending fault is embedded in the uniform triaxial compression field produced by principal deviatoric stress $\Delta\sigma_1$,

$\Delta\sigma_2$ and $\Delta\sigma_3$ with $\Delta\sigma_1 > \Delta\sigma_2 > \Delta\sigma_3$. Here we take compression to be positive. The fault consists of a primary planar part on the $x_1 x_2$ plane, where the applied shear stress has a maximum value, and a secondary bending part off the $x_1 x_2$ plane. Both the direction in which the fault bends and the maximum shear stress loads are taken to be parallel to the x_1 axis, and the fault shape is invariant toward the x_2-axis.

Now we introduce a local coordinate system (ξ_1, ξ_2, ξ_3) fixed on the bending fault, taking the ξ_1 axis along the direction of bending, the ξ_3 axis normal to the local fault, and the ξ_2 axis perpendicular to them, namely, parallel to the x_2 axis. For simplicity, the direction of fault slip is presumed to be parallel to the local ξ_1 axis, that is, the bending direction of the fault. Then the shear stress driving the fault slip is the τ'_{31} component defined in the local coordinate system. In the present situation (uniform triaxial compression), the applied $\tau^{0\prime}_{31}$ component on the fault can be written as

$$\tau^{0\prime}_{31} = \frac{1}{2}(\Delta\sigma_1 - \Delta\sigma_3)\cos 2\theta \tag{12}$$

where θ is the angle of inclined fault measured from the primary planar fault ($x_1 x_2$ plane) counter-clockwise. Equation(12) gives the initial stress condition for numerical simulation of dynamic rupture propagation.

We introduce a parameter S that relates the slip-weakening features and applied shear stress. S is defined by $(\tau_p - \tau^{0\prime}_{31})/(\tau^{0\prime}_{31} - \tau_r)$ following DAS and AKI (1977) and DAY (1982), where the peak strength is defined by $\tau_p \equiv \tau_r + \Delta\tau_b$. This parameter can vary on the fault and controls the velocity of rupture propagation. In this paper we take $S = 1/3$ on the primary plane ($\theta = 0$). This value of S is slightly larger than the critical value for which rupture propagation is initiated under the circumstances we gave.

For each simulation, we assume that the values of τ_r and τ_p are constant over the fault except for the region of a given initial crack. Then only the applied shear stress $\tau^{0\prime}_{31}$ varies along the curved ξ_1 axis following equation (12). We indicate the relation between the absolute stress level τ_r and the applied shear stress $\tau^{0\prime}_{31}$ in Figure 3 for three different cases of residual stress, (a) $\tau_r = 5\Delta\tau_b$, (b) $2\Delta\tau_b$ and (c) 0. $\tau^{0\prime}_{31}$ decreases as the inclined angle θ through equation (12). In the case of $\tau_r = 2\Delta\tau_b$, the applied shear stress (thick curved line) given by equation (12) crosses the residual stress level (dotted straight line). On the primary plane ($\theta = 0$) including an initial crack, the initial stress is certainly loaded at a level between τ_r and τ_p, that is, $0 \leq S \leq 1$. However, as the angle of inclined fault θ increases, the applied shear stress decreases below the residual stress level. In such a region, the shear stress would not be released but be stored, if rupture occurred. In the case of $\tau_r = 0$, which corresponds to the case of tensile rupture or sliding without friction, the applied shear stress is

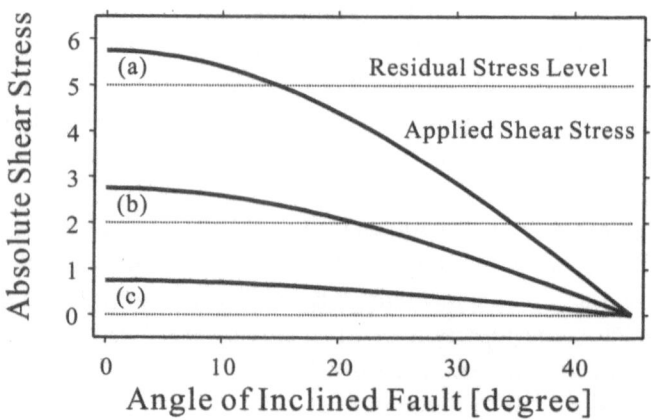

Figure 3

Relation between the absolute residual stress level τ_r and the applied shear stress $\tau_{31}^0{}'$ as a function of the angle of the inclined fault θ. The vertical axis is normalized by the breakdown strength drop $\Delta\tau_b$. Thick lines are applied shear stress τ_{31}^0 from equation (12) and dotted lines are assumed residual stresses (a) $\tau_r = 5\Delta\tau_b$, (b) $2\Delta\tau_b$ and (c) 0, respectively. In all cases, we take $S = 1/3$ on the plane of $\theta = 0°$.

always above the residual stress level unless $|\theta| \geq \pi/4$, and therefore a positive stress drop occurs to accelerate rupture propagation. When the residual stress level increases, the positive stress drop region becomes narrow, consequently it is expected to become more difficult for the rupture to propagate beyond the fault bend. This suggests that the absolute stress level would be a strong constraint for dynamic rupture propagation on a bending fault.

Furthermore, we should consider the effect of dynamic stress change due to the propagation of elastic waves, which is one of the most important purposes of this study. Perturbation of stress is carried by the P wave first and by the following S wave, and stress remains at static level after the dynamic disturbance finishes. This effect will be examined in detail through numerical simulations in the following section.

In Figure 4, we show an example of a bending fault model in which the left panel represents the applied shear stress field $\tau_{31}^0{}'$ and the right panel is the strength distribution τ_p at the beginning of the simulation ($t = 0$). Both are normalized by the breakdown strength drop $\Delta\tau_b$. Annotations on the spatial axes are normalized by grid size Δs, and their origin are located at the center of the initial crack. The initial crack diameter is taken to be 5 grids, that is, $L = 5\Delta s$. Inside the initial crack, the peak strength is lower than the applied initial stress and the breakdown strength drop is half of that outside. The fault bend starts at the 11th grid along the ξ_1 axis and is approximated with an arc whose radius is about $20\Delta s$, though we also approximate it with a series of small planar patches, that is, 10 grids for $\theta = \pi/6$, for example. We also consider a kinked fault which abruptly bends at 11th grid

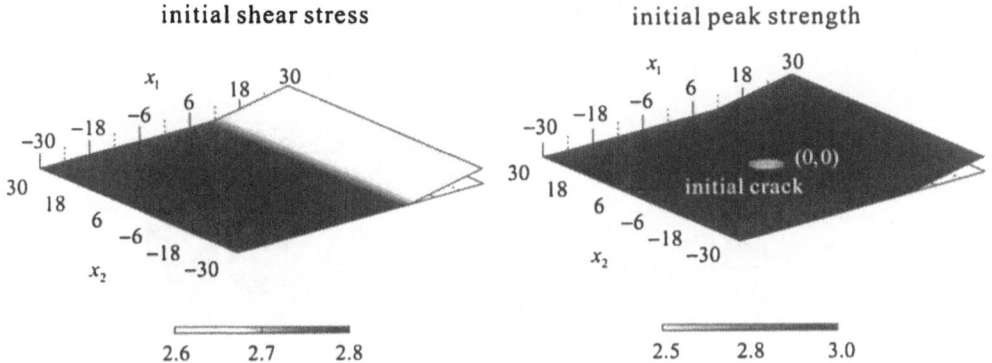

Figure 4

An example of a bending fault. The fault bends around $x_1 = 10.5$ and the bending direction finally changes ($\theta = \pi/18$) from the primary plane. Coordinate is normalized by grid size Δs. The left figure is applied shear stress $\tau_{31}^{0}{}'$ and the right one is assumed peak strength distribution σ for residual stress level $\tau_r = 2\Delta\tau_b$. We assume a circular crack, weak zone, at the center, to initiate rupture propagation. This is the initial condition of the simulation shown in Figure 5(a).

instead of the curved fault. Its final direction of the inclined fault (θ) is changed from $\pi/18$ to $\pi/2$.

Results of Numerical Simulation

Snapshots of Rupture Propagation

A series of snapshots in Figure 5 shows the evolution of slip velocity V_1' (the first row of each group) and shear stress τ_{31}' (the second row) for two bending fault models with different final angles of the inclined secondary fault (a) $\theta = \pi/18$, corresponding to the model as shown in Figures 4, and (b) $\theta = \pi/6$, respectively. The numbers along both axes represent the grid number on the $\xi_1\xi_2$ plane and fault starts bending at $\xi_1 = 10.5$. Ruptures propagate symmetrically on the primary planar fault until the rupture front reaches the bending portion. In the case of a small angle of bending ($\theta = \pi/18$), the rupture continues expanding beyond the bending portion, although expansion of the rupture area in the ξ_2 direction is decelerated by the bend, compared to that on the primary planar fault. When the angle of bending is large, on the other hand, the rupture is nearly arrested at the bending portion. At time step $t \sim 100$, the rupture front reaches the left edge of the computation region. However, its disturbance does not arrive at the opposite side of the crack tip during the calculation.

Figure 6 shows the slip velocities as a function of time step at 3 points, $(\xi_1, \xi_2) = (5, 5)$, $(15, 0)$ and $(12, 12)$, on the fault models of (a) $\theta = \pi/18$ and (b) $\pi/6$, corresponding to Figures 5(a) and (b), respectively. $(\xi_1, \xi_2) = (5, 5)$ is on the

Snapshots of rupture propagation

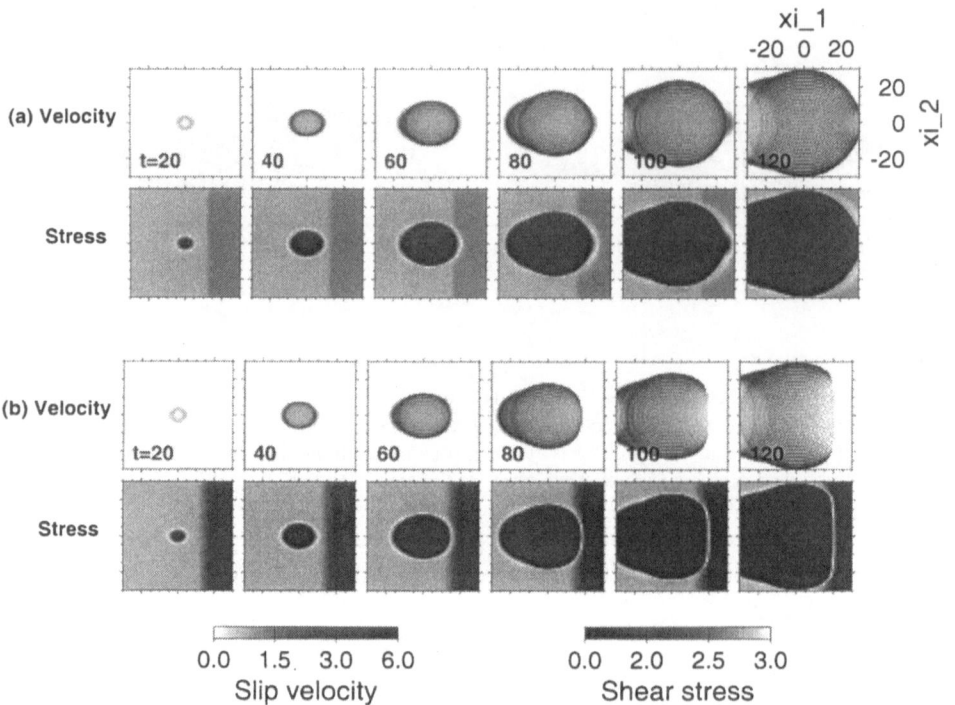

Figure 5
Two computations with different angles of the inclined fault (a) $\theta = \pi/18$ and (b) $\pi/6$. Upper and lower columns of each panel show slip velocity and shear stress, respectively.

primary fault plane near the initial crack, $(15, 0)$ and $(12, 12)$ are on the curved portion. We observe that the slip velocities are disturbed for each points. However, the magnitude of the disturbance is relatively small compared to the maximum slip velocity which appears just after the passages of the rupture front. That is why the disturbance may not affect the rupture propagation. When the disturbance becomes too large to further continue calculation with BIEM, we should re-scale the slip-weakening distance (D_c) in order to make D_c large enough comparing to spatial grid interval (FUKUYAMA and MADARIAGA, 2000; LAPUSTA *et al.*, 2000). Moreover a numerical experiment using a twice-higher resolution for both space and time leads to the same results as those presented here. This indicates that the numerical solution converges adequately and the numerical oscillations do no affect the ultimate results.

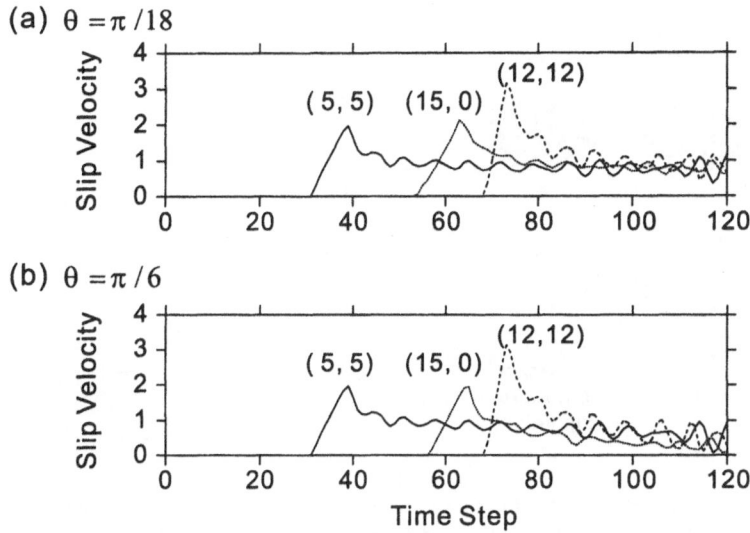

Figure 6
Slip velocities as a function of time step on various points on the curved fault. Cases (a) and (b) have the inclined angle $\theta = \pi/18$ and (b) $\pi/6$, which correspond to Figures 5(a) and (b), respectively.

Simulations of Kinked Fault

In order to examine the geometrical effect of the bend on dynamic rupture propagation, we show the difference of rupture propagation for various final angles of the secondary inclined fault θ, i.e., assuming, contrary to the previous case, that a uniform shear stress $\tau_{31}^{\prime 0}$ applies upon the whole fault. In other words, the value of S is constant everywhere on the fault ($S = 1/3$). In Figure 7, we display snapshots of slip velocity on the fault at time step $t = 80$ (left four panels) and the total amount of slip at that time (right panel). The fault bends abruptly at $\xi_1 = 10.5$ with the angle (a) $\theta = \pi/18$ (b) $\pi/6$ (c) $\pi/3$ and (d) $\pi/2$, respectively. The numerical results show that rupture continues to propagate beyond the kinked portion with the same rupture velocity even though the fault bends perpendicularly. The effect of bending is more apparent on the slip distribution as shown in Figure 7. As the angle of bend becomes steeper, a lower slip region appears around the bending portion. This indicates that the bending portion acts as a kind of barrier in the rupture process. However, the results depend on the position of fault bending. When the bending is located near the initial crack, for example $\xi_1 = 5.5$, rupture does not propagate across the bend for the case of $\theta = \pi/2$, because of insufficient stress accumulation on the secondary bending fault.

TADA and YAMASHITA (1996) have pointed out the paradox that the opposite sign of normal stress appears on the fault for the analytical expression of the smooth bending fault and its numerical approximation with many small planar patches because of artificial kinks. However, our result shows that an artificial

Snapshots at t=80

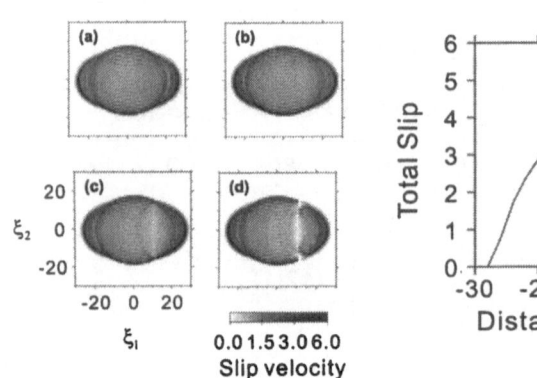

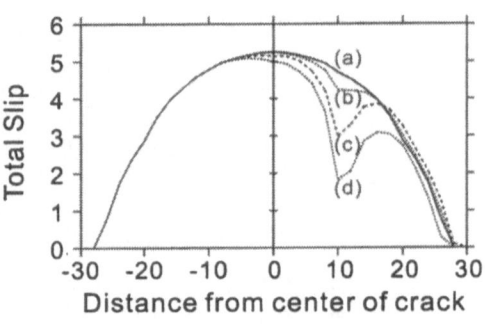

Slip velocity

Figure 7

Difference in rupture propagation due to various final angles of the inclined fault. The fault bends abruptly at $\xi_1 = 10.5$ with the angle (a) $\theta = \pi/18$, (b) $\pi/6$, (c) $\pi/3$ and (d) $\pi/2$, respectively. The left figures show snapshots of slip velocity on the fault at time step $t = 80$, and the right one is the distribution of total amount of slip at the same time. Each rupture velocity of the crack front is only slightly different, although slip velocity around the bend is decelerated by the steep angle of the bend.

structure in which two planar grids connect with a finite angle produces no opposite polarity of stress field ahead of it, since rupture continues propagating without decreasing its rupture velocity for any angle of bend within $0 \leq \theta < \pi/2$. Here we must notice that the rupture growth is perfectly determined by the shear stress field on the fault in our simulations. Thus a series of planar small patches would be a proper approximation of a curved fault in the shear rupture model as long as we consider the fracture criterion which depends not on the normal traction but on the shear stress. We shall compare the models with different way of bending later.

Effect of Fault Angle

Now we return to the case of rupture propagation under the triaxial compression field. Figure 8 shows rupture fronts propagating in the direction of both the positive and negative ξ_1 directions at the cross section $\xi_2 = 0$ on various bending faults at the same residual stress level $\tau_r = 0$. Each geometry is shown at the top of the figure and the relation between the initial shear stress and the angle of the inclined fault is also shown in the case (c) of Figure 3. In the hypothetical situation where $\tau_{31}^{0\prime}$ is assumed to be constant everywhere as shown in Figure 7, we see no difference of rupture velocities for the angle of the secondary inclined fault, since energy balance between stored shear stress and consumed fracture energy around the crack tip is invariant, regardless of the bending angle. In the uniform triaxial compression field, however, we can find significant differences in rupture velocity

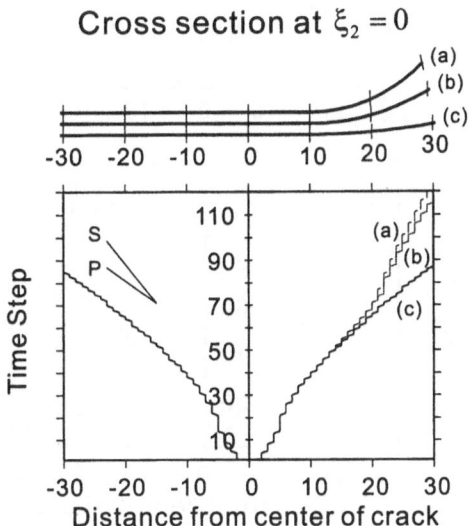

Figure 8

Rupture propagation for various final angles of the inclined fault in the case of $\tau_r = 0$. At the top we show each geometry of the faults with the inclined angle (a) $\theta = \pi/4$, (b) $\pi/6$ and (c) $\pi/18$, respectively. Rupture velocity decreases with the initial stress level on the inclined fault.

between the three fault models, because the value of S depends on the final angle of the inclined fault θ. Rupture propagates with super-shear velocity for the case of $\theta = \pi/18$ which leads to $S = 0.42$, while rupture velocity remains below shear wave speed for above $\theta = \pi/6$ which corresponds to $S = 1.67$. Our result is consistent with DAS and AKI (1977) who reported that sub-shear velocity of rupture propagation requires S larger than 1.6 for 2-D in-plane problems.

Absolute Level of Residual Stress

Next we examine the effects of absolute level of residual stress on rupture propagation. We fix the final angle of the inclined fault θ to $\pi/6$ and change the residual stress level (a) $\tau_r = 5\Delta\tau_b$, (b) $\tau_r = 2\Delta\tau_b$, and (c) $\tau_r = 0$ as shown in Figure 9. For case (b) we have presented snapshots of rupture propagation in Figure 5(b). As absolute stress level increases, we observe that the rupture propagation beyond the fault bend becomes more difficult. In the case of $\tau_r \neq 0$, the rupture almost stops around the bending portion, while opposite side of the rupture front propagates further at a super-shear wave velocity. In fact, Figure 3 indicates that the initial shear stress becomes less than the residual stress level for an angle larger than about 15° in case (a), and 22° in case (b), respectively. These angles correspond to distances of about 15 and 18 from the center of the crack. For both cases, rupture proceeds slightly in the region where negative stress drop is brought about and then is arrested.

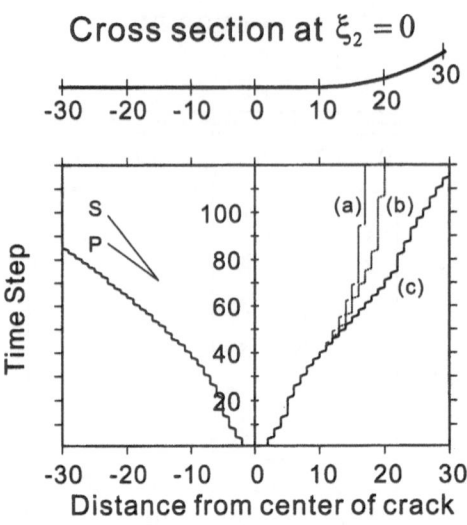

Figure 9

Rupture propagation for absolute stress level on the same bending fault, with final angle of the bending fault $\theta = \pi/6$ as shown at the top. Each case corresponds to (a) $\tau_r = 5\Delta\tau_b$, (b) $\tau_r = 2\Delta\tau_b$ and (c) $\tau_r = 0$, respectively. Rupture is nearly arrested around the bending for high residual stress level.

Kinked Fault and Smooth Curved Fault

Finally, we investigate the difference caused by the shape of bending. Figure 7 enables us to estimate that an artificial abrupt kink would produce no different result concerning rupture propagation, since the rupture velocity does not change significantly with the angle of the inclined fault. We consider a kinked fault which consists of two plane faults and a curved bending fault which we approximate with many small planar patches in order to understand its role. Figure 10 exhibits the difference in rupture propagation between two models for the case of $\tau_r = 5\Delta\tau_b$ (a, c) and $\tau_r = 0$ (b, d). Comparing (b) and (d), we can see that the terminal rupture velocities are almost the same in both cases, though the rupture propagates faster on the smooth curved fault (d), around the bending portion, than on the kinked fault (b). This phenomena can be interpreted in terms of the value of S. In the case of (d), the value of S increases gradually from 0.33 to 1.67. Conversely, it changes abruptly on the kinked fault (b). Similarly, the rupture propagates more smoothly for the curved fault (c) than for the kinked fault (a), which is considered to be caused by the pre-applied shear stress. For the inclined fault with angle $\pi/6$, the rupture necessarily produces a negative stress drop, as shown in Figure 3. Thus the rupture stops suddenly at the bending in case (a). We do not think that neither the effect of discretization nor approximation in modeling with many small planar patches produces any artificial result.

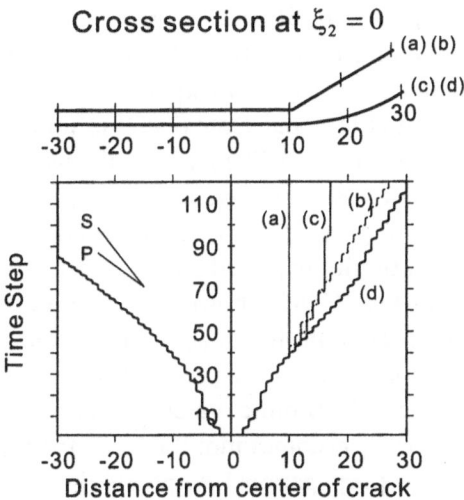

Figure 10

Difference of rupture propagation between (a, b) kinked fault and (c, d) curved fault. Their geometries are drawn at the top of the figure and they have the same final angle of the bending fault $\theta = \pi/6$. We put $\tau_r = 5\Delta\tau_b$ for cases (a) and (c), and $\tau_r = 0$ for (b) and (d). The pre-loaded shear stress on the inclined fault produces the difference of the rupture propagation between that on the kinked and that on the smooth curved faults.

Discussion and Conclusion

In this paper we have constructed a new simulation scheme of spontaneous dynamic rupture propagation on non-planar faults in 3-D elastic medium. We removed all hyper-singularities in BIEs following the regularization method proposed by FUKUYAMA and MADARIAGA (1995, 1998), and also analytically evaluated all the BIEs for a basic box-like discrete source. This enables one to evaluate stress fields accurately and to simulate dynamic rupture propagation stably. This method is also applicable to rupture propagation on unknown paths in intact material, as KAME and YAMASHITA (1999a,b) simulated growths of a crack bending spontaneously in 2-D medium. Our BIE agrees with special case of the general formulation derived by TADA et al. (2000), and our scheme of simulation becomes the same as those proposed by FUKUYAMA and MADARIAGA (1995, 1998) for the case of a planar fault in 3-D.

As an example of numerical simulations, we considered a simple physical situation in which a bending fault is embedded in a 3-D elastic medium subjected to a uniform triaxial compression field. The simulation results demonstrate that rupture propagation is decelerated or arrested for some inclined angle of the bending fault and absolute value of the fault strength. We also observed that the effect of dynamic stress change caused by rupture propagation itself was not strong enough to stop the rupture propagation. The decrease of rupture velocity due to bending also has been reported by BOUCHON and STREIFF (1997) in 2-D in-plane simulations. The velocity of rupture front depends mainly upon the pre-loaded initial stress on the fault with various directions of the inclined fault, that is, the parameter S. DAS and AKI (1977) and DAY (1982) examined the effect of S in 2-D anti- and in-plane cases and for a 3-D planar fault, respectively. Our results are consistent with theirs. For the special case of $S < 0$ in which negative stress drop occurs, we note that bending plays the role of a barrier strong enough to stop rupture propagation as naturally expected. In the field observation, the inversion result by WALD and HEATON (1995) shows that rupture velocity clearly changes with the strike direction of the fault, for example. From our results, we surmise that it is probably caused by the variation of pre-loaded shear stress on the faults.

We discussed the role of absolute value of shear stress level on rupture propagation. As it increases, it becomes more difficult for rupture to proceed on an inclined fault. The rupture propagation beyond the bend requires a low absolute level compared to breakdown strength drop $\Delta \tau_b$. From the inversion analysis by BOUCHON *et al.* (1998a) of the 1992 Landers earthquake, there exists a portion in which stress accumulation before rupture exceeds 20 MPa, although static stress does not change at all before and after earthquake on the fault with unfavorable strike. Assuming $\Delta \tau_b = 20$ MPa in the simulations we presented, for example, it would be difficult for the rupture to grow on a 30° inclined fault in the case of an absolute residual stress τ_r larger than 40 MPa (see Fig. 9). At the depth of 10 km, continuing pressure reaches 300 MPa, which leads to a frictional coefficient μ_f of 0.1–0.2, considerably smaller than that observed in the laboratory. This discussion, however, is constructed under the assumption that the fault system is embedded in such a homogeneous triaxial compression field as that in the simulations we demonstrated, and that the constitutive parameters are also homogeneous over the fault. The Landers earthquake might not be in such a simple state. AOCHI (1999) simulated the rupture propagation on the non-planar faults for that earthquake, considering a tectonic loading system, and discussed the physical background in detail.

In order to understand the physical process of earthquake generation, a forward modeling of dynamic rupture propagation on off-plane faults must be very efficient to reveal characteristics of the realistic field in which earthquakes occur. For this purpose we should model a more realistic condition using the scheme we have just constructed, and discussed in detail the effect of the complex geometrical structure of faults on earthquake generation.

Acknowledgements

We thank Dr. Taku Tada at *Laboratoire de Géologie, École Normale Supérieure*, Paris (now at Earthquake Research Institute, University of Tokyo, Japan). He provided us many valuable comments and suggestions for simplifying the formulations. We also express appreciation to Dr. Alain Cochard who reviewed and offered many valuable comments. For numerical study, we have used the computer system of Earthquake Information Center of Earthquake Research Institute, University of Tokyo, Japan.

Appendix A. Regularized Boundary Integral Equation

A Green's function in 3-D homogeneous, isotropic, infinite elastic medium (the Stokes tensor) is

$$
G_{ij}(\vec{r}, t) = \frac{\beta^2}{4\pi\mu} \frac{3\gamma_i\gamma_j - \delta_{ij}}{r^3} \int_{r/\alpha}^{r/\beta} t'\delta(t - t')\, dt'
$$

$$
+ \frac{p^2}{4\pi\mu} \frac{\gamma_i\gamma_j}{r} \delta\left(t - \frac{r}{\alpha}\right) - \frac{1}{4\pi\mu} \frac{\gamma_i\gamma_j - \delta_{ij}}{r} \delta\left(t - \frac{r}{\beta}\right), \tag{A-1}
$$

where λ and μ are Lamé's constants, α and β are P- and S-wave velocities, respectively, $p = \beta/\alpha$, and $\delta(\cdot)$ is the delta function. The Laplace form of the integration appearing in equation (A-1)

$$
\bar{I} = \int_0^\infty dt\, e^{-st} \frac{\beta^2}{r^2} \int_{r/\alpha}^{r/\beta} t'\delta(t - t')\, dt' = \frac{\beta^2}{r^2 s^2} \left\{ \left(1 + \frac{sr}{\alpha}\right) e^{-st/\alpha} - \left(1 + \frac{sr}{\beta}\right) e^{-sr/\beta} \right\} \tag{A-2}
$$

is useful for rewriting equation (2). For example, to obtain the off-plane expression ($x_3 \neq 0$) of the 3η component, we continue with equation (C8) in FUKUYAMA and MADARIAGA (1998), referred to as FM98 hereafter, without taking the limit $x_3 \to 0$. We rewrite equation (C9) and (C11) in FM98 for $x_3 \neq 0$, and then obtain the final expression instead of equation (C20) in FM98

$$
\bar{\tau}_{3\eta}(\vec{x}, s) = -\frac{\mu}{4\pi} \int \frac{\gamma_\eta}{r^2} \Delta\bar{u}_{\zeta,\zeta}(\vec{\xi}, s) \left[12\bar{I} + 4p^2 e^{-sr/\alpha} - 5 e^{-sr/\beta} - \frac{sr}{\beta} e^{-sr/\beta} \right] d\Sigma
$$

$$
- \frac{\mu}{4\pi\beta^2} \int \frac{1}{r} s^2 \Delta\bar{u}_\eta(\vec{\xi}, s) e^{-sr/\beta}\, d\Sigma - \frac{\mu}{4\pi\beta} \int \frac{\gamma_\zeta}{r} s\Delta\bar{u}_{\eta,\zeta}(\vec{\xi}, s) e^{-sr/\beta}\, d\Sigma
$$

$$
- \frac{\mu}{4\pi} \int \frac{\gamma_\zeta}{r^2} \Delta\bar{u}_{\eta,\zeta}(\vec{\xi}, s) e^{-sr/\beta}\, d\Sigma + x_3^2 \frac{\mu}{\pi} \int \frac{\gamma_\eta}{r^4} \Delta\bar{u}_{\zeta,\zeta}(\vec{\xi}, s)
$$

$$
\times \left[15\bar{I} + 6p^2 e^{-sr/\alpha} - 6 e^{-sr/\beta} + \frac{sr}{\alpha} p^2 e^{-sr/\alpha} - \frac{sr}{\beta} e^{-sr/\beta} \right] d\Sigma, \tag{A-3}
$$

where $\bar{I}$ we defined here is (C10) in FM98 multiplied by (-1). When we take the limit $x_3 \to 0$, we can extract the instantaneous term from the third term in equation

(A-3) using (C17) in FM98, and then the above expression agrees with on-plane expression (C20) in FM98. Finally we derive the expression in time domain

$$
\begin{aligned}
\tau_{3\eta}(\vec{x}, t) = \frac{\mu}{4\pi} \int \frac{\gamma_\eta}{r^2} \Bigg[& -12 \frac{\beta^2}{r^2} \int_{r/\alpha}^{r/\beta} t' \Delta u_{\zeta,\zeta}(\vec{\xi}, \|t - t'\|)\, dt' - 4p^2 \Delta u_{\zeta,\zeta}\left(\vec{\xi}, \left\|t - \frac{r}{\alpha}\right\|\right) \\
& + 5\Delta u_{\zeta,\zeta}\left(\vec{\xi}, \left\|t - \frac{r}{\beta}\right\|\right) + \frac{r}{\beta} \Delta \dot{u}_{\zeta,\zeta}\left(\vec{\xi}, \left\|t - \frac{r}{\beta}\right\|\right) \Bigg] d\Sigma \\
& - \frac{\mu}{4\pi\beta^2} \int \frac{1}{r} \Delta \ddot{u}_\eta \left(\vec{\xi}, \left\|t - \frac{r}{\beta}\right\|\right) d\Sigma - \frac{\mu}{4\pi\beta} \int \frac{\gamma_\zeta}{r} \Delta \dot{u}_{\eta,\zeta}\left(\vec{\xi}, \left\|t - \frac{r}{\beta}\right\|\right) d\Sigma \\
& - \frac{\mu}{4\pi} \int \frac{\gamma_\zeta}{r^2} \Delta u_{\eta,\zeta}\left(\vec{\xi}, \left\|t - \frac{r}{\beta}\right\|\right) d\Sigma + \frac{\mu}{\pi} \int \frac{\gamma_\zeta^2 \gamma_\eta}{r^2} \\
& \times \Bigg[15 \frac{\beta^2}{r^2} \int_{r/\alpha}^{r/\beta} t' \Delta u_{\zeta,\zeta}(\vec{\xi}, \|t - t'\|)\, dt' + 6p^2 \Delta u_{\zeta,\zeta}\left(\vec{\xi}, \left\|t - \frac{r}{\alpha}\right\|\right) \\
& - 6\Delta u_{\zeta,\zeta}\left(\vec{\xi}, \left\|t - \frac{r}{\beta}\right\|\right) + p^2 \frac{r}{\alpha} \Delta \dot{u}_{\zeta,\zeta}\left(\vec{\xi}, \left\|t - \frac{r}{\alpha}\right\|\right) \\
& - \frac{r}{\beta} \Delta \dot{u}_{\zeta,\zeta}\left(\vec{\xi}, \left\|t - \frac{r}{\beta}\right\|\right) \Bigg] d\Sigma,
\end{aligned}
\tag{A-4}
$$

in which the slip functions $\Delta u_i(\|a\|)$ are only evaluated for positive values of a. In the limit $x_3 \to 0$ on equation (A-4), the instantaneous term ($\Delta \dot{u}$) is extracted from the 2nd term and the 5th off-plane term disappears.

Similarly we can derive the other components,

$$
\begin{aligned}
\tau_{33}(\vec{x}, t) = \frac{\mu}{2\pi} \int \frac{\gamma_3}{r^2} \Bigg[& -18 \frac{\beta^2}{r^2} \int_{r/\alpha}^{r/\beta} t' \Delta u_{\zeta,\zeta}(\vec{\xi}, \|t - t'\|)\, dt' + (1 - 8p^2) \Delta u_{\zeta,\zeta}\left(\vec{\xi}, \left\|t - \frac{r}{\alpha}\right\|\right) \\
& + 7\Delta u_{\zeta,\zeta}\left(\vec{\xi}, \left\|t - \frac{r}{\beta}\right\|\right) + (1 - 2p^2) \frac{r}{\alpha} \Delta \dot{u}_{\zeta,\zeta}\left(\vec{\xi}, \left\|t - \frac{r}{\alpha}\right\|\right) \\
& + \frac{r}{\beta} \Delta \dot{u}_{\zeta,\zeta}\left(\vec{\xi}, \left\|t - \frac{r}{\beta}\right\|\right) \Bigg] d\Sigma + \frac{\mu}{\pi} \int \frac{\gamma_3^3}{r^2} \\
& \times \Bigg[15 \frac{\beta^2}{r^2} \int_{r/\alpha}^{r/\beta} t' \Delta u_{\zeta,\zeta}(\vec{\xi}, \|t - t'\|)\, dt' + 6p^2 \Delta u_{\zeta,\zeta}\left(\vec{\xi}, \left\|t - \frac{r}{\alpha}\right\|\right) \\
& - 6\Delta u_{\zeta,\zeta}\left(\vec{\xi}, \left\|t - \frac{r}{\beta}\right\|\right) + p^2 \frac{r}{\alpha} \Delta \dot{u}_{\zeta,\zeta}\left(\vec{\xi}, \left\|t - \frac{r}{\alpha}\right\|\right) - \frac{r}{\beta} \Delta \dot{u}_{\zeta,\zeta}\left(\vec{\xi}, \left\|t - \frac{r}{\beta}\right\|\right) \Bigg] d\Sigma
\end{aligned}
\tag{A-5}
$$

$$
\tau_{\eta\eta}(\vec{x}, t) = \frac{\mu}{2\pi} \int \frac{\gamma_\eta^3}{r^2} \Bigg[-6 \frac{\beta^2}{r^2} \int_{r/\alpha}^{r/\beta} t' \Delta u_{\zeta,\zeta}(\vec{\xi}, \|t - t'\|)\, dt' + (1 - 4p^2) \Delta u_{\zeta,\zeta}\left(\vec{\xi}, \left\|t - \frac{r}{\alpha}\right\|\right)
$$

$$+ 2\Delta u_{\zeta,\zeta}\left(\vec{\xi}, \left\|t - \frac{r}{\beta}\right\|\right) + (1 - 2p^2)\frac{r}{\alpha}\Delta \dot{u}_{\zeta,\zeta}\left(\vec{\xi}, \left\|t - \frac{r}{\alpha}\right\|\right)\Bigg]d\Sigma$$

$$+ \frac{\mu}{\pi}\int\frac{\gamma_\eta^2\gamma_3}{r^2}\Bigg[15\frac{\beta^2}{r^2}\int_{r/\alpha}^{r/\beta} t'\Delta u_{\zeta,\zeta}(\vec{\xi}, \|t - t'\|)\,dt' + 6p^2\Delta u_{\zeta,\zeta}\left(\vec{\xi}, \left\|t - \frac{r}{\alpha}\right\|\right)$$

$$- 6\Delta u_{\zeta,\zeta}\left(\vec{\xi}, \left\|t - \frac{r}{\beta}\right\|\right) + p^2\frac{r}{\alpha}\Delta\dot{u}_{\zeta,\zeta}\left(\vec{\xi}, \left\|t - \frac{r}{\alpha}\right\|\right) - \frac{r}{\beta}\Delta\dot{u}_{\zeta,\zeta}\left(\vec{\xi}, \left\|t - \frac{r}{\beta}\right\|\right)\Bigg]d\Sigma$$

$$+ \frac{\mu}{2\pi}\int\frac{\gamma_3}{r^2}\Bigg[\Delta u_{\eta,\eta}\left(\vec{\xi}, \left\|t - \frac{r}{\beta}\right\|\right) + \frac{r}{\beta}\Delta\dot{u}_{\eta,\eta}\left(\vec{\xi}, \left\|t - \frac{r}{\beta}\right\|\right)\Bigg]d\Sigma, \tag{A-6}$$

and

$$\tau_{12}(\vec{x}, t) = \frac{\mu}{\pi}\int\frac{\gamma_1\gamma_2\gamma_3}{r^2}\Bigg[15\frac{\beta^2}{r^2}\int_{r/\alpha}^{r/\beta} t'\Delta u_{\zeta,\zeta}(\vec{\xi}, \|t - t'\|)\,dt' + 6p^2\Delta u_{\zeta,\zeta}\left(\vec{\xi}, \left\|t - \frac{r}{\alpha}\right\|\right)$$

$$- 6\Delta u_{\zeta,\zeta}\left(\vec{\xi}, \left\|t - \frac{r}{\beta}\right\|\right) + p^2\frac{r}{\alpha}\Delta\dot{u}_{\zeta,\zeta}\left(\vec{\xi}, \left\|t - \frac{r}{\alpha}\right\|\right)$$

$$- \frac{r}{\beta}\Delta\dot{u}_{\zeta,\zeta}\left(\vec{\xi}, \left\|t - \frac{r}{\beta}\right\|\right)\Bigg]d\Sigma$$

$$+ \frac{\mu}{4\pi}\int\frac{\gamma_3}{r^2}\Bigg[\Delta u_{\zeta,\bar{\zeta}}\left(\vec{\xi}, \left\|t - \frac{r}{\beta}\right\|\right) + \frac{r}{\beta}\Delta\dot{u}_{\zeta,\bar{\zeta}}\left(\vec{\xi}, \left\|t - \frac{r}{\beta}\right\|\right)\Bigg]d\Sigma, \tag{A-7}$$

where we define $\bar{\zeta} = 3 - \zeta$.

Appendix B. Discretized Kernel for a Semi-infinite Slip Velocity Field

We obtain the discretized kernels for the basic shear source given by equation (5). In this case, we can analytically evaluate all the integrals in equations (A-4)–(A-6) and (A-7), for discrete BIE without using any numerical calculation. We will introduce notations before summarizing discrete kernels,

$$\chi_\zeta'^2 = x_\zeta^2 + x_3^2, \quad X_\zeta(c) = \sqrt{c^2t^2 - \chi_\zeta'^2}, \quad \text{and} \quad \chi^2 = x_1^2 + x_2^2 + x_3^2 \tag{B-1}$$

and

$$I_{i/\zeta} = \frac{x_i}{\chi_\zeta'^2}, \quad J_{ijk/\zeta} = \frac{x_i x_j x_k}{\chi_\zeta'^2}, \quad K_\zeta(c) = \frac{ct}{\chi_\zeta'}, \quad L_\zeta(c) = \frac{ctx_\zeta}{\chi}, \quad \text{and} \quad M(c) = \frac{ct}{\chi} \tag{B-2}$$

and define three cases which depend on the observation point $\vec{x}$ and time t

$$
\begin{cases}
(\mathrm{I}-c)_{\bar{\zeta}} & x_{\bar{\zeta}} < -X_{\zeta}(c) \quad \text{for} \quad X_{\zeta}(c)^2 > 0, \quad \text{or} \quad X_{\zeta}(c)^2 \le 0 \\
(\mathrm{II}-c) & \chi^2 < c^2 t^2 \\
(\mathrm{III}-c)_{\bar{\zeta}} & x_{\bar{\zeta}} > X_{\zeta}(c) \quad \text{for} \quad X_{\zeta}(c)^2 > 0
\end{cases}
\tag{B-3}
$$

With the above notations and definitions, we create the final expressions of the discretized kernels

$$
\tau_{31}(\check{x}, t) = \frac{\mu}{4\pi\beta} \times [\, -V_1 E_{II}(\beta)
$$

$$
+ V_1(-Q_{11}(\beta) + Q_{11}(\alpha) + \mathscr{B}'_{11}(\beta) + R_{3311}(\beta) - R_{3311}(\alpha))
$$

$$
+ V_2(-Q_{12}(\beta) + Q_{12}(\alpha) + \mathscr{B}'_{12}(\beta) + R_{3312}(\beta) - R_{3312}(\alpha)], \tag{B-4}
$$

$$
\tau_{33}(\check{x}, t) = \frac{\mu}{4\pi\beta} \times [V_1(-3(Q_{31}(\beta) - Q_{31}(\alpha)) + 2(\mathscr{B}'_{31}(\beta) - p(2p^2-1)\mathscr{B}'_{31}(\alpha))
$$

$$
+ R_{3331}(\beta) - R_{3331}(\alpha)) + V_2(-3(Q_{32}(\beta) - Q_{32}(\alpha))
$$

$$
+ 2(\mathscr{B}'_{32}(\beta) - p(2p^2-1)\mathscr{B}'_{32}(\alpha)) + R_{3332}(\beta) - R_{3332}(\alpha))], \tag{B-5}
$$

$$
\tau_{11}(\check{x}, t) = \frac{\mu}{4\pi\beta} \times [V_1(-Q_{31}(\beta) + Q_{31}(\alpha) + 2(\mathscr{B}'_{31}(\beta) + p(1-2p^2)\mathscr{B}'_{31}(\alpha)) + R_{3111}(\beta)
$$

$$
- R_{3111}(\alpha)) + V_2(-Q_{32}(\beta) + Q_{32}(\alpha) + 2p(1-2p^2)\mathscr{B}'_{32}(\alpha)
$$

$$
+ R_{3112}(\beta) - R_{3112}(\alpha))], \tag{B-6}
$$

$$
\tau_{12}(\check{x}, t) = \frac{\mu}{4\pi\beta} \times [V_1(-Q_{32}(\beta) + Q_{32}(\alpha) + \mathscr{B}'_{32}(\beta) + R_{3112}(\beta) - R_{3112}(\alpha))
$$

$$
+ V_2(-Q_{31}(\beta) + Q_{31}(\alpha) + \mathscr{B}'_{31}(\beta) + R_{3221}(\beta) - R_{3221}(\alpha))], \tag{B-7}
$$

where we use the following 4 terms for the basic Heaviside function:

$$
\mathscr{B}_{i\zeta}(c) \overset{i \ne \bar{\zeta}}{\longrightarrow}
\begin{cases}
0 & \text{for } (\mathrm{I}-c)_{\bar{\zeta}} \\
I_{i/\zeta}(X_{\zeta}(c) + L_{\bar{\zeta}}(c)) & \text{for } (\mathrm{II}-c) \\
2I_{i/\zeta}X_{\zeta}(c) & \text{for } (\mathrm{III}-c)_{\bar{\zeta}}
\end{cases}
\tag{B-8}
$$

$$
\overset{i=\bar{\zeta}}{\longrightarrow}
\begin{cases}
1 - M(c) & \text{for } (\mathrm{II}-c) \\
0 & \text{otherwise}
\end{cases}
\tag{B-9}
$$

$$Q_{i\zeta}(c) \overset{i \neq \bar{\zeta}}{\to}$$

$$\begin{cases} 0 & \text{for } (\mathrm{I} - c)_{\bar{\zeta}} \\[2mm] 2\dfrac{\beta^3}{c^3} I_{i/\zeta}\left[\dfrac{2}{3} X_\zeta(c)(1 - K_\zeta(c)^2) + L_{\bar{\zeta}}(c)\left(1 - \dfrac{1}{3}(M(c)^2 + 2K_\zeta(c)^2)\right)\right] & \text{for } (\mathrm{II} - c) \\[2mm] \dfrac{8}{3}\dfrac{\beta^3}{c^3} I_{i/\zeta} X_\zeta(c)(1 - K_\zeta(c)^2) & \text{for } (\mathrm{III} - c)_{\bar{\zeta}} \end{cases}$$

$$\text{(B-10)}$$

$$\overset{i = \bar{\zeta}}{\to}\begin{cases} 2\dfrac{\beta^3}{c^3}\left(\dfrac{2}{3} - M(c)\left(1 - \dfrac{1}{3}M(c)^2\right)\right) & \text{for } (\mathrm{II} - c) \\[2mm] 0 & \text{otherwise} \end{cases} \qquad \text{(B-11)}$$

$$R_{3ij\zeta}(c) \overset{i,j \neq \bar{\zeta}}{\to}$$

$$\begin{cases} 0 & \text{for } (\mathrm{I} - c)_{\bar{\zeta}} \\[2mm] 2\dfrac{\beta^3}{c^3} J_{3ij/\zeta}\left[\dfrac{2}{3}\dfrac{X_\zeta(c)}{\chi_\zeta'^2}(1 - 4K_\zeta(c)^2) + \dfrac{L_{\bar{\zeta}}(c)}{c^2 t^2}\left\{(M(c)^2 + 2K_\zeta(c)^2)\left(1 - \dfrac{4}{3}K_\zeta(c)^2\right) - M(c)^4\right\}\right] & \text{for } (\mathrm{II} - c) \\[2mm] \dfrac{8}{3}\dfrac{\beta^3}{c^3} J_{3ij/\zeta}\dfrac{X_\zeta(c)}{\chi_\zeta'^2}(1 - 4K_\zeta(c)^2) & \text{for } (\mathrm{III} - c)_{\bar{\zeta}} \end{cases}$$

$$\text{(B-12)}$$

$$\overset{i = 3, j = \bar{\zeta}}{\to}\begin{cases} -2\dfrac{\beta^3}{c^3}\dfrac{x_3^2}{\chi^2} M(c)(1 - M(c)^2) & \text{for } (\mathrm{II} - c) \\[2mm] 0 & \text{otherwise} \end{cases} \qquad \text{(B-13)}$$

$$\overset{i,j = \bar{\zeta}}{\to}\begin{cases} 0 & \text{for } (\mathrm{I} - c)_{\bar{\zeta}} \\[2mm] 2\dfrac{\beta^3}{c^3} I_{3/\zeta}\left[\dfrac{2}{3} X_\zeta(c)(1 - K_\zeta(c)^2) + \dfrac{x_\zeta^2}{\chi^2} L_\zeta(c)\left(1 - \dfrac{2}{3}K_\zeta(c)^2 - M(c)^2\right)\right] & \text{for } (\mathrm{II} - c) \\[2mm] \dfrac{8}{3}\dfrac{\beta^3}{c^3} I_{3/\zeta} X_\zeta(c)(1 - K_\zeta(c)^2) & \text{for } (\mathrm{III} - c)_{\bar{\zeta}} \end{cases} \text{,}$$

$$\text{(B-14)}$$

$$E_{II}(\beta) = 2\pi H(x_1)H(x_2)H(\beta t - |x_3|)$$

$$+ \sum_{\zeta=1}^{2} \begin{cases} 0 & \text{for} \quad (\text{I} - \beta)_{\bar{\zeta}} \\ I_{\zeta/\zeta}(X_\zeta(\beta) + L_{\bar{\zeta}}(\beta)) - \arctan \dfrac{X_\zeta(\beta)}{x_\zeta} - \arctan \dfrac{x_{\bar{\zeta}}}{x_\zeta} & \text{for} \quad (\text{II} - \beta) \\ 2\left(I_{\zeta/\zeta}X_\zeta(\beta) - \arctan \dfrac{X_\zeta(\beta)}{x_\zeta} \right) & \text{for} \quad (\text{III} - \beta)_{\bar{\zeta}} \end{cases} \qquad \text{(B-15)}$$

Here $E_{II}(\beta)$ includes the instantaneous term.

All the discrete kernels derived in this appendix are the functions of the distance from the origin to an observation point. Substituting $(x_1 - \xi_1^l + \Delta s/2, x_2 - \xi_2^m + \Delta s/2; t - \tau + e_t\Delta t)$ for $(x_1, x_2; t)$ etc. we finally obtain the discrete kernels $P_{pq/\zeta}(\vec{r})$ for a piecewise source $\Delta \dot{u}_\zeta = S(\xi_1, \xi_2, \tau; \xi_1^l, \xi_2^m, \tau^n)$ that appears in equation (6).

REFERENCES

AKI, K., and RICHARDS, P. G., *Quantitative Seismology: Theory and Methods* (Freeman and Co., San Francisco 1980).

AOCHI, H. (1999), *Theoretical Studies on Dynamic Rupture Propagation along a 3-D Non-planar Fault System*, D.Sc. Thesis, University of Tokyo, Japan.

AYDIN, A., and DU, Y. (1995), *Surface Rupture at a Fault Bend: The 28 June 1992 Landers, California, Earthquake*, Bull. Seismol. Soc. Am. *85*, 111–128.

BOUCHON, M., CAMPILLO, M., and COTTON, F. (1998a), *Stress Field Associated with the Rupture of the 1992 Landers, California, Earthquake and its Implications Concerning the Fault Strength at the Onset of the Earthquake*, J. Geophys. Res. *103*, 21,091–21,097.

BOUCHON, M., SEKIGUCHI, H., IRIKURA, K., and IWATA, T. (1998b), *Some Characteristics of the Stress Field of the 1995 Hyogo-ken Nanbu (Kobe) Earthquake*, J. Geophys. Res. *103*, 24,271–24,282.

BOUCHON, M., and STREIFF, D. (1997), *Propagation of a Shear Crack on a Nonplanar Fault: A Method of Calculation*, Bull. Seismol. Soc. Am. *87*, 61–66.

COCHARD, A., and MADARIAGA, R. (1994), *Dynamic Faulting under Rate-dependent Friction*, Pure appl. geophys. *142*, 419–445.

DAS, S., and AKI, K. (1977), *A Numerical Study of Two-dimensional Spontaneous Rupture Propagation*, Geophys. J. R. astr. Soc. *50*, 643–668.

DAY, S. M. (1982), *Three-dimensional Simulation of Spontaneous Rupture: The Effect of Nonuniform Prestress*, Bull. Seismol. Soc. Am. *72*, 1881–1902.

ELLSWORTH, W. L., and BEROZA, G. C. (1995), *Seismic Evidence for an Earthquake Nucleation Phase*, Science *268*, 851–855.

FUKUYAMA, E., and MADARIAGA, R. (1995), *Integral Equation Method for Plane Crack with Arbitrary Shape in 3-D Elastic Medium*, Bull. Seismol. Soc. Am. *85*, 614–628.

FUKUYAMA, E., and MADARIAGA, R. (1998), *Rupture Dynamics of a Planar Fault in a 3-D Elastic Medium: Rate- and Slip-weakening Friction*, Bull. Seismol. Soc. Am. *88*, 1–17.

FUKUYAMA, E., and MADARIAGA, R. (2000), *Dynamic Rupture Propagation and Interaction of a Rupture Front on a Planar Fault*, Pure appl. geophys. *157*, 1959–1979.

HARRIS, R. A., and DAY, S. M. (1993), *Dynamics of Fault Interaction: Parallel Strike-slip Faults*, J. Geophys. Res. *98*, 4461–4472.

HARRIS, R. A., and DAY, S. M. (1999), *Dynamic 3-D Simulations of Earthquakes on en echelon Faults*, Geophys. Res. Lett. *26*, 2089–2092.

IDA, Y. (1972), *Cohesive Force across the Tip of a Longitudinal-shear Crack and Griffith's Specific Surface Energy*, J. Geophys. Res. *77*, 3796–3805.

IDE, S., and TAKEO, M. (1997), *Determination of Constitutive Relations of Fault Slip Based on Seismic Wave Analysis*, J. Geophys. Res. *102*, 27,379–27,391.

IRIKURA, K., IWATA, T., SEKIGUCHI, H., PITARKA, A., and KAMAE, K. (1996), *Lesson from the 1995 Hyogoken Nanbu Earthquake: Why Were such Destructive Motions Generated to Buildings?* J. Nat. Disaster Sci. *17*, 99–127.

KAME, N., and YAMASHITA, T. (1997), *Dynamic Nucleation Process of Shallow Earthquake Faulting in a Fault Zone*, Geophys. J. Int. *128*, 204–216.

KAME, N., and YAMASHITA, T. (1999a), *A New Light on Arresting Mechanism of Dynamic Earthquake Faulting*, Geophys. Res. Lett. *26*, 1997–2000.

KAME, N., and YAMASHITA, T. (1999b), *Simulation of the Spontaneous Growth of a Dynamic Crack without Constraints on the Crack Tip Path*, Geophys. J. Int. *139*, 345–358.

KASE, Y., and KUGE, K. (1998), *Numerical Simulation of Spontaneous Rupture Processes on two Non-coplanar Faults: The Effect of Geometry on Fault Interaction*, Geophys. J. Int. *135*, 911–922.

KOLLER, M. G., BONNET, M., and MADARIAGA, R. (1992), *Modeling of Dynamical Crack Propagation Using Time-domain Boundary Integral Equations*, Wave Motion. *16*, 339–366.

LAPUSTA, N., RICE, J. R., BEN-ZION, Y., and ZHENG, G.-T. (2000), *Elastodynamic Analysis for Slow Tectonic Loading with Spontaneous Rupture Episodes on Faults with Rate- and State-dependent Friction*, J. Geophys. Res., submitted.

MAGISTRALE, H., and DAY, S. M. (1999), *3-D Simulations of Multi-segment Thrust Fault Rupture*, Geophys. Res. Lett. *26*, 2093–2096.

MATSU'URA, M., KATAOKA, H., and SHIBAZAKI, B. (1992), *Slip-dependent Friction Law and Nucleation Processes in Earthquake Rupture*, Tectonophysics *211*, 135–148.

OKUBO, P. G., and DIETERICH, J. H. (1984), *Effects of Physical Fault Properties on Frictional Instabilities Produced on Simulated Fault*, J. Geophys. Res. *89*, 5817–5827.

OHNAKA, M. (1996), *Nonuniformity of the Constitutive Law Parameters for Shear Rupture and Quasistatic Nucleation to Dynamic Rupture: A Physical Model of Earthquake Generation Processes*, Proc. Natl. Acad. Sci. USA *93*, 3795–3802.

OHNAKA, M., KUWAHARA, Y., and YAMAMOTO, K. (1987), *Constitutive Relations between Dynamic Physical Parameters near a Tip of the Propagating Slip Zone during Stick-slip Shear Failure*, Tectonophysics *144*, 109–125.

PALMER, A. C., and RICE, J. R. (1973), *The Growth of Slip Surface in the Progressive Failure of Over-consolidated Clay*, Proc. Roy. Soc. Lond. A. *332*, 527–548.

SEELIG, TH., and GROSS, D. (1997), *Analysis of Dynamic Crack Propagation Using a Time-domain Boundary Integral Equation Method*, Int. J. Solids Struc. *34*, 2087–2103.

SEELIG, TH., and GROSS, D. (1999a), *On the Stress Wave Induced Curving of Fast Running Cracks—A Numerical Study by a Time-domain Boundary Element Method*, Acta Mechanica *132*, 47–61.

SEELIG, TH., and GROSS, D. (1999b), *On the Interaction and Branching of Fast Running Cracks—A Numerical Investigation*, J. Mech. Phys. Solids *47*, 935–952.

SEKIGUCHI, H., IRIKURA, K., IWATA, T., KAKEHI, Y., and HOSHIBA, M. (1996), *Minute Locating of Faulting beneath Kobe and the Waveform Inversion of the Source Process during the 1995 Hyogo-ken Nanbu, Japan, Earthquake Using Strong Ground Motion Records*, J. Phys. Earth *44*, 473–487.

TADA, T., FUKUYAMA, E., and MADARIAGA, R. (2000), *Non-hypersingular Boundary Integral Equations for 3-D Non-planar Crack Dynamics*, Comput. Mech. *25*(6), 613–626.

TADA, T., and MADARIAGA, R. (2000), *Dynamic modeling of the flat 2-D Crack by a Semi-analytic BIEM Scheme*, Int. J. Num. Meth. Eng., submitted.

TADA, T., and YAMASHITA, T. (1996), *The Paradox of Smooth and Abrupt Bends in Two-dimensional In-plane Shear-crack Mechanics*, Geophys. J. Int. *127*, 795–800.

TADA, T., and YAMASHITA, T. (1997), *Non-hypersingular Boundary Integral Equations for Two-dimensional Non-planar Crack Analysis*, Geophys. J. Int. *130*, 269–282.

WALD, D. J., and HEATON, T. H. (1995), *Spatial and Temporal Distribution of Slip for the 1992 Landers, California, Earthquake*, Bull. Seismol. Soc. Am. *84*, 668–691.

YOSHIDA, S., KOKETSU, K., SHIBAZAKI, B., SAGIYA, T., KATO, T., and YOSHIDA, Y. (1996), *Joint Inversion of Near- and Far-field Waveforms and Geodetic Data for the Rupture Process of the 1995 Kobe Earthquake*, J. Phys. Earth. *44*, 437–454.

(Received July 30, 1999, revised April 5, 2000, accepted April 7, 2000)

Pure appl. geophys. 157 (2000) 2029–2046
0033–4553/00/122029–18 $ 1.50 + 0.20/0

Constraints on Stress and Friction from Dynamic Rupture Models of the 1994 Northridge, California, Earthquake

STEFAN B. NIELSEN[1,2] and KIM B. OLSEN[3]

Abstract—We have simulated several scenarios of dynamic rupture propagation for the 1994 Northridge, California, earthquake, using a three-dimensional finite-difference method. The simulations use a rate- and slip-weakening friction law, starting from a range of initial conditions of stress and frictional parameters. A critical balance between initial conditions and friction parameters must be met in order to obtain a moment as well as a final slip distribution in agreement with kinematic slip inversion results. We find that the rupture process is strongly controlled by the average stress and connectivity of high-stress patches on the fault. In particular, a strong connectivity of the high-stress patches is required in order to promote the rupture propagation from the initial nucleation to the remaining part of the fault. Moreover, we find that a small amount of rate-weakening is needed in order to obtain a level of inhomogeneity in the final slip, similar to that obtained in the kinematic inversion results. However, when the amount of rate-weakening is increased, the overall moment drops dramatically unless the average prestress is raised to unrealistic levels. A velocity-weakening parameter on the order of 10 cm per second is found to be adequate for an average prestress of about a hundred bars. The presence of the free surface and of the uppermost low-impedance layers in the model are found to have negligible influence on the rupture dynamics itself, because the top of the fault is at a depth of several kilometers. The 0.1–0.5 Hz radiated waves from the dynamic simulation provides a good fit to strong motion data at sites NWH and SSA. Underprediction of the recorded peak amplitude at JFP is likely due to omission of near-surface low velocity and 3-D basin effects in the simulations.

Key words: Earthquake source, dynamic rupture, rate-weakening friction, Northridge earthquake, stress distribution, rupture history.

Introduction

The principal aim of this study is to outline constraints on the fundamental dynamic parameters involved in fault rupture (e.g., the prestress distribution and the frictional parameters), based on observations from the 1994 Northridge, California, earthquake. This is indeed an ambitious task, since dynamic parameters can only be inferred from the source kinematics (i.e., the irregular growth of

[1] Institute for Crustal Studies and Materials Research Laboratory, University of California, Santa Barbara, California, U.S.A.

[2] Currently at Istituto Nazionale di Geofisica, Dipartimento di Scienze Fisiche, Università degli Studi di Napoli, Federico II, Via Cintia 45, 80126 Napoli, Italy. E-mail: snielsen@na.infn.it.

[3] Institute for Crustal Studies, University of California, Santa Barbara, California, U.S.A.

rupture and onset of dislocation on the fault as a function of time). In addition, our knowledge of the source kinematics is itself retrieved through an indirect inversion process, with a sparse amount of data limiting both the spatial and temporal resolution. For most well-studied earthquakes, the spatial and temporal resolutions are at best about 3–4 km and 1 s, respectively (see, for example, the inversion results of WALD et al., 1996 for the Northridge earthquake; HARTZELL and HEATON, 1983 and ARCHULETA, 1984 for the 1979 Imperial Valley earthquake; COTTON and CAMPILLO, 1995a and WALD and HEATON, 1994 for the 1992 Landers earthquake; YOSHIDA et al., 1996, IDE et al., 1996 for the 1995 Kobe earthquake). Despite this lack of resolution, constraints have been derived on parameters such as the absolute level of stress and the characteristic slip-weakening distance in the friction law. Using observed and inferred rake rotation during the 1995 Kobe earthquake, SPUDICH et al. (1998) inferred a relatively low tectonic shear stress on the fault, confirming the results obtained by COTTON and CAMPILLO (1995b) for the Landers earthquake, also based on rake rotation. IDE and TAKEO (1997) inferred a slip-weakening distance on the order of 0.5–1 m for the friction law, based on observations of the Kobe earthquake of 1995. OLSEN et al. (1997) modeled the 1992 Landers earthquake as the propagation of a spontaneous rupture controlled by a heterogeneous prestress distribution using a three-dimensional finite-difference method (MADARIAGA et al., 1998). They computed the prestress distribution based on a slip distribution obtained from kinematic slip inversion. Their results, using a simple slip-weakening friction law, showed rupture propagation along a complex path with highly variable speed and rise time, changing the magnitude and pattern of the stress significantly. Moreover, they found that the rupture speed and healing of the fault were critically determined by the level of the yield stress and the slip-weakening distance in the friction law. For example, if the slip-weakening distance was chosen less than about 60–80 cm, the rupture duration and therefore rise times were considerably shorter than those obtained from kinematic inversion, while larger values introduced rupture resistance at a level that prevented the rupture from propagating at all. The Landers study suggested that earthquakes are critical phenomena, occurring only for a limited range of rupture resistance. The idea of criticality was furthermore tested by MADARIAGA and OLSEN (2000) (this volume), who identified a non-dimensional parameter κ that controlled rupture.

In this paper, we study dynamic rupture propagation for the 1994 Northridge earthquake. In particular, we investigate the effects of rate weakening in the friction law and the connectivity of the high-stress patches on the rupture propagation. As for the OLSEN et al. (1997) Landers model, we assume a planar fault and slip occurring only along the fault plane. However, due to the dip of the causative fault for the Northridge earthquake, we introduce the interaction between the fault-normal stress and the free surface, an effect which is absent for vertical fault models such as that used by OLSEN et al. (1997). We quantify this interaction by comparison between dynamic models with and without a free surface.

Fault Boundary Condition and Friction Law

We use the mixed boundary conditions by NIELSEN and OLSEN (2000) in our dynamic simulations of the Northridge earthquake. This mixed dislocation-traction boundary condition allows us to impose the slip rate on the fault instead of the shear stress, thus bypassing the inertial stage introduced by the finite node size. This considerably reduces spurious oscillations in the dynamic rupture propagation. The implementation of this boundary condition has two further advantages: it allows the consistent use of higher-order spatial operators on a staggered grid; both friction and dislocation are defined on the same nodes of the grid, exactly on the fault surface. The boundary condition can be expressed as

$$\frac{\mu}{2\beta} \Delta \dot{u}(\mathbf{x}, t) = T(\mathbf{x}, t) - \sigma_e(\mathbf{x}, t), \tag{1}$$

where the T is the fault traction, the slip rate is $\Delta \dot{u}$, μ is the shear stiffness of the medium and β the shear-wave velocity. Here, σ_e represents the cumulative nonlocal contribution to the stress due to radiation arriving from other regions of the fault; its explicit form is a convolution integral within the elastic wave causality cone. Specific forms of σ_e have been derived inside a homogeneous elastic space in the 2-D anti-plane configuration (COCHARD and MADARIAGA, 1994) and in 3-D (COCHARD and RICE, 1997; FUKUYAMA and MADARIAGA, 1998). However, no analytical expression for σ_e is known inside an inhomogeneous medium; in our case σ_e is computed numerically by the finite-difference scheme and then substituted into expression (1).

During active slip we use a rate- and slip-weakening friction law whose value is given by the following bimodal expression:

$$T = \mu_d \sigma_n + (\sigma_y + \mu_s \sigma_n) \, \text{Max} \left[\frac{1}{1 + \Delta \dot{u}/v}, \frac{1}{1 + \Delta u/\delta} \right], \tag{2}$$

where σ_y is the yield stress in the absence of normal stress perturbations, σ_n is the stress normal to the fault in excess of the lithostatic stress and δ and v are the slip- and rate-weakening parameters, respectively. The static and the dynamic friction coefficients take the values $\mu_s = 0.7$ and $\mu_d = 0.3$, respectively. The fracture criterion is met whenever the shear traction $T = \sigma_{zx}$ on the fault exceeds the sum $\sigma_y + \mu_s \sigma_n$, where z is the fault normal and x is along the dip of the fault. Healing occurs when the shear traction on the fault exceeds the radiated stress (σ_e in expression (1)), which is equivalent to a reversal in the sign of the slip rate.

Fault Plane and Geological Model

The 1994 M 6.7 Northridge earthquake occurred on a blind thrust fault below the San Fernando Valley in southern California. We use the fault model for the Northridge earthquake proposed by WALD et al. (1996) with a planar fault plane, with the top of the fault at 5 km depth. Rupture propagation in most of our simulations is contained within the slip region defined by WALD et al. (1996) i.e., 18 km along strike and 24 km along dip (see Fig. 1), although we allow fracture to expand to a slightly larger area (20 km along strike and 26 km along dip). The fault

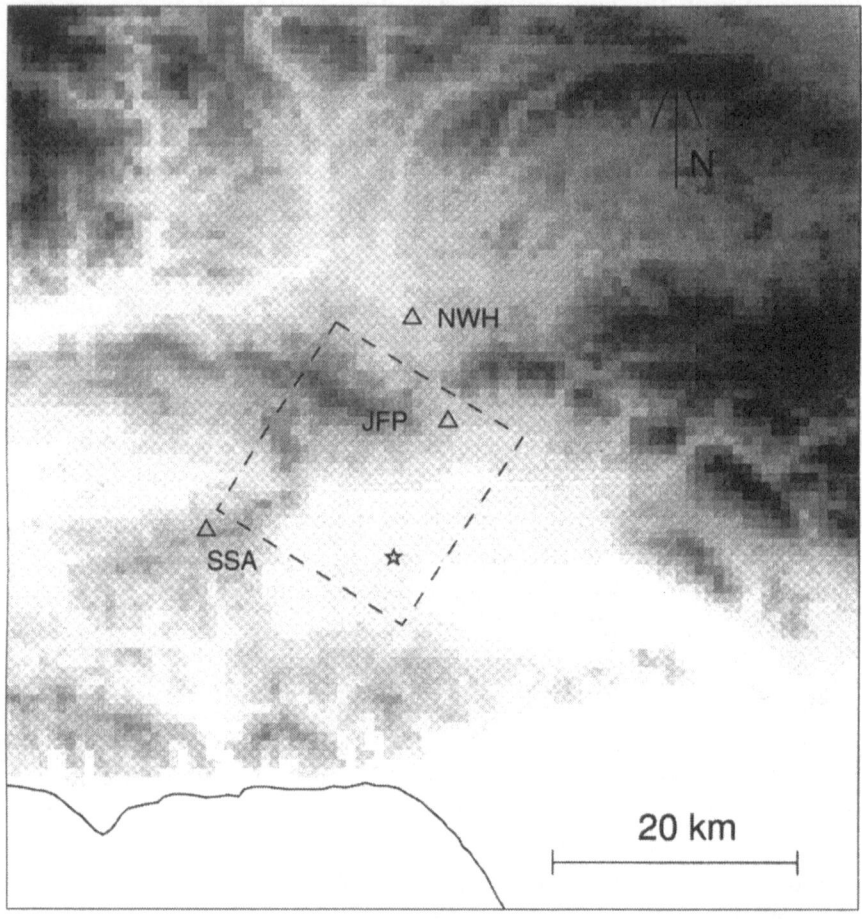

Figure 1
Topographic map of the San Fernando Valley area. The rectangle depicts the surface projection of the causative fault for the Northridge earthquake. The triangles depict the location of the three strong motion stations NWH, JFP, and SSA, where we compare radiated waves from our dynamic simulations to data. The star depicts the epicenter of the Northridge earthquake. Grey shading represents the local topography.

Table 1

Layered Model Parameters

V_p (km/s)	V_s (km/s)	ρ (g/cm³)	Top (km)
1.9	1.0	2.1	0.0
4.0	2.0	2.4	0.5
5.5	3.20	2.7	1.5
6.3	3.6	2.8	4.0

is confined to a single geological layer of constant properties, located below surficial layers, as listed in Table 1. However, simulations in a homogeneous whole space were also carried out to test the combined effect of the near-surface layers and the free surface. We define the numerical grid parallel to the fault plane to avoid cumbersome numerical interpolations that would be necessary if the fault were at any angle $\theta \neq \eta\pi/2$ with the numerical grid. On the other hand, it is straightforward to define the free surface at any angle θ with respect to the numerical grid by introducing a vacuum layer, a method which is widespread for finite-difference simulations in the presence of irregular topography (e.g., OLSEN *et al.*, 1996). Similarly, the top layers are tilted θ with respect to the grid, where θ is the fault dip. The model and its mapping onto the numerical grid is schematically illustrated in Figure 2. The numerical grid used is a cube of $276 \cdot 191 \cdot 130$ nodes with space and time sampling $dx = 200$ m and $dt = 0.015$ s, respectively.

Friction Parameters and Initial Stress

All the simulations presented in this study are conducted under the assumption that both the frictional parameters δ and v and the yield stress are constant across the fault for a given simulation. The heterogeneity generated in the dynamic rupture thus originates either in the initial stress with which the fault is loaded prior to rupture, in the presence of the free surface and the geological stratification in the uppermost part of the model, or near the unbreakable boundaries of the fault.

Given the assumption of a heterogeneous prestress, we now need to define the initial stress distribution on the fault. We used the finite-difference method, in a fashion similar to that of OLSEN *et al.* (1997), to estimate the stress variation $\Delta\sigma$ induced by an arbitrary dislocation on the fault plane. Starting from the combined slip distribution obtained by WALD *et al.* (1996) in their kinematic modeling of the 1994 Northridge earthquake, we computed a static stress variation on the fault. Subsequently, the assumptions are made that (1) the static and the dynamic onset of dislocation generate a similar stress drop and (2) the final stress level after rupture is close to homogeneous on the fault, so that the stress drop and the initial stress can be identified. With assumptions (1) and (2), we can use $-\Delta\sigma$ as an

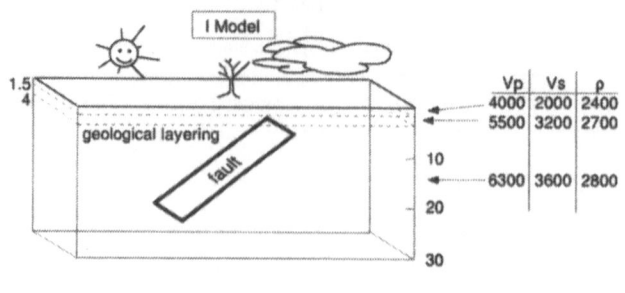

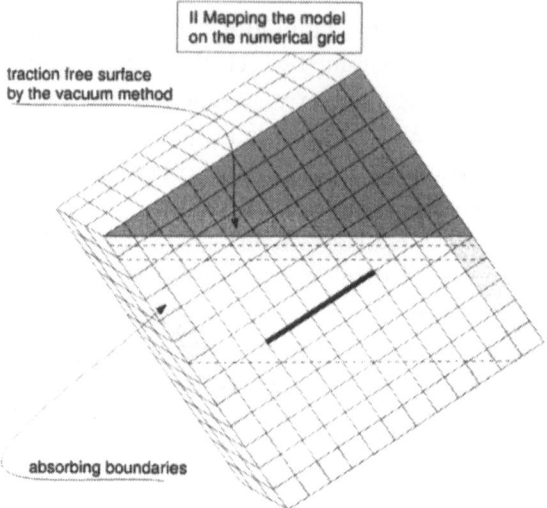

Figure 2
Schematic illustration of the geological model, including the dipping fault (top) and its mapping onto the numerical grid (bottom). The fault is parallel to the grid to maximize the numerical accuracy on the fault interface. As a consequence, the free surface is at an angle with the grid. The free surface condition is implemented by the vacuum method, which yields satisfactory results when modeling body waves for wavelengths not smaller than ≈ 10 grid steps (in our case wavelengths of about 2 Km); besides, surface waves are absent or negligible in this near-source configuration.

approximation of the prestress distribution prior to seismic rupture. It is clear that the dynamics of the rupture process can generate highly variable results for a fixed initial stress distribution (for example, by varying the frictional parameters), one of the main points illustrated by the results of this study. However, given the lack of constraints on the stress distribution on the fault, the initial stress field obtained through this method is not an unreasonable starting point.

We then use trial-and-error modeling to refine the stress distribution and friction parameters in order to optimize the fit between the dynamic and kinematic rupture history as well as between the dynamic radiation and observed strong motion seismograms. The trial-and-error modeling includes various transforma-

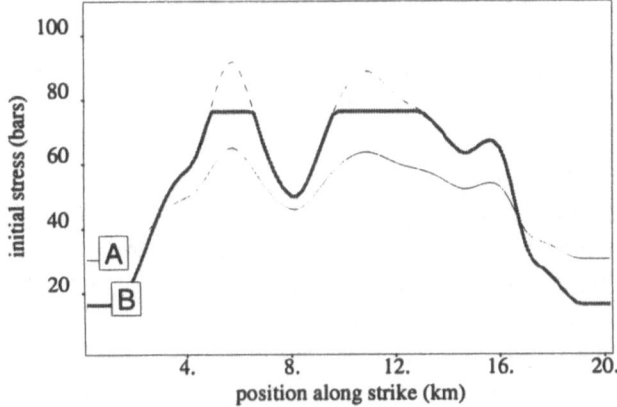

Figure 3
Illustration of the scaling procedure of the initial stress distribution used in this study. The curves show two initial stress distributions, (A) and (B), along a horizontal section of the fault at a depth of 14 km. Starting from distribution (A), the prestress is rescaled, shifted and truncated above an arbitrary value in order to obtain a new distribution (B). Depending on the amount of rescaling, shifting and truncation, the new stress distribution can be characterized, among other parameters, by its new mean, range and skewness. The most severe truncations will result in larger and more connected high stress patches, associated with a relatively large value of the normalized skewness value ζ for the distribution (see definition (5) and further details in text).

tions of the initial stress $\sigma_0(\mathbf{x})$ as explained below. We rescale and shift the original $\sigma_0(\mathbf{x})$ so that it spans a given interval. In most simulations, we truncate the values of the stress field that are above a threshold value between 90% and 99% of the yield stress to avoid rupture initiation from several points on the fault. An example of this procedure for altering the prestress field is illustrated in Figure 3. These operations generate various distributions that can be characterized by means of several arbitrary parameters, such as the mean $(\overline{\sigma_0})$, the range $\langle \sigma_0 \rangle$ and the skewness $\xi(\sigma_0)$. Skewness describes the amount of asymmetry in the distribution; in our case it also provides an arbitrary measure of the connectivity between the high-stress patches on the fault. We choose to use ζ, a renormalized estimate of the skewness that we define in Appendix (equation (5)). For a given mean stress, larger values of ζ indicate an abundance of high, interconnected stress patches surrounded by fewer areas of much lower stress.

Analysis of Dynamic Rupture Propagation

Using the slip- and rate-weakening friction law described earlier, we performed trial-and-error modeling with values of δ within the range [0.05, 1] m and of v within the range [0.0, 0.53] m/s. Approximately 30 simulations were performed in total, with user-estimated values for the parameters. The yield stress used as a criterion for fracture was varied within the range [100, 550] bars. The prestress values were also highly variable, within the range [30, 550] bars.

Table 2 states initial values and final slip for five examples among the models that were tested. We observe essentially three categories of rupture propagations. Ruptures in the first category do not propagate much further than the area immediately around the nucleation (see, for example, model O in Table 2 and Fig. 4-O). Conditions under which an initial crack will either stop or continue to propagate, when subjected to a slip-weakening friction and homogeneous prestress, have been investigated (e.g., ANDREWS, 1976; SCHOLZ, 1990; MADARIAGA and OLSEN (2000), who introduced the role of the finite fault width). In our case, the propagation of the initial nucleation is also subject to strong heterogeneity in the prestress and to rate-weakening in the friction. For such case, all the relevant model parameters are listed in the Appendix and combined to form three dimensionless parameters. Two of these parameters are based on fault rheology and average prestress, and the third parameter describes properties of the initial stress distribution. In essence, the dimensional analysis discussed in the Appendix demonstrates that large values of rate weakening and slip weakening in the friction as well as a high yield stress hinder the propagation of an initial nucleation. On the contrary, a

Table 2

Properties of some of the models discussed. $\overline{\sigma_0}$ and $\langle\sigma_0\rangle$ are the average prestress and the variation of the prestress, respectively, σ_y is the yield stress, δ and v are the slip- and rate-weakening parameters, respectively, $\zeta(\sigma_0)$ is the connectivity of the prestress, and Δu_m and $\overline{\Delta u}$ are the maximum and average slip, respectively. M_0 is the total moment, defined as the areal integral of slip times the shear stiffness. For the computation of the average slip $\overline{\Delta u}$, the effective area is limited to that of the actual fracture for each model, as opposed to the total area of the brittle fault. The values for simulations M, I2 and L have been rescaled, including the frictional parameters δ and v, in order to yield a moment of $1.4 \ 10^{19} N$ m, but not for the two extreme models O and R. M is the "preferred" model, i.e., the model generating a rupture sequence in closest agreement with kinematic results (WALD et al., 1996) in terms of duration, final distribution of slip, and synthetic seismograms. Note that a change in the value of the rate-weakening parameter v in models M and R generated extremely different rupture propagation. The effect of connectivity between high-stress patches is illustrated in the difference between models M and O. These models share the same average stress, yield stress and friction parameters, however they differ in the normalized skewness ζ of the prestress distribution, depicting the relative abundance of high stress patches and their connectivity (see definition in equation (5)). A decrease of about 13% in ζ suffices to prevent the propagation of the initial nucleation in model O.

#	Rupture style	$\overline{\sigma_0}$ bars	$\langle\sigma_0\rangle$ Bars	σ_y bars	δ m	v m s^{-1}	$\zeta(\sigma_0)$ n/a	Δu_m m	$\overline{\Delta u}$ m	M_0 N m
O	aborting nucleation	97	72–216	240	0.15	0.092	0.28	0.93	0.2	$1.33 \ 10^{17}$
M	Preferred model	97	55–183	240	0.15	0.092	0.32	2.44	1.03	$1.4 \ 10^{19}$
R	crack-like smooth slip	97	55–183	240	0.15	0.001	0.32	7.14	3.74	$7.2 \ 10^{19}$
$I2$	wide pulse	84	60–120	132	0.24	0.11	0.38	1.78	0.87	$1.4 \ 10^{19}$
L	narrow pulse slow rupture	276	158–522	527	0.53	0.527	0.32	4.22	1.72	$1.4 \ 10^{19}$

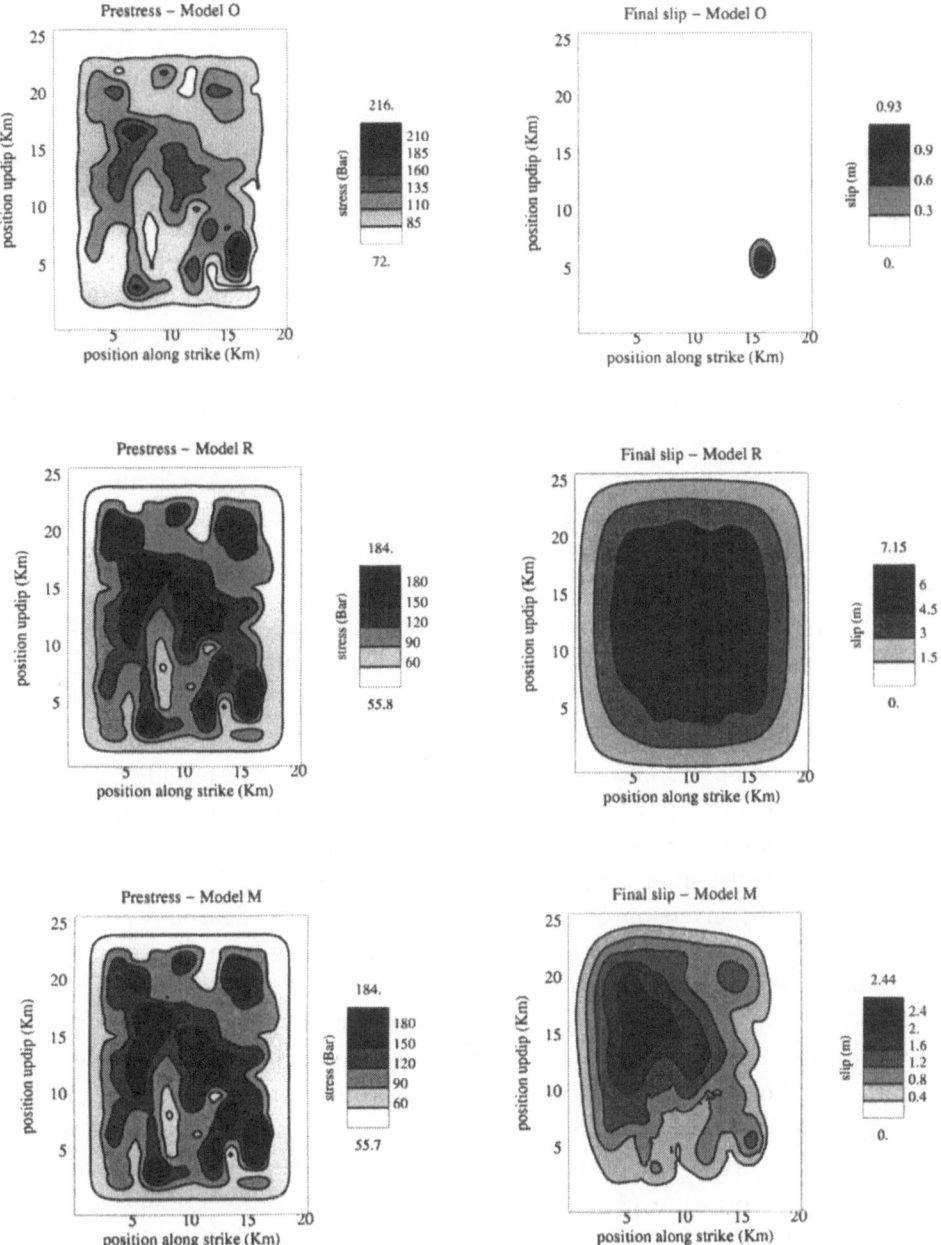

Figure 4
Illustration of the three categories of rupture propagation obtained by trial-and-error modeling of the Northridge earthquake. Model *O* fails to propagate outside the initial nucleation area, while rupture for model *R* yields an excessively smooth slip distribution. Model *M* generates rupture propagation and a final slip in closer agreement with kinematic inversion results. See Table 2 and the text for further details.

high, strongly connected prestress field and a large nucleation diameter promote rupture propagation. A strong connectivity of the high-stress patches is required to promote the rupture propagation of an initial nucleation to the rest of the fault. Figure 4-O shows how the rupture fails to grow beyond a small circular area, due to an insufficient amount of dynamic rupture energy generated within the isolated high-stress patch where nucleation occurred.

If the average stress is increased or the rate-weakening friction is omitted in model O, one obtains nucleations that develop into large size ruptures. However, in most cases the resulting slip distribution is extremely smooth and most characteristics of the initial heterogeneity in the prestress field are lost. These ruptures generate an ellipsoidal slip distribution typical of a crack-like rupture (see model R in Table 2 and in Fig. 4-R, where propagation is facilitated by a lower rate-weakening parameter v). For this reason, the moment corresponding to the peak slip deduced from the kinematic slip results (≈ 3 m) is unrealistically large. A velocity-weakening parameter on the order of a few centimeters per second is found to be adequate for an average prestress of about a hundred bars. We classify this type of events—resulting in a crack-like rupture—in a second category.

Most events generated by our trial-and-error simulations fall within these two extreme categories. Obviously, we seek models belonging to a third category that produces both a propagating rupture and preserves some level of inhomogeneity in the final slip distribution, in agreement with the observations and kinematic results (WALD et al., 1996). We find that this third family of events can only be generated in models that have a particular balance of parameters to be discussed. A representative model from the third category is illustrated by example M (Table 2). The final slip distribution (Fig. 4-M) and rupture history (Fig. 5) are similar to those obtained by inversion of combined teleseismic, near-field and geodetic data (WALD et al., 1996). The overall duration of rupture is also in agreement with kinematic results (about 8 seconds).

Rupture propagation within this third category meets the two criteria noted for categories one and two, namely a high level of connectivity and the presence of a minimum level of rate weakening (as discussed further, between 0.07 and 0.5 m.s^{-1}, depending on the model). The high level of connectivity is necessary in the prestress field, so that rupture can progress through an almost uninterrupted high-stress path, or, in the presence of a localized low-stress patch, by leaping across to a nearby high-stress patch, a behavior that is also known as a tunneling effect (see snapshot at 3 s in Fig. 5 for an illustration of the leaping rupture front). The minimum amount of rate weakening should be included in the friction, to allow formation of a relatively narrow rupture pulse. We find that the rate weakening is necessary for preserving heterogeneity in the final slip distribution.

However, slip and the moment are considerably reduced when rate weakening is further increased, unless the prestress is raised to unrealistic values (see model L in

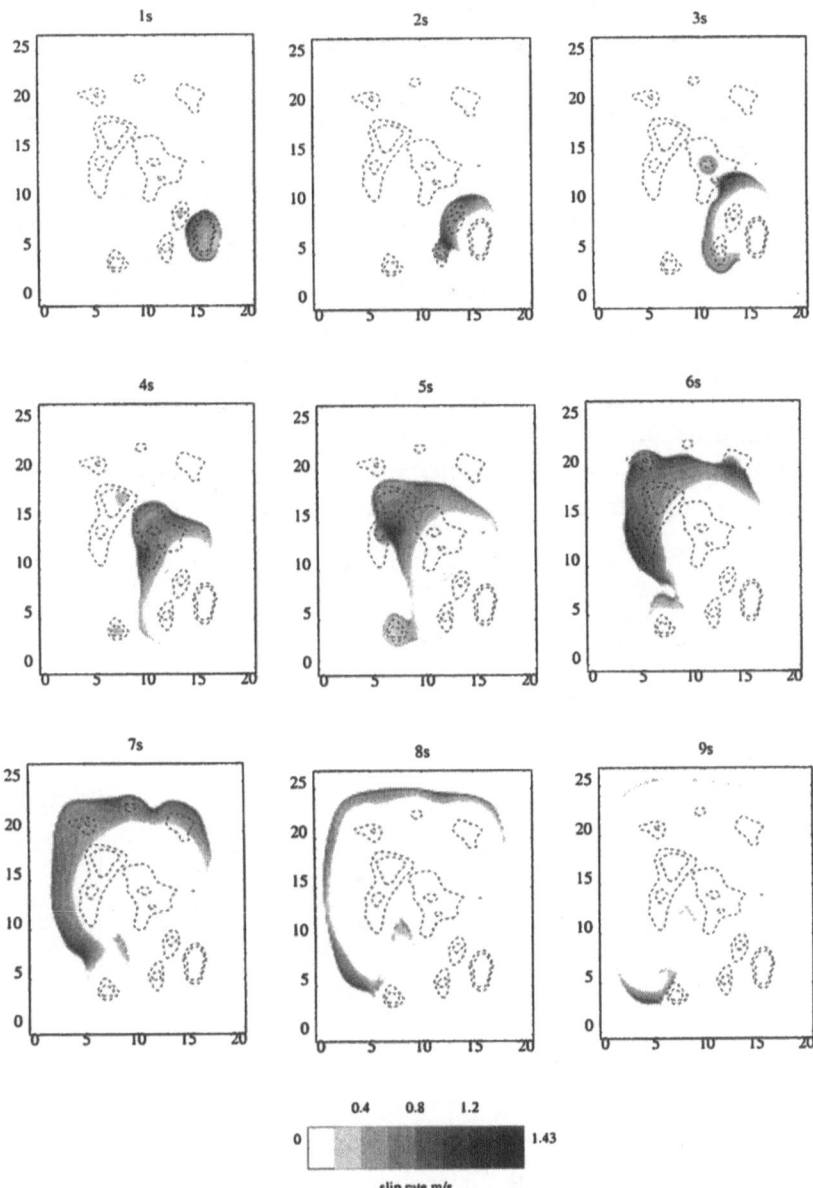

Figure 5

Snapshots of the sliprate for the dynamic simulation of the Northridge earthquake, using an initial stress distribution and friction parameters as defined in Table 2, model M. The rupture follows a path between the patches of high stress, depicted by the dashed contours (150 and 170 bars). Note that the rupture jumps into a high stress patch at 3–4 s, subsequently breaking the interconnecting low-stress regions.

Table 2). As a general rule, a higher dynamic stress drop is required to obtain the same amount to slip within a narrower rupture pulse. Besides, the rupture pulse becomes narrower when the rate weakening is increased (see Appendix for the parameters controlling the rupture pulse width) and the final slip distribution is more heterogeneous.

In synthesis, Figure 6 presents domains of parameters generating the three main families of events described above. Only eight examples are shown (among nearly thirty trials in total), because most parameters corresponding to unsatifactory models were discarded in the course of the simulations. The graph represents connectivity ζ versus normalized rate weakening H, as defined in Appendix. It is concluded that the successful models are found in a roughly delimited domain of parameters, with connectivity ζ above 0.3 and normalized rate weakening H at least above 0.1 (in the successful experiments H actually lies in the interval 0.26–0.52).

Assuming standard elastic parameters for crustal rocks such as those of Table 1, bottom layer, a prestress $\sigma_0 = 100$ bars and a yield stress $\sigma_y = 240$ bars, a value

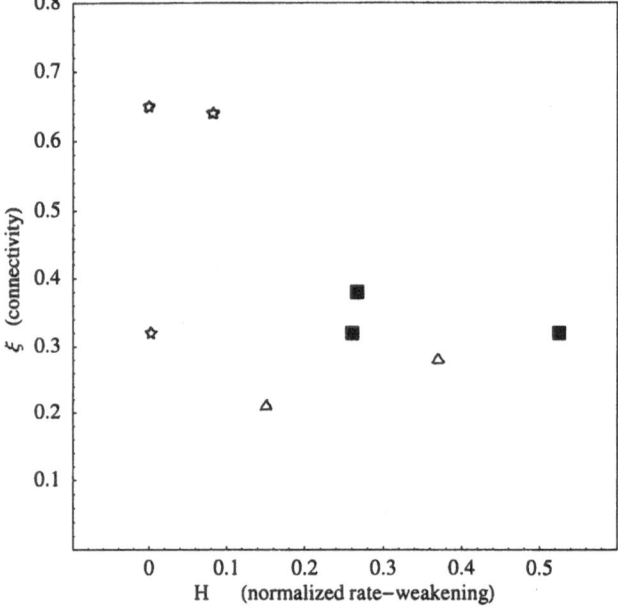

Figure 6

Three families of models and their parameters in terms of stress field connectivity ζ and normalized rate weakening H. Triangles represent models that yield aborting nucleations. Stars represent models yielding crack-like ruptures with smooth slip distributions. Finally, squares represent successful models generating pulse-like ruptures and heterogeneous slip distributions closest to those inferred for the Northridge earthquake. The domain of parameters for successful models is found at connectivities above 0.3 and renormalized rate weakening H in the interval 0.26–0.52. For rock properties and stress conditions as described in the text, such values of H correspond to rate-weakening values on the order of 0.07–0.5 m.s^{-1} in the friction law.

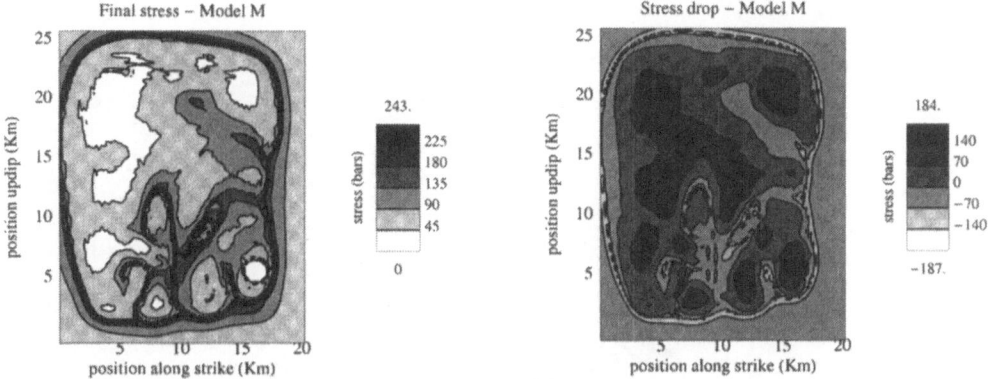

Figure 7
Final stress (left) and stress drop (right) for model *M* (see Table 2 and Fig. 5 and 4-*M*). See discussion in the text.

$H = 0.26$ yields a rate weakening on the order of $v = 0.07$ m.s^{-1}. Alternatively, a value $H = 0.5$ yields $v \approx 0.5$ m.s^{-1}.

The final stress for model *M* has large areas of almost homogeneous low stress, surrounded by narrow regions of high stress along the perimeter of the ruptured fault area (Fig. 7, left). Most of these high stress bands are located at the perimeter of the global rupture area, but some are also found within the fracture area, owing to local complex sequences of rupture arrest (see rupture sequence of Fig. 5). The stress drop (Fig. 7, right) is reminiscent of the initial stress (Fig. 4-*M*), as assumed in the procedure used to generate the initial stress field (described earlier), except for the regions of increased stress at the edges. The average stress drop within the rupture area is 35 bars.

The presence of the free surface and of the uppermost low impedance layers in the model is found to have a negligible influence on the rupture dynamics itself, because the top of the fault is at a depth of several kilometers. The negligible difference between the rupture propagation with and without the free surface is in agreement with the findings of OGLESBY *et al.* (1998) for a fault that is buried more than a couple of kilometers.

Comparison of Dynamic Radiation and Observed Seismograms

Figure 8 compares 0.1–0.5 Hz radiation from the dynamic simulation of the Northridge earthquake in Figure 5 to strong motion data recorded at three sites in the San Fernando Valley area: two alluvium sites, Newhall Firestation (NWH) and Jensen Filtration Plant (JFP), and a rock site, Santa Susanna (SSA). Owing to the finite size of the crustal block that we simulated, motion at further monitoring sites

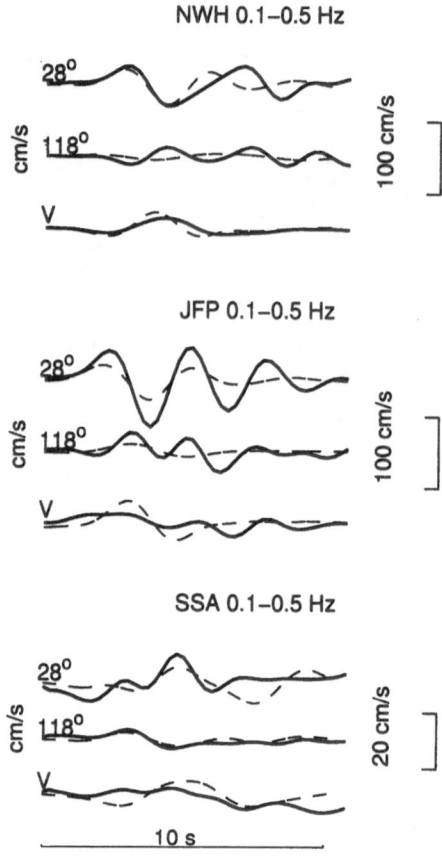

Figure 8

Comparison of 0.1–0.5 Hz radiated waves from the dynamic simulation *M* (dashed traces, with rupture history shown in Fig. 5) to data (solid traces) recorded by the three strong motion sites Newhall Firestation (NWH), Jensen Filtration Plant (JFP), and Santa Susanna (SSA) (see Fig. 1).

was not computed and could not be compared. The topmost layer (see Table 1) was included in the simulation generating the synthetics for the two alluvium sites, but omitted for SSA. The match between synthetics and data is best for NWH and SSA. At JFP, the synthetics underpredict the peak velocity by about a factor of three. It is possible that part of this discrepancy is due to an underprediction of the stress drop on the northeastern part of the fault in the simulation. However, JFP is located immediately above low-impedance sediments, extending to a depth of several kilometers (MAGISTRALE et al., 1998). Kinematic finite-fault simulations of the Northridge earthquake, which cannot be shown here, indicates that the basin structure amplifies the long-period waves at JFP by more than 50%.

Discussion and Conclusions

Our 3-D simulations of dynamic rupture propagation for the Northridge earthquake show that only a critical combination of initial conditions and friction parameters produce a seismic moment as well as a final slip distribution in general agreement with kinematic slip inversion results (WALD *et al.*, 1996). The rupture process is strongly controlled by the stress level and connectivity of high-stress patches on the fault. Unless a strong connectivity of the stress distribution is present, the rupture propagation from the initial asperity nucleation to the remaining part of the fault is inhibited. This is in agreement with dynamic rupture modeling of the 1992 Landers earthquake (OLSEN *et al.*, 1997), which also used a realistic, heterogeneous initial stress distribution derived from kinematic slip inversion results.

Unless a small amount of rate weakening (a few cm/s) is included in the dynamic simulations of Northridge, the level of inhomogeneity in the final slip is significantly smaller than that obtained in the kinematic inversion results. However, when the amount of rate weakening is increased, the overall moment drops dramatically unless the average prestress is raised to unrealistic levels. The necessity of including rate weakening to preserve heterogeneity in the Northridge rupture is different from the results of OLSEN *et al.* (1997), who were able to generate a complex dislocation distribution on the Landers fault surface with simple slip-weakening friction. However, this difference may be due to the different aspect ratios of the causative faults for the Northridge and Landers earthquakes. For example, NIELSEN *et al.* (2000) showed that long narrow faults can maintain a level of complexity and heterogeneity in the stress and slip fields through rupture, even for cases with negligible amounts of rate weakening. Indeed, the finite width of the fault promotes the formation of a pulse by confining the rupture laterally, preventing the development of a crack-like rupture. Thus, it appears that the generation of rupture heterogeneity on a fault with large aspect ratio (i.e., 0.9 for Northridge) requires a certain level of rate-weakening friction, while that is not the case for rupture propagation on faults with smaller aspect ratios (i.e., 0.2 for Landers). The presence of the free surface and of the uppermost low impedance layers in the model are found to have negligible influence on the rupture dynamics itself, in agreement with the dynamic results by OGLESBY *et al.* (1998).

The radiated waves for the preferred dynamic simulation correlate reasonably well with data recorded at three strong motion sites in the San Fernando Valley area. The match between synthetics and data is best at the alluvium site Newhall Firestation (NWH) and reasonable for rock site Santa Susanna (SSA). At the soft alluvium site Jensen Filtration Plant (JFP), the synthetics underpredict the peak velocity by about a factor of three. This discrepancy may be partly due to an insufficient stress drop on the northeastern part of the fault in the simulation, partly to the lack of near-surface layers of lower impedance and 3-D amplification effects in the model.

Appendix: Parameters Controlling Rupture Propagation

Previous studies have discussed the conditions under which a crack will propagate, based on the balance of the fracture energy necessary for the creation of new crack surface, versus the energy provided by the elastic stress drop during the crack propagation. For example, based on rupture energy considerations, ANDREWS (1976) identified the critical diameter for the propagation of a Mode II crack as:

$$L_c = \frac{\mu}{\pi} \frac{\sigma_y - \sigma_f}{(\sigma_0 - \sigma_f)^2} \delta \tag{3}$$

where σ_y is the yield stress, σ_f is the dynamic sliding friction, σ_0 is the prestress, δ is the slip-weakening distance and μ the shear stiffness. Recently, MADARIAGA and OLSEN (this issue) proposed use of a similar criterion to describe the propagation of rupture along a narrow fault, by replacing L_c by the fault width in expression (3).

In our models of the Northridge earthquake we are furthermore dealing with rate-weakening friction and length scales smaller than the fault width. The rate weakening allows premature healing of the rupture and promotes the abortion of the nucleation in a way that is not quantified by (3). For a homogeneous prestress, a parameter describing the ability of self-healing or rupture pulse formation was identified by NIELSEN et al. (2000) and by NIELSEN and CARLSON (1999),

$$H = \frac{\mu v}{2\beta(2\sigma_y - \sigma_0 - 2\sqrt{\sigma_y(\sigma_y - \sigma_0)})}, \tag{4}$$

where lower values of H promote self-healing (a reference value $\sigma_f = 0$ is assumed or can be otherwise subtracted from each stress variable above). As pointed out by NIELSEN and CARLSON (1999) self-healing also facilitates the arrest of rupture propagation, so that low values of H will hinder the propagation of the initial nucleation. On the other hand, for a fixed nucleation geometry (i.e., circular or quasi-circular) the criterion (3) for propagation in the absence of rate weakening can be rewritten with L_c as the nucleation diameter.

Since the initial stress field for our Northridge models is heterogeneous, no equivalent value for σ_0 can be easily identified. Here, we propose to replace σ_0 by the mean stress, and to introduce a descriptive parameter for the distribution of prestress. By introducing the normalized skewness ζ as

$$\zeta(\sigma_0) = \frac{1}{2}\left(1 - \frac{\xi(\sigma_0)}{1 + |\xi(\sigma_0)|}\right) \tag{5}$$

we obtain a parameter in the interval [0,], where ξ is the skewness, calculated by dividing the third central moment by the cube of the standard deviation. Like ξ, ζ describes the amount of asymmetry in the distribution. Larger values of ζ indicate a distribution with relatively high stress and a long tail in the low values, i.e., a fault surface with a majority of high-stress patches and a few very low-stress regions. On the contrary, low values of ζ indicate a minority of high-stress patches in the distribution. Moreover, we arbitrarily associate the normalized skewness $\zeta(\sigma_0)$ to the "connectivity" of high-stress patches. Indeed, in our specific case, ζ is also an indicator of how close and interconnected the high-stress patches are on the fault. We do not claim any generality of such measure of the connectivity, as it is only meaningful for comparing the modifications of a single original distribution. Given the relevance of both the statistical and the geometrical distribution of stress on the fault, more appropriate and specific measures of these features may need to be defined in the future, as more quantitative and systematic investigations are conducted under inhomogeneous stress conditions.

In summary, the parameters affecting rupture for the Northridge simulations are the yield stress σ_y, the dynamic friction σ_f (usually with a 0 reference level), the prestress distribution (σ_0 for the homogeneous case), shear stiffness μ and density ρ, a characteristic slip-weakening distance δ, a characteristic rate weakening v, and the shape of the initial nucleation. Large rate-weakening and slip-weakening friction as well as high yield stress hinder the propagation of an initial nucleation. On the contrary a high, strongly connected prestress field and a large nucleation diameter will promote rupture propagation (for a fixed shear stiffness and shear-wave velocity β). This can be quantified by the three dimensionless parameters in equations (3), (4) and (5).

Acknowledgements

The computations in this study were carried out on the SGI Origin 2000 at University of California, Santa Barbara (NSF grant CDA96-01954). This is Institute for Crustal Studies contribution ICS # 0357-100EQ.

References

ANDREWS, J. (1976), *Rupture Velocity of Plane Strain Shear Cracks*, J. Geophys. Res. *81*, 5679–5687.

ARCHULETA, R. (1984), *A Faulting Model for the 1979 Imperial Valley Earthquake*, J. Geophys. Res. *89*, 4559–4585.

COCHARD, A., and MADARIAGA, R. (1994), *Dynamic Faulting Under Rate-dependent Friction*, Pure appl. geophys. *142*, 419–445.

COCHARD, A., and RICE, J. (1997), *A Spectral Method for Numerical Elastodynamic Fracture Analysis without Spatial Replication of the Rupture Event*, J. Mech. Phys. Solids *45*, 1393–1418.

COTTON, F., and CAMPILLO, M. (1995a), *Frequency Domain Inversion of Strong Motions: Application to the 1992 Landers Earthquake*, J. Geophys. Res. *100*, 3961–3975.

COTTON, F., and CAMPILLO, M. (1995b), *Stability of the Rake during the 1992 Landers Earthquake; An Indication for a Small Stress Release?* Geophys. Res. Lett. *22*, 1921–1924.

FUKUYAMA, E., and MADARIAGA, R. (1998), *Rupture Dynamics of a Planar, Fault in a 3-D Elastic Medium: Rate- and Slip-weakening Friction*, Bull. Seismol. Soc. Am. *88*, 1–17.

HARTZELL, S. H., and HEATON, T. H. (1983), *Inversion of Strong Ground Motion and Teleseismic Waveform Data for the Fault Rupture History of the 1979 Imperial Valley, California, Earthquake*, Bull. Seismol. Soc. Am. *73*, 1553–1583.

IDE, S., and TAKEO, M. (1997), *Determination of Constitutive Relations of Fault Slip Based on Seismic Wave Analysis*, J. Geophys. Res. *102*, 27,379–27,391.

IDE, S., TAKEO, M., and YOSHIDA, Y. (1996), *Source Process of the 1995 Kobe Earthquake: Determination of Spatio-temporal Slip Distribution by Bayesian Modeling*, Bull. Seismol. Soc. Am. *86*, 547–566.

MADARIAGA, R., OLSEN, K. B., and ARCHULETA, R. (1998), *Modeling Dynamic Rupture in a 3-D Earthquake Model*, Bull. Seismol. Soc. Am. *88*, 1182–1197.

MADARIAGA, R., and OLSEN, K. B. (2000), *Criticality of Rupture Dynamics in 3-D*, Pure appl. geophys. *157*, 1981–2001.

MAGISTRALE, H., GRAVES, R. W., and Clayton, R. (1998), *A standard Three-dimensional Seismic Velocity Model for Southern California: Version 1*, EOS Trans. AGU *79*, F605.

NIELSEN, S. B., CARLSON, J. M. and OLSEN, K. B. (2000), *Friction, Fault Geometry, and Earthquake Rupture*, J. Geophys. Res. *105*, B3, 6069–6088.

NIELSEN, S. B., and OLSEN, K. B. (2000), *Mixed Boundary Condition for Rupture Dynamics*, in preparation.

NIELSEN, S. B., and CARLSON, J. M. (1999), *Rupture Pulse Characterization: Convergence to a Self-similar Steady-state*, accepted for publication in Bull. Seismol. Soc. Am., 2000.

OGLESBY, D. D., ARCHULETA, R. J., and NIELSEN, S. B. (1998), *Earthquakes on Dipping Faults: The Effects of Broken Symmetry*, Science *280*, 1055–1059.

OLSEN, K. B., PECHMANN, J. C., and SCHUSTER, G. T. (1996), *An Analysis of Simulated and Observed Blast Records in the Salt Lake Basin*, Bull. Seismol. Soc. Am. *86*, 1061–1076.

OLSEN, K. B., MADARIAGA, R., and ARCHULETA, R. (1997), *Three-dimensional Dynamic Simulation of the 1992 Landers Earthquake*, Science *278*, 834–838.

SCHOLZ, C., *The Mechanics of Earthquake Faulting* (Cambridge University Press 1990) pp. 166–177.

SPUDICH, P., GUATTERI, M., KENSHIRO, O., and JUN, M. (1998), *Use of Fault Striations and Dislocation Models to Infer Tectonic Shear Stress during the 1995 Hyogo-ken Nandu (Kobe) Earthquake*, Bull. Seismol. Soc. Am. *88*, 413–427.

WALD, D., and HEATON, T. (1994), *Spatial and Temporal Distribution of Slip for the 1992 Landers, California Earthquake*, Bull. Seismol. Soc. Am. *84*, 668–691.

WALD, D., HEATON, T., and HUDNUT, K. W. (1996), *The Slip History of the 1994 Northridge, California, Earthquake Determined from Strong Motion, Teleseismic, GPS, and Levelling Data*, Bull. Seismol. Soc. Am. *86*, S49–S70.

YOSHIDA, S., KOKETSU, K., SHIBAZAKI, B., SAGIYA, T., KATO, T., and YOSHIDA, Y. (1996), *Joint Inversions of Near- and Far-field Waveforms and Geodetic Data for the Rupture Process of the 1995 Kobe Earthquake*, J. Phys. Earth *44*, 437–454.

(Received October 22, 1999, revised March 20, 2000, accepted April 5, 2000)

To access this journal online:
http://www.birkhauser.ch

Pure appl. geophys. 157 (2000) 2047–2062
0033–4553/00/122047–16 $ 1.50 + 0.20/0

▌Pure and Applied Geophysics

Parallel 3-D Simulation of Ground Motion for the 1995 Kobe Earthquake: The Component Decomposition Approach

TAKASHI FURUMURA[1,*] and KAZUKI KOKETSU[2]

Abstract—A new approach to parallel pseudospectral simulation of 3-D seismic wave propagation is developed based on component decomposition of the wavefield. Field quantities and equations of motion are distributed over three processors according to their relationship to the x-, y- and z-coordinates, and computation is carried out concurrently on each processor with inter-processor communications. The efficiency of this approach is evaluated by a theoretical estimate and actual benchmark computations. We then conduct a 3-D simulation of strong ground motion for the 1995 Kobe (Hyogo-ken Nanbu) earthquake in order to show the feasibility of the parallel pseudospectral method. The results of the simulation demonstrate that the complex 3-D structures of the subsurface medium and source fault greatly affect the strong ground motion on the surface.

Key words: 1995 Kobe earthquake, parallel computing, pseudospectral method, strong ground motion, 3-D simulation.

Introduction

For large-scale 3-D modeling of seismic wave propagation, RESHEF *et al.* (1998) first attempted parallel Fourier pseudospectral computation using a CRAY X-MP supercomputer, and he has been followed by SATO *et al.* (1995), FURUMURA *et al.* (1998) and HUNG and FORSYTH (1998). In these studies the whole 3-D space is divided into subdomains, and the field quantities in each subdomain are assigned to each processor for concurrent computing. Instead of this domain partition approach, we can alternatively separate the x- y- and z-components of the field quantities and assign them to three processors separately.

We will present here a new parallel scheme of the pseudospectral method based on the component decomposition, with a brief overview of the 3-D wavefield calculation by the pseudospectral method. The efficiency of this parallel scheme will

[1] Hokkaido University of Education, Midorigaoka 2-34-1, Iwamizawa, 068-8642, Japan. E-mail: furumura@iwa.hokkyodai.ac.jp

[2] Earthquake Research Institute, University of Tokyo, Yayoi 1-1-1, Bunkyo-ku, 113-0032, Japan. E-mail: inquiry@taro.eri.u-tokyo.ac.jp

* Present address: Earthquake Research Institute, University of Tokyo, Yayoi 1-1-1, Bunkyo-ku, 113-0032, Japan. E-mail: furumura@eri.u-tokyo.ac.jp

be predicted by theoretical experiments, and complemented by actual computations. We then conduct a large-scale parallel 3-D simulation of strong ground motion for the 1995 Kobe (Hyogo-ken Nanbu) earthquake to demonstrate the feasibility of the component decomposition approach.

Pseudospectral Simulation of 3-D Wave Propagation

In a 3-D Cartesian coordinate system with the z-axis taken position downward, the equations of motion are represented as

$$\rho \ddot{u}_p = \frac{\partial \sigma_{xp}}{\partial x} + \frac{\partial \sigma_{yp}}{\partial y} + \frac{\partial \sigma_{zp}}{\partial z} + f_p, \quad (p = x, y, z), \tag{1}$$

where σ_{pq}, f_p and ρ stand for a stress, body force and density, respectively. $\ddot{u}_p$ is a particle acceleration, namely the second-order time derivative of a displacement. The stresses in an isotropic medium are given by

$$\sigma_{pq} = \lambda(e_{xx} + e_{yy} + e_{zz})\delta_{pq} + 2\mu e_{pq}, \quad (p, q = x, y, z), \tag{2}$$

with the Lamé's constants λ, μ and the strains defined by

$$e_{pq} = \frac{1}{2}\left(\frac{\partial u_p}{\partial q} + \frac{\partial u_q}{\partial p}\right), \quad (p, q = x, y, z). \tag{3}$$

Equation (1) is numerically integrated with the time step Δt as

$$\dot{u}_p^{n+1/2} = \dot{u}_p^{n-1/2} + \frac{1}{\rho}\left(\frac{\partial \sigma_{px}^n}{\partial x} + \frac{\partial \sigma_{py}^n}{\partial y} + \frac{\partial \sigma_{pz}^n}{\partial z} + f_p^n\right)\Delta t, \quad (p = x, y, z). \tag{4}$$

$\dot{u}_p^{n \pm 1/2}$ is a particle velocity at the time of $t = (n \pm \frac{1}{2})\Delta t$. The spatial derivatives in Equation (4) are calculated analytically in the wavenumber domain by use of the fast Fourier transform (FFT) (see e.g., FURUMURA et al., 1998). Other spectral expansions based on the Chebyshev polynomials (e.g., KOSLOFF et al., 1990) or the Legendre polynomials (e.g., FACCIOLI et al., 1997; KOMATITSCH and VILOTTE, 1998) can be also used, that are superior in convergence properties and the handling of boundary conditions for highly complex underground structures.

The differentiation of Equations (2) and (3) with respect to time yields

$$\sigma_{pq}^{n+1} = \sigma_{pq}^n + \left[\lambda\left(\frac{\partial \dot{u}_x^{n+1/2}}{\partial x} + \frac{\partial \dot{u}_y^{n+1/2}}{\partial y} + \frac{\partial \dot{u}_z^{n+1/2}}{\partial z}\right)\delta_{pq} + \mu\left(\frac{\partial \dot{u}_p^{n+1/2}}{\partial q} + \frac{\partial \dot{u}_q^{n+1/2}}{\partial p}\right)\right]\Delta t,$$
$$(p, q = x, y, z). \tag{5}$$

The spatial derivatives in Equation (5) are again calculated in the wavenumber domain. The stresses at the next time step $t = (n + 1)\Delta t$ are then calculated by Equation (5) and the particle velocities resulting from Equation (4).

Component Decomposition

For parallel pseudospectral simulation of 3-D seismic wavefields, the domain partition approach, which divides the whole model into 2^N subregions and assigns them to processors, has usually been used (e.g., SATO et al., 1995; FURUMURA et al., 1998; HUNG and FORSYTH, 1998). However, if multiples of three processors are available, it is straightforward to partition the 3-D wavefield into three parts according to the closeness to the x-, y- and z-coordinates.

We first separate the nine equations of 3-D in Equations (4) and (5) into three groups. The x-component group consists of

$$\dot{u}_x^{n+1/2} = \dot{u}_x^{n-1/2} + \frac{1}{\rho}\left(\frac{\partial \sigma_{xx}^n}{\partial x} + \frac{\partial \sigma_{xy}^n}{\partial y} + \frac{\partial \sigma_{xz}^n}{\partial z}\right)\Delta t, \tag{6}$$

$$\sigma_{xx}^{n+1} = \sigma_{xx}^n + \left[(\lambda+2\mu)\frac{\partial \dot{u}_x^{n+1/2}}{\partial x} + \lambda\left(\frac{\partial \dot{u}_y^{n+1/2}}{\partial y} + \frac{\partial \dot{u}_z^{n+1/2}}{\partial z}\right)\right]\Delta t, \tag{7}$$

$$\sigma_{xy}^{n+1} = \sigma_{xy}^n + \mu\left(\frac{\partial \dot{u}_x^{n+1/2}}{\partial y} + \frac{\partial \dot{u}_y^{n+1/2}}{\partial x}\right)\Delta t, \tag{8}$$

and the y-, z-component groups are

$$\dot{u}_y^{n+1/2} = \dot{u}_y^{n-1/2} + \frac{1}{\rho}\left(\frac{\partial \sigma_{xy}^n}{\partial x} + \frac{\partial \sigma_{yy}^n}{\partial y} + \frac{\partial \sigma_{yz}^n}{\partial z}\right)\Delta t, \tag{9}$$

$$\sigma_{yy}^{n+1} = \sigma_{yy}^n + \left[(\lambda+2\mu)\frac{\partial \dot{u}_y^{n+1/2}}{\partial y} + \lambda\left(\frac{\partial \dot{u}_x^{n+1/2}}{\partial x} + \frac{\partial \dot{u}_z^{n+1/2}}{\partial z}\right)\right]\Delta t, \tag{10}$$

$$\sigma_{yz}^{n+1} = \sigma_{yz}^n + \mu\left(\frac{\partial \dot{u}_y^{n+1/2}}{\partial z} + \frac{\partial \dot{u}_z^{n+1/2}}{\partial y}\right)\Delta t, \tag{11}$$

and

$$\dot{u}_z^{n+1/2} = \dot{u}_z^{n-1/2} + \frac{1}{\rho}\left(\frac{\partial \sigma_{xz}^n}{\partial x} + \frac{\partial \sigma_{yz}^n}{\partial y} + \frac{\partial \sigma_{zz}^n}{\partial z}\right)\Delta t, \tag{12}$$

$$\sigma_{zz}^{n+1} = \sigma_{zz}^n + \left[(\lambda+2\mu)\frac{\partial \dot{u}_z^{n+1/2}}{\partial z} + \lambda\left(\frac{\partial \dot{u}_x^{n+1/2}}{\partial x} + \frac{\partial \dot{u}_y^{n+1/2}}{\partial y}\right)\right]\Delta t. \tag{13}$$

$$\sigma_{xz}^{n+1} = \sigma_{xz}^n + \mu\left(\frac{\partial \dot{u}_x^{n+1/2}}{\partial z} + \frac{\partial \dot{u}_z^{n+1/2}}{\partial x}\right)\Delta t, \tag{14}$$

respectively.

If three processors called CPU-x, CPU-y, and CPU-z are available, these equation groups are assigned to them separately. The field quantities are also distributed over the three processors as $\dot{u}_x$, $\partial \dot{u}_x/\partial x$, σ_{xx}, and σ_{xy} are assigned to CPU-x; $\dot{u}_y$, $\partial \dot{u}_y/\partial y$, σ_{yy}, and σ_{yz} to CPU-y, and $\dot{u}_z$, $\partial \dot{u}_z/\partial y$, σ_{zz}, and σ_{xz} to CPU-z. Each processor is responsible for the calculation of the corresponding equations, as

well as performing interprocessor communications to obtain necessary quantities on
the other processors. The material parameters λ, μ, ρ, and Q are commonly stored
in the local memories of the three processor in a compressed form. They are
arranged to be fit for a 3-D array of one-byte data, and retrieved directly through
an index table.

Figure 1 illustrates how parallel computing is performed on a cluster of three
processors by the component decomposition approach mentioned above. In the first
stage of CPU-x, the derivatives $\partial \sigma_{xx}^n / \partial x$, and $\partial \sigma_{xy}^n / \partial y$ are calculated directly from
quantities stored in the local memory. Message passing is then carried out to obtain
σ_{xz}^n from CPU-z for computing $\partial \sigma_{xz}^n / \partial z$. The particle velocity $\dot{u}_x$ is then updated
from $\dot{u}_x^{n-1/2}$ to $\dot{u}_x^{n+1/2}$ following Equation (6). In the second stage, CPU-x first
calculates $\partial \dot{u}_x^{n+1/2} / \partial x$ using $\dot{u}_x^{n+1/2}$ in local memory. The processor then obtains
$\partial \dot{u}_y^{n+1/2} / \partial y$ and $\partial \dot{u}_z^{n+1/2} / \partial z$ which have been calculated by CPU-y and CPU-z
simultaneously, and updates the stress σ_{xx} from σ_{xx}^n to σ_{xx}^{n+1} following Equation
(7). CPU-x finally receives $\dot{u}_y^{n+1/2}$ from CPU-y by message passing, and updates the
stress σ_{xy} from σ_{xy}^n to σ_{xy}^{n+1} following Equation (8). Since we implement the above
procedure by introducing standard Message Passing Interface (MPI; GROPP et al.,
1995), our code can run on a variety of platforms with few modifications.

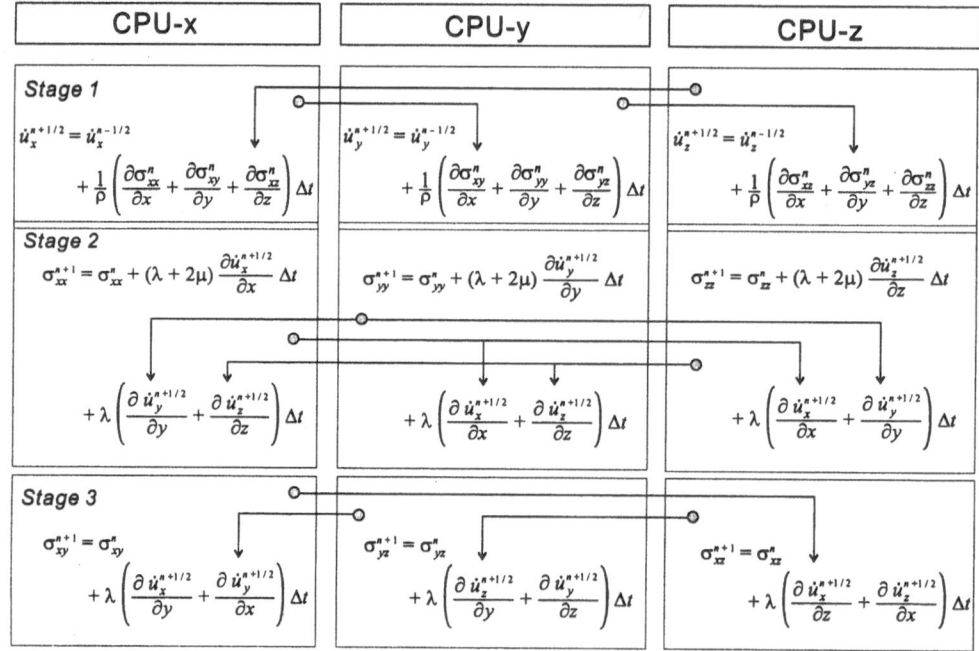

Figure 1
Diagram of the component decomposition for the parallel pseudospectral computing by the processors
CPU-x, CPU-y, CPU-z. Each processor concurrently updates its corresponding particle velocity and
stresses, and inter-processor communications indicated by arrows are required at each stage of the
computation due to the coupling of the equations of motion over components. MPI is used for the
inter-processor communications.

Efficiency of the Parallel Computation

If three processors are connected by an ideally fast network, the CPU time at each processor should be reduced to 1/3, and the speed-up rate by the parallel computation should approach 3. However, actual computer networks perform the communication only at a limited speed, compared with the computation speed of processors. We assume that the computation is carried out entirely in single-precision arithmetic for a 3-D model with $N_x \times N_y \times N_z$ grid points, and thus all the field quantities and their spatial derivatives occupy $N_x \times N_y \times N_z \times 4$ bytes. We further assume that the communication between each processor pair has the same communication speed V_c bytes/s, and no collisions occur during communications. The component decomposition approach requires twelve transmissions of quantities as shown in Figure 1. Therefore, the total communication time at each processor is given by

$$T_c = \frac{N_x \times N_y \times N_z \times 4}{V_c} \times 4, \tag{15}$$

if three transmissions can be performed concurrently by the three processors.

We define here the computation speed V_d as

$$V_d = \frac{N_x \times N_y \times N_z \times 4}{T_d}, \tag{16}$$

using the total CPU time T_d for the computation on a single processor. For a three-processor parallel system, the CPU time at each processor must be $T_d/3$, and thus we can roughly estimate the speed-up rate by

$$R = \frac{T_d}{T_d/3 + T_c} = \frac{3}{1 + 12V_d/V_c}. \tag{17}$$

We now implement the conventional pseudospectral code on a workstation with a DEC Alpha 21164A processor of 500 MHz and 1 Gbyte memory, and measure V_d for various model sizes. We then connected three such workstations using 100BaseT Ethernet, and measured V_c through a simple communication performance test of the MPI library. The results of these experiments showed that a cluster of workstations runs at around $V_d = 0.18$ Mbyte/s and $V_c = 6.2$ Mbyte/s. By inserting these values into Equation (17), we expect a peak speed-up rate of $R = 2.2$ for parallel computing using the cluster of three workstations. We note that R will approach to $2.7 \sim 2.8$ if we employ a much faster computer network such as 1000BaseT Ethernet which is at least three or four times faster than the 100BaseT system.

To compliment the theoretical estimate we next implement the component decomposition pseudospectral code on the three-workstation cluster, and measure

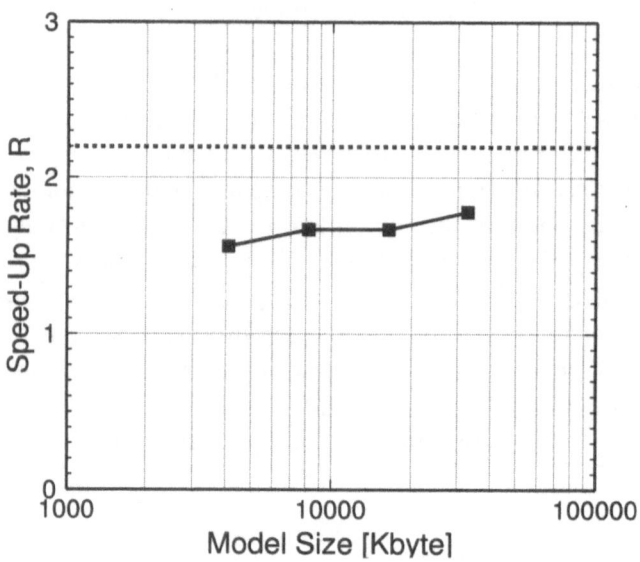

Figure 2
Experimental efficiency of the parallel 3-D pseudospectral computing using a cluster of three workstations with various model sizes. The dashed line indicates the theoretical peak performance rate of
$R = 2.2$.

the actual speed-up rate relative to the single processor computation time for
various model sizes. The result in Figure 2 indicates that the rate gradually
increases from $R = 1.5$ to 1.8 as the model size increases. Although the rates are
slightly lower than the theoretical peak rate, probably because of the overhead of
MPI with small data size transmissions (GROPP et al., 1995), we expect good
performance rates for large-scale simulations.

The 1995 Kobe Earthquake

The 1995 Kobe earthquake (M_w 6.9) is the most damaging in recent Japanese
history after the 1923 Kanto earthquake (M_s 8.2), and its notable feature is that
most of the damage and over 6000 casualties occurred in the narrow zone called
"damage belt." This 25 km-long belt extends through the City of Kobe with a
width of about 2 km (gray zone in Fig. 3), and a source fault was assumed just
beneath it at first (SHIMAMOTO, 1995), However, no fault with evidence of recent
rupture has been found there by any geophysical survey, but the distribution of
aftershocks and static displacements indicates the active strike-slip faults northwest
of the belt (thick lines in Fig. 3) to be the source fault system of the earthquake
(e.g., YOSHIDA et al., 1996; SEKIGUCHI et al., 1996).

Usually a fault-rupture propagation generates strong ground motions along the direction of propagation, and this directivity effect should be significant in fault-normal ground motions above the source fault system and beyond its leading end (KOKETSU, 1996; INOUE and MIYATAKE, 1997). The strong motion records in Figure 3 show the effect clearly. They mainly consist of two large, long-period (1 ~ 2 s) pulses, since two asperities of large slip occurred in the source fault system as recovered by YOSHIDA et al. (1996) (Fig. 4). However, the strong ground motion causing the severe damage is migrated from the source fault traces into the center of Kobe city as mentioned above, and consequently this migration should be due to the complex subsurface structure in Kobe. Simulations of ground motion have already been carried out by FURUMURA and KOKETSU (1998), PITARKA et al. (1998) and others (see chap. 5 IRIKURA et al., 1999) for reproducing the migration.

Kobe forms the western half of the Osaka basin together with the Osaka bay. The basin is bounded on the north by the Rokko mountains, where the granite bedrock is exposed, and this boundary corresponds to the source fault system. The

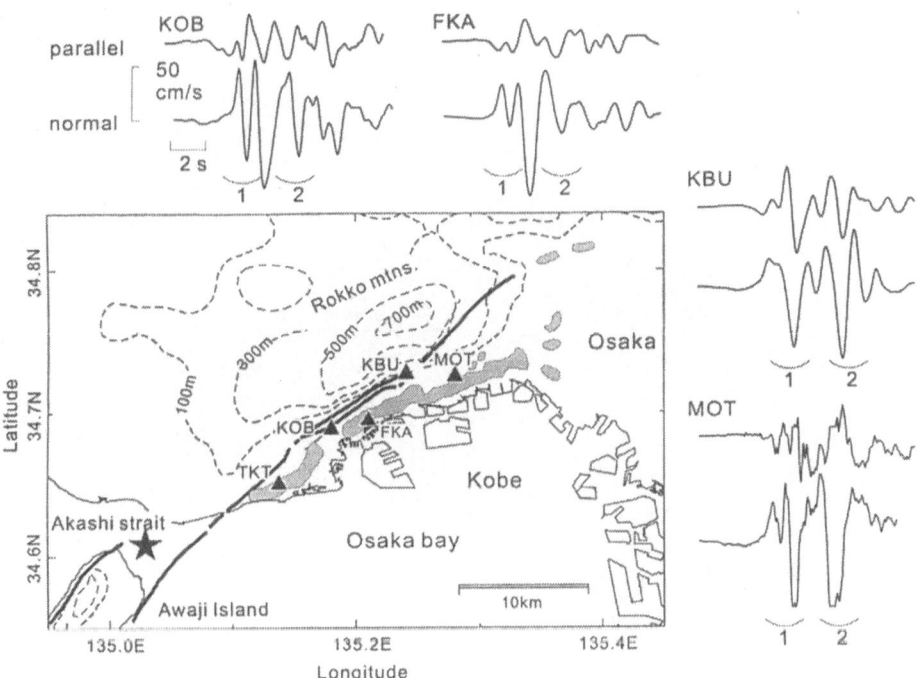

Figure 3

Index map of the 1995 Kobe (Hyogo-ken Nanbu) earthquake. The star symbol, gray zone and thick lines indicate the epicenter, damage belt and source faults, respectively. Dashed lines denote ground elevation. The fault rupture was initiated beneath the Akashi strait and propagated bilaterally towards Kobe and the Awaji island along known active fault traces. The seismograms on the top and right sides show the fault-parallel and fault-normal ground velocities recorded at the stations near the fault (note that the seismograms of MOT are clipped). The two large long-period pulses are numbered 1 and 2.

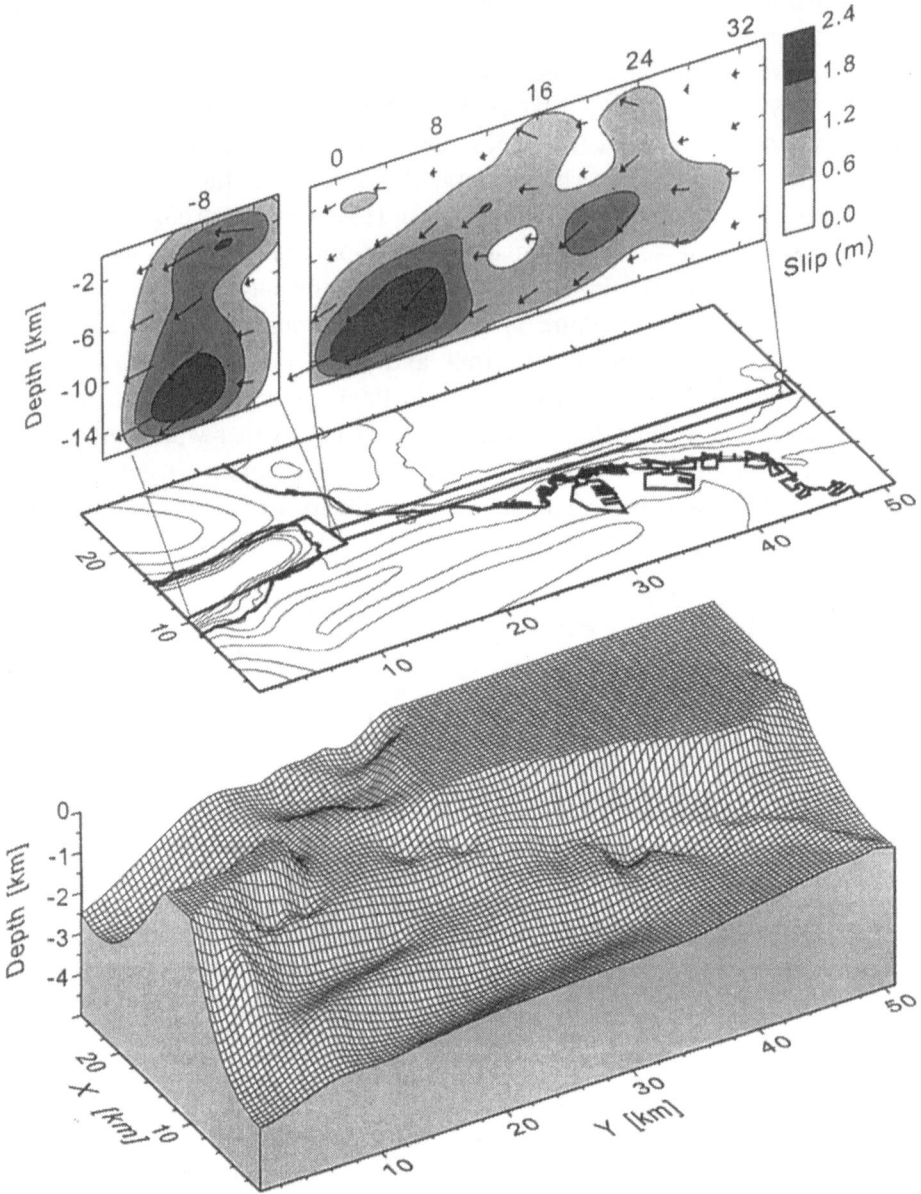

Figure 4
The topography of sediment/bedrock interface by NAKAGAWA et al. (1996) is shown by tthe contour drawing (middle) and in a bird's-eye view (bottom). The surface projections of the fault planes are also shown in the middle diagram. The top diagram represents the slip distribution recovered by YOSHIDA et al. (1996).

Table 1

Velocity model by KAGAWA et al. (1993)

	V_p (km/s)	V_s (km/s)	ρ (t/m^3)	Q
Sediment 1	1.6	0.35	1.7	60
Sediment 2	1.8	0.55	1.8	100
Sediment 3	2.5	1.0	2.1	150
Bedrock 1	5.4	3.2	2.7	300
Bedrock 2	6.0	3.46	2.8	500

thickness of the sediments rapidly changes from the mountains to the shoreline of the Osaka bay. A gentle slope then extends to the center of the bay, where the thickness is over 2500 m (Fig. 4; NAKAGAWA et al., 1996). KAGAWA et al. (1993) derived very low S-wave velocities 0.35 ~ 1.0 km/s for the sediments from refraction experiments (Table 1). The contrast between them and the velocity in the bedrock (3.2 km/s) should result in high amplification of seismic motions in the sediments. However, the previous simulations of the ground motion did not include the near-surface low-velocity layer of $V_s = 0.35$ km/s due to limitations of computer memory. We therefore carry out the simulation once again with this low-velocity layer by using the parallel pseudospectral method mentioned in the preceding sections.

3-D Ground Motion Simulation in Kobe

We employ a 51.2 × 25.6 × 40 km model, which is discretized by a uniform grid spacing of 0.1 km in the horizontal directions. Variable grid spacing is applied in the vertical direction with intervals of 0.1 to 0.4 km by using the mapping technique of FORNBERG (1998). The area covered by this 3-D model is almost the same as that of our previous study (FURUMURA and KOKETSU, 1998), however the finer spacing results in 16.77 million grid points and the pseudospectral simulation with them requires 1.8 Gbyte memory, which is four times larger than that of the previous study. Even though we introduce the shallowest layer with $V_s = 0.35$ km/s, which was omitted in the previous study, the fine spacing allows us to simulate frequency components up to 1.75 Hz with at least 3.5 grid points per shortest S-wave wavelength.

In order to suppress artificial reflections and wraparound phases, the absorbing buffer zones of CERJAN et al. (1985) are applied along the outer limits with a width of twenty grid points. The boundary condition at the free surface is incorporated simply by introducing a zone of $\lambda = \mu = 0$ over the surface ("vacuum formulation"; see e.g., GRAVES, 1996). Since the vacuum formulation often causes the Gibbs phenomenon in case of a seismic source near the free surface, we use the "symmet-

ric differentiation" technique (see FURUMURA and TAKENAKA, 1992 and FURU-
MURA et al., 1998 for details).

We adopt the 3-D topography of the sediment/bedrock interface derived by
NAKAGAWA et al. (1996) from gravity and reflection surveys, and the velocity
model of KAGAWA et al. (1993) shown in Table 1. We assume the Q factors to be
$60 \sim 50$ and $300 \sim 500$ in the sediments and bedrock, respectively. The anelastic
attenuation is incorporated in the time integration procedure of Equations (4) and
(5) by multiplying the stresses and velocities by the damping coefficients corresponding to the Q factors (GRAVES, 1996).

For the earthquake source, we adopt the fault model by YOSHIDA et al. (1996),
whose slip distribution is shown in Figure 4 with the topography of the sediment/
bedrock interface. Their model is represented by point dislocations at the centers of
4 by 4 km subfaults, and the fault slip is allowed to occur at each subfault in three
successive time windows of length 1 s. The slip rate functions are approximated by
HERRMANN's (1979) pseudo-delta function with a corner frequency of 1 Hz. To
avoid spatial aliasing, the interval between the point sources must be shorter than
the minimum wavelength of S-waves, so that we further divide each subfault into
four smaller pieces, assigning linearly interpolated slip rate functions.

The 3-D parallel simulation takes 160 hours to complete the 22.5 s (5000 time
steps) of ground motion using the cluster of three DEC Alpha workstations. Figure
5 displays the result of this simulation by snapshots of horizontal ground motion.
In the first panel ($T = 6.3$ s), the first directivity pulse with amplitude over 50 cm/s
is being generated from the asperity below western Kobe. The second pulse then
begins to emerge from the second asperity below central Kobe in the $T = 8.6$ s
panel. These two pulses are well developed and propagating to the east in the third
panel of $T = 10.4$ s. The lower-right panel of Figure 5 illustrates the distribution of
peak ground velocities of the horizontal motion during the entire duration of the
simulation. This distribution reproduces the narrow belt of ground velocity larger
than 60 cm/s. This zone extends over 20 km along the basin edge and nearly
coincides with the damage belt in Figure 3.

Interpretation of the Simulation Results

Figure 6 compares the simulated fault-normal ground velocities with the observed seismograms at the near-fault stations TKT, KOB, FKA, KBU, and MOT
(see Fig. 3). A band-pass filter has been applied to both the synthetics and
observations with a frequency range between 0.1 to 1.75 Hz except for the clipped
observation at MOT. Compared with the previous study (FURUMURA and
KOKETSU, 1998), the two directivity pulses are reproduced more clearly by this
simulation. In particular, the peak ground velocities at the basin stations (TKT,
FKA) are considerably improved by the introduction of the 0.35 km/s layer.

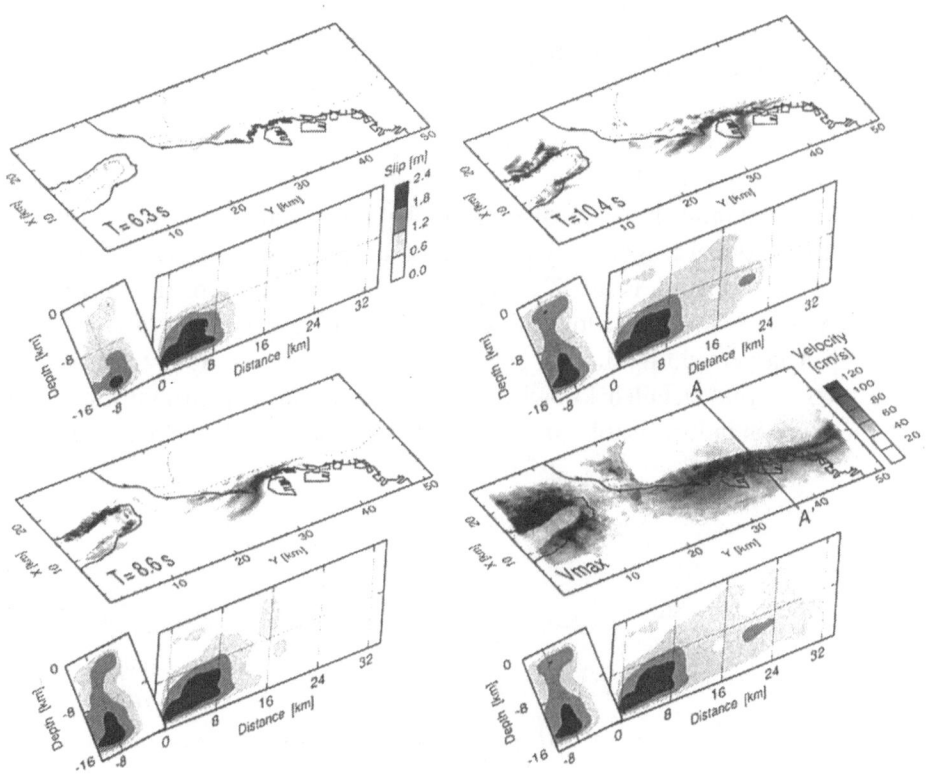

Figure 5
Snapshots of simulated horizontal ground velocities. The bottom-right panel V_{max} displays the distribution of peak ground velocities during the 22.5 s of the simulation. The dashed lines denote the basin edge. The progress of the fault rupture is also shown by snapshots.

However, they are still $60 \sim 80\%$ of the observed values, probably because the source model used does not include frequency components higher than 1 Hz. The discrepancy in the waveforms may be due to the variation of the shallowmost subsurface structure.

Figure 7 illustrates the simulated ground motions along the line A-A' in Figure 5. The distribution of horizontal peak velocities along this line is also displayed in the uppermost panel of the figure. The two directivity pulses are significantly amplified by the propagation through the low-velocity sediments, and the second one produces large surface waves at the basin edge. We can see in the basin that the constructive interference between these surface waves and the direct S-wave propagating upward from the basin bottom enhances the ground velocity in the zone slightly away from the basin edge. This is called the "basin edge effect" (KAWASE,

1996; PITARKA et al., 1996, 1998). The width of the damage belt and its distance from the basin edge is quite sensitive to the structure of the basin edge; wall angle, depth to the basin bottom, and the velocity contrast at the sediment/bedrock interface (KAWASE et al., 1998).

We note that the damage belt can also appear even when the source fault is distant from the basin edge. Figure 8(a) illustrates the peak horizontal velocities calculated for the fault that is located in the Rokko mountains 2 km from the basin edge. In the figure we again see the belt of large ground velocity in the basin, though the peak velocity is much smaller than that in Figure 6. In contrast, when the source fault is located below the basin, the zone appears just above the fault trace with a larger width, shorter length and higher amplitude (Fig. 8(b)). These results indicate that detailed knowledge of a 3-D subsurface structure and source fault geometry is indispensable to understand ground motions near a source fault.

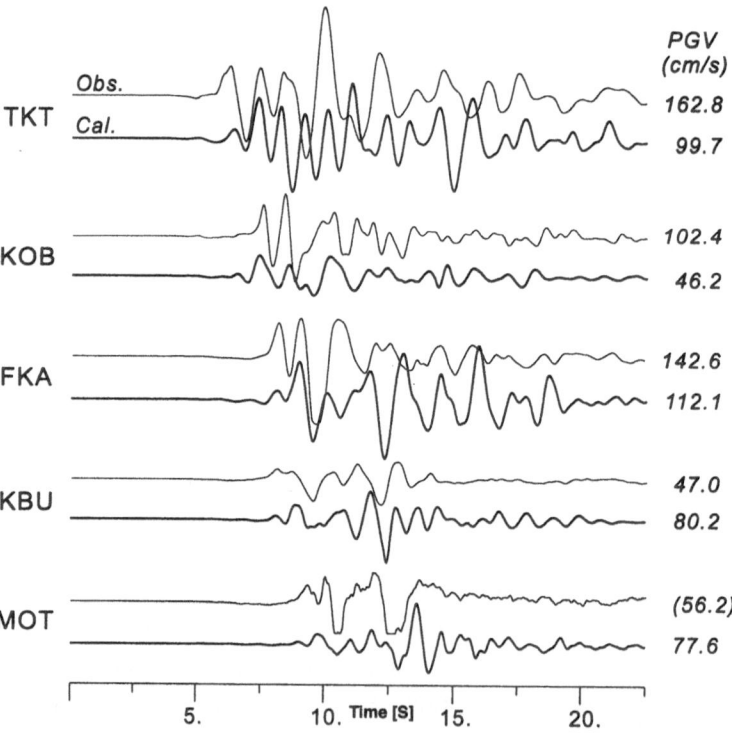

Figure 6

Comparison of the simulated (thick lines) and observed fault-normal seismograms (thin lines) at the five stations in Figure 3. The maximum amplitudes are shown on the right sides of the traces. A band-pass filter of 1.0–1.75 Hz has been applied to the seismograms except for the clipped observation at MOT.

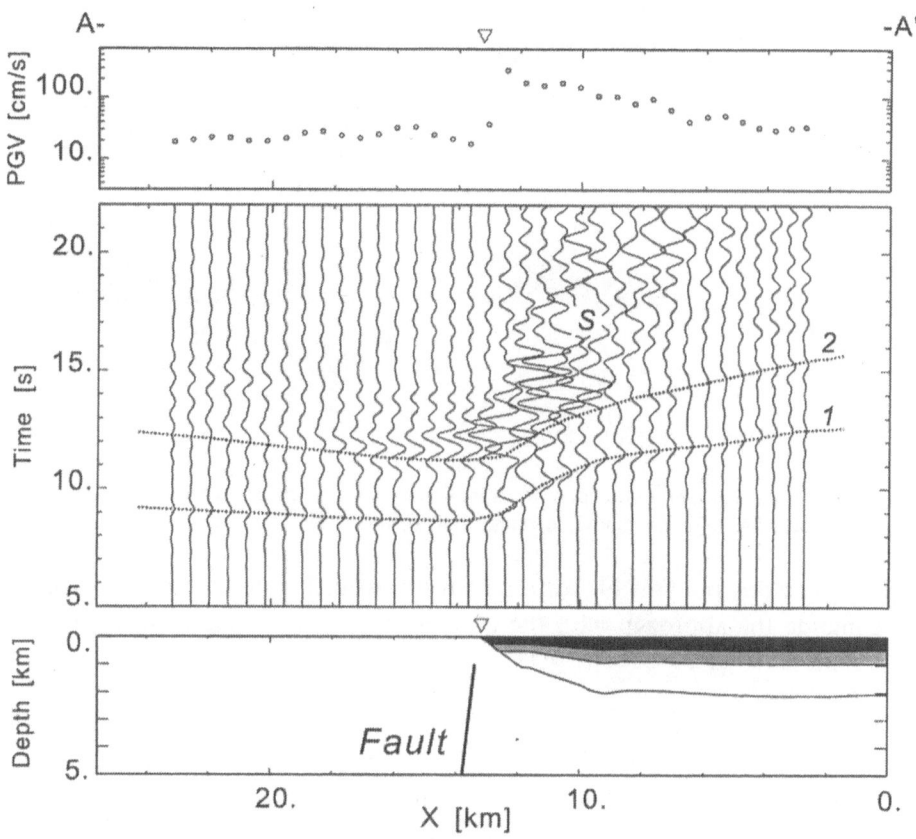

Figure 7
Sections of synthetic fault-normal motions (middle) and peak ground velocities (top) along the line A-A' in Figure 5. The vertical profiles of the velocity model and source fault are also shown at the bottom. The direct S-wave fronts for the two directivity pulses are marked by the dashed lines and numbers 1, 2. S indicates surface waves generated at the basin edge.

Conclusions

We developed the component decomposition approach to the pseudospectral simulation of 3-D seismic wavefield, and applied it to the strong motion simulation of the 1995 Kobe earthquake with a realistic subsurface model in Kobe. We have successfully reproduced the strong ground shaking during the earthquake.

Numerical modeling of 3-D seismic wavefield has long been expensive and its application is restricted to rather long-period components of the wavefield due to the limitation of computer memory, except for a few pioneering studies using massively parallel computers. However, the component decomposition approach to the parallel pseudospectral simulation has realized a practical 3-D simulation by using only desktop workstations. This approach is not restricted to a system of

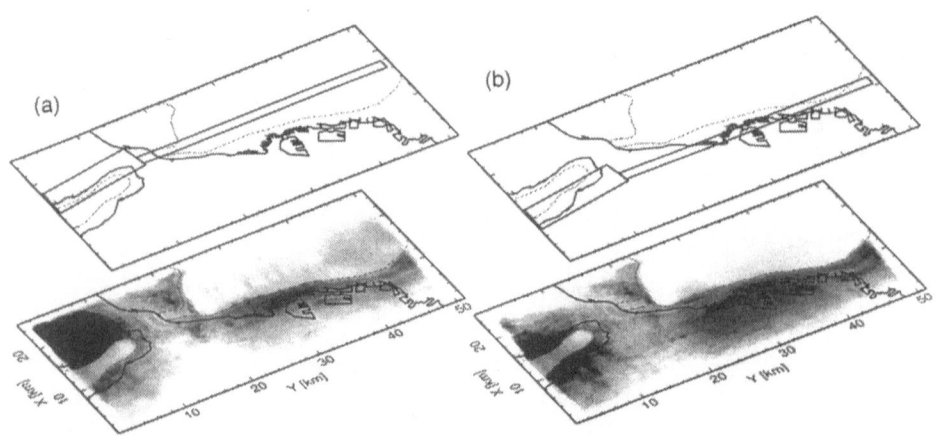

Figure 8

Distribution of peak ground velocities of horizontal motion for (a) the source fault 2 km from the actual location (Fig. 4) to the mountain range, and (b) that 2 km from the center of the basin.

three processors, but feasible for a massively parallel system of 3×2^N processors if we combine the approach with the domain partition scheme (e.g., FURUMURA et al., 1998).

Acknowledgements

This study was supported by grants from the Japanese Ministry of Education, Science and Culture (Nos. 11115202 and 10740213) and the Earth Simulator Project of the Science and Technology Agency of Japan. We thank the Japan Meteorological Agency, Committee of Earthquake Observation Research in Kansai, Osaka Gas and Japan Railway (NAKAMURA et al., 1996; No. R-036) for providing us with strong motion data. We gratefully acknowledge constructive comments by R. Madariaga, P. Mora and two anonymous reviewers.

REFERENCES

CERJAN, C., KOSLOFF, D., KOSLOFF, R., and RESHEF, M. (1985), *A Nonreflecting Boundary Condition for Discrete Acoustic and Elastic Wave Equations*, Geophysics 50, 705–708.

FACCIOLI, E., MAGGIO, F., PAOLUCCI, R., and QUARTERONI, A. (1997), *2D and 3D Elastic Wave Propagation by a Pseudospectral Domain Decomposition Method*, J. Seismol. 1, 237–251.

FORNBERG, B. (1998), *The Pseudospectral Method: Accurate Representation of Interfaces in Elastic Wave Calculations*, Geophysics 53, 625–637.

FURUMURA, T., and TAKENAKA, H. (1992), *A Stable Method for Numerical Differentiation of Data with Discontinuities at End-points by Means of Fourier Transform-symmetric Differentiation*, Butsuri-tansa (J. SEGJ) 45, 303–309 (in Japanese).

FURUMURA, T., KENNETT, B. L. N., and TAKENAKA, H. (1998), *Parallel 3-D Pseudospectral Simulation of Seismic Wave Propagation*, Geophysics *63*, 279–288.

FURUMURA, T., and KOKETSU, K. (1998), *Specific Distribution of Ground Motion During the 1995 Kobe Earthquake and its Generation Mechanism*, Geophys. Res. Lett. *25*, 785–788.

GRAVES, R. W. (1996), *Simulating Seismic Wave Propagation in 3D Elastic Media Using Staggered-grid Finite Differences*, Bull. Seismol. Soc. Am. *86*, 1091–1106.

GROPP, W., LUSK, E., and SKJELLUM, A., *Using MPI, Portable Parallel Programming with the Message-passing Interface* (The MIT Press, London 1995).

HERRMANN, R. B. (1979), *SH-wave Generation by Dislocation Source—A Numerical Study*, Bull. Seismol. Soc. Am. *69*, 1–15.

HUNG, S.-H., and FORSYTH, D. W. (1998), *Modeling Anisotropic Wave Propagation in Oceanic Inhomogeneous Structures Using the Parallel Multidomain Pseudospectral Method*, Geophys. J. Int. *133*, 726–740.

INOUE, T., and MIYATAKE, T. (1997), *3-D Simulation of Near-field Strong Motion: Basin Edge Effect Derived from Rupture Directivity*, Geophys. Res. Lett. *24*, 905–908.

IRIKURA, K., KUDO, K., OKADA, H., and SASATANI, T., *The Effects of Surface Geology on Seismic Motion, Volume 3* (A. A. Balkema, Rotterdam 1999).

KAGAWA, T., SAWADA, S., IWASAKI, Y., and NANJO, A. (1993), *Modeling of Deep Sedimentary Structure of the Osaka Basin*, Proc. 22th Earthq. Eng. Symp., 192–202 (in Japanese).

KAWASE, H. (1996), *The Cause of the Damage Belt in Kobe: "The Basin-edge Effect," Constructive Interference of the Direct S-wave with the Basin-induced Diffracted/Rayleigh Waves*, Seism. Res. Lett. *67*, 25–34.

KAWASE, H., MATSUSHIMA, S., GRAVES, R. W., and SOMERVILLE, P. G. (1998), *Three-dimensional Wave Propagation Analysis of Simple Two-dimensional Basin Structure with Special Reference to "the Basin-edge Effect"—The Cause of the Damage Belt during the Hyogo-ken Nanbu Earthquake*, Zisin *50*, 431–449 (in Japanese).

KOKETSU, K. (1996), *Damaging Californian Earthquakes and the 1995 Hyogo-ken Nanbu Earthquake*, Kagaku *66*, 93–97 (in Japanese).

KOMATITSCH, D., and VILOTTE, J.-P. (1998), *The Spectral Element Method: An Efficient Tool to Simulate the Seismic Response of 2D and 3D Geological Structures*, Bull. Seismol. Soc. Am. *88*, 368–392.

KOSLOFF, D., KESSLER, D., FILHO, A. Q., TESSMER, E., BEHLE, A., and STRAHILEVITZ, R. (1990), *Solution of the Equations of Dynamic Elasticity by a Chebyshev Spectral Method*, Geophysics *55*, 734–748.

NAKAGAWA, K., SHIONO, K., INOUE, N., and SANO, M. (1996), *Geological Characteristics and Problems in and Around Osaka Basin as a Basis for Assessment of Seismic Hazards*, Soil and Foundations, Special Issue, 15–28.

NAKAMURA, Y., UEHAN, F., and INOUE, H. (1996), *Waveform and its Analysis of the 1995 Hyogo-Ken-Nanbu Earthquake (II)*, JR Earthquake Information *23d*, Railway Technical Research Institute, 112 pp. (in Japanese).

PITARKA, A., IRIKURA, K., IWATA, T., and KAGAWA, K. (1996), *Basin Structure Effects in the Kober Area Inferred from the Modeling of Ground Motions from two Aftercocks of the January 17, 1995 Hyogoken-nanbu Earthquake*, J. Phys. Earth *44*, 563–576.

PITARKA, A., IRIKURA, K., IWATA, T., and SEKIGUCHI, H. (1998), *Three-dimensional Simulation of the Near-fault Ground Motion for the 1995 Hyogo-Ken Nanbu (Kobe), Japan, Earthquake*, Bull. Seismol. Soc. Am. *88*, 428–440.

RESHEF, M., KOSLOFF, D., EDWARDS, M., and HSIUNG, C. (1998), *Three-dimensional Elastic Modeling by the Fourier Method*, Geophysics *53*, 1184–1193.

SATO, T., MATSUOKA, T., and TSURA, T. (1995), Wavefield Modeling by Pseudospectral Method on a Parallel Computer (2) Seismic Data Simulation and Processing, Proc. SEGJ Conference *92*, 392–396 (in Japanese).

SEKIGUCHI, H., IRUKURA, K., IWATA, T., KAKEHI, Y., and HOSHIBA, M. (1996), *Minute Locating of Faulting beneath Kobe and the Waveform Inversion of the Source Process during the 1995 Hyogo-ken Nanbu, Japan, Earthquake Using Strong Ground Motion Records*, J. Phys. Earth *44*, 473–487.

SHIMAMOTO, T. (1995), *Earthquake Disasters in Kobe and Presumed Earthquake—Generating Faults*, Kagaku *65*, 195–198 (in Japanese).

YOSHIDA, S., KOKETSU, SHIBAZAKI, B., SAGIYA, T., KATO, T., and YOSHIDA, Y. (1996), *Joint Inversion of Near- and Far-field Waveforms and Geodetic Data for the Rupture Process of the 1995 Kobe Earthquake*, J. Phys. Earth *44*, 437–454.

(Received August 14, 1999, revised March 20, 2000, accepted April 10, 2000)

To access this journal online:
http://www.birkhauser.ch

Pure appl. geophys. 157 (2000) 2063–2081
0033–4553/00/122063–19 $ 1.50 + 0.20/0

❙Pure and Applied Geophysics

Computer Simulation of Strong Ground Motion near a Fault Using Dynamic Fault Rupture Modeling: Spatial Distribution of the Peak Ground Velocity Vectors

TAKASHI MIYATAKE[1]

Abstract—Computer simulation was used to study the nature of the strong ground motion near a strike-slip fault. The faulting process was modeled by stress release with fixed rupture velocity in a uniform elastic half-space or layered half-space. The fourth-order 3-D finite-difference method with staggered grids was employed to compute both ground motions and slip histories on the fault. The fault rupture was assumed to start from a point and propagate circularly with 0.8 times shear-wave velocity. In the present paper, we focused on the spatial pattern of ground velocity vectors, i.e., the direction of strong motions. In the case of bilateral rupture propagation, the strong fault parallel ground motion appeared near the center of the fault. The fault normal motions of ground velocity appeared near the edges of the fault. In the case of unilateral rupture, the fault parallel motion appeared near the starting point however, the amplitude was lower than that for the bilateral rupture case. The fault normal motion was predominant near the terminal point of the rupture. The results were applied to the earthquake damage data, especially the directions that simple bodies overturned and wooden houses collapsed, caused by the 1927 Tango, the 1930 Kita-Izu, and the 1948 Fukui earthquakes. The spatial distributions of the direction data were found to reflect the strong ground motions generated from the earthquake source process.

Key words: Strong ground motion, source process.

Introduction

The 1995 Kobe earthquake occurred beneath an urban city, causing severe damage. Research focused on near source strong ground motion simulations has been conducted since this event (e.g., PITARKA *et al.*, 1998; FURUMURA and KOKETSU, 1998). Studies involving the prediction of strong ground motion generated from a potential fault have also been conducted, using finite-difference simulations (e.g., GRAVES, 1998; OLSEN and ARCHULETA, 1996). It is, however, important to study the physical mechanism that controls ground motion. INOUE and MIYATAKE (1998) conducted one such study in 3-D simulation. In their study, the effect of source processes on ground motion near the fault was investigated.

---

[1] Earthquake Research Institute, University of Tokyo, Yayoi, 1-1-1, Bunkyo, Tokyo, 113-0032, Japan. E-mail: miyatake@eri.u-tokyo.ac.jp

They mainly focused on the spatial distribution of peak ground velocity near the fault. The direction of ground motion, however, is another important property. Sometimes wooden houses, grave stones, trees, and other objects in a local area fall in the same direction due to strong ground motion (e.g., SHIMAMOTO et al., 1996). These data certainly reflect not only the site effect and the earth's structure, but also the source process. In this present paper, we study the direction of ground motion near the fault.

The Method of Computation

Figure 1 shows our modeling space for the strike-slip fault. The fourth-order finite-difference method with staggered grid (GRAVES, 1996) is employed in the present study. Absorbing boundary conditions (CLAYTON and ENGQUIST, 1977) are adopted on all sides of the calculating space, except on the free surface. In the numerical computation we calculate only a quarterspace, using a symmetry condition on the $y = 0$ (fault plane); $V_x(x, -y, z) = -V_x(x, y, z)$, $V_y(x, -y, z) = V_y(x, y, z)$, $V_z(x, -y, z) = -V_z(x, y, z)$. We also added an artificial viscosity to the finite-difference equations in order to dissipate spurious high-frequency oscillation in the numerical solution generated by abrupt stress drop on the crack (VIRIEUX and MADARIAGA, 1982). We represent an earthquake as a propagating stress relaxation over a finite fault plane embedded within a half-space. This is dynamic modeling (e.g., ANDREWS, 1976; DAY, 1982a,b; IDA, 1972; VIRIEUX and

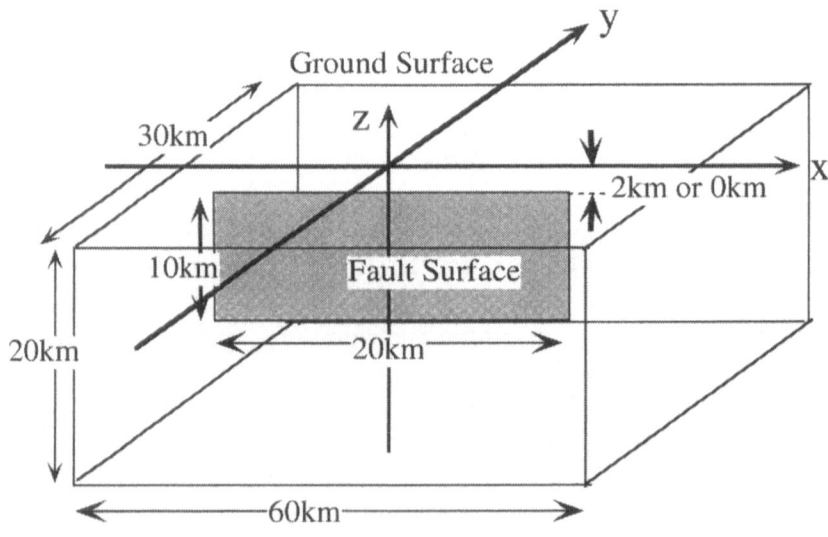

Figure 1
Schematic figure of the fault model in the present paper.

MADARIAGA, 1982) with constant rupture velocity and is more realistic than kinematic modeling.

Our models have been constructed as simply as possible in order to clarify the fundamental relationships between strong motion and various rupture model parameters. We consider a rectangular-shaped strike-slip fault existing in the zx-plane as a causative fault. The length and width of the fault is 20 km and 10 km, respectively. The uppermost depth of the fault is assumed to be 2 km for main calculations. The calculation for the depth of 0 km is also carried out in order to investigate the effect of breakout of the fault. Grid spacing is 0.1 km in all directions and the time step is 0.005 sec. A rupture is assumed to initiate from a point on the fault and propagate circularly outward at a given velocity V_r. When the rupture front arrives at a point on the fault plane, shear stress drops at that same point to a prescribed frictional stress. The stress drop is assumed to be uniform and 10 MPa on the whole fault plane. The rupture propagation suddenly stops when it reaches the boundaries of the fault. The structure parameters used in the present paper are V_p (P-wave velocity) of 5.2 km/s, V_s (S-wave velocity) of 3.0 km/s and density of 3.0 g/cm^3.

The Spatial Pattern of Peak Ground Velocities Near the Fault

Figure 2 shows our source and structure model. Firstly, I investigate the effect of rupture propagation mode (unilateral or bilateral rupture propagation). The waveforms for unilateral rupture propagation (Case U_0) are shown in Figure 3. The waveforms were 0.5 Hz-lowpass filtered to avoid the effect of numerical dispersion. The particle motions, particle velocity orbits, and peak horizontal ground velocity vectors for Case U_0 are plotted in Figure 4. The thick line indicates the fault trace and the rupture propagates from the left-hand side to the right-hand side on the fault in Figure 4. The shape of the particle-motion orbit at most points on the ground is like a flat ellipse or a line except the area located around points a 45-degree direction apart from the fault edge. The fault normal motions appear along the fault trace except the area around the starting point of rupture where fault parallel motions appear. The fault normal motion becomes strongest around the terminal point of the rupture. These are known as the forward directivity effect and have already been pointed out by several seismologists (e.g., AKI, 1968; BOUCHON, 1980; INOUE and MIYATAKE, 1998).

In the case of bilateral rupture propagation, the fault normal motion is predominant at both edges of the fault due to forward directivity. The fault parallel motion is predominant near the center of the fault trace where the starting point of rupture is located.

We also investigate the effect of the depth of the rupture starting point on the ground velocity. The peak ground velocity vectors for unilateral rupture propaga-

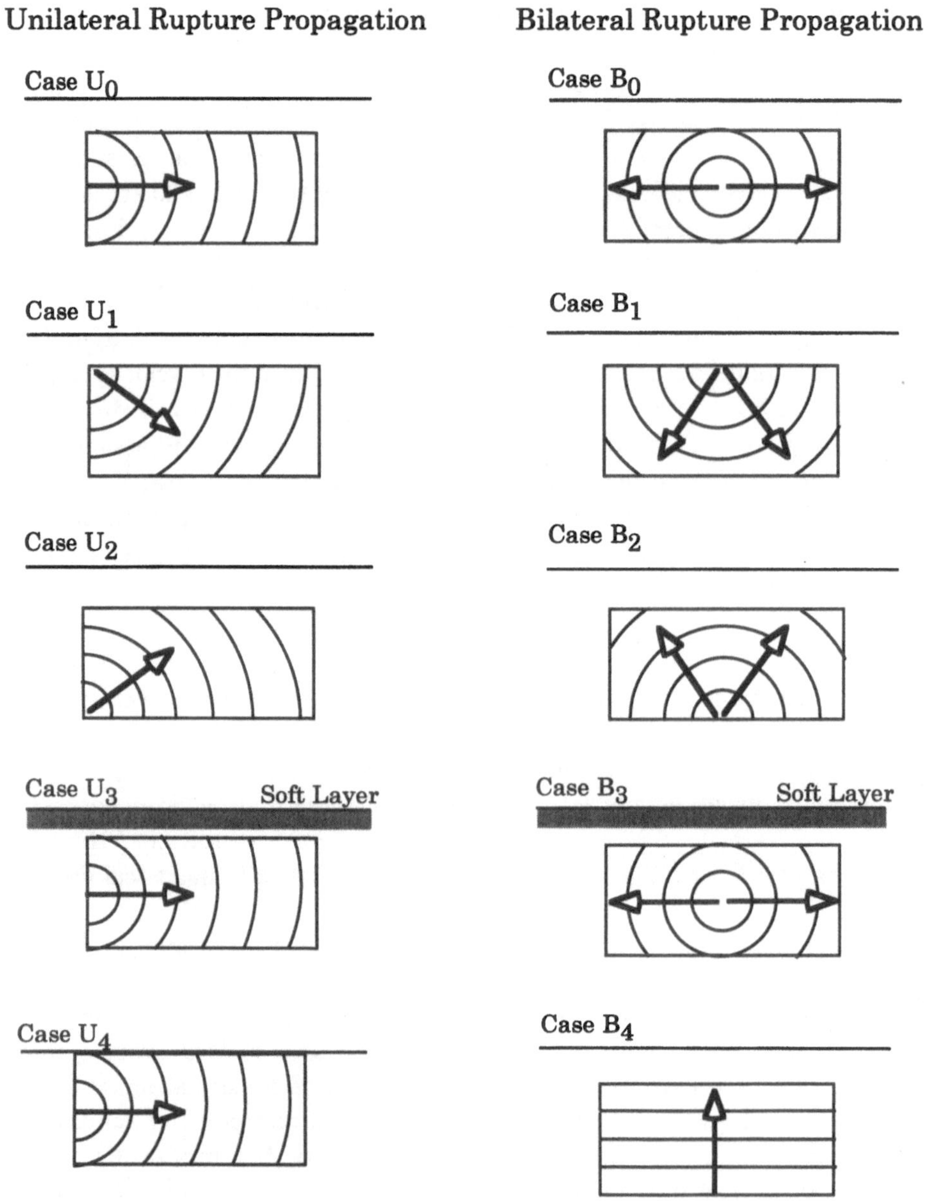

Unilateral Rupture Propagation   Bilateral Rupture Propagation

Case U_0

Case B_0

Case U_1

Case B_1

Case U_2

Case B_2

Case U_3 Soft Layer

Case B_3 Soft Layer

Case U_4

Case B_4

Figure 2

Schematic figure of the fault and structure models in the present paper. The horizontal lines indicate ground surface. The contours indicate rupture fronts. The uppermost depth of the fault is assumed 2 km except Case U_4.

tion are shown in Figure 5a. The starting point for cases U_0, U_1 and U_2 are the middle point, the shallowest point, and the deepest point of the left margin of the fault, respectively. In Figure 5b, those for bilateral rupture are shown. The starting

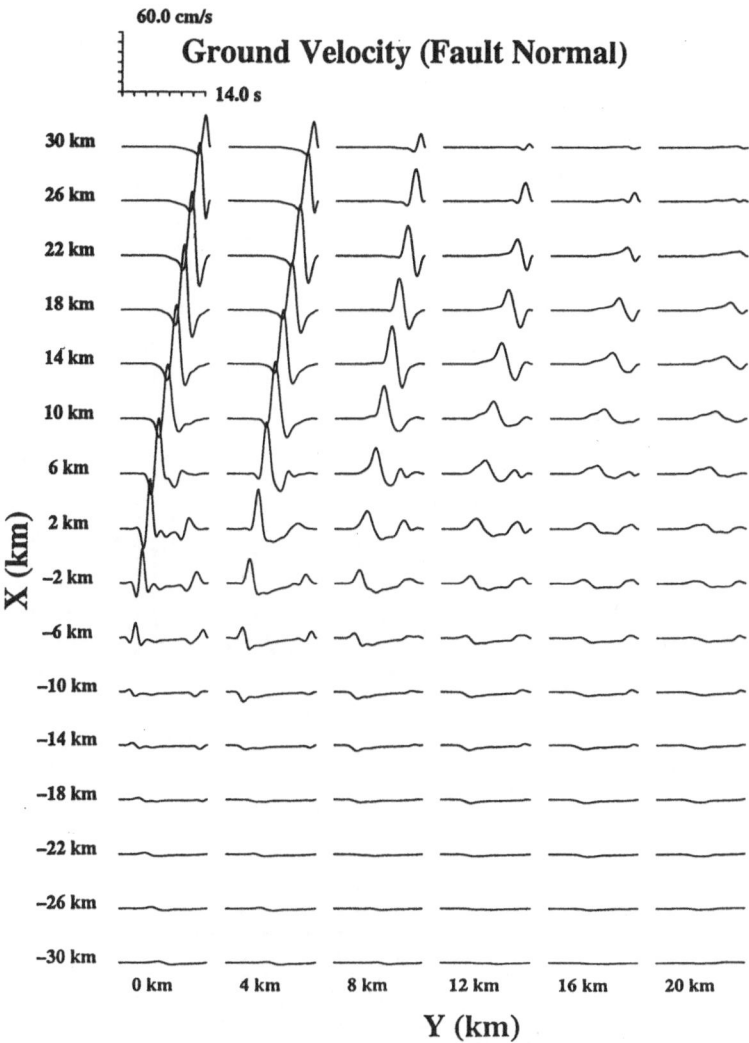

Figure 3

The 0.5 Hz-low-pass-filtered ground velocity wave form for Case U_0. The waveforms are plotted at every 4 km.

point for cases B_0, B_1 and B_2 are the middle point, the shallowest point, and the deepest point of the left margin of the fault, respectively. The spatial pattern of peak ground velocity vectors depends on whether the rupture propagates unilaterally or bilaterally as mentioned previously, but is not affected by the depth of the starting point, although the amplitude of peak ground velocity is enhanced by the depth of the rupture starting point, which is considered as the directivity effect of rupture propagation toward the ground surface.

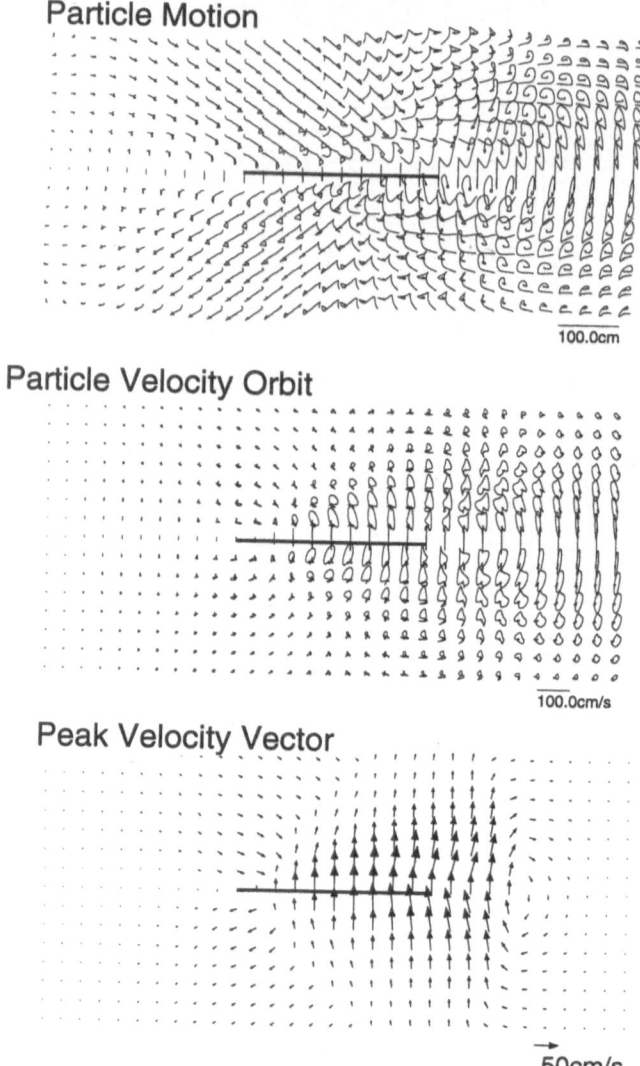

Figure 4
Particle motion, ground velocity orbits, and peak ground velocity vectors for Case U_0 are plotted. The thick line indicates the fault trace.

In the above computations, we assumed the uppermost depth of the fault to be 2 km. However, the breakout of the fault surface has often been observed. Thus, I calculate the case of breakout of the fault (0 km depth) to the ground surface. Figure 6 shows the particle motion, particle velocity orbits, and peak ground velocity. The fault normal motion is predominant around the terminal point of rupture, and the fault parallel motion appeared around the rupture starting point. These features are the same as shown in Figure 5a. The evidence that the fault normal velocity was considerably higher than the fault parallel velocity near the breakout fault surface for the 1995 Neftegorsk earthquake, has also been shown in SHIMAMOTO *et al.* (1996) or INOUE and MIYATAKE (1998).

Finally, the effect of soft layer is investigated (Case U_3). The thickness of soft layer is 1 km and the uppermost depth of the fault is 2 km. The P-wave (2.6 km/s) and S-wave speeds (1.5 km/s) are assumed half of those in the basement. The density of soft layer is 2.0 g/cm^3. Figure 7 shows the ground velocity waveforms of the fault normal component. Although the ground velocity waveform in a uniform

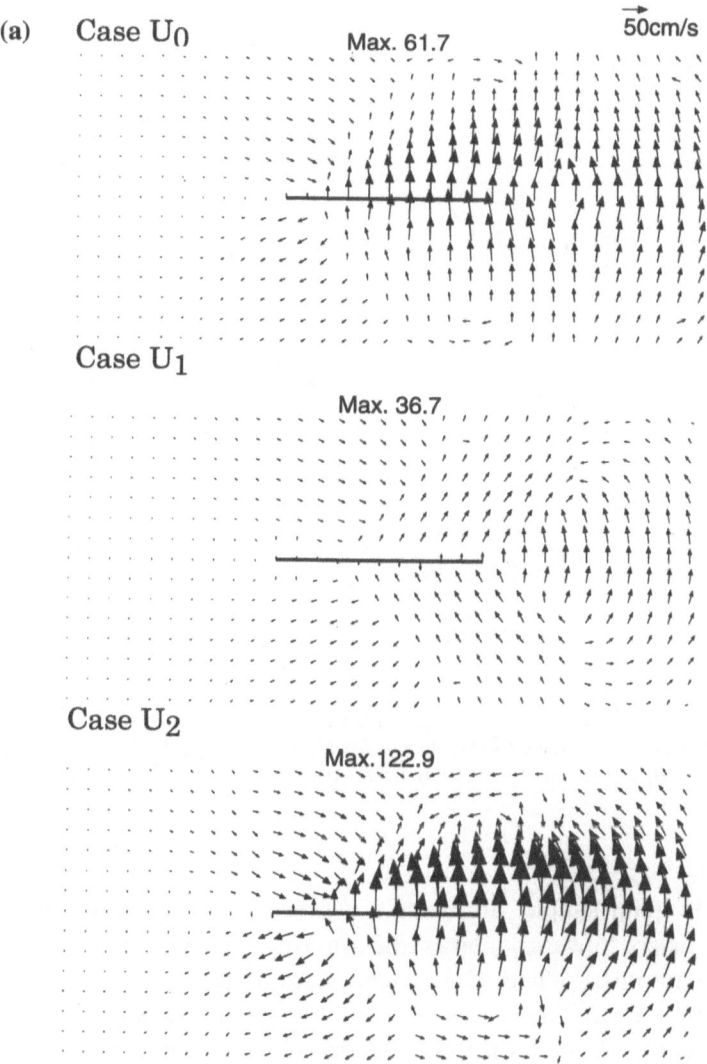

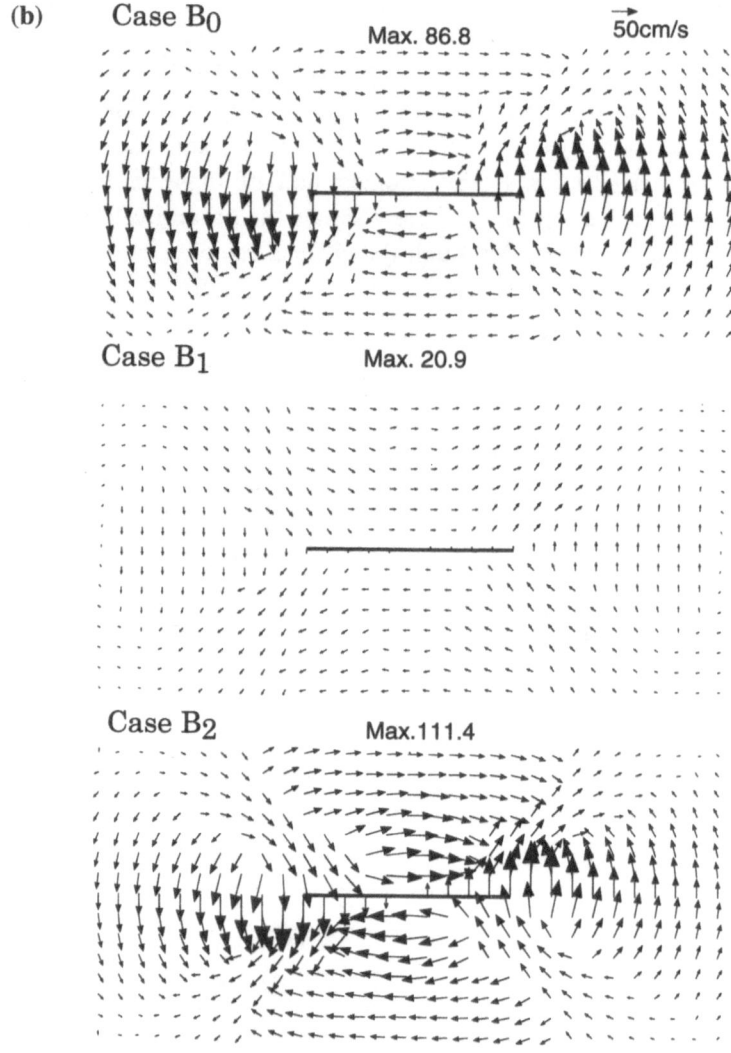

Figure 5
(a) The distribution peak ground velocity vectors for U_0, U_1, and U_2 (b). The same figures for B_0, B_1 and B_2 as (a). The thick line indicates the fault trace. See also Figure 2.

half-space is quite simple, and is like a single pulse as shown in Figure 3, the sedimental layer, however, makes the waveforms complex, amplifying the back swing and later phases. The peak ground motion vectors for Case U_3 are shown in Figure 8, and it seems they are a quite similar spatial pattern to those in Figure 5a.

Thus the distribution of PGV vectors has the characteristic feature shown schematically in Figure 9.

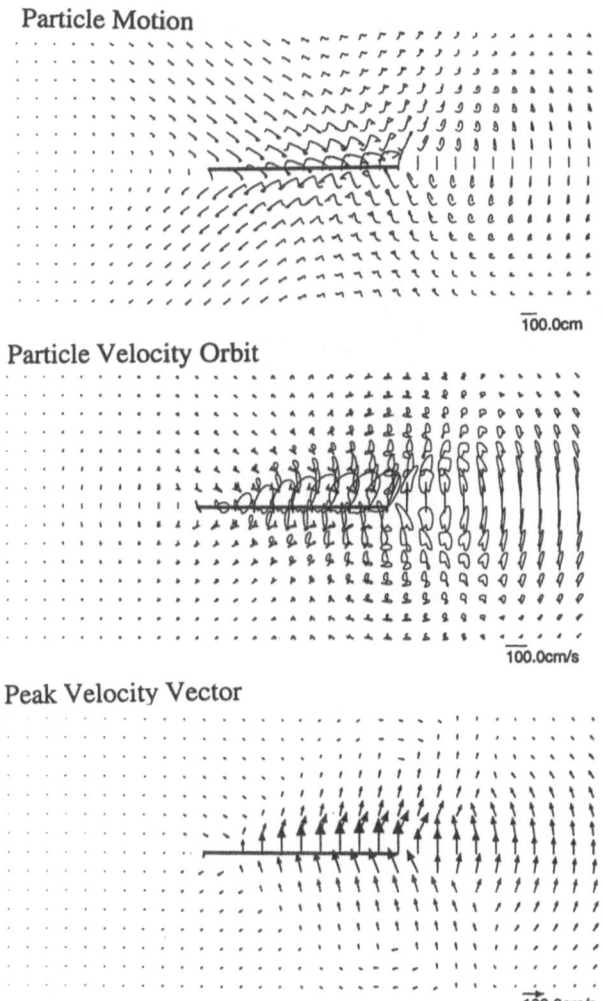

Figure 6
Particle motion, ground velocity orbits, and peak ground velocity vectors for the case of breakout of the fault (Case U_3) are plotted.

The results were obtained for a single event. Many earthquakes, however, are not single events but often multiple shocks. We often observe the segmented faults. Before I apply these results to the earthquake damage data, I discuss how to apply to the data, especially to data generated from the multiple shock. The nearfield waveforms from multiple shock are a summation of waves from each subfault. On the other hand, peak value of any component of ground velocity waveform at some observation point from multiple shock are not a summation of that from each

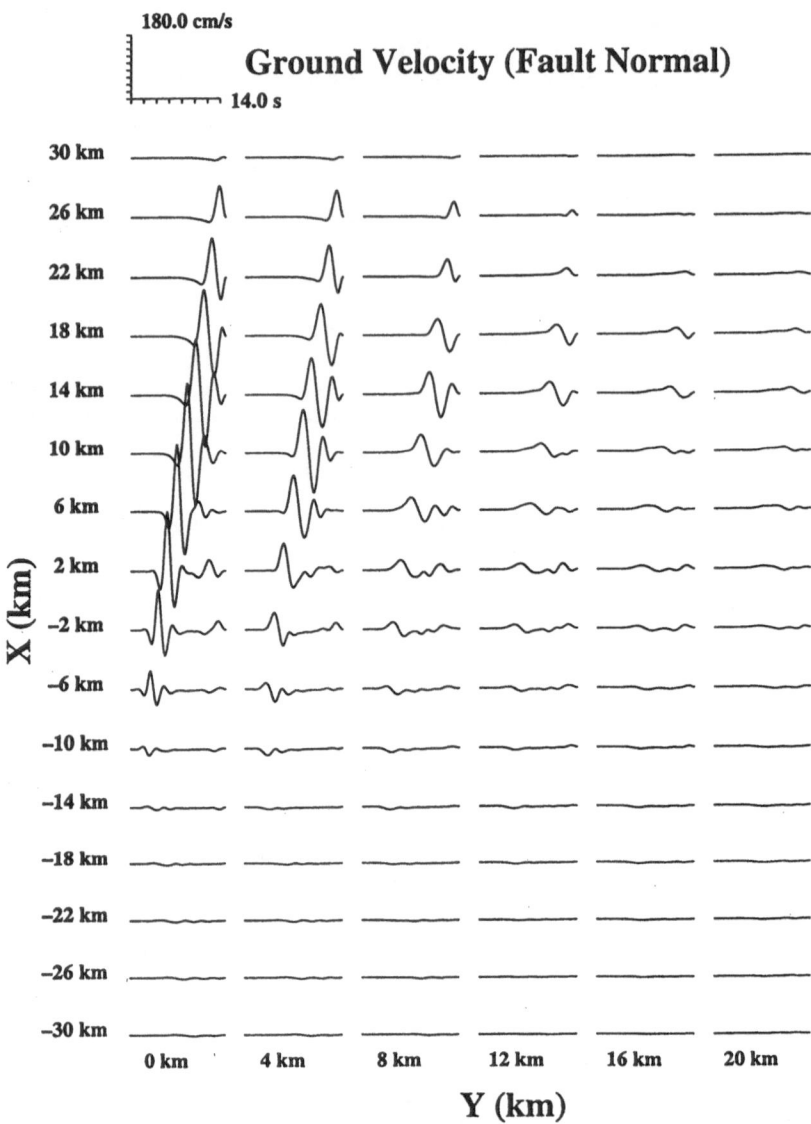

Figure 7
The 0.5 Hz-low-pass-filtered ground velocity waveform for Case U_3. The waveforms are plotted at every 2 km.

subfault, because of the time difference of peak value of each waveform. Hence, for a rough estimation of the spatial pattern of peak ground motion, or the spatial pattern of the intensity and direction of strong ground motion, it would be reasonable to consider the spatial pattern of peak ground velocity vector from multiple shock as the largest of each peak ground velocity vector pattern from each subfault because of the arrival time difference of waveform from each subfault.

(a) Unilateral Rupture Propagation

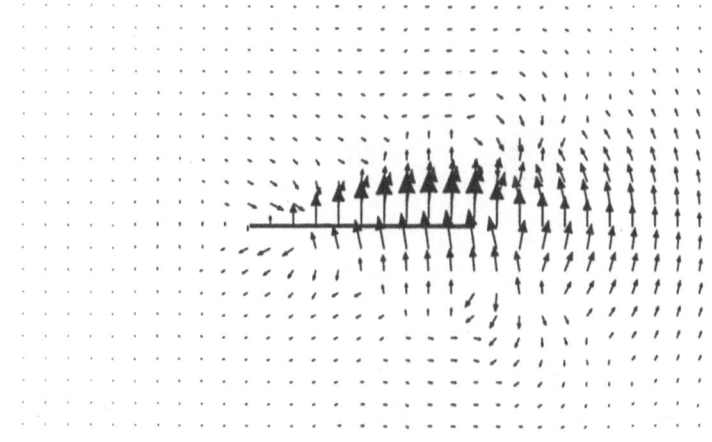

(b) Bilateral Rupture Propagation

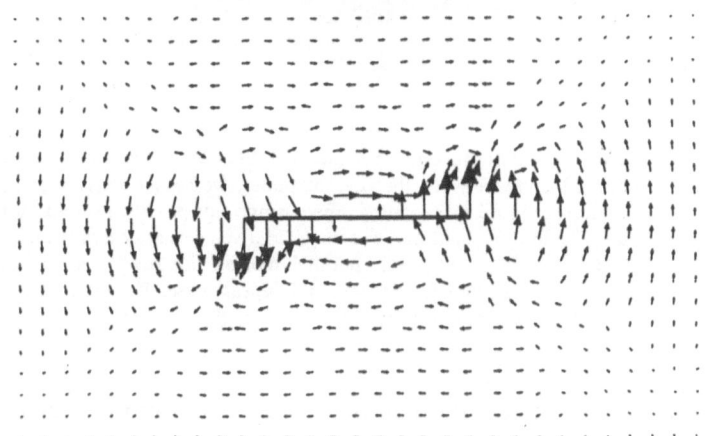

Figure 8

The distribution of peak ground velocity for (a) Case U_3, and (b) Case B_3.

Earthquake Damage Data

In Japan there are numerous documents detailing earthquake damages (e.g., USAMI, 1987; TAKEMURA et al., 1998). Damage reports for individual earthquakes have been published subsequent to each earthquake. These reports often include data detailing the direction of buildings' collapse and the direction of simple bodies' overturned, such as gravestones, monuments, chimneys, and trees. These direction

Unilateral Rupture Propagation

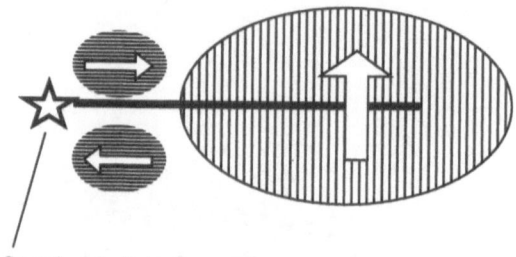

Starting point of rupture

Bilateral Rupture Propagation

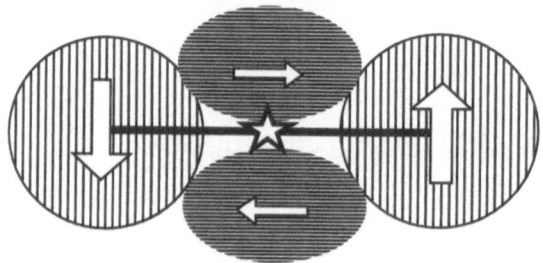

Figure 9

Schematic ground velocity vector patterns for right lateral strike-slip event. For left lateral slip, the polarity of the vector changed oppositely. In the present paper, we only discuss the direction not its polarity. In the case of bilateral rupture propagation, the strong fault parallel ground motion appeared near the center of the fault. The fault normal motions of ground velocity appeared near the edges of the fault. In the case of unilateral rupture, the fault parallel motion appeared near the starting point, however the amplitude was lower than that for the bilateral rupture case. The fault normal motion was predominant near the terminal point of the rupture.

data have not been well studied, though they certainly reflect the source process. In this present paper, we interpret these data using the previous findings. As our calculations are for strike-slip faults, we have chosen data involving strike-slip events. The direction of building collapse includes a difficult problem. The breakdown of buildings and gravestones is a very complex nonlinear process in which the response of buildings and gravestones must be considered. Square bodies, such as houses or gravestones, do not easily fall in the direction of their own diagonal line but easily fall in a direction perpendicular to their side. This means the direction data have an uncertainty of 45 degrees in the worst case and may cause apparent scattering of patterns in the direction data. As shown in the previous section, the ground motion near the fault is quite simple and similar to a single pulse (Fig. 3). However, soft sediments make it complex at a point apart from the fault (Fig. 7). If the ground motion is strong enough or if the buildings are weak enough, the attack of the first strong motion may cause a collapse in the same direction as that of the ground motion. However, if the ground motion is not strong enough or if the houses are not weak enough, the collapse may occur at the time of the second or later swing of the ground motion. Since it will have an ambiguity of 180 degrees, we do not discuss its polarity.

(1) The 1927 Tango Earthquake

The 1927 Tango earthquake occurred on the Japan Sea coast. Two seismic faults, i.e., the Gomura and the Yamada, appeared on the ground surface as a conjugate system. KANAMORI (1973) studied the seismic waveform recorded at Tokyo, located at a distance of 430 km, and modeled the source process. His model, indicates (1) that the moment release on the Gomura fault was about 9 times larger than that on the Yamada, (2) that the rupture of the Gomura fault extended toward the seaside and consequently, the length of the Gomura fault became twice as long as that of the fault on the land sides, (3) that the rupture started from a point which has not yet been well estimated but is considered to be located near the crossing point between the Gomura fault and the coast line, and (4) that the Yamada fault subsequently ruptured. The spatial distribution of collapsed wooden houses and their falling directions are plotted in Figure 10 (TAKEMURA et al., 1998). The rupture processes of the Gomura fault caused a strong forward-directivity effect and the fault normal motion, i.e., east-west component, became strong, as shown in Figure 10. The motion along the Yamada fault may also have been caused by the high moment release of the Gomura fault.

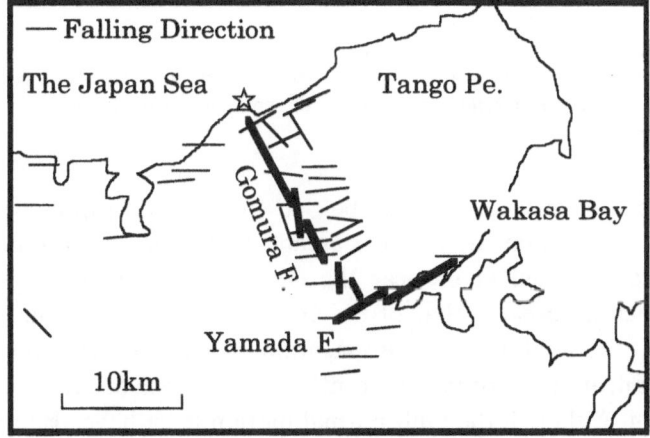

Figure 10
The direction of collapsed wooden houses for the 1927 Tango earthquake after TAKEMURA et al. (1998). The star symbol ☆ indicates the epicenter of the event. Thick lines mean fault traces.

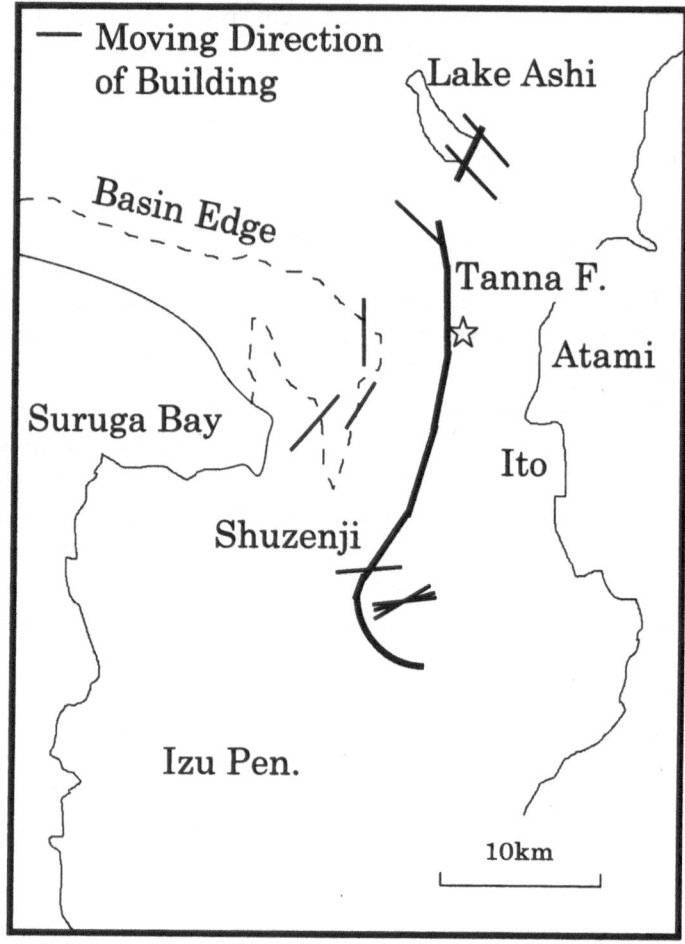

Figure 11
Directions of movement of buildings for the 1930 Kita-Izu earthquake after TAKEMURA *et al.* (1998).
The star symbol ☆ indicates the epicenter of the event.

Another possible explanation may be that the bilateral rupture of the Yamada fault triggered the motion along the Yamada fault. The rupture of the Yamada fault should have started at or near the crossing point between two faults and propagate bilaterally because that of the Gomura derives from the rupture of the Yamada fault. Bilateral rupture propagation generated the fault, creating the east-west ground motion around the center of the Yamada fault as shown in the previous section. Although the fault normal motion must have also appeared along the two edges of the Yamada fault, such motion unfortunately cannot be seen in the figure. This could be caused by stronger motion generated from the Gomura fault than that from the Yamada fault, as mentioned previously.

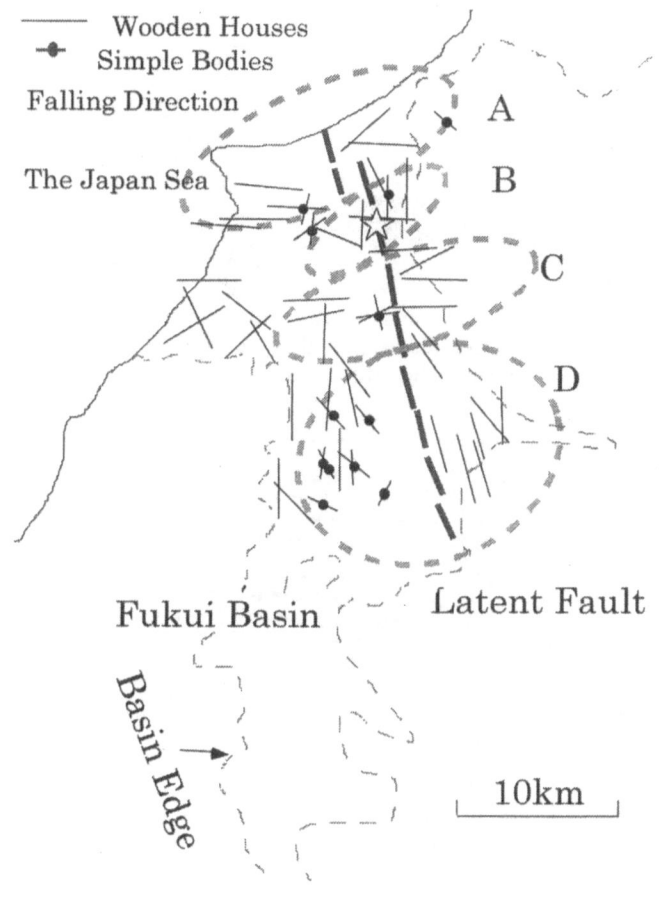

Figure 12

Directions of movement of buildings for the 1948 Kita-Izu earthquake after TAKEMURA *et al.* (1998) with modifications. A and C indicate the area in which fault normal motion is predominant. On the other hand, in B and C, fault parallel motion is predominant.

(2) The 1930 Kita-Izu Earthquake

Several remarkable surface faults appeared, as shown in Figure 11. As the epicenter is located at a point slightly north of the center of the fault, the rupture is considered to have propagated bilaterally. The fault normal motion therefore is expected to be around the northern and the southern parts of the fault, and the fault parallel motion is expected to be central but apart from the fault. The falling directions of wooden houses are shown in Figure 11 (TAKEMURA *et al.*, 1998). The fault parallel motion appears in proximity west of the central part of the fault. In addition, the fault normal motion appears at the southern edge. The direction of movement near the northern end seems to be rotated.

The 1948 Fukui Earthquake

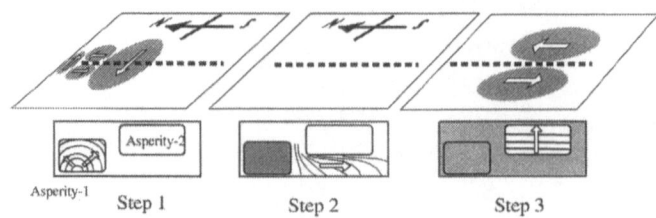

Figure 13
The scenario of the 1948 Fukui earthquake inferred from the spatial distribution and the location of the epicenter after MIYATAKE (1999).

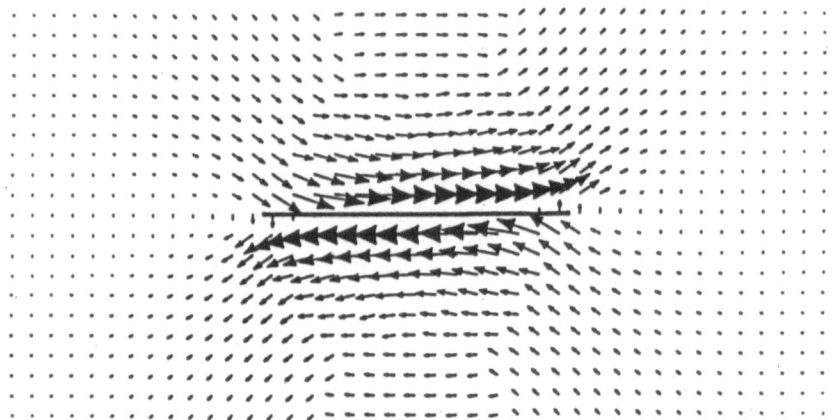

Figure 14
The maximum ground velocity vector for upward rupture propagation (Case B_4).

(3) 1948 Fukui Earthquake

The direction distribution of simple bodies overturned and wooden houses collapsed is shown in Figure 12 (TAKEMURA *et al.*, 1998). The fault normal motion is at the northern end of the fault (A in Fig. 12) and at the center of the fault (denoted by C), the fault parallel motion at B and D is apparent. KANAMORI (1973) analyzed the seismogram recorded at Abuyama station located at a distance of 155 km and concluded that the earthquake consisted of two subevents with a time difference of 9 seconds, the moment release of the second subevent being exceedingly larger than that of the first. The locations of the subevents were not resolved in Kanamori's study. The epicenter has been determined and located around the northern part of the fault (HAMADA, 1987). This means that the rupture mode is almost unilateral. Recently, KIKUCHI *et al.* (1999) modeled the rupture process using moment-rate functions at grid points on the fault with a spacing of 10 km, and inverted the waveforms recorded at four near-field stations. This revealed that the rupture initiated at a depth of about 10 km and propagated mainly to the south. The high moment release of the southern part corresponds to KANAMORI's second subevent (1973). The direction data can be explained by the following rupture process (MIYATAKE, 1999). If the subevent rupture first started at a deeper part and propagated circularly, the fault normal ground motion in areas A and C and the fault parallel motion in area B, were generated by the first subevent (Step-1 in Fig. 14). The widely apparent fault parallel motion (D in Fig. 13) around the central to the southern part of the fault can be explained, if the rupture process inside the second subevent is mainly an upward rupture propagation (Case B_4, see Fig. 14). The relatively flat rupture front in the second subfault is possible if the surrounding part near the subfault ruptured during the 9 seconds after the rupture of first subevent (DAS and KOSTROV, 1983). Considering the time difference of 9 seconds between the first and second event, the rupture may have propagated to a deeper part of the fault, silently without stress drop. The above scenario is inferred from the direction data with constraint of the following two facts:

1. The epicenter is located at the northern part of the fault.
2. The event consisted of two subevents, the second subevent being larger than the first (KANAMORI, 1973).

While we do not take into account the results of KIKUCHI et al. (1999), our scenario is consistent with their model. The inversion of the earthquake has a resolution of more than 10 km, thus it is impossible to detect the rupture process at less than 10 km. However, combining the damage data with the waveform inversion results, we obtained new information and deeper understanding of the rupture process than previously available, or inferred by the rupture process involved in historical earthquakes. Even for the recent earthquake, the strong motion direction data near the fault inferred by a building's collapse may provide constraints necessary for the rupture process to be inferred by waveform inversion.

Conclusion

The nature of strong ground motion generated from a strike-slip fault is, (1) strong fault-normal-motion around the terminal part of the fault, (2) fault parallel ground motion also appears around the starting point of rupture. The above features are demonstrated by the earthquake damage data of the 1930 Kita-Izu, the 1927 Tango, and the 1948 Fukui earthquakes. The damage data were interpreted in relationship with the source process, and new results for the source process have been obtained in this process. The present paper demonstrates the possibility of estimating the source process from the damage data, especially the earthquakes which occurred from the 1890s to 1960s in which there had been insufficient seismic network coverage for source inversion in Japan although the damage of buildings had been reported in detail.

REFERENCES

AKI, K. (1968), *Seismic Displacement Near a Fault*, J. Geophys. Res. *73*, 5359–5376.

ANDREWS, D. J. (1976), *Rupture Propagation with Finite Stress in Antiplane Strain*, J. Geophys. Res. *81*, 3575–3582.

BOUCHON, M. (1980), *The Motion of the Ground during an Earthquake. 1. The Case of a Strike-slip Fault; 2. the Case of a Dip Slip Fault*, J. Geophys. Res. *85*, 356–375.

CLAYTON, R., and ENGQUIST, B. (1977), *Absorbing Boundary Conditions for Acoustic and Elastic Wave Equations*, Bull. Seismol. Soc. Am. *67*, 1529–1540.

DAS, S., and KOSTROV, B. V. (1983), *Breaking of a Single Asperity: Rupture Process and Seismic Radiation*, J. Geophys. Res. *88*, 4277–4288.

DAY, S. M. (1982a), *Three-dimensional Finite-difference Simulation of Fault Dynamics: Rectangular Faults with Fixed Rupture Velocity*, Bull. Seismol. Soc. Am. *72*, 705–727.

DAY, S. M. (1982b), *Three-dimensional Simulation of Spontaneous Rupture: The Effect of Nonuniform Prestress*, Bull. Seismol. Soc. Am. *72*, 1881–1902.

FURUMURA, T., and KOKETSU, K. (1998), *Specific Distribution of Ground Motion during the 1995 Kobe Earthquake and its Generation Mechanism*, Geophys. Res. Lett. *25*, 785–788.

GRAVES, R. W. (1996), *Simulating Seismic Wave Propagation in 3-D Elastic Media Using Staggered-grid Finite Differences*, Bull. Seismol. Soc. Am. *86*, 1091–1106.

GRAVES, R. W. (1998), *Three-dimensional Finite-difference Modeling of the San Andreas Fault: Source Parameterization and Ground-motion Levels*, Bull. Seismol. Soc. Am. *88*, 881–897.

HAMADA, N. (1987), *Re-examination of Seismicity Associated with Destructive Inland Earthquakes of Japan and its Seismological Significance*, Papers in Meteorol. Geophys. *38*, 77–156.

IDA, Y. (1972), *Cohesive Force across the Tip of a Longitudinal-shear Crack and Griffith's Specific Surface Energy*, J. Geophys. Res. *77*, 3796–3805.

INOUE, T., and MIYATAKE, T. (1998), *3-D Simulation of Near-field Strong Ground Motion*, Bull. Seismol. Soc. Am. *88*, 1445–1456.

KANAMORI, N. (1973), *Mode of Strain Release Associated with Major Earthquakes in Japan*, Ann. Rev. Earth Planet. Sci. *1*, 212–239.

KIKUCHI, M., NAKAMURA, M., YAMADA, M., FUSHIMI, M., TATSUMI, Y., and YOSHIKAWA, K. (1999), *Source Parameters of the 1948 Fukui Earthquake Inferred from Low-gain Strong Motion Records*, Zisin II *52*, 121–128 (in Japanese).

MIYATAKE, T. (1999), *Strong Ground Motion Generated from the 1948 Fukui Earthquake—Direction of Simple Bodies' Overturn and Buildings' Collapse, and the Source Process*, Zisin *52*, 151–160 (in Japanese).

PITARKA, A, IRIKURA, K., IWATA, T., and SEKIGUCHI, H. (1998), *Three-dimensional Simulation of the Near-fault Ground Motion for the 1995 Hyogo-ken Nanbu (Kobe), Japan, earthquake*, Bull. Seismol. Soc. Am. *88*, 428–440.

SHIMAMOTO, T., WATANABE, M., and SUZUKI, Y. (1996), *Surface faults associated with the Neftegorsk earthquake. In* Special Issue *The Neftegorsk Earthquake of May 27(28), 1995*, FSSN (Russian's Federal System of Seismological Networks and Earthquake Prediction, Information and Analytical Bulletin) pp. 99–114.

TAKEMURA, M., MOROI, T., and YASHIRO, K. (1998), *Characteristics of Strong Ground Motions as Deduced from Spatial Distributions of Damages due to the Destructive Inland Earthquakes from 1891 to 1995 in Japan*, Zisin *50*, 485–505 (in Japanese).

USAMI, T. (1987), *Destructive Catalogue of Disaster Earthquakes in Japan*, University of Tokyo Press, Tokyo (in Japanese).

VIRIEUX, J., and MADARIAGA, R. (1982), *Dynamic Faulting Studied by a Finite-difference Method*, Bull. Seismol. Soc. Am. *72*, 345–369.

(Received September 30, 1999, revised March 15, 2000, accepted April 6, 2000)

 To access this journal online:
http://www.birkhauser.ch

Pure appl. geophys. 157 (2000) 2083–2104
0033–4553/00/122083–22 $ 1.50 + 0.20/0

Pure and Applied Geophysics

Numerical Simulation of Dynamic Process of the Tangshan Earthquake by a New Method—LDDA

YONGEN CAI,[1] TAO HE[2] and REN WANG[3]

Abstract—LDDA (Lagrangian Discontinuous Deformation Analysis) is a new numerical analysis method to deal with the problems of discontinuous deformation in elastic block systems. The method is based on DDA (Discontinuous Deformation Analysis) and the domain decomposition algorithm. Due to the use of the contact criteria of the DDA, it is not necessary to define a slide line in advance for contact problems, which is needed in the classical FEM. The method prevails over the penalty method in the satisfaction of constrain conditions. By using the domain decomposition algorithm, the efficiency in solving equations is improved greatly.

The process of solving a multi-elastic body system by LDDA is as follows: 1) to find total contact points (Lagrange multiplier points) among the elastic bodies according to the contact criteria of the DDA; 2) to solve the contact forces (Lagrange multipliers) by domain decomposition method; and 3) to calculate the displacement and stress caused by the contact forces and other loads by the FEM for each elastic body.

In this paper, the method is used to model the dynamic process of the Tangshan earthquake ($M_s = 7.8$) of 28 July, 1976 and to directly obtain the quasi-static and dynamic dislocations, shear stress drop, and rupture velocity of the earthquake fault.

The simulation shows that the method can be used to efficiently solve the dynamic problems of earthquakes. It is also applicable for solving rock-engineering problems as a multi-elastic body system.

Key words: Multi-elastic body system, discontinuous deformation, dynamic analysis method, modeling earthquake process.

Introduction

Earthquakes, in mechanics, can be considered as a phenomenon in which rock breaks suddenly and strain energy stored in the seismic source region is released rapidly. Due to inadequate knowledge of the material properties and geometry of seismic sources at depth, it is very difficult to model its dynamic process. The analytical or numerical methods to address the problem are still limited.

[1] Department of Geophysics, Geodynamic Research Center of Peking University, Bejing, 100871, China. E-mail: yongen@pku.edu.cn
[2] Department of Geology, Peking University, Beijing, 100871, China.
[3] Department of Mechanics and Engineering Science, Peking University, Beijing, 100871, China.

In seismology, the finite difference method is widely used to model the dynamic process of earthquakes (ANDREWS, 1973; DAY, 1982; FRANKEL, 1993; OLSEN and ARCHULETA, 1996). It is very suitable and efficient for uniform and homogeneous media as well as simple geometries of seismic sources and boundaries. Although the method has been improved continuously (MADARIAGA, 1976; LEVANDER, 1988; OHMINOTO and CHOUET, 1997), it is still difficult to handle the nonlinearities both in material and geometry. The stability of the solution is conditional. The boundary element method based on Green's functions has been used in dealing with the rupture dynamics of faults (KIM and PAPAGEORGIOU, 1993; FUKUYAMA and MADARIAGA, 1998). It overcomes the numerical difficulties arising from boundaries in the finite difference method, and the solution to be solved is one dimension lower than that obtained by other methods. The limitations of the method, on the one hand, are the same as those in the finite difference method for linear and homogeneous problems, on the other hand, in the non-existence of the fundamental solution. Spectral methods frequently used in fluid dynamics have been introduced in seismology for elastic waves (GAZDAG, 1981; KOSLOFF et al., 1990; CARCIONE, 1991). However, the method is still limited to simple geometry and homogeneous material properties. KOMATITSCH and VILOTTE (1998) proposed a practical spectral element method based on the variational form of the governing equations and Legendre polynomials and Gause-Lobatto Legendre quadrature. This method can deal with inhomogeneous media and complex boundary geometries, however difficulties still exist in modeling the discontinuous interfaces in the solution domain. The finite element method is used widely in mechanics with continuous media. It cannot be directly used for a system with joints or faults across which the displacement can be discontinuous. The joint element was first proposed (GOODMAN et al., 1968) to handle such a system. This is a kind of spring element in mechanics and penalty element in mathematics. It connects the contact surfaces of joints or faults together. The finite element method with joint elements can solve the dynamic problems involving faults. The method implements easy in programming, however the solution so obtained is strongly dependent on the choice of penalty parameter. The finite element method with perturbed Lagrangian multipliers is also used in solving such a problem (SIMO et al., 1985). The method improves the matrix properties of FEM equations but decreases the precision of computed contact forces and increases the number of unknown variables. Usually it must define a slide line or surface in advance. For the contact problems, the displacement and contact force are solved together.

The dynamic problems of the multi-elastic body system with friction are very difficult. In mathematics, it can be considered as solving the problems with variational inequality; in mechanics, it can be treated as contact problems with friction. LDDA (Lagrangian Discontinuous Deformation Analysis) is a new numerical method especially suitable to study contact and impact problems. It is developed from the work of HILBERT et al. (1994) which was based on the DDA

(Discontinuous Deformation Analysis, SHI and GOODMAN, 1985) and the domain decomposition method (PRINTICE and FIKANI, 1992; LIANG and HE, 1993). Three types of contact between elastic bodies are considered in the DDA method for plane problems. They are line-to-line, angle-to-line and angle-to-angle contacts. The first one can be transformed to two angle-to-line contacts. In numerical calculations, the third type also can be treated as angle-to-line contact. In LDDA, contacting judgement criteria in DDA is used to determine the contacting elastic bodies, and the domain decomposition method is employed to seek the solution of the system. Lagrangian multipliers are introduced to represent contact forces on contact surfaces and are unknown variables. Obviously the method prevails over penalty methods, as used in DDA and DEM (Distinct Element Method, CUNDALL, 1988) in the satisfaction of constraint conditions. The method need not define slide lines or surfaces in advance as in FEM. The contact force obtained by the method is precise. The method has been successfully used to analyze the impact problems and sliding of rock slopes (CAI et al., 1996, 1998).

This study is the further improvement and development of our previous work in LDDA. The main purpose of the paper is to introduce this new numerical method and its application to seismology. The method can deal with the dynamic problems with complex geological structure and material distributions. In this paper, as an example of its application in seismology, it is used to model the dynamic process of the Tangshan earthquake ($M_s = 7.8$) of 28 July, 1976, in China. The quasi-static and dynamic dislocations, shear stress drops, and the rupture velocity of the earthquake fault can be obtained directly.

Introduction to LDDA Method

Consider a generic body α in a multi-elastic body system in two dimensions (see Fig. 1). Assume the displacement vector, $\mathbf{d}^\alpha = [d_x^\alpha \, d_y^\alpha]^T$, and traction, $\mathbf{q}^\alpha = [q_x^\alpha \, q_y^\alpha]^T$, are prescribed on the boundaries Γ_a^α and Γ_q^α, respectively, Γ_c^α, Γ_c^β and $\Gamma_c^{\alpha\beta}$ are contact boundaries and possible contact position or interface in the process of motion between the bodies α and β, respectively. $\mathbf{b}^\alpha = [b_x^\alpha \, b_y^\alpha]^T$ is the body force. Assume $\mathbf{u}_0^\alpha$ and $\mathbf{v}_0^\alpha$ are the initial displacement and velocity fields, respectively, then the dynamic problem of the system can be attributed to the initial-boundary value problem as follows.

Conservation of linear momentum for body α implies a governing equation of the form

$$\mathbf{L}^T \boldsymbol{\sigma}^\alpha + \mathbf{b}^\alpha = \rho^\alpha \mathbf{a}^\alpha + \eta^\alpha \mathbf{v}^\alpha \quad (\alpha = 1, \ldots, N), \tag{1}$$

where $\mathbf{v}$ and $\mathbf{a}$ are velocity and acceleration vectors, respectively, and ρ^α and η^α, the density and damping parameter, respectively. Internal stresses $\boldsymbol{\sigma}^\alpha$ are related linearly to internal strains $\boldsymbol{\varepsilon}^\alpha$ via the Hooke's tensor, $\mathbf{D}^\alpha$

$$\sigma^\alpha = \mathbf{D}^\alpha \varepsilon^\alpha. \tag{2}$$

Internal strains may be computed from displacements using the differential operator, $\mathbf{L}$, i.e.,

$$\varepsilon^\alpha = \mathbf{L}\mathbf{u}^\alpha. \tag{3}$$

The boundary conditions for body α are

$$\mathbf{u}^\alpha|\Gamma_u^\alpha = \mathbf{d}^\alpha \tag{4}$$

$$\mathbf{n}^\alpha \sigma^\alpha|\Gamma_c^\alpha = \mathbf{p}^{\alpha\beta} \tag{5}$$

$$\mathbf{n}^\alpha \sigma^\alpha|\Gamma_q^\alpha = \mathbf{q}^\alpha, \tag{6}$$

and the initial conditions are given by

$$\mathbf{u}^\alpha|_{t=0} = \mathbf{u}_0^\alpha \tag{7}$$

$$\mathbf{v}^\alpha|_{t=0} = \mathbf{v}_0^\alpha. \tag{8}$$

Consider the possible contact position ($\Gamma_c^{\alpha\beta}$) between bodies α and β. Let $g_{0_P}^{\alpha\beta} = [g_{P_n}^{\alpha\beta} \ g_{P_s}^{\alpha\beta}]^T$ and $g_P^{\alpha\beta} = [g_n^{\alpha\beta} \ g_s^{\alpha\beta}]^T$ be the initial and current distances or gaps respectively, and $P_P^\alpha = [p_n^\alpha \ p_s^\alpha]^T$ be the contact force between the bodies. The subscripts n and s indicate the normal and tangential components of the vectors on the contact surface.

In the dynamic contact problems, there are three possible contact states or constraints:

1) *Sticking state*

$$g_n^{\alpha\beta} \equiv u_n^\beta - u_n^\alpha + g_{P_n}^{\alpha\beta} = 0, \quad g_s^\alpha \equiv u_s^\beta - u_s^\alpha + g_{P_s}^{\alpha\beta} = 0, \tag{9}$$

$$p_n^\beta = -p_n^\alpha \le 0, \quad p_s^\beta = -p_s^\alpha. \tag{10}$$

$g_n^{\alpha\beta} = 0$ means no penetration and no opening in the normal direction, and $g_s^{\alpha\beta} = 0$, no sliding. Formula (10) expresses Newton's third Law.

2) *Coulomb's sliding state*

$$u_n^\beta - u_n^\alpha + g_{P_n}^{\alpha\beta} = 0, \tag{11}$$

$$v_s \equiv v_s^\beta - v_s^\alpha = -\kappa p_s^\alpha, \tag{12}$$

$$p_n^\beta = -p_n^\alpha \le 0, \tag{13}$$

$$p_s^\alpha = -\mu_d p_n^\alpha v_s / |v_s|. \tag{14}$$

In this case, the normal constraint is as before but sliding can occur in the tangential direction, and the relative sliding v_s is assumed to depend on the mechanic parameter κ and the shear contact force p_s^α which obeys the Coulomb's friction law.

3) *Non-contact state*

$$\mathbf{p}^\beta = -\mathbf{p}^\alpha = 0, \quad g_n^{\alpha\beta} > 0. \tag{15}$$

The contact point is open in the normal direction. In this case there are no forces on the contact surface.

The principle of virtual work can be used to find the solution of the initial-boundary value problem with N elastic bodies and with the constraints mentioned above. At time t it gives (FARHAT and CRIVELLI, 1994)

$$\sum_{\alpha=1}^N (W_e^\alpha - W_c^\alpha) = 0, \tag{16}$$

where

$$W_e^\alpha = \int_\Omega (\delta\varepsilon^\alpha)^T \boldsymbol{\sigma}^\alpha \, d\Omega - \int_\Omega (\delta\mathbf{u}^\alpha)^T [\mathbf{b}^\alpha - \rho^\alpha \mathbf{a}^\alpha - \eta^\alpha \mathbf{v}^\alpha] \, d\Omega - \int_{\Gamma_q^\alpha} (\delta\mathbf{u}^\alpha)^T \mathbf{q}^\alpha \, d\Gamma_q = 0,$$

is the total sum of the virtual work done by external forces acting on the body α and the virtual strain energy within it.

$$W_c^\alpha = W_n^\alpha + W_s^\alpha = \int_{\Gamma_c^\alpha} (\delta u_n^\alpha)^T p_n^\alpha \, d\Gamma_c^\alpha + \int_{\Gamma_c^\alpha} (\delta u_s^\alpha)^T p_s^\alpha \, d\Gamma_c^\alpha$$

is the virtual work done by the contact force $\mathbf{p}^\alpha$ on the boundary Γ_c^α of the body. W_n^α and W_s^α are the virtual works done by the normal contact force p_n^α and tangential contact force p_s^α, respectively.

By using the finite element method, the displacement field $\mathbf{u}_l^\alpha$ in the local coordinate system (n, s) of the body can be approximated by the nodal displacements $\mathbf{u}_e^\alpha$ in the global coordinate system (x, y) of the element

$$\mathbf{u}_l^\alpha = [u_n^p \; u_s^p]^T = \mathbf{N}_P^\alpha \mathbf{H}_e^\alpha \mathbf{S}_e^\alpha \mathbf{U}^\alpha$$

where $\mathbf{N}_P^\alpha$ is the coordinate transposition matrix of the normal and tangential displacements at point P from local (n, s) to global (x, y) coordinates. $\mathbf{H}_e^\alpha$ and $\mathbf{U}^\alpha$ are the shape function matrix of the element and nodal displacement vector of the body α, $\mathbf{S}_e^\alpha$ is an assembly matrix which builds up the connections between $\mathbf{u}_e^\alpha$ and $\mathbf{U}^\alpha$.

To express the distributed force $\mathbf{p}^\alpha$ on the possible contact boundary Γ_c^α (Fig. 1) in terms of the equivalent concentrated force vectors λ_p ($p = 1, \ldots, M_\alpha$, the number of contact points on the body α) at the discrete contact points, the Lagrangian mesh is introduced. The nodal points, or Lagrangian multiplier points, on the boundary can either be the same as or independent of the nodal points of the finite element related to the contact boundary. Then formulae (14) and (15) can be replaced by

$$\lambda_s^p = -\mu_d \lambda_n^p v_s / |v_s| \tag{17}$$

$$\lambda_p^\beta = -\lambda_p^\alpha \quad \text{or} \quad \mathbf{L}_P^\beta \Lambda^\beta = -\mathbf{L}_P^\alpha \Lambda^\alpha \tag{18}$$

where

$$\lambda_p^i = [\lambda_n^i \ \lambda_s^i]^T = \mathbf{L}_P^i \Lambda^i, \quad i = \alpha, \beta.$$

The vector Λ^i is the contact force vector of block i and λ_n^i and λ_s^i are the normal and tangential components at point p on the contact boundary of block i. $\mathbf{L}_P^i$ is the assembly matrix of the contact point p for the body i, which connects λ_p^i and Λ^i.

If the total number of contact points is M_α on the boundary of the body α, then the virtual work done by the contact forces P can be expressed by the nodal displacement vector $\mathbf{U}^\alpha$ and contact force vector Λ^α of the body as (for the contact point without sliding)

$$\begin{aligned}
W_c^\alpha &= \sum_{p=1}^{M_\alpha} (\delta u_n^p \lambda_n^p + \delta u_s^p \lambda_s^p) \\
&= (\delta \mathbf{U}^\alpha)^T \sum_{p=1}^{M_\alpha} \mathbf{G}_P^\alpha \mathbf{S}_\lambda^\alpha \Lambda^\alpha
\end{aligned} \tag{19}$$

$$\mathbf{G}_P^\alpha = (\mathbf{N}_P^\alpha \mathbf{H}_e^\alpha \mathbf{S}_e^\alpha)^T, \quad \mathbf{S}_\lambda^\alpha = \mathbf{L}_P^\alpha.$$

For a contact point undergoing Coulomb sliding, by introducing an assembly matrix $\mathbf{l}_P^\alpha$ and using formula (17), the normal contact force λ_n^α and $\mathbf{G}_P^\alpha$ can be expressed as

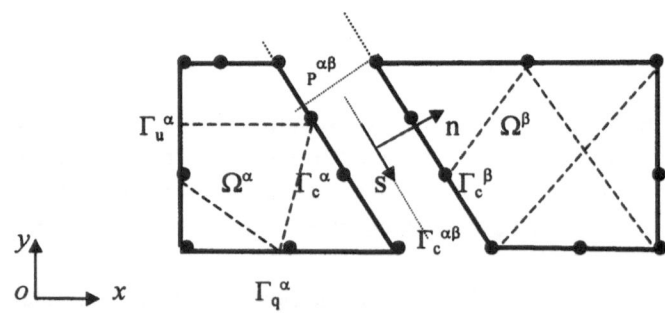

Figure 1
The boundary conditions and meshes of the bodies α and β. Each elastic body is a Lagrangian mesh. The solid, black points on the boundaries are possible contact force points or so-called Lagrangian points. The meshes marked by dashed lines in each body (Lagrangian mesh) are FE meshes.

$$\lambda_n^\alpha = l_P^\alpha \Lambda^\alpha \equiv l_P^{\alpha\beta} \Lambda^{\alpha\beta}$$

$$\mathbf{G}_p^\alpha = (\mathbf{N}_P^\alpha \mathbf{H}_e^\alpha \mathbf{S}_e^\alpha)^T [1 - \mu_d v_s / |v_s|]^T,$$

$$\mathbf{S}_\lambda^\alpha = \mathbf{l}_P^\alpha,$$

and for the non-contact state

$$W_c^\alpha = 0. \tag{20}$$

After employing the finite element method to the body α and using Newmark's time integration scheme in the time interval $[t, t + \tau]$ (BATHE, 1982), we can obtain the virtual work of the body at time $t + \tau$:

$$W^\alpha = W_e^\alpha + W_c^\alpha = (\delta \mathbf{U}^\alpha)^T (\mathbf{K}^\alpha \mathbf{U}^\alpha - \mathbf{G}^\alpha \Lambda^\alpha - \mathbf{F}^\alpha) = 0, \tag{21}$$

where $\mathbf{K}^\alpha$, $\mathbf{U}^\alpha$ and $\mathbf{F}^\alpha$ are the stiffness matrix, nodal displacement vector and equivalent nodal loads of the body respectively, and

$$\mathbf{G}^\alpha = \sum_{P=1}^{M_\alpha} \mathbf{G}_P^\alpha \mathbf{S}_\lambda^\alpha.$$

From formula (21), we can obtain the equilibrium equations with nodal displacements and contact forces of the body, i.e.,

$$\mathbf{K}^\alpha \mathbf{U}^\alpha - \mathbf{G}^\alpha \Lambda^\alpha - \mathbf{F}^\alpha = 0. \tag{22}$$

The displacement vector $\mathbf{U}^\alpha$ of the body α and the contact point force Λ^α between the body α and the body β can be expressed in terms of the nodal displacement vector $\mathbf{U}$ and contact force vector Λ of the multi-elastic body system by introducing the assembly matrices $\mathbf{S}^\alpha$ and $\mathbf{L}^{\alpha\beta}$, that is

$$\mathbf{U}^\alpha = \mathbf{S}^\alpha \mathbf{U}, \quad \Lambda^{\alpha\beta} = \mathbf{L}^{\alpha\beta} \Lambda.$$

Substituting them into formula (21) and using formula (18), we can obtain the equilibrium equations at the nodal points of the multi-elastic body system:

$$\mathbf{KU} + \mathbf{G}\Lambda = \mathbf{F} \tag{23}$$

where $\mathbf{K}$ and $\mathbf{F}$ are the global stiffness matrix and equivalent nodal load vector, with

$$\mathbf{K} = \sum_{\alpha=1}^{N} (\mathbf{S}^\alpha)^T \mathbf{K}^\alpha \mathbf{S}^\alpha,$$

$$\mathbf{F} = \sum_{\alpha=1}^{N} (\mathbf{S}^\alpha)^T \mathbf{F}^\alpha,$$

$$\mathbf{G} = \sum_{\alpha=1}^{N} \sum_{p=\alpha+1}^{N} \sum_{p=1}^{M_{\alpha\beta}} \mathbf{G}_P^{\alpha\beta},$$

where $M^{\alpha\beta}$ is the number of the contact points between bodies α and β, and

$$G_P^{\alpha\beta} = [(\mathbf{S}^\beta)^T \mathbf{G}_P^\beta - (\mathbf{S}^\alpha)^T \mathbf{G}_P^\alpha] \cdot \mathbf{L}_P^{\alpha\beta} \mathbf{L}^{\alpha\beta}.$$

For the sliding contact point,

$$G_P^{\alpha\beta} = [(\mathbf{S}^\beta)^T \mathbf{G}_p^\beta - (\mathbf{S}^\alpha)^T \mathbf{G}_P^\alpha][1 - \mu_d v_s/|v_s|]^T \mathbf{l}_P^{\alpha\beta} \mathbf{L}^{\alpha\beta}.$$

Since the number of unknowns is more than that of the equations (23), additional equations are needed. The constraint conditions (9) can be used to solve the problem. Putting the discretized displacement vector on the contact boundary into (9), we can obtain

$$\mathbf{BU} + \mathbf{D}_0 = 0 \tag{24}$$

where

$$\mathbf{D}_0 = \sum_{\alpha=1}^N \sum_{p=\alpha+1}^N \sum_{p=1}^{M_{\alpha\beta}} (\mathbf{l}_P^{\alpha\beta} \mathbf{L}^{\alpha\beta})^T g_{P_n}^{\alpha\beta},$$

$$\mathbf{B} = \sum_{\alpha=1}^N \sum_{\beta=\alpha+1}^N \sum_{p=1}^{M_{\alpha\beta}} \mathbf{B}_P^{\alpha\beta}.$$

For the contact point without sliding, $(\mathbf{B}_P^{\alpha\beta})^T = \mathbf{G}_P^{\alpha\beta}$. For the sliding contact point, $(\mathbf{B}_P^{\alpha\beta})^T \neq \mathbf{G}_P^{\alpha\beta}$. From the non-penetration condition of formula (9), we can obtain

$$\mathbf{B}_P^{\alpha\beta} = (\mathbf{L}_P^{\alpha\beta} \mathbf{L}^{\alpha\beta})^T [\mathbf{B}_P^\beta \mathbf{S}^\beta - \mathbf{B}_P^\alpha \mathbf{S}^\alpha],$$

$$\mathbf{B}_P^i = \mathbf{n}_P^i \mathbf{H}_e^i \mathbf{S}_e^i, \quad i = \alpha, \beta$$

where $\mathbf{n}_P^i$ is the coordinate transposition matrix of the normal displacement at point P from local (n, s) to global (x, y) coordinates. Finally, we can obtain the displacement U and the contact force Λ by solving formulas (23) and (24). For a system with many elastic bodies, it is not a good choice to solve them all together. The domain decomposition method is used here to solve the equations in LDDA which is very suitable for parallel computation.

Express U in formula (23) in terms of Λ and put it into formula (24), the contact force equation at time $t + \tau$ can be obtained

$$\mathfrak{N}\Lambda = \mathfrak{R} \tag{25}$$

where

$$\mathfrak{N} = \mathbf{B}^T \mathbf{K}^{-1} \mathbf{G} \quad \text{and} \quad \mathfrak{R} = \mathbf{D}_0^T + \mathbf{B}^T \mathbf{K}^{-1} \mathbf{F}.$$

For the sticking state with no sliding point, $\mathfrak{N}$ is a symmetric matrix. It is no longer symmetric when sliding occurs. Note, $\mathfrak{N}$ and $\mathfrak{R}$ should not be calculated directly from the above formulas in LDDA, since calculating $\mathbf{K}^{-1}$ will take excessive CPU time and needs considerable storage space for a huge system with many elastic bodies. They are calculated and assembled in the procedure of forming the stiffness of each elastic body, the amount of calculation is limited because it is related only to the points on the contact surface. From (22) we can obtain

$$\mathbf{U}^\alpha = \mathbf{X}^\alpha \mathbf{\Lambda}^\alpha + \mathbf{Y}^\alpha, \tag{26}$$

where

$$\mathbf{X}^\alpha = (\mathbf{K}^\alpha)^{-1} \mathbf{G}^\alpha, \quad \mathbf{Y}^\alpha = (\mathbf{K}^\alpha)^{-1} \mathbf{F}^\alpha$$

that is

$$\mathbf{K}^\alpha \mathbf{X}^\alpha = \mathbf{G}^\alpha,$$
$$\mathbf{K}^\alpha \mathbf{Y}^\alpha = \mathbf{F}^\alpha.$$

Since $\mathbf{K}^\alpha$ has been decomposed into a triangular matrix, it is easy to compute the vectors $\mathbf{X}^\alpha$ and $\mathbf{Y}^\alpha$, i.e., $(\mathbf{K}^\alpha)^{-1} \mathbf{G}^\alpha$ and $(\mathbf{K}^\alpha)^{-1} \mathbf{F}^\alpha$. They will be used in constructing $\mathfrak{N}$ and $\mathfrak{R}$. After the contact force Λ is obtained from formula (25), the displacement vector can be obtained from (26) for each elastic body. Then the velocity and acceleration vectors as well the stress are easily calculated by the procedure of the finite element method. The efficiency of this solving process is obviously better than that in classical algorithms, especially in modeling the dynamic problems with many elastic bodies and a multitude of nodal points in each body.

To sum up, the process of solving the dynamic problems of multi-elastic body system by using the LDDA is as follows:

Assume the displacement $\mathbf{U}$, velocity $\mathbf{V}$ and acceleration $\mathbf{A}$ are known at time t, then the contact force Λ, $\mathbf{U}$, $\mathbf{V}$ and $\mathbf{A}$ at time $t + \tau$ can be obtained by following the steps:

1. Loop for all the elastic bodies to form $\mathbf{K}^{t+\tau}$ and $\mathbf{F}^{t+\tau}$.
2. Loop for all the elastic bodies and find the total number of contact points in the system by employing the DDA method.
3. Loop for all the elastic bodies.
   – calculate $\mathbf{G}^\alpha$, $\mathbf{B}^\alpha$ and $\mathbf{D}_0^\alpha$ and
   – assemble them into $\mathfrak{N}$ and $\mathfrak{R}$.
4. Solve Λ to match the contact constraint conditions from (25).
5. Loop for all the elastic bodies and solve for $\mathbf{U}^\alpha$ from (24) and then calculate velocity and acceleration vectors and the stress in the elastic body.

Modeling Tangshan Earthquake by the LDDA

1. The Procedures of Modeling the Dynamic Process of Earthquakes

In mechanics, earthquake occurrence can be considered as the result of the shear stress on the fault and overcomes the frictional resistance of the fault. From this point of view, WANG et al. (1983) simulated the earthquake sequence in North China in the last 700 years by a static elasto-plastic finite element method. The earthquakes in their study were modeled by reducing the friction coefficient of faults. We will take the idea to model the dynamic process of the Tangshan earthquake ($M_s = 7.8$), which occurred on July, 28, 1976, in China, by means of the LDDA method. Suppose the tectonic field far from Tangshan is stable throughout the duration of the dynamic process of the earthquake. The modeling is separated into two steps: the first step is to build up the initial stress field according to the far stress field, and the second step is to artificially reduce the friction coefficient at the middle segment of the fault to trigger the Tangshan earthquake. Subsequently the rupture development of the fault obeys Coulomb's friction law.

2. The Geometry and Mechanic Models of the Tangshan Earthquake

The geometry model used to simulate the Tangshan earthquake is shown in Figure 2. Its area is 250 km × 240 km, the region is divided into 25 elastic subregions or Lagrangian meshes. The total number of nodal points of the multi elastic body system is 10,909. The length of the Tangshan earthquake fault is taken as 60 km and is constructed of 24 finite element boundaries (25 nodal points),

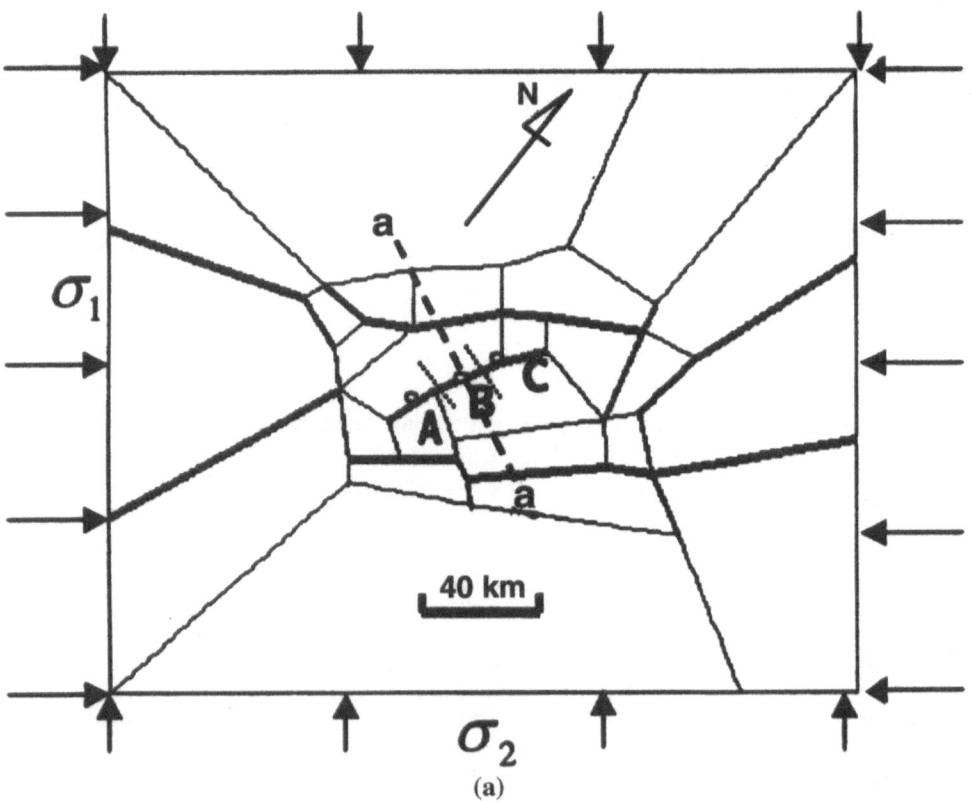

(a)

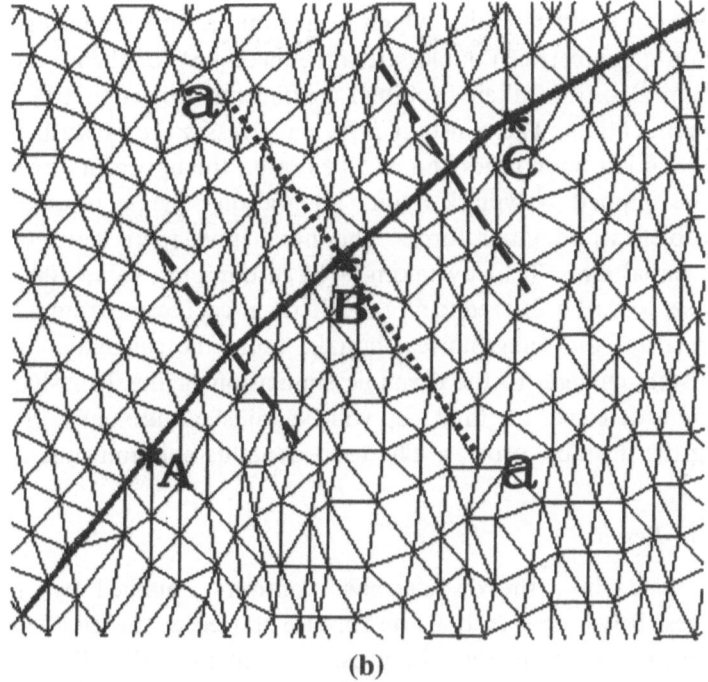

(b)

Figure 2

a) Model geometry and boundary conditions used to simulate the Tangshan earthquake (EARTHQUAKE EDITORIAL GROUP OF THE 1976 TANGSHAN EARTHQUAKE, SSB, 1982). The solid lines are faults marked by the letters A, B and C. b) Local FE meshes in the area surrounding the Tangshan earthquake fault.

which is marked by A, B and C, with the location of point B as the epicenter. The fault also is divided into three segments by dashed lines, of which the middle segment length is 20 km. The aim in doing so is to simulate the different frictional strengths of the fault.

As the code of the LDDA is limited to two-dimensional problems in this study, the small deformation, plane stress and homogeneous material model is adopted. The Poisson's ratio v and density ρ are taken as 0.25 and 2700 kg/m^3, respectively. The average velocity of P wave in this area is taken as 6.56 km/s. The Young's modulus $E = 9.72 \times 10^4$ MPa is obtained from the well-known formula for the Poisson's ratio

$$V_p^2 = 3E/2\rho(1 + v).$$

The faults in the area are simulated by the contact surface with Coulomb's friction law, and across which the displacement can be discontinuous when an earthquake occurs. The frictional coefficient of the middle and the remaining two segments on the Tangshan fault are taken as 0.8 and 0.6, respectively when it is in the sticking state. The frictional coefficients on the other faults are set to 0.5. They will automatically drop to 0.2 when the shear stress on the faults is greater than the frictional resistance of the faults, which means an earthquake occurs. The contact segments, not faults, between bodies are introduced just for the need of the domain decomposition algorithm, their frictional coefficients are set to 10^{20}, a large value used to prevent sliding. The constant principal stresses are prescribed as 100 MPa on the right and left boundaries and as 50 MPa on the upper and lower boundaries (Fig. 2). The implicit Newmark's time integration scheme is used to solve the dynamic problem with contact forces. The initial time step is set to 0.5 s to form a quasi-static initial stress field by taking a large damping value 100, and then the time step is set to 0.01 s. By reducing the frictional coefficient from 0.6 to 0.2 at the middle segment of the Tangshan earthquake fault, the calculation state from the quasi-static state changes to the dynamic simulation of the Tangshan earthquake. When sliding is detected, the time step will be adjusted automatically for the needs of modeling the unstable process. The total dynamic solution time is 30 s.

3. Simulation Results

Figures 3a–d show the displacement contours obtained by the LDDA at different times. It can be seen that the rupture started at the middle segment (Fig. 3a) and then propagated to the southwest segment first (Fig. 3b) and later to the northeast (Fig. 3c). The displacement distributions of the P and S waves are basically consistent in magnitude with that predicted by the seismic dislocation theory. The maximum shear stress contour at 17.5 s after the earthquake is shown in Figure 3e. Usually the stress is calculated at the Gauss integration points of the element in the general FEM. In the LDDA the stress is given at the nodal points of the element. For a typical element with three nodal points, say element e, it is calculated by constructing the following functional

$$I = \int_{\Omega_e} (\sigma_P^\alpha)^T \sigma_P^\alpha \, d\Omega_e^\alpha. \tag{27}$$

Express the stress σ_P^α and strain ε_P^α at point P in the body α in terms of the unknown nodal stress vector σ_c^α and the known nodal displacement vector $\mathbf{U}_c^\alpha$ respectively, as follows:

$$\sigma_P^\alpha = \mathbf{H}_e \sigma_e^\alpha, \quad \varepsilon_P^\alpha = \mathbf{B}_e \mathbf{U}_e^\alpha,$$

$$\sigma_e^\alpha = [\sigma_x^1 \sigma_y^1 \sigma_{xy}^1 \ \sigma_x^2 \sigma_y^2 \sigma_{xy}^2 \ \sigma_x^3 \sigma_y^3 \sigma_{xy}^3]^T,$$

$$\mathbf{U}_e^\alpha = [u_x^1 u_y^1 \ u_x^2 u_y^2 \ u_x^3 u_y^3]^T,$$

where $\mathbf{B}_e$ is the strain-displacement connecting matrix. The superscripts in σ_c^α and $\mathbf{U}_c^\alpha$ indicate the nodal point number. Substituting them into formula (27) and taking its variation, we can obtain.

$$\mathbf{M}_e^\alpha \sigma_e^\alpha = \mathbf{R}_e, \tag{28}$$

where

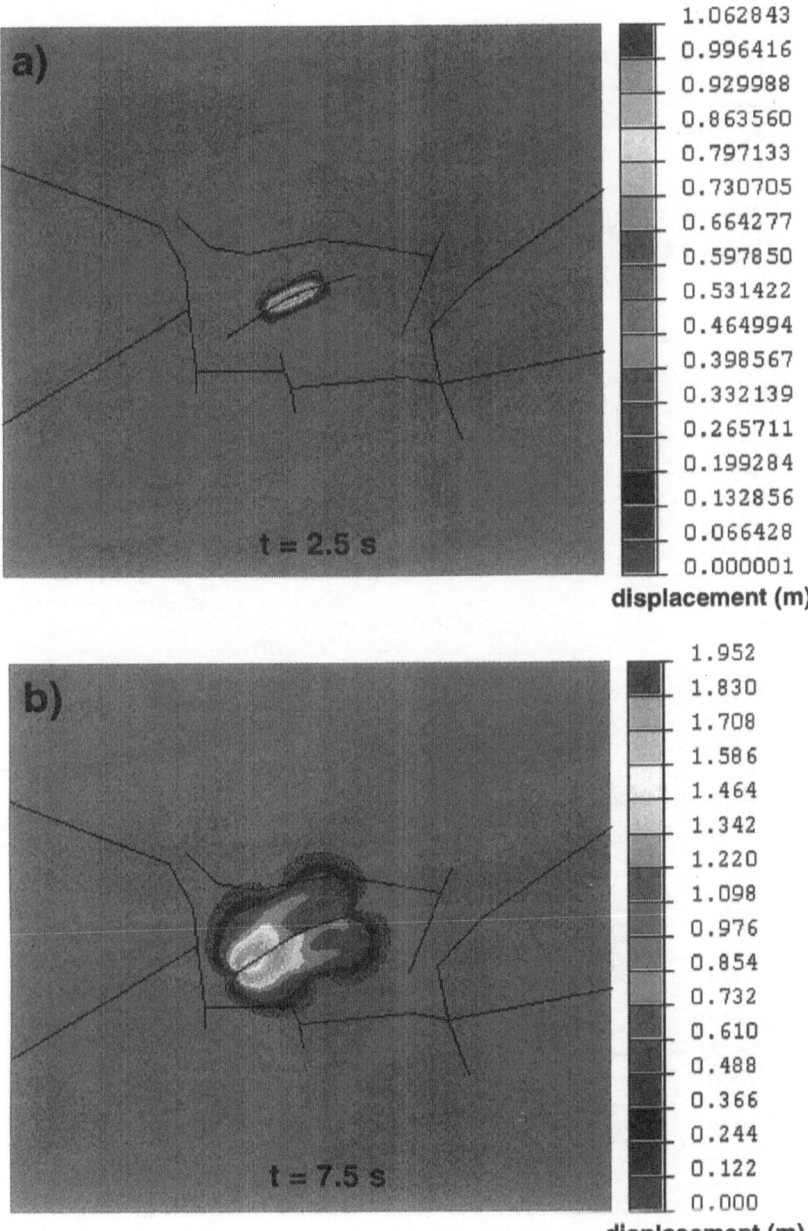

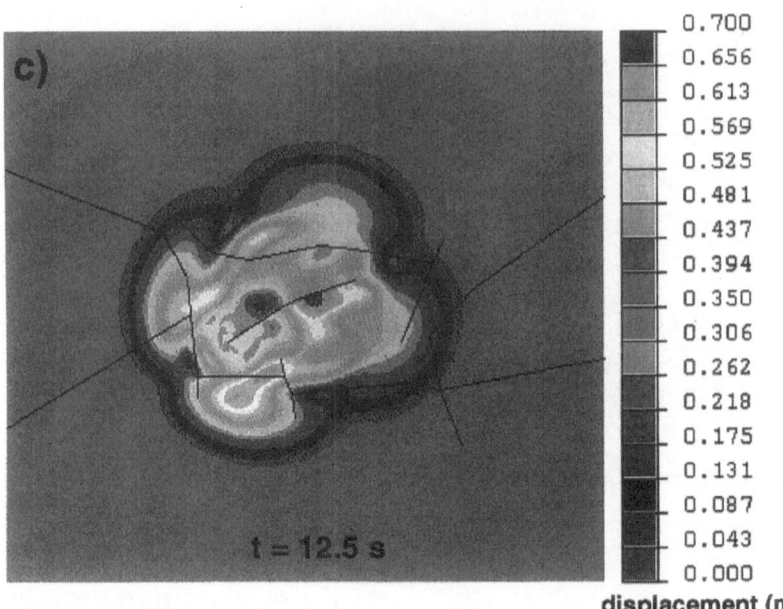

displacement (m)

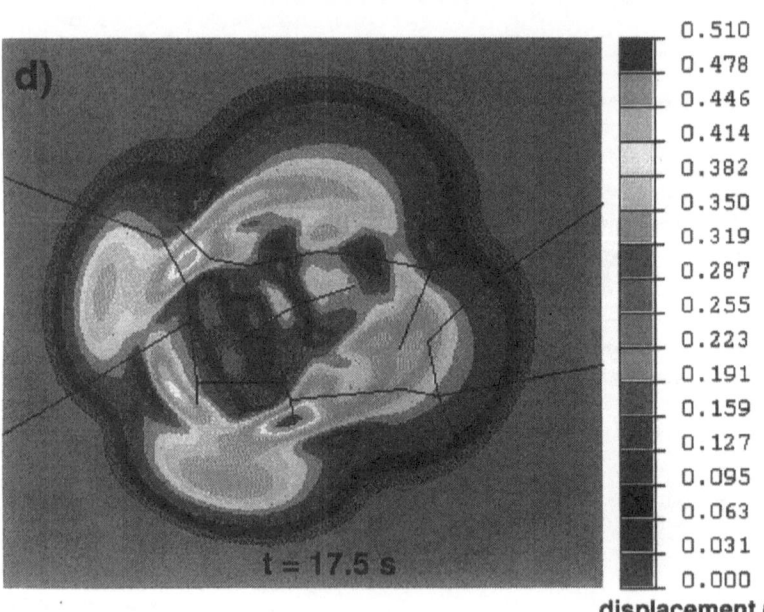

displacement (m)

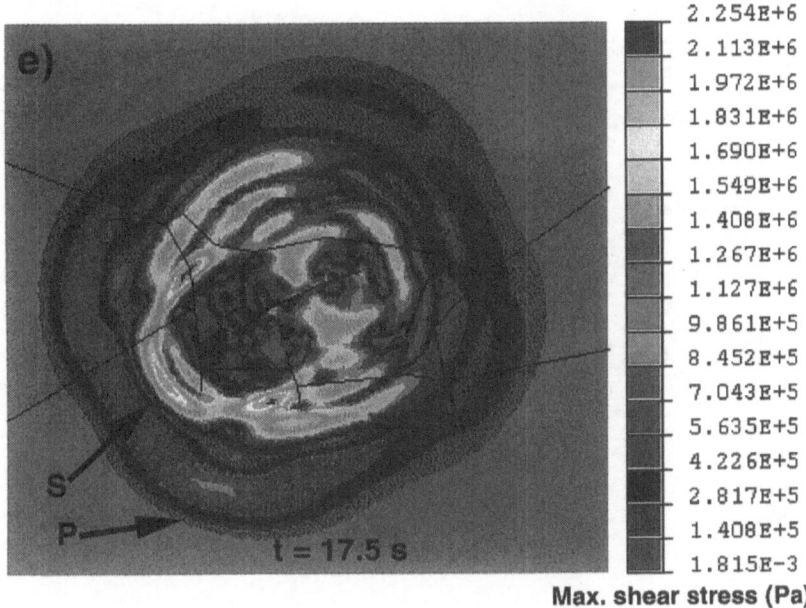

Max. shear stress (Pa)

Figure 3
(a)–(d) simulated displacements in the duration of the Tangshan earthquake. (e) Maximum shear stress contour in the duration of the Tangshan earthquake.

$$\mathbf{M}_e^\alpha = \int_{\Omega_e} \mathbf{H}_e^T \mathbf{H}_e \, d\Omega_e,$$

$$\mathbf{R}_e^\alpha = \int_{\Omega_e} \mathbf{H}_e^T \mathbf{D}_e \mathbf{B}_e \mathbf{u}_e^\alpha \, d\Omega_e.$$

The stress can then be solved from formula (28). The maximum shear stress at the nodal point can be easily obtained using the classical elasticity theory.

We can see that the distribution of the dynamic maximum shear stress is very different in the area, the magnitude in the southwest region is greater than that in the northeast region. The result can be used to explain why the intensity of the Tangshan earthquake in the northeast region attenuated faster than that in the southwest (EARTHQUAKE EDITORIAL GROUP OF THE 1976 TANGSHAN EARTH-QUAKE, SSB, 1982). The P and S waves can be easily identified from their travel times and speeds. We have prescribed the speed of P wave as 6.6 km/s, from this we compute the speed of the S wave as 3.8 km/s. The travel time is 17.5 s, therefore we know that they should travel 115.5 km and 66.5 km, respectively during the time interval. The identified P and S waves are marked by the black arrows in Figure 3e. The distances from the epicenter, or point B, to the places pointed at by the arrows are 114.9 km and 65.2 km, respectively. From the results we can see the contribution of contours in the process of the Tangshan earthquake.

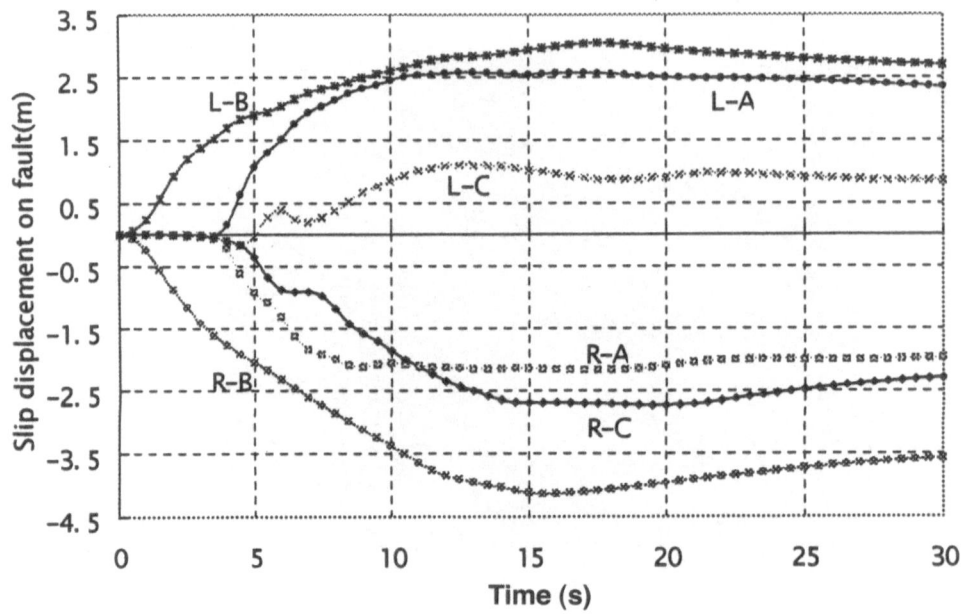

Figure 4
Simulated slip displacements at the Tanshang earthquake fault.

Figure 4 shows the tangential displacement versus time at the points A, B and C on the two sides of the Tangshan earthquake fault during the Tangshan earthquake. The letters L (left) and R (right) represent the northwest side and the southeast side of the fault respectively. These curves are the "seismic source time functions," which are usually prescribed in the dislocation model for modeling the seismic process. It shows that the slip on the two sides of the fault is asymmetric. The maximum dynamic and quasi-static relative displacements are 7.1 m and 6.2 m respectively. The average quasi-static relative displacement after 30 s is about 4.5 m. We can also see that the average for the concave side of the fault is greater than that for the convex side of the fault. At the same side of the fault, the slips near the one end in tension exceed that near the other end in pressure. The curves also show the time at which the rupture begins (the word "rupture" used below means sliding). We know that the time is the travel time of the propagating rupture. The time from the middle segment to the southwest end (point A) and the northeast and (point C) of the Tangshan earthquake fault is 3.5 s and 4.5 s, respectively. The distance from each point to its initial rupture source is 10,770 m and 5320 m,

Figure 5
a) Simulated displacement vector field after the Tangshan earthquake. b) Observed displacement vector field after the Tangshan Earthquake (EDITORIAL GROUP OF THE 1976 TANGSHAN EARTHQUAKE, SSB, 1982). 1. Seismic fault from the crustal deformation analysis after the earthquake. 2. Seismic fault inverted from the geodesy data. 3. Horizontal displacement vector. 4. Displacement vector scale.

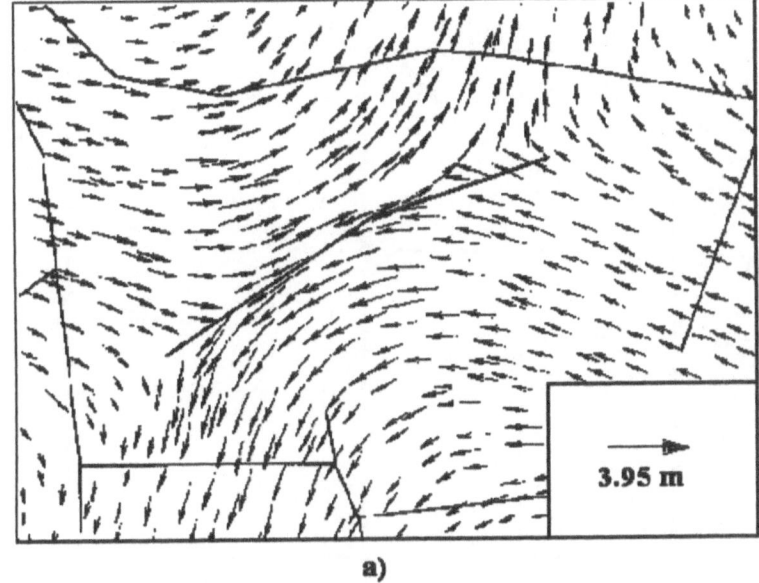

a)

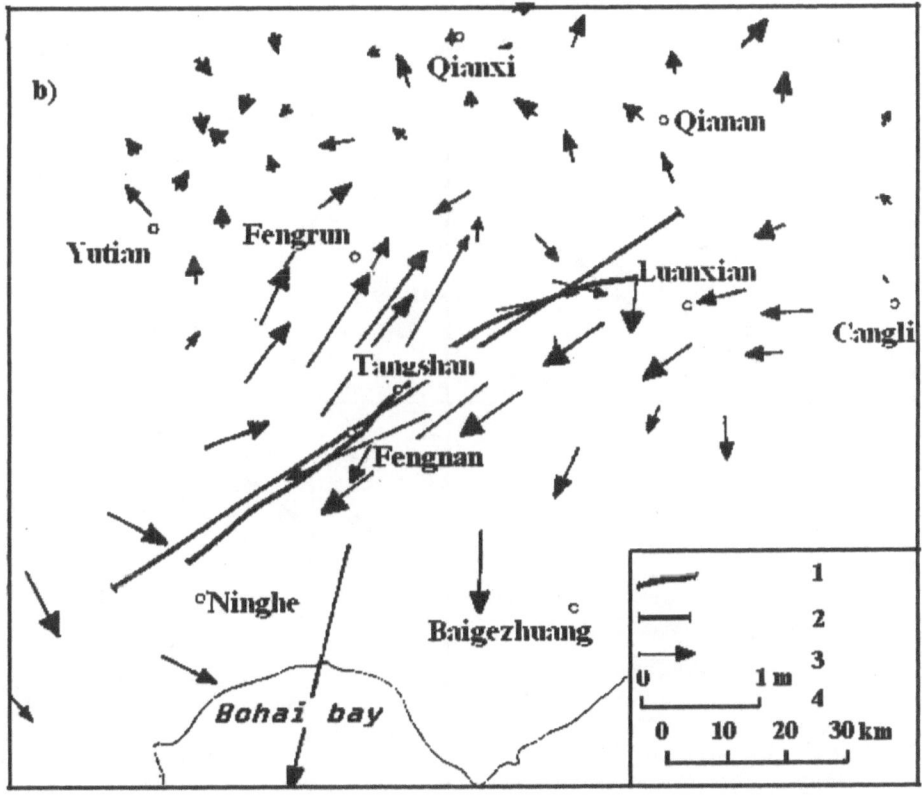

b)

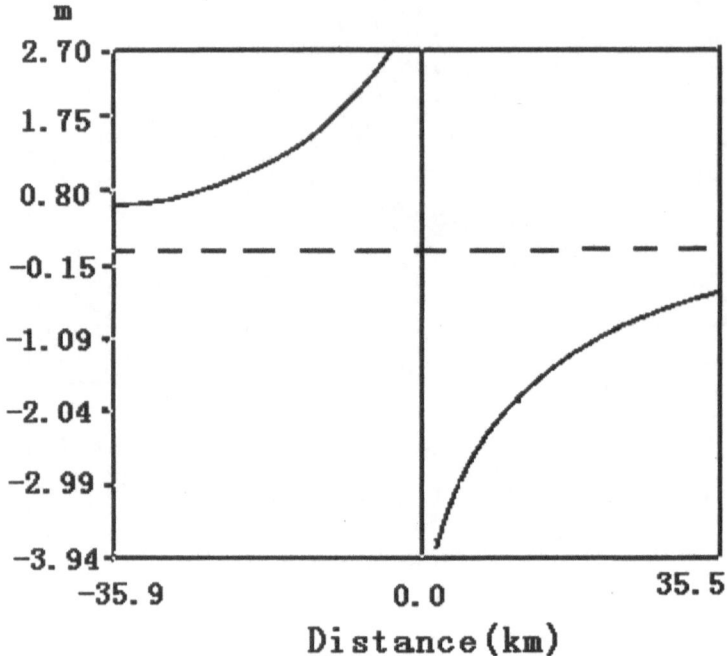

a)

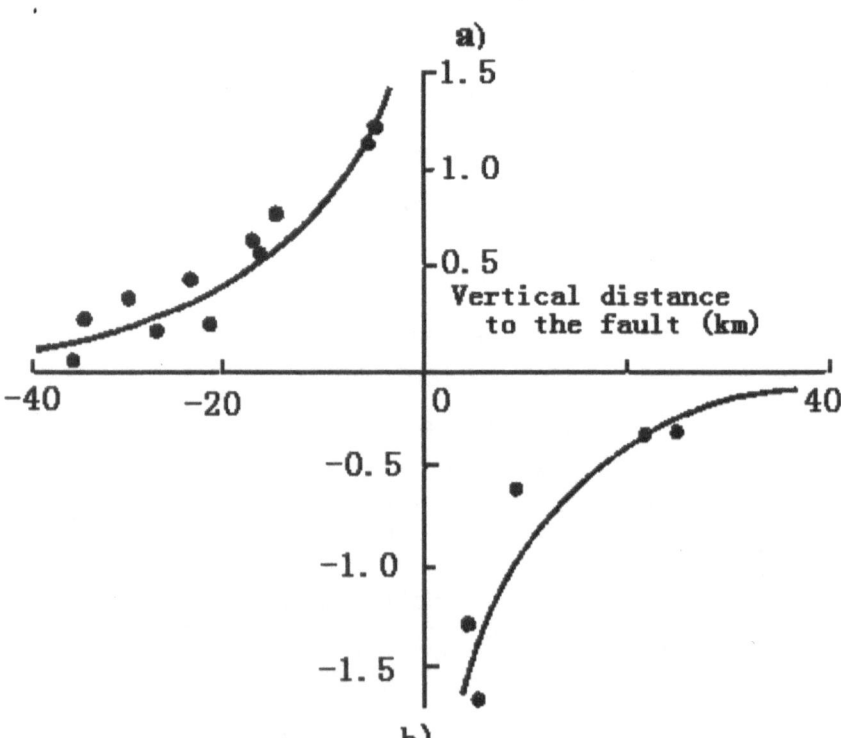

b)

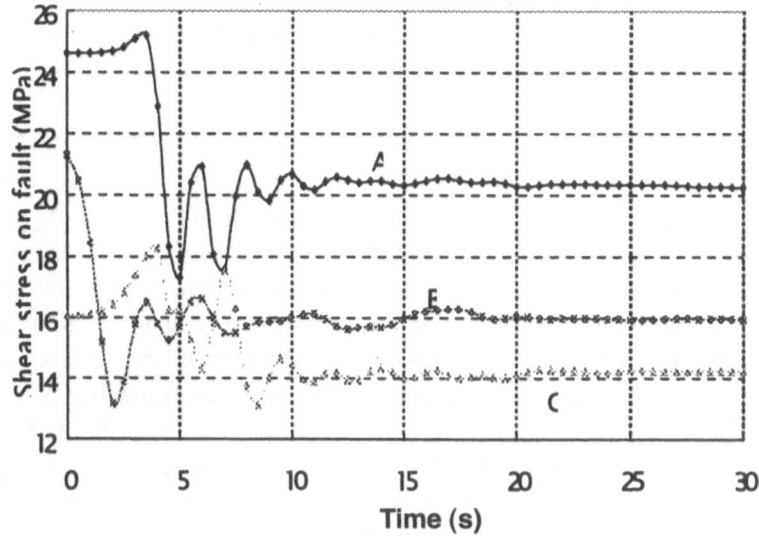

Figure 7
Simulated dynamic shear stress along the Tangshan earthquake fault.

respectively. Using the travel time and the distance, we can deduce that the rupture velocity of the fault is 3.08 km/s in the southwest and 1.18 km/s in the northeast, respectively. The average is 2.13 km/s, which is consistent with 2.5 km/s derived by BUTLER *et al.* (1979) by surface wave inversion, using a symmetric finite moving source model. We also find that the "overshoot" is different on either side of the fault.

Figure 5a shows the simulated displacement vector field in a quasi-static state after the Tangshan earthquake. We can see that the displacement vectors near the Tangshan earthquake fault are greater than those distant from the fault, and that the vectors at the concave side of the fault are greater than those at the convex side of the fault. This may be caused by the nonplanar shape of the fault surface. We also can see that the displacement vector is greater at the middle segment and attenuates to the ends of the Tangshan earthquake fault. The features of the distribution agree fairly we with the observed data (Fig. 5b) after the earthquake (EARTHQUAKE EDITORIAL GROUP OF THE 1976 TANGSHAN EARTHQUAKE, SSB, 1982).

Figure 6 displays the displacement distribution perpendicular to the profile, a-a in Figure 2. The "elastic rebound" can be clearly seen. The displacements are asymmetric about the Tangshan earthquake fault. The most extensive deformation is confined to within 35 km of the fault, which is also consistent with observations after the earthquake (EARTHQUAKE EDITORIAL GROUP OF THE 1976 TANGSHAN EARTHQUAKE, SSB, 1982).

Figure 6
a) Simulated "elastic rebound" of the Tangshan earthquake, the displacements along the perpendicular a–a. b) Observed displacements along the perpendicular a–a (EARTHQUAKE EDITORIAL GROUP OF THE 1976 TANGSHAN EARTHQUAKE, SSB, 1982).

Figure 7 indicates the simulated dynamic shear stresses at points A, B and C on the Tangshan earthquake fault. The quasi-static shear stress can be calculated by

$$\sigma = (\sigma_1 - \sigma_2) \sin \theta / 2.$$

Taking $\sigma_1 = 100$ MPa, $\sigma_2 = 50$ MPa and $\theta_A = 40°$, $\theta_B = 30°$, $\theta_C = 20°$ in the formula, we can derive the shear stresses as 24.6 MPa, 21.6 MPa, 16.0 MPa, respectively, to match the results in Figure 7 at $t = 0.0$ s. The results reveal that the stress distribution is not uniform on the fault. It can be seen that the shear stresses diverge sharply with time at these points. The severe changes occur during about the first 10 s of the earthquake. The shear stress drops can be directly obtained from the figure. The maximum dynamic shear stress drop, the difference of the maximum and minimum shear stresses, is 8.1 MPa at point B reached at 2 seconds after the earthquake starts; the drops at points A and C are 7.3 MPa and 3.0 MPa, and reached at 5.0 s and 8.0 s, respectively. The average dynamic shear stress drop on the fault is 6.1 MPa. The quasi-static shear stress drops, the difference of the shear stress at the beginning and the end of the earthquake, are about 4.3 MPa, 5.4 MPa and 1.9 MPa, respectively. The average quasi-static stress drop on the fault is 3.9 MPa, which is consistent with 3.0 MPa derived from the seismic surface wave by BUTLER *et al.* (1979). Also we can see that the stress wave arrives at point A earlier than at point C. It is very interesting to see that the shear stress at the two points increases before a sharp shear stress drop when ruptures begin. It seems that the rupture velocity is related to the initial stress field, i.e., the stronger the initial stress at the fault, the faster the rupture speed of the fault.

Conclusions

The preliminary results demonstrate that the LDDA method can be used to model the process of earthquakes. The dislocation, seismic source time function, shear stress drop and rupture velocity of an earthquake fault can be obtained directly by controlling the change of the frictional coefficient of the faults. The method can handle inhomogeneous material distributions and irregular, complex geometry of seismic sources and boundaries. It can also be used to solve contact and impact problems as well as earthquake engineering obstacles.

The contact force obtained by the LDDA is precise. This is very important in modeling the process of earthquakes. It is favorable to parallel computation, achieving a higher efficiency than that of other methods.

The modeling results obtained by the LDDA method are very preliminary. The computer used for the simulation is a Pentium II PC with 64 Mb RAM and 2 Gb of disk space. Due to the limitations of this machine, the FE mesh is not fine enough to accurately resolve high frequency waves. Assuming a mesh spacing of only 2.5 km and that 10 nodal points are required to resolve the shortest wavelengths, P- and S-wave periods larger than about 4 s and 6 s can be identified in the simulation. The periods used in studying the source process may be acceptable. However, for the higher frequency waves emitted from the rupture, the mesh spacing is insufficient, and considerably finer resolutions are required to accurately describe the rupture process at say 1 Hz. The corresponding mesh spacing required is 600 m and 400 m for P and S waves, respectively. The frequency for structural engineering needs is about 4 Hz (BIELAK and GHATTAS, 1999), requiring a far finer mesh. In this case a parallel supercomputer is required to solve the simultaneous equations with millions of degrees of freedom. The inaccurate simulation of high frequency waves arises from the limited mesh size not from the LDDA method. This problem may be easily overcome by using a high performance computer. The displacement field is not affected substantially by higher frequency waves. Therefore, the simulated displacement field for the dynamic (Figs. 3a–d) and the quasi-static states (Fig. 5) reflects the main characteristics of displacements excited by the earthquake.

The aim of the paper is to introduce a new numerical method which can deal with the dynamic problems that have complex geological structure and material distributions. Modeling the dynamic process of the Tangshan earthquake is an example of its newly attempted application to seismology. Therefore the modeling results presented here are rather preliminary. Further improvements are needed not only to the theory but also in the implementation. Further work entails use of the rate- and state-dependent friction law and extention of the LDDA code to 3-D.

Acknowledgements

We wish to thank Chen Yuntai for his discussions, and appreciation is extended to Peter Mora for his supportive suggestions to the paper and Dion Weatherley for correcting errors resulting from transportation of the manuscript file. We are grateful also to the reviewers for their comments and suggestions. This work was supported by SSTC (95-S-05), NSFC (58779013), and JESF (197008).

REFERENCES

ANDREWS, D. J. (1973), *Rupture Velocity of Plane Strain Shear Cracks*, J. Geophys. Res. *81*, 5679–5687.
BATHE, K. J., *Finite Element Procedures in Engineering Analysis* (Prentice-Hall, Inc. Englewood Cliffs, New Jersey 1982).
BIELAK, J., and GHATTAS, O., *Computational challenges in seismology*, APEC Cooperation for Earthquake Simulation, 1st ACES Workshop Proceedings (ed. P. Mora) (GOPRINT, Brisbane 1999).
BUTLER, R. G., STEWART, S., and KANAMORI, H. (1979), *The July 27, 1976 Tangshan, China Earthquake—A Complex Sequence of Intraplate Events*, Bull. Seismol. Soc. Am. *69*, 207–220.
CAI, Y., LIANG, G. P. SHI, G. H., and COOK, N. G. W., *Studing an impact problem by using LDDA*. In *Discontinuous Deformation Analysis (DDA) and Simulation of Discontinuous Media* (eds. Salami, M. R., and Banks, D.) (TSI 1996).

CAI, Y., LIANG, G. P., and YIN, Y. (1998). *A New Numerical Method Modeling Dynamic Process of Rockslides*, Numer. Anal. Meth. in Rock and Soil Mech., STG, Guangdong.

CARCIONE, J. M. (1991), *Domain Decomposition for Wave Propagation Problem*, J. Sci. Comp. *6*, 453–482.

CUNDALL, P. A. (1988), *Formulation of a Three-dimensional Distinct Element Model, Part I*, Int. J. Rock Mech. Sci. and Geomech. *25*, 107–116.

DAY, S. M. (1982), *Three-dimensional Simulation of Spontaneous Rupture: The Effect of Nonuniform Prestress*, Bull. Seismol. Soc. Am. *72*, 1881–1902.

FARHAT, C., and CRIVELLI, L. (1994), *A Transient FETI Methodology for Large-scale Parallel Implicit Computations in Structural Mechanics*, Int. J. Numer. Meth. Eng. *37*, 1945–1975.

FRANKEL, A. (1993), *Three-dimensional Simulations of Ground Motions in the San Bernardino Valley, California, for Hypothetical Earthquakes on the San Andreas Fault*, Bull. Seismol. Soc. Am. *83*, 1020–1041.

FUKUYAMA, E., and MADARIAGA, R. (1998), *Rupture Dynamics of a Planar Fault in 3-D Elastic Medium: Rate- and Slip-weakening Friction*, Bull. Seismol. Soc. Am. *88*, 1–17.

EARTHQUAKE EDITORIAL GROUP OF THE 1976 TANGSHAN EARTHQUAKE (ed. by the State Seismology Bureau of China) (Earthquake Press 1982).

GAZDAG, J. (1981), *Modeling of the Acoustic Wave Equation with Transform Methods*, Geophys. *46*, 854–859.

GOODMAN, R. E., TAYLOR, R. L., and BREKKE, T. L. (1968), *A Model for Mechanics of Jointed Rock*, Proc. Am. Soc. Civ. Engr. *94*, SM3, 637–659.

HILBERT, L. B., Jr., YI, W., COOK, N. G. W., CAI, Y., and LIANG, G. P. (1994), *A New Discontinuous Finite Element Method for Interaction of Many Deformable Bodies in Geomechanics*, Proceed. 8th Int. Conf. Comp. Meth. Adv. Geomech., 931–836.

KIM, J., and PAPAGEORGIOU, A. (1993), *Discrete Wavenumber Boundary Element Method for 3-D Scattering Problem*, J. Eng. Mech., ASCE *119*, 603–624.

KOSLOFF, D., KESSLER, A., FILHO, Q., TESSMER, E., BEHLE, A., and STRAHILEVITZ, R. (1990), *Solution of the Equations of Dynamic Elasticity by a Chebyshey Spectral Method*, Geophysics *55*, 748–754.

KOMATITSCH, D., and VILOTTE, J. P. (1998), *The Spectral Element Method: An Efficient Tool to Simulate the Seismic Response of 2-D and 3-D Geological Structures*, Bull. Seismol. Soc. Am. *88*, 368–392.

LEVANDER, A. R. (1988), *Fourth-order Finite-difference P-SV Seismograms*, Geophysics *53*, 1425–1436.

LIANG, G. P., and HE, J. (1993), *The Non-conforming Domain Decomposition Method for Elliptic Problems with Lagrangian Multipliers*, Chinese J. Num. Math. and Appl. *15*, 8–19.

MADARIAGA, R. (1976), *Dynamics of an Expanding Circular Fault*, Bull. Seismol. Soc. Am. *65*, 163–182.

OHMINOTO, T., and CHOUET, A. B. (1997), *A Free Surface Boundary Condition for Including 3-D Topography in the Finite Difference Method*, Bull. Seismol. Soc. Am. *87*, 494–515.

OLSEN, K. B., and ARCHULETA, R. J. (1996), *B-D Simulation of Earthquakes on the Los Angeles Fault System*, Bull. Seismol. Soc. Am. *86*, 575–596.

PRINTICE, J. K., and FIKANI, M. M., *A Domain Decomposition Technique for Computational Solid Dynamics*, Fifth Intern. Symp. on *Domain Decomposition Methods for Partial Differential Equations* (SIAM, Philadelphia 1992) pp. 541–544.

SHI, GENG-HUA, and GOODMAN, R. E. (1985), *Two-dimensional Discontinuous Deformation Analysis*, Int. J. Numer. Anal. Meth. Geomech. *9*, 541–556.

SIMO, J. C., WRIGGERS, P., and TAYLOR, R. L. (1985), *A Perturbed Lagrangian Formulation for the Finite Element Solution of Contact Problem*, Comp. Meth. Appl. Mech. Engr. *50*, 163–180.

WANG, R., XUNYING, S., and CAI, Y. (1983), *A Mathematical Simulation of Earthquake Sequence in North China in Last 700 Years*, Scientia Sinica (in English) *26* (1), 32–42.

(Received July 30, 1999, revised March 29, 2000, accepted April 7, 2000)

Pure appl. geophys. 157 (2000) 2105–2124
0033–4553/00/122105–20 $ 1.50 + 0.20/0

| Pure and Applied Geophysics

Nonlinear Structural Subsystem of GeoFEM for Fault Zone Analysis

Mikio Iizuka,[1] Hiroshi Okuda[2] and Genki Yagawa[3]

Abstract—GeoFEM (Iizuka *et al.*, 1999) is a parallel finite element analysis system intended for multi-physics, multi-scale problems of solid earth field phenomena. Very large linear elastic problems have already been resolved by parallel computation with GeoFEM. The next stage is to examine large-scale nonlinear problems using GeoFEM. The analysis of large-scale contact problems for fault zones is particularly important in the development of models that simulate the occurrence and cycle of earthquakes.

This paper proposes a parallel FEM using an iterative solver with the augmented Lagrange method for solving large-scale contact problems. Direct solvers are presently applied to contact problem analysis because the matrix for such problems is ill-conditioned. However, direct solvers are not suitable for large-scale matrices. The augmented Lagrange method can improve the matrix conditions. The present study evaluates a parallel FEM using an iterative solver with the augmented Lagrange method. Analysis of a contact problem with the augmented Lagrange method revealed that an optimal penalty parameter exists and that large-scale parallel contact analysis using the iterative solver with localized preconditioning is promising.

Key words: Fault zone, contact problem, augmented Lagrange method, iterative solver, GeoFEM.

1. Introduction

Solid earth simulations have recently been developed (Rundle *et al.*, 1999; Bielak and Ghattas, 1999; Zienkiewicz *et al.*, 1993; Sivathasan *et al.*, 1998; Zhao *et al.*, 1998) to address issues such as natural disasters, global environmental destruction and the conservation of natural resources. The simulation of solid earth phenomena involves the analysis of complex structures including strata, faults, and heterogeneous material properties. Because complex phenomena such as multi-phases and chemical reactions must be addressed, simulations require much greater computing capacity than that which is currently available.

---

[1] Research Organization for Information Science & Technology-RIST, 1-18-16 Hamamatsucho Minato-ku Tokyo, 105-0013, Japan. E-mail: iizuka@tokyo.rist.or.jp

[2] Department of Mechanical Engineering and Materials Science, Faculty of Engineering, Yokohama National University, 79-5 Jobandai Hodogaya-ku Yokohama City, 240-0067 Japan. E-mail: okuda@typhoon.cm.me.ynu.ac.jp

[3] Department of Quantum Engineering and Systems Science, The University of Tokyo, 7-3-1 Hongo Bunkyo-ku Tokyo, 113-0033 Japan. E-mail: yagawa@garlic.q.t.u-tokyo.ac.jp

The Science and Technology Agency have therefore promoted "the earth simulator project," in an effort to predict geological changes. One of the large-scale software programs developed through this project is the parallel finite element analysis system called GeoFEM (see also http://geofem.tokyo.rist.or.jp), which is designed to handle solid earth phenomena.

GeoFEM has conventionally used an iterative solver with the ICCG method (MEIJERINK and VAN DER VORST, 1977). This is considered to be the most suitable means of solving symmetric definite matrices and produces outstanding results in the field of large-scale linear elastic analysis (GARATANI *et al.*, 1999). A current challenge is to develop nonlinear analysis methods based on these results.

Simulation of the generation and cycle of earthquakes is particularly important in the nonlinear analysis of solid earth phenomena, however such a simulation requires analysis of a complex fault network (MATSU'URA and SATO, 1997). To perform such analyses, contact problems must be solved using the large-scale finite element method in which parallel computation is essential for such large-scale finite element analysis to be practical. Using a direct solver in large-scale parallel computation is difficult because it requires a huge memory capacity and extensive communication between processors.

However, iterative solvers are not yet sufficiently versatile to be used for all structural analysis problems. To deal with the contact problem by imposing contact constraints, the penalty (BELYTSCHKO and NEAL, 1991) and Lagrange multiplier methods (BATHE and CHAUDHARY, 1985) are being applied, usually with a direct solver because the matrix is ill-conditioned and iterative solvers are not applicable.

This study presents an effective way to analyze a large-scale parallel contact problem using GeoFEM, with the iterative solver and the augmented Lagrange method (LANDERS and TAYLOR, 1985, 1986; HEEGAARD and CURNIER, 1993) to improve matrix conditions. We also explain the application of parallel computation. This method initially applied to simple, small-scale problems, and solver convergence is examined. Finally, the paper will show an example of large-scale parallel contact problem analysis of simulated faults that traverse the Japanese islands.

2. Formulation of Contact Problem Analysis Using the Augmented Lagrange Method

2.1 Formulation of Contact Problem Analysis

This section outlines the formulation of frictionless elastic contact problem analysis. Figure 1 supplies a description of the contact system using a plate boundary as an example. Here, Ω, Γ_σ, $\Gamma_{\sigma c}$ and p are the domain, domain boundary of force, contact body boundary and the domain number, respectively. In the contact problem, several domain Ω are in contact at the boundaries $\Gamma_{\sigma c}$. The formula is given by the following virtual work and added conditions:

$$\sum_p \left[\int_{\Omega_p} \lfloor \delta\varepsilon \rfloor \{\sigma\} \, dV - \int_{\Gamma_{\sigma p}} \lfloor \delta u \rfloor \{f_o\} \, dS - \int_{\Omega_p} \lfloor \delta u \rfloor \{r_o\} \, dV \right]$$
$$= -\sum_{kl} \left[\int_{\Gamma_{\sigma ckl}} \lfloor \delta(\bar{\Delta}u) \rfloor \{f_{oc}\} \, dS \right] \tag{1}$$

$$\sum_{kl} \int_{\Gamma_{\sigma ckl}} \lfloor \delta(f_{oc}) \rfloor \, g \, dS = 0 \tag{2}$$

where, Equation (1) shows the balance of force and Equation (2) a kinematic contact constraint as an added condition. A kinematic contact constraint means that domains in contact along a contact surface have no penetration. The symbol kl represents a pair of contact boundaries k and l, whereas $\{\varepsilon\}$, $\{u\}$, $\{\sigma\}$, $\{f_o\}$, $\{r_o\}$, $\{f_{oc}\}$, $\{g\}$ and $\{\bar{\Delta}u\}$ are strain, displacement, stress, external force, body force, contact force, contact boundary gap, and relative displacement, respectively.

The first term of Equation (1) shows the internal force, the second term shows the traction force, and the third term shows the volume force. In the formulation of structural analysis without a contact surface, the right-hand side becomes zero and Equation (2) was not required. Therefore, it is a feature of the contact problem to have the contact force term in the formulation and to add the geometrical condition (no penetration) of the contact surface. This added condition causes considerable difficulty in solving the contact problem.

2.2 Contact Constraint Method

The Lagrange, penalty and augmented Lagrange methods can be used to solve problems with added conditions. Each method has unique features. However, we used the augmented Lagrange method because that method is more suitable for the iterative solver.

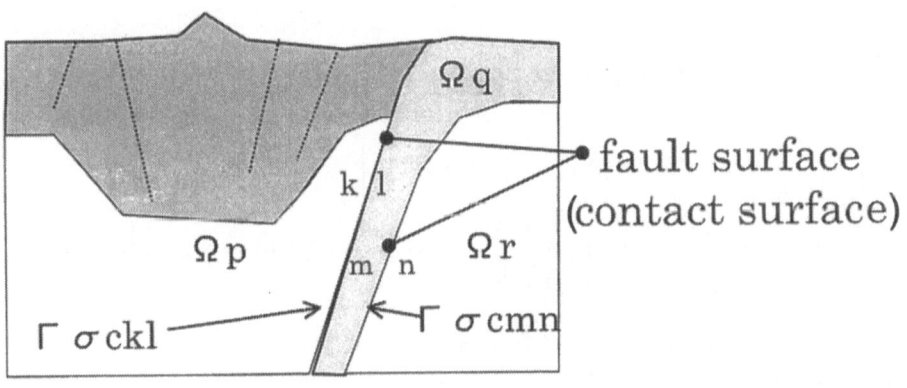

Figure 1
Contact system.

Analysis of a contact problem using the Lagrange method can be formulated by adding Equation (2) as a Lagrange multiplier term to Equation (1) and by applying the incremental and Newton-Raphson methods.

$$
\sum_p \left[\int_{\Omega_p} \lfloor \delta\varepsilon \rfloor^{(n+1,q+1)} \{d(\Delta\sigma)\} \, dV \right] = -\left[\sum_p \left[\int_{\Omega_p} \lfloor \delta\varepsilon \rfloor \, ({}^{(n+1,q)}\{\Delta\sigma\} + {}^{(n)}\{\sigma\}) \, dV \right. \right.
$$
$$
\left. - \int_{\Gamma_{\sigma p}} \lfloor \delta u \rfloor^{(n+1)} \{f_o\} \, dS - \int_{\Omega_p} \lfloor \delta u \rfloor^{(n+1)} \{r_o\} \, dV \right]
$$
$$
\left. + \sum_{kl} \left[\int_{\Gamma_{\sigma ckl}} \lfloor \delta(\bar{\Delta}u) \rfloor \, ({}^{(n)}\{f_{oc}\} + {}^{(n+1,q)}\{\Delta f_{oc}\}) \, dS \right] \right] - \sum_{kl} \int_{\Gamma_{\sigma ckl}} \lfloor \delta(f_{oc}) \rfloor \, g \, dS \quad (3)
$$

where (n) is the load step and (q) is the iteration number. The fourth term of the right-hand side is the contact force, and it represents the Lagrange multiplier. The equation which has the modified displacement increment and contact force $\{d(\Delta u)\}$, $\{d(\Delta f_{oc})\}$ as unknown variables, is derived from Equation (3). Since the matrix diagonal components with the added condition are zero in the Lagrange multiplier method, the iterative solver cannot be used. Uzawa's method (ARROW *et al.*, 1958) can be used to solve the problem using the iterative form of mixed problems, but it is not efficient.

The contact problem can be formulated using the penalty method (ZIENKIEWICZ, 1974). First, the contact force is obtained by multiplying the contact boundary gap by the penalty parameter, then substituting it for Equation (3). The following relation is obtained by applying the incremental and Newton-Raphson methods.

$$
{}^{(n+1,q)}\{f_{oc}^n\} = \{\alpha^{(n+1,q)}g\} \tag{4}
$$

$$
\{d(\Delta f_{oc}^n)\}_{PNL} = \{\alpha d(\Delta g)\} \tag{5}
$$

$$
\sum_p \left[\int_{\Omega_p} \lfloor \delta\varepsilon \rfloor^{(n+1,q+1)} \{d(\Delta\sigma)\} \, dV \right] + \sum_{kl} \left[\int_{\Gamma_{\sigma ckl}} \lfloor \delta(\bar{\Delta}u) \rfloor \, dS^{(n+1,q+1)} \{d(\Delta f_{oc}^n)\}_{PNL} \right]
$$
$$
= -\left[\sum_p \left[\int_{\Omega_p} \lfloor \delta\varepsilon \rfloor \, ({}^{(n+1,q)}\{\Delta\sigma\} + {}^{(n)}\{\sigma\}) \, dV - \int_{\Gamma_{\sigma p}} \lfloor \delta u \rfloor^{(n+1)} \{f_o\} \, dS \right. \right.
$$
$$
\left. \left. - \int_{\Omega_p} \lfloor \delta u \rfloor^{(n+1)} \{r_o\} \, dV \right] + \sum_{kl} \left[\int_{\Gamma_{\sigma ckl}} \lfloor \delta(\bar{\Delta}u) \rfloor^{(n+1,q)} \{f_{oc}^n\} \, dS \right] \right] \tag{6}
$$

where $\{f_{oc}^n\}$ is the normal contact force and α the penalty parameter. The diagonal components of the stiffness matrix from the left-hand side of Equation (6) are not zero because of the second term in the left-hand side of Equation (6) due to the penalty method. However, the penalty parameter needs a large value which worsens the matrix condition and causes difficulties in using the iterative solver. We therefore employed the augmented Lagrange method as shown below to improve the matrix condition for large-scale parallel contact analysis.

The augmented Lagrange method can be formulated as follows by applying the incremental and Newton-Raphson methods and combining a modified increment with the penalty term $()_{PNL}$ and the augmented Lagrange term $()_{ALM}$.

$$^{(n+1,q+1)}\{\Delta f^n_{oc}\} = {}^{(n+1,q)}\{\Delta f^n_{oc}\} + {}^{(n+1,q+1)}\{d(\Delta f^n_{oc})\}_{PNL} + {}^{(n+1,q+1)}\{d(\Delta f^n_{oc})\}_{ALM} \tag{7}$$

$$\{d(\Delta f^n_{oc})\}_{ALM} = \{\alpha g\}. \tag{8}$$

Contact problem analysis using the augmented Lagrange method is formulated as follows:

$$\sum_P \left[\int_{\Omega_p} \lfloor \delta\varepsilon \rfloor \, {}^{(n+1,q+1)}\{d(\Delta\sigma)\} \, dV \right] + \sum_{kl} \left[\int_{\Gamma_{\sigma ckl}} \lfloor \delta(\bar{\Delta}u) \rfloor \, dS^{(n+1,q+1)}\{d(\Delta f^n_{oc})\}_{PNL} \right]$$

$$= -\left[\sum_P \left[\int_{\Omega_p} \lfloor \delta\varepsilon \rfloor \, ({}^{(n+1,q)}\{\Delta\sigma\} + {}^{(n)}\{\sigma\}) \, dV - \int_{\Gamma_{\sigma p}} \lfloor \delta u \rfloor \, {}^{(n+1)}\{f_o\} \, dS \right.\right.$$

$$\left.- \int_{\Omega_p} \lfloor \delta u \rfloor \, {}^{(n+1)}\{r_o\} \, dV \right] + \sum_{kl} \left[\int_{\Gamma_{\sigma ckl}} \lfloor \delta(\bar{\Delta}u) \rfloor \, dS(^{(n)}\{f^n_{oc}\} \right.$$

$$\left.\left.+ {}^{(n+1,q)}\{\Delta f^n_{oc}\} + {}^{(n+1,q)}\{d(\Delta f^n_{oc})\}_{ALM}) \, dS \right]\right] \tag{9}$$

$$\{d(\Delta f^n_{oc})\}_{PNL} = \{\alpha d(\Delta g)\} \tag{10}$$

$$\{d(\Delta f^n_{oc})\}_{ALM} = \{\alpha g\}. \tag{11}$$

Modification is repeated on the right side (augmented Lagrange term) until the gap g on the contact boundaries becomes zero. The penalty term on the left-hand side makes the matrix non-singular and convergence can be achieved over a wide range of penalty value. When the penalty value is large, in particular, the gap g of the contact boundaries converges rapidly.

2.3 Iterative Algorithm of the Augmented Lagrange Method

The above formulation creates the following iterative algorithm of the augmented Lagrange method.
1. Contact force modification $\{d(\Delta f^n_{oc})\}_{PNL}$ is obtained by Equation (5) from the modified increment of displacement after which the contact force is corrected by Equation (7).

2. Gap g of the contact boundary is obtained and the contact condition is renewed from the new gap and the contact force.
3. The estimated increase in the modified contact force $\{d(\Delta f_{oc}^n)\}_{ALM}$ is obtained from the contact boundary gap by Equation (8) and then the contact force is renewed by Equation (7).
4. The equation is solved using the modified contact force to obtain the modified increment of displacement, and the displacement is renewed.

These procedures are repeated until the contact force gap g falls within the convergence tolerance range. During this iterative procedure, the sign of the contact force and the gap amount are combined to judge whether or not there is contact and the contact condition has converged. Once the contact boundary gap g has converged, the equation becomes equivalent to the Lagrange multiplier method.

2.4 Finite Element Models of Contact Boundaries

Node-to-node and node-to-segment contact models as shown in Figure 2 are used as the finite element models of contact boundaries. The node-to-node model is easier to handle, but it can only be applied to small-slip contact problems. Therefore, the node-to-segment model should be used for more general problems. For this reason, the node-to-segment model was used in this study.

3. Parallel Computation Method for Contact Problems in GeoFEM

The parallel computation method described here has been developed for parallel computers with distributed memories. To optimize the method for use on this type of computer, the whole model is divided into smaller regions, each of which is then

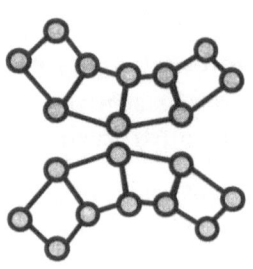

Node-to-node contact model

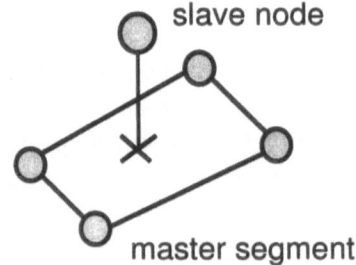

slave node

master segment

Node-to-segment contact model

Figure 2
Finite element models of contact boundaries.

allocated to a processing element. There are two ways to perform this domain decomposition. One approach is the use of an iterative solver to deal with the overall degrees of freedom, and the other is the use of the iterative method for solving the degrees of freedom condensed at the partitioned domain boundary by eliminating the inner degree of freedom in each domain. The former is used for GeoFEM because of the stability of the solver and because of the flexibility of its application to various problems (several iterative solvers that have been developed can be used according to the type of problem). The iterative method (currently, the ICCG method) is used for the solver (NAKAJIMA and OKUDA, 1998; NAKAJIMA, 1998).

The domain decomposition method for contact problem analysis applied in the present study is introduced below. Contact problems can be solved in two ways; one gathers contact boundaries within a single region for processing, whereas the other divides and allocates the contact boundaries to each region. The former does not require matching of contact conditions between regions, thus simplifying the parallel processing algorithms. On the other hand, it creates bad boundary divisions and therefore worsens the convergence of iterative solvers. The latter requires matching of contact conditions between regions, and algorithms for parallel processing become more complex as a result. One advantage is that domain decomposition can be performed flexibly and the quality of domain division is good. This study used the latter, considering the flexibility and quality of domain decomposition. To ease the contact point search, we also used a method that has overlapping information about nodes with contact potential within the designated distance. The domain decomposition method for contact problems and communication during parallel computation is explained below.

1. Figure 3(a) shows that contact boundaries are set for the master body and slave body for which contact is expected. The node-to-segment model is used as a finite element model and therefore the study focused on contact between the contact boundary element surfaces at the contact boundaries of the master body (master segment) and the nodes at the contact boundaries of the slave body (slave node).

2. Overlapping information pertaining to nodes with contact potential inside the designated distance within the domain boundary is necessary. Before dividing the domain, the contact potential distance (CPD) is considered to select contact potential elements (CPE), as shown in Figure 3(b). When this domain decomposes along with these CPEs, they are shared as overlapping elements in each region, leading to sharing of the node data necessary for contact problem analysis.

3. First, as shown in Figure 3(c), domain decomposition for parallel computation is achieved by edge cutting inside the continuous domains to determine regions and overlapping areas. From this information regarding the overlapping areas,

we set the external and boundary points of the division data of the continuous domains.

4. Next, when dividing the domain by edge cutting (Fig. 3(d)), the new nodes that are not included in the overlapping elements of the continuous domain are generated as external, boundary points with contact potential. These nodes are called contact potential external points (CPEP) and contact potential boundary points (CPBP). For the domain decomposition method used here, contact boundaries are divided among the regions, the data for contact problem analysis are automatically shared at the boundaries, and inter-region communication to search for contact points is no longer needed.

5. During parallel computation, communication occurs only between the external point and the boundary point if no contact problem analysis takes place. If contact problem analysis proceeds, CPEP and CPBP are added to the nodes for inter-region communication (Fig. 4).

Because only slight distortion was handled in this study, CPD is sufficient by the length of one element. Even if large slips are to be handled, this method should be feasible by setting the slip-potential distance at CPD.

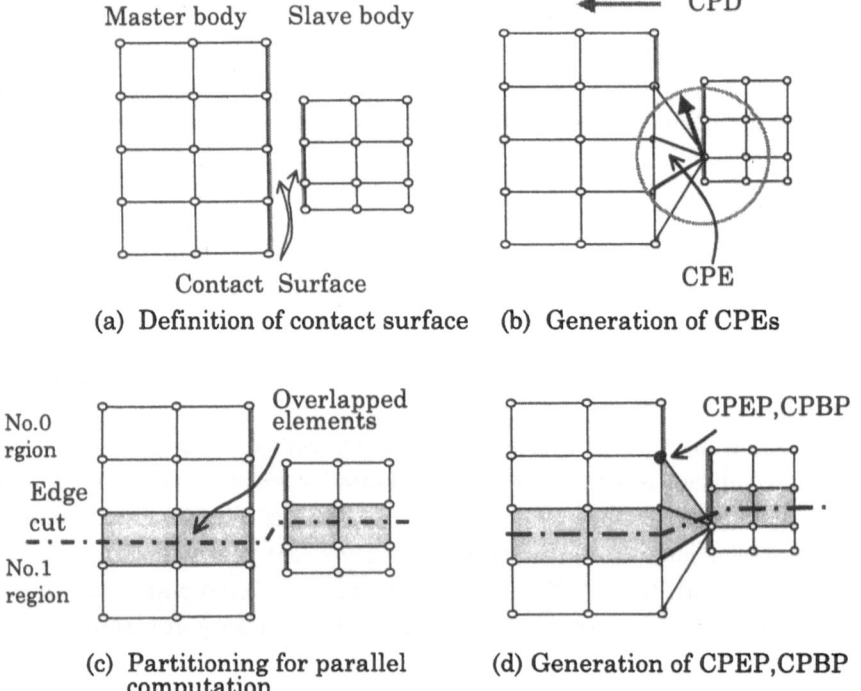

Figure 3
Partition of contact problem.

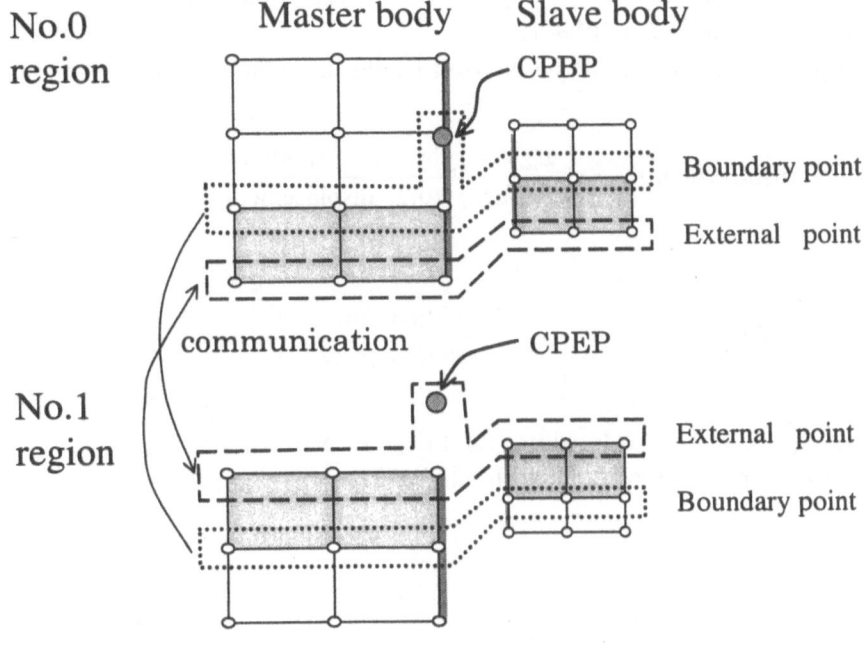

Figure 4
Communication of contact problem.

4. Application of Augmented Lagrange Method to the Iterative Solver in the Solution of Contact Analysis Problems

4.1 Example for Evaluation

To evaluate the contact problem analysis method based on the augmented Lagrange method, we analyzed a two-dimensional problem involving a half-infinite body compressed by an elastic plane. This requires stress analysis of a half-infinite body compressed by an elastic plane measuring 2B in width and height (see Okubo's theoretical solution (OKUBO, 1952)). Figure 5 shows our model (4B-wide, 4B-deep) of this half-infinite body and we used half of the plane for analysis. Solutions were nondimensionally adjusted using the ratio of Young's modulus of the half-infinite body and the elastic plane. We used the specific example shown in Figure 5. For plane strain analysis, the entire Z direction was fixed, Poisson's ratio was 0.3, Youngs modulus of the half-infinite body and the elastic plane was 1.0, and the friction coefficient between contact boundaries was 0.0. The following examples of mesh division were analyzed: when node positions facing on the contact boundaries matched and when they did not. There were 675 elements and 1504 nodes.

Analysis conditions for the augmented Lagrange method were as follows:
- Penalty parameter (α) $10^4 \sim 10^{16}$
- Convergence tolerance of the Newton-Raphson method ε_u, $\varepsilon_f < 10^{-5}$

  - For displacement $\varepsilon_u = \left[\dfrac{\Sigma_{idof}^{ndof}(\text{displacement modification}_{idof})^2}{\Sigma_{idof}^{ndof}(\text{total displacement}_{idof})^2} \right]^{1/2}$

  - For residual force $\varepsilon_f = \left[\dfrac{\Sigma_{idof}^{ndof}(\text{residual force}_{idof})^2}{\Sigma_{idof}^{ndof}(\text{external force}_{idof})^2} \right]^{1/2}$

idof, ndof degree of freedom, total degrees of freedom.
The convergence tolerance of the iterative solver was $\varepsilon_{sol} < 10^{-8}$.

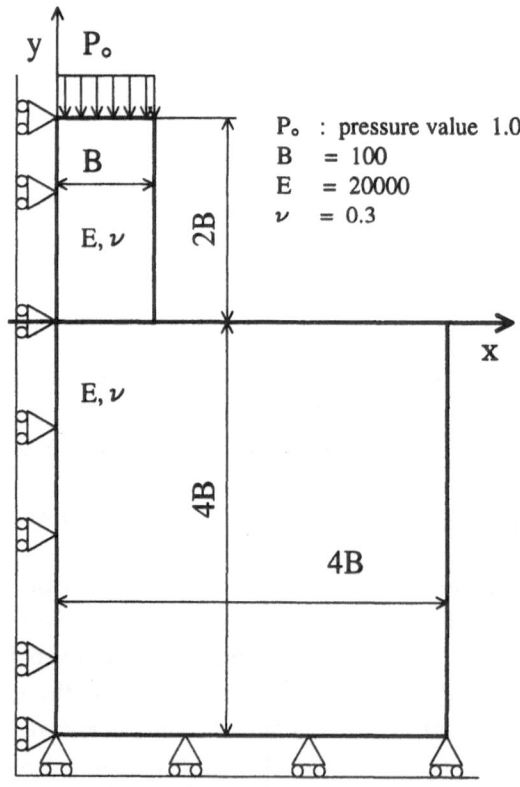

Figure 5
Two-dimensional problem of semi-infinite elastic body compressed by an elastic plane.

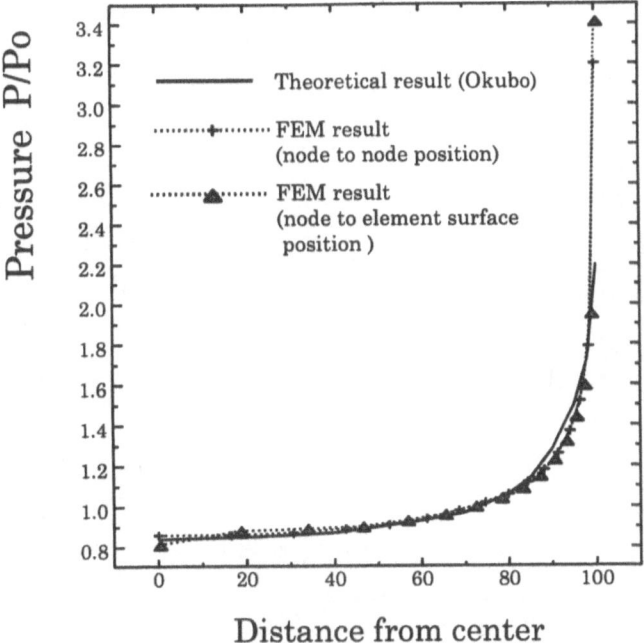

Figure 6
Pressure distribution on contact surface.

4.2 Contact Pressure Distribution of Analysis Results at the Contact Boundaries

The pressure distribution at the contact boundaries is expressed by x, the distance from the center. The contact pressure is normalized as $-p/p_o$. Figure 6 shows the contact pressure and its theoretical solution. Simulation results compare well with the analytical solution.

4.3 Applicability of the Augmented Lagrange Method to Iterative Solver

In the above analysis, the penalty parameter of the augmented Lagrange method was changed to evaluate the applicability of this method to the iterative solver in the analysis of contact problems. The facing node positions matched in the mesh on the contact boundary were used in the evaluation.

4.3.1 Penalty parameter and computation time

Figure 7 shows the relationship between the penalty parameter and the computation time. The computation time is the sum of all the iterations made until convergence by the Newton-Raphson method. The minimal value was obtained when the penalty parameter was 10^7, indicating that an optimal penalty parameter exists in the augmented Lagrange method. The following section explains this.

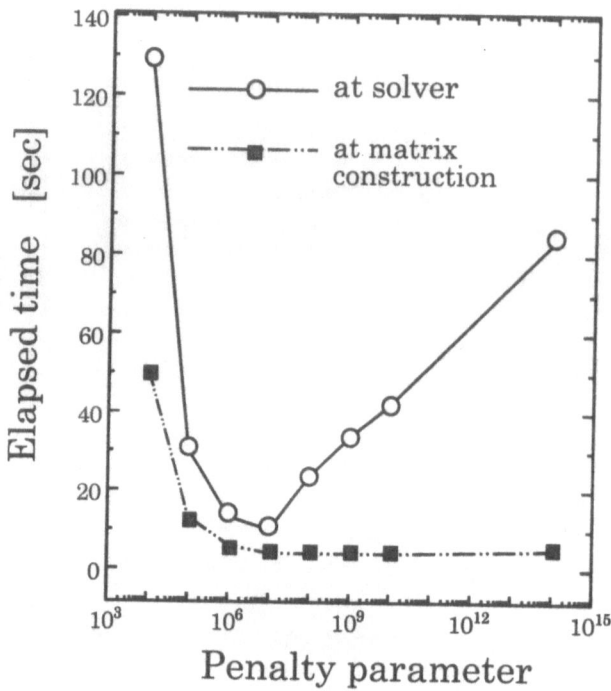

Figure 7

Elapsed time required in solver and matrix construction.

4.3.2 Penalty parameter and iterative number

Figure 8 shows the relationship between the penalty parameter and the number of iterations. From the figure the following is clear: (a) Using the Newton-Raphson method, the number of Newton-Raphson iterations does not change after the penalty parameter has increased to a specific level. This is explained as follows. The iteration number of the Newton-Raphson method decreases because of a large modified value, but at a larger penalty value, the iteration number of the Newton-Raphson method does not decrease because of numerical error. (b) The number of solver iterations per Newton-Raphson iteration becomes the smallest at a certain penalty parameter; the number of iterations slightly increases when the parameter is small, but considerably increases when it is large. We surmise that the penalty parameter affects the matrix condition and that the number of solver iterations changes significantly because of iterative solver convergency that strongly depends on the matrix condition. Therefore, the minimal number of solver iterations can be obtained using the optimal penalty parameter. (c) The number of solver iterations over the entire Newton-Raphson procedure, as deduced from the above changes, becomes the smallest at a certain penalty parameter. Iterations increase when the penalty parameter is either very large or very small. These explain why the minimal computation time exists as the penalty parameters change.

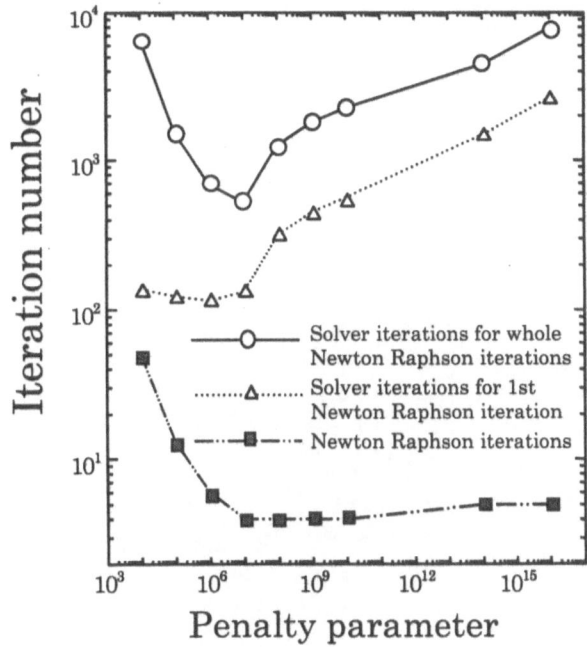

Figure 8
Number of iterations required in solver and Newton-Raphson iterations.

Therefore, when using the augmented Lagrange method, the choice of the penalty parameter is very important taking into account the possibility of iterative solver convergence and the Newton-Raphson method convergence. In addition, contact problems can be analyzed using iterative solvers over a wide range of penalty parameters. This means that the augmented Lagrange method provides a stable method which is easily applied to contact problems. Therefore, the augmented Lagrange method with iterative solvers is suited to solving large-scale contact problems.

5. Example of Small-scale Parallel Contact Computation

We examined solver convergence in parallel computation. When the boundaries of the divided domain cross contact surfaces, localized preconditioning is affected by parallel computations in the matrix with a penalty. This is likely to adversely affect solver convergence. We made two and four region calculations to address the two-dimensional problem of a semi-infinite body compressed by an elastic plane, then we checked the solver convergence. Figure 9 shows the divisions of the four regions.

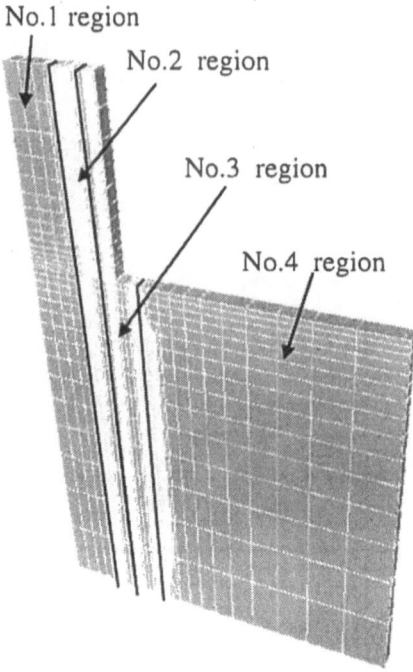

Figure 9
Partitioning of 4 regions for semi-infinite elastic body and elastic plane.

Table 1 ((%) means increment of iteration number) shows the number of solver iterations each time the Newton-Raphson method is repeated. The number of iterations increases considerably from one to two regions during parallel computation, but the increase is less than that from two to four regions. In GeoFEM where localized preconditioning is carried out for the iterative solver, solver convergence is likely to worsen as the number of domains increases. However, solver convergence will likely be affected only to a limited degree by an increase in the number of regions.

Table 1

Number of solver iterations

Number of Newton-Raphson iterations	1 region	2 regions	4 regions
1st	163	234 (43.6%)	237 (45.4%)
2nd	121	186 (53.7%)	196 (62.0%)
3rd	117	178 (52.1%)	188 (60.7%)
4th	123	171 (39.0%)	176 (43.1%)

(%) shows the increment of iteration number for one region.

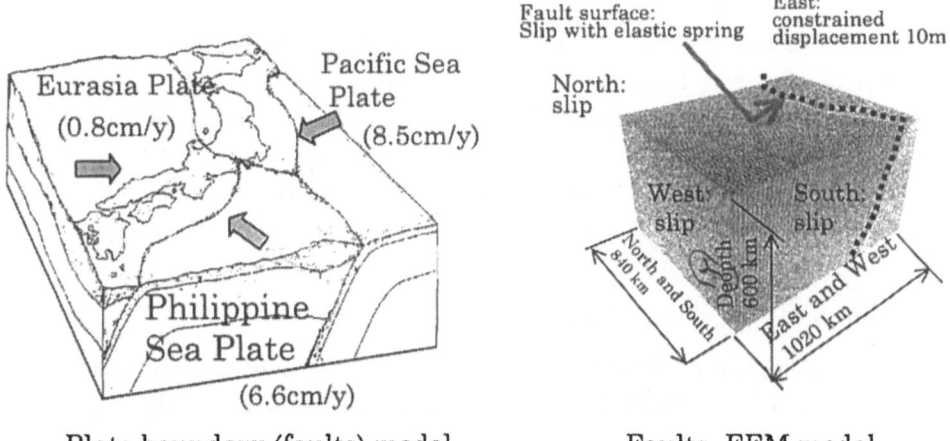

Plate boundary (faults) model   Faults FEM model

Figure 10
Fault analysis around Japan islands.

6. Example of Large-scale Parallel Contact Analysis

6.1 Analysis Model

We simulated faults in the Japanese islands for large-scale parallel contact analysis, to demonstrate the validity of the proposed analytical method in large-scale computations. Figure 10 shows how fault surfaces were obtained by simulating the colliding surfaces of the Eurasia, Philippine and pacific Sea plates. The area for analysis measures 1020 km × 840 km × 600 km and the boundary conditions were as follows: boundaries running north-south, on the west side, and on the bottom are slip boundaries, and the east side is a 10 m constrained displacement boundary. The load was given in one step, and the nonlinear equation was solved using the Newton-Raphson method. We analyzed small-scale (21,660 DOFs, 7220 nodes, 5832 elements), medium-scale (156,066 DOFs, 52,022 nodes, 46,656 elements), and large-scale models (1,183,038 DOFs, 394,346 nodes, 373,248 elements). The small-scale model was divided into 16 regions, the medium-scale model into 16 and 32 regions, and the large-scale model into 32 regions. In each model, Young's modulus is set to 50 GPa and Poisson's ratio is 0.3.

The analysis conditions of the augmented Lagrange method were as follows:
- Penalty parameter (α) $10^9 \sim 10^{14}$
- Newton-Raphson method convergence tolerance $\varepsilon_u < 10^{-5}$, $\varepsilon_f < 10^{-3}$

Iterative solver convergence tolerance is $\varepsilon_{sol} < 10^{-8}$.

6.2 Analysis Scale and Analysis Time

Figure 11 shows the relationship between the scale and time of analysis. The penalty parameter was 10^{10}. The SR2201 computer installed at the University of Tokyo completed the computations within about 2.5 hours for 1.18×10^6 DOFs model, which represented the largest scale. These results indicate that large-scale parallel contact analysis using the iterative solver with the augmented Lagrange method is promising.

6.3 Convergence Plot of Gap Errors

Figure 12 shows the relationship between gap errors and iterative numbers by the Newton-Raphson method. These gap errors were obtained by adding up all gap errors at the contact points and dividing the sum by the total number of contact points. The gap errors will rapidly converge, using the augmented Lagrange method.

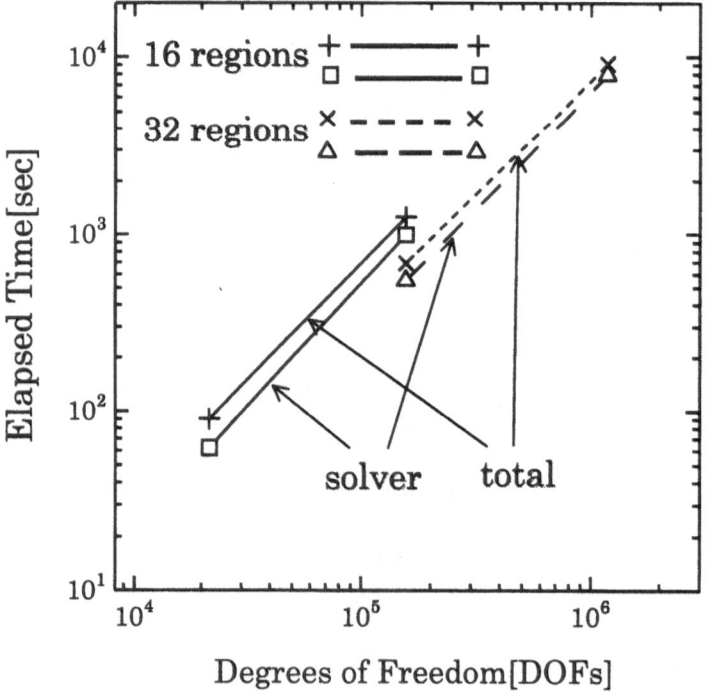

Figure 11
Relation between degrees of freedom and elapsed time.

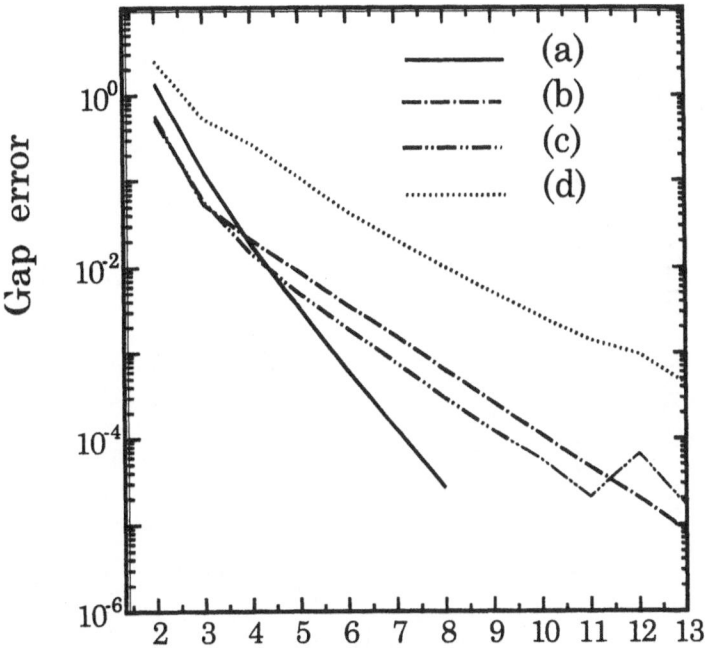

Number of Newton-Raphson iterations

Figure 12

Relationship between number of Newton-Raphson iterations and gap error. (a) Gap error for small-scale model; (b) gap error for No. 1 region of large-scale model; (c) gap error for No. 16 region of large-scale model; (d) gap error for No. 32 region of large-scale model.

6.4 Penalty Parameter and Iterative Number

Figure 13 shows the relationship between the penalty parameter and the number of iterations. We can see from this relationship that by selecting a proper penalty parameter, the conditions for calculation can be optimized even in large-scale parallel contact problem analysis. In this computation, the iterative solvers did not converge when the penalty parameter was 10^{14}, thus confirming the validity of the augmented Lagrange method for solving large-scale contact problems.

6.5 Contour of Normal Contact Force on Fault Surface

Figure 14 delineates the contour of normal contact force on the fault surface. As this study focused on the analyzing large-scale parallel contact problems, we used idealized boundary conditions and frictionless models of the fault surface. The normal contact force on the fault surface therefore has no geophysical meaning, but the contours distribution of the analysis results for the small-scale and large-scale models is similar, demonstrating that parallel contact problems can be accurately analyzed on a large scale.

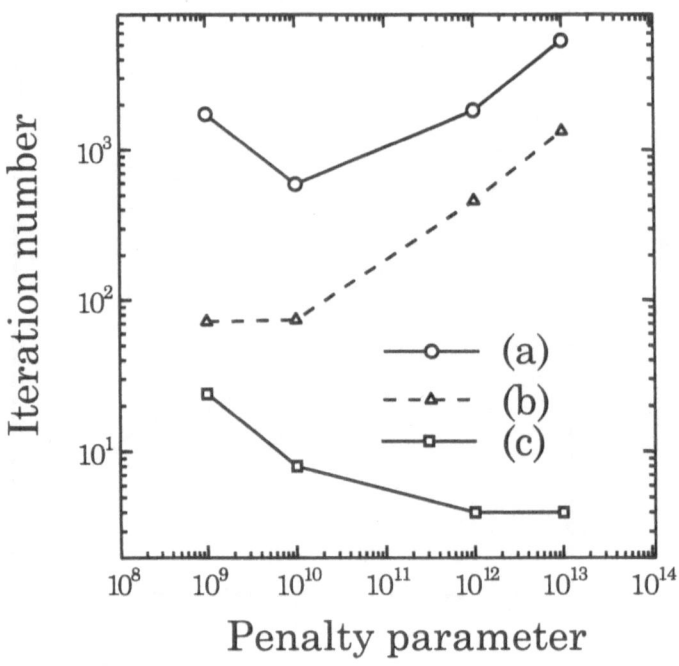

Figure 13

Relationship between penalty parameter and iteration number. (a) Solver iteration for entire Newton-Raphson iteration; (b) solver iteration for one Newton-Raphson iteration; (c) Newton-Raphson iteration.

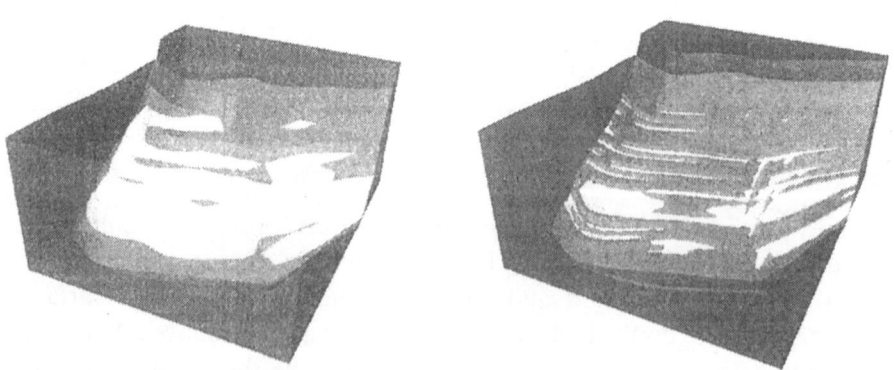

Figure 14

Contour of normal contact force at fault surface.

7. Summary

This research has implications for the generation and frequency of earthquakes. We studied parallel FEM using an iterative solver with the augmented Lagrange method to solve large-scale contact problems for faults. The results were as follows:

(1) The augmented Lagrange method allows contact problems to be analyzed using iterative solvers for a wide range of penalty parameters. This means that the analysis of contact problems by the augmented Lagrange method is relatively stable and simple to perform. The method is therefore suited to the analysis of large-scale contact problems using iterative solvers.

(2) Optimal computation conditions can be obtained through the selection of a proper penalty parameter.

(3) It is expected that, for the iterative solver using localized preconditioning, there will be only limited deterioration in solver iterations due to the increase in the number of regions, suggesting that parallel performance will be high.

(4) It has been confirmed that the augmented Lagrange method is suited to analysis of large-scale parallel contact problems using iterative solvers.

REFERENCES

ARROW, K. J., HURWICZ, L., and UZAWA, H., *Studies in Non-Linear Programming* (Stanford University Press, Stanford 1958).

BATHE, K. L., and CHAUDHARY, A. (1985), *A Solution Method for Planar and Axisymmetric Contact Problems*, Int. J. Numer. Methods Eng. *21*, 65–88.

BELYTSCHKO, T., and NEAL, M. O. (1991), *Contact-Impact by the Pinball Algorithm with Penalty and Lagrangian Methods*, Int. J. Numer. Methods Eng. *31*, 547–572.

BIELAK, JACOBO, and GHATTAS, OMAR, *Computational challenges in seismology*, 1-st ACES Workshop Proceedings (ed. Mora, P.) (The APEC Cooperation for Earthquake Simulation, Brisbane, Australia 1999) pp. 325–328.

GARATANI, K., NAKAMURA, H., OKUDA, H., and YAGAWA, G. (1999), *GeoFEM: High Performance Parallel FEM for Solid Earth*. Proceedings of 7th High-performance Computing and Networking (HPCN Europe '99), LNCS-1593, 133–140.

HEEGAARD, J.-H., and CURNIER, A. (1993), *An Augmented Lagrangian Method for Discrete Large-slip Contact Problems*, Int. J. Numer. Meth. Eng. *36*, 569–593.

IIZUKA, MIKIO, GARATANI, KAZUTERU, NAKAJIMA, KENGO, NAKAMURA, HISASHI, OKUDA, HIROSHI, and YAGAWA, GENKI (1999), *GeoFEM: High-performance Parallel FEM Geophysical Applications*, ISHPC99, Second International Symposium Proceedings, High Performance Computing, Lecture Notes in Computer Science 1615, 292–303.

LANDERS, J. A., and TAYLOR, R. L. (1985), *An Augmented Lagrangian Formulation for the Finite Element Solution of Contact Problems*, Rep. No. UCB/SESM-85/09, University of California, Berkeley.

LANDERS, J. A., and TAYLOR, R. L. (1986), *An Augmented Lagrangian Formulation for the Finite Element Solution of Contact Problems*, Rep. No. AD-A166 649, University of California, Berkeley.

MATSU'URA, M., and SATO, T. (1997), *Loading Mechanism and Scaling Relations of Large Interplate Earthquakes*, Tectonophysics *277*, 189–1998.

MEIJERINK, J. A., and VAN DER VORST, H. A. (1977), *Iterative Solution Method for Linear Systems of which the Coefficient Matrix is Asymmetric M-matrix*, Math. Comp. *61*, 148–162.

NAKAJIMA, K. (1998), *GeoFEM: Multi-purpose Parallel FEM for Solid Earth (2) Large-scale Parallel Iterative Solvers*, Proceedings of the Conference on Computational Engineering and Science, JSCES *3-1*, 93–94.

NAKAJIMA, K., and OKUDA, H. (1998), *Parallel iterative solvers with localized ILU preconditioning for unstructured grids on workstation Cluster*, 4th Japan-US Symposium on *FEM in Large-scale Computational Fluid Dynamics Proceedings*, pp. 25–30.

OKUBO, H. (1952), *On the Two-dimensional Problem of a Semi-infinite Elastic Body Compressed by an Elastic Plane*, Trans. Jap. Soc. Mech. Eng. *18–65*, 58–62.

RUNDLE, JOHN B., HENYEY, TOM, MINSTER, J.-BERNARD, and FOX, GEOFFREY, *General earthquake models*, 1-st ACES Workshop Proceedings (ed. Mora, P.) (The APEC Cooperation for Earthquake Simulation, Brisbane, Australia 1999) pp. 281–287.

SIVATHASAN, K., PAULINO, G. G., LI, K. S., and ARULANANDAN, K., *Validation of Site Characterization Method for the Study of Dynamic Pre-pressure Response*, Geotechnical Special Publication No. 75, Volume one, Geotechnical Earthquake Engineering and Solid Dynamics III (ASCE, Seattle, Washington 1998).

ZHAO, C., HOBBS, B. E., and MÜHLHAUS, H. B. (1998), *Finite Element Modelling of Temperature Gradient Driven Rock Alteration and Mineralization in Porous Rock Masses*, Comp. Meth. Appl. Mech. Eng. *165*, 175–187.

ZIENKIEWICZ, O. C., *Constrained variational principles and penalty function methods in the finite element analysis*. In *Lecture Notes in Mathematics* (Springer-Verlag, Heidelberg 1974).

ZIENKIEWICZ, O. C., HUANG, M., and PASTOR, M., *Numerical prediction for Model No 1*. In *Verification of Numerical Procedures for the Analysis of Soil Liquefaction Problems 1* (ed. Arulanandan Scott) (Balkema, Rotterdam. 1993) pp. 259–274.

(Received September 23, 1999, revised March 30, 2000, accepted April 15, 2000)

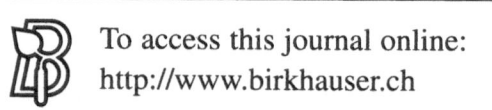

To access this journal online:
http://www.birkhauser.ch

Part III
Macroscopic Simulation: Long Time Scale Phenomena (Earthquake Cycle)

Pure appl. geophys. 157 (2000) 2125–2147
0033–4553/00/122125–23 $ 1.50 + 0.20/0

❘Pure and Applied Geophysics

3-D Physical Modelling of Stress Accumulation Processes at Transcurrent Plate Boundaries

CHIHIRO HASHIMOTO[1] and MITSUHIRO MATSU'URA[2]

Abstract—We constructed a 3-D physical model of tectonic loading at transcurrent plate boundaries by considering viscoelastic stress relaxation in the asthenosphere and spatial variation in frictional properties (peak strength and critical weakening displacement) of faults. With this model we simulated the process of stress accumulation and release at a seismogenic region with relatively high strength on the plate interface. In low strength regions surrounding the seismogenic region, quasi-static fault slip gradually proceeds with the progress of relative plate motion. The increase of slip deficits in the seismogenic region brings about stress concentration at its margin. The stress accumulation rate is roughly proportional to the inverse of the effective fault length. The accumulated stress is released by unstable dynamic rupture if the critical weakening displacement D_c is small, and by stable fault slip if D_c is very large. When a fault system consists of two adjacent seismogenic regions, sudden stress release in one region accelerates the stress accumulation process in another region through transient viscoelastic stress transfer as well as instantaneous elastic stress transfer. This indicates the importance of elastic and viscoelastic interaction between adjacent seismic faults even in stress accumulation processes.

Key words: Tectonic loading, viscoelastic stress relaxation, fault constitutive law, slip deficit.

1. Introduction

Generation of large interplate earthquakes can be regarded as the process of tectonic stress accumulation and release, resulting from relative plate motion. The stress accumulation process essentially controls the subsequent dynamic rupture process in seismogenic regions because the stress released at the time of dynamic rupture is nothing but the tectonic stress that has been accumulated through the interseismic period.

To date, in many earthquake simulation studies, it has been simply assumed that the tectonic loading uniformly proceeds both in space and time. The assumption of uniform loading is clearly contradictory to geodetic observations at and around the San Andreas Fault system in California. For example, LISOWSKI *et al.*

---
[1] Department of Earth and Planetary Science, University of Tokyo, Hongo 7-3-1, Bunkyo-ku, Tokyo 113-0033, Japan. E-mail: hashi@solid.eps.s.u-tokyo.ac.jp
[2] Department of Earth and Planetary Science, University of Tokyo, Hongo 7-3-1, Bunkyo-ku, Tokyo 113-0033, Japan. E-mail: matsuura@eps.s.u-tokyo.ac.jp

(1991) have reported that interseismic strain buildup concentrates within an approximately 80-km wide zone parallel to the San Andreas Fault in the San Francisco Bay area, where the great San Francisco earthquake occurred in 1906. On the other hand, THATCHER (1983) has pointed out the temporal change in near-fault shear strain rate during one earthquake cycle, indicating significant effects of viscoelastic stress relaxation in the asthenosphere.

Recently, MATSU'URA and SATO (1997) conducted a numerical simulation for tectonic loading at transcurrent plate boundaries with a lithosphere-asthenosphere coupling model subjected to steady relative plate motion, and elucidated that stress accumulation on a seismic fault is partly due to the base loading (viscous drag at the base of the lithosphere) and partly due to the edge loading (dislocation pile-ups at horizontal edges of the fault). In their model, however, the seismic fault area is *a priori* set as a completely locked portion on the plate interface. Therefore, we cannot discuss the details of stress accumulation and release processes on the plate interface by using this model.

On the other hand, there have been many simulation studies dealing with the spatio-temporal change in fault slip (and/or shear stress) during one earthquake cycle on a plate interface (e.g., TSE and RICE, 1986; STUART, 1988; KATO and HIRASAWA, 1997) in a two-dimensional framework. In these studies, the rate- and state-dependent type of constitutive laws, proposed by DIETERICH (1979) and RUINA (1983), is coupled with a slip response function for an elastic plate or an elastic half-space as a basic equation governing the fault slip process. Namely, the effects of transient viscoelastic stress relaxation in the asthenosphere are completely ignored in these models.

In the present study we construct a 3-D physical model of tectonic loading at transcurrent plate boundaries, and examine the process of stress accumulation and release resulting from relative plate motion through numerical simulations. In section 2, we show that the physical process of tectonic loading can be quantitatively described by coupled nonlinear equations, consisting of a viscoelastic slip response function, a fault constitutive relation, and a steady relative plate motion as driving force. In section 3, we derive the concrete expressions of the viscoelastic slip response function by applying a propagator matrix method. In section 4, a new algorithm for numerical computation is developed. In section 5, after briefly discussing the fault constitutive relation, we present the results of numerical simulation for stress accumulation and release processes in four representative cases.

2. Physical Model of Tectonic Loading

Given the rheological structure of the lithosphere-asthenosphere system and the geometry of a plate interface, we can describe the physical process of tectonic loading by coupled nonlinear equations, consisting of a viscoelastic slip response function that relates fault slip with shear stress changes on the plate interface, a friction law that prescribes the constitutive relation between shear stress and fault slip (and/or slip velocity), and a steady relative plate motion to drive the system.

In the present study we model the lithosphere-asthenosphere system by an elastic surface layer overlying a Maxwellian viscoelastic half-space. The constitutive equation of the elastic surface layer is given by

$$\sigma_{ij} = \lambda^{(1)}\varepsilon_{kk}\delta_{ij} + 2\mu^{(1)}\varepsilon_{ij} \tag{1}$$

and that of the underlying viscoelastic half-space by

$$\dot{\sigma}_{ij} + \frac{\mu^{(2)}}{\eta}\left(\sigma_{ij} - \frac{1}{3}\sigma_{kk}\delta_{ij}\right) = \lambda^{(2)}\dot{\varepsilon}_{kk}\delta_{ij} + 2\mu^{(2)}\dot{\varepsilon}_{ij} \tag{2}$$

where σ_{ij}, ε_{ij} and δ_{ij} are the stress tensor, the strain tensor, and the unit diagonal tensor, respectively. The dot indicates differentiation with respect to time, $\lambda^{(i)}$ and $\mu^{(i)}$ ($i = 1, 2$) is the Lamé elastic constants of each medium, and η is the viscosity of the underlying half-space.

We introduce an infinitely long, vertical interface Σ that divides the elastic surface layer into two plates being in a relative horizontal motion, and take Cartesian coordinates (x, y, z) and cylindrical coordinates (r, φ, z) as shown in Figure 1. Interaction between these two plates is rationally represented by the increase of tangential displacement discontinuity parallel to the x direction across the interface Σ (MATSU'URA and SATO, 1997). The displacement discontinuity (dislocation) is mathematically equivalent to the force system of a double couple without moment (MARUYAMA, 1963; BURRIDGE and KNOPOFF, 1964), which has

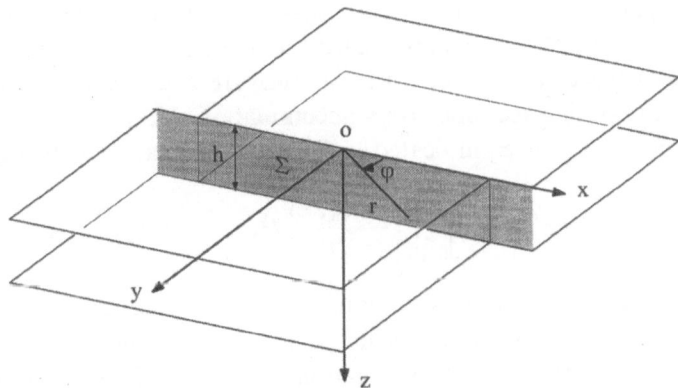

Figure 1

A structural model and two coordinate systems. The lithosphere-asthenosphere system is modeled by an elastic surface layer with a thickness h overlying a Maxwellian viscoelastic half-space. An infinitely long, vertical interface Σ divides the elastic surface layer into two plates which are in relative horizontal motion.

no net force and no net torque. Such a property must be satisfied for any force system acting on plate interfaces, since it is the internal force produced by a dynamic process within the earth.

Given a viscoelastic stress response, $H_{ij}(x, t; \xi, \tau)$ to a unit step slip on the plate interface, we can generally calculate the internal stress field $\sigma_{ij}(x, t)$ due to the fault slip motion $w(x, t)$ by applying the technique of hereditary integral as

$$\sigma_{ij}(x, t) = \int_{-\infty}^{t} \int_{\Sigma} \frac{\partial w(\xi, \tau)}{\partial \tau} H_{ij}(x, t - \tau; \xi, 0) \, d\xi \, d\tau. \tag{3}$$

In the present problem, where pure strike-slip motion on a vertical fault is treated, only the yx component of the stress tensor is needed. For simplicity we use σ and H instead of σ_{yx} and H_{yx} hereafter. The concrete expression of the viscoelastic slip response function H will be given in the next section and Appendix.

Now we decompose the total fault slip motion w into the steady plate motion at v_{pl} and its perturbation u_s as

$$w(x, t) = v_{pl}t + u_s(x, t) \tag{4}$$

and substitute it into Eq. (3). Then we obtain

$$\sigma(x, t) = v_{pl} \int_{-\infty}^{t} \int_{\Sigma} H(x, t - \tau; \xi, 0) \, d\xi \, d\tau + \int_{-\infty}^{t} \int_{\Sigma} \frac{\partial u_s(\xi, \tau)}{\partial \tau} H(x, t - \tau; \xi, 0) \, d\xi \, d\tau \tag{5}$$

where the first and second terms on the right-hand side indicate the contributions from the steady plate motion and the slip perturbation, respectively. As regards transcurrent plate boundaries, the steady-state plate motion is merely horizontal block motion parallel to the strike of the plate boundary, which causes no change in tectonic stress. Therefore, we may regard the first term as a constant (σ^0) in time, and take it as a basic stress level when we measure stress changes due to the slip perturbation. Supposing tectonic stress accumulation due to slip deficits begins at $t = 0$ in the model region Σ_s indicated in Figure 2, we can rewrite Eq. (5) as

$$\sigma(x, t) = \sigma^0(x) + \int_{0}^{t} \int_{\Sigma_s} \frac{\partial u_s(\xi, \tau)}{\partial \tau} H(x, t - \tau; \xi, 0) \, d\xi \, d\tau. \tag{6}$$

In the present problem the fault slip motion w in the model region Σ_s is unknown. The fault slip velocity outside Σ_s is taken to be the same as the relative plate velocity v_{pl}. We know the constitutive relation between the fault slip w and the shear stress σ, that defines frictional properties of the fault. As to the fault constitutive relation we take a slip-weakening type of law proposed by MATSU'URA et al. (1992), which is written in the following functional form;

$$\sigma(x, t) = f[w(x, t); x]. \tag{7}$$

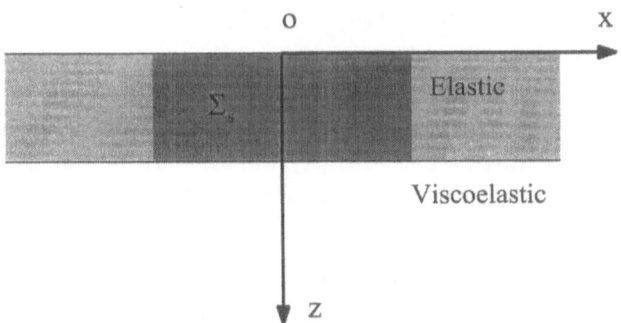

Figure 2
The model region Σ_s on the vertical plate interface. The boundary conditions on Σ_s are given by the position-dependent constitutive relations in Eq. (7) of the text. On the plate interface outside Σ_s we impose the uniform fault slip at a constant rate v_{pl}.

The physical process of stress accumulation on the transcurrent plate boundary is essentially governed by the coupled nonlinear system of Eqs. (4), (6) and (7).

3. Derivation of Viscoelastic Slip Response Functions

Mathematical expressions for surface deformation fields due to a point dislocation source in an elastic-viscoelastic composite medium have been obtained by many investigators (RUNDLE and JACKSON, 1977; RUNDLE, 1978, 1982; COHEN, 1980; MATSU'URA et al., 1981; IWASAKI and MATSU'URA, 1981). Considering a hypothetical horizontal interface at a depth in the elastic surface layer and applying Thomson-Haskel's forward propagator matrix method, MATSU'URA and SATO (1997) derived the expressions of internal deformation fields for an elastic-viscoelastic two-layered model. Their expressions, however, contain a term showing apparent singularity when the depth z of calculation points is deeper than a source depth d. To remove this singularity we need a very complicated technique in numerical computation.

Recently we have succeeded in obtaining the complete expressions of internal deformation fields due to a point dislocation source in an elastic-viscoelastic composite medium by applying the forward propagator matrix method for $0 \leq z < d$ and the backward propagator matrix method for $d < z$ (MATSU'URA et al., unpublished paper). The derivation of these expressions is basically accomplished in the cylindrical coordinate system (r, φ, z) shown in Figure 1. Given the expressions of stress components in the cylindrical coordinate system, we can calculate the yx component of stress tensor H_{yx} that we need by applying the rule of coordinate transformation;

$$H_{yx} = \frac{1}{2}(H_{rr} - H_{\varphi\varphi}) \sin 2\varphi + H_{r\varphi} \cos 2\varphi. \tag{8}$$

In general, according to the correspondence principle of linear viscoelasticity (LEE, 1955; RADOK, 1957), the solution of a viscoelastic problem in the Laplace transformed domain (s domain) can be directly obtained from the associated elastic problem by replacing the Lamé elastic constants with the corresponding Laplace operators and a source time function with its Laplace transform; that is, in the present case, $\lambda^{(2)}$ and $\mu^{(2)}$ with

$$\hat{\lambda}^{(2)}(s) = \frac{\lambda^{(2)}s + \mu^{(2)}(\lambda^{(2)} + 2\mu^{(2)}/3)/\eta}{s + \mu^{(2)}/\eta} \quad \text{and} \quad \hat{\mu}^{(2)}(s) = \frac{\mu^{(2)}s}{s + \mu^{(2)}/\eta} \tag{9}$$

and the step-type source time function $\Theta(t)$ with $1/s$. Note that the constitutive equation of the elastic surface layer in Eq. (1) does not change in the Laplace transformation. The expression of the viscoelastic solution in the physical domain (time domain) is obtained by applying the inverse Laplace transformation to this solution. As to the practical method of the inverse Laplace transformation in the present problem, refer to MATSU'URA et al. (1981).

In the following part of this section and Appendix, we present the concrete expressions of internal displacement components G_i ($i = r, \varphi, z$) and stress components H_{ij} ($i, j = r, \varphi, z$) due to a left-lateral unit step slip at $(0, 0, d)$ in the two-layered elastic half-space. In either case of $0 \le z < d$ and $d < z \le h$, the displacement components G_i ($i = r, \varphi, z$) and stress components H_{zj} ($j = r, \varphi, z$) can be written in the following form of semi-infinite wavenumber integrals:

$$G_r(r, \varphi, z, t) = \frac{1}{8\pi} \Theta(t) \sin 2\varphi \int_0^\infty \left[U_1(z; \xi) \frac{\partial J_2(\xi r)}{\partial r} - 4U_3(z; \xi) \frac{J_2(\xi r)}{r} \right] d\xi \tag{10}$$

$$G_\varphi(r, \varphi, z, t) = \frac{1}{8\pi} \Theta(t) \cos 2\varphi \int_0^\infty \left[2U_1(z; \xi) \frac{J_2(\xi r)}{r} - 2U_3(z, \xi) \frac{\partial J_2(\xi r)}{\partial r} \right] d\xi \tag{11}$$

$$G_z(r, \varphi, z, t) = \frac{1}{8\pi} \Theta(t) \sin 2\varphi \int_0^\infty \xi U_2(z; \xi) J_2(\xi r) \, d\xi \tag{12}$$

$$H_{zr}(r, \varphi, z, t) = \frac{1}{8\pi} \Theta(t) \sin 2\varphi \int_0^\infty \xi \left[2T_1(z; \xi) \frac{\partial J_2(\xi r)}{\partial r} - 4T_3(z; \xi) \frac{J_2(\xi r)}{r} \right] d\xi \tag{13}$$

$$H_{z\varphi}(r, \varphi, z, t) = \frac{1}{8\pi} \Theta(t) \cos 2\varphi \int_0^\infty \xi \left[4T_1(z; \xi) \frac{J_2(\xi r)}{r} - 2T_3(z; \xi) \frac{\partial J_2(\xi r)}{\partial r} \right] d\xi \tag{14}$$

$$H_{zz}(r, \varphi, z, t) = \frac{1}{8\pi} \Theta(t) \sin 2\varphi \int_0^\infty 2\xi^2 T_2(z; \xi) J_2(\xi r) \, d\xi \tag{15}$$

where $\Theta(t)$ and J_2 indicate the Heaviside step function and the Bessel function of order 2, respectively. By applying the forward propagator matrix method for $0 \leq z < d$ and the backward propagator matrix method for $d < z \leq h$, the expressions of the kernels of the integrals,

$$Y^T \equiv [U_1(z) \; U_2(z) \; T_1(z)/\mu^{(1)} \; T_2(z)/\mu^{(1)}] \quad \text{and} \quad Y'^T \equiv [U_3(z) \; T_3(z)/\mu^{(1)}] \quad (16)$$

are obtained in matrix form as

$$Y(z; 0 \leq z < d) = C(z)F_1(z)GY^0 \tag{17}$$

$$Y'(z; 0 \leq z < d) = C(z)F_1'(z)Y'^0 \tag{18}$$

and

$$Y(z; d < z \leq h) = C(z-h)F_1(z-h)D_1^{-1}E_2A_2 \tag{19}$$

$$Y'(z; d < z \leq h) = C(z-h)F_1'(z-h)D_1'^{-1}A_2' \tag{20}$$

with

$$C(z) = \cosh \xi z. \tag{21}$$

The concrete expressions of the vectors, Y^0, Y'^0, A_2 and A_2', and the matrices, F_1, F_1', D_1^{-1}, $D_1'^{-1}$, E_2 and G, in the above equations are given in the Appendix.

Given the displacement components G_i $(i = r, \varphi, z)$, we can easily obtain the expressions of H_{rr}, $H_{\varphi\varphi}$ and $H_{r\varphi}$, needed for the calculation for the yx component of stress tensor H_{yx}, from the constitutive equations.

$$H_{rr}(r, \varphi, z) = \lambda^{(1)}[E_{rr}(r, \varphi, z) + E_{\varphi\varphi}(r, \varphi, z) + E_{zz}(r, \varphi, z)] + 2\mu^{(1)}E_{rr}(r, \varphi, z) \tag{22}$$

$$H_{\varphi\varphi}(r, \varphi, z) = \lambda^{(1)}[E_{rr}(r, \varphi, z) + E_{\varphi\varphi}(r, \varphi, z) + E_{zz}(r, \varphi, z)] + 2\mu^{(1)}E_{\varphi\varphi}(r, \varphi, z) \tag{23}$$

$$H_{r\varphi}(r, \varphi, z) = 2\mu^{(1)}E_{r\varphi}(r, \varphi, z) \tag{24}$$

and the definition of the strain,

$$E_{rr}(r, \varphi, z) = \frac{\partial G_r(r, \varphi, z)}{\partial r} \tag{25}$$

$$E_{\varphi\varphi}(r, \varphi, z) = \frac{1}{r}G_r(r, \varphi, z) + \frac{1}{r}\frac{\partial G_\varphi(r, \varphi, z)}{\partial \varphi} \tag{26}$$

$$E_{zz}(r, \varphi, z) = \frac{\partial G_z(r, \varphi, z)}{\partial z} \tag{27}$$

$$E_{r\varphi}(r, \varphi, z) = \frac{1}{2}\left(\frac{1}{r}\frac{\partial G_r(r, \varphi, z)}{\partial \varphi} + \frac{\partial G_\varphi(r, \varphi, z)}{\partial r} - \frac{1}{r}G_\varphi(r, \varphi, z)\right). \tag{28}$$

4. Numerical Computation Algorithm

In order to solve the coupled nonlinear system of Eqs. (4), (6) and (7) numerically, first we rewrite these equations in discrete form both in space and time. We represent the slip perturbation u_s by the superposition of a finite number $(K \times L)$ of basis functions M_{kl}:

$$u_s(\boldsymbol{x}, t) = \sum_{k=1}^{K} \sum_{l=1}^{L} a_{kl}(t) M_{kl}(\boldsymbol{x}). \tag{29}$$

As the basis functions we take the bicubic B-spline functions defined by

$$M_{kl}(\boldsymbol{x}) = N_{4,k}(x) N_{4,l}(z) \tag{30}$$

where $N_{4,j}(s)$ is the B-spline function of order 4 (degree 3) with an equally spaced local support $(s_j - 4\Delta s \leq s < s_j)$, defined by the following de Boor-Cox recurrence formula (DE BOOR, 1972; COX, 1972):

$$N_{r,j}(s) = \frac{(s - s_{j-r}) N_{r-1,j-1}(s) + (s_j - s) N_{r-1,j}(s)}{s_j - s_{j-r}} \tag{31}$$

with

$$N_{1,j}(s) = \begin{cases} 1/(s_j - s_{j-1}) & (s_{j-1} \leq s < s_j) \\ 0 & \text{(otherwise)} \end{cases} \tag{32}$$

Substituting Eq. (29) into Eq. (6), we obtain

$$\sigma(\boldsymbol{x}, t) = \sigma^0(\boldsymbol{x}) + \int_0^t \sum_{k=1}^{K} \sum_{l=1}^{L} \frac{\partial a_{kl}(\tau)}{\partial \tau} S_{kl}(\boldsymbol{x}, t - \tau) \, d\tau \tag{33}$$

with

$$S_{kl}(\boldsymbol{x}, t - \tau) = \int_{\Sigma_s} M_{kl}(\boldsymbol{\xi}) H(\boldsymbol{x}, t - \tau; \boldsymbol{\xi}, 0) \, d\boldsymbol{\xi}. \tag{34}$$

The integration with respect to time in Eq. (33) is carried out stepwise at each time step τ_i $(i = 0, 1, \ldots)$ as

$$\sigma(\boldsymbol{x}, \tau_j) = \sigma^0(\boldsymbol{x}) + \sum_{i=0}^{j} \sum_{k=1}^{K} \sum_{l=1}^{L} \Delta a_{kl}(\tau_i) S_{kl}(\boldsymbol{x}, \tau_j - \tau_i) \tag{35}$$

where $\Delta a_{kl}(\tau_i)$ indicates the increase of a_{kl} at the i-th time step τ_i. We set all Δa_{kl} to zero at $t = \tau_0$. Similarly, the total fault slip w in Eq. (4) can be written as

$$w(\boldsymbol{x}, \tau_j) = v_{pl} \tau_j + \sum_{i=0}^{j} \sum_{k=1}^{K} \sum_{l=1}^{L} \Delta a_{kl}(\tau_i) M_{kl}(\boldsymbol{x}). \tag{36}$$

Consequently, our problem of solving the coupled nonlinear equations can be reduced to the problem of determining the values of $\Delta a_{kl}(\tau_i)$ stepwise in time so that the shear stress σ in Eq. (35) and the fault slip w in Eq. (36) satisfy the given constitutive relation in Eq. (7). In the following part of this section, we briefly show a practical technique to determine the unknown parameters Δa_{kl} at each time step.

We select N sampling points within the model region Σ_s. The fault slip and the shear stress at these sampling points are denoted by w_n ($n = 1, 2, \ldots, N$) and σ_n ($n = 1, 2, \ldots, N$), respectively. Furthermore, we arrange the parameters Δa_{kl} in some order, and made a M ($= K \times L$)-dimensional column vector $\Delta a = [\Delta a_m]$. Then we define the column vectors w and σ by one-dimensional arrays of w_n and σ_n, respectively. Thus we may rewrite Eqs. (36) and (35) as

$$w_n(\tau_j) = v_{pl}\tau_j + \sum_{i=0}^{j} \sum_{m=1}^{M} \Delta a_m(\tau_i) M_{nm} \tag{37}$$

and

$$\sigma_n(\tau_j) = \sigma_n^0 + \sum_{i=0}^{j} \sum_{m=1}^{M} \Delta a_m(\tau_i) S_{nm}(\tau_j - \tau_i). \tag{38}$$

Similarly, we may rewrite the constitutive relation in Eq. (7) as

$$\sigma_n(\tau_j) = f_n[w_n(\tau_j)]. \tag{39}$$

These equations are written in matrix form as

$$w(\tau_j) = v_{pl}\tau_j \mathbf{1} + \sum_{i=0}^{j} M \, \Delta a(\tau_i) \tag{40}$$

$$\sigma(\tau_j) = \sigma^0 + \sum_{i=0}^{j} S(\tau_j - \tau_i) \, \Delta a(\tau_i) \tag{41}$$

$$\sigma(\tau_j) = f[w(\tau_j)] \tag{42}$$

with

$$w^T(\tau_j) = [w_1(\tau_j), \ldots, w_n(\tau_j), \ldots, w_N(\tau_j)] \tag{43}$$

$$\sigma^T(\tau_j) = [\sigma_1(\tau_j), \ldots, \sigma_n(\tau_j), \ldots, \sigma_N(\tau_j)] \tag{44}$$

$$f^T[w(\tau_j)] = [f_1[w_1(\tau_j)], \ldots, f_n[w_n(\tau_j)], \ldots, f_N[w_N(\tau_j)]] \tag{45}$$

$$\mathbf{1}^T = [1, \ldots, 1, \ldots, 1] \tag{46}$$

$$\sigma^{0T} = [\sigma_1^0, \ldots, \sigma_n^0, \ldots, \sigma_N^0] \tag{47}$$

$$\Delta a^T(\tau_j) = [\Delta a_1(\tau_j), \ldots, \Delta a_m(\tau_j), \ldots, \Delta a_M(\tau_j)] \tag{48}$$

and

$$M = [M_{nm}], \quad S(\tau_j - \tau_i) = [S_{nm}(\tau_j - \tau_i)]. \tag{49}$$

Given the complete slip history from the beginning (τ_0) to the present (τ_j), namely, the values of Δa at all the time steps from τ_0 to τ_j, we can calculate the total slip w and the stress state σ at the present from Eqs. (40) and (41), respectively. Next, to determine the slip perturbation $\Delta a(\tau_{j+1})$ realized at the next step, we define a misfit function,

$$\Phi[\Delta a(\tau_{j+1})] \equiv \|\sigma(\tau_{j+1}) - f[w(\tau_{j+1})]\|^2$$

$$= \left\| \sigma^0 + \sum_{i=0}^{j+1} S(\tau_{j+1} - \tau_i)\,\Delta a(\tau_i) - f\left[v_{pl}\tau_{j+1}\mathbf{1} + \sum_{i=0}^{j+1} M\,\Delta a(\tau_i) \right] \right\|^2 \qquad (50)$$

and minimize it for $\Delta a(\tau_{j+1})$. The misfit function gives a measure of discrepancy between the shear stress directly calculated from Eq. (41) and that indirectly calculated from the total slip in Eq. (40) with the constitutive relation in Eq. (42). In order to find the optimal values of $\Delta a(\tau_{j+1})$, we can use the Levenberg-Marquardt algorithm for nonlinear least-squares (LEVENBERG, 1944; MARQUARDT, 1963).

5. Results of Numerical Simulation

For numerical simulation we need to give a concrete form of the constitutive relation in Eq. (7). In the present study we take a slip-weakening type of constitutive relation,

$$f(\chi) = \begin{cases} \sigma^0 + \Delta\sigma_p \chi(3 - \chi^2)/2 & (0 \leq \chi < 1) \\ \sigma^0 + \Delta\sigma_p \chi^v \exp[v(1 - \chi)] & (1 \leq \chi) \end{cases} \qquad (51)$$

with

$$\chi = w/w_0 \quad \text{and} \quad v = w_0/w_c. \qquad (52)$$

Here, σ^0 is the constant residual stress level, $\Delta\sigma_p$ is the position-dependent breakdown strength drop (peak strength measured from σ^0), and w_c and w_0 are also the position-dependent parameters, which define the critical weakening displacement D_c as

$$D_c \simeq w_0 + 5w_c. \qquad (53)$$

The slip-weakening type of constitutive relation has been confirmed by OKUBO and DIETERICH (1984) and OHNAKA et al. (1987) through laboratory experiments of stick-slip, and theoretically explained by MATSU'URA et al. (1992) through physical modelling of frictional sliding. Many laboratory experiments suggest linear scale-dependence of D_c and scale-independence of $\Delta\sigma_p$ (e.g., OHNAKA, 1996). On the basis of the comparative study of laboratory experiments and seismological observations with a theoretical model of rupture nucleation, SHIBAZAKI and MATSU-'URA (1998) have demonstrated that the fracture surface energy G_c, the critical nucleation-zone size L_c, and the dimension of seismic faults L_f scale with D_c. And furthermore, on the basis of laboratory experiments, OHNAKA (1992) has concluded that D_c is insensitive to temperature T in the range of $T < 300°C$, but has significant temperature dependence in the range of $T > 300°C$. In the case of vertical transcurrent plate boundaries, the critical temperature of 300°C corresponds to about 12 km in depth, where brittle-ductile transition occurs (SIBSON, 1984; MARONE and SCHOLZ, 1988). The temperature threshold over which brittle-ductile transition occurs might be stress sensitive, but we have insufficient data regarding it. Therefore we do not consider the stress dependence to the temperature sensitivity of D_c in the present simulation model. Integrating these different types of knowledge, we obtain a basic pattern of spatial variation in $\Delta\sigma_p$ and D_c used for numerical simulations.

In the following part of this section we show the results of numerical simulation for four representative cases. In the first and second cases with the effective fault length of 40 km, the patterns of spatial variation in $\Delta\sigma_p$ and D_c are identical, however the absolute values of D_c are taken to be different. In the third and fourth cases with the effective fault length of 100 km, the patterns of spatial variation in D_c are the same, but those in $\Delta\sigma_p$ are different in the fault strike direction. As for the structural parameters, we take the same values, which are listed in Table 1, in every case. The relative plate velocity v_{pl} is fixed to 5 cm/yr through the computation.

In the first case, we set an approximately 40-km long seismogenic region with relatively high strength in the 60×30 km model region Σ_s. The depth variation of constitutive relations at the center of the seismogenic region is shown in Figure 3, where the functional form of the constitutive relation at each depth is given by Eq. (51). The value of $\Delta\sigma_p$ gradually increases from 4 MPa to 5 MPa with depth in the brittle zone ($0 \le z < 12$ km), and rapidly decreases in the ductile zone ($z > 12$ km). In the direction of fault strike the value of $\Delta\sigma_p$ is taken to be constant and tapered off at both ends of the seismogenic region. On the other hand, the value of D_c is nearly constant (0.3 m) in the brittle zone, and rapidly increases in the ductile zone. Figure 4 is a series of snapshots showing the evolution of a shear stress field on the

Table 1

The structural parameters used in numerical computation

	ρ [kg/m^3]	λ [GPa]	μ [GPa]	η [Pa s]
Lithosphere	3000	40	40	∞
Asthenosphere	3400	90	60	10^{19}

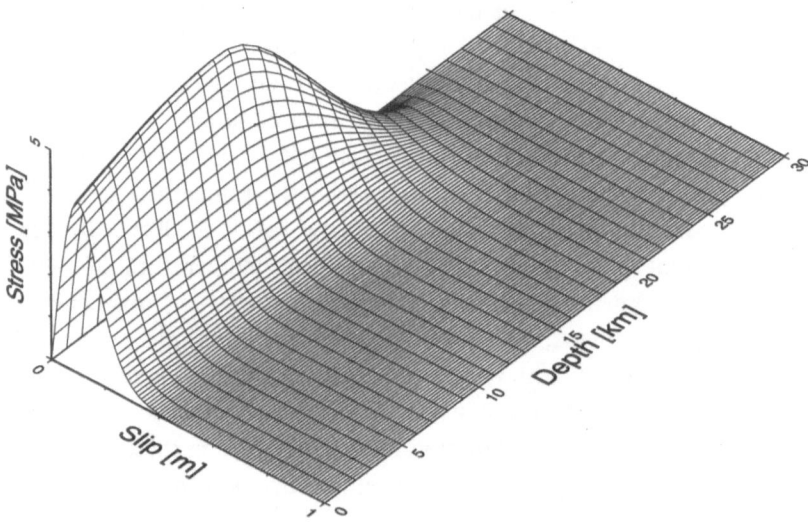

Figure 3
The depth-variation of constitutive relations at the center of the seismogenic region. The functional form
of constitutive relation is given in Eq. (51) of the text. The value of D_c in the brittle zone is taken to be
0.3 m.

plate interface. The stress accumulation due to slip deficits gradually proceeds in the seismogenic region until $t = 19$ yr. The stress concentration at the margin of the seismogenic region is caused by dislocation pile-ups there. At the next step ($t = 20$ yr) the system becomes unstable, indicating the occurrence of earthquake rupture. The process of dynamic rupture propagation cannot be treated in the quasi-static simulation algorithm. Thus we permit a jump in fault slip and shear stress and search a quasi-static solution at the next step. In the present case the process ends with the nearly complete release of the shear stress stored in the seismogenic region. The average stress drop of this seismic event is evaluated from the stress distribution at $t = 19$ yr as about 1.5 MPa. This value is consistent with obervational results for large interplate strike-slip earthquakes in California and North Anatolia (e.g., FUJII and MATSU'URA, 2000).

In the second case, we consider the same model setting as in the first case except for the absolute value of D_c. Namely, we assume the value of D_c in the brittle zone to be 2.5 m, which is about ten times as large as that in the first case. In this case, as can be seen from a series of snapshots in Figure 5, the stress accumulation gradually proceeds in the same manner as in the first case in the early stage, but the sudden stress release does not occur even after $t = 20$ yr. The gradual stress increase continues in the central part of the seismogenic region until the shear stress reaches the peak strength there. Then the process turns to stable stress release. This process continues until the stored stress is completely released.

In the third case, we set a considerably longer (about 100 km) seismogenic region in the 150 km × 30 km model region. The depth variation of constitutive relations at the center of the seismogenic region is shown in Figure 6, where the value of D_c in the brittle zone is taken to be 1.0 m. As to $\Delta\sigma_p$, we take the same

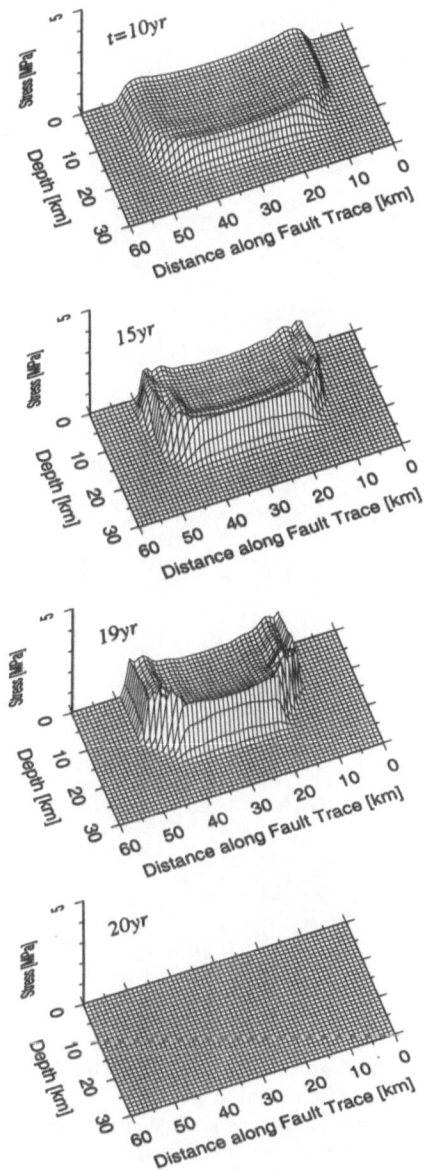

Figure 4

A series of snapshots showing the stress accumulation process in the 40-km long seismogenic region with $D_c = 0.3$ m. The stress concentration at the margin of the seismogenic region is caused by dislocation pile-ups there. At $t = 20$ yr the system becomes unstable, and sudden stress release occurs.

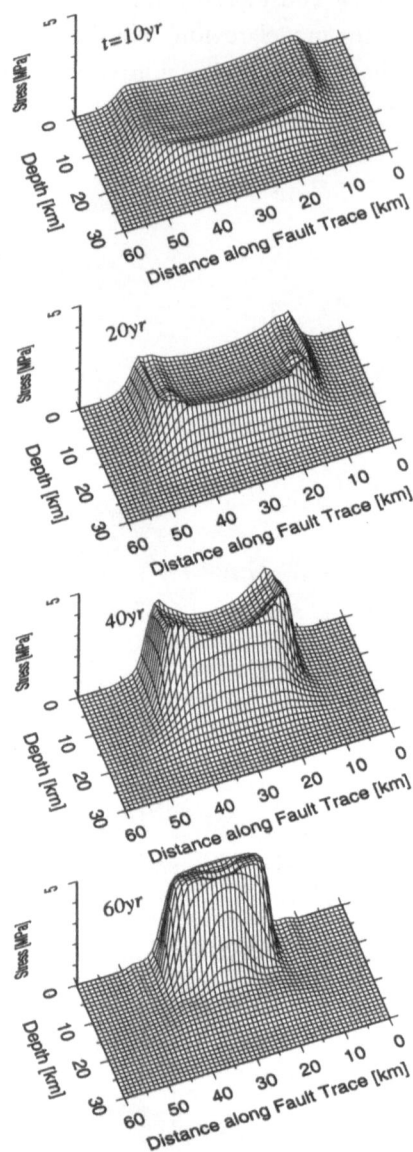

Figure 5

A series of snapshots showing the stress accumulation process in the 40-km long seismogenic region with $D_c = 2.5$ m. The model setting is the same as that in the case of Figure 4 except for the value of D_c. In this case sudden stress release does not occur. The gradual stress increase continues in the central part of the seismogenic region until the shear stress reaches the peak strength there. Then the process turns to stable stress release.

value as in the previous cases. From a series of snapshots in Figure 7, we can see that the stress accumulation gradually proceeds in the same manner as in the first case except for its rate. The stress accumulation rate in the central part of the seismogenic region is about one half of that in the first case. The sudden stress release occurs just after $t = 40$ yr. This means that the recurrence time of 100-km long earthquake rupture is about twice as long as that of a 40-km long earthquake rupture. On the other hand, the average stress drop of this seismic event, evaluated from the stress distribution at $t = 40$ yr, is nearly the same as that in the first case.

In the fourth case, we consider the same model setting as in the third case except that the 100-km long seismogenic region is divided into two subregions by a narrow weak zone with $\Delta\sigma_p \simeq 1$ MPa. The effective length of the greater subregion is about 70 km, and that of the smaller subregion is about 20 km. In Figure 8 we show the process of stress accumulation and release by a series of snapshots. In the very early stage the stress accumulation gradually proceeds at nearly the same rate as in the third case. Then the stress is rapidly concentrated in the smaller subregion, and sudden stress release occurs there at $t = 35$ yr. The occurrence of sudden stress release in the smaller subregion accelerates stress accumulation in the greater subregion, and induces the occurrence of sudden stress release there at just after $t = 37$ yr. This result suggests the importance of interaction between adjacent seismic faults through transient viscoelastic stress transfer as well as instantaneous elastic stress transfer.

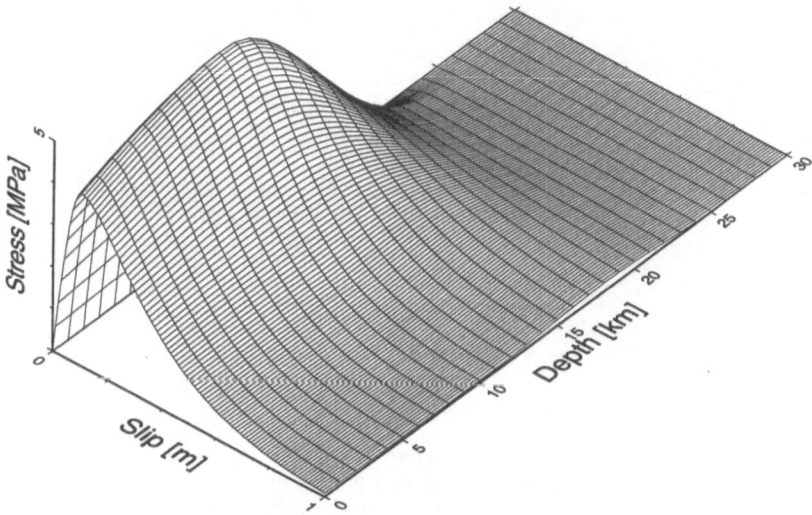

Figure 6

The depth variation of constitutive relations at the center of the seismogenic region. The value of D_c in the brittle zone is taken to be 1.0 m.

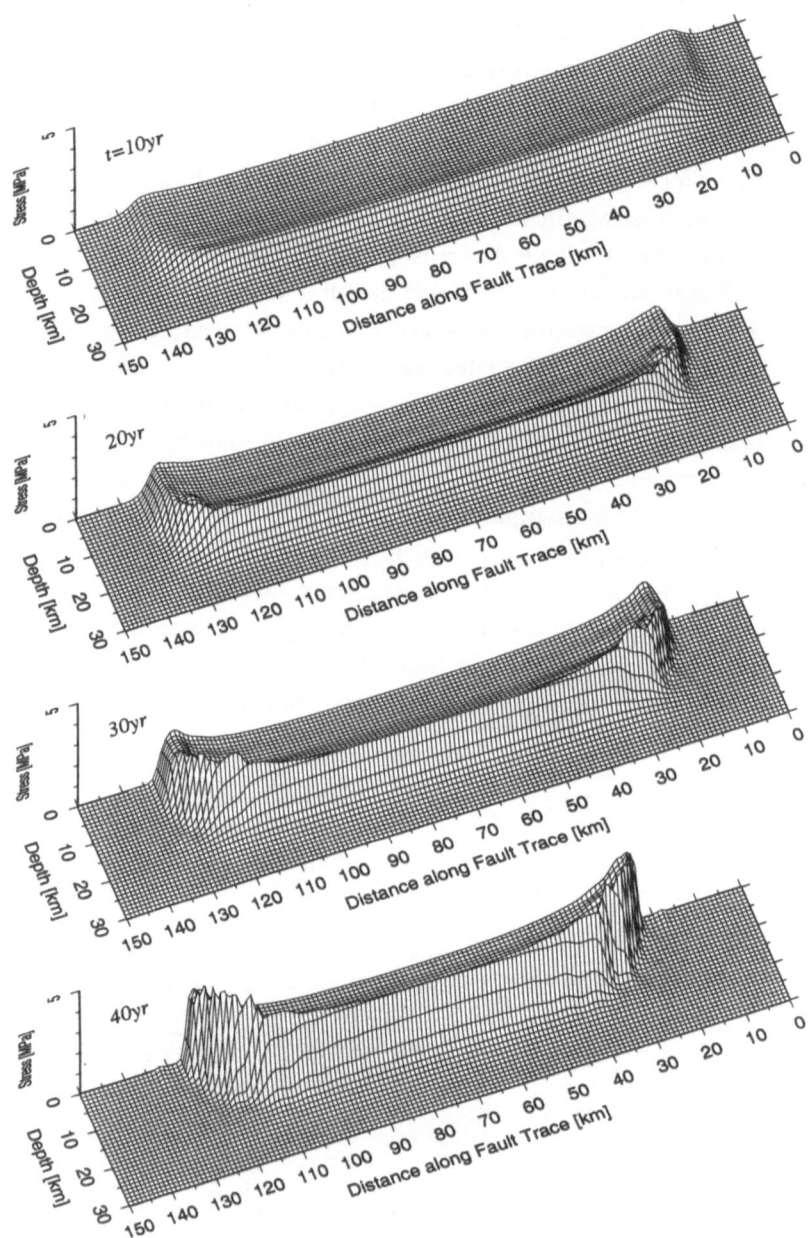

Figure 7
A series of snapshots showing the stress accumulation process in the 100-km long seismogenic region with $D_c = 1.0$ m. Just after $t = 40$ yr the system becomes unstable, and sudden stress release occurs.

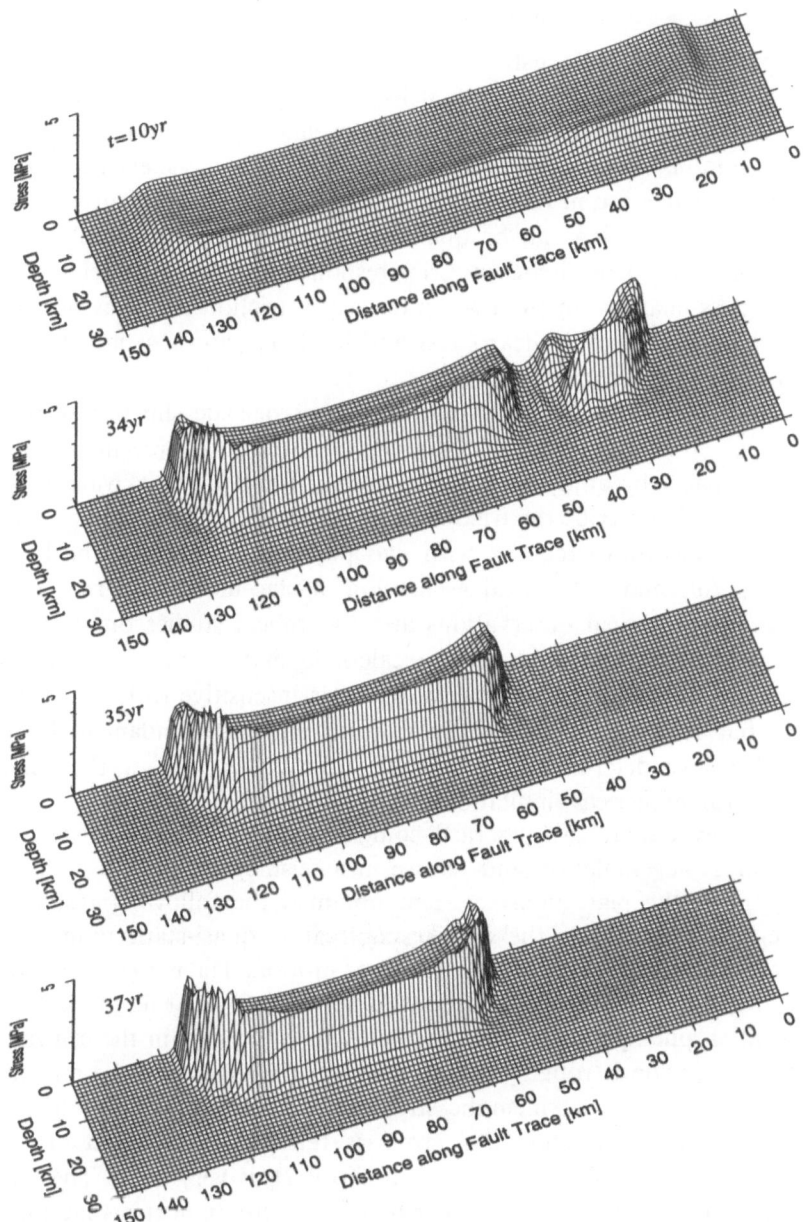

Figure 8

A series of snapshots showing the process of stress accumulation and release in the 100-km long seismogenic region with a narrow weak zone. The model setting is the same as that in the case of Figure 7 except for the narrow weak zone. The stress is rapidly concentrated in the smaller subregion, and sudden stress release occurs there at $t = 35$ yr. This event accelerates stress accumulation in the greater subregion, and induces the occurrence of sudden stress release there at just after $t = 37$ yr.

6. Discussion and Conclusions

In the present study we constructed a 3-D physical model of tectonic loading at transcurrent plate boundaries. We modeled the lithosphere-asthenosphere system by an elastic surface layer overlying a Maxwellian viscoelastic half-space, and introduced an infinitely long, vertical interface that divides the elastic surface layer into two plates which are in relative horizontal motion. Subsequently the physical process of tectonic loading can be quantitatively described by coupled nonlinear equations, consisting of a viscoelastic slip response function that relates fault slip with shear stress changes on the plate interface, a friction law that prescribes the constitutive relation between shear stress and fault slip, and a steady relative plate motion to drive the system.

We derived the concrete expressions of the viscoelastic slip response function (internal shear stress change due to a point dislocation source in an elastic-viscoelastic half-space) by applying both the forward and the backward propagator matrix methods. To describe the frictional properties of faults we took a slip-weakening type of constitutive relation with two physical parameters: the breakdown strength drop $\Delta\sigma_p$ and the critical weakening displacement D_c. From laboratory experiments, seismological observations and theoretical studies on rupture nucleation, it has been revealed that $\Delta\sigma_p$ is scale-independent, but D_c is inherently a scale-dependent quantity. And furthermore, D_c is insensitive to temperature T for $T < 300°C$, but sensitive for $T > 300°C$. Considering these fundamental properties of $\Delta\sigma_p$ and D_c we determined the spatial variation pattern of the constitutive relation used for numerical simulations.

With a newly developed numerical computation algorithm, we simulated the process of stress accumulation and release in a seismogenic region with relatively high strength on the plate interface, and obtained the following results. In low strength regions surrounding the seismogenic region, quasi-static fault slip gradually proceeds with the progress of relative plate motion. The increase of slip deficits in the seismogenic region brings about stress concentration at its margin, resulting from dislocation pile-ups there. The stress accumulation rate in the central part of the seismogenic region is roughly proportional to the inverse of the effective fault length, as expected from a kinematic stress accumulation model by MATSU'URA and SATO (1997). The accumulated stress is released by an unstable dynamic rupture if D_c is small, and by a stable fault slip if D_c is very large. The latter case may correspond to a kind of silent earthquakes. This indicates that the critical weakening displacement D_c is an essential parameter governing the process of stress accumulation and release. In numerical simulations, if we take the reasonable values of D_c on the basis of the scaling relation, $D_c \simeq 10^{-5} L_f$, proposed by SHIBAZAKI and MATSU'URA (1998), the recurrence time of seismic events, which corresponds to the time needed for the occurrence of sudden stress release in the present case, is roughly proportional to the effective fault length, but the coseismic average stress drop is independent of it.

We examined the process of stress accumulation and release for a fault system consisting of two adjacent seismogenic regions, and found that sudden stress release in one region accelerates stress accumulation in another region through transient viscoelastic stress transfer as well as instantaneous elastic stress transfer. This suggests the importance of elastic and viscoelastic interaction between adjacent seismic faults in furthering the understanding of complexity and diversity in seismic activity at transcurrent plate boundaries.

Appendix

In this Appendix we give the concrete expressions of structural matrices, F_1, F'_1, D_1^{-1}, D'_1^{-1}, E_2 and G, the surface displacement vectors, Y^0 and Y'^0, and the coefficient vectors, A_2 and A'_2, of potential functions in the substratum.

The expressions of the structural matrices, F_1, F'_1, D_1^{-1}, D'_1^{-1}, E_2 and G, are given by

$$F_1(z) = \begin{bmatrix} 1 + \gamma^{(1)}\xi z T(z) & -(1-\gamma^{(1)})T(z) - \gamma^{(1)}\xi z \\ -(1-\gamma^{(1)})T(z) + \gamma^{(1)}\xi z & 1 - \gamma^{(1)}\xi z T(z) \\ \gamma^{(1)}T(z) + \gamma^{(1)}\xi z & -\gamma^{(1)}\xi z T(z) \\ \gamma^{(1)}\xi z T(z) & \gamma^{(1)}T(z) - \gamma^{(1)}\xi z \end{bmatrix}$$

$$\begin{bmatrix} (2-\gamma^{(1)})T(z) + \gamma^{(1)}\xi z & -\gamma^{(1)}\xi z T(z) \\ \gamma^{(1)}\xi z T(z) & (2-\gamma^{(1)})T(z) - \gamma^{(1)}\xi z \\ 1 + \gamma^{(1)}\xi z T(z) & (1-\gamma^{(1)})T(z) - \gamma^{(1)}\xi z \\ (1-\gamma^{(1)})T(z) + \gamma^{(1)}\xi z & 1 - \gamma^{(1)}\xi z T(z) \end{bmatrix} \quad (A.1)$$

$$F'_1(z) = \begin{pmatrix} 1 & T(z) \\ T(z) & 1 \end{pmatrix} \quad (A.2)$$

$$D_1 = \begin{bmatrix} 1 & 0 & 0 & 0 \\ 0 & 1 & 0 & 0 \\ 0 & 0 & \dfrac{\mu^{(1)}}{\mu^{(2)}} & 0 \\ 0 & 0 & 0 & \dfrac{\mu^{(1)}}{\mu^{(2)}} \end{bmatrix}, \quad D'_1 = \begin{bmatrix} 1 & 0 \\ 0 & \dfrac{\mu^{(1)}}{\mu^{(2)}} \end{bmatrix} \quad (A.3)$$

$$
E_2 = \begin{bmatrix} 1 & 0 & 0 & 0 \\ 0 & 2-\gamma^{(2)} & 1 & 0 \\ 0 & 1-\gamma^{(2)} & 1 & 0 \\ 1 & 0 & 0 & 1 \end{bmatrix}, \quad
G = \begin{bmatrix} 1 & 0 & 0 & 0 \\ 0 & 1 & 0 & 0 \\ 0 & 0 & 1 & 0 \\ 0 & \dfrac{\rho^{(1)}g}{2\mu^{(1)}\xi} & 0 & 1 \end{bmatrix} \tag{A.4}
$$

where

$$
T(z) = \tanh \xi z, \quad \gamma^{(i)} = \frac{\lambda^{(i)} + \mu^{(i)}}{\lambda^{(i)} + 2\mu^{(i)}} \quad (i = 1, 2) \tag{A.5}
$$

and g is the acceleration of gravity at the earth's surface.

The surface displacement vectors, Y^0 and Y'^0, are given by

$$
Y^{0T} = (U_1^0 \; U_2^0 \; 0 \; 0) \quad \text{and} \quad Y'^{0T} = (U_3^0 \; 0) \tag{A.6}
$$

where

$$
U_1^0 = \frac{1}{C(h)} \frac{1}{\Delta} (X_1 \; X_2 \; X_3 \; X_4) \begin{bmatrix} (P_{22} + P_{42})(Q_{11} + Q_{31}) - (P_{12} + P_{32})(Q_{21} + Q_{41}) \\ (P_{22} + P_{42})(Q_{12} + Q_{32}) - (P_{12} + P_{32})(Q_{22} + Q_{42}) \\ (P_{22} + P_{42})(Q_{13} + Q_{33}) - (P_{12} + P_{32})(Q_{23} + Q_{43}) \\ (P_{22} + P_{42})(Q_{14} + Q_{34}) - (P_{12} + P_{32})(Q_{24} + Q_{44}) \end{bmatrix}
$$

$$\tag{A.7}$$

$$
U_2^0 = \frac{1}{C(h)} \frac{1}{\Delta} (X_1 \; X_2 \; X_3 \; X_4) \begin{bmatrix} (P_{11} + P_{31})(Q_{21} + Q_{41}) - (P_{21} + P_{41})(Q_{11} + Q_{31}) \\ (P_{11} + P_{31})(Q_{22} + Q_{42}) - (P_{21} + P_{41})(Q_{12} + Q_{32}) \\ (P_{11} + P_{31})(Q_{23} + Q_{43}) - (P_{21} + P_{41})(Q_{13} + Q_{33}) \\ (P_{11} + P_{31})(Q_{24} + Q_{44}) - (P_{21} + P_{41})(Q_{14} + Q_{34}) \end{bmatrix}
$$

$$\tag{A.8}$$

with

$$
\Delta = (P_{11} + P_{31})(Q_{22} + Q_{42}) - (P_{12} + P_{32})(Q_{21} + Q_{41}) \tag{A.9}
$$

and

$$
U_3^0 = \frac{1}{C(h)} \frac{1}{\Delta'} (X_1' \; X_2') \begin{pmatrix} Q_{11}' + Q_{21}' \\ Q_{12}' + Q_{22}' \end{pmatrix} \tag{A.10}
$$

$$
\Delta' = P_{11}' + P_{21}' \tag{A.11}
$$

P_{ij}, Q_{ij}, P_{ij}' and Q_{ij}' are the ij-elements of the matrices P, Q, P' and Q' defined by

$$
P = E_2^{-1} D_1 F_1(h) G, \quad Q = E_2^{-1} D_1 \tag{A.12}
$$

$$P' = D'_1 F'_1(h), \quad Q' = D'_1 \tag{A.13}$$

and X_i and X'_i are the i-elements of the vectors X and X' defined by

$$X = \begin{pmatrix} X_1 \\ X_2 \\ X_3 \\ X_4 \end{pmatrix} = e^{(h-d)\xi}(1 + e^{-2(h-d)\xi}) \left[\begin{pmatrix} 2T(h-d) \\ 0 \\ 1 \\ T(h-d) \end{pmatrix} \right.$$

$$\left. + \gamma^{(1)} \begin{pmatrix} -T(h-d) + (h-d)\xi \\ (h-d)\xi T(h-d) \\ (h-d)\xi T(h-d) \\ -T(h-d) + (h-d)\xi \end{pmatrix} \right] \tag{A.14}$$

$$X' = \begin{pmatrix} X'_1 \\ X'_2 \end{pmatrix} = e^{(h-d)\xi}(1 + e^{-2(h-d)\xi}) \begin{pmatrix} -T(h-d) \\ -1 \end{pmatrix}. \tag{A.15}$$

The coefficient vectors, A_2 and A'_2, of potential functions in the substratum are given by

$$A_2^T = (A_2^- \ B_2^- \ -A_2^- \ -B_2^-) \quad \text{and} \quad A_2'^T = (C_2^- \ -C_2^-) \tag{A.16}$$

where

$$\begin{pmatrix} A_2^- \\ B_2^- \end{pmatrix} = \frac{1}{C(h)} \frac{1}{\tilde{\Delta}} \begin{pmatrix} \tilde{P}_{44} - \tilde{P}_{42} & \tilde{P}_{32} - \tilde{P}_{34} \\ \tilde{P}_{41} - \tilde{P}_{43} & \tilde{P}_{33} - \tilde{P}_{31} \end{pmatrix} \begin{pmatrix} \tilde{X}_3 \\ -(\rho^{(1)}g/2\mu^{(1)}\xi)\tilde{X}_2 + \tilde{X}_4 \end{pmatrix} \tag{A.17}$$

with

$$\tilde{\Delta} = (\tilde{P}_{33} - \tilde{P}_{31})(\tilde{P}_{44} - \tilde{P}_{42}) - (\tilde{P}_{34} - P_{32})(\tilde{P}_{43} - \tilde{P}_{41}) \tag{A.18}$$

and

$$C_2^- = \frac{1}{C(h)} \frac{1}{\tilde{\Delta}'} \tilde{X}'_2 \tag{A.19}$$

with

$$\tilde{\Delta}' = \tilde{P}'_{22} - \tilde{P}'_{21} \tag{A.20}$$

$\tilde{P}_{ij}, \tilde{Q}_{ij}, \tilde{P}'_{ij}$ and $\tilde{Q}'_{ij}$ are the ij-elements of the matrices $\tilde{P}, \tilde{Q}, \tilde{P}'$ and $\tilde{Q}'$ defined by

$$\tilde{P} = G^{-1}F_1(-h)D_1^{-1}E_2, \quad \tilde{Q} = G^{-1} \tag{A.21}$$

$$\tilde{P}' = F'_1(-h)D_1'^{-1}, \quad Q' = I \tag{A.22}$$

and $\tilde{X}_i$ and $\tilde{X}'_1$ are the i-elements of the vectors $\tilde{X}$ and $\tilde{X}'$ defined by

$$\tilde{X} = \begin{pmatrix} \tilde{X}_1 \\ \tilde{X}_2 \\ \tilde{X}_3 \\ \tilde{X}_4 \end{pmatrix} = e^{d\xi}(1 + e^{-2d\xi}) \left[\begin{pmatrix} -2T(d) \\ 0 \\ 1 \\ -T(d) \end{pmatrix} + \gamma^{(1)} \begin{pmatrix} -T(d) - d\xi \\ d\xi T(d) \\ d\xi T(d) \\ -T(d) - d\xi \end{pmatrix} \right] \qquad \text{(A.23)}$$

$$\tilde{X}' = \begin{pmatrix} \tilde{X}'_1 \\ \tilde{X}'_2 \end{pmatrix} = e^{d\xi}(1 + e^{-2d\xi}) \begin{pmatrix} T(d) \\ -1 \end{pmatrix}. \qquad \text{(A.24)}$$

Acknowledgements

We are grateful to Bruce Shaw for his constructive review.

REFERENCES

BURRIDGE, R., and KNOPOFF, L. (1964), *Body Force Equivalents for Seismic Dislocations*, Bull. Seismol. Soc. Am. *54*, 1875–1888.

COHEN, S. C. (1980), *Postseismic Viscoelastic Surface Deformation and Stress, 1. Theoretical Considerations, Displacement, and Strain Calculations*, J. Geophys. Res. *85*, 3131–3150.

COX, M. G. (1972), *The Numerical Evaluation of B-splines*, J. Inst. Math. Appl. *10*, 134–149.

DE BOOR, C. (1972), *On Calculating with B-splines*, J. Approx. Theory *6*, 50–62.

DIETERICH, J. H. (1979), *Modeling of Rock Friction 1. Experimental Results and Constitutive Equations*, J. Geophys. Res. *84*, 2161–2168.

FUJII, Y., and MATSU'URA, M. (2000), *Regional Difference in Scaling Laws for Large Earthquakes and its Tectonic Implication*, Pure appl. geophys. *157*, 2283–2301.

IWASAKI, T., and MATSU'URA, M. (1981), *Quasi-static Strain and Tilt due to Faulting in the Layered Half-space with an Intervenient Viscoelastic Layer*, J. Phys. Earth. *29*, 499–518.

KATO, N., and HIRASAWA, T. (1997), *A Numerical Study on Seismic Coupling along Subduction Zones Using a Laboratory-derived Friction Law*, Phys. Earth Planet. Inter. *29*, 499–518.

LEE, E. H. (1955), *Stress Analysis in Visco-elastic Bodies*, Q. Appl. Math. *13*, 183–190.

LEVENBERG, K. (1944), *A Method for the Solution of Certain Nonlinear Problems in Least-squares*, Q. Appl. Math. *2*, 164–168.

LISOWSKI, M., SAVAGE, J. C., and PRESCOTT, W. H. (1991), *The Velocity Field along the San Andreas Fault in Central and Southern California*, J. Geophys. Res. *96*, 8369–8389.

MARONE, C., and SCHOLZ, C. H. (1988), *The Depth of Seismic Faulting and the Upper Transition from Stable to Unstable Slip Regimes*, Geophys. J. Lett. *15*, 621–624.

MARQUARDT, D. W. (1963), *An Algorithm for Least-squares Estimation of Nonlinear Parameters*, Indust. Appl. Math. *11*, 431–441.

MARUYAMA, T. (1963), *On the Force Equivalents of Dynamical Elastic Dislocations with Reference to the Earthquake Mechanism*, Bull. Earthq. Res. Inst. Tokyo Univ. *41*, 467–486.

MATSU'URA, M., KATAOKA, H., and SHIBAZAKI, B. (1992), *Slip-dependent Friction Law and Nucleation Processes in Earthquake Rupture*, Tectonophysics *211*, 135–148.

MATSU'URA, M., and SATO, T. (1997), *Loading Mechanism and Scaling Relation of Large Interplate Earthquakes*, Tectonophysics *277*, 189–198.

MATSU'URA, M., TANIMOTO, T., and IWASAKI, T. (1981), *Quasi-static Displacements due to Faulting in a Layered Half-space with an Intervenient Viscoelastic Layer*, J. Phys. Earth. *29*, 23–54.

OHNAKA, M. (1992), *Earthquake Source Nucleation: A Physical Model for Short-term Precursors*, Tectonophysics *211*, 149–178.

OHNAKA, M. (1996), *Nonuniformity of the Constitutive Law Parameters for Shear Fracture and Quasi-static Nucleation to Dynamic Rupture: A Physical Model of Earthquake Generation Process*, Proc. Natl. Acad. Sci. USA *93*, 3795–3802.

OHNAKA, M., KUWAHARA, Y., and YAMAMOTO, K. (1987), *Constitutive Relations between Dynamic Physical Parameters near a Tip of the Propagating Slip Zone during Stick-slip Shear Failure*, Tectonophysics *144*, 109–125.

OKUBO, P. G., and DIETERICH, J. H. (1984), *Effects of Physical Fault Properties on Frictional Instabilities Produced on Simulated Faults*, J. Geophys. Res. *88*, 5817–5827.

RADOK, J. R. M. (1957), *Visco-elastic Stress Analysis*, Q. Appl. Math. *15*, 198–202.

RUINA, A. (1983), *Slip Instability and State Variable Friction Law*, J. Geophys. Res. *88*, 10,359–10,370.

RUNDLE, J. B. (1978), *Viscoelastic Crustal Deformation by Finite Quasi-static Sources*, J. Geophys. Res. *83*, 5937–5945.

RUNDLE, J. B. (1982), *Viscoelastic-gravitational Deformation by a Rectangular Thrust Fault in a Layered Earth*, J. Geophys. Res. *87*, 7787–7796.

RUNDLE, J. B., and JACKSON, D. D. (1977), *A three-dimensional Viscoelastic Model of a Strike Slip Fault*, Geophys. J. R. Astron. Soc. *49*, 575–591.

SHIBAZAKI, B., and MATSU'URA, M. (1998), *Transition Process Nucleation to High-speed Rupture Propagation: Scaling from Stick-slip Experiments to Natural Earthquakes*, Geophys. J. Int. *132*, 14–30.

SIBSON, R. H. (1984), *Roughness at the Base of the Seismogenic Zone: Contributing Factors*, J. Geophys. Res. *89*, 5791–5799.

STUART, W. D. (1988), *Forecast Model for Great Earthquakes at the Nankai Trough Subduction Zone*, Pure appl. geophys. *126*, 619–641.

THATCHER, W. (1983), *Nonlinear Strain Buildup and Earthquake Cycle on the San Andreas Fault*, J. Geophys. Res. *88*, 5893–5902.

TSE, S. T., and RICE, J. R. (1986), *Crustal Earthquake Instability in Relation to the Depth Variation of Frictional Slip Properties*, J. Geophys. Res. *91*, 9452–9472.

(Received August 6, 1999, revised March 8, 2000, accepted April 6, 2000)

 To access this journal online:
http://www.birkhauser.ch

Pure appl. geophys. 157 (2000) 2149–2164
0033–4553/00/122149–16 $ 1.50 + 0.20/0

⎾Pure and Applied Geophysics

The Edges of Large Earthquakes and the Epicenters of Future Earthquakes: Stress-induced Correlations in Elastodynamic Fault Models

BRUCE E. SHAW[1]

Abstract—Fault models can generate complex sequences of events from frictional instabilities, even when the material properties are completely uniform along the fault. These complex sequences arise from the heterogeneous stress and strain fields which are produced through the dynamics of repeated ruptures on the fault. Visual inspection of the patterns of events produced in these models shows a striking and ubiquitous feature: future events tend to occur near the edges of where large events died out. In this paper, we explore this feature more deeply. First, using long catalogues generated by the model, we quantify the effect. We show, interestingly, that it is an even larger effect for future small events than it is for future large events. Then, using our ability to directly measure all aspects of the model, we find a physical explanation for our observations by examining the stress fields associated with large events. Looking at the average stress field we see a large stress concentration left at the edge of the large events, out of which the future events emerge. Further, we see the smearing out of the stress concentration as small events occur. This indicates why the epicenters of future small events are more correlated with the edges of large events than are the epicenters of future large events. Finally, we discuss how results from our simple model may be relevant to the more complicated case of the earth.

Key words: Earthquake dynamics, stress interactions, seismology, spatial correlations.

1. Introduction

Earthquake prediction has remained a long sought, but highly elusive goal. Despite decades of effort, a debate still rages on whether or not it is even possible. Implicit in most of this discussion is the goal of predicting when earthquakes will happen. If, instead, we think about other scientific and useful predictions that can be made, we see that this is only part of the story. Thus, even if the goal of when future earthquakes will occur remains unachieved, it would nevertheless be very useful to be able to predict what will occur and where it will occur. In this paper, we present measurements in simple dynamical earthquake models which show strong correlations between past large model events and the locations of epicenters of future model events. Specifically, we find that the edges of where large events

[1] Lamont-Doherty Earth Observatory, Columbia University, Palisades, NY, 10964, U.S.A. E-mail: shaw@ldeo.columbia.edu

died out produce large stress concentrations, which then tend to be the most stressed regions and therefore the epicenters of events which occur in the future. We discuss as well, the applicability of these results from this simple model to the more complex case of the earth.

Previous work on earthquakes has examined a number of the issues we discuss in this paper. PEREZ and SCHOLZ (1997) inspected the epicenters of events before and after large events for numerous cases globally. They found an increase in activity around the ends of future large events both preceding and following the large events. They also pointed out the stress concentrations left at the edges of large events. Availability of data, however, limited the discussion to mainly qualitative answers. OLSEN et al. (1997) noted large stress concentrations left at edges of the inferred 1992 Landers M7.4 earthquake, and referred to them as likely nucleation centers of future seismicity.

There is extensive literature detailing that stress perturbations from large earthquakes trigger future earthquakes (HARRIS and SIMPSON, 1992; JAUME and SYKES, 1992; STEIN et al., 1992; KING et al., 1994; HARRIS et al., 1995). Good agreement between calculated stress perturbations and the numbers of aftershocks as a function of location has been demonstrated (DIETERICH, 1994). Long-term correlations of stress increments and future seismicity have also been shown (DENG and SYKES, 1996). Collectively, this work has solidified the central role that stress plays in earthquake mechanics. In all these cases, however, there remains an ambiguity of what the absolute stress levels are, since the initial stress to which the stress increments are added remains unknown. Further, there are many more locations where we have insufficient information about past large events to make sufficiently accurate estimates of stress perturbations to be used for future seismicity forecasts. Thus, finding general patterns which relate past earthquakes to future seismicity could be quite useful.

We have studied a number of very simple scalar elastodynamic models of earthquake faults, and found what seems to be a quite general pattern emerging from the long catalogues generated in these models. Specifically, we find that the edges of where past large events die out tend to be the locations where epicenters of future events occur. By investigating the guts of the model, we find a physical origin of the effect in terms of the stress field. Understood in this way, the results become less surprising; they can be used, however, to assist our means of extrapolation from these simple models to the more complex earth. The rest of the paper is organized as follows. In the forthcoming Section 2 we describe the simple models. We then present in Section 3 the new results for the correlations and the origin of the correlations. Finally, in Section 4, we discuss how these results might apply to the more complicated case of the earth.

2. The Model

Elastodynamic models of earthquake faults have been shown to produce remarkably vast complexity, with a wide distribution of sizes of events, even with completely uniform properties along the fault (CARLSON and LANGER, 1989). Complex solutions are a legitimate outcome of the continuum of elastodynamic equations (SHAW and RICE, 2000), although the richest population of events occurs only over restricted regions of frictional parameter space (SHAW and RICE, 2000). These models spontaneously generate a heterogeneous stress distribution from repeated ruptures along the fault, and sequences of events on this complex attractor produce chaotic, complex behavior. While complex, the behavior is clearly not random: we see quite systematic aspects to it, one of which we will focus on in this paper. First, however, we present the model we will use to exhibit this systematic feature.

We focus our attention on the dynamics of a single fault, and a very simplified fault at that. There are three main reasons for studying such a simplified system. First, the simplest systems generally allow the most minimal parameterizations, thus the fewest features must be specified and the parameter space can be most fully explored. Secondly, our simplifications, particularly the low dimensionality of the fault we study, are computationally very fast, and allow for long sequences on large faults to be studied. Thirdly, we want to understand why our system is behaving as it does, and the simpler systems make this understanding easier to achieve.

The model we study here is a two-dimensional model of a fault (SHAW, 1997; SHAW and RICE, 2000). The two-dimensional fault has the advantage over a one-dimensional fault (BURRIDGE and KNOPOFF, 1967) of preserving long-range elastic interactions, and the advantage over a full three-dimensional fault of being vastly less expensive numerically. We consider only scalar motions, consequently we have the wave equation in the bulk. The coupling to the stably sliding lower fault, which provides the main driving both in our model and in real faults, is represented by a stiff boundary sliding at a constant creep rate. A long strip of the wave equation connects this stably sliding lower fault to the unstably sliding seismogenic fault; on the seismogenic fault we have a frictional boundary condition, and sliding occurs in episodic stick-slip events. We study a system with uniform material properties, so that all of the irregularities which develop are dynamic in origin.

In equations, we have, for the bulk, the wave equation for the scalar displacement field U:

$$\frac{\partial^2 U}{\partial t^2} = \nabla^2 U \tag{1}$$

where t is time, and ∇^2 is the two-dimensional Laplacian operator

$$\nabla^2 = \frac{\partial^2}{\partial x^2} + \frac{\partial^2}{\partial y^2}$$

for the directions x, which we take to be along the fault, and y which is then the direction perpendicular to the fault. Along the fault, located at $y = 0$, we have the boundary condition that the strain equals the traction, here the friction Φ on it:

$$\left.\frac{\partial U}{\partial y}\right|_{y=1} = \Phi. \tag{2}$$

We will return to a discussion of Φ shortly; first let us specify the other boundary conditions. Away from the fault, a distance of a crust depth (normalized to unity) away, the displacement creeps along at a constant steady slow rate v, the plate velocity:

$$\left.\frac{\partial U}{\partial t}\right|_{y=1} = v \tag{3}$$

with $v \ll 1$. Along the fault, we use periodic boundary conditions:

$$U(x + L_x) = U(x). \tag{4}$$

To complete the description of the model, we must specify the friction Φ. All of the nonlinearity in our model is contained in the friction. We specify the friction through a constitutive relation between motions on the fault, and the traction there which resists the motion.

The Friction

We present here a somewhat detailed discussion of the physical basis of the friction we use, to motivate the resulting constitutive equations. The results we present in this paper are not specific to the physical mechanism of frictional weakening discussed. Rather, they arise more generally as a consequence of dynamic stress heterogeneities. Nevertheless, we find it useful to see the physical basis of the friction we use. The reader less interested in the specifics of the friction can skip to the next part which treats numerics.

The friction we use is motivated by a physical idea which reverts to SIBSON (1973). Simple estimates of the heat generated by sliding at usual values of friction suggest that the fault would melt (MCKENZIE and BRUNE, 1972); since this is not generally observed in exhumed faults, something else must be transpiring. SIBSON (1973) noted an interesting feedback mechanism: if there are pore fluids present, then the heat generated would raise the temperature and pressure of the fluids, thereby reducing the effective normal stress and thus the friction. A simple mathematical quantification of this physical idea was presented by SHAW (1995) as follows. The friction Φ is the product of the coefficient of friction μ and the effective normal stress N:

$$\Phi = \mu N. \tag{5}$$

The effective normal stress is reduced by the pore fluid pressure, which increases with temperature:

$$N = N_0 - \alpha Q, \tag{6}$$

where N_0 is the ambient normal stress, α is a proportionality constant, and Q is the heat. The heat Q evolves in two ways. It is produced by frictional sliding and, in our approximation, is dissipated over some timescale:

$$\frac{\partial Q}{\partial t} = -\gamma Q + \Phi V, \tag{7}$$

where V is the velocity on the fault and γ is some dissipation inverse timescale. Two limiting cases were found for this friction; in the case where the dissipation was slow compared to the rupture timescale, $\gamma \ll 1$, slip-weakening results, a case LACHENBRUCH (1980) had examined long ago. In the opposite limit, when the dissipation is fast compared to the rupture timescale, when $\gamma \gg 1$, velocity-weakening results (SHAW, 1995). In between, some mixture of slip- and velocity-weakening occurs.

One physical process neglected in this friction is the possibility of hydrofracturing, which can occur at high pore fluid pressures, and would limit the drop in friction to some nonzero value. SHAW (1995) considered various possibilities of how this might affect the friction, and examined how different nonlinear saturations might affect the dynamics. The conclusion was that the initial linear regime was the most important feature to the resulting complexity, and the nonlinear saturation was much less important. Therefore we use a simplified friction presented by SHAW (1997) which retains the basic heat-weakening physics, and saturates at a finite value. We use:

$$\Phi = \phi(V(t'), t' \leq t)H(V) - \eta \nabla_{\parallel}^2 V. \tag{8}$$

Here ϕ depends on the past history of slip. The function H is the antisymmetric step function, with

$$H = \begin{cases} \dfrac{\partial \hat{S}}{\partial t} & \dfrac{\partial S}{\partial t} \neq 0; \\[2mm] |H| < 1 & \dfrac{\partial S}{\partial t} = 0, \end{cases} \tag{9}$$

where $\widehat{\partial S / \partial t}$ is the unit vector in the sliding direction. Thus H represents the stick-slip nature of the friction, being multivalued at zero slip rate.

The parameter η is the strength of the viscous-like boundary dissipation, with $\nabla_{\parallel}^2 = \partial^2 / \partial x^2$ being the fault-parallel Laplacian operator. This term usefully provides stability to the smallest lengthscales (LANGER and NAKANISHI, 1993; SHAW, 1997).

The history dependent ϕ we use in this paper is given by

$$\phi = \Phi_0 - \frac{\alpha Q}{1 + \alpha Q} - \Sigma. \tag{10}$$

The first term Φ_0 is a constant, the threshold value of sticking friction, which, as long as it is large compared to the maximum friction drop, becomes an irrelevant parameter in the problem. The second term in (10) generates the same initial linear decrease with heat Q as in the full nonlinear case (Eqs. (5)–(7)), but here has a different nonlinear saturation; a difference expected to be unimportant. It is normalized so that it drops at most a stress of unity. The last term in (10) Σ represents the drop in friction as we go from sticking to slipping; it corresponds with what the standard rate and state friction (DIETERICH, 1979; RUINA, 1983) would contribute. Here we simplify and consider a Σ which depends instead only on time:

$$\Sigma(t) = \begin{cases} \sigma \dfrac{t - t_s}{\tau} & t - t_s < \tau; \\[2mm] \sigma & t - t_s \geq \tau, \end{cases} \tag{11}$$

where τ is a timescale for the drop to occur, and t_s is the time since last sticking. With this nucleation mechanism there is instantaneous healing upon sticking. This simplification of the friction allows for a dramatic speed-up of the numerics. Fortunately, these two-dimensional models with the friction we consider are, interestingly, quite insensitive to many of the details of the nucleation term Σ; for example, a slip-weakening σ and this time dependent σ produce the same large-scale results (SHAW and RICE, 2000). Thus, while our Σ is a drastic simplification of what occurs, the results are quite insensitive to it, and it manifest a tremendous speedup.

Numerics

In the simulations, beginning from any nonuniform initial condition, the system settles down to an attractor which has a statistically steady state. Two general types of attractors are seen, depending on the friction parameter values—principally the heat-weakening parameter α. For α below a critical value, a simple attractor with only large events which span the whole system size is seen. For α above the critical value, a complex attractor with a distribution of large events is seen. Close to the critical value, a rich population of small events occurs, when σ is not too large compared to unity. This population of small events exhibits power-law distribution of sizes and displays a remarkable richness of behavior; while it is a legitimate outcome of the continuum equations (SHAW and RICE, 2000), it does, however, only occur over a quite restricted range of parameter space (SHAW and RICE, 2000). Because we are interested in how the small events correlate with past large events, we will focus our attention on the restricted parameter range where the small events occur.

We solve the equations with a second-order explicit finite difference technique. The displacement and velocity are integrated forward in time on a non-staggered grid. This technique can provide some long-term drift relative to the continuum, however numerical evidence suggests we are shadowing the attractor, and thus the drift is principally along the attractor, rather than away from it. We have confirmed that an alternative formulation which follows velocity and stress (VIRIEUX and MADARIAGA, 1982) and the formulation used here produces the same statistics on their shadowed attractors (SHAW and RICE, 2000). Grid independence of the large scales is a principal signature of the continuum. Dispersion of the waves in the bulk at small wavelengths, an inevitable consequence of the finite difference (ALFORD et al., 1974), interestingly does not appear to significantly affect the fault attractor, which seems to be relatively insensitive to the dispersion in the bulk away from the fault (SHAW and RICE, 2000).

What does the model behavior look like? Figures 1 and 2 show examples of the attractors for two different sets of parameters, both in the range where numerous small events are seen. Two different views of the attractor are shown in each picture. At the top, we plot the times at which various parts of the fault have slipped. On the bottom, we plot the net slip at each point of the fault. Because the small events slip considerably less than the large events, they are difficult to see on the bottom plots, but are much more apparent on the top plots. The detailed sequence of events is in fact chaotic (CRISANTI et al., 1992; DE SOUSA VIEIRA, 1999; SYKES et al., 1999), and thus is highly sensitive to the initial conditions. The statistics of the attractor, however is stationary, and we see long-term regularities in the patterns. Visually, we see considerable order and structure to the patterns, amidst this chaos. It is our purpose in this paper to discuss one such general, striking regularity of these patterns: how the edges of large events influence where future events occur.

3. Results

Inspection of Figures 1 and 2 confirms a remarkable regularity of the model behavior: observing where most events initiate, we see that the vast majority occur near the edges of past large events. This is most easily seen by looking at the small events in (a) of the figures, and mentally tracing backwards in time at the locations where they occurred: most frequently they line up very close to an edge of a previous large event. (Despite the epicenters not being directly indicated in the figure, this effect can be easily seen with the small events because of their limited spatial extent which bounds the epicenter location.) This basic effect, which seems apparent from visual inspection, transcends many aspects of the models, including dimensionality of the models (it is also true in the one-dimensional models), geometry of the two-dimensional models (alternative coupling to the stably sliding layer, e.g., through the bulk which gives a Klein-Gordon equation (MYERS *et al.*,

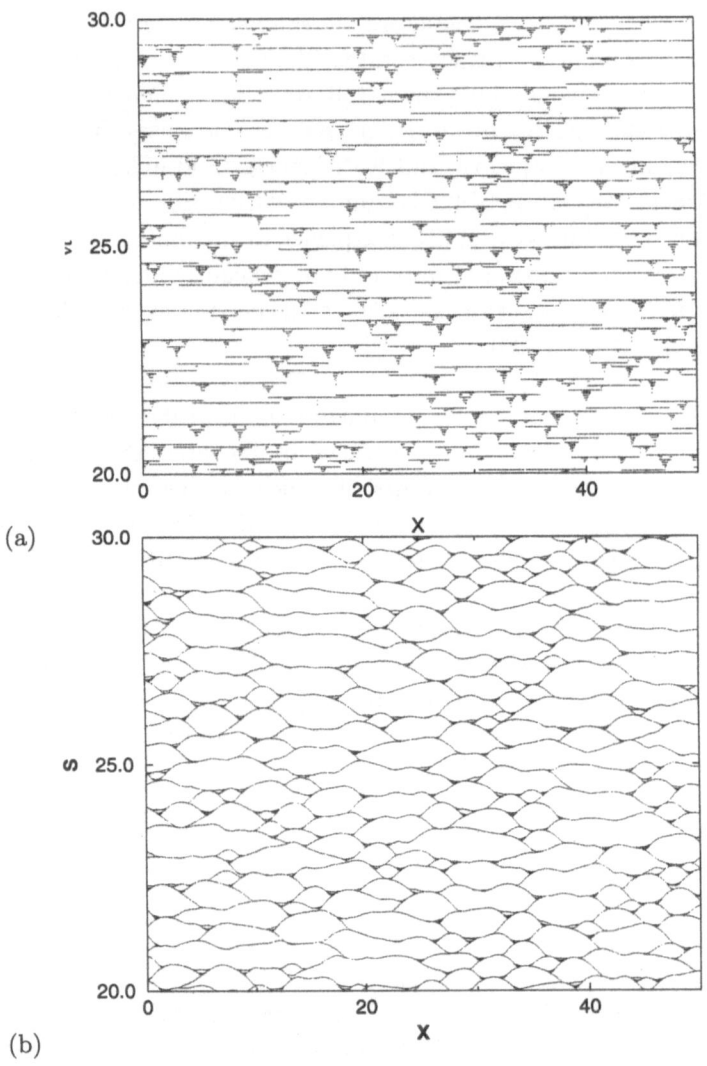

(a)

(b)

Figure 1

Two different representations of the attractor produced by the model for the parameter range where numerous small events occur. The horizontal axis is distance along the fault, measured in units of the brittle crust depth. The material properties are uniform along the fault. (a) We plot the times at which various parts of the fault break—a standard space-time plot in seismology, only here for very many loading cycles, whereas only a fraction of a loading cycle is generally available for real data. (b) We plot the cumulative slip along the fault following the event, for the same set of events as in (a). The friction parameters used are $\alpha = 3$, $\sigma = .05$, $\gamma = 1$, $\tau = .1$, and $\eta = .00001$.

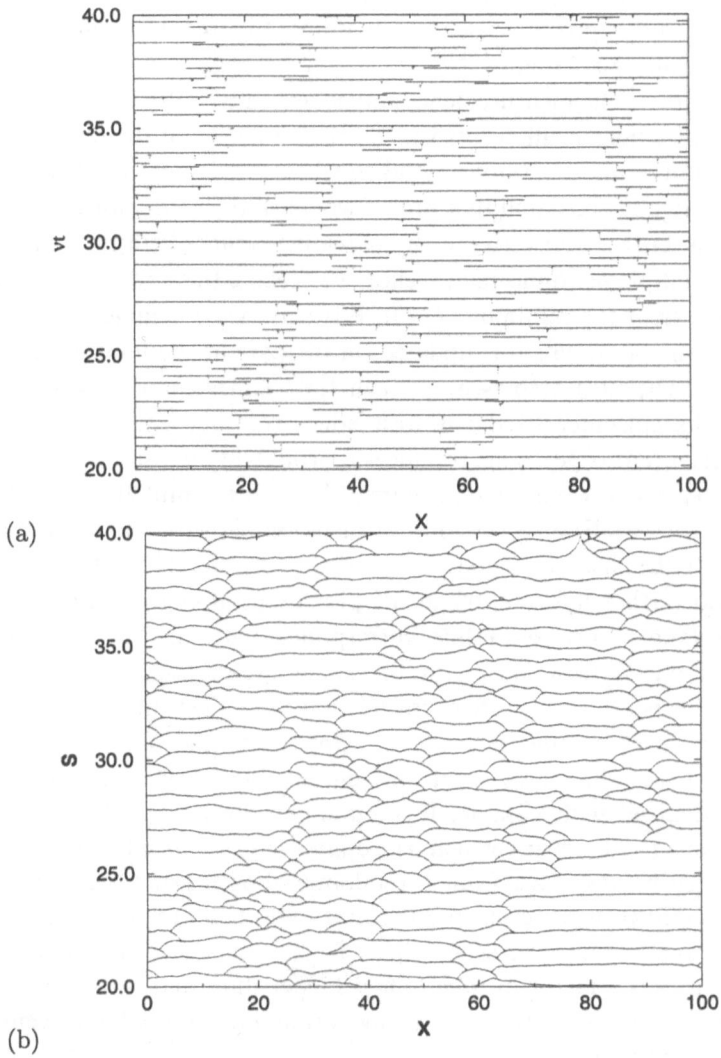

Figure 2
Two different representations of the attractor produced by the model for a somewhat larger value of the weakening parameter. The same type of plots as in Figure 1 is shown, although with the axes rescaled somewhat. (a) The times at which various parts of the fault break. (b) The cumulative slip along the fault following an event, for the same set of events as in (a). The same set of parameters as in Figure 1 are used, except now for $\alpha = 8$.

1996), also shows this effect), and a variety of frictional instabilities (slip-weakening and velocity-weakening to name two). It is a basic feature of the dynamic complexity set-up by the uniform fault models: it is the edges of where the large events die out which create the largest strain heterogeneities in the model, and are the main sites where future events initiate.

Let us explore this effect more quantitatively with the model, and assure we are not being misled by apparent visual correlations. Figure 3 quantifies this effect, and affirms its significance. Figure 3a uses events of Figure 1, while Figure 3b uses events of Figure 2. The measurement is made as follows. We begin by defining large events, which we take as all events which break a length of the fault which is larger than some cutoff length L_0 (typically taken to be a crust depth in length, so $L_0 = 1$). Then, at the edge of the patch that broke, we mark the spot, along the fault. Within any patch that broke, we overwrite any previously marked edges. If an event breaks the fault in multiple disconnected patches, we execute this procedure with any patch that is longer than the minimum L_0. In this way, we conclude with a list of edge marks along the fault, which is updated on the part of the fault where a new large event occurs. The next step is easy: each time an event occurs, we measure the distance of its epicenter to the nearest edge from the list we just described of previous large event edges. This produces a probability density function of distance from epicenters to previous edges. Figure 3 shows the cumulative of this probability density function (PDF) as a function of distance, showing on the vertical axis the probability of the distance being less than the distance plotted on the horizontal axis. We add further information to the plot by grouping events according to the size of the events. The solid line corresponds with the large events. The top short-dashed line corresponds with the small events. Interestingly, we see that small-sized events are even more correlated with the edges of past large events than are the large events. One last curve helps show the size of the effect. The long-dashed curve illustrates what this cumulative PDF would be if the event epicenters occurred with completely randomness in space. We calculate this by, every time an event occurs which is counted by the solid curve, calculating a distance to the nearest edge from a randomly chosen point along the fault. This generates a control distribution for the null hypothesis of uncorrelated epicenters with respect to previous large event edges. We see that the clustering is a substantial effect: close in it is a factor of many times higher than that of the cumulative PDF for the random case. Thus we have strong correlations of the large event edges with future event epicenters.

How do the results change with parameter values? Using larger values of the weakening parameter α only increases the correlation. Comparing (a) where $\alpha = 3$ with (b) where $\alpha = 8$, both the small event curve and the large event curve move up with increasing α, being even more closely clustered at the large event edges, while the control curve moves down, as the large event lengths elongates. Thus larger values of α correspond with stronger correlations. Changing other parameters has only relatively small effects. In all cases, the correlations are strong.

Why do these correlations occur? Thinking in terms of stress, we gain a clear physical understanding. Large events tend to die out in regions of low stress and while propagating into regions that are even less stressed ahead. The propagating ruptures carry stress concentrations with them, which are able to break through

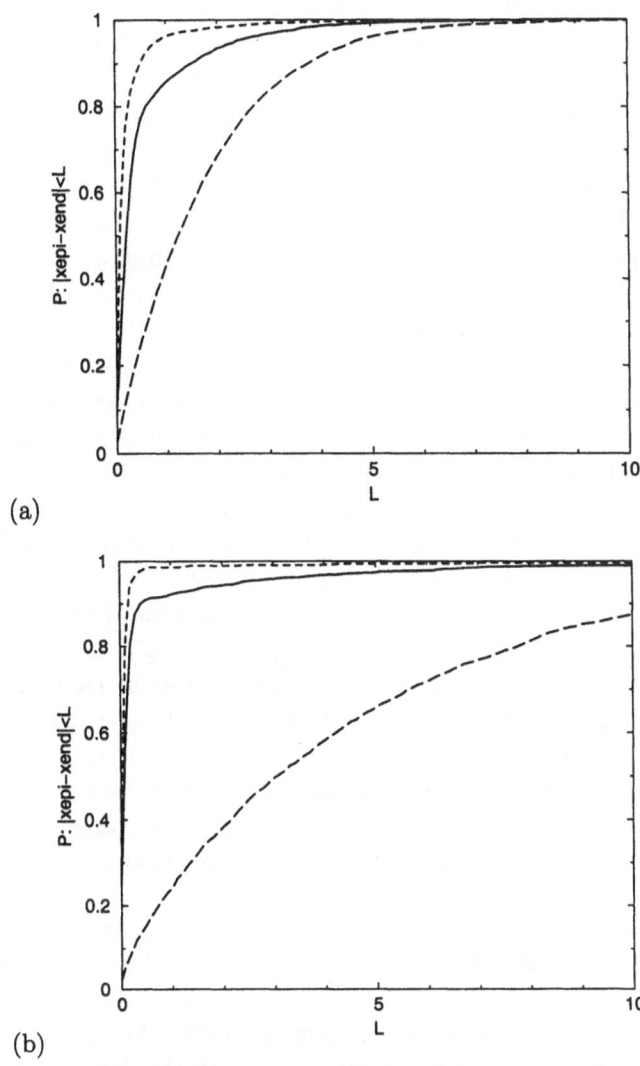

Figure 3

Epicenter correlations with large event edges. The vertical axis is the cumulative probability density function of the distance from an epicenter to a previous large event edge being less than the distance on the horizontal axis. Distances are in units of the brittle crust depth. Within each figure three curves are shown. The upper small dashed curve is the epicenters of the small and medium events. The middle solid curve is the epicenters of the large events. The lower long-dashed curve is the control for the case where the epicenters are randomly distributed. The curves lying far above this control case show clearly the epicenters are closely clustered near the large event edges. Note that the small and medium epicenter curve is even more clustered than the large event epicenter curve. The two different figures correspond to two different values of the weakening mechanism, with $\alpha = 3$ in (a), using the events of Figure 1, and $\alpha = 8$ in (b), using the events of Figure 2.

somewhat understressed regions, but not overly understressed regions. The regions they leave behind are even less stressed after they have broken. Thus we end up with a stress minimum behind the ruptures, a stress concentration at the edge of the rupture, and a less stressed region ahead of the rupture; all of which leaves the edge of the event where the rupture died out as the highest stressed region locally. Subsequently, when the system is uniformly reloaded, this region is the closest to threshold, and the first to break. Figure 4 shows this effect quantitatively where we plot the average stress before and after a stack of large events. There are four curves in each plot, all showing the average stress as a function of distance from the edge of a large event, averaged over many events. A fifth vertical thin solid line demarcates where the large events died out, at $x = 0$, about which all the stress stacks are made. The thick long-dashed curve delineates the average stress before the large events occurred, events which then break through $x < 0$ and die at $x = 0$. The thick solid curve shows the average stress after the large events occurred. Both of these thick curves show all of the effects we just described above: the events died propagating into less stressed regions; they left behind destressed regions; and added large stress concentrations at their edges. The last two curves in the plots, the two thin-dashed curves, indicate additional information. They show the average stress just before the next large event ruptures through, measured at the same points where the thick solid line is taken, though now at later times. The upper thin-dashed curve shows the stress just before the next large event ruptures through. The lower dashed thin curve shows the same data, but with a constant stress subtracted from it; this is done so as to allow comparison with the thick solid curve, which shows the stress at the same locations at an earlier time. We see something interesting here: the stress concentration at the large event edges has been lessened and smeared out onto the neighboring regions.

This is a result of the small events which occurred at the stress concentration. They have two effects: they lower the stress where they break, and they also shunt this stress onto the neighboring regions. The net effect is a smearing of the stress concentration. This provides an explanation as to why the epicenters of the future small events in Figure 3 were more correlated with the edges of the large events than were the epicenters of the future large events: by the time the fault has reloaded enough for a large event to occur again, the stress concentration has been somewhat eroded and smeared out by the smaller events.

Completing our examination of Figure 4, looking to the sides of the stress concentration we see to the left the ruptured interior of the last large event, the shifted thick-dashed curve overlays the original thick solid curve, indicating this least stressed previously broken region has generally not rebroken yet. Again, stress levels are consistent with activity, with more stressed regions being more active.

The effect of changing parameters reappears in the stress field, and again explains the correlations. Two weakening values, again $\alpha = 3$ in (a), and $\alpha = 8$ in (b), illustrate this. In Figure 3 we noted that larger values of α increased the

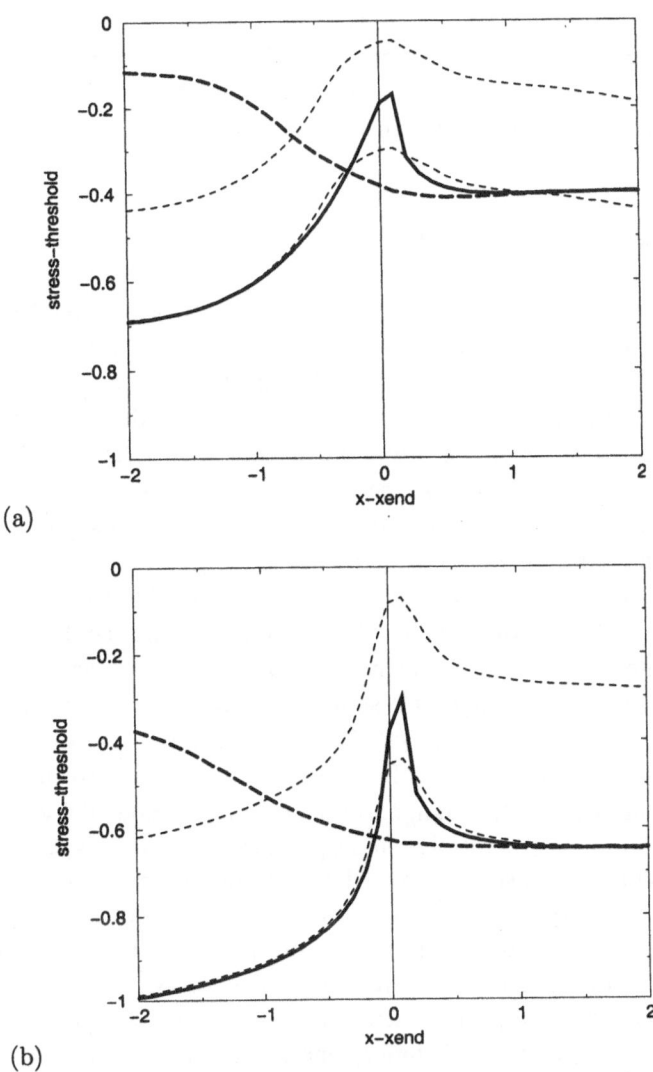

(a)

(b)

Figure 4

Stress averages. The stress state relative to the edges of large events, averaged over many events. The stack is made with respect to the end of a large event which ruptured $x < x_{end}$ and died out at $x - x_{end} = 0$. The thin solid line shows this edge. Within each figure four curves are shown. The long-dashed curve shows the stress before the large event. The thick solid curve shows the stress after the large event ruptured. The upper thin-dashed line shows the stress just before the next large event ruptures through the previous edge; the lower thin-dashed line shows this same stress but shifted down by a constant, to compare with the thick solid line. Note the stress concentration left by the large event peak in the thick solid line at the edge. Note further the smearing of the stress concentration with time, evidenced by the smearing of the peak in the thin-dashed line. The two different figures correspond to two different values of the weakening mechanism, with $\alpha = 3$ in (a), using the events of Figure 1, and $\alpha = 8$ in (b), using the events of Figure 2.

correlations. Examining the stress fields in Figure 4 demonstrate why: larger values of α lead to larger stress concentrations, with the peak at $x = 0$ being higher and the averages off to the sides of the peak being lower. Qualitatively, the picture is the same: quantitatively, we see and can understand the differences.

4. Discussion and Conclusion: Extrapolating Results to Real Earthquakes

While the models show the large event edges to be a dominant effect on future events, we are most interested in how these results might apply to the case of real earthquakes. Here, there are a number of issues to consider.

First, one significant approximation of the model is that we have considered a one-dimensional fault, as opposed to a more realistic two-dimensional fault surface. Thus, actual fault edges are not merely points, but lines, and potentially quite complicated lines at that. Further, the bottom of the seismogenic depth will not have broken in a large earthquake, so the closest edge will be at most a crust depth away, if we include vertical distance. At a minimum, this would have substantial quantitative effects on the results. Nevertheless, if we confine our results to qualitative statements, we can still make certain observations. To reiterate, first, large event edges have the highest residual stress concentrations, and consequently are the first locations to be reactivated. Second, small events smear out the stress concentrations. Thus with time a broader area becomes activated. Eventually, a large event comes ripping through, and the cycle is repeated. These statements seem quite general. Thus, the qualitative underpinning of our results seem to carry over, even if the quantitative aspects would be substantially modified.

A second geometry complication is that faults exist not as isolated individuals faults, as considered in our simple model, but as parts of systems of faults. Here, if one thinks in the language of stress, which involves, in the elastic earth case, considerations of tensor aspects of stress and failure, the basic ideas still carry over. However, rather than simple distance, one must consider direction as well. Still, again, the qualitative ideas ought to transfer. In this context, DENG and SYKES (1996) have made some interesting observations of seismicity in Southern California. They noted, what they have euphemistically called a "great wall," marking a change in the level of seismic activity where the large M8.1 1857 Fort Tejon earthquake was believed to have died out. Below this end the activity is much higher than behind, in the region estimated to be shadowed by the stress relief from the events. This, along with the observations of PEREZ and SCHOLZ (1997), provide some of the strongest observational evidence for the qualitative picture being presented here (though the suggestion of PEREZ and SCHOLZ (1997) of enhanced activity at the edges preceding the large events raises other questions).

An additional complication of real earthquakes not present in the simple model is the existence of aftershocks and foreshocks for earthquakes. In the simple frictions we have considered, we do not have this intermediate timescale. The re-emergence of future events in the model is more akin to what would be called "preshocks" in seismology, which occur over fractions of the loading cycle, and indicate an emergence out of the low stress regime following a large event. If the model were to have an aftershock timescale, however, some of this signal would likely appear as aftershocks as well. Thus, to look for the signature of the effect being described here, counting aftershocks would be appropriate. The fact that aftershocks are the most numerous around the edges of the rupture support the picture presented here.

The fact that earthquakes eventually stop requires heterogeneities to exist in the Earth. A central question which remains unanswered is the role of relatively fixed material and geometrical heterogeneities and the role of changing, dynamic, stress heterogeneities. In the simple models we have examined here, we have completely neglected any fixed heterogeneities and studied a completely uniform fault. All of our heterogeneities are stress heterogeneities. In this way, we have provided a baseline context with which to compare the data. Deviations from, or similarities with, the simple picture emerging from these models would speak to these heterogeneity questions: to what extent does seismicity concentrate on the edges of large events? To what extent is seismicity tied to an absolute location (fixed in space) as opposed to a relative location (relative to previous events)?

This work suggests immediate practical prescriptions: to heighten monitoring at the edges of known previous large ruptures, and to identify and locate more accurately the previous large event edges. Even with all the other additional complications of the real earth, on long faults where great earthquakes occur there should be an enhanced probability of events occurring at the edges of previous large events; what remains to be determined is the strength of this enhancement.

Acknowledgements

Chris Scholz provided a useful suggestion which added to the plot of Figure 4. This work was supported by NSF grant EAR-99-09287 and USGS grant 1434-HQ-97-GR-O3074.

REFERENCES

ALFORD, R. M., KELLY, K. R., and BOORE, D. M. (1974), *Accuracy of Finite Difference Modeling of the Acoustic Wave Equation*, Geophysics *36*, 834.

BURRIDGE, R., and KNOPOFF, L. (1967), *Model and Theoretical Seismicity*, Bull Seismol. Soc. Am. *57*, 341.

CARLSON, J. M., and LANGER, J. S. (1989), *Mechanical Model of an Earthquake Fault*, Phys. Rev. A *40*, 6470.

CRISANTI, A., JENSEN, M. H., VULPIANI, A., and PALADIN, G. (1992), *Strongly Intermittent Chaos and Scaling in an Earthquake Model Fault*, Phys. Rev. A *46*, R7363.

DE SOUSA VIEIRA (1999), *Chaos and Synchronized Chaos in an Earthquake Model*, Phys. Rev. Lett. *82*, 201.

DENG, J., and SYKES, L. R. (1996), *Evolution of the Stress Field in Southern California and Triggering of Moderate-size Earthquakes: A 200 Year Perspective*, J. Geophys. Res. *102*, 9859.

DIETERICH, J. H. (1979), *Modeling of Rock Friction: 1. Experimental Results and Constitutive Equations*, J. Geophys. Res. *84*, 2161.

DIETERICH, J. H. (1994), *A Constitutive Law for the Rate of Earthquake Production and its Application to Earthquake Clustering*, J. Geophys. Res. *99*, 2601.

HARRIS, R. A., and SIMPSON, R. W. (1992), *Changes in Static Stress on Southerna California Faults after the 1992 Landers Earthquake*, Nature *360*, 251.

HARRIS, R. A., SIMPSON, R. W., and REASENBERG, P. A. (1995), *Influence of Static Stress Changes on Earthquake Locations in Southern California*, Nature *375*, 221.

JAUME, S. C., and SYKES, L. R. (1992), *Changes in the State of Stress on the Southern San Andreas Fault resulting from the California Earthquake Sequence of April to June 1992*, Science *258*, 1325.

KING, G. C. P., STEIN, R. S., and LIN, J. (1994), *Static Stress Changes and the Triggering of Earthquakes*, Bull. Seismol. Soc. Am. *84*, 935.

LACHENBRUCH, A. (1980), *Frictional Heating, Fluid Pressure, and the Resistance to Fault Motion*, J. Geophys. Res. *85*, 6097.

LANGER, J. S., and NAKANISHI, H. (1993), *Models of Crack Propagation II: Two-dimensional Model with Dissipation on the Fracture Surface*, Phys. Rev. E *48*, 439.

McKENZINE, D. P., and BRUNE, J. N. (1972), *Melting on Fault Planes during Large Earthquakes*, Geophys. J. Roy. Astron. Soc. *29*, 65.

MYERS, C. R., SHAW, B. E., and LANGER, J. S. (1996), *Slip Complexity in a Two-dimensional Crustal Plane Model*, Phys. Rev. Lett. *77*, 972.

OLSEN, K. B., MADARIAGE, R., and ARCHULETA, R. J. (1997), *Three-dimensional Dynamic Simulation of the 1992 Landers Earthquake*, Science *278*, 834.

PEREZ, O. J., and SCHOLZ, C. H. (1997), *Long-term Seismic Behaviour of the Focal and Adjacent Regions of Great Earthquakes during the Time between two Successive Shocks*, J. Geophys. Res. *102*, 8203.

RUINA, A. (1983), *Slip Instability and State Variable Friction Laws*, J. Geophys. Res. *88*, 10359.

SHAW, B. E. (1995), *Frictional Weakening and Slip Complexity on Earthquake Faults*, J. Geophys. Res. *100*, 18239.

SHAW, B. E. (1997), *Modelquakes in the Two-dimensional Wave Equation*, J. Geophys. Res. *102*, 27367.

SHAW, B. E., and RICE, J. R. (2000) *Existence of Continuum Complexity in the Elastodynamics of Repeated Fault Ruptures*, J. Geophys. Res., in press.

SIBSON, R. H. (1973), *Interactions between Temperature and Pore Fluid Pressure during Earthquake Faulting and a Mechanism for Partial or Total Stress Relief*, Nature *243*, 66.

STEIN, R. S., KING, G. C. P., and LIN, J. (1992), *Change in Failure Stress on the Southern San Andreas Fault System Caused by the 1992 Magnitude = 7.4 Landers Earthquake*, Science *258*, 1328.

SYKES, L. R., SHAW, B. E., and SCHOLZ, C. H. (1999), *Rethinking Earthquake Prediction*, Pure appl. geophys. *155*, 207.

VIRIEUX, J., and MADARIAGA, R. (1982), *Dynamic Faulting Studied by a Finite Difference Method*, Bull. Seismol. Soc. Am. *72*, 345.

(Received August 28, 1999, revised April 14, 2000, accepted April 28, 2000)

Pure appl. geophys. 157 (2000) 2165–2182
0033–4553/00/122165–18 $ 1.50 + 0.20/0

Pure and Applied Geophysics

Precursory Seismic Activation and Critical-point Phenomena

JOHN B. RUNDLE,[1] WILLIAM KLEIN,[2] DONALD L. TURCOTTE[3] and
BRUCE D. MALAMUD[3]

Abstract—In this paper we relate the behavior of seismicity prior to a characteristic earthquake to
the excitation in proximity to a spinodal instability. We illustrate the spinodal instability as the upper
limit of superheated water prior to a steam explosion. We draw an analogy between the steam explosion
and a characteristic earthquake, and show that the power-law activation associated with the spinodal
instability is essentially identical to the power-law increase in Benioff strain observed prior to characteristic earthquakes. We find that theory and actual data give very similar results.

Key words: Seismicity, spinodal, critical-point phenomena, characteristic earthquake, Benioff strain.

1. Introduction

There is increasing observational evidence that seismic activation occurs prior to
some major earthquakes. Qualitatively the number of intermediate-sized earthquakes systematically increases prior to the major or characteristic earthquake in a
region. Quantitatively this activation has been associated with a power-law increase
in the cumulative Benioff strain prior to the characteristic earthquake (BUFE and
VARNES, 1993; BOWMAN *et al.*, 1998). In this paper we will show that this increase
can be explained by the standard theory for phase transitions.

BOWMAN *et al.* (1998) carried out a systematic study of the size of the region
associated with seismic activation. They found that seismic activation occurs over a
region with a characteristic length of about one order of magnitude larger than the
rupture length of the subsequent major earthquake. Thus this activation cannot be
explained by the classical theories of fracture. These theories would restrict activation to a length scale that approximates the rupture length. However, the large

[1] Department of Physics and Colorado Center for Chaos and Complexity, Cooperative Institute for
Research in Environment Sciences, University of Colorado, Boulder, CO 80309, U.S.A. E-mail:
rundle@cires.colorado.edu
[2] Department of Physics and Center for Computational Science, Boston University, Boston, MA
02215, U.S.A. E-mail: klein@buphyc.bu.edu
[3] Department of Geological Sciences, Cornell University, Snee Hall, Ithaca, NY 14853-1504, U.S.A.
E-mails: turcotte@geology.cornell.edu and bruce@malamud.com

region of activation is consistent with the statistical physics of phase changes. The correlation length ξ, the scale over which events are correlated, approaches infinity as the critical point is approached.

2. The Ising Model

In order to relate critical-point concepts to the earthquake problem, we first present a brief discussion of a standard model with a second-order phase change, the Ising model (DEBENEDETTI, 1996) followed by a discussion of the metastable behavior associated with a first-order transition. The Ising model is an approximation to a magnetic material; a square grid of particles is considered. Each particle has either a $(+)$ up spin or a $(-)$ down spin, and force interactions between particles are restricted to the nearest neighbors. The magnetic polarization M of the system is the sum of the spins. The behavior of this model as a function of temperature T without an applied magnetic field is illustrated schematically in Figure 1a. At zero temperature $T = 0$ all the particles have either $(+)$ spins $(M = M_0)$ or $(-)$ spins $(M = -M_0)$. As the temperature increases, a mixture of particle spins develops until the critical point T_c is reached. For $T > T_c$ the system is completely random and there is no net magnetization.

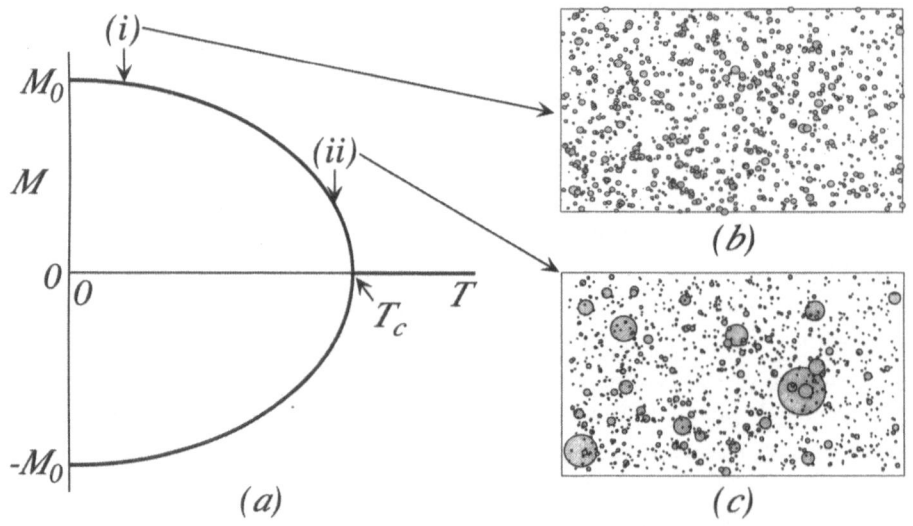

Figure 1

(a) Dependence of the magnetic moment M on temperature T for the Ising model. (b) Exponential frequency-area distribution of patches of negative magnetization at point (i) in (a). (c) Power-law frequency area distribution of patches of negative magnetization at point (ii) in (a) as the second-order critical point at $T = T_c$ is approached.

For T in the critical region just below T_c (see Fig. 1a), the magnetic polarization behaves as a power law,

$$M \sim (T_c - T)^\alpha \tag{1}$$

where $\alpha > 0$ is one of a family of characteristic critical exponents and $T < T_c$. The value of α is determined by a variety of factors including the exact nature of the model, such as whether interactions among spins are strictly nearest neighbor, or are of longer range. The magnetic polarization $M \to 0$ as $T \to T_c$.

There is a second-order critical point at $T = T_c$. Assume that initially $T = 0$ and all particles have $(+)$ spins. Assume that the temperature increases with time. At a low but finite temperature, point (i) in Figure 1a, patches of particles with $(-)$ spins form randomly. There are no spatial correlations and there is an exponential frequency-area distribution of patches. This distribution is illustrated in Figure 1b. As the temperature T approaches the critical temperature T_c, point (ii) in Figure 1a, the patches of particles with $(-)$ spin become correlated and there is a power-law (fractal) frequency-area distribution of patches. This distribution is illustrated in Figure 1c. The radius of the region over which patches are correlated is the correlation length ξ. For T in the critical region just below T_c, the correlation length also behaves as a power-law

$$\xi \sim (T_c - T)^{-\gamma} \tag{2}$$

where $\gamma > 0$ is another critical exponent related to α. Thus $\xi \to \infty$ as $T \to T_c$.

3. Spinodals

In the discussion above we have described the physics of a second-order phase transition. There is also a less well known, but similarly important class of critical phenomena associated with the more familiar first-order phase transitions (DEBENEDETTI, 1996). An example of a first-order phase transition is the boiling of water. A schematic temperature-volume diagram for the transition of water-to-water vapor (steam) is shown in Figure 2. A saturated mixture of liquid water and water vapor coexists in equilibrium beneath the domal region between the saturated liquid line (SL) and the saturated vapor line (SV). Liquid water is found to the left of the saturated liquid line (SL) and superheated vapor (steam) to the right of the saturated vapor line (SV).

In Figure 2, water initially at point "a" in the liquid phase is heated at constant pressure until it reaches the boiling temperature T_b (and volume v_b) at point "b". This is the usual coexistence point, the point at which liquid water and water vapor (steam) can coexist in stable equilibrium. In thermodynamic equilibrium, boiling occurs at the boiling temperature T_b until all the water is converted to saturated steam at point b'. Further heating to point c' produces superheated steam.

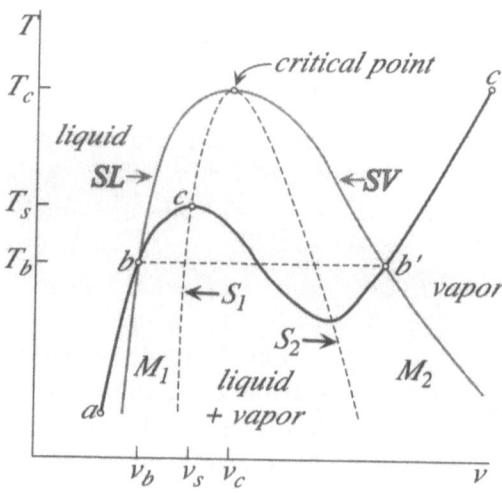

Figure 2

Schematic temperature–volume diagram for the transition of water-to-water vapor. A saturated mixture of liquid water and water vapor coexists in equilibrium beneath the domal region between the saturated liquid line (*SL*) and the saturated vapor line (*SV*). Liquid water is found to the left of *SL* and superheated vapor (steam) to the right of *SV*. Metastable superheated water can be produced in the region M_1, which occurs between the saturated liquid line (*SL*) and the dashed spinodal line S_1. Metastable supercooled steam can be produced in the region M_2, which occurs between the saturated vapor line (*SV*) and the dashed spinodal line S_2. The two spinodal lines S_1 and S_2 converge on the critical point at temperature $T = T_c$ and volume $v = v_c$. The constants T_b, T_s, and T_c represent the boiling, theoretical maximum, and critical temperatures; associated volumes are v_b, v_s, and v_c. The points a, b, b', c, c', are discussed in the text.

However, if the temperature is increased with sufficient carefulness, a superheated state of water can be achieved, in which the water remains in the liquid state even though the temperature T is above the nominal boiling temperature, $T > T_b$. This superheated condition is a state of metastable equilibrium, because if the experimenter waits long enough, random thermal or pressure fluctuations will occur that cause the system to nucleate bubbles of steam and the water will suddenly boil.

Many years ago, it was shown by van der Waals (1873) that there is a theoretical maximum temperature T_s beyond which the superheated state can no longer exist, and the liquid state must transform to vapor. In Figure 2, water can be heated (in principle) as a metastable liquid from point "b" to point "c". Point "c" is on the spinodal line S_1 that is the limit of the metastable heating. At point "c", catastrophic boiling must occur.

According to current understanding, the spinodal temperature can only be closely approached in systems with comparatively long-range interactions, which act to stabilize the system against small fluctuations. In theoretical studies of the nature of the spinodal line, it has been found that the spinodal behaves as a critical point similar to that characterizing second-order phase transitions. The critical point for water is shown in Figure 2; this occurs at critical temperature T_c and volume v_c and is a point where a second-order phase transition takes place. The two dashed lines are two spinodal lines· for the water system; S_1 represents the upper limit of superheated water and S_2 represents the lower limit of supercooled water vapor (steam). The two spinodal lines converge on the critical point.

In addition to the water-steam system, magnetic systems such as the Ising model discussed above also have spinodal points. Experimentally, it is found that the power-law scaling relations, (1) and (2), are satisfied in the vicinity of the spinodal; however, values of the scaling exponents differ from those valid near the second-order phase change. Thus as the spinodal is approached, the system moves more deeply into the metastable regime and the correlation length ξ increases dramatically. Without the presence of a spinodal, the entire field of paleomagnetism would be impossible and the discovery of plate tectonics would have been delayed. The metastable condition associated with the magnetic spinodal is what permits sea floor basalts with a "reversed," frozen in, magnetic polarization to exist in an era of "normal" magnetic field polarization. Metastability is also responsible for the existence of the high-pressure phase of carbon, diamond, at the earth's surface.

4. Fracture and Earthquakes

With respect to rocks, HIRATA et al. (1987) have used acoustic emissions and other observations to conclude that the approach to fracture in pristine rock is essentially identical to the second-order phase change in the Ising model. The stress on the rock is analogous to the temperature in the Ising model. As stress is increased, microcracks occur randomly and are uncorrelated. As the fracture stress is approached, the microcracks become correlated and satisfy a power-law (fractal) distribution. Many authors have associated the fracture of pristine materials with second-order critical points (HERRMANN, 1991).

While there are important similarities between the fracture of pristine rocks and an earthquake rupture, there are also important differences. The fracture of pristine rock is an irreversible process. However, earthquakes on a fault represent a cyclic process. Crustal faults experience many earthquakes. Between earthquakes faults heal. If the earth's crust, prior to a major earthquake, behaved like the fracture of pristine rocks, there would be no problem in predicting earthquakes. There would be a systematic increase in seismic activity prior to an earthquake. This is clearly not the case. Thus the direct application of critical point analyses is not adequate to explain the behavior of the earth's crust. Crustal seismicity is a repetitive process and resembles the cycling of water through points "a", "b", and "c" as illustrated in Figure 2. This analogy is illustrated in Figure 3, where during an earthquake cycle the stress on a fault zone, σ, is given as a function of the slip deficit, ϕ. After an earthquake, the stress on the fault is σ_0 (in analogy to point "a" in Figure 2).

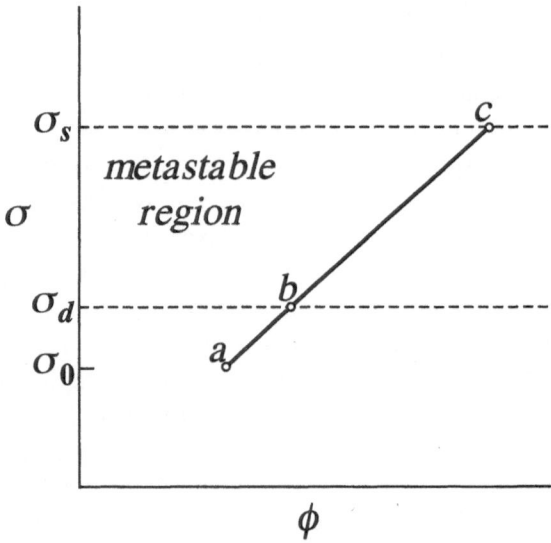

Figure 3
Stress on a fault zone σ as a function of the slip deficit ϕ during an earthquake cycle. The low stress after a characteristic earthquake is at point "*a*", tectonic forces increase the stress into the metastable region at "*b*" and to the rupture point "*c*" where the earthquake occurs. The cycle then repeats.

The fault is locked so that the slip deficit ϕ and stress σ increase with time, due to the tectonic displacements, until the stress is equal to the "dynamic friction" stress σ_d (in analogy to point "*b*" in Figure 2). For $\sigma_0 < \sigma < \sigma_d$, the fault is stable and an earthquake cannot be initiated. The stress increases further until the stress is equal to the "static friction" stress σ_s where an earthquake is catastrophically triggered (in analogy to point "*c*" in Figure 2). For $\sigma_s > \sigma > \sigma_d$, the fault is in a metastable state and an earthquake rupture can propagate if initiated. The stress σ in Figure 3 is analogous to the temperature T in Figure 2.

5. *Self-organized Critically*

Theories describing second-order transitions have been proposed to explain earthquake phenomena, however these do not capture all of the important physics. The concept of self-organized criticality was introduced to explain the sandpile model (BAK *et al.*, 1988). In this model, particles are randomly dropped onto a square grid of boxes. When a box accumulates four particles they are redistributed to the four adjacent boxes or lost off the edge of the grid. Redistributions can lead to further instabilities with the possibility of more particles being lost from the grid, contributing to the size of each "avalanche." These model "avalanches" satisfied a power-law frequency-area distribution with a slope near unity. In self-organized criticality the "input" to a complex system is constant, whereas the "output" is a series of events or "avalanches" that follow a power-law (fractal) frequency-size distribution. For example, in the case of the sandpile model, the input is the steady addition of sand grains, and the output is the sand avalanches.

Slider-block models have long been recognized as simple analog models for earthquakes (BURRIDGE and KNOPOFF, 1967). Multiple slider-block models exhibit essentially the same behavior as the sandpile model (RUNDLE and JACKSON, 1977; CARLSON and LANGER, 1989). Thus it was suggested that earthquakes are an example of self-organized critical behavior.

TURCOTTE (1999) and TURCOTTE et al. (1999) introduced an inverse-cascade model to explain self-organized critical behavior. In the slider-block model there are metastable patches of slider blocks that will slip once a slip event is triggered. As the stress on the slider blocks is increased, the metastable patches of blocks coalesce in an inverse cascade from small patches to larger patches. Slip events sample the distribution of patches but significant numbers of metastable blocks are lost only from the large slip events that terminate the cascade. The cascade process leads directly to a power-law frequency-area distribution of metastable clusters.

6. Self-organized Spinodal (SOS) Behavior

RUNDLE et al. (1996, 1999) have argued that slider block and other models for earthquakes are consistent with systems undergoing a repetitive series of first-order phase transitions. The limit of stability (spinodal) corresponds to the failure threshold of the fault. In first-order systems, sudden spontaneous changes in macroscopic properties of the fault are possible, whereas in second-order systems, only gradual changes can occur. Thus characteristic earthquakes can occur for a system with first-order transitions, but not for a system with only second-order transitions. Sufficiently near the spinodal, a power-law scaling regime is possible for a first-order system, similar to the power-law statistics that exist near a second-order transition. Metastability is possible in first-order systems, but not in second-order systems.

We suggest that the behavior of the earth's crust can be understood in terms of the physics of a first-order phase transition. For want of a better term, we call this self-organizing spinodal (SOS) behavior. This type of behavior is similar to the intermittent criticality model of SORNETTE and SAMMIS (1995). We consider the time evolution of seismicity in the earth's crust prior to a major earthquake, which is a consequence of the fault reaching its limit of stability, or spinodal. We refer to this earthquake as the characteristic earthquake in a region.

We divide the precursory seismicity in the region into small- and intermediate-sized events. We hypothesize that the characteristic earthquake is a first-order phase transition. The intermediate-sized earthquakes prior to the first-order phase transition are associated with an activation and increase in the correlation length as

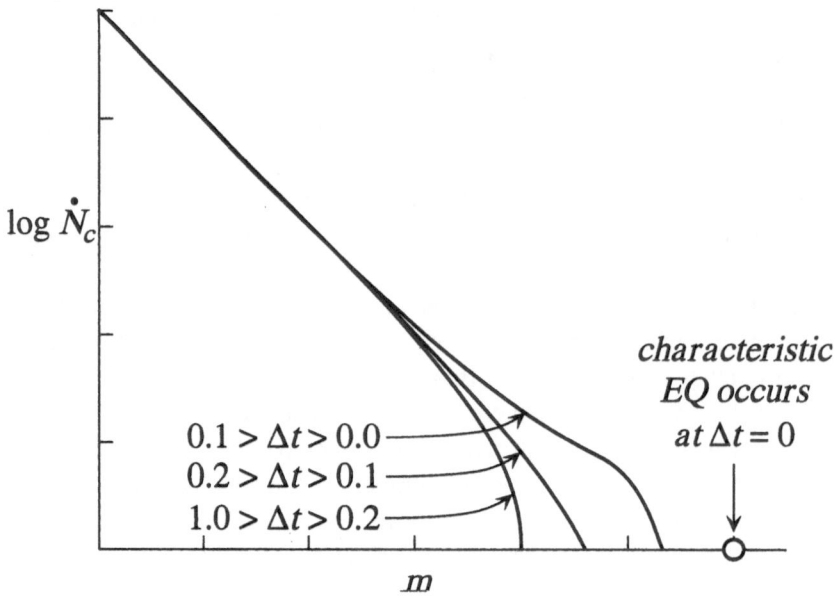

Figure 4

Gutenberg–Richter illustration of the cumulative number of earthquake occurring per unit time $\dot{N}_c$ with magnitudes greater than m as a function of m. Data for different time periods during the earthquake cycle $1 \geq \Delta t \geq 0$ prior to the occurrence of a characteristic earthquake at $\Delta t = 0$. Small earthquakes occur at a constant rate but there is an activation of intermediate-sized earthquakes prior to the occurrence of the characteristic earthquake.

illustrated in Figure 1. However, the small earthquakes represent a cascade in the SOS system. The rates at which small earthquakes occur during a seismic cycle is independent of time, thus they cannot be used to forecast the occurrence of a characteristic earthquake. Nonetheless there is an increase in the rate of occurrence of intermediate-sized earthquakes prior to the occurrence of the characteristic earthquake (TRIEP and SYKES, 1997; JAUMÉ and SYKES, 1999).

This seismic activation hypothesis is illustrated in Figure 4. The cumulative frequency-magnitude distribution of earthquakes is shown for three time periods during the earthquake cycle, $1.0 > \Delta t > 0.2$, $0.2 > \Delta t > 0.1$, and $0.1 > \Delta t > 0.0$, where

$$\Delta t = 1 - \frac{t}{t_0} \tag{3}$$

is the nondimensional time remaining until the next characteristic earthquake, t is the time measured forward from the previous characteristic earthquake and t_0 is the time interval between characteristic earthquakes. Just after a characteristic earthquake we have $\Delta t = 1$; the next characteristic earthquake occurs at $\Delta t = 0$.

Small earthquakes occur uniformly at all times but early on in the earthquake cycle, $1.0 > \Delta t > 0.2$; there is a systematic lack of intermediate-sized earthquakes. In the time period $0.2 > \Delta t > 0.1$, there is an increase in the number of intermediate-sized events. In the precursory time period, $0.1 > \Delta t > 0.0$, there is seismic activation and a further increase in the numbers of intermediate-sized events. Finally, at the SOS critical point (spinodal), $\Delta t = 0$, the characteristic earthquake occurs.

We further suggest that the behavior described above is nested. The 1992 Landers and 1994 Northridge earthquakes generated precursory activation that will be described in the next section. But it is likely that the Landers and Northridge earthquakes represent precursory activation that is associated with the next great (characteristic) earthquake on the San Andreas Fault in southern California.

7. Seismic Activation

We begin by discussing the accumulating evidence that there may be an activation of intermediate-sized earthquakes prior to a great earthquake. The occurrence of a relatively large number of intermediate-sized earthquakes in northern California prior to the 1906 San Francisco earthquake has been noted by SYKES and JAUMÉ (1990). These authors also pointed out increases in intermediate-sized events before the 1868 Hayward and 1989 Loma Prieta events in northern California. ELLSWORTH et al. (1981) first pointed out that the rate of intermediate-sized events in the San Francisco Bay region started increasing about 1955 from its post-1906 low. It has also been proposed that there is a power-law increase in seismicity prior to a major earthquake. This was the first proposed by BUFE and VARNES (1993). They considered the cumulative amount of Benioff strain (square-root of seismic energy) in a specified region. They showed that an accurate retrospective prediction of the Loma Prieta earthquake could be made by assuming a power-law increase in Benioff strain prior to the earthquake.

Systematic increases in intermediate-level seismicity prior to a large earthquake have been proposed by several authors (VARNES, 1989; BUFE et al., 1994; KNOPOFF et al. 1996; VARNES and BUFE, 1996; BREHM and BRAILE, 1998, 1999). A systematic study of the optimal spatial region and magnitude range to obtain the power-law seismic activation has been carried out by BOWMAN et al. (1998). Four examples of their results are given in Figure 5, where the cumulative Benioff strain $\varepsilon[t]$ is given as a function of time. Clear increases in seismic activity prior to the 1952 Kern County, 1989 Loma Prieta, 1992 Landers, and 1983 Coalinga earthquakes are illustrated.

In each of the four examples (BOWMAN et al., 1998) given in Figure 5 the data has been correlated (solid line) with the relation

$$\varepsilon[t] = \varepsilon_0 - \beta (\Delta t)^S \tag{4}$$

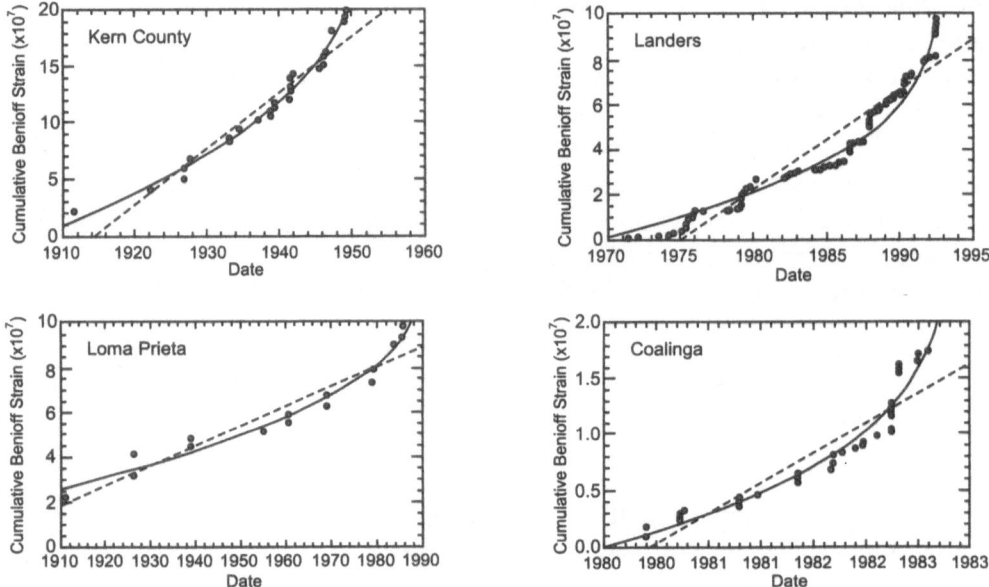

Figure 5

Power-law increases in the cumulative Benioff strains prior to four major earthquakes in California (BOWMAN *et al.*, 1998). Data points are cumulative Benioff strains $\varepsilon[t]$ determined from (5) prior to four major earthquakes in California (BOWMAN *et al.*, 1998). Clear increases on seismic activity prior to the 1952 Kern County, 1989 Loma Prieta, 1992 Landers, and 1983 Coalinga earthquakes are illustrated. In each of the four examples the data has been correlated (solid lines) with the power-law relation given in (4). The values of the power-law exponent s used in (4) are given in Table 1. Dashed straight lines represent a best-fit constant rate of seismicity.

where the cumulative Benioff strain $\varepsilon[t]$ is defined by

$$\varepsilon[t] = \sum_{i=1}^{N[t]} E_i^{1/2}. \tag{5}$$

In (5), E_i is the seismic energy release of the ith precursory earthquake and is proportional to M_i the seismic moment of the ith precursory earthquake, and $N[t]$ is the number of earthquakes considered up until time t. In (4), ε_0 is the cumulative Benioff strain when the characteristic earthquake occurs at $t = t_0$, Δt is nondimensional time remaining until the next characteristic earthquake and is defined in (3), and β and s are constants used to fit the data.

In obtaining the excitation data given in Figure 5, BOWMAN *et al.* (1998) investigated circular regions about each of the characteristic earthquakes they considered. They determined the optimal radius for activation. SORNETTE and SAMMIS (1995) associated this precursory activation with the approach to a critical phase transition. SALEUR *et al.* (1996a,b) interpreted this behavior in terms of correlation lengths that increase prior to the characteristic earthquake. Thus we will refer to this radius of the optimal activation region as a correlation length.

Table 1

Power-law exponents s and correlation lengths ξ for twelve major earthquakes, as given by BOWMAN et al. (1998)

Earthquake	Date	Magnitude	ξ, km	s
Assam	August 15,1950	8.6	900 ± 175	0.22
San Francisco	April 18, 1906	7.7	575 ± 240	0.49
Kern Country	July 21, 1952	7.5	325 ± 75	0.30
Landers	June 28, 1992	7.3	150 ± 15	0.18
Loma Prieta	October 18, 1989	7.0	200 ± 30	0.28
Coalinga	May 2, 1983	6.7	175 ± 10	0.18
Northridge	January 17, 1994	6.7	73 ± 17	0.10
San Fernando	February 9, 1971	6.6	100 ± 20	0.13
Superstition Hills	November 24, 1987	6.6	275 ± 95	0.43
Borrego Mountain	April 8, 1968	6.5	240 ± 60	0.55
Palm Springs	July 8, 1986	5.6	40 ± 5	0.12
Virgin Islands	February 14, 1980	4.8	24 ± 2	0.11

The power-law exponents s and the correlation length ξ (see (2)) for the twelve earthquakes studied by BOWMAN *et al.* (1998) are given in Table 1. The mean value of s for these earthquakes is $\bar{s} = 0.26 \pm 0.15$ (error given is one standard deviation). The correlation lengths are given as a function of earthquake magnitude in

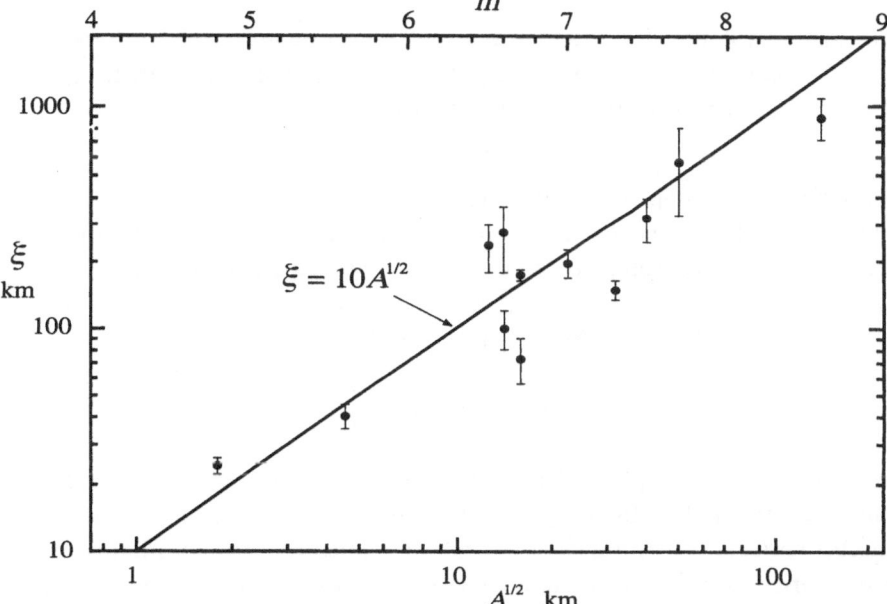

Figure 6

Correlation lengths ξ for precursory seismic activation are given as a function of the square-root of the rupture area $A^{1/2}$ and the magnitude m for twelve major earthquakes (BOWMAN *et al.*, 1998) as given in Table 1.

Figure 6. The dependence on the square-root of rupture area, A, is also shown, and to a good approximation

$$\xi = 10A^{1/2}. \tag{6}$$

DOBROVOLSKY *et al.* (1979) and KEILIS-BOROK and KOSSOBOKOV (1990) reported a similar scaling for the maximum distance between an earthquake and its precursors using pattern recognition techniques.

8. Activation Near A Self-organizing Spinodal

We will now demonstrate that the power-law increase in Benioff strain prior to a characteristic earthquake can be obtained from the analysis of a first-order transition. The basic equation that governs the approach to the critical point associated with the occurrence of the characteristic earthquake in the spinodal approach is (RUNDLE *et al.*, 1997, eq. (54))

$$- K_C \nabla^2 \psi + K_L v \Delta t - \gamma_c \kappa^3 \{ \sin \left[\kappa(\phi_0 + vt_0) \right] \} [\psi^2 - \psi v \Delta t + (v \Delta t)^2] = 0 \tag{7}$$

where Δt has been defined in (3) and v is the tectonic loading velocity. The independent variable $\psi[x, y, \Delta t]$ is a measure of the slip deficit on the ensemble of faults in a seismic zone. The slip deficit $\phi[x, y, t]$ is given by

$$\phi[x, y, t] = vt - S[x, y, t] \tag{8}$$

where $S[x, y, t]$ is the slip, x and y are spatial coordinates, and t is time. The slip deficit is expanded about its value ϕ_0 at the time of the characteristic earthquake (the time of approach to the spinodal) t_0 with the result

$$\psi[x, y, \Delta t] = \phi_0 - \phi[x, y, t] \tag{9}$$

which defines the independent variable ψ.

The constants K_C and K_L in (7) are defined to be moments of the stress Greens function, $T[x, y]$, according to (KLEIN *et al.* 1996, eqs. (6) and (7))

$$K_L = - \int \int T[x, y] \, dx \, dy \tag{10}$$

$$K_C = \frac{1}{2} \int \int T[x, y](x^2 + y^2) \, dx \, dy. \tag{11}$$

When applied to multiple slider-block models K_L can be related to the spring constant of the puller (loader) springs and K_C to the spring constant of the connector springs. In writing (7) it has been assumed that the solution is dominated by the Fourier term with wave number κ. Finally, the constant γ_c in (7) is a measure of the frictional resistance.

Our objective is to obtain scaling relations from (7) rather than solutions. In the application of slider-block models to seismicity, it is appropriate to consider a "stiff" system with $K_C \gg K_L$. In this limit we can approximate (7) with

$$-\nabla^2\psi - C(\psi^2 - \psi v \Delta t + (v\Delta t)^2) = 0, \tag{12}$$

where C is a constant. Our analysis will be restricted to obtaining scaling relations from (12). In order that the three terms within the bracket remain of the same order in the limit $\Delta t \to 0$, we require that

$$\psi \sim \Delta t. \tag{13}$$

The characteristic length in a spinodal analysis is the correlation length ξ, thus we conclude that

$$\nabla^2 \psi \sim \xi^{-2}\psi. \tag{14}$$

In order for the first term in (12) to be of the same order as the other terms in the limit $\Delta t \to 0$, we also require

$$\xi \sim (\Delta t)^{-1/2}. \tag{15}$$

As expected, the correlation length ξ approaches infinity (STAUFFER and AHARONY, 1994) as the time to failure approaches zero.

We also need to know the rate, dN/dt, at which precursory events (earthquakes) occur. This rate of occurrence can be related to the power-law (fractal) distribution of unstable regions in the vicinity of the spinodal. The total number of events (foreshocks) of all magnitudes per unit time is given by

$$\frac{dN}{dt} = \frac{C'(1 - \exp(-k\Delta t A_0))}{\Delta t} \tag{16}$$

where A_0 is the rupture area of the characteristic earthquake and C' and k are constants. RUNDLE et al. (1997) showed that this relation gives Omori's law for the falloff of aftershocks with time after a main shock. In the limit $\Delta t \to 0$ the exponent in (16) can be expanded to give

$$\frac{dN}{dt} = \text{constant}. \tag{17}$$

This is the result that we will use. The rate at which precursory events occur is independent of time. This is consistent with observations of earthquakes. If there were a systematic increase in the rate of occurrence of precursory earthquakes prior to a main event, they could be used for earthquake prediction.

We now use the results given above to obtain the power-law increase in Benioff strain prior to a characteristic earthquake. The mean (indicated by $\langle \, \rangle$) seismic moment of the precursory earthquakes, $\langle M \rangle$, is related to shear modulus, μ, the mean of the slips, $\langle \delta_s \rangle$, and the mean of the rupture areas, $\langle A \rangle$, by

$$\langle M \rangle = \mu \langle \delta_s \rangle \langle A \rangle. \tag{18}$$

We assume that the rate at which precursory earthquakes generate moment dM/dt is given by

$$\frac{dM}{dt} = \mu \langle \delta_s \rangle \langle A \rangle \frac{dN}{dt} \tag{19}$$

where dN/dt is the rate of occurrence of the earthquakes.

Under a wide variety of conditions it is found that the earthquake slip scales with the square-root of the rupture area (KANAMORI and ANDERSON, 1975)

$$\langle \delta_s \rangle \sim \langle A \rangle^{1/2}. \tag{20}$$

Since the mean seismic energy release (see (5)) is proportional to the mean seismic moment, $\langle E \rangle \sim \langle M \rangle$, substitution of (20) into (18) gives

$$\langle E \rangle \sim \langle A \rangle^{3/2} \tag{21}$$

and substitution of (17) and (20) into (19) gives

$$\frac{dE}{dt} \sim \langle A \rangle^{3/2}. \tag{22}$$

The Benioff strain ε is defined to be the square-root of the seismic energy release E so that we have

$$\frac{d\varepsilon}{dt} \sim \frac{dE^{1/2}}{dt} \sim \frac{1}{\langle E \rangle^{1/2}} \frac{dE}{dt}. \tag{23}$$

Substitution of (21) and (22) into (23) gives

$$\frac{d\varepsilon}{dt} \sim \langle A \rangle^{3/4}. \tag{24}$$

The independent variable ψ has been defined to be a measure of the slip deficit on the ensemble of faults in an active seismic zone. However, $\langle \delta_s \rangle$ is the mean slip on individual faults. We assume that

$$\langle \psi \rangle \sim \frac{\langle \delta_s \rangle}{\langle V \rangle} \tag{25}$$

where $\langle V \rangle$ is the mean of the activated volumes of individual faults so that

$$\langle V \rangle \sim \langle A \rangle^{3/2}. \tag{26}$$

Combining (20), (25) and (26) gives

$$\langle \psi \rangle \sim \frac{1}{\langle A \rangle}. \tag{27}$$

Substitution of (13) into (27) gives

$$\langle A \rangle \sim \frac{1}{\Delta t}. \tag{28}$$

Although the number of earthquakes does not increase as $\Delta t \to 0$, the mean rupture areas of these earthquakes does increase. This is the basic observation of seismic activation.

Combining (15) and (28) gives

$$\xi \sim \langle A \rangle^{1/2}. \tag{29}$$

The correlation length ξ scales with the square-root of the mean rupture area. This is identical to the observed scaling given in (6). From (24) and (28) the increase in Benioff strain is given by

$$\frac{d\varepsilon}{dt} \sim \Delta t^{-3/4}. \tag{30}$$

The cumulative Benioff strain as a function of Δt is given by

$$\varepsilon[\Delta t] = \varepsilon_0 - \int_\varepsilon^{\varepsilon_0} d\varepsilon = \varepsilon_0 - \int_0^{\Delta t} \frac{d\varepsilon}{dt} d(\Delta t) = \varepsilon_0 - C'' \int_0^{\Delta t} \frac{d(\Delta t)}{(\Delta t)^{3/4}} = \varepsilon_0 - \beta (\Delta t)^{1/4} \tag{31}$$

where C'' and β are constants. Comparing (24) to (4), we see that the two equations are identical if we take $s = 0.25$. Previously we found from the twelve actual earthquakes given in Table 1, a mean value of $\bar{s} = 0.26 \pm 0.15$. Thus excellent agreement is found between the spinodal theory and actual earthquake data.

9. Discussions and Conclusions

There is considerable observational evidence for seismic activation before at least some major earthquakes. This evidence was discussed in section 7 of this paper. Independent support for seismic activation comes from the pattern recognition algorithms for intermediate-range earthquake prediction developed at the *International Institute of Earthquake Prediction Theory and Mathematical Geophysics* in Moscow (KEILIS-BOROK, 1990; KEILIS-BOROK and ROTWAIN, 1990; KEILIS-BOROK and KOSSOBOKOV, 1990). The first algorithm, $M8$, was developed to make intermediate term predictions of the largest earthquakes ($m > 8$). This method utilizes overlapping circles of seismicity with diameters of 384, 560, 854 and 1333 km for earthquakes, with magnitudes 6.5, 7.0, 7.5 and 8.0, respectively. Within each circle four quantities are determined. The first three are measures of increase in intermediate levels of seismicity and the fourth is a measure of aftershock activity. This approach was used to successfully predict the Loma Prieta earthquake. This approach to earthquake prediction utilizes the concept of precursory activation that is the primary focus of this paper.

The principal object of this paper has been to provide an explanation for the seismic activation in terms of observed scaling laws. We have related the earthquake cycle to thermodynamic cycle involving a spinodal line. The thermodynamic cycle is illustrated in Figure 2. It consists of the following:

1. Water at point "*a*" is heated until it reaches the metastable region at point "*b*".
2. The superheated water is heated further until it reaches the spinodal line at point "*c*", where a steam explosion must occur. This steam explosion can do work.
3. The resulting steam and water mixture is cooled until it is again at point "*a*" and the cycle repeats.

The related earthquake cycle is illustrated in Figure 3. It consists of the following:

1. After an earthquake, the stress is in the stable region at point "*a*". This slip deficit is increased until it reaches the metastable region at point "*b*".
2. This slip deficit is increased further until the spindol or failure line at point "*c*" is reached and an earthquake occurs.
3. The system returns to point "*a*" and the cycle repeats.

The illustration given above demonstrates how earthquakes can be related to spinodal explosions. In both cases a systematic "excitation" occurs prior to the explosion (i.e., the characteristic earthquake). A power-law increase in the Benioff strain prior to a characteristic earthquake is now well documented and from real data appears to satisfy (4) with an exponent $\bar{s} = 0.26 \pm 0.15$. We have shown in this paper that the excitation near a spinodal instability also satisfies this relation with a theoretical value $s = 0.25$. In addition, our analysis predicts that the correlation length ζ scales with the square-root of rupture area, which is also true for observed seismic activation.

Acknowledgements

The authors wish to acknowledge valuable discussions with Charlie Sammis and Robert Shcherbakov. This manuscript benefited significantly from the comments and suggestions of two anonymous reviewers. JBR received support from DOE grant DE-FG03-95ER14499, WK received support from DOE grant DE-FG02-95ER14498, and DLT and BDM received support from NSF grant EAR 9804859.

REFERENCES

BAK, P., TANG, C., and WIESENFELD, K. (1988), *Self-organized Criticality*, Phys. Rev. *A38*, 364–374.
BOWMAN, D. D., OUILLON, G., SAMMIS, C. G., SORNETTE, A., and SORNETTE, D. (1998), *An Observational Test of the Critical Earthquake Concept*, J. Geophys. Res. *103*, 24,359–24,372.

BREHM, D. J., and BRAILE, L. W. (1998), *Intermediate-term Earthquake Prediction Using Precursory Events in the New Madrid Seismic Zone*, Seis. Soc. Am. Bull. *88*, 564–580.

BREHM, D. J., and BRAILE, L. W. (1999), *Intermediate-term Earthquake Prediction Using the Modified Time-to-failure Method in Southern California*, Seis. Soc. Am Bull. *89*, 275–293.

BUFE, C. G., NISHENKO, S. P., and VARNES, D. J (1994), *Seismicity Trends and Potential for Large Earthquakes in the Alaska–Aleutian Region*, Pure appl. geophys. *142*, 83–99.

BUFE, C. G., and VARNES, D. J. (1993), *Predictive Modelling of the Seismic Cycle of the Greater San Francisco Bay Region*, J. Geophys. Res. *98*, 9871–9883.

BURRIDGE, R., and KNOPOFF, L. (1967), *Model and Theoretical Seismicity*, Seis. Soc. Am. Bull. *57*, 341–371.

CARLSON, J. M., and LANGER, J. S. (1989), *Mechanical Model of an Earthquake Fault*, Phys. Rev. *A40*, 6470–6484.

DEBENEDETTI, P. G., *Metastable Liquids: Concepts and Principles* (Princeton University Press, Princeton, NJ 1996).

DOBROVOLSKY, I. R., ZUBKOV, S. I., and MIACHKIN, V. I. (1979), *Estimation of the Size of Earthquake Preparation Zones*, Pure appl. geophys. *117*, 1025–1044.

ELLSWORTH, W. L., LINDH, A. G., PRESCOTT, W. H., and HERD, D. G., *The 1906 San Francisco earthquake and the seismic cycle*. In *Earthquake Prediction: An International Review* (eds. Simpson, D. W., and Richards, P. G.) (AGU, Washington D.C. 1981) pp 126–140.

HERRMANN, H. J., *Fractures*. In *Fractals and Disordered Systems* (eds. Bunde, A., and Havlin, S.) (Springer-Verlag, Berlin 1991) pp. 175–205.

HIRATA, H. J., SATOH, T., and ITO, K. (1987), *Fractal Structure of Spatial Distribution of Microfracturing in Rock*, Geophys. J. Roy. Astron. Soc. *90*, 369–374.

JAUMÉ, S. C., and SYKES, L. R. (1999), *Evolving Towards a Critical Point: A Review of Accelerating Seismic Moment/Energy Release Prior to Large and Great Earthquakes*, Pure Appl. Geophys. *155*, 279–305.

KANAMORI, H., and ANDERSON, D. L. (1975), *Theoretical Basis of Some Empirical Relations in Seismology*, Seis. Soc. Am. Bull. *65*, 1073–1096.

KEILIS-BOROK, V. I. (1990), *The Lithosphere of the Earth as a Nonlinear System with Implications for Earthquake Prediction*, Rev. Geophys. *28*, 19–34.

KEILIS-BOROK, V. I., and KOSSOBOKOV, V. G. (1990), *Premonitory Activation of Earthquake Flow: Algorithm M8*, Phys. Earth Planet. Int. *61*, 73–83.

KEILIS-BOROK, V. I., and ROTWAIN, I. M. (1990), *Diagnosis of Time of Increased Probability of Strong Earthquakes in Different Regions of the World: Algorithm CN*, Phys. Earth Planet. Int. *61*, 57–72.

KLEIN, W., FERGUSON, C. D., and RUNDLE, J. B., *Spinodals and scaling in slider-block models*. In *Reduction and Predictability of Natural Disasters* (eds. Rundle, J. B., Turcotte, D. L., and Klein, W.) (Addison–Wesley, Reading 1996) pp. 223–242.

KNOPOFF, L., LEVSHINA, T., KEILIS-BOROK, V. I., and MATTONI, C. (1996), *Increased Long-range Intermediate-Magnitude Earthquake Activity Prior to Strong Earthquakes in California*, J. Geophys. Res. *101*, 5779–5796.

RUNDLE, J. B., KLEIN, W., GROSS, S., and FERGUSON, C. D. (1997), *Travelling Density Wave Models for Earthquakes and Driven Threshold Systems*, Phys. Rev. E*56*, 293–307.

RUNDLE, J. B., and JACKSON, D. D. (1977), *Numerical Simulation of Earthquake Sequences*, Seis. Soc. Am. Bull. *67*, 1363–1377.

RUNDLE, J. B., KLEIN, W., and GROSS, S. (1996), *Dynamics of a Travelling Density Wave Model for Earthquakes*, Phys. Rev. Lett. *76*, 4285–4288.

RUNDLE, J. B., KLEIN, W., and GROSS, S. (1999), *Physical Basis for Statistical Patterns in Complex Earthquake Populations: Models, Predictions and Tests*, Pure appl. geophys. *155*, 575–607.

SALEUR, H., SAMMIS, C. G., and SORNETTE, D. (1996a), *Renormalization Group Theory of Earthquakes*, Nonlinear Proc. Geophys. *3*, 102–109.

SALEUR, H., SAMMIS, C. G., and SORNETTE, D. (1996b), *Discrete Scale Invariance, Complex Fractal Dimensions, and Log-periodic Fluctuations in Seismicity*, J. Geophys. Res. *101*, 17,661–17,677.

SORNETTE, D., and SAMMIS, C. G. (1995), *Complex Critical Exponents from Renormalization Group Theory of Earthquakes: Implications for Earthquakes Predictions*, J. Phys. 1 France *5*, 607–619.

STAUFFER, D., and AHARONY, A., *Introduction to Percolation Theory*, 2nd ed. (Taylor and Francis, London 1994).

SYKES, L. R., and JAUMÉ, S. C. (1990), *Seismic Activity on Neighbouring Faults as a Long-term Precursor to Large Earthquakes in the San Francisco Bay Area*, Nature *348*, 595–599.

TRIEP, E. G., and SYKES, L. R. (1997), *Frequency of Occurrence of Moderate to Great Earthquakes in Intracontinental Regions: Implications for Changes in Stress. Earthquake Prediction, and Hazards Assessments*, J. Geophys. Rev. *102*, 9923–9948.

TURCOTTE, D. L. (1999), *Seismicity and Self-organised Criticality*, Phys. Earth Planet Int. *111*, 275–293.

TURCOTTE, D. L., MALAMUD, B. D., MOREIN, G., and NEWMAN, W. I. (1999), *An Inverse Cascade Model for Self-organized Critical Behavior*, Physica *A268*, 629–643.

VAN DER WAALS, J. D., *De Continuitet van den Gas-en Vloeistoftoestand*, Ph.D. thesis (University of Leiden, Holland 1873).

VARNES, D. J. (1989), *Predicting Earthquakes by Analyzing Accelerating Precursory Seismic Activity*, Pure appl. geophys. *130*, 661–686.

VARNES, D. J., and BUFE, C. G. (1996), *The Cyclic and Fractal Seismic Series Preceding an m_b 4.8 Earthquake on 1980 February 14 near the Virgin Islands*, Geophys. J. Int. *124*, 149–158.

(Received October 13, 1999, revised January 25, 2000, accepted April 5, 2000)

 To access this journal online:
http://www.birkhauser.ch

Pure appl. geophys. 157 (2000) 2183–2207
0033–4553/00/122183–25 $ 1.50 + 0.20/0

❙Pure and Applied Geophysics

Evolution of Stress Deficit and Changing Rates of Seismicity in Cellular Automaton Models of Earthquake Faults

DION WEATHERLEY,[1,2] STEVEN C. JAUMÉ[1,3] and PETER MORA[1,4]

Abstract—We investigate the internal dynamics of two cellular automaton models with heterogeneous strength fields and differing nearest neighbour laws. One model is a crack-like automaton, transferring all stress from a rupture zone to the surroundings. The other automaton is a partial stress drop automaton, transferring only a fraction of the stress within a rupture zone to the surroundings. To study evolution of stress, the mean spectral density $\mathscr{S}(k_r)$ of a stress deficit field is examined prior to, and immediately following ruptures in both models. Both models display a power-law relationship between $\mathscr{S}(k_r)$ and spatial wavenumber (k_r) of the form $\mathscr{S}(k_r) \sim k_r^{-\beta}$. In the crack model, the evolution of stress deficit is consistent with cyclic approach to, and retreat from a critical state in which large events occur. The approach to criticality is driven by tectonic loading. Short-range stress transfer in the model does not affect the approach to criticality of broad regions in the model. The evolution of stress deficit in the partial stress drop model is consistent with small fluctuations about a mean state of high stress, behaviour indicative of a self-organised critical system. Despite statistics similar to natural earthquakes these simplified models lack a physical basis. Physically motivated models of earthquakes also display dynamical complexity similar to that of a critical point system. Studies of dynamical complexity in physical models of earthquakes may lead to advancement towards a physical theory for earthquakes.

Key words: Cellular automata, self-organised criticality, critical point hypothesis, stress correlations.

Introduction

In recent years, theoretical research in seismology has been boosted by the infusion of ideas from statistical physics, particularly the concepts of self-organised criticality and critical point phenomena (BÅK and TANG, 1989; SORNETTE and SAMMIS, 1995). Self-organised critical systems are spatially extended, driven dynamical systems which spontaneously self-organise into a critical state where they remain (BÅK *et al.*, 1987, 1988). In the critical state, cascades of energy release of any size may occur at any time. Critical point systems are closely related to

---
[1] Queensland University Advanced Centre for Earthquake Studies, The University of Queensland, Brisbane 4072 Queensland, Australia.
[2] E-mail: weatherley@shake.earthsciences.uq.edu.au
[3] E-mail: jaume@shake2.earthsciences.uq.edu.au
[4] E-mail: mora@shake2.earthsciences.uq.edu.au

self-organised critical systems. Rather than remaining perpetually near a critical state, critical point systems progressively approach and retreat from a critical state in which large-scale cascades of energy release may occur (SORNETTE and SAMMIS, 1995).

In simple critical point systems, the approach to criticality corresponds with the growth of long-range spatial correlations of a physical property of the system (e.g., stress). A theory for seismogenesis based upon the physics of critical point systems called the critical point hypothesis, has been proposed as a theory for regional seismicity. According to this theory, the action of tectonic stresses and small earthquakes result in the growth of long-range stress correlations within a region consisting of many faults.

As longer-range stress correlations form, larger events become progressively more likely. The region reaches a critical state when stress correlations exist up to the size of the region. In the critical state, a large earthquake is possible. A large earthquake fails a considerable portion of the region, destroying long-range stress correlations and returning the system to a state far-from-failure (SORNETTE and SAMMIS, 1995; SALEUR *et al.*, 1996). The cycle then repeats, culminating in another large event. Cumulative energy release prior to large events follows a power-law time-to-failure relation.

Seismologists have found evidence for accelerating seismic moment release prior to large or great earthquakes from a variety of tectonic settings (see JAUMÉ and SYKES, 1999 for a review). Large or great earthquakes are often preceded by a period of increased activity in moderate-sized events and in a few cases, there is an increase in the rate of events of all magnitudes (JAUMÉ, 1999). Cumulative Benioff strain (the square-root of seismic energy release) prior to large events follows a power-law time-to-failure relation (VARNES, 1989; BUFE and VARNES, 1993). However, accelerating moment release is not universally observed (JAUMÉ and SYKES, 1999) and there is currently no direct means of detecting the predicted growth and destruction of stress correlations held responsible for the accelerating sequences by the critical point hypothesis.

Dynamical complexity such as critical point behaviour and self-organised criticality can be simulated using highly simplified models of spatially extended dynamical systems called cellular automata (BÅK and TANG, 1989; SAMMIS and SMITH, 1998). A cellular automaton consists of an array of cells, each of which is assigned a dynamical variable which might represent seismic energy, stress, or accumulated strain. In this paper, we will assume the dynamical variable represents stress. Cells are also assigned a failure condition (or failure strength); when the stress of a given cell exceeds its strength, the cell fails. A failed cell redistributes its stress to neighbouring cells according to a pre-defined nearest neighbour law. The system is driven by repeatedly incrementing the stress of cells by a small amount.

Cascades of cell failings of any size may occur in cellular automata. In a highly stressed state, failure of one cell may trigger neighbouring cells to fail. In this manner, a rupture may cascade through the model. As cell fail, stress is concentrated on the rupture front. A rupture will grow if the stress concentrations are sufficient to fail the cells adjacent to the rupture front. If the stress concentrations are insufficient to fail these cells, the rupture arrests leaving behind a boundary of cells with high stress.

Although highly simplified, cellular automata display power-law statistics and fractal clustering of events, features of natural earthquake sequences. A number of researchers have studied earthquake-like automata with differing properties (LOMNITZ-ADLER, 1993; SAMMIS and SMITH, 1998; STEACY et al., 1996; STEACY and McCLOSKEY, 1998, 1999). Models with different nearest neighbour laws or model geometry produce different types of dynamical behaviour. In this paper, we compare two cellular automata with identical model geometry but differing nearest neighbour interactions.

Natural fault zones have a heterogeneous strength distribution (BROWN and SCHOLZ, 1985) and it is likely that strength heterogeneity plays a role in rupture nucleation (HUANG and TURCOTTE, 1988). The effects of strength heterogeneity can be simulated in automaton models by defining a fractal distribution of cell strengths (STEACY et al., 1996). We examine two such heterogeneous automata in this paper. The models studied are reproductions of models by STEACY and McCLOSKEY (1998, 1999).

One model defines a crack-like automaton in which all stress from a failed region is transferred to the surroundings. The other model, a partial stress drop model, only partially decreases the stress level within a rupture zone. The next section is a detailed description of the properties of the two models. Following the description, we present results which illustrate the internal dynamics of the models. In particular, we develop a method for measuring the evolution of stress. In one model, the dynamics progressively approach and retreat from a critical state. The approach to criticality is primarily driven by tectonic loading while the occurrence of large events substantially reduces the stress in the model. The dynamics of the second model are consistent with small fluctuations about a critical state. Further study of cellular automata and physical models displaying dynamical complexity may lead to an improved understanding of seismogenesis.

Properties of the Models

In this section we describe the properties of the two models. The only significant difference between the models is their nearest neighbour law. As we show, the differing nearest neighbour laws give rise to quite different dynamics. The following features are common to both models:

1. A 2-D statistical fractal distribution of strengths is defined on a grid of $N_x \times N_y$ cells. The fractal dimension of the distribution is $D = 2.3$ and cell strengths lie in the range (0.01, 1.00). The strength field was generated using a Fourier filtering technique (see TURCOTTE, 1997).

2. The stress of each cell, which is initially zero, is incremented uniformly. The stress increment is computed each loading time as the minimum stress increment necessary to fail at least one cell.

3. A cell fails when its stress exceeds its strength. The stress from a failed cell is redistributed to at most $N = 8$ neighbouring cells via a nearest law described below.

4. Model boundaries are dissipative. Stress transferred across model boundaries is removed from the model.

5. The size of an event is defined as the sum of the stress drops of all cells which fail in response to a single stress increment.

6. A simulation consists of 5×10^5 stress increments. There is an initial transient period after which the dynamics of the model settles into a statistical steady state. Model data is only recorded after the initial transient is completed, typically after 2×10^4 stress increments.

Each model has a different nearest neighbour law. One law redistributes the stress from failed cells only to cells which have not failed since the previous stress increment. This defines a crack-like automaton model (LOMNITZ-ADLER, 1993); stress within the rupture zone drops to zero. The nearest neighbour law of the other model redistributes stress preferentially to unfailed neighbours giving a lesser proportion to previously failed cells. This defines a partial stress drop (PSD) automaton (LOMNITZ-ADLER, 1993); the stress within the rupture zone remains high after the passage of a rupture. In the following, the first model will be referred to as the crack model and the second will be the PSD model.

The crack model is a reproduction of a model examined by STEACY and MCCLOSKEY (1998). By the nature of its nearest neighbour law, all stress within a rupture zone is concentrated on the rupture front leaving behind a region with zero stress. Subsequent to a rupture, stress within the rupture zone increases until another rupture fails the region. This results in cyclic dynamical behaviour. Regions cycle between a state of low stress (far-from-failure) to a state of high stress (close-to-failure).

Let σ_f represent the stress of a failed cell. Then, the stress redistributed to nearest neighbours in the crack model is given by

$$\Delta\sigma_u = \frac{\gamma\sigma_f}{(N - n_b)},$$

(1)

$$\Delta\sigma_b = 0.0,$$

where n_b is the number of broken nearest neighbours and $\Delta\sigma_u$ and $\Delta\sigma_b$ is the stress redistributed to unbroken and broken nearest neighbours, respectively. Stress transfer is non-conservative in this model; only 75% of the stress of a failed cell is redistributed to nearest neighbours (i.e., the damping factor $\gamma = 0.75$). The remaining stress is removed from the model. Non-conservative stress transfer is necessary in this model to prevent the occurrence of global ruptures. A global rupture would reset the stress of all cells to zero i.e., the system would reset to the initial state. A factory $\gamma = 0.75$ was chosen so that the maximum event sizes in the crack model were of comparable size to those of the PSD model.

The nearest neighbour law of the crack model leads to stress concentrations that scale with rupture size, higher than the square-root of rupture size predicted by linear elastic fracture mechanics. However the law may be physically justified by assuming that failed cells are yet to heal and thus cannot support stress (STEACY and MCCLOSKEY, 1998).

The PSD model was introduced by STEACY and MCCLOSKEY (1999). Unlike the crack model, some stress in redistributed to cells within the rupture zone. As a result, the stress level within rupture zones only partially drops. In general, each cell within the rupture zone receives stress from multiple cells on the rupture front. Consequently, the stress level within the rupture zone is not constant and cells maintain a high level of stress after a rupture. Once every cell in the model has failed at least once, the entire model remains in a state of high stress.

Stress from failed cells in the PSD model, is redistributed to nearest neighbours according to

$$\Delta\sigma_u = \frac{\gamma\sigma_f}{(N+1) - \sqrt{n_b + 1}}, \tag{2}$$

$$\Delta\sigma_b = \frac{\gamma\sigma_f - \sum_{i=1}^{N-n_b} \Delta\sigma_u}{n_b}; \quad n_b \neq 0.$$

A nearest neighbour law of this form leads to stress concentrations that, for a circular rupture in a homogeneous medium, initially scale with the square-root of rupture radius then asymptotically approach a constant value (STEACY and MC-CLOSKEY, 1999). Unlike the crack model, stress transfer in this model is conservative ($\gamma = 1.00$). By virtue of the nearest neighbour law, global ruptures do not occur in this model.

Cellular automata are driven systems. The models in this paper are driven by incrementing the stress of all cells by an amount $\Delta\sigma$. The automata of STEACY and MCCLOSKEY (1998, 1999) used a constant stress increment $\Delta\sigma$ which was chosen to be small enough that approximately one cell failed in response to a single stress increment. The models in this paper employ a variable stress increment. Each loading time we compute the minimum stress necessary to fail at least one cell and increment all cells by this amount. This method approximates slow tectonic loading in the models.

SAMMIS and SMITH (1998) employed a similar method for their uniform and hierarchical fractal automata. Such a method warrants the introduction of a model time. The time scale for loading is set by the assumption of a uniform rate of loading. Thus, time increments in the models are proportional to stress i.e., $\Delta t = \kappa \Delta \sigma$. Since the constant of proportionality κ is arbitrary, we choose $\kappa = 1$ here.

The stress increment at each timestep is the stress necessary to fail at least one cell. The distribution of stress increment sizes is a measure of the minimum stress required to fail cells in the models. The interval stress increment size-distributions for both models are given in Figure 1. Stress increments are higher in general in the crack model. Maximum stress increments in the crack model are of order $\Delta \sigma_m \sim$ 0.0055 compared with the PSD model where $\Delta \sigma_m \sim 0.00125$. Subsequent to the initial transient period, stress increments are always less than the minimum cell strengths (0.01) in the models. Smaller stress increments in the PSD model imply that cells are closer-to-failure on average than cells in the crack model.

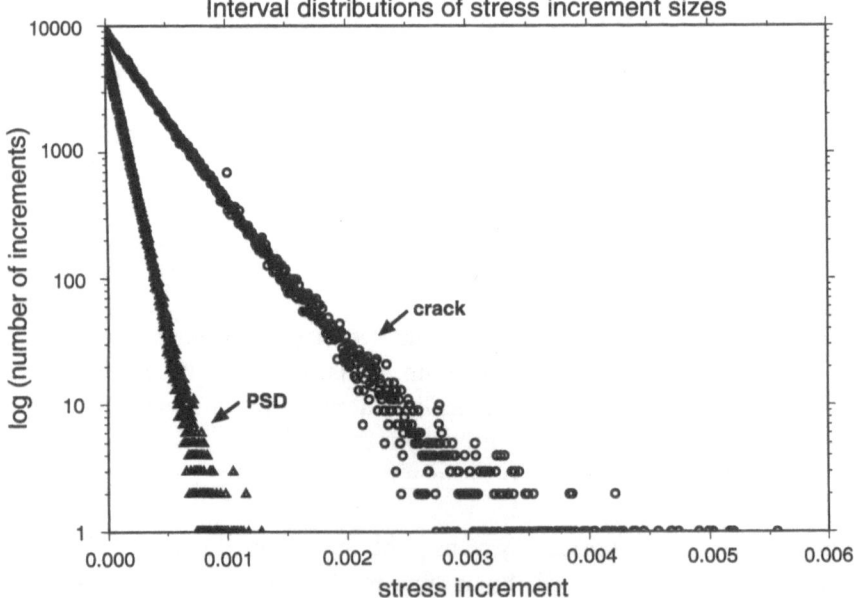

Figure 1
Interval distribution of stress increment sizes for both models. The PSD model (triangles) has lower stress increments than the crack model (circles). Stress increments are always much lower than the minimum cell strength (0.01) in the models.

Internal Dynamics of the Models

In these automata, two processes are responsible for altering the state of the system, namely the loading mechanism and the rupture dynamics. Tectonic loading in heterogeneous automata causes regions of differing strength to fail at different times. All cells are driven towards failure at the same rate. However, a patch of cells with a relatively high strength will take longer to approach failure than a similar patch of weaker cells. Thus different regions of the model rupture at different times. Rupture dynamics also play an important role in the evolution of the system.

Ruptures alter the stress level of cells within the rupture zone and adjacent to the rupture boundary. In the crack model, the stress level drops to zero within a rupture zone. All the stress from within a rupture zone is concentrated on the rupture front. Thus when a rupture arrests, it leaves behind a boundary of cells with high residual stress concentrations. Cells within a rupture zone in the PSD, model have a non-zero stress level due to the redistribution of stress from the propagating rupture front. As a consequence, stress concentrations at rupture boundaries are considerably smaller in the PSD model.

Event Statistics

Through cooperative failing of large numbers of cells, ruptures of any size up to the size of the system may occur. We examine the interval event size-distribution for both models in Figure 2. Unless otherwise stated, all results in this paper were obtained from models consisting of 64 × 64 cells. The PSD model has a power-law, or Gutenberg-Richter size-distribution over approximately three orders of magnitude.

The crack model has a characteristic earthquake-like distribution. The over-abundance of large events is due to the high stress concentrations on rupture fronts in the model (STEACY and MCCLOSKEY, 1998). As a rupture grows, the stress transferred to unbroken cells becomes considerably larger than the stress necessary to fail those cells. Consequently, ruptures in the crack model can propagate through regions of relatively low stress level, resulting in a large event. Both distributions in Figure 2 display a roll-off in event numbers at the smallest and largest scales. This is a finite size effect. Maximum event sizes are constrained by the number of cells in the model while minimum event sizes depend upon the strengths of individual cells.

For the rupture of a single cell to initiate a large event, the stress level within the future rupture zone must be sufficiently high to permit the propagation of the rupture front through the region. STEACY and MCCLOSKEY (1998) introduced normalised stress as a mean-field measure of the stress level prior to events. The normalised stress $\zeta(t)$ is simply the mean of the stresses $\sigma(x, y, t)$ of all cells normalised by their individual cell strengths $\tau(x, y)$ i.e.,

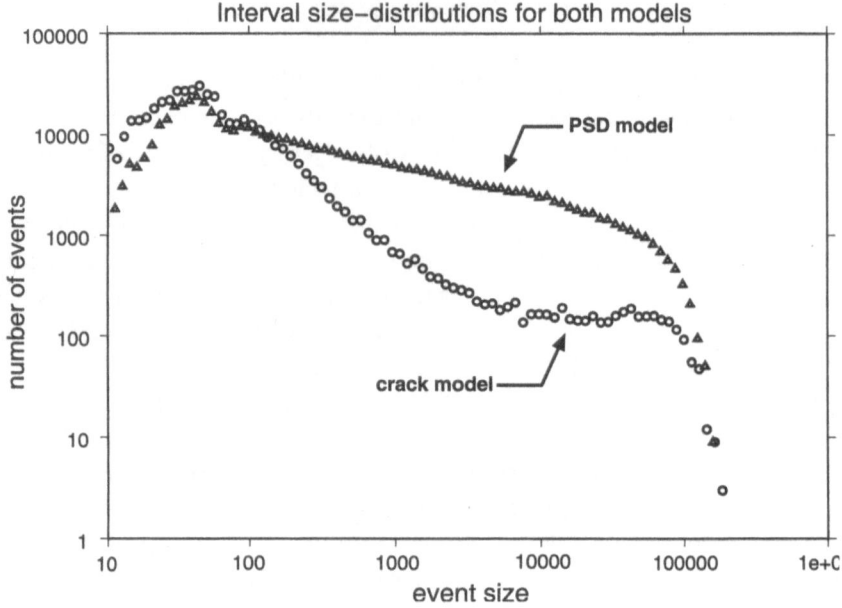

Figure 2

The interval event size distributions of both models. The crack model (circles) has a characteristic earthquake-like distribution while the PSD model (triangles) has a power-law event size-distribution. Roll-off in event numbers at small and large event sizes is due to finite size effects.

$$\zeta(t) = \frac{1}{N_c} \sum_{x=0}^{N_x-1} \sum_{y=0}^{N_y-1} \frac{\sigma(x, y, t)}{\tau(x, y)},$$

where $N_c = N_x N_y$ is the total number of cells and the sum is taken over all cells (x, y). It may be seen from the definition that a high value of normalised stress implies that many cells are close-to-failure. STEACY and MCCLOSKEY (1998) found that large events in the crack model only occur when normalised stress is high.

We have sorted events according to the normalised stress prior to each event and computed the event size-distribution for subsets of events with differing levels of normalised stress (Fig. 3). We interpret these distributions in terms of probability density functions for earthquake occurrence. In both models the probability of a large event is higher for higher normalised stress. In the crack model there is a three-fold increase in the maximum event size as normalised stress increases in the range $0.05 < \zeta(t) < 0.45$. Normalised stress is near-constant in the PSD model $(0.545 < \zeta(t) < 0.575)$ and the probability of large events increases by a lesser amount than the crack model. Large variations in normalised stress in the crack model implies that the model is not always in a state of high stress.

JAUMÉ *et al.* (1999) examined the evolution of event time and size statistics in these models. Large events in the crack model were found to be preceded by a period of accelerating seismic energy release more often than a randomised cata-

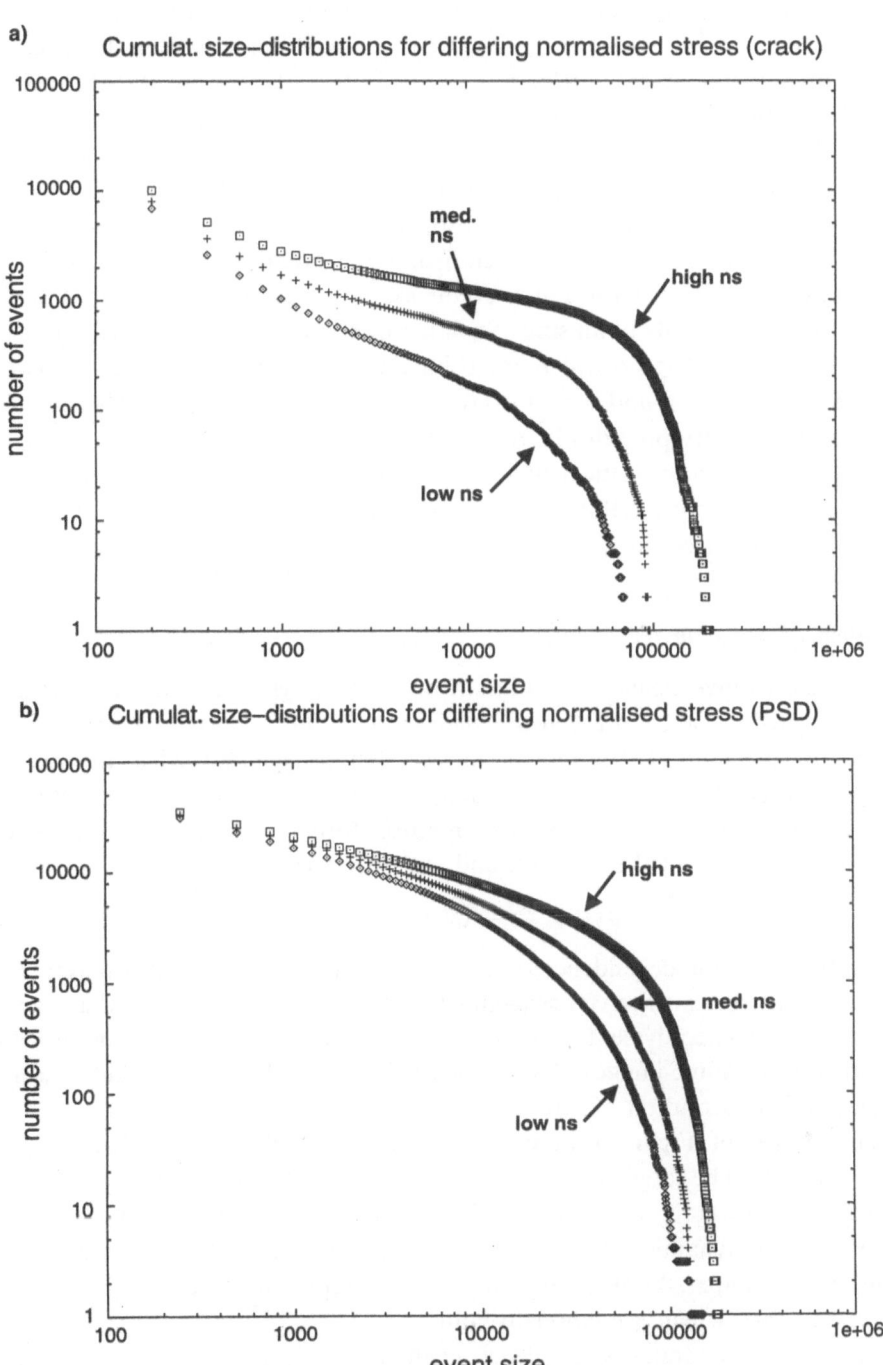

Figure 3
Cumulative event size-distributions for low, medium, and high normalised stress in a) the crack model and b) the PSD model. Larger events are more likely for higher normalised stress in both models, although the increase in maximum event size in much smaller in the PSD model.

logue. The accelerating sequences in the crack model consist of an increase in the rate of events of all sizes, consistent with a small number of natural cases of accelerating seismic energy release (JAUMÉ, 1999). However, this behaviour is inconsistent with a larger number of cases where there is an increase in the rate of only moderate-sized events (JAUMÉ, 1999). On average, no increase in the rate of events of any size is observed prior to large events in the PSD model.

These results highlight emergent properties of the internal dynamics in the models. The crack model has characteristic large events and displays large variations in the mean stress level. Large events are preceded by a precursory interval of increasing rate of events of all sizes. Such features suggest that the crack model may be a system displaying critical point behaviour. The PSD model has a near-constant, high stress level and a power-law distribution of event sizes. No systematic precursory seismicity precedes large events in this model. Such features are characteristic of self-organised critical system. To confirm that the crack model is a critical point system and that the PSD model self-organises into a critical state, we examine the evolution of stress in the models.

Evolution of Stress in the Models

We seek to investigate the evolution of the two systems from one state to the next. As noted earlier, both tectonic loading and rupture dynamics are responsible for altering the state of stress in the models. Rupture initiate when the stress of a cell exceeds its strength. For this reason, it is desirable to study the evolution of stress deficit rather than the evolution of stress. Stress deficit $\delta(x, y, t)$ is defined as the difference between the strength and stress of a cell (x, y) i.e.,

$$\delta(x, y, t) = \tau(x, y) - \sigma(x, y, t). \tag{3}$$

Notice that the time-dependence of stress deficit arises solely from the time-dependence of cell stresses since cell strengths remain constant. Thus, the time evolution of stress deficit is equivalent to the time evolution of stress in the models. In the following we examine images of stress deficit in the models and present a technique to quantify the evolution of stress deficit.

The images of Figure 4 were obtained from simulations involving a grid of 256×256 cells. The larger grid size was chosen to ensure the features of the model are easily resolved in the images. The dynamics illustrated in these images is identical to the dynamics of the smaller (64×64) models used to generate the other results in this paper. Darker greys in Figure 4 represent cells with a lower stress deficit, i.e., cells which are closer-to-failure.

Figure 4a is the initial stress deficit in the two models. This is equivalent to the strength distribution in the models since all cells initially have zero stress. A broad stronger (lighter) region roughly bisects the model with weaker (darker) regions either side. We have already noted that regions of differing strength will rupture at different times. Thus in general, different regions will have differing stress deficit i.e., some regions will be closer-to-failure than others.

Consider an image of stress deficit in the crack model (Fig. 4b) taken after the initial transient period. The model is divided into regions of differing sizes separated by crack-like boundaries. The crack-like structures are the boundaries of previous ruptures. High stress concentrates along such boundaries when a rupture arrests. Due to the high stress concentrations, rupture boundaries are often the sites for nucleation of subsequent events. Crack geometry evolves over time. At a later time, some of the larger regions in Figure 4b may subdivide into several smaller regions which fail almost independently. Later still, a rupture may propagate through a number of these smaller regions producing another large event.

To see how this may arise, consider two adjacent regions with differing strengths. For the sake of discussion, we define the repeat time for each region as the time between two consecutive ruptures of the region. Since the two regions have

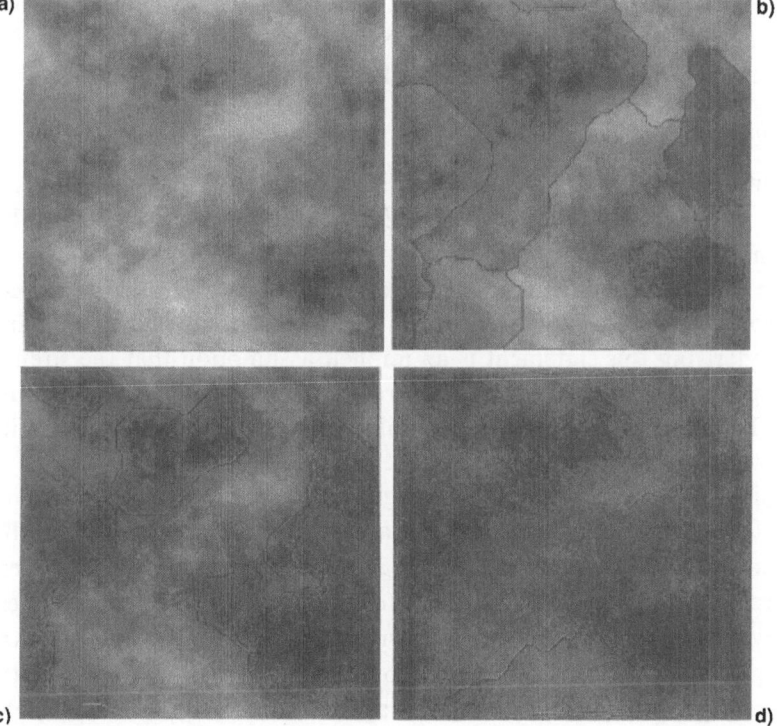

Figure 4

a) The strength distribution for models consisting of 256 × 256 cells. Lighter greys represent higher strengths. b) Stress deficit for the crack model (darker cells are closer-to-failure). Crack-like structures are the boundaries of previous ruptures. They are the sites of high residual stress concentration. c) Stress deficit in the PSD model prior to failure of all cells. Regions which have failed have a lower stress deficit on average than unfailed cells. d) Stress deficit in the PSD model after all cells failed. All cells have a low stress deficit due to previous ruptures and the model is everywhere close-to-failure.

differing strengths, they also have different repeat times. Initially, the two regions will fail independently; one region will fail while the other is still approaching failure and *vice versa*. However, at some stage the two regions will approach failure simultaneously. A rupture nucleating in either region will propagate to the other region producing an event which is larger than the previous events which failed only one or the other of the two smaller regions. The stress deficit of adjacent regions may randomly correlate in time producing quasi-periodic large events.

Ruptures in the PSD model redistribute stress inside the rupture zone. As a result, the stress level in the rupture zone remains high after a rupture has occurred. In Figure 4c we examine the stress deficit in the PSD model before every cell in the model has failed. Regions which have failed have a much higher stress level than regions which are yet to fail. Once every cell has failed at least once (Fig. 4d) the stress deficit is small across the entire model. Since a greater proportion of stress is redistributed outside the rupture zone, there are small stress concentrations along rupture boundaries in the PSD model. However, these are transient features because subsequent ruptures easily propagate across previous rupture boundaries. Random correlations in stress deficit are much less important to the dynamics of the PSD model; the stress level of all cells remains high at all times.

Quantifying Evolution of Stress Deficit

The images of Figure 4 provide only a qualitative measure of stress deficit evolution in the models. Here, we quantify evolution of stress deficit in the models. It is often stated that the critical point hypothesis predicts a growth in correlation lengths in the stress field prior to the largest events in a given region. As correlation length grows, larger events become likely (SALEUR *et al.*, 1996; JAUMÉ and SYKES, 1999). Underlying this statement is an implicit assumption that the strength within the region is homogeneous. As stress correlation lengths increase in a homogeneous medium, larger areas become close to failing thereby increasing the likelihood that ruptures will run-away into large events.

The picture is not that simple if strength within the region is heterogeneous. For ruptures to grow into large events, large areas within the region need to be close to failing. In a heterogeneous medium, a region is close to failure if the stress is close to the strength everywhere, i.e., if the difference between strength and stress is close to zero everywhere. Thus, for a heterogeneous medium we would not expect there to be a growth in stress field correlations prior to a large event. Rather, there should be a growth in correlations of a stress deficit field $\delta(x, y, t)$ defined by Equation (3).

We have devised a technique to quantify evolution of stress deficit correlations in the models. The technique involves the computation of the mean spectral density of the stress deficit field at each loading time. The mean spectral density is a measure of the spatial variations in stress deficit at different wavelengths in the models.

The mean spectral density is computed by first calculating the Fourier transform $\Delta(k_x, k_y, t)$ of the stress deficit $\delta(x, y, t)$:

$$\Delta(k_x, k_y, t) = \frac{1}{N_x N_y} \sum_{x=0}^{N_x-1} \sum_{y=0}^{N_y-1} \delta(x, y, t) \exp\left[-2\pi i\left(\frac{xk_x}{N_x} + \frac{yk_y}{N_y}\right)\right].$$

$\Delta(k_x, k_y, t)$ is the Fourier coefficient for a component with wavenumbers (k_x, k_y). We introduce the radial wavenumber $k_r = \sqrt{k_x^2 + k_y^2}$ and compute the mean spectral density $\mathscr{S}(k_r)$ for each radial wavenumber i.e.,

$$\mathscr{S}(k_r) = \frac{1}{N_j} \sum_{i=1}^{N_j} |\Delta(k_x, k_y)|^2, \quad \sqrt{k_x^2 + k_x^2} = k_r,$$

where N_j is the number of Fourier coefficients with radial wavenumber k_r (TUR-COTTE, 1997).

Figure 5 is a log-log plot of the mean spectral density of the stress deficit field versus radial wavenumber for one timestep in each model. For both models, there is a power-law relationship between the mean spectral density $\mathscr{S}(k_r)$ and radial wavenumber k_r of the form $\mathscr{S}(k_r) \sim k_r^{-\beta}$. This is evidenced by the linear trend of data points in the figure. As we show later, the exponent is lower in general in the PSD model.

The exponent β may be calculated from the slope of $\log \mathscr{S}(k_r)$ versus $\log k_r$. A simple least-squares line fit giving equal weighting to each wavenumber introduces a bias towards high wavenumbers because there are more data points at high wavenumbers (see Fig. 5). The critical point hypothesis predicts a growth in long-range correlations. These correspond to short wavenumbers where data points are sparse. To overcome the implicit bias towards high wavenumbers it is preferable to use a weighted least-squares fit to the data. The following algorithm describes the calculation of the weighting for each wavenumber:

1. Compute the number $N(\log k_r)$ of wavenumbers between $\log k_r - \Delta \log k_r$ and $\log k_r$, where $\Delta \log k_r$ is a small constant.
2. Assign an initial weighting $W(\log k_r) = 1/N(\log k_r)$ and compute the minimum weighting W_o. Small wavenumbers will have a higher weighting than large wavenumbers.
3. Specify a maximum weighting (here $W_m = 10\ W_o$) and assign this weighting to all wavenumbers for which $W(\log k_r) > W_m$. This ensures small wavenumbers are not too highly weighted relative to large wavenumbers.
4. Except for the smallest wavenumbers, the weighting relation $W(\log k_r)$ should be a decreasing function of $\log k_r$. Due to the finite size of the model, the very largest wavenumbers may have a weighting $W(\log k_r) > W_o$. These wavenumbers are assigned a new weighting W_o to decrease statistical fluctuations at the highest wavenumbers.

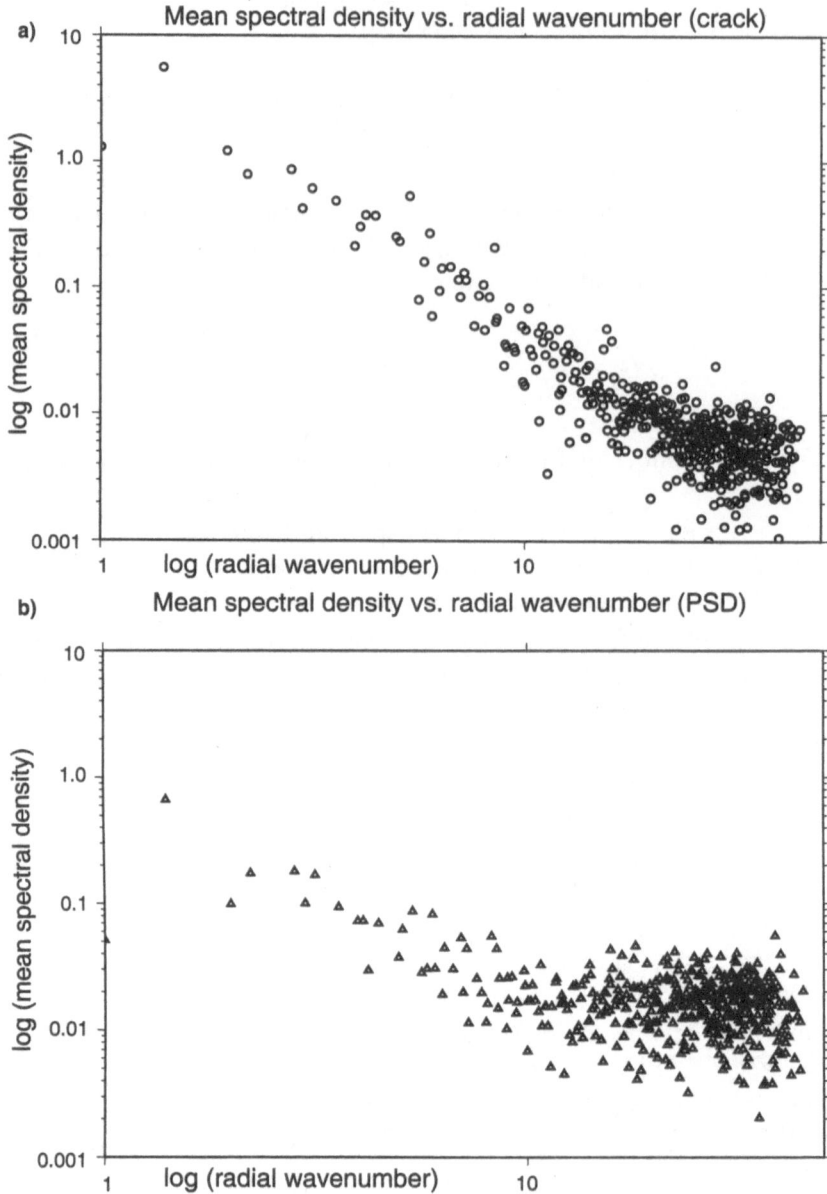

Figure 5

a) The logarithm of the mean spectral density of the stress deficit field $\mathscr{S}(k_r)$ versus the logarithm of radial wavenumber k_r at some time t for the crack model. b) Mean spectral density of the stress deficit field for one timestep of the PSD model. The linear trend of data points in each plot identifies the mean spectral density as a power-law of the form $\mathscr{S}(k_r) \sim k_r^{-\beta}$.

It is important to discuss the relationship between the value of the exponent (β) and the stress deficit in the models prior to presenting results. The exponent controls the scaling of spectral density with wavelength. We compare the mean spectral density at large wavelengths for two different values of the exponent β in the crack model (Fig. 6). In general, the mean spectral density is lower at larger wavelengths when the exponent β is smaller. Recall from the definition (Equation (3)) that high stress deficit implies a cell is far-from-failure and that stress deficit approaches zero as a cell approaches failure. Thus, if the spectral density is high at large wavelengths then ruptures are unlikely to propagate to large scales. A rupture may propagate to large scales only if the variation in stress deficit is small for all scales up to the final size of the rupture.

To illustrate this concept we have sketched (Fig. 7) the stress deficit near a failed cell for three values of the exponent β. Each curve consists of three sinusoids with wavelengths (long, intermediate, and short wavelengths). The sinusoid with the shortest wavelength has a constant amplitude. We scale the amplitudes of the other sinusoids in each curve to emulate the differing variation in stress deficit associated with differing values of the exponent β (cf. inset of Fig. 7). If the exponent is large, stress deficit increases rapidly with increasing distance from the failed cell at $x = 0$ (the epicentre). A growing rupture will not propagate far before the stress deficit of neighbouring cells is higher than the stress concentrated on the rupture front, causing the rupture to arrest. As the value of the exponent decreases, stress deficit increases much less rapidly with distance from the epicentre. Thus, ruptures are more likely propagate to larger distances.

Based upon this argument we may conclude that ruptures will grow to large scales only when the exponent β is small. If the exponent is small, the mean spectral density is near-constant at all wavelengths; the stress deficit has a near-white spectrum. Conversely, ruptures arrest if the mean spectral density is large at long wavelengths; the stress deficit has a red spectrum. A similar result was obtained by RUNDLE et al. (1998) who derived a stochastic Griffith crack theory in which ruptures arrest (grow) if the stress deficit field has a red (blue) spectrum. The maximum value of the exponent is constrained by the strength distribution. If all cells have zero stress, then the stress deficit is equal to the strength distribution. For a fractal dimension of $D = 2.3$, the maximum value of the exponent of the mean spectral density is $\beta = 2.4$ according to the relation (TURCOTTE, 1997):

$$D = \frac{7 - \beta}{2}.$$

Tectonic loading causes the stress deficit of all cells to decrease on average. This will decrease the mean spectral density of stress deficit at all wavelengths thereby decreasing the value of the exponent and increasing the likelihood of large events. Small and moderate events fail only a small portion of the model so we expect little

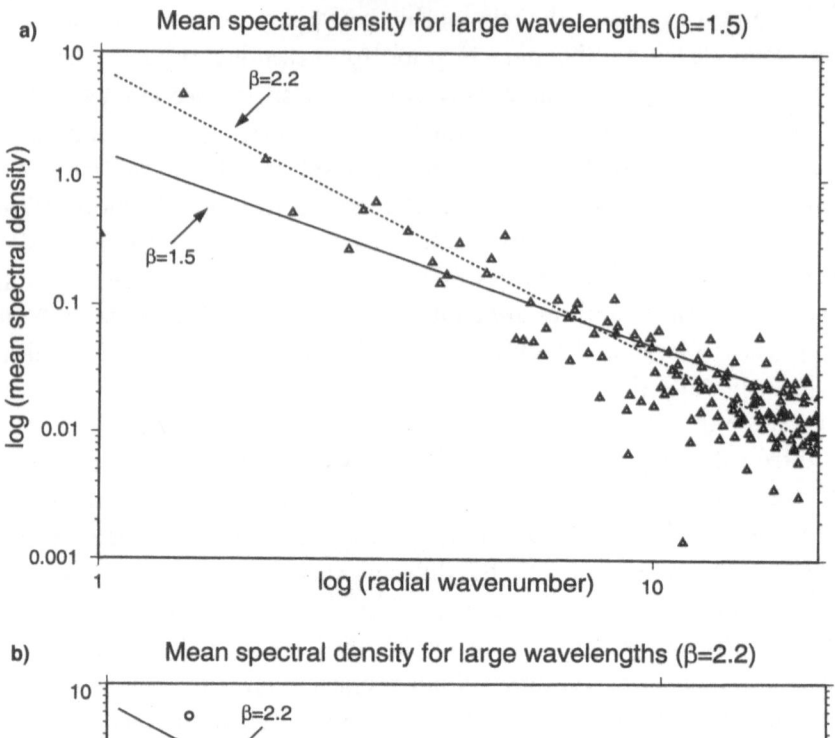

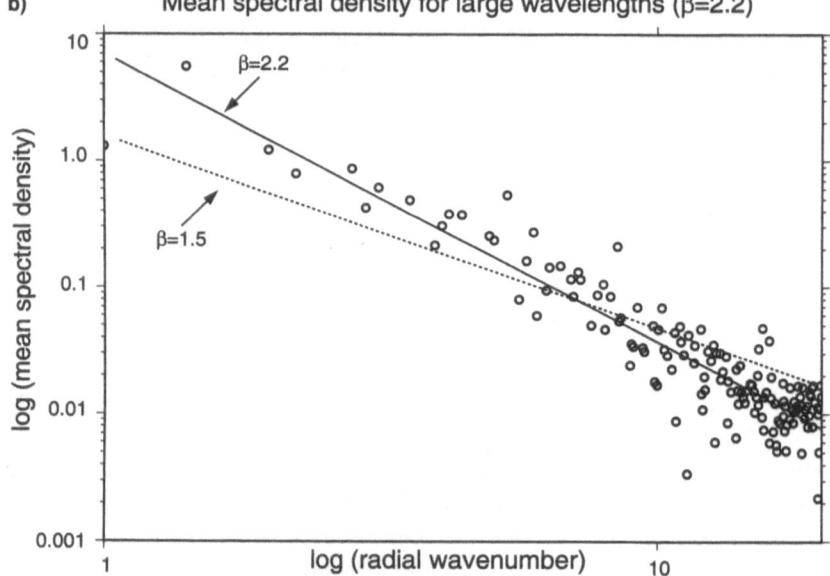

Figure 6

Log-log plot of the mean spectral density at large wavenumbers for two values of the exponent a) $\beta = 1.50$ and b) $\beta = 2.20$ for the crack model. As the exponent decreases, the mean spectral density at large wavelengths decreases indicating that larger regions are close-to-failure. The straight lines in each figure are the weighted least-squares fit to each dataset.

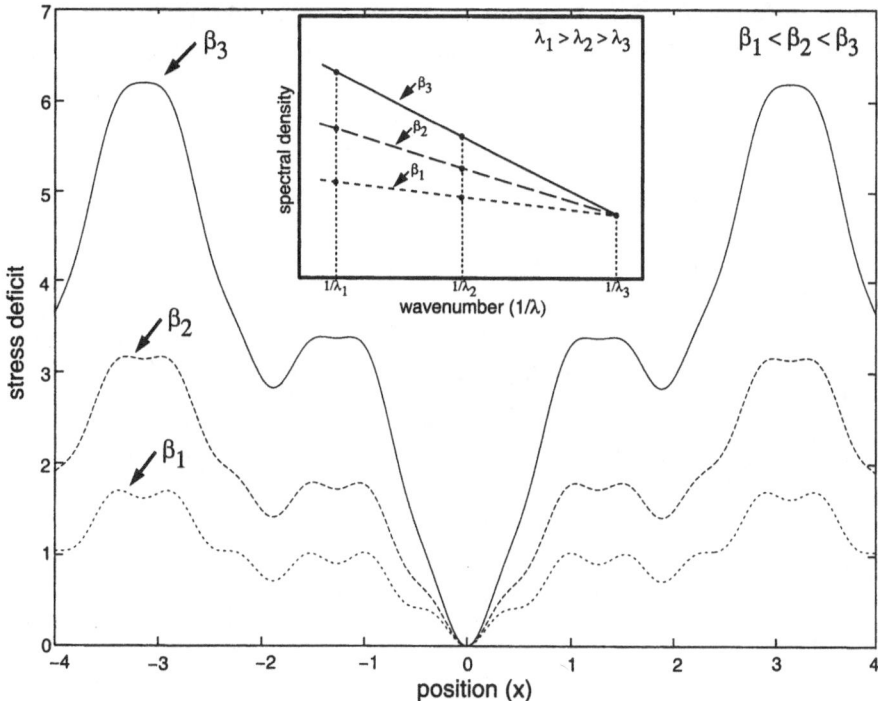

Figure 7

A conceptual illustration of the stress deficit near a failed cell (at $x = 0$) for three different values of the mean spectral density exponent $\beta_1 < \beta_2 < \beta_3$. As the exponent decreases, ruptures are more likely to propagate to large scales because barriers of high stress deficit decrease in size.

change in the exponent due to small events. Large events fail a considerable portion of the model increasing the stress deficit of a broad region. This will considerably increase the mean spectral density at large wavelengths, causing a sudden rise in the value of the exponent.

Evolution of Stress Deficit and Model Dynamics

In Figures 8 and 9 we examine the time evolution of the exponent β in the crack and PSD models. Overlaying each graph is a plot of cumulative energy release for the same interval. This provides a measure of the model dynamics associated with the evolution of stress deficit. We have examined the evolution of the exponent for three time intervals of differing length in each model. This illustrates the evolution of stress deficit at different timescales.

Figure 8a is the evolution of the exponent for a time interval consisting of approximately 30,000 events in the crack model. Cumulative energy release during this interval is linear, i.e., there is a constant rate of energy release. The exponent shows large variations about a mean of $\beta_m = 1.87 \pm 0.115$, remaining less than the maximum value of $\beta = 2.4$. A similar time interval from the PSD model is shown in Figure 9b. The exponent varies about a mean of $\beta_m \sim 0.667 \pm 0.0364$. The

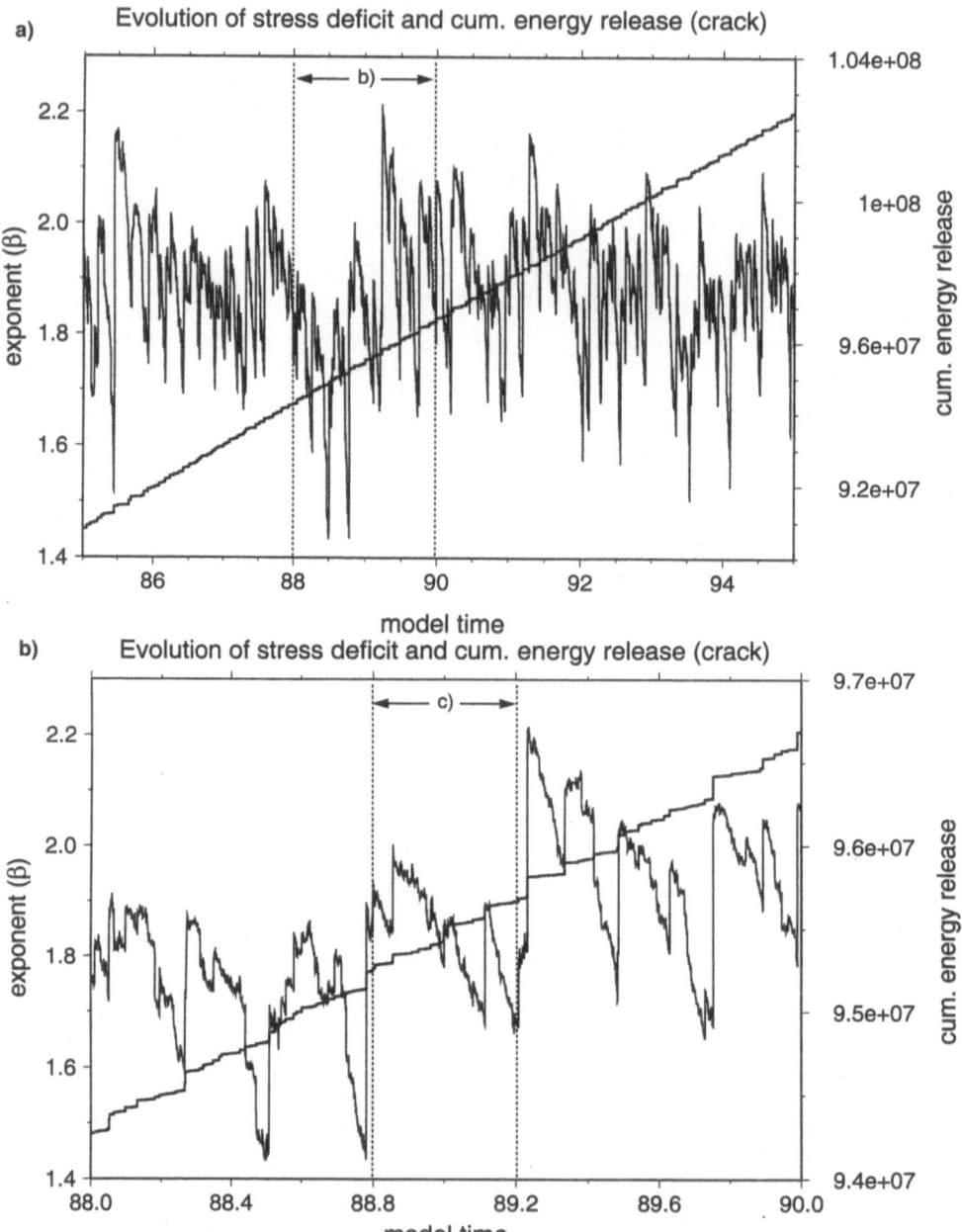

a) Evolution of stress deficit and cum. energy release (crack)

b) Evolution of stress deficit and cum. energy release (crack)

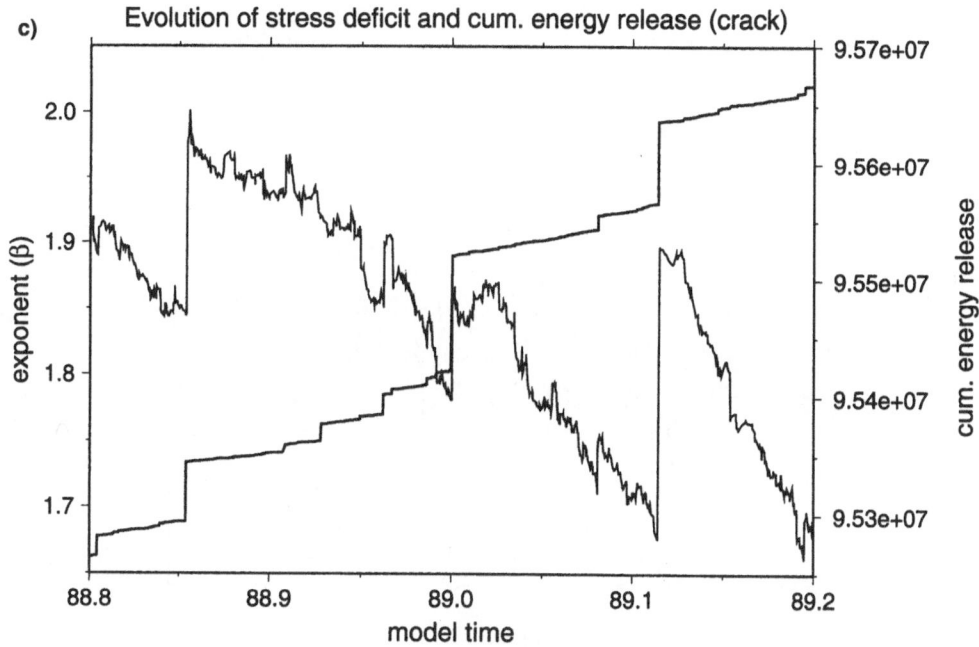

Figure 8
Graphs of the evolution of the exponent β as a function of model time for the crack model. Each graph has been overlain with a plot of the cumulative energy release for the same interval. a) An interval consisting of approximately 30,000 events. Long-period variations in the exponent are evident. b). A subset of 5000 events. Large events are preceded by periods in which the exponent decreases on average culminating in a sudden increase after the large event. c) A few cycles of energy release consisting of 1000 events. Small and moderate events occur during periods of decreasing exponent.

exponent is smaller in general in the PSD model than the crack model, indicating that the PSD model remains much closer-to-failure. A subset from these long intervals is examined in Figure 8b and Figure 9b.

During a shorter interval, cumulative energy release is linear on average, punctuated by sharp jumps in energy release associated with the largest events in the model. In the crack model, there is a clear correlation between the evolution of the exponent and the occurrence of large events. Prior to large events, the exponent decreases on average due to tectonic loading. As expected, large events correspond with sudden increases in the value of the exponent. Subsequent to a large event the exponent continues to decrease on average until another large event occurs. Such cyclic behaviour is typical of the crack model. Exponent fluctuations are more variable in the PSD model. Both periods in which the exponent increases and periods of decreasing exponent are evident.

In Figure 8c we examine a few cycles from the crack model consisting of approximately 1000 events. Prior to each large event, the exponent decreases on average, rising sharply once the event has occurred. During periods of decreasing

a)

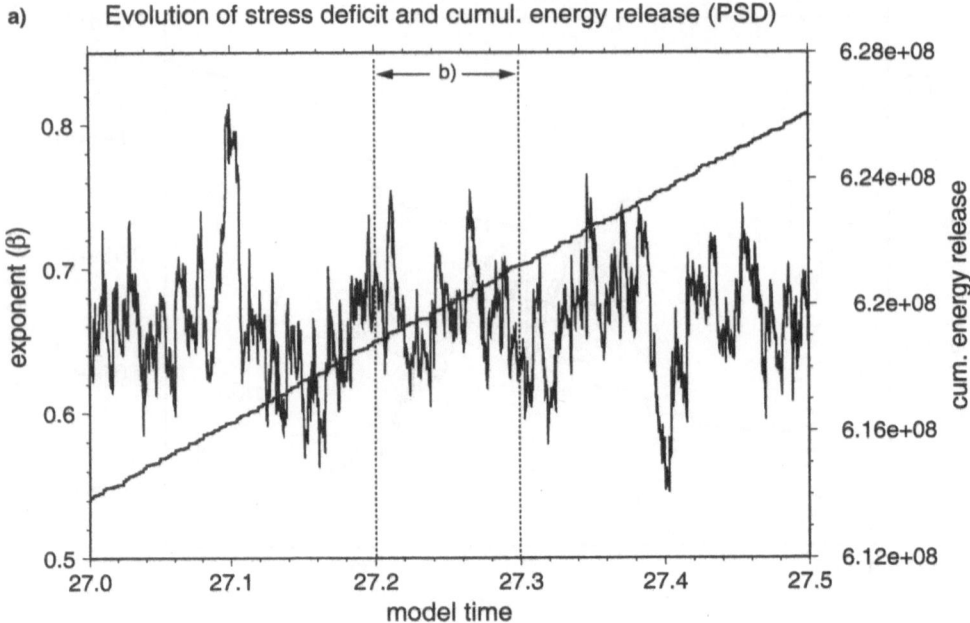

b)

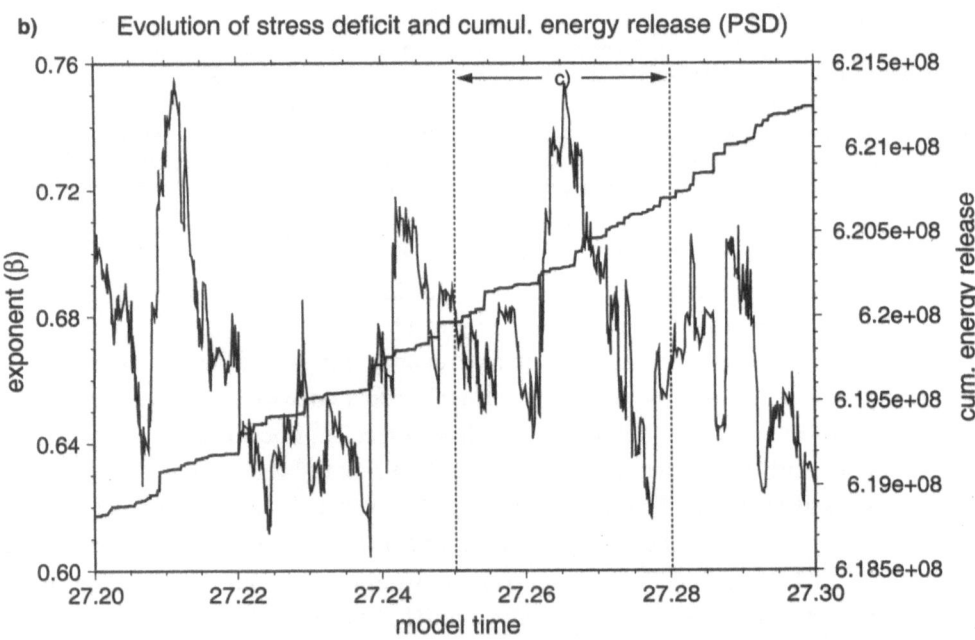

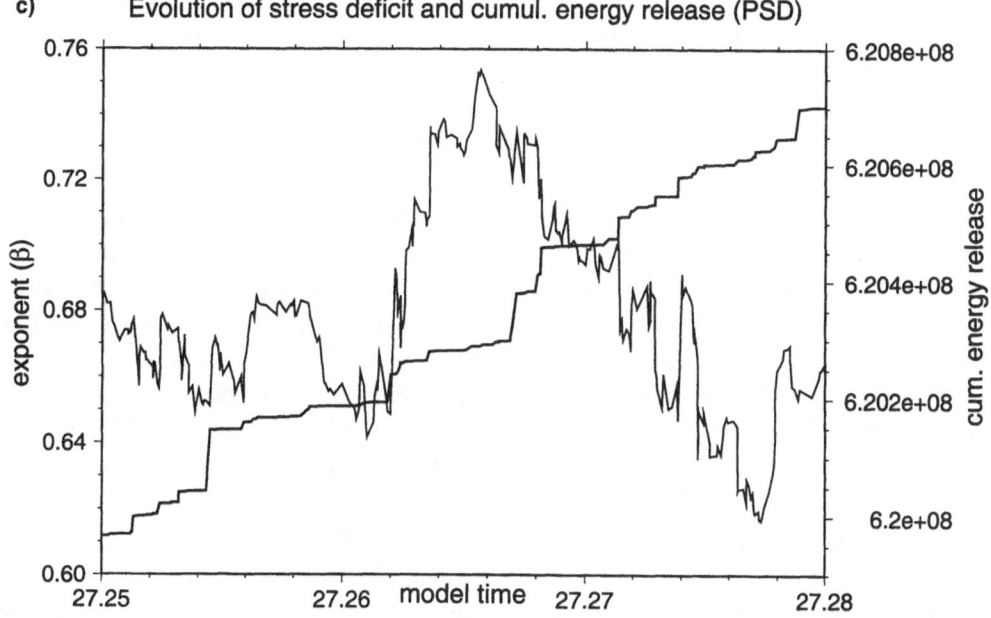

Figure 9

Graphs of the evolution of stress deficit in the PSD model. a) The exponent varies randomly about a means of $\beta_m \sim 0.667$ over long time intervals. b) Shorter time intervals reveal long periods in which the exponent decreases and short periods where the exponent rises sharply. c) A short interval consisting of 300 events illustrating the behaviour typical of the PSD model. The dynamics are consistent with small fluctuations about a mean-state of high stress.

exponent, small to moderate-sized events may occur. Some of these may occur within the rupture zone of the subsequent mainshock. The occurrence of these events ensures that stress deficit is correlated over the entire region which fails during the large event. Notice that the largest event (at $t \sim 89.0$) produces a smaller change in the exponent than the smaller event at $t \sim 89.12$. This is because the largest events do not necessarily rupture the largest area. An event in a large, weak region may have a smaller size than an event in a smaller, stronger region.

Figure 9c illustrates exponent fluctuations during short periods in the PSD model. Commencing at a model time $t \sim 27.26$, the exponent rises sharply from a value of $\beta \sim 0.66$ to a maximum value of $b \sim 0.76$. During this period several moderate to large events occur; the rate of energy release is high. Subsequent to this sudden increase, the exponent decreases on average to a minimum of $\beta \sim 0.62$. Energy release during this interval occurs at a lower rate than the previous period. This type of behaviour is typical of the PSD model. We find that periods of increased activity in larger events are initiated when the exponent drops below the mean value ($\beta < \beta_m$). Conversely, periods with a decreased rate of energy release commence once the exponent rises above the mean ($\beta > \beta_m$). Evolution of the stress deficit is consistent with dynamical fluctuations which maintain the system near a critical state close-to-failure. Such behaviour is expected of self-organised critical systems (MAIN, 1996).

Discussion and Conclusions

The crack model progressively approaches and retreats from a critical state in which large events occur. During the approach to criticality, the rate of events of all sizes increases and the stress deficit of small adjacent regions correlates. This provides the conditions necessary for a large event in the crack model. Large events occur almost at random in the PSD model. Fluctuations in stress deficit in the model are consistent with small fluctuations about a mean-state of high stress. Event sizes follow a power-law distribution. These features identify the PSD model as a self-organised critical system; a system which remains perpetually close-to-failure in large events.

The differing dynamics of the two models in this paper arise solely from differing nearest neighbour laws. The nearest neighbour law determines the fraction of stress which remains in the rupture zone after an event (and the fraction of stress which is transferred to the surroundings). The crack model transfers all stress to the surroundings leaving behind a rupture zone devoid of stress; stress drops in the crack model are large. The PSD model maintains a high stress level within the rupture zone; stress drops in the PSD model are small. Since stress is only transferred to nearest neighbours, interactions during rupture are short-range in both models. Stress transfer during rupture does not affect the stress level of regions spatially separated from the rupture zone.

Earthquakes have a relatively constant and relatively small stress drop over a broad a range of scales. Stress drops are typically of order 3 MPa compared with tectonic stresses in the crust of order 10–100 MPa (MAIN, 1996; ABERCROMBIE and LEARY, 1993; ABERCROMBIE, 1995). Such observations tend to support the PSD model over the crack model as an analogue for seismogenesis. However, power-law acceleration in energy release prior to large events in the crack model (JAUMÉ *et al.*, 1999) are similar to accelerations in Benioff strain release prior to a few cases of natural earthquakes (JAUMÉ, 1999).

Coulomb Failure analysis using models of dislocations in an elastic halfspace have suggested that static stress changes during a given earthquake may either promote or inhibit the occurrence of earthquakes on nearby faults (KING *et al.*, 1994; JAUMÉ and SYKES, 1996). Dynamic stress changes during one earthquake may trigger other events via fault zone interactions (BELARDINELLI *et al.*, 1999). These results suggest that longer-range stress interactions may be important factors determining rupture nucleation. Since interactions are explicitly short-range in cellular automata, such longer range effects cannot be simulated in these simplified models.

It is doubtful that the dynamics of either of these models is a good representation of natural seismicity. However many of the features of these models resemble processes thought to occur within the earth's crust. In this paper we have shown that power-law accelerations in energy release and cyclic occurrence of large events may arise from a model which is driven towards instability primarily by tectonic loading. Random correlation in the stress level of adjacent regions is responsible for preparing the system for a large event. Stress transfer during rupture has very little affect upon the occurrence of large events. A model with relatively small stress drops during events remains perpetually close to failure and there are no systematic patterns of energy release prior to large events.

In reality, earthquake dynamics may lie somewhere between the dynamics of the two automaton models. Stress drops may be larger than those of the PSD model but insufficiently large to drain a region of all its stress. If stress drops are larger than the PSD model, one might expect that cyclic behaviour similar to that of the crack model might begin to play an important role in the dynamics. Also, if stress within rupture zones is not reduced to zero but remains above a certain minimum stress level, then the rate of small earthquakes may be unaffected by larger ruptures. Changing occurrence rates may then be confined only to moderate and large events, in agreement with the majority of cases of accelerating Benioff strain release identified in the literature (JAUMÉ and SYKES, 1999; BUFE and VARNES, 1993; BREHM and BRAILE, 1998).

Stress transfer during ruptures may be an important factor in models of earthquakes with longer-range interactions. Physically motivated models of earthquakes such as Burridge-Knopoff and finite element models display accelerating sequences of events which may be related to stress field correlations (WINTER, 1999). Mora *et al.* (1999) report evidence for accelerating cumulative Benioff strain release prior to large events in a numerical model of earthquakes which simulates the elasto-dynamics of a system representing a simplified fracture zone (PLACE and MORA, 1999). This model is a good candidate for future studies of long-range stress transfer and its influence upon the occurrence of large events.

Much may be learned about the emergence of dynamical complexity in spatially extended, driven dynamical systems by studying the dynamics of simple cellular automaton models. Two models examined in this paper display quite different dynamics which arise solely from differing interaction laws. These models represent only two members of a class of automata whose rupture dynamics result in a variety of dynamical behaviour. Further study of simplified rupture models and the types of behaviour they display may be greatly beneficial in the pursuit of a comprehensive theory for seismogenesis.

Acknowledgements

This work was supported by the University of Queensland and the sponsors of QUAKES. We wish to thank S. Steacy and J. McCloskey for sharing their automaton code. The authors wish to thank colleagues at QUAKES for insightful discussion and advise in the preparation of this manuscript.

REFERENCES

ABERCROMBIE, R. (1995), *Earthquake Source Scaling Relationships from (1 to 5 ML using Seismograms Recorded at 2.5 km Depth*, J. Geophys. Res. *100*, 24,015–24,036.

ABERCROMBIE, R., and LEARY, P. (1993), *Source Parameters of Small Earthquakes Recorded at 2.5 km Depth, Cajon Pass, California; Implications for Earthquake Scaling*, Geophys. Res. Lett. *20*, 1511–1514.

BÁK, P., TANG, C., and WIESENFELD, K. (1987), *Self-organised Criticality: An Explanation of 1/f Noise*, Phys. Review Lett. *59* (4), 381–384.

BÁK, P., TANG, C., and WIESENFELD, K. (1988), *Self-organised Criticality*, Physical Rev. *A38* (1), 364–374.

BÁK, P., and TANG, C. (1989), *Earthquakes as a Self-organised Critical Phenomenon*, J. Geophys. Res. *94* (B11), 15,635–15,637.

BELARDINELLI, M. E., COCCO, M., COUTANT, O., and COTTON, F. (1999), *Redistribution of Dynamic Stress During Coseismic Ruptures: Evidence for Fault Interaction and Earthquake Triggering*, J. Geophys. Res. *104* (B), 14,925–14,945.

BREHM, D. J., and BRAILE, L. W. (1998), *Intermediate-term Earthquake Prediction Using Precursory Events in the New Madrid Seismic Zone*, Bull. Seismol. Soc. Am. *88* (2), 564–580.

BROWN, S. R., and SCHOLZ, C. H. (1985), *Broad Bandwidth Study of the Topography of Natural Rock Surface*, J. Geophys. Res. *90* (B14), 12,575–12,582.

BUFE, C. G., and VARNES, D. J. (1993), *Predictive Modelling of the Seismic Cycle of the Greater San Francisco Bay Region*, J. Geophys. Res. *98* (B6), 9871–9883.

HUANG, J., and TURCOTTE, D. L. (1988), *Fractal Distributions of Stress and Strength and Variations of b-valve*, Earth and Planet. Sci. Lett. *91*, 223–230.

JAUMÉ, S. C. (1999), *Changes in earthquake size-frequency distributions underlying accelerating seismic moment/energy release*. In *The Physics of Earthquakes* (eds. Rundle, J. B., Turcotte, D. L., and Klein, W.) (Am. Geophys. Union, Washington), in press.

JAUMÉ, S. C., and SYKES, L. R. (1996), *Evolution of Moderate Seismicity in the San Francisco Bay Region, 1850 to 1993: Seismicity Changes Related to the Occurrence of Large and Great Earthquakes*, J. Geophys. Res. *101* (B), 765–789.

JAUMÉ, S. C., and SYKES, L. R. (1999), *Evolving Towards a Critical Points: A Review of Accelerating Seismic Moment/Energy Release Prior to Large and Great Earthquakes*, Pure appl. geophys. *155*, 278–305.

JAUMÉ, S. C., WEATHERLEY, D., and MORA, P. (1999), *Accelerating Seismic Energy Release and Evolution of Event Time and Size Statistics: Comparison of Models and Observations*, Pure appl. geophys. *157*, 2209–2226.

KING, G. C. P., STEIN, R. S., and LIN, J. (1994), *Static Stress Changes and the Triggering of Earthquakes*, Bull. Seismol. Am. *84* (3), 935–953.

LOMNITZ-ADLER, J. (1993), *Automaton Models of Seismic Fracture: Constraints Imposed by the Magnitude-frequency Relation*, J. Geophys. Res. *98* (B10), 17,745–17,756.

MAIN, I. (1996), *Statistical Physics, Seismogenesis, and Seismic Hazard*, Rev. Geophys. *34* (4), 433–462.

PLACE, D., and MORA, P. (1999), *The Lattice Solid Model to Simulate the Physics of Rocks and Earthquakes: Incorporation of Friction*, J. Comp. *150*, 332–372.

RUNDLE, J., PRESTON, E., MCGINNIS, S., and KLEIN, W. (1998), *Why Earthquakes Stop: Growth and Arrest in Stochastic Fields*, Phys. Rev. Lett. *80* (25), 5698–5701.

SALEUR, H., SAMMIS, G. G., and SORNETTE, D. (1996), *Discrete Scale Invariance, Complex Fractal Dimensions, and Log-periodic Fluctuations in Seismicity*, J. Geophys. Res. *101* (B8), 17,661–17,677.

SAMMIS, C. G., and SMITH, S. W. (1998), *Seismic Cycles and the Evolution of Stress Correlation in Cellular Automaton Models of Finite Fault Networks*, Pure appl. geophys. *155*, 307–334.

SORNETTE, D., and SAMMIS, C. G. (1995), *Complex Critical Exponents from Renormalisation Group Theory of Earthquakes: Implications for Earthquake Predictions*, J. Phys. I France *5*, 607–619.

STEACY, S. J., MCCLOSKEY, J., BEAN, C. J., and REN, J. (1996), *Heterogeneity in a Self-organised Critical Earthquake Model*, Geophys. Res. Lett. *23* (4), 383–386.

STEACY, S. J., and MCCLOSKEY, J. (1998), *What Controls an Earthquake's size? Results from a Heterogeneous Cellular Automaton*, Geophys. J. Int. *133*, F11–F14.

STEACY, S. J., and MCCLOSKEY, J. (1999), *Heterogeneity and the Earthquake Magnitude-frequency Distribution*, Geophys. Res. Lett. *26* (7), 899–902.

TURCOTTE, D. L., *Fractals in Geology and Geophysics* (Cambridge Univ. Press, New York, 2nd ed. 1997).

VARNES, D. J. (1989), *Predicting Earthquakes by Analysing Accelerating Precursory Seismic Activity*, Pure appl. geophys. *130*, 661–686.

WINTER, M. E. (1999), *The Plausibility of Long-period Stress Correlation or Stress Magnitude as a Mechanism for Precursory Seismicity: Results from Two Simple Elastic Models*, Pure appl. geophys. *157*, 2227–2248.

(Received October 3, 1999, revised April 5, 2000, accepted April 7, 2000)

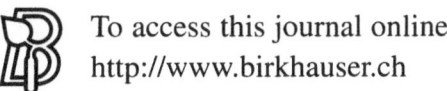

To access this journal online:
http://www.birkhauser.ch

Pure appl. geophys. 157 (2000) 2209–2226
0033–4553/00/122209–18 $ 1.50 + 0.20/0

© Birkhäuser Verlag, Basel, 2000

⎮Pure and Applied Geophysics

Accelerating Seismic Energy Release and Evolution of Event Time and Size Statistics: Results from Two Heterogeneous Cellular Automaton Models

STEVEN C. JAUMÉ,[1] DION WEATHERLEY[1] and PETER MORA[1]

Abstract—The evolution of event time and size statistics in two heterogeneous cellular automaton models of earthquake behavior are studied and compared to the evolution of these quantities during observed periods of accelerating seismic energy release prior to large earthquakes. The two automata have different nearest neighbor laws, one of which produces self-organized critical (SOC) behavior (PSD model) and the other which produces quasi-periodic large events (crack model). In the PSD model periods of accelerating energy release before large events are rare. In the crack model, many large events are preceded by periods of accelerating energy release. When compared to randomized event catalogs, accelerating energy release before large events occurs more often than random in the crack model but less often than random in the PSD model; it is easier to tell the crack and PSD model results apart from each other than to tell either model apart from a random catalog. The evolution of event sizes during the accelerating energy release sequences in all models is compared to that of observed sequences. The accelerating energy release sequences in the crack model consist of an increase in the rate of events of all sizes, consistent with observations from a small number of natural cases, however inconsistent with a larger number of cases in which there is an increase in the rate of only moderate-sized events. On average, no increase in the rate of events of any size is seen before large events in the PSD model.

Key words: Accelerating seismic energy, heterogeneous cellular automaton, self-organized criticality, critical point hypothesis.

Introduction

There is at present considerable controversy in seismology as to whether the earthquake process is best described as being in a state of continuous self-organized criticality (BAK and TANG, 1989; ITO and MATSUZAKI, 1990; GELLER *et al.*, 1997) or whether regions of the earth's seismogenic crust cyclicly approach and retreat from a state of criticality (SAMMIS and SMITH, 1999; SYKES *et al.*, 1999). The answer to this question has important implications for a number of areas of seismology, including the practice of earthquake hazard assessment (MAIN, 1996)

---

[1] Queensland University Advanced Centre for Earthquake Studies, The University of Queensland, Brisbane 4072, Queensland, Australia. E-mail: jaume@earthsciences.uq.edu.au; weatherley@earthsciences.uq.edu.au; mora@earthsciences.uq.edu.au

and the feasibility of earthquake prediction (GELLER *et al.*, 1997; SYKES *et al.*, 1999).

The observed power-law scaling of earthquake size-frequency statistics and other statistical features of seismicity are often used as evidence to support the self-organized criticality (SOC) hypothesis (BAK and TANG, 1989; KAGAN, 1997). Observations of changes in the rate of occurrence of moderate earthquakes and acceleration in the rate of seismic energy release leading up to large earthquakes are often used as evidence for an approach and retreat from criticality (critical point hypothesis; SORNETTE and SAMMIS, 1995). Some authors even argue that these two concepts can co-exist, with their applicability depending upon the time and space scales observed (HUANG *et al.*, 1998; SYKES *et al.*, 1999).

In this paper we examine synthetic earthquake catalogs output by two heterogeneous cellular automaton models with different nearest-neighbor energy redistribution laws, originally studied by STEACY and MCCLOSKEY (1998, 1999). We choose these two models because the different nearest-neighbor laws lead to very different dynamics; i.e., one model shows classic SOC behavior (STEACY and MCCLOSKEY, 1999) but the other has quasi-periodic large events (STEACY and MCCLOSKEY, 1998). In this paper we focus on the output of the two different models; i.e., the time history of earthquake events. The internal dynamics of these two models are studied in a companion paper (WEATHERLEY *et al.*, this volume).

The observation most often cited as support for the critical point hypothesis is the presence of accelerating seismic moment/energy release (ASR) before the occurrence of many large earthquakes (BOWMAN *et al.*, 1998; SORNETTE and SAMMIS, 1995). In this work we examine the output of the two models in this context, to determine to what degree ASR characterizes seismicity prior to large model events. We also examine "randomized" versions of the output of these two models, to gauge to what degree the patterns shown by the two models can occur by chance.

In the body of the paper we first describe the two heterogeneous cellular automaton models, followed by a description of the algorithm used to analyze the model output. We then compare/contrast the two models (and their randomized versions) as a function of how well they fit the ASR model. Finally, we compare these results to relevant observations of natural seismicity.

Models and Analysis

Heterogeneous Cellular Automaton Models

The details of the models from which the synthetic earthquake catalogs are derived are described in WEATHERLEY *et al.* (this volume); therefore here we will only outline the two models. Both models consist of a grid of 64 by 64 cells with the strength of each cell defined by a two-dimensional fractal distribution, with a fractal dimension $D = 2.3$ and cell strengths in the range (0.01, 1.00). The stress in the model is uniformly incremented until the stress in at least one cell equals its strength; that cell then fails and its stress is redistributed to up to 8 nearest neighbor cells (see details of stress redistribution laws below). If the stress on the neighboring cells equals or exceeds their strength, these cells also fail and their stress is subsequently redistributed. This cascade continues until no more cells fail as a result of stress redistribution. The entire model is then uniformly loaded to the level where the next cell failure occurs and the stress redistribution cycle is repeated. The size of any individual event is defined as the sum of the strengths of all cells that fail during a single loading step.

The major difference between these two models lies in their stress redistribution laws and degree of energy conservative. The first model uses the stress redistribution law of STEACY and MCCLOSKEY (1998). Let σ_f represent the stress on a failed cell. The stress redistributed to the nearest neighbors is given by:

$$\Delta\sigma_u = \frac{\gamma\sigma_f}{(N - n_b)},\tag{1}$$

$$\Delta\sigma_b = 0.0,$$

where n_b is the number of broken cells, and $\sigma_u(\sigma_b)$ is the stress redistributed to unbroken (broken) cells. Thus in this model stress is only redistributed to unbroken cells and stress in the rupture zone drops to zero, leading to crack-like behavior (hereafter called the crack model). An event in this model concentrates all the stress from the rupture zone at the rupture front, leaving behind a region with zero stress. Only 75% of the stress of a failed cell is redistributed in this model ($\gamma = 0.75$), in order to prevent ruptures from failing the entire model and resetting the model to the starting condition. Note that this makes the crack model a non-conservative cellular automaton.

The second model uses the stress redistribution law of STEACY and MC-CLOSKEY (1999). The stress redistributed to the 8 nearest neighbors in this model is given by:

$$\Delta\sigma_u = \frac{\gamma\sigma_f}{(N + 1) - \sqrt{n_b + 1}},\tag{2}$$

$$\Delta\sigma_b = \frac{\gamma\sigma_f - \sum_{i=1}^{N - n_b}\Delta\sigma_u}{n_b}, \quad n_b \neq 0.$$

This model preferentially redistributes stress to unbroken cells, but still distributes a substantial amount to previously broken cells, resulting in a partial stress drop (PSD) in the rupture zone. An event in the PSD model concentrates stress at the rupture front, but to a lesser degree than in the crack model. More importantly, an event in the PSD model leaves behind a relatively high level of stress in the rupture zone. The PSD model features internal energy conservation (i.e., $\gamma = 1.0$). In both models energy is lost if stress is transferred across the model boundaries.

As described in detail by WEATHERLEY *et al.* (this volume), the different stress redistribution laws in these two models lead to different event-size distributions and very different internal dynamics. The crack model produces a characteristic-like event size distribution while the PSD model produces a power-law event size distribution. More importantly, a large event in the crack model significantly **perturbs the stress state of the system whereas large events do not** change the stress state in the PSD model. This suggests that the crack model is more likely to produce critical-point-like behavior than the PSD model.

Model Data Analysis

Each model described above was run for 500,000 model stress increments (i.e., a total of 500,000 events). The first 20,000 events were deleted from the catalog before analysis, to remove any effect of the initial loading period, and the remaining 480,000 model earthquakes were subject to analysis.

In addition to the event catalogs produced by the two cellular automata models, we also created two new catalogs with random (Poisson) event times. This was done by randomly reassigning new times to each event in the catalogs from the cellular automaton models, mapped onto the same model time interval. This produces two new catalogs with the same event-size statistics as those they were derived from but with any temporal relationships between events destroyed. The mean interevent times of the **two randomized catalogs are very similar (to within 1%) to the interevent times of the** catalogs from which they were derived.

Since the critical point hypothesis described in the Introduction pertains to variations in seismicity preceding the largest events within a region, we decided to focus our analysis on the largest 0.1% (i.e., largest 480 events) of our synthetic earthquake catalogs. As an initial step in our analysis, we examined the interevent time statistics for the largest 480 events from the cellular automaton models, comparing them to the interevent statistics of the largest 480 events in the randomized catalogs (Fig. 1). The large events in the crack model tend to occur quasi-periodically (Fig. 1a) but large events in the PSD model (Fig. 1b) are more clustered in time, similar to large events in the randomized catalogs.

We then took the four catalogs and ran them through an algorithm that searches for "best-fitting" ASR sequences preceding large events. This algorithm takes the cumulative energy release preceding a large event and fits it to a power-law time-to-failure relationship of the form (BUFE and VARNES, 1993; VARNES, 1989):

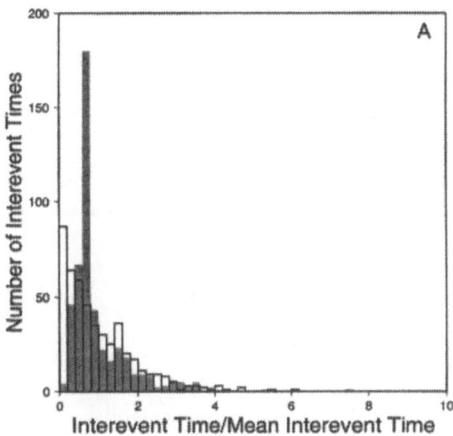 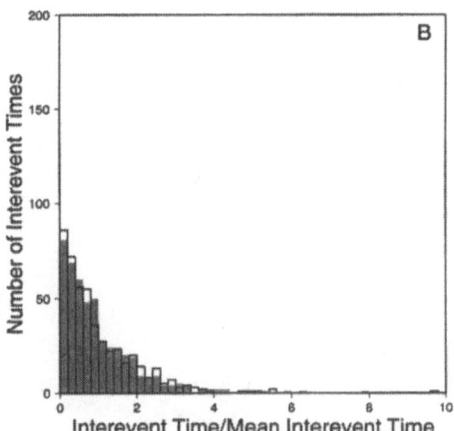

Figure 1

Histograms of interevent times for the largest 480 events from the two models (gray shade) and their randomized counterparts (open boxes), normalized by the mean interevent time: A) the crack model, and B) the PSD model.

$$\Sigma\Omega(t) = A + B(t_f - t)^m, \tag{3}$$

where Ω is a measure of seismic energy release to be summed over time, t_f is the time of the mainshock event, t the time at which Ω is calculated, m is an exponent that defines the curvature of the power-law acceleration and A and B are constants to be determined. The quality of the fit to Equation (3) is determined in the same manner as BOWMAN et al. (1998), who defined a curvature parameter $\mathscr{C}$:

$$\mathscr{C} = \frac{\text{power law fit RMS error}}{\text{linear fit RMS error}}. \tag{4}$$

The A, B, t_f and m in Equation (3) that produce a minimum value of $\mathscr{C}$ is defined as the best fit.

For the purposes of our study, it was necessary to constrain a number of parameters in Equation (3). First, we fixed t_f as the time of the mainshock event. Also, we restricted our search for power-law accelerations to time periods before the largest 480 events in each catalog, as described above. We limited the time periods over which we searched for fits to Equation (3) to be between 0.1 and 10.0 times the mean interevent time for the largest 480 events. In the case where one of the 480 large events is preceded by an even larger event within the minimum time interval, that event was rejected from further analysis. We restricted the allowed range of the exponent m in Equation (3) to be between 0.0 and 0.8. As noted by BOWMAN et al. (1998), Equation (3) is nearly indistinguishable from a straight line for $0.8 < m < 1.0$; restricting m requires that the fit has a minimum degree of curvature. In the results reported here we calculate $\Sigma\Omega(t)$ from the square root of

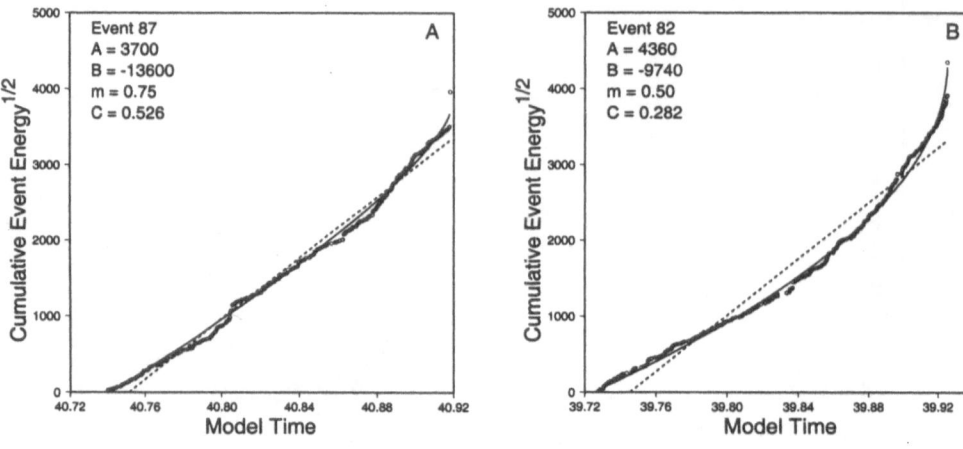

Figure 2

Power-law (solid line) and linear (dashed line) fits to cumulative square-root energy release prior to two large events from the crack model. A) Fits to the largest event and B) best-fitting event among the ten largest events. The parameters from Equation (3) and the $\mathscr{C}$ value are also shown.

the event energies (i.e., Benioff strain). Finally, in calculating $\Sigma\Omega(t)$ we only used events with energies within 3 or 4 orders of magnitude of the mainshock energy, in accordance with observational results suggesting that only earthquakes with magnitudes within 2.0 to 3.0 units of the mainshock magnitude (i.e., 3 to 4.5 orders of magnitude in energy/moment release) participate in accelerating energy release sequences (BREHM and BRAILE, 1999a; JAUMÉ and SYKES, 1999). The output of the algorithm includes the values of A, B, and m for the best fit to Equation (3) for each analyzed sequence, the value of $\mathscr{C}$ for that fit, and the time period (in model time) over which the best-fitting sequence occurs.

Examples of the types of fits achieved are shown in Figures 2 to 5. For each of the four catalogs analyzed, the fit to Equation (3) for both the largest event in each catalog plus the fit to the event with the smallest $\mathscr{C}$ from among the largest 10 events is shown, together with the corresponding linear fit. Periods of ASR occur before all of the largest events in the crack model (Fig. 2); in some cases there is substantial curvature in the cumulative Benioff strain (Fig. 2B). Good fits to Equation (3) can also often be found before large events in the randomized catalogs (Figs. 3 and 5), although not as often as for the crack model (see next section). Equation (3) poorly describes the cumulative Benioff strain before large events in the PSD model; the best fits tend to be for very short sequences (Fig. 4).

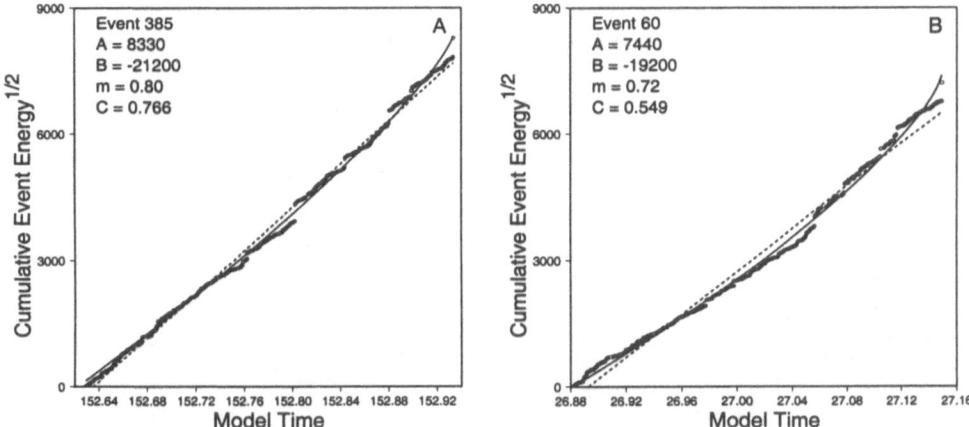

Figure 3

Power-law (solid line) and linear (dashed line) fits to cumulative square-root energy release prior to two large events from the randomized crack model catalog. A) Fits to the largest event and B) best-fitting event among the ten largest events. The parameters from Equation (3) and the $\mathscr{C}$ value are also shown.

Results

The main result of this work is an estimate of how often seismicity preceding large events produced by the two cellular automaton models fits predictions of the critical point hypothesis; i.e., how many large events are preceded by ASR. Figure 6 displays histograms of $\mathscr{C}$-values from the analysis of the four catalogs. For the

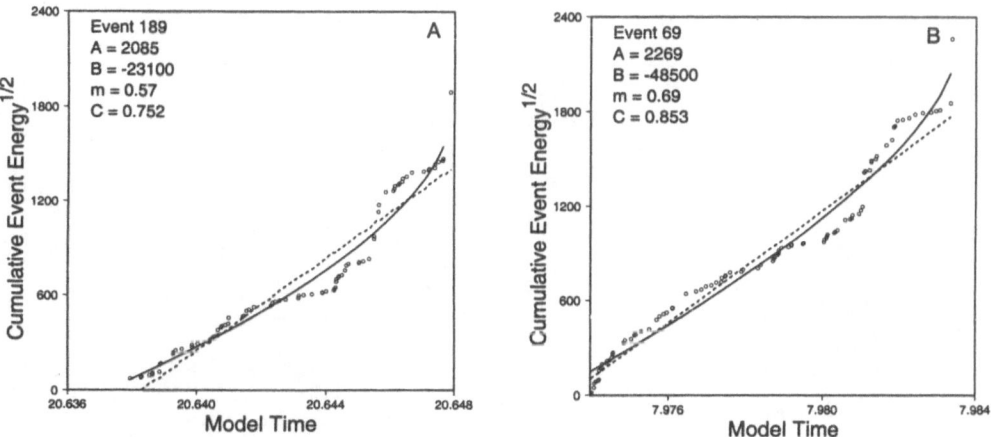

Figure 4

Power-law (solid line) and linear (dashed line) fits to cumulative square-root energy release prior to the two largest events from the PSD model catalog. A) Fits to the largest event and B) second largest event among the ten largest events. Note that the largest event also has the lowest $\mathscr{C}$ value. The parameters from Equation (3) and the $\mathscr{C}$ value are also shown.

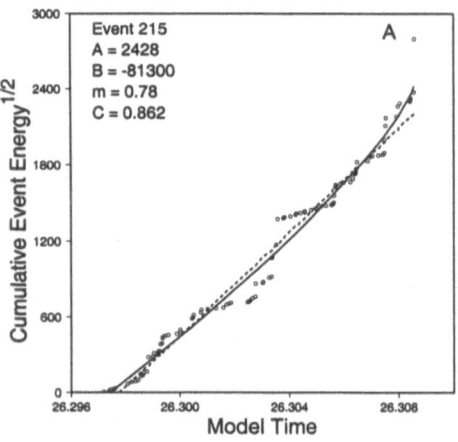
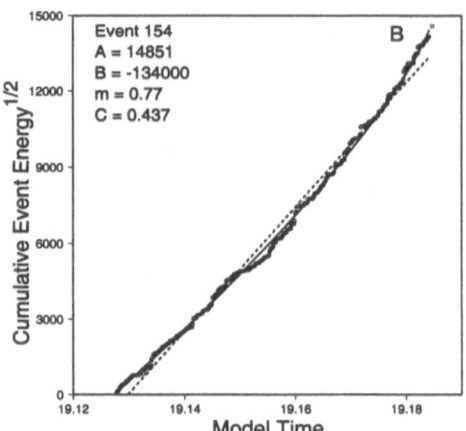

Figure 5

Power-law (solid line) and linear (dashed line) fits to cumulative square-root energy release prior to two large events from the randomized PSD model catalog. A) Fits to the largest event and B) best-fitting event among the ten largest events. The parameters from Equation (3) and the $\mathscr{C}$ value are also shown.

crack model (Fig. 6A), we find that most of the 480 large events are preceded by a time period in which the energy release is better described by Equation (3) than by a linear energy release model. This is not true for the PSD model (Fig. 6B), in which only about half the large events are better fit by Equation (3). Surprisingly, we find that while the randomized catalogs produce fewer good fits than the crack model, they actually produce more good fits than the PSD model. This suggests

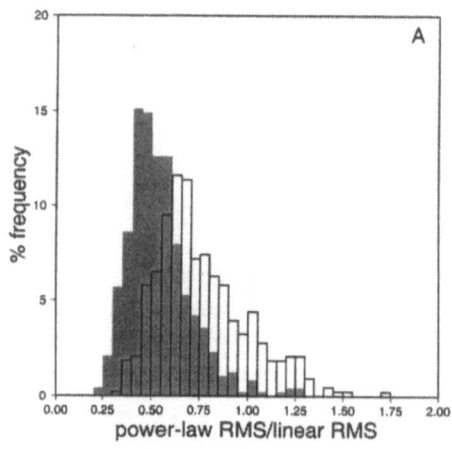
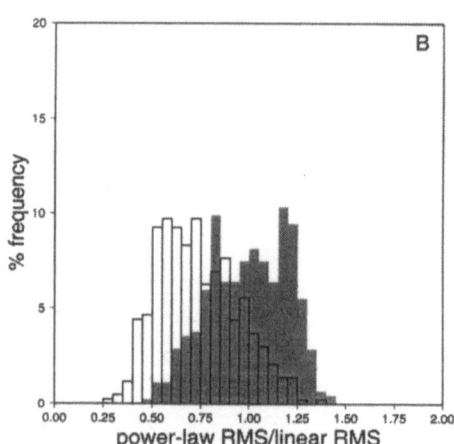

Figure 6

Frequency of fit quality (as measured by $\mathscr{C}$) to Equation (3) for cumulative Benioff strain before large events in the model catalogs. A) The crack model (grey-shaded histogram) and its randomized equivalent (open histogram), and B) the PSD model (grey-shaded histogram) and its randomized equivalent (open histogram).

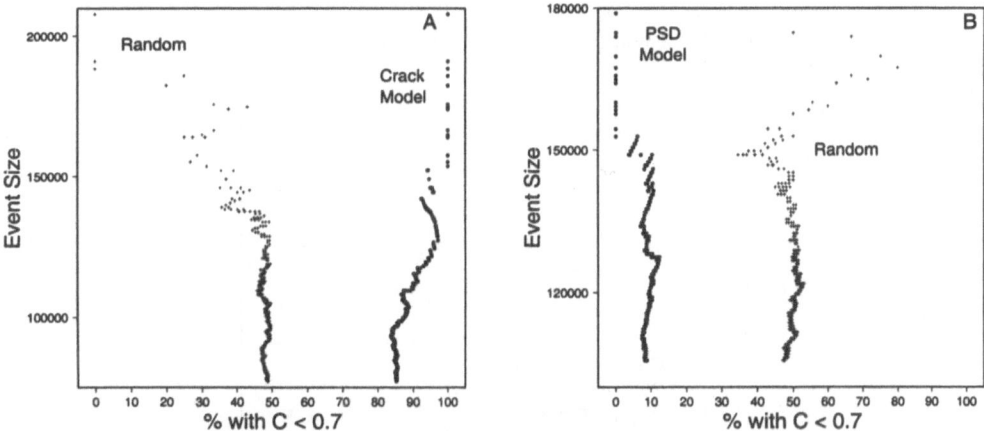

Figure 7

Frequency of fits with $\mathscr{C} < 0.7$ as a function of event size. A) The crack model (solid circles) and its randomized equivalent (crosses), and B) the PSD model (solid circles) and its randomized equivalent (crosses).

that it may be easier to distinguish between critical point and self-organized critical models of regional seismicity than it is to distinguish between either model and randomly generated seismicity.

Another prediction of the critical point model is an expectation that accelerating energy release will be seen most often for the largest events in a system. We test this prediction by examining what percentage of events have fits below a certain $\mathscr{C}$-value (here chosen to be 0.7; i.e., same cutoff used by BOWMAN et al., 1998) as a function of their size (Fig. 7). The crack model fits the critical point hypothesis best, with the 14 largest events all having $\mathscr{C}$ less than 0.7 and a decreasing percentage of good fits to smaller events (Fig. 7A). Just the opposite occurs for events from the PSD model (Fig. 7B). Again, the behavior of the two random catalogs falls in between. We note that for the random catalogs the percentage of good fits changes very rapidly when only looking at a small number of very large events. Again, this may make it difficult to tell random seismicity apart from the two proposed hypotheses for seismicity (i.e., self-organized criticality and critical point), particularly when only a small number of events are examined.

As noted above, the algorithm we use will not produce a fit to Equation (3) before one of the 480 large events if there is an even larger event within the minimum time period allowed. Based upon the statistics of the interevent times between the largest events (Fig. 1), we suspected that this may be one way to differentiate between the randomized catalogs and the two model catalogs. To test this idea we re-ran our analysis using different values for the minimum time period over which $\Sigma\Omega(t)$ is calculated, and plotted the percentage of events fit by the algorithm as a function of the minimum fitting time (Fig. 8). The crack model

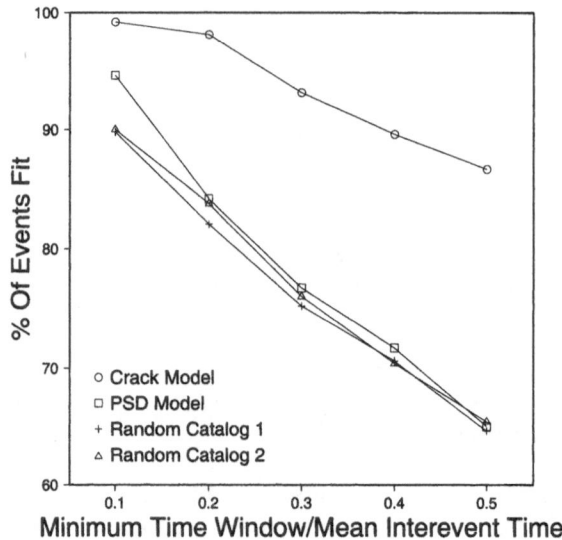

Figure 8
Frequency of fits to Equation (3) as a function of minimum allowed cycle length for the four catalogs.

always has the largest percentage of fitted events, with 476 (over 99%) of the largest 480 events yield a fit. The percentage of events yielding fits to either the PSD model or the two randomized catalogs is always less than for the crack model. As the minimum time window is expanded, the percentage of fitted events drops, least rapidly for the crack model and more rapidly for the other catalogs. When the minimum time window reaches 0.2 times the mean interevent time, the PSD model is indistinguishable from the two randomized catalogs.

Finally, we examine the results for the four catalogs in terms of changes in the distribution of event sizes with time during the fitted accelerating energy release sequences. In an analysis of coarse grained models of the earthquake process, RUNDLE *et al.* (1999) point out that the critical point hypothesis implies that ASR is underlain by a change in the earthquake frequency-magnitude distribution with time, and that this change is greatest at the large magnitude end of the distribution. JAUMÉ and SYKES (1999) tested this idea for a small number of observed cases of ASR by comparing the frequency-magnitude distributions during the first and second halves of the ASR sequences, and found that three of the four cases examined fit the RUNDLE *et al.* (1999) hypothesis.

JAUMÉ (2000) points out that the traditional Gutenberg-Ritcher magnitude-frequency distribution is ill-suited to compare changes at the large magnitude end of earthquake size distribution. Therefore, JAUMÉ (2000) introduced a graphical method based upon rank-ordering statistics to examine changes in the event size distribution with time. The rank-ordering technique emphasizes the extreme event tail of a distribution (GUMBEL, 1958) and has been used to estimate large

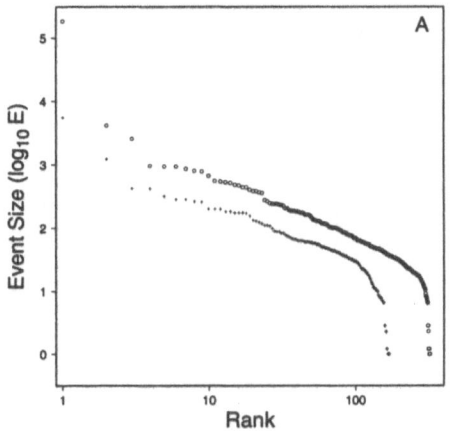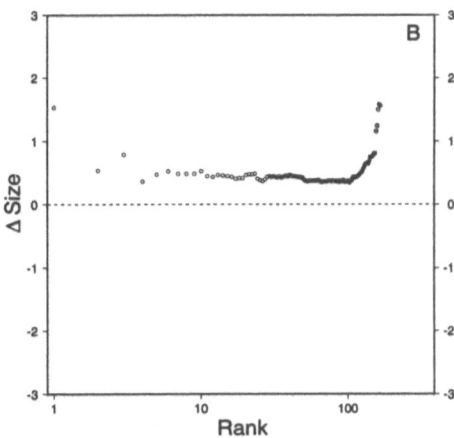

Figure 9

Illustration of the graphical method of JAUMÉ (1999) for determining event size differences as a function of frequency. A) Rank-ordering plot of event-size distribution from the ASR sequence preceding event 82 in the crack model (Fig. 2B). Crosses are the events occurring during the first half of the sequence and circles are events occurring during the second half (the mainshock included). In the rank-ordering method, the largest event is in the first rank, the second largest is in the second rank, etc., until the smallest event in the distribution is reached. This is equivalent to reversing the axes on the traditional frequency-magnitude graph. Note that $\log_{10} E$ is used instead of magnitude as in JAUMÉ (1999). B) The *magnitude/energy difference* graph for the same data. This is formed by differencing the size of the events at each rank; i.e., subtracting the size of the event in the first half of the ASR sequence (crosses in A) from the size of the events in the second half of the sequence (circles in A).

earthquake recurrence times in eastern North America (HOWELL, 1993, 1994) and test for changes in scaling of earthquake populations at large magnitudes (SOR-NETTE *et al.*, 1996). JAUMÉ (2000) finds that, for 15 of 17 observed cases of ASR before large earthquakes, the ASR can be explained by a model in which the allowed maximum earthquake size increases with time. For the remaining two cases, the change in event-size statistics was better explained by an overall increase in the rate of earthquake events (JAUMÉ, 2000).

To illustrate the methodology of JAUMÉ (2000), we examine the change in the event-size distributions during the ASR sequence before event 82 (Fig. 2B) from the crack model (Fig. 9). First, we use the results of the algorithm described in the previous section to define the time period over which the ASR sequence takes place. We extract the events that fall within that time period from the catalog and then split it into two sections, covering the 1st and 2nd halves of the accelerating sequence. We then plot the event sizes as a function of rank (Fig. 9A), putting the largest event in the first rank, the second largest event in the second rank, etc. This is equivalent to taking the more common cumulative frequency-magnitude distribution and switching axes. One advantage of this display is that it emphasizes the large event tail of the distribution. Another advantage is that one can easily take the difference between the two distributions and plot them as a function of rank

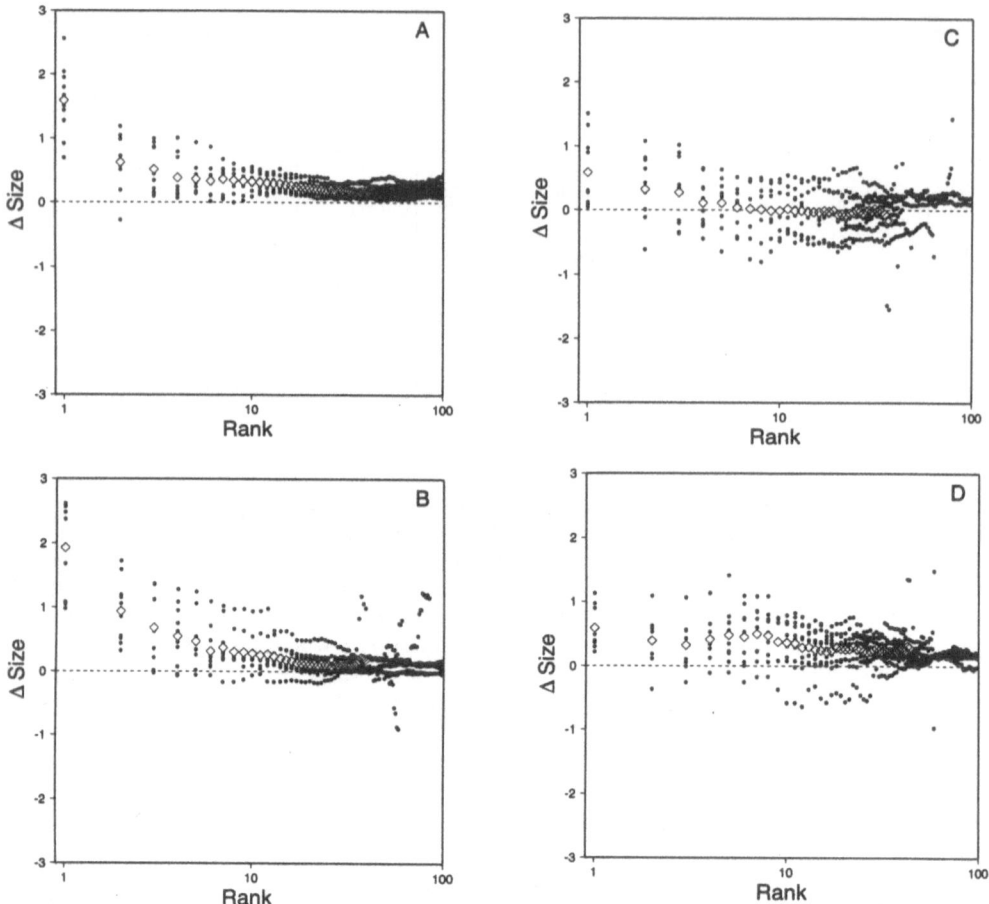

Figure 10
Magnitude difference as a function of rank for the best-fitting ASR sequences before the ten largest events (circles) from A) the crack model, B) the randomized crack model event catalog, C) the PSD model, and D) the randomized PSD model event catalog. The average magnitude difference for the ten largest events is also shown (diamonds); note that the average can only be defined up to largest rank at which the magnitude difference is defined for all 10 sequences. The mainshock event is included in the distribution for the second half of the ASR sequence on these plots. Only the first 100 ranks are shown.

(Fig. 9B). JAUMÉ (2000) has called this a *magnitude difference* graph. When the magnitude difference (here we use $\log_{10}$ energy, E, instead of magnitude) is greater than zero the size of the event at that rank is greater in the second half of the accelerating sequence than the size of the event at the same rank from the first half of the sequence. A magnitude difference that remains above zero for all ranks (as in Fig. 9) represents an overall higher rate of event occurrence during the second half of the accelerating sequence than during the first half.

In Figures 10 and 11 we show magnitude difference graphs for the ten largest events in each of the four catalogs. As noted by JAUMÉ (2000), it is unclear whether

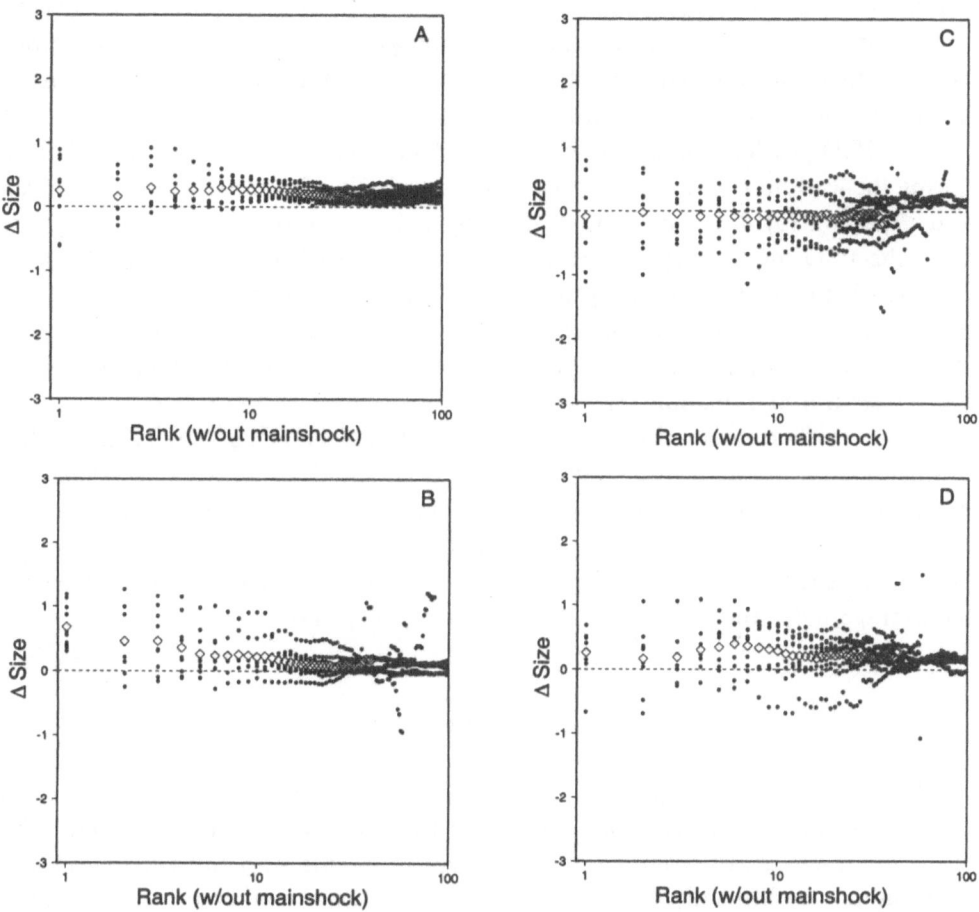

Figure 11

Magnitude difference as a function of rank for the best-fitting ASR sequences before the ten largest events (circles), but with the mainshock excluded from the distribution for the second half of the ASR sequence. A) The crack model, B) the randomized crack model event catalog, C) the PSD model, and D) the randomized PSD model event catalog. The average magnitude difference for the ten largest events is also shown (diamonds). Only the first 100 ranks are shown.

or not the mainshock should be included when forming the magnitude difference; therefore we construct magnitude difference graphs both with (Fig. 10) and without the mainshock (Fig. 11) in the event size distribution for the second half of the accelerating sequence. One thing that is clear from Figure 10 is that $\Delta \log_{10} E$ will always be positive at rank 1 when the largest event in the distribution (i.e., the mainshock) is included. This is not necessarily true when the mainshock is excluded (Fig. 11).

Figure 11A shows, in general, that larger events occur during the second half of accelerating sequences in the crack model than during the first half. As pointed out above, this could be due to either an overall increase in event size or an overall increase in the rate of occurrence. Closer examination of the event-size distributions indicates that, for most sequences from the crack model, there is an overall increase in the rate of event occurrence leading up to the large events.

The PSD model exhibits little overall change in the event size distribution during the lead-up to the ten largest events (Fig. 11C). This is not surprising, given that Equation (3) does not fit the seismicity before any of the ten largest events from the PSD model very well (Fig. 7). There is also a much greater degree of variation between the individual magnitude differences for each sequence than in the results from the crack model (Fig. 11A). This is due in part because the fitting algorithm generally prefers short sequences when the seismicity preceding a large event does not fit Equation (3) well.

Again, the two randomized catalogs (Figs. 11B and 11D) illustrate behavior that is intermediate between the crack and the PSD models. There is considerable variation between the individual sequences, similar to the PSD model (Fig. 11B). For both randomized catalogs the fitting algorithm appears to be doing a good job of picking sequences which yield more larger events near the end of sequence; i.e., the average magnitude difference is above zero for all ranks.

Discussion

Perhaps the most striking result of this work is that it appears to be easier to define measures of seismicity that can tell the two models apart than it would be to tell either model apart from a catalog of random events with the same event-size distribution. Traditionally, a random (Poisson) process is used as the null hypothesis when attempting to evaluate the probability that seismicity follows some other distribution (e.g., GROSS and RUNDLE, 1998). Our results suggest that a more profitable approach may be to define two or more hypotheses/models that yield specific predictions for different measures of seismicity, and then compare these models to equivalent measures of natural events. Such an approach would likely yield varying confidence limits on the different hypotheses, allowing some to be clearly rejected and others to be provisionally accepted (i.e., until the next set of hypothesis/model tests).

There have been a number of studies in recent years that have addressed the question of how often large earthquakes are preceded by ASR. The results of these studies have been mixed, with some authors reporting a high success rate searching for ASR sequences (BOWMAN *et al.*, 1998; BREHM and BRAILE, 1998, 1999a,b) while other report that a Poisson model better fits regional seismicity (GROSS and RUNDLE, 1998). BOWMAN *et al.* (1998) found that all eight $M > 6.5$ earthquakes between 1952 and 1998 along the San Andreas Fault system in California appear to be preceded by ASR. BREHM and BRAILE (1998) found ASR sequences prior to

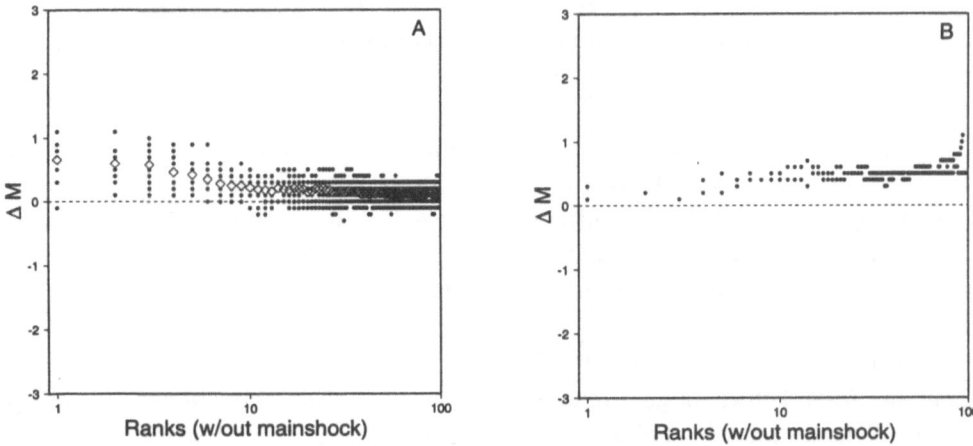

Figure 12
Magnitude difference as a function of rank for the natural ASR sequences studied by JAUMÉ (1999). A)
Fifteen cases showing increases in the magnitude difference at small rank (average of 11 cases with more
than 100 ranks shown as open diamonds), and B) two cases where no increase in magnitude difference
with rank is seen. Modified from JAUMÉ (1999).

mainshocks as small as $M = 3.5$ in the New Madrid Seismic Zone; however, they
also found that ASR could not be defined before a substantial number of
mainshocks due to other large events nearby in space and time. BREHM and
BRAILE (1999a,b) obtained similar results studying $M > 5.5$ events in southern
California and $M > 6.5$ events in the western USA. GROSS and RUNDLE (1998)
modeled cumulative Benioff strain release from 1960 through 1985 in North and
Central America and tested for the success of predictions using either Equation (3)
or a modified version including log-periodic corrections (SORNETTE and SAMMIS,
1995). They found that a Poisson model was more effective at predicting future
(1986–1995) large earthquakes than either time-to-failure model. However, JAUMÉ
and SYKES (1999) criticized GROSS and RUNDLE's (1998) use of m_b as a measure of
an earthquake's size, noting that the saturation of m_b at larger magnitudes would
lead to an underestimate of the Benioff strain release in the largest events.

Based on our results, we suspect it is unlikely a sufficiently large set of real
earthquakes have yet been analyzed to differentiate between a model of random
earthquake occurrence and critical point behavior. However, the considerable
number of cases in which ASR appears to fit seismicity preceding large earthquakes
argues that self-organized criticality is not a good model for the evolution of
natural seismicity.

Another comparison between natural ASR sequences and our model results that
can be made regards the evolution of the event-size distribution as the time of the
mainshock is approached. In Figure 12 we present magnitude difference graphs for
seismicity during several known ASR sequences (JAUMÉ, 2000). For most natural

cases of ASR there is a larger ΔM at small rank than at large rank (Fig. 12A), indicating that there are larger events in the size distribution during the second half of the sequence. For a smaller number of events ΔM is similar at all ranks (Fig. 12B), indicating an overall increase in event numbers without a change in the distribution. Comparing this to our results (Fig. 11), we see that the crack model is most similar to the natural cases with an overall event rate increase. Interestingly, the average magnitude (size) difference curve for the randomized crack model catalog is most similar to the natural cases of presumed critical point behavior, although the amount of scatter in the random catalog results is substantially greater.

Besides the tests on natural seismicity, there have been other studies that examine ASR before large events in computer models. HUANG *et al.* (1998) analyzed the output of a cellular automaton with fractal cell sizes (BARRIERE and TURCOTTE, 1994). They found that the time of 95% of the largest events (i.e., failures of the largest cell) could be predicted with high accuracy using the log-periodic formulation of Equation (3) (SORNETTE and SAMMIS, 1995). Failures of the second largest cells were more difficult to predict. SAMMIS and SMITH (1999) examined both conservative and non-conservative uniform cellular automata plus the fractal cell size automaton. They found that all models have a "scaling region" as the critical state is reached; in this region the energy release increases as a power-law with time. The uniform conservative cellular automaton remains in a critical state forever, however large events in both the non-conservative automaton and the fractal automaton move the system out of a critical state. Power-law increases in energy release with time occur as the system returns to a critical state.

Our results generally agree with those of other workers who have studied cellular automata models of seismicity. The PSD model behaves much like the classical SOC automaton and, after an initial run-in period, does not display ASR with time before large events. The crack model is a non-conservative model, and, like the simpler non-conservative model studied by SAMMIS and SMITH (1999), also commonly shows ASR before large events. To our knowledge, however, none of the results of these models has been compared to random event catalogs as ours have.

One limitation of using cellular automaton models to simulate fault systems is that stress transfer between cells is normally restricted to the nearest neighbor interactions (as in the models studied here). Yet considerable evidence exists for stress transfer effecting the timing of earthquake occurrence across considerable distances (see HARRIS, 1998 and references therein). MORA *et al.* (2000) have studied the evolution of synthetic earthquake occurrence in a lattice solid model (LSM) representation of a complex 2-D fracture zone (i.e., representing a simplified interacting fault system). Unlike the cellular automaton models studied in this paper, stress transfer from a slip event along any given rupture surface in the model propagates throughout the model via elastic wave radiation. MORA *et al.* (2000) find that large events in that model are frequently preceded by ASR, similar to our results with the crack model. They also found that a random event catalog does not produce the same number and quality of fits, both to Equation (3) and the power-law time-to-failure function of SORNETTE and SAMMIS (1995). In most cases the ASR sequences are a result of an increase in the rate of moderate-sized events (i.e., similar to Fig. 12A), but an increase in the rate of events of all sizes is seen in a smaller number of cases (i.e., similar to Figs. 11A and 12B).

Conclusions

We have analyzed two catalogs of 480,000 events produced by two different cellular automaton models of seismicity plus randomized versions of the same catalogs. We find that a nonconservative automaton model with a crack-like stress redistribution law commonly manifests periods of accelerating energy release prior to large model events. Conversely, a conservative automaton model that behaves much like the classical SOC model rarely shows this behavior; accelerating energy release before large events occurs more often in random event catalogs than in this model. In the crack model, it appears that an increase in the rate of events of all sizes underlies the acceleration in energy release. This behavior is seen in a minority of natural cases; the majority of natural cases of accelerating seismic energy release show an increase in the rate of only larger magnitude events.

Acknowledgements

This work was supported by the Australian Research Council and the sponsors of QUAKES. We thank S. Steacy for sharing their cellular automaton code. We also thank M. Matsu'ura and an anonymous reviewer for their helpful comments. The GMT system was used to produce the figures in this paper (WESSEL and SMITH, 1991).

REFERENCES

BAK, P., and TANG, C. (1989), *Earthquakes as a Self-organized Critical Phenomenon*, J. Geophys. Res. *94*, 15,635–15,637.

BARRIERE, B., and TURCOTTE, D. L. (1994), *Seismicity and Self-organized Criticality*, Phys. Rev. E *49*, 1151–1160.

BOWMAN, D. D., OUILLON, G., SAMMIS, C. G., SORNETTE, D., and SORNETTE, A. (1998), *An Observational Test of the Critical Earthquake Concept*, J. Geophys. Res. *103*, 24,359–24,372.

BREHM, D. J., and BRAILE, L. W. (1998), *Intermediate-term Prediction Using Precursory Events in the New Madrid Seismic Zone*, Bull. Seismol. Soc. Am. *88*, 564–580.

BREHM, D. J., and BRAILE, L. W. (1999a), *Intermediate-term Earthquake Prediction Using the Modified Time-to-failure Method in Southern California*, Bull. Seismol. Soc. Am. *89*, 275–293.

BREHM, D. J., and BRAILE, L. W. (1999b), *Refinement of the Modified Time-to-failure Method for Intermediate-term Earthquake Prediction*, J. Seismol. *3*, 121–138.

BUFE, C. G., and VARNES, D. J. (1993), *Predictive Modeling of the Seismic Cycle in the Greater San Francisco Bay Region*, J. Geophys. Res. *98*, 9871–9983.

GELLER, R. S., JACKSON, D. D., KAGAN, Y. Y., and MULARGIA, F. (1997), *Earthquakes Cannot be Predicted*, Science *275*, 1616–1617.

GROSS, S., and RUNDLE, J. (1998), *A Systematic Test of Time-to-failure Analysis*, Geophys. J. Int. *133*, 57–64.

GUMBEL, E. J., *Statistics of Extremes* (Columbia University Press, New York 1958).

HARRIS, R. A. (1998), *Introduction to Special Section: Stress Triggers, Stress Shadows, and Implications for Seismic Hazard*, J. Geophys. Res. *103*, 24,347–24,358.

HOWELL, B. F., JR. (1993), *Recurrence Expectation for Earthquakes in Eastern North America South of 50° Latitude*, Seism. Res. Lett. *64*, 139–147.

HOWELL, B. F., JR. (1994), *Earthquake Recurrence Rates in the Central Atlantic United States*, Seism. Res. Lett. *65*, 149–156.

HUANG, Y., SALEUR, H., SAMMIS, C., and SORNETTE, D. (1998), *Precursors, Aftershocks, Criticality and Self-organized Criticality*, Europhys. Lett. *41*, 43–48.

ITO, K., and MATSUZAKI, M. (1990), *Earthquakes as Self-organized Critical Phenomena*, J. Geophys. Res. *95*, 6853–6860.

JAUMÉ, S. C., *Changes in earthquake size-frequency distributions underlying accelerating seismic moment/energy release.* In *Physics of Earthquakes* (eds. Rundle, J. B., Turcotte, D. L., and Klein, W.) (AGU, Washington, D.C. 2000), pp. 199–210.

JAUMÉ, S. C., and SYKES, L. R. (1999), *Evolving Towards a Critical Point: A Review of Accelerating Seismic Moment/Energy Release Prior to Large and Great Earthquakes*, Pure appl. geophys. *155*, 279–305.

KAGAN, Y. Y. (1997), *Are Earthquakes Predictable?*, Geophys. J. Int. *131*, 505–525.

MAIN, I. (1996), *Statistical Physics, Seismogenesis, and Seismic Hazard*, Rev. Geophys. *34*, 433–462.

MORA, P., PLACE, D., ABE, S., and JAUMÉ, S., *Lattice solid simulation of the physics of fault zones and earthquakes.* In *Physics of Earthquakes* (eds. Rundle, J. B., Turcotte, D. L., and Klein, W.) (AGU, Washington, D.C. 2000), 105–125.

RUNDLE, J. B., KLEIN, W., and GROSS, S. (1999), *Physical Basis for Statistical Patterns in Complex Earthquake Populations: Models, Predictions, and Tests*, Pure appl. geophys. *155*, 575–607.

SAMMIS, C. G., and SMITH, S. W. (1999), *Seismic Cycles and the Evolution of Stress Correlation in Cellular Automaton Models of Finite Fault Networks*, Pure appl. geophys. *155*, 307–334.

SORNETTE, D., and SAMMIS, C. G. (1995), *Complex Critical Exponents from Renormalization Group Theory of Earthquakes: Implications for Earthquake Predictions*, J. Phys. I France *5*, 607–619.

SORNETTE, D., KNOPOFF, L., KAGAN, Y. Y., and VANESTE, C. (1996), *Rank-ordering Statistics of Extreme Events: Application to the Distribution of Large Earthquakes*, J. Geophys. Res. *101*, 13,883–13,893.

STEACY, S. J., and MCCLOSKEY, J. (1998), *What Controls an Earthquake Size? Results from a Heterogeneous Cellular Automaton*, Geophys. J. Int. *133*, F11–F14.

STEACY, S. J., and MCCLOSKEY, J. (1999), *Heterogeneity and the Earthquake Magnitude-frequency Distribution*, Geophys. Res. Lett. *26*, 899–902.

SYKES, L. R., SHAW, B. E., and SCHOLZ, C. H. (1999), *Rethinking Earthquake Prediction*, Pure appl. geophys. *155*, 207–232.

VARNES, D. J. (1989), *Predicting Earthquakes by Analyzing Accelerating Precursory Seismic Activity*, Pure appl. geophys. *130*, 661–686.

WESSEL, P., and SMITH, W. H. F. (1991), *Free Software Helps Map and Display Data*, EOS, Trans. AGU *72*, 445–446.

WEATHERLEY, D., JAUMÉ, S., and MORA, P. (2000), *Evolution of Stress Deficit and Changing Rates of Seismicity in Cellular Automaton Models of Faults*, Pure appl. geophys. *157*, 2183–2207.

(Received October 3, 1999, revised April 5, 2000, accepted April 7, 2000)

Pure appl. geophys. 157 (2000) 2227–2248
0033–4553/00/122227–22 $ 1.50 + 0.20/0

The Plausibility of Long-wavelength Stress Correlation or Stress Magnitude as a Mechanism for Precursory Seismicity: Results from Two Simple Elastic Models

MICHAEL E. WINTER[1]

Abstract—Observations of accelerating seismic activity prior to large earthquakes in natural fault systems have raised hopes for intermediate-term earthquake forecasting. If this phenomena does exist, then what causes it to occur? Recent theoretical work suggests that the accelerating seismic release sequence is a symptom of increasing long-wavelength stress correlation in the fault region. A more traditional explanation, based on Reid's elastic rebound theory, argues that an accelerating sequence of seismic energy release could be a consequence of increasing stress in a fault system whose stress moment release is dominated by large events. Both of these theories are examined using two discrete models of seismicity: a Burridge-Knopoff block-slider model and an elastic continuum based model. Both models display an accelerating release of seismic energy prior to large simulated earthquakes. In both models there is a correlation between the rate of seismic energy release with the total root-mean-squared stress and the level of long-wavelength stress correlation. Furthermore, both models exhibit a systematic increase in the number of large events at high stress and high long-wavelength stress correlation levels. These results suggest that either explanation is plausible for the accelerating moment release in the models examined. A statistical model based on the Burridge-Knopoff block-slider is constructed which indicates that stress alone is sufficient to produce accelerating release of seismic energy with time prior to a large earthquake.

Key words: Precursory seismicity, stress correlation.

Introduction

A generally accepted view of earthquakes argues that the process is inherently random, a consequence of a "self-organized critical" system (BÅK and TANG, 1989). In this model the earth's crust exists in a critical steady state with stationary event distributions. As a consequence of the model, earthquakes have been argued to be unpredictable (GELLER *et al.*, 1997). Self-organized criticality has been observed in a variety of models of seismicity including block-slider (MCCLOSKEY and BEAN, 1994) and cellular automata models (STEACY *et al.*, 1996; BÅK and TANG, 1989).

[1] QUAKES, Department of Earth Sciences, University of Queensland, Australia. Now at: Hawaii Institute of Geophysics and Planetology, University of Hawaii at Manoa, U.S.A. E-mail: winter@quakes.earthsciences.uq.edu.au

This view of seismicity may be simplistic. The conclusion that the earth's crust exists in a state of self-organized criticality rests on the observation of a power-law scaling of the magnitude-frequency distribution of earthquakes—the Gutenberg-Richter distribution (GUTENBERG and RICHTER, 1956). The manner in which this distribution is obtained may mask large deviations from a power-law scaling due to a smoothing out of local variation (SCHOLZ, 1968). An alternative view, the "critical-point theory," is that on smaller spatial scales a fault system approaches and retreats from a critical point (SORNETTE and SORNETTE, 1990), resulting in a change in likelihood of a large event over a seismic cycle.

Recent observations of accelerating seismic moment release (AMR) prior to large earthquakes (KNOPOFF et al., 1996; VARNES, 1989; SYKES and JAUMÉ, 1990; BUFE and VARNES, 1993; BREHM and BRAILE, 1999) support the critical point theory. These observations show an increasing seismic energy release prior to a large earthquake. A large earthquake serves to move a region from a critically stressed state to an understressed state, resulting in increasing seismic energy release as the system returns to criticality. These observations have raised the hope of intermediate-term earthquake prediction.

If the phenomenon of increased precursory seismicity does indeed exist, what then causes it to manifest itself? If anything, the process must be statistical, with some internal configuration of the fault system resulting in an increase in seismic energy release and an increase in the likelihood of a large event. BUFE and VARNES (1993) suggest inelastic phenomena such as crack propagation or progressive rock damage are responsible. This would seem to be incompatible with the observation of precursory seismicity in perfectly elastic systems, such as Burridge-Knopoff block-slider models (SHAW et al., 1992). More recently a statistical physics-based explanation was offered by RUNDLE et al. (1999). They suggest that an increasing degree of spatial correlation will lead to an increase in the number and magnitude of larger earthquakes in the system. This sort of effect was also observed in cellular automaton models by SAMMIS and SMITH (1999) and by STEACY et al. (1996).

Nonetheless is a new explanation for the phenomena necessary? Changes in earthquake distribution are often observed in nature and are generally argued to be due to increases in stress in the region surrounding the fault. That the state of stress on faults influences seismicity is implicit even in the early elastic rebound theory of REID (1910). Seismically active faults exist in a state of constant elastic loading. For a typical fault, 70 percent of the total seismic moment is released within a magnitude of the largest event (Fig. 1). Central to the theory of elastic rebound is the idea that seismic energy release correlates with increasing stress. In this case, a fault system will accumulate strain and consequently manifest an increased rate of seismicity. This increasing rate of seismicity would be terminated when a large event occurs, releasing a significant portion of the accumulated strain. This concept has been shown to be consistent with seismicity in the San Francisco region and the New Madrid fault zone by HOWELL (1997). Howell notes an increase in the size of the largest event with time since the last great earthquake on a fault segment. A strong correlation was found between calculated coseismic stress increase and future earthquake locations by KING *et al.* (1994), and aftershock distributions often lie predominantly in areas where a mainshock has increased stress (for example see STIEN *et al.*, 1994). In a study of rock microfracture, SCHOLZ (1968) observed a systematic decrease of b-value micro-earthquakes when stress was increased. STEACY and MCCLOSKEY (1998) have also observed this effect in simple cellular automata models. Constant stress increases coupled with a stress-induced increase in seismicity, terminated by a large event releasing stress, would result in observations similar to the heightened levels of seismicity prior to large earthquakes seen in nature.

The objective of this paper is to investigate whether or not stress or long-wavelength stress correlation are plausible mechanisms for heightened precursory seismicity. Here, changes in event distribution as a function of both stress correlation and fault stress are analyzed in two conceptual models of seismicity: a one-dimensional block-slider model, and a two-dimensional model with full inertial dynamic. The two different models act as end-member cases for stress concentration. The block-slider model demonstrates the case in which all stress is concentrated on the fault surface, while in the continuum model stress is distributed throughout the solid. The goal of this analysis is to ascertain whether an accelerating release of seismic energy prior to a large earthquake can be described as an inherent statistical property of either a stressed elastic system or an elastic system with high levels of long-wavelength spatial correlation of stress. As fault systems represent complex

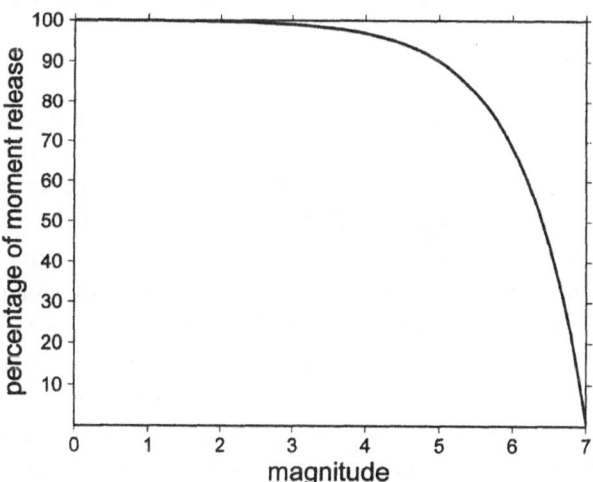

Figure 1

Percentage of seismic release due to earthquakes above a given magnitude for a fault with a *b*-value of one and a maximum earthquake of $M = 7.0$. Seventy percent of moment release is due to events within one magnitude of the largest event.

nonlinear dynamic systems the processes cannot be individually analyzed. If either process is responsible for heightened precursory seismicity then it should correlate with seismicity, indicating a systematic increase in seismic energy release. Here changes in Gutenberg-Richter distributions are examined to see if this correlation exists. Demonstrating a cause-effect relationship between either process and precursory seismicity is a considerably more ambitious goal and is not attempted here.

Models

In order to understand phenomenon of heightened precursory seismicity, it is beneficial to construct model faults. These models are either mathematical or computational approximations to the processes responsible for stress accumulation, transfer, and seismic rupture. They are inherently limited in the scope of physical phenomena that can be included, due to constraints in computation time and model resolution. Nonetheless, high levels of precursory activity have been observed in several simplified models of seismic rupture: a cellular automata model (SAMMIS and SMITH, 1999), a Burridge-Knopoff "block-slider" based model (SHAW et al., 1992), and an elastic model which includes a realistic fault gouge layer (MORA et al., 2000a). Since all of these simple models display this behavior, by implication the phenomena responsible must be the simplest set of physical phenomena modeled by all three. The phenomena must therefore be encapsulated by the processes of elastic stress, strain transfer, and rupture. Here the phenomena will be examined in both a Burridge-Knopoff block-slider model like that used by SHAW et al. (1992), and in a finite element fault model based on the equations of an elastic continuum.

Burridge-Knopoff Model

The first model used in this discussion is the Burridge-Knopoff (BK) block-slider model (BURRIDGE and KNOPOFF, 1967). This model consists of a series of rigid blocks connected by springs. The blocks are connected to a driving plate at the top by a set of leaf springs, and rest on a stationary surface (Fig. 2). The BK model has been widely used in seismology as it represents the simplest possible model of dynamic elastic rupture. BK models represent deterministic chaotic systems that spontaneously generate complex seismicity with a Gutenberg-Richter-like distribution (BURRIDGE and KNOPOFF, 1967; LANGER et al., 1996; SHAW, 1995). They also mimic other more specific properties of faults such as narrow slip pulses (LANGER et al., 1996). While the BK model produces earthquake statistics similar to those seen for real faults, it is not entirely representative of the processes of seismicity. Its nature is intrinsically discrete and one-dimensional, with stress only being transferred between adjoining blocks. Nonetheless, the BK model encompasses the kernel of the basic processes of seismicity, stress accumulation, basic stress transfer, and seismic rupture.

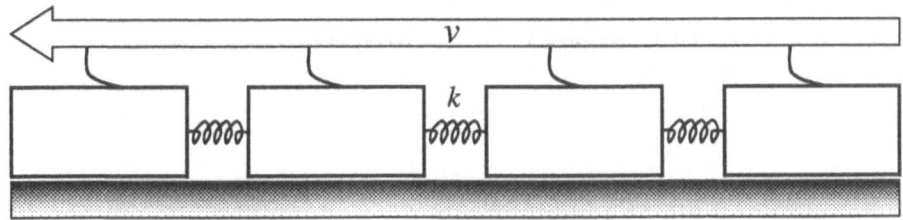

Figure 2
The Burridge-Knopoff (BK) model represents a fault system as a series of blocks, coupled by linear springs. The system is stressed elastically by a driving plate moving at velocity v.

For this discussion, the most interesting observation made using this model is that it exhibits behavior consistent with accelerating seismic energy release prior to a large earthquake. A variation of this model was used by SHAW et al. (1992) to examine seismicity prior to large simulated events. They found that seismicity preceding a large event clusters near it, and that the rate of earthquake occurrence increases prior to a large event. These observations are remarkably consistent with the generally accepted manifestations of precursory seismicity (cf., JAUMÉ and SYKES, 1999), specifically the acceleration of seismicity prior to a large event and the correlation of precursory activity with mainshock location.

The variation on the BK model used here is based on that used by SHAW et al. (1992). Dimensionless quantities were used wherever possible. For a given block i, the acceleration due to elastic force is given by:

$$\ddot{U}_i = k(U_{i+1} - 2U_i + U_{i-1}) + vt - U_i, \tag{1}$$

where U_i and $\ddot{U}_i$ are the displacement and acceleration of block i, k is the interblock spring constant (0.75 is used here), v is the velocity of the driving plate, and t is time. At equilibrium ($vt = U_i$) the equation of motion amounts to a finite-difference form of the one-dimensional wave equation. Blocks in motion have frictional interaction with the base of the model. The friction law used here is slightly modified from the velocity-weakening law used by SHAW et al. (1992):

$$\mu_i = \frac{1}{1 + 2\alpha|\dot{U}_i|}, \tag{2}$$

where α controls the rate of decay of friction with block velocity, $\dot{U}_i$. A value of 2.5 was used for α consistent with the value used by SHAW et al. (1992).

In order to obtain the greatest possible accuracy in terms of determining event distributions, the model was stressed statically. Here the approximation is used that the time scale of tectonic loading is much greater than the time scale of elastic rupture ($v \approx 0$). The onset of an earthquake was found by iteratively solving the equations for frictional traction and elastic stress to find the point where a block begins to slide. An event was terminated when there were no more blocks in motion. The event energy was obtained by subtracting the post-event internal energy from the pre-event internal energy, where internal energy for the model is given by:

$$E = \sum_{i=1}^{N} \left(\frac{1}{2}(U_i - vt)^2 + \frac{k}{2}(U_i - U_{i-1})^2 \right). \tag{3}$$

The magnitude of a simulated earthquake is defined in terms of earthquake energy release. BURRIDGE and KNOPOFF (1967) define magnitude for this model as proportional as the logarithm of earthquake energy:

$$M = \log_{10}(\Delta E), \tag{4}$$

where ΔE is the decrease in model internal energy due to an event.

The simplicity of this model allows large numbers of simulated earthquakes to be gathered. For the analysis presented here a 250-block system was run until 70,000 earthquakes were collected. The first few thousand events were then discarded to ensure that transient effects related to the initialization of the model were no longer apparent.

The seismicity produced by Burridge-Knopoff models like this one has been extensively studied. These models tend to produce complex, power-law distributed seismicity. This model produced approximately Gutenberg-Richter seismicity (GUTENBERG and RICHTER, 1956) (Fig. 3) over just under two orders of magni-

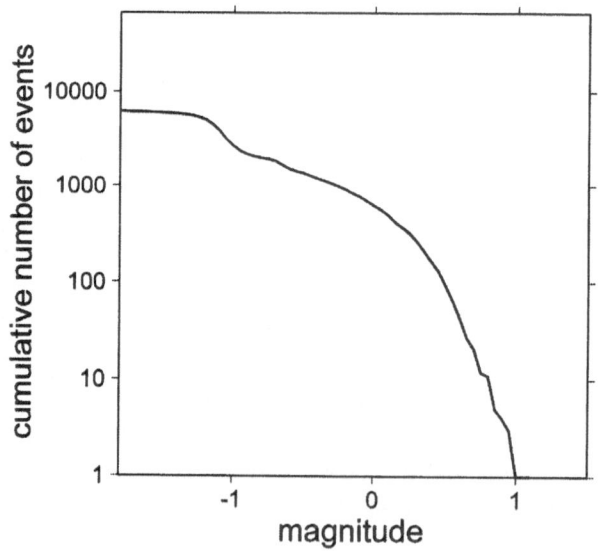

Figure 3
Cumulative magnitude-frequency plot for the block-slider system.

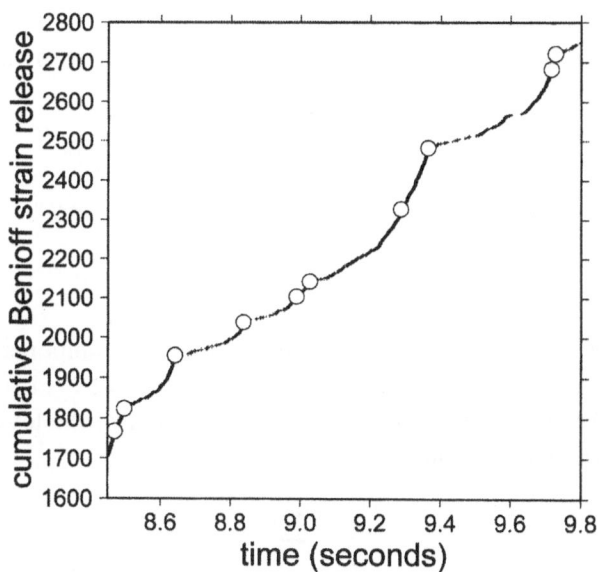

Figure 4
Cumulative seismic Benioff strain release for the BK model. Large events are shown as circles. The BK model shows accelerating release of seismic energy prior to the large events.

tude. The resolution of the magnitude-frequency relationship is limited by the number of blocks (250) and fact that several blocks are required to reproduce the elastic wave equation. The model shows accelerating seismic energy release (as measured by cumulative Benioff strain release) prior to the largest two percent of all events (Fig. 4).

Dependence of Seismicity on Stress and Long-wavelength Stress Correlation

What causes the heightened precursory seismicity in the block-slider model? Typically, the cause of precursory seismicity is described as an increasing correlation in the elastic stress field surrounding the fault (SORNETTE and SAMMIS, 1995; RUNDLE et al., 1999). Alternatively, elastic rebound theory suggests that a general increase in the rate of seismic energy release should be observed as the fault system becomes more stressed (HOWELL, 1997). The idea that stress correlation grows prior to a large earthquake is in some ways counter-intuitive. One would expect that the stress field would be most correlated *after* a large event, where the stress has been reduced near the epicenter, followed by a slow decrease in the correlation as small events increase stress heterogeneity.

Both these theories can be easily tested on the block-slider model. The term "long-period stress correlation" is somewhat ill defined. Here, the degree of spatial similarity of the stress field is used as a proxy. Two quantities are defined to describe the internal state of stress of the block-slider system. A stress magnitude measurement ($\bar{\sigma}$) and a long-wavelength stress correlation score (σ_C). The stress magnitude is defined as the Euclidean norm of the model stress field:

$$\bar{\sigma} = \sqrt{\sum_{1}^{N} \frac{\ddot{U}_i^2}{N}}. \tag{5}$$

A stress variation field is then defined as:

$$\sigma_i' = \frac{\ddot{U}_i}{\bar{\sigma}}. \tag{6}$$

This serves to separate the stress field into two components: the absolute stress magnitude ($\bar{\sigma}$) and the spatial variation of the stress field (σ'). The long-period stress correlation score is defined as the simple sum of the lower order Fourier components of the stress variation field:

$$\sigma_C = \sum_{i=0}^{m} \tilde{\sigma}_i', \tag{7}$$

where $\tilde{\sigma}'$ is the Fourier transform of σ', and m is the number of Fourier components to include. Here the first eight Fourier components were used, as this measures the stress correlation at approximately the spatial scale of the largest events observed in the model. This sum is intended to measure the long-wavelength stress correlation of the block-slider system with the effects of the absolute stress magnitude removed. Figure 5 indicates that these quantities can be analyzed separately, as they do not correlate with each other.

Figure 6 shows the time evolution of these two quantities in the BK model with large events plotted as triangles. The long-wavelength stress correlation score decreases in the leadup to large events, as smaller earthquakes add more and more high frequencies to the stress field, and jumps after each large event. This behavior is opposite to that expected of critical point systems, and therefore suggests that the block-slider system is not a critical point system. In spite of this, the block-slider displays the largest events which tend to occur in periods of high correlation. The plot of RMS stress shows predictable behavior, as the total model stress decreases after each large event. The RMS stress graph also shows that the largest events seem to occur at the highest stresses. The observations of stress correlation evolution here which is inconsistent to that of critical point systems are contrary to some models of two-dimensional earthquake fault systems (e.g., MORA et al., 2000a,b), and may indicate that the block-slider system has dissimilar behavior in some respects to two-dimensional fault systems.

Figure 7a shows the cumulative magnitude-frequency plots for the BK model divided into two populations; high pre-event stress and low pre-event stress. An event was assigned to a population based on whether the model RMS stress was greater than or less than the median score for the entire model run. Larger events tend to occur at high stresses, consistent with elastic rebound theory.

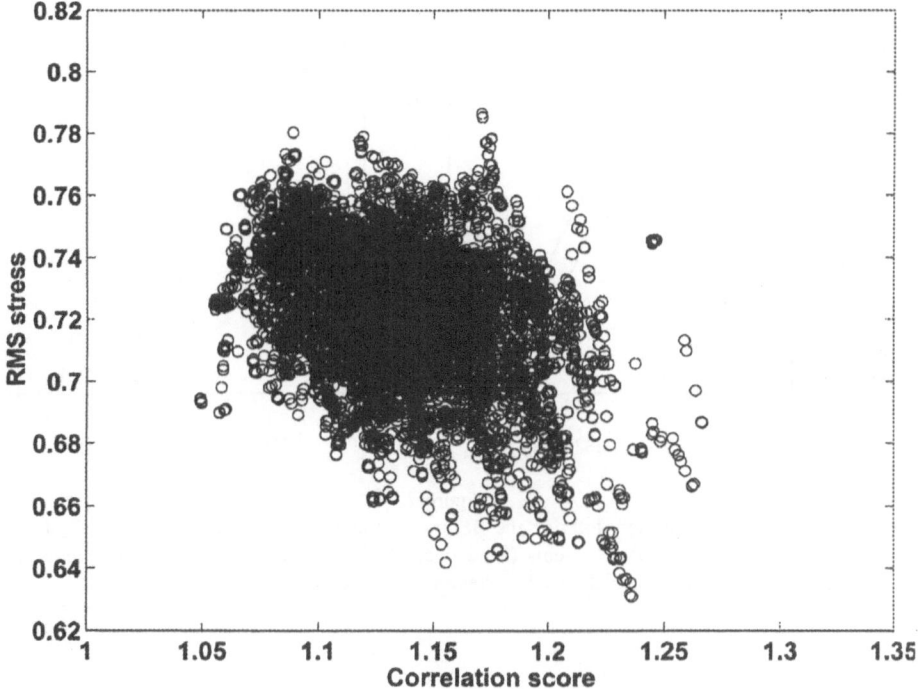

Figure 5
RMS stress versus correlation score in the BK model for 80,000 events. There is no correlation between
the two quantities.

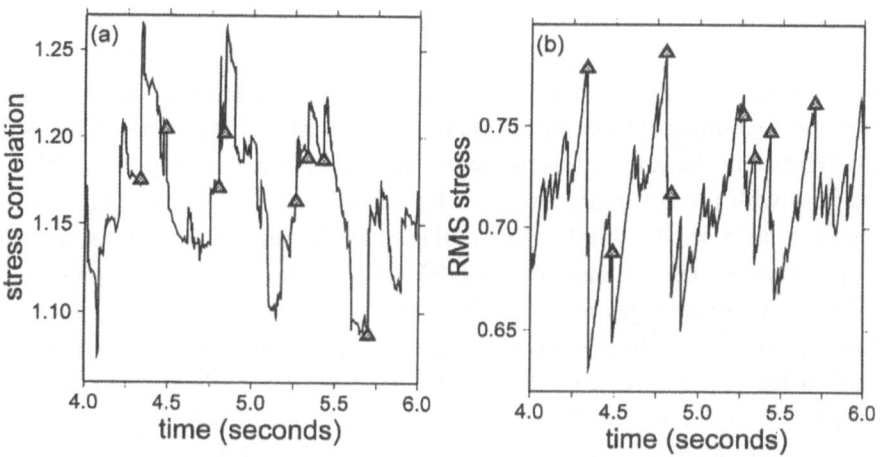

Figure 6
Evolution of the state of stress in the BK model: (a) long-wavelength stress correlation score σ_C; (b)
root-mean-squared model stress $\bar{\sigma}$. The largest events ($M > 0.7$) are denoted by triangles.

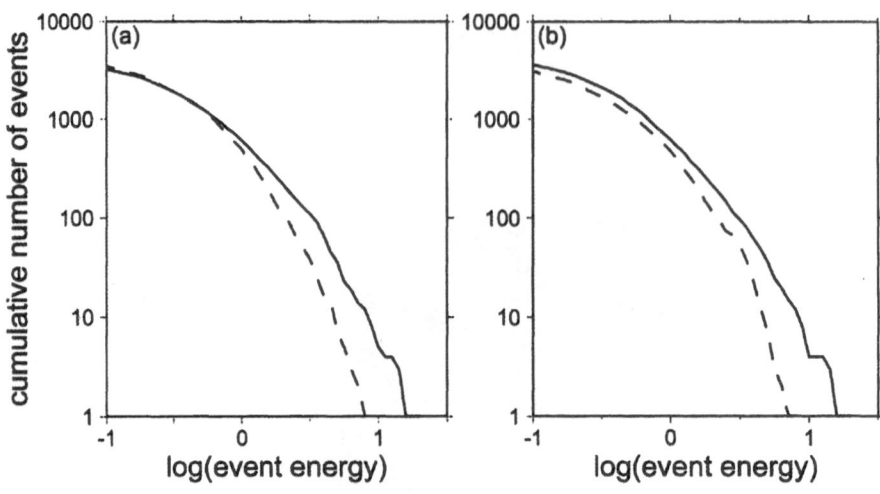

Figure 7
Cumulative magnitude-frequency plots for the BK model under different conditions: (a) for low stress events (dashed line) and high stress events (solid line); (b) low stress correlation score events (dashed line) and high stress correlation score events (solid line).

A similar plot is shown in Figure 7b, where events were divided into two populations based on whether the pre-event correlation score was above or below the median value. In a theoretical model of accelerating moment release developed by RUNDLE et al. (1999), the magnitude-frequency distribution of earthquakes shifts, with larger magnitude events tending to occur at high levels of stress correlation. In the model presented here, seismicity appears to correlate well with the long-wavelength stress correlation score, showing an increase in event magnitude for the medium and large sized events. In contrast to Figure 7a there is a slight overall increase in the rate of seismicity for the highly correlated system. This result is consistent with the theoretical work of RUNDLE et al. (1999). The magnitude-frequency plots also appear similar to the theoretically derived magnitude-frequency plots presented by JAUMÉ and SYKES (1999) which show a change in earthquake distribution due to changes in stress correlation.

Both stress magnitude and long-wavelength stress correlation offer a plausible mechanism for precursory seismicity in the BK model. The highly stressed block-slider system shows an overall increase in seismic moment release through both a greater number of events and an increase in the size of the medium and large events. High stress correlation in the BK model leads to an increasing energy release due to greater numbers of medium and large events.

Continuum Model

A similar study can be performed with a more physically complete model. Here a fault is represented as a discontinuity between two elastic blocks. For perfectly elastic blocks the continuum obeys Cauchy's equation of motion:

$$\rho \ddot{\vec{U}} = \vec{\nabla}\mathbf{T}, \tag{8}$$

with ρ representing material density. $\mathbf{T}$ represents the stress tensor:

$$\mathbf{T} = \lambda \vec{\nabla} \cdot \vec{U}\mathbf{I} + 2\mu\mathbf{E}, \tag{9}$$

with λ and μ being the Lamé constants, $\mathbf{I}$ the rank three identity matrix and $\mathbf{E}$ the strain tensor. For this model the density and Lamé constants were chosen to be consistent with those for granite. These equations were formulated using the finite element method (ZIENKIEWICZ and TAYLOR, 1994), resulting in two elastic blocks. The model uses periodic boundary conditions on the x boundary and an absorbing layer at the extremes of the y-axis (see Fig. 8).

Central to any attempt at modeling earthquake rupture is an accurate representation of the physical processes of friction. Friction at the interface was modeled using a simple finite difference law, where normal stress was transmitted and tangential stress obeyed a prescribed frictional law. A velocity (rate) dependent friction law is applied here. For consistency, the form used here is similar to that utilized in the block-slider model:

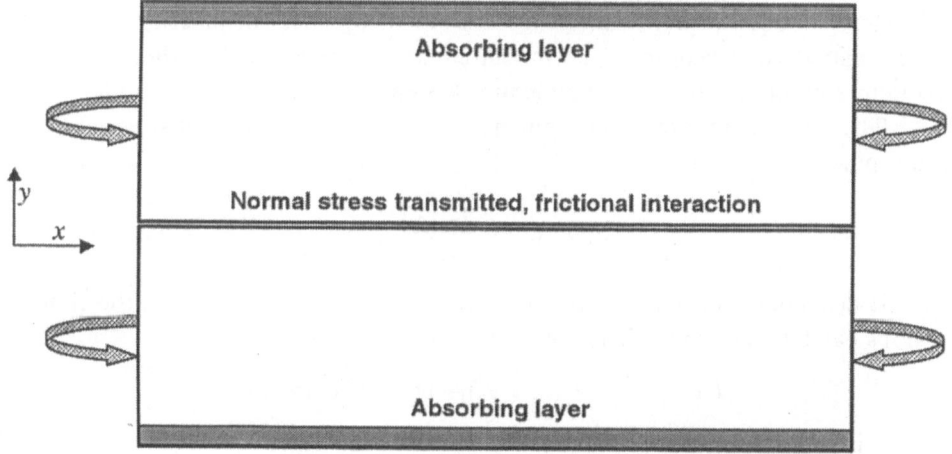

Figure 8
The finite element continuum model. A fault zone is modeled as two elastic blocks pressed together, with absorbing boundary conditions parallel to the fault and periodic boundary conditions perpendicular.

$$\mu(x) = \frac{\mu_0(x)}{1 + 2\alpha|V(x)|},$$ (10)

where $\mu(x)$ is the friction at point x on the fault surface, $\mu_0(x)$ is the value of friction before slip occurs, $V(x)$ is the relative velocity at a point on the fault interface, and α controls the rate of decay of friction with velocity. A value of 25 was chosen for α which results in a drop of approximately 4 percent in friction during slip. Perfectly elastic models such as this one are not capable of reproducing complex seismicity without some sort of spatial inhomogeneity (BEN-ZION and RICE, 1995; RICE, 1993; UMEDA et al., 1996). In order to obtain an event distribution that approximates a Gutenberg-Richter power law, asperities have been added to the model by modulating the value of the frictional coefficient $\mu_0(x)$. Initially, a constant value of 0.6 was used for $\mu_0(x)$. This is similar to the coefficient of friction observed in granite by TULLIS and WEEKS (1986). Added to this was distribution of values mimicking a power-law (fractal) distribution similar to that observed in nature. This distribution was intentionally biased towards strong local heterogeneity in order to promote a large range of earthquake sizes (BEN-ZION and RICE, 1995) and to reproduce the distribution of asperities inferred from seismic studies (NADEAU and JOHNSON, 1998). The net result is a fault with small patches of high strength. These strong patches account for approximately ten percent of the fault area. Slight stress smoothing accomplished by low-pass filtering the stress field at just below its Nyquist frequency was employed at the fault interface to dampen spurious interface waves. The result of the addition of asperities to the continuum model results in a model that is intrinsically discrete. As a result, this model can be best visualized as an update of the BK model to include two-dimensionality and full inertial dynamics.

Like the BK model, the continuum model employed a static stress accumulation cycle. Here, an equilibrium strain field for a one meter shear displacement ($\vec{U}_s(x, y)$) was calculated via relaxation. The model could be pre-stressed to the point of the next failure by iteratively solving an equation relating frictional traction and normal stress. The point of failure is determined when traction and normal stress are equal at any point on the fault:

$$\sigma_{yy}(x) = \frac{\sigma_{xy}(x)}{\mu(x)}.$$ (11)

Stress is dependent on the deformation, which evolves statically from the time the previous earthquake concluded (t_0):

$$\vec{U}(x, y, t) = \vec{U}(x, y, t_0) + v(t - t_0)\vec{U}_s(x, y),$$ (12)

where v is the driving plate velocity.

As with the block-slider model, magnitude is defined in terms of earthquake energy release. Again, the definition of BURRIDGE and KNOPOFF (1967) is adopted, defining magnitude as proportional to the logarithm of earthquake energy:

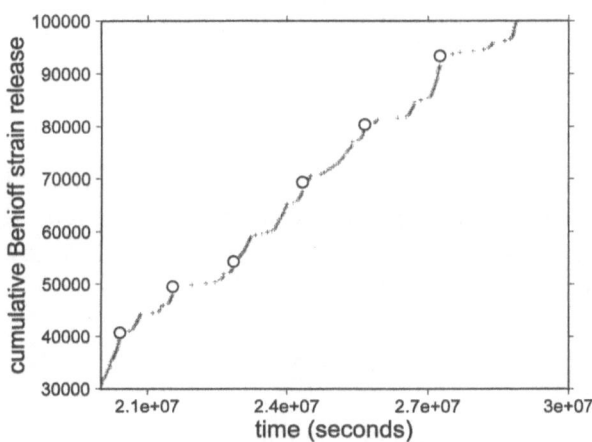

Figure 9
Cumulative seismic Benioff strain release for the continuum model. Large events are shown as circles.
The continuum model shows accelerating seismic energy releases prior to the largest events.

$$M = \log_{10}(\Delta E),\tag{13}$$

where ΔE is the decrease in model internal energy due to a simulated earthquake.

The need to include heterogeneity implicitly limits the resolution of the model, restricting the range of realistic earthquake behavior. Earthquakes in this model generally occur as quanta of the regions between asperities which, given a large enough model, will produce seismicity with a power-law distribution. The resulting simulated fault system is semi-periodic with large earthquakes occurring in regular intervals. Figure 9 shows cumulative Benioff strain release for a portion of a model run, illustrating an accelerating rate of seismicity prior to the largest events.

Dependence of Seismicity on Stress and Long-wavelength Stress Correlation

The reproduction of classic seismicity patterns in the continuum model is less perfect than for the BK model. It does however manifest the same statistical deviations from true random seismicity. A similar analysis to that performed on the BK model was applied to here. The stress magnitude measurement ($\bar{\sigma}$) and a long-wavelength stress correlation score (σ_C) are calculated similar to those for the BK model, with the exception that field variables are treated as continuous functions in finite element analysis. For the continuum model, stress magnitude is defined as:

$$\bar{\sigma} = \sqrt{\int \frac{\tau_x^2(x)}{L}\,dx},\tag{14}$$

where $\tau_x(x)$ is the total traction parallel to the fault at the fault interface. The integral is taken over the entire fault. A stress variation field is then defined as:

$$\tau'_x(x) = \frac{\tau_x(x)}{\bar{\sigma}}. \tag{15}$$

This serves to separate the stress field into two components: an absolute stress magnitude field ($\bar{\sigma}$) and the spatial variation of the stress field ($\tau'_x(x)$). The long-period stress correlation score is defined as the sum of the lower order Fourier components of the stress variation field:

$$\sigma_C = \sum_{i=0}^{m} \tilde{\tau}'_{xi}, \tag{16}$$

where $\tilde{\tau}'_{xi}$ is the Fourier transform of $\tau'_x(x)$, and m is the wave number of the largest Fourier component to include. Here Fourier components with wavelengths of at least an eighth of the model were used, as this measures the stress correlation at approximately the spatial scale of the largest events observed. As with the BK model, these scores are intended to separate the effects of long-wavelength stress correlation and stress magnitude. Figure 11 shows that, although there is a correlation between the two values over individual loading cycles, there is no general correlation between the two values.

Figure 10 delineates the time evolution of these two quantities in the continuum model with large events plotted as triangles. Contrary to the BK model, the continuum model shows stress correlation dropping after each large event, in line with two-dimensional models of fault systems (MORA et al., 2000a,b). As with the

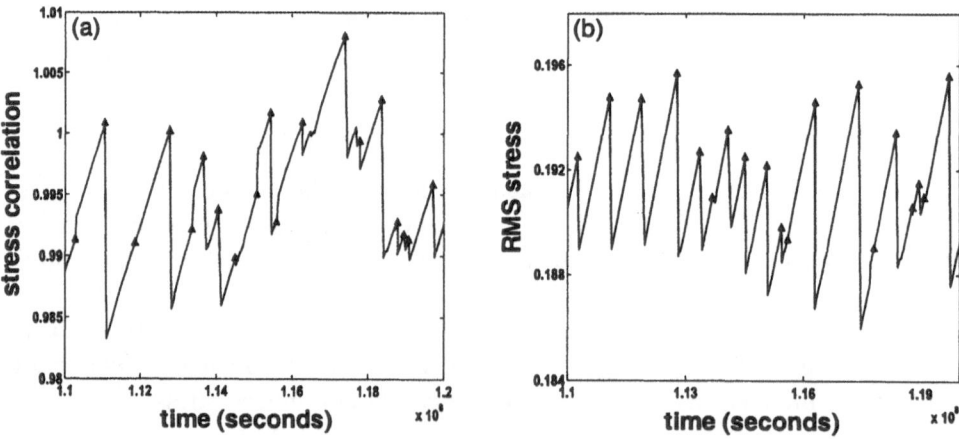

Figure 10

Evolution of the state of stress in the continuum model: (a) long-wavelength stress correlation score σ_C; (b) root-mean-squared model stress $\bar{\sigma}$. The largest events ($M > 0.7$) are denoted by triangles.

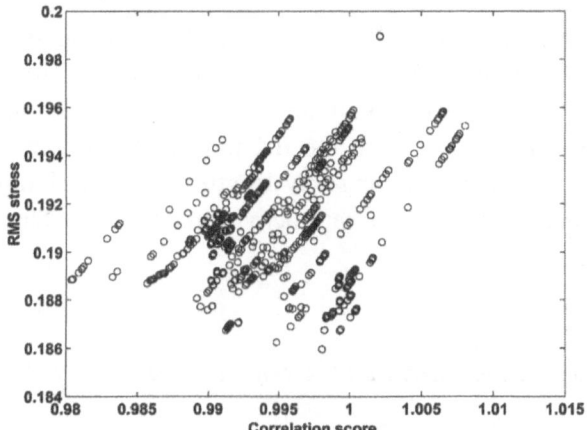

Figure 11

RMS stress versus correlation score in the continuum model for over 1000 events. There is a correlation between the two values over a single loading cycle (shown by the positive linear trends), however there is at best only a weak correlation between the two quantities overall.

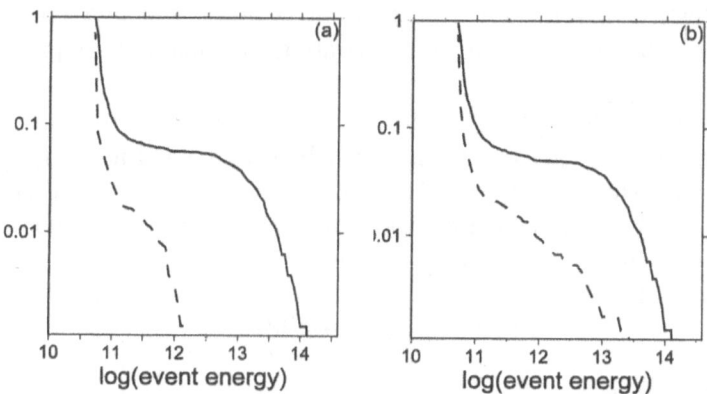

Figure 12

Cumulative magnitude-frequency plots for the continuum model under different conditions: (a) for low stress events (dashed line) and high stress events (solid line); (b) low stress correlation score events (dashed line) and high stress correlation score events (solid line). The continuum model shows heightened levels of seismicity at high levels of long-wavelength stress correlation and high levels of RMS stress.

BK model, the plot of RMS stress exhibits predictable behavior, as the total model stress decreases after each large event. The RMS stress graph also indicates that the largest events seem to occur at the highest stresses.

The change in seismicity due to either high stress or high long-wavelength stress correlation is similar to that observed in the BK model (see Fig. 12). The seismicity produced when the model is in a highly stressed state shows a large increase in the number and magnitude of the largest events, up to 1.5 magnitude units versus the less stressed model. High levels of stress correlation also boost the frequency and magnitude of the large events with an additional increase in the numbers of medium-sized earthquakes. As with the BK model, the continuum model demonstrates that both high levels of long-wavelength stress correlation and high stress magnitude produce deviations in seismicity capable of producing the observed heightened levels of seismicity.

Discussion: Stress and Accelerating Moment Release

Is stress alone a plausible mechanism for heightened precursory activity? Natural fault systems tend to release a majority of accumulated moment with the largest few events. If this is the case then the fault system will accumulate strain due to tectonic loading until a large event occurs. The earthquake models presented show an increase in seismic energy release at high stresses, as do observations of faults in nature. If this is the case then the increasing stress will result in an increase in seismic energy release until a large event occurs, "resetting" the strain. This would result in an observation of accelerating moment release prior to a large event.

To demonstrate the viability of this theory, a simple statistical seismicity model based on an energy balance argument will be examined. Energy is a convenient measure of the state of stress of an elastic system since it is approximately proportional to the mean-squared stress field. Here the block-slider system described above will be considered, as it is better statistically behaved than the continuum model.

Accelerating Sequence

Consider a block-slider system under continuous loading. For this system the rate of energy increase due to tectonic processes at high-loading stress levels is approximately proportional to the square-root of energy (BURRIDGE and KNOPOFF, 1967):

$$\frac{dE^+}{dt} \approx v\sqrt{2EN}, \tag{17}$$

where v is the velocity of the loading plate, N is the total number of blocks, and E is the internal energy of the block-slider system. This relationship is approximate as it neglects energy due to interblock forces. At typical stress levels seen in the model presented here this amounts to an error of no greater than two percent.

Observations of natural fault systems (HOWELL, 1997) and rock microfracture (SCHOLZ, 1990) suggest that highly stressed systems show greater seismic energy release. The block-slider model expresses behavior consistent with this. In order to more systematically study the change in seismic energy release as a function of stress, model energy was divided into 18 regions spanning the total range of energies occupied by the model. The energy of the model prior to each event was then used to bin the earthquakes into these regions. From this seismic energy, release for each energy division was calculated by dividing the total energy released in the division by the total amount of time the block-slider system occupied that energy range. The "great" earthquakes, defined as the largest 0.1 percent of all earthquakes ($M > 0.7$), were removed from the statistical analysis in order to study the rate of energy release in the period preceding large events. The largest earthquakes are implicitly removed from an analysis of precursory seismicity as the precursory seismicity is defined as being prior to a large event. Since large earthquakes are rare, this has the effect of removing them from the analysis. Figure 13 shows the rate of energy accumulation and release for this case. It indicates that the block-slider system with the largest events removed is fundamentally unstable, as it is always in a state of increasing energy. For this system the rate of energy release appears to be linearly related to total internal energy:

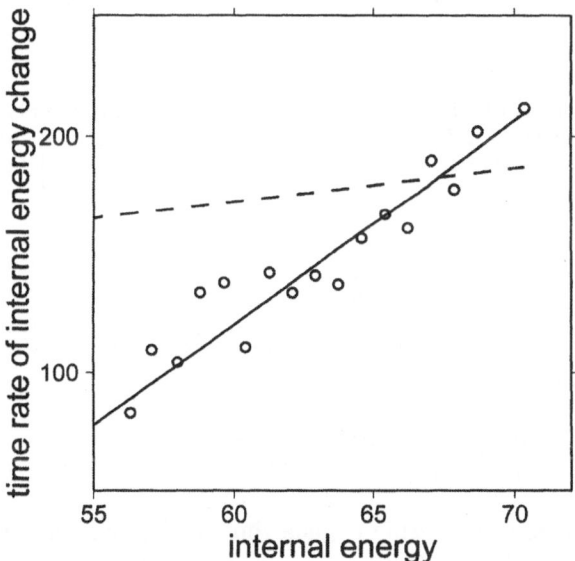

Figure 13

Rates of energy change as a function of the energy (approximately proportional to the mean-squared stress field) with catastrophic events removed for the BK model. Points indicate derived rates on energy release, the line represents a least-squares linear fit of the rate of energy release as a function of energy, and the dashed line shows the rate of energy increase due to loading as a function of energy.

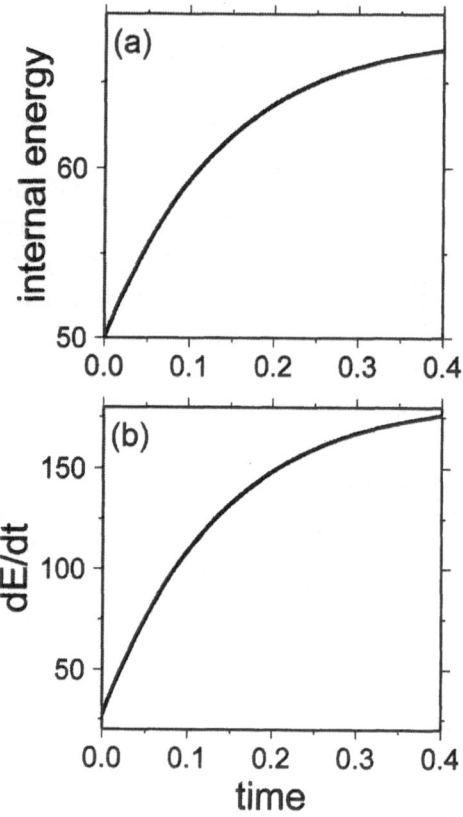

Figure 14
Statistical simulation of a block-slider system under stress prior to a large event based on Equation (19):
(a) total internal energy (E); (b) time rate of energy release (dE^-/dt). The statistically derived rate of
energy release shows an accelerating release of energy with time.

$$\frac{dE^-}{dt} \approx c_0 + c_1 E. \tag{18}$$

The rate of energy increase in the fault system is simply the difference between the
two rates (Equations (17) and (18)):

$$\frac{dE}{dt} = \frac{dE^+}{dt} - \frac{dE^-}{dt} = v\sqrt{2EN} - c_0 - c_1 E. \tag{19}$$

Equation (19) can be solved using explicit finite-difference time integration to give
energy as a function of time. Here c_0, c_1 and a typical initial energy were based on
derived values for the BK model. The rate of seismic energy release as a function
of time prior to a large event is shown in Figure 14. This shows a monotonically
increasing release of seismic energy as the system becomes more stressed, consistent
with elastic rebound theory. The degree of accelerating moment release observed on

a plot of cumulative moment release depends on the internal energy of the model at the point of the terminating large event. Cumulative moment release will be approximately parabolic with time at low initial internal energy levels. Traditionally accelerating sequences are fit with a power-law time to failure function on the basis of a material failure model (VOIGHT, 1989) or a statistical physics argument (RUNDLE et al., 1999). While the derived behavior of this statistical model does not produce the same behavior, the result is of a similar basic form: a monotonic increase in energy release with time to earthquake rupture. For a system with a Gutenberg-Richter magnitude-frequency distribution, a plot of rate of event occurrence as a function of time prior to a large event would be simply proportional to the rate of energy release plot shown in Figure 14b.

The energy of the system will eventually reach an equilibrium value at high stress when

$$\frac{dE^+}{dt} = \frac{dE^-}{dt}.$$ (20)

In this model, the internal energy will never decrease without the occurrence of a large earthquake.

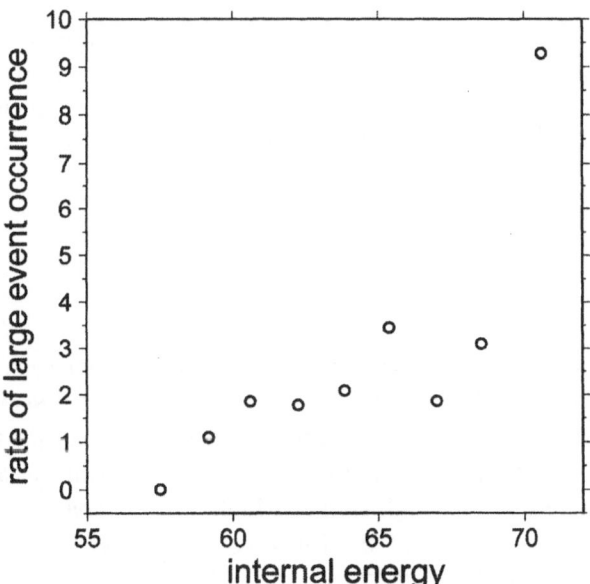

Figure 15

Rate of large event ($M > 0.7$) occurrence as a function of model internal energy. The model shows a sharp increase in the rate of large events occurrence at the highest energy (stress) levels.

Catastrophic Events

As the stress in either the block-slider or continuum system increases, a catastrophic event becomes increasingly likely. This is a consequence of the general bias in the distribution of events towards larger earthquakes at high stress levels. For the block-slider system, the rate of large event occurrence at the highest model energies is roughly four times that at moderate or low energies (see Fig. 15). A statistical change in event distribution was demonstrated for both the block-slider system and the more realistic elastic finite element model. In both models, larger events tend to occur at high stress (see Fig. 7a for the BK model and Fig. 12a for the continuum model). This is similar to the results observed by SCHOLZ (1990) in an analysis of rock microfracture where a decrease in b-value was observed with increasing stress. STEACY and MCCLOSKEY (1998) observed event distributions dependent on a normalized stress in a cellular automata model, with larger events occurring more frequently at higher stresses. This effect is most likely due to high stress levels allowing slip pulses to overcome strong portions of the model fault. For the BK model, this strength is inherent due to the implicit discreteness of the model (BEN-ZION and RICE, 1995; RICE, 1993), while for the continuum model presented here and cellular automata models it is artificially added in the form of a fractal asperity distribution.

This statistical model shows the manifestations of precursory seismicity due solely to increasing stress. This is coupled with a dramatically increasing likelihood of a large event occurring as the system becomes increasingly stressed. Indeed, the stress level of the block-slider model will never decrease without the occurrence of a great earthquake. An accelerating release of seismic energy prior to a large event in the model presented can be described as a statistical consequence of a stressed elastic system. This sequence will be inevitably terminated by a large earthquake, which relieves stress.

Conclusion

Two simple elastic systems were presented: a block-slider system based on the model proposed by Burridge and Knopoff (BURRIDGE and KNOPOFF, 1967), and an elastic continuum fault model. In both models statistical deviations are seen which can lead to the phenomena of accelerating moment release. When both models are in a state of either high stress or high stress correlation, they exhibit an increase in seismicity generally of the form of an increase in the numbers of the largest events. A statistical system based on observations of the block-slider model indicates that stress alone is sufficient to produce an observation of accelerating seismic energy release prior to a large earthquake. There is a limit to the extent to which these results may be extrapolated to natural fault systems because of the simplicity of the one-dimensional models presented here. Indeed, fully two- or three-dimensional fault systems may behave substantially different. Nonetheless, the results from these two models suggest that both mechanisms are at least a plausible explanation for the observations of precursory seismicity in nature.

Acknowledgements

This research was supported by the Australian Research Council and the University of Queensland. Simulations were carried out on QUAKES Origin 2000 supercomputer. The comments of two anonymous reviewers greatly enhanced this article.

REFERENCES

BÅK, P., and TANG, CHAO (1989), *Earthquakes as a Self-organized Critical Phenomenon*, J. Geophys. Res. *94*, 15,635–15,637.

BEN-ZION, Y., and RICE, JAMES R. (1995), *Slip Patterns and Earthquake Populations along Different Classes of Faults in Elastic Solids*, J. Geophys. Res. *100*, 12,959–12,983.

BUFE, C. G., and VARNES, D. J. (1993), *Predictive Modeling of the Seismic Cycle in the Greater San Francisco Bay Region*, J. Geophys. Res. *98*, 9871–9883.

BREHM, D. J., and BRAILE, L. W. (1999), *Intermediate-term Earthquake Prediction Using the Modified Time-to-failure Method in Southern California*, Bull. Seismol. Soc. Am. *89*, 275–293.

BURRIDGE, R., and KNOPOFF, L. (1967), *Model and Theoretical Seismicity*, Bull. Seismol. Soc. Am. *57*, 341–371.

GELLER, R. J., JACKSON, D. D., KAGAN, Y. Y., and MULARGIA, F. (1997), *Earthquakes Cannot be Predicted*, Science *275*, 1616–1617.

GUTENBERG, B., and RICHTER, C. F. (1956), *Magnitude and Energy of Earthquakes*, Ann. Geofis. *9*, 1.

HOWELL, B. F.. JR. (1997), *Patterns of Seismic Activity after Three Great Earthquakes in the Light of Reid's Elastic Rebound Theory*, Bull. Seismol. Soc. Am. *87*, 50–60.

JAUMÉ, STEVEN C., and SYKES, LYNN R. (1999), *Evolving Towards a Critical Point: A Review of Accelerating Seismic Moment/Energy Release Prior to Large and Great Earthquakes*, Pure appl. geophys. *15*, 279–309.

KING, G. C. P., STEIN, R. S., and LIN, J. (1994), *Static Stress Changes and the Triggering of Earthquakes*, Bull. Seismol. Soc. Am. *84*, 935–953.

KNOPOFF, L., LEVSHINA, T., KEILIS-BOROK, V. I., and MATTONL, C. (1996), *Increased Long-range Intermediate-magnitude Earthquake Activity Prior to Strong Earthquakes in California*, J. Geophys. Res. *101*, 5779–5796.

LANGER, J. S., CARLSON, J. M., MYERS, CHRISTOPHER R., and SHAW, BRUCE E. (1996), *Slip Complexity in Dynamic Models of Earthquake Faults*, Proc. Natl. Acad Sci USA *93*, 3825–3829.

McCLOSKEY, JOHN, and BEAN, C. J. (1994), *Temporally Unstable Recurrence of Earthquakes Due to Breaks in Fractal Scaling*, Science *266*, 410–412.

MORA, PETER, PLACE, D., ABE, S., and JAUMÉ, S. (2000a), *Lattice solid simulation of the physics of fault zones and earthquakes: the model, results and directions*. In *Geocomplexity and the Physics of Earthquakes*, John B. Rundle, Donald L. Turcotte, and William Klein, editors, Geophysical Monograph Series, Vol. 120, American Geophysical Union.

MORA, PETER, PLACE, DAVID, ABE, STEFFEN, and JAUMÉ, STEVEN (2000b), *Lattice solid simulation: Thermal effects on the earthquake dynamics, stress evolution, and the earthquake cycle*. In Proc. Int. Workshop on *Solid Earth Simulation* and ACES WG Meeting (Jan. 17–21, 2000, The University of Tokyo, Tokyo, Japan).

NADEAU, ROBERT M., and JOHNSON, LANE R. (1998), *Seismological Studies at Parkfield VI: Moment Release Rates and Estimates of Source Parameters for Small Repeating Earthquakes*, Bull. Seismol. Soc. Am. *88*, 790–814.

REID, H. F. (1910), *The Mechanics of the Earthquake*, Rept. State Earthquake Inv. Comm., The California Earthquake of April 18, 1906 (Washington, D.C.: Carnegie Inst.)

RICE, JAMES R. (1993), *Spatio-temporal Complexity of Slip on a Fault*, J. Geophys. Res. *98*, 9885–9907.

RUNDLE, JOHN B., KLEIN, W., and GROSS, SUSANNA (1999), *Physical Basis for Statistical Patterns in Complex Earthquake Populations; Models, Predictions and Tests*, Pure appl. geophys. *155*, 575–608.

SAMMIS, CHARLES G., and SMITH, STEWERT W. (1999), *Seismic Cycles and the Evolution of Stress Correlation in Cellular Automaton Models of Finite Fault Networks*, Pure appl. geophys. *155*, 307–334.

SCHOLZ, C. H., *The Mechanics of Earthquakes and Faulting* (Cambridge Univ. Press, New York, 1990).

SHAW, BRUCE E., CARLSON, J. M., and LANGER, J. S. (1992), *Patterns of Seismic Activity Preceding Large Earthquakes*, J. Geophys. Res. *97*, 479–488.

SHAW, BRUCE E. (1995), *Frictional Weakening and Slip Complexity in Earthquake Faults*, J. Geophys. Res. *100*, 18,239–18,251.

SCHOLZ, C. H. (1968), *The Frequency-magnitude Relation of Microfracturing in Rock and its Relation to Earthquakes*, Bull. Seismol. Soc. Am. *58*, 399–415.

SORNETTE, A., and SORNETTE, D. (1990), *Earthquake Rupture as a Critical Point: Consequences for Telluric Precursors*, Tectonophysics *179*, 327–334.

SORNETTE, D., and SAMMIS, C. G. (1995), *Complex Critical Exponents from Renormalization Group Theory of Earthquakes; Implications for Earthquake Predictions*, J. Phys. I France *5*, 607–619.

STEACY, SANDRA, MCCLOSKEY, J., BEAN, C. J., and REN, J. (1996), *Heterogeneity in a Self-organized Critical Earthquake Model*, Geophys. Res. Lett. *23*, 383–386.

STEACY, SANDRA, and MCCLOSKEY, JOHN (1998), *What Controls and Earthquake's size? Results from a Heterogeneous Cellular Automaton*, Geophys. J. Int. *133*, F11–F14.

STIEN, ROSS S., KING, GEOFFREY C. P., and LIN, JIAN (1994), *Stress Triggering of the 1994 M = 6.7 Northridge, California, Earthquake by its Predecessors*, Science *265*, 1432–1435.

SYKES, L. R., and JAUMÉ, S. C. (1990), *Seismic Activity on Neighboring Faults as a Long-term Precursor to Large Earthquakes in the San Francisco Bay Region*, Nature *348*, 595.

TULLIS, TERRY E., and WEEKS, JOHN D. (1986), *Constitutive Behavior and Stability of Frictional Sliding of Granite*, Pure appl. geophys. *124*, 383–414.

UMEDA, YASUHIRO, YAMASHITA, TERUO, TADA, TAKU, and KAME, NOBUKI (1996), *Possible Mechanisms of Dynamic Nucleation and Arresting of Shallow Earthquake Faulting*, Tectonophysics *261*, 179–192.

VARNES, D. J. (1989), *Predicting Earthquakes by Analyzing Accelerating Precursory Seismic Activity*, Pure appl. geophys. *130*, 661–686.

VOIGHT, BARRY (1989), *A Relation to Describe Rate-dependent Material Failure*, Science *243*, 200–203.

ZIENKIEWICZ, O. C., and TAYLOR, R. L., *The Finite Element Method* (McGraw-Hill, London 1994).

(Received October 3, 1999, revised May 5, 2000, accepted May 5, 2000)

Part IV
Scaling, Data Assimilation and Forecasting

Pure appl. geophys. 157 (2000) 2249–2258
0033–4553/00/122249–10 $ 1.50 + 0.20/0

▌Pure and Applied Geophysics

Scale-dependence in Earthquake Processes and Seismogenic Structures

KEIITI AKI[1]

Abstract—We provide a positive answer to the question whether large earthquakes can be separated from small ones using various observations on the earthquake processes and seismogenic structures. For California, the boundary of separation is at $M = 5$. This conclusion leads to the possibility of deterministic modeling of earthquake occurrence for prediction. We recognize, however, that the system of earthquake occurrence involves nonlinear physics with many degrees of freedom. We suggest that the geometrical barriers and asperities may be a good starting point for the modeling. We also discuss the implication of our finding on the numerical simulation of earthquake rupture propagation, and conclude that the model fault zone must be made thicker for a more realistic simulation of major earthquakes.

Key words: Scaling, earthquake, fault, prediction, modeling and rupture.

Introduction

A fundamental question regarding the deterministic modeling of earthquake processes is whether we can separate large earthquakes from small ones. As MATSU'URA (1999) stated in the beginning of the ACES Inaugural Workshop, if the answer is negative, there will be no earthquake cycle, no use for deterministic modeling and no hope for earthquake prediction. The purpose of the present paper is to state a positive answer to this question, based on observations of earthquake processes and fault zone structures in California.

The strongest evidence for the positive answer comes from a series of works by Li and his coworkers on the seismic guided waves trapped in the fault zone (LI *et al.*, 1990, 1994a,b, 1997a,b, 1998a,b; LI and VIDAL, 1996). In particular, the Landers, California, earthquake of 1992 offered a wealth of data for studying various aspects of fault zone trapped modes, including their spatial variations and temporal changes. By comparing the observed waveform with the synthetic waveform for a low-velocity, low-Q zone, LI *et al.* (1994a,b) estimated a fault zone width around 180 m, with shear velocity of 2.0–2.2 km/s and a Q value of about 50.

---

[1] Observatoire Volcanologique du Piton de la Fournaise, 14 Route Nationale 3, 97418 La plaine des Cafres, La Reunion, France. E-mail: aki@iremia.univ-reunion.fr

A similar estimate of fault zone width was made by an entirely different method at the same sites where the trapped modes were observed. From a detailed study of tension cracks on the surface, JOHNSON et al. (1994) concluded that the Landers fault rupture is not a distinct slip across a fault plane but rather a belt of localized shearing spread over a width of 50–200 m. We identify this shear zone with the low-velocity, low Q zone found from the trapped modes because their widths are virtually the same at the same location on the fault. Since the trapped modes were observed from aftershocks with focal depths greater than 10 km, we conclude that the shear zone found by JOHNSON et al. extends to the depth.

It is natural to identify this zone with the breakdown zone of the slip-weakening model of the shear rupture process. Ohnaka and his colleagues (OHNAKA and YAMASHITA, 1989; OHNAKA, 1996; OHNAKA and SHEN, 1999) have made a thorough study of nucleation and propagation of shear rupture by laboratory experiments on the basis of a slip-dependent constitutive law. They studied the dependence of various physical quantities inherent in shear rupture on the critical slip displacement D_c, and extrapolated their results to earthquakes, explaining the observations on strong ground motion by PAPAGEORGIOU and AKI (1983) and on the nucleation phase of earthquake by ELLSWORTH and BEROZA (1995).

In the following we shall revisit the analysis of the strong ground motion, based on the specific barrier model (PAPAGEORGIOU and AKI, 1983) in which the slip-weakening model was used for interpreting the observed upper limit frequency of the strong motion spectra. We shall find that the slip distribution and the fault zone width found from the trapped mode for the Landers earthquake fit well to the systematic magnitude dependence of the barrier model parameters found by AKI (1992) for major California earthquakes. This systematic magnitude dependence will lead to the conclusion that the magnitude of the minimum earthquake which can occur on major faults in California is 5. In other words, large and small earthquakes in California are separated at $M = 5$. This result agrees with the conclusion of KNOPOFF and LEE (1999) based on spatial and temporal distribution of aftershocks of the Kern County earthquake of 1952 and the Landers earthquake of 1992. It is also consistent with JAUMÉ's (1999) observation on the magnitude distribution of the pre-shocks of the Landers earthquake presented at the same workshop.

Magnitude Dependence of the Barrier Model Parameters for Major Earthquakes in Southern California

PAPAGEORGIOU and AKI (1983) studied the strong motion acceleration spectra obtained for major California earthquakes, using the specific barrier model in which high-frequency radiation is generated by the random rupture of circular cracks with equal radii filling a rectangular fault plane, and furthermore the high-frequency limit of the seismic excitation is attributed to a slip-weakening friction law. The idea behind the model had been developed from both theoretical and observational studies on earthquake rupture propagation (DAS and AKI, 1977; AKI, 1979). The model parameters are (1) the breakdown stress, (2) the local stress drop (the stress drop in the circular crack), (3) the barrier interval (the diameter of the circular crack), (4) Griffith's specific fracture energy, (5) the critical weakening slip, and (6) the size of the breakdown zone. The original values published by PAPAGEORGIOU and AKI (1983) were approximately corrected for the recording site effect by AKI and PAPAGEORGIOU (1989), using the amplification factor obtained by PHILLIPS and AKI (1986) and were shown in AKI (1992), where an additional result for the Loma Prieta earthquake of 1989 obtained by CHIN and AKI (1991) is included. These parameters indicated a remarkably systematic dependence on magnitude which ranges from 6.1 to 7.5. The breakdown stress increases with magnitude from 200 to 700 bars; the local stress drop increases from 100 to slightly less than 200 bars; the barrier interval increases from 2 km to just over 20 km; Griffith's energy increases from 100 to 5000 joule per cm^2, the critical weakening slip increases from 20 cm to a little over 1 m; and the size of the breakdown zone is nearly constant lying within 300 to 500 m.

Although the relative constancy of local stress drop has been known universally (HANKS and McGUIRE, 1981), the sharp increase in the barrier interval (subevent size) with the increase in magnitude is peculiar to major earthquakes in California. This occurs because the increase in magnitude is predominantly due to the increase in fault slip, and consequently the barrier interval must increase with the increasing slip to keep the local stress drop nearly constant (as required to explain the observed strong motion acceleration).

The barrier interval for the Landers earthquake of 1992 may be estimated from the variation of fault slip along the fault as determined from surface measurements (SIEH et al., 1993) and from seismic inversion (WALD and HEATON, 1995). The resultant value around 10 km for $M = 7.2$ fits well with the empirical relation mentioned above.

The relative constancy of the size of the breakdown zone, on the other hand, is consistent with the relative constancy of the fault zone width determined by the use of the trapped modes. For example, according to Li and his coworkers, the fault zone width was estimated as 180 m for the fault zone of the Landers earthquake ($M = 7.2$) and as 160 m for that of the Parkfield earthquake ($M = 6.1$). The size of the breakdown zone was estimated by PAPAGEORGIOU and AKI (1983) from the upperlimit frequency, assuming that the latter is equal to the rupture speed divided by the former. Although this relation was verified in laboratory experiments by OHNAKA and YAMASHITA (1989), the use of the upper limit frequency, so called f_{max}, for the estimation of the size of the breakdown zone has been clouded by the controversy of whether the f_{max} is due to the propagation path effect (as originally proposed by HANKS, 1982), the source effect (PAPAGEORGIOU and AKI, 1983) or the recording site effect (ANDERSON and HOUGH, 1984). In fact the recording site effect on the strong ground motion, involving the nonlinear soil response pervasively, proved to be a considerably more complicated problem than that which seismologists previously thought (AKI, 1988, 1993).

Although we still believe that the f_{max} effect on strong motion spectra from major earthquakes is primarily controlled by the earthquake source because, for example, we did not recognize any systematic difference in f_{max} between rock and soil sites for the San Fernando earthquake of 1971, we consider them as a more uncertain estimator of the breakdown zone size than the low-velocity, low Q zone revealed from the trapped modes.

An interesting consequence of the systematic magnitude dependence of the barrier model parameter, as pointed out by AKI (1992), is that if we extrapolate the relation toward smaller magnitudes, the barrier interval, or the diameter of the subevent crack, becomes comparable to the size of the breakdown zone at $M = 5$. Since this is a physically impossible situation as a rupture model, we conclude that the minimum size of earthquakes which can occur on the major faults in California is $M = 5$. AKI (1992) also drew attention to an interesting fact that the magnitude of the characteristic earthquake in the creeping zone of the San Andreas Fault in Central California is about 5. In terms of the barrier model, the creeping zone corresponds to the most heterogeneous (the shortest barrier intervals) part of the San Andreas Fault. The locked segment of the fault would correspond to the smoothest part with the longest barrier interval, roughly corresponding to the thickness of the brittle zone of the lithosphere. (See AKI, 1979 for the discussion of the repeat of the great California earthquake of 1857.)

The boundary magnitude of 5 between large and small earthquakes in California derived above is consistent with two recent studies of Southern California earthquakes. One is authored by KNOPOFF and LEE (1999) and the other by JAUMÉ (1999). Knopoff and Lee recognized that the aftershocks of the Kern County earthquake of 1952 with $M > 4.7$ occurred only in the first six months. They also found a discontinuity in the magnitude-frequency relation at about $M = 5$ for earthquakes in Southern California. They also noted that the width of the aftershock zone of the Landers earthquake of 1992 is about 3 km, which corresponds to the linear dimension of an earthquake with $M = 5$. Combining these observations, they concluded that earthquakes smaller than $M = 5$ are self-similar generated by a self-organized critical process, however those with larger magnitudes are not. Jaumé pointed out an anomalous increase in the frequency of earthquakes with $M > 5$ before the Landers earthquake, and that the fault slips in these preshocks increased the Coulomb stress on the fault plane of the Landers earthquake. These studies suggest that we need only consider earthquakes greater than $M = 5$ for a deterministic modeling of earthquake processes in California.

In order to develop a deterministic modeling of earthquake processes, it is important to know how heterogeneous the fault plane (zone) is. Our barrier concept was proposed to describe the heterogeneity, and we now conclude that the barrier interval (subevent size) on the major faults of California ranges from the fault zone width (a few hundred meters) to the brittle zone thickness (about 15 km). This type of heterogeneity may affect the propagation of seismic waves either as an attenuator or a scatterer. We shall now turn to the evidence from seismic scattering and attenuation in support of the reality of such a heterogeneity in seismically active regions.

Frequency Dependence of Q of S Waves in the Lithosphere between 0.1 and 100 Hz

The lithosphere in a seismic region containing earthquake faults may scatter and absorb seismic waves propagating through it, and the frequency dependence of the seismic attenuation may reveal the nature of the heterogeneity related to the fault. The first clear indication of the strong attenuation of seismic waves in a tectonically active region was found in the temporal decay of the coda of local earthquakes in California and Japan, which was more than ten times as rapid as those in stable areas such as the central United States and Scandinavia (AKI and CHOUET, 1975; SINGH and HERRMANN, 1983). It was also found that the Q factor of attenuation in tectonically active regions inferred from the coda decay is strongly frequency dependent, namely, Q increases nearly proportionally with frequency ranging from 1 to 10 Hz.

The frequency dependence of seismic attenuation is a difficult subject to study for the same reason that the f_{max} mentioned in the preceding section is a difficult one. The amplitudes of seismic waves are intricately combined results of the source radiation effect, the propagation path effect and the recording site effect, and it is not easy to separate one effect from the others. Most studies addressing the frequency dependence of seismic attenuation have been unsatisfactory because the spectral shapes of either the source and/or the site effect are somewhat arbitrarily assumed, until AKI (1980) found a method isolating the propagation path effect on S waves by the use of coda waves which share the same source and path effects on S waves. The method is called "the coda normalization method" by SATO and FEHLER (1998), who reviewed its development from Aki's study of Q of S waves in Japan to the multiple window analysis by FEHLER *et al.* (1992) which offered an effective method of separating the scattering Q and the intrinsic (absorptive) Q based on the Monte Carlo simulation of the envelope of local earthquake seismogram by HOSHIBA *et al.* (1991). JIN *et al.* (1994) summarized the results obtained by the application of the multiple time window analysis, and found empirically that the total Q due to both scattering and absorption effects is close to the coda Q obtained from the decay rate of seismic amplitude of local earthquake by the method of AKI and CHOUET (1975), including the strong frequency dependence between 1 and 10 Hz mentioned earlier.

This frequency dependence of Q and S waves in the lithosphere that is nearly proportional to frequency cannot be extended to frequencies lower than 1 Hz, because we know that the Q is very high at the period around 20 seconds, gained from the global measurements on surface waves (see TSAI and AKI (1969) for the first simultaneous determination of the source and propagation path effects on Rayleigh waves and MITCHEL and ROMANOVICZ (1998) for the latest review on Q of the earth). For this reason AKI (1980) proposed that there must be a peak in the reciprocal of Q (Q inverse) at a frequency slightly below 1 Hz. Some indication of such a peak has been occasionally reported, however, the well known high seismic background noise in the frequency range of 0.1 to 1 Hz, made attaining a definitive conclusion regarding the peak frequency difficult.

Recent studies of Q for frequencies higher than 10 Hz using borehole data, are bringing new insight to the frequency dependence of Q. Both ADAMS and ABERCROMBIE (1998) and YOSHIMOTO et al. (1998) found that the Q of S waves obtained from the borehole recordings of local earthquakes in Southern California and Central Japan tend to become constant for frequencies above roundly 10 to about 100 Hz.

Combining the above three pieces of observations, namely, (1) the near proportionality of Q with frequency from 1 to 10 Hz, (2) the high Q at frequencies below 0.1 Hz and (3) the nearly constant Q for frequencies above 10 Hz, we see a broad peak in the Q inverse of S waves in the lithosphere of tectonically active regions between 0.1 and 10 Hz. The corresponding range of wave length of S waves would be from a few tens of km to a few hundreds of meters. This range agrees with the range of the heterogeneity scale length of the fault plane inferred from various observations of the earthquake processes and fault zone structures in the preceding section. The reason for the upper limit was attributed to the thickness of the brittle zone in the lithosphere and that for the lower limit to the thickness (width) of the major fault zone.

We now revert to the question raised by MATSU'URA (1999), mentioned in the introduction. Our answer to his question is positive. We can separate large and small earthquakes. However the heterogeneity of the fault plane to be parameterized for a deterministic modeling may be in the scale length range of a few hundred meters to a few tens of km. We now address that which this range of heterogeneity scale implies in the deterministic modeling of earthquake processes.

Deterministic Modeling of Earthquake Processes for Prediction

Our conclusion regarding the distinction between large and small earthquakes in California leads to the general acceptance of the idea of an earthquake cycle and the plausibility of deterministic modeling of earthquake processes for prediction. We found, however, that there exists a rather broad range (2 orders of magnitude from a few hundred meters to a few tens of km) of heterogeneity scale length along the fault zone to be considered in the modeling. Since the system to be modeled is the plate boundary with the scale of several hundred km, a large number of parameters are needed to characterize it. Since the system is governed by nonlinear physics of the friction law, we are facing a problem of nonlinear physics of a system with many degrees of freedom. This is a formidable problem to solve, nonetheless the recognition of a problem may facilitate its solution.

The deterministic modeling was lacking in the prediction research program in most countries. Its absence is understandable because of difficulties with the nonlinearity of physics involved and the multitude of parameters needed to specify the system. The deterministic modeling, however, is essential for anchoring the science of earthquake prediction.

The absence of modeling in the past made the prediction research heavily dependent on the so-called "precursory signals." These signals can be generated by many different factors and should only be considered as something to constrain the model rather than objectively leading to an earthquake. The earthquake is a considerably more complicated process than that assumed in this kind of "precursory signal" approach.

We need a physical model for prediction. The Coulomb stress calculation for the effect of an earthquake on the neighbor fault segments is a good start in this direction. Although, of course, it tells only a small part of the story. We need a model that can relate to many other observed phenomena, such as foreshocks, quiescence, b-value, coda Q, etc. This belongs to the domain of nonlinear physics, which can be studied only by time-consuming computer simulations, because there is no analytic solution of the problem. In the past we tried to find a panacea which would tell us where and when an earthquake will occur. These precursors are being abandoned because none of them proved to be a panacea. I think that many of these precursors were actual physical expressions of the earthquake process, reflecting the conditions of the seismogenic structure, but it is not correct to expect any of them to be a reliable predictor from the viewpoint of the complex nonlinear physics of the earthquake system. We need to construct a model, via computer simulation, that can be constrained by these precursory observations. Our first model could consist of geometrical barriers (AKI, 1979, and subsequent papers). Once such a model is constructed, with the loading information from the GPS data, we may have some idea of the temporal and spatial distribution of stress under a seismic region that may be correlated with the occurrence of major earthquakes.

Meanwhile, our recognition of scale-dependent earthquake phenomena must be included in numerical simulations of dynamic ruptures. For example, MADARIAGA et al. (1998) made a finite difference simulation of a spontaneous propagation of rupture with the slip-weakening and rate-dependent friction law, and found that for a stable simulation, the characteristic weakening slip must more than 4 times exceed the slip unit defined by the yield stress and the grid interval which is taken to be the thickness of the fault zone.

Applying the fault zone parameters discussed earlier allows us to determine what this condition means for an actual fault. For example, for the Landers, California, earthquake of 1992, we estimate the characteristic weakening slip to be about 0.8 m, the yield (cohesive) stress to be about 650 bars, and the thickness of the fault zone to be about 180 m. These parameters effect the slip unit of Madariaga *et al.* to be about 0.8 m. This result illustrates that the condition for a stable simulation is not met if the fault zone thickness is taken to be the grid interval, as was done by MADARIAGA *et al.* (1998). The fault zone thickness must be some 4 times the grid interval for the Landers earthquake.

Conclusions

A positive answer provided the question of whether large earthquakes can be separated from small ones using various observations of the earthquake processes and seismogenic structures. For California, the boundary between them is $M = 5$. This conclusion leads to the possibility of deterministic modeling of earthquake occurrence for prediction. We recognize, however, that the system of earthquake occurrence involves nonlinear physics with many degrees of freedom. We suggest that the geometrical barriers and asperities may be a good starting point for the modeling. We also discussed the implication of our finding on the numerical simulation of earthquake rupture propagation, and conclude that the model fault zone must be made thicker for a more realistic simulation of major earthquakes.

REFERENCES

ADAMS, D. A., and ABERCROMBIE, R. E. (1998), *Seismic Attenuation above 10 Hz in Southern California from Coda Waves Recorded in the Cajon Path Borehole*, J. Geophys. Res. *103*, 24,257–24,266.
AKI, K. (1979), *Characterization of Barriers on an Earthquake Fault*, J. Geophys. Res. *84*, 6140–6148.
AKI, K. (1980), *Attenuation of Shear Waves in the Lithosphere for Frequencies from 0.05 to 25 Hz*, Phys. Earth Planet. Inter. *21*, 50–60.
AKI, K. (1988), *Local site effect on ground motion*. In *Earthquake Engineering and Soil Dynamics II: Recent Advances Ground Motion Evaluation* (J. L. von Thun, ed.), Am. Soc. Civil Eng. Geotechnical Spec. Publ. 20, 103–115.
AKI, K. (1992), *Higher-order Interrelations between Seismogenic Structures and Earthquake Processes*, Tectonophysics *211*, 1–12.
AKI, K. (1993), *Local Site Effects on Weak and Strong Ground Motion*, Tectonophysics *218*, 93–111.
AKI, K., and CHOUET, B. (1975), *Origin of Coda Waves: Source, Attenuation and Scattering Effects*, J. Geophys. Res. *80*, 3322–3342.
AKI, K., and PAPAGEORGIOU, A. S. (1989), *Separation of Source and Site Effects in Acceleration Power Spectra of Major California Earthquakes*, Proc. 9th World Conf. Earthquake Engineering *8*, 163–167.
ANDERSON, J. G., and HOUGH, S. E. (1984), *A Model for the Shape of the Fourier Amplitude Spectrum of Acceleration at High Frequencies*, Bull. Seismol. Soc. Am. *74*, 1969–1994.
CHIN, B. H., and AKI, K. (1991), *Simultaneous Determination of Source, Path and Recording Site Effects on Strong Motion during the Loma Prieta Earthquake; A Preliminary Result on Pervasive Nonlinear Site Effect*, Bull. Seismol. Soc. Am. *81*, 1859–1884.

DAS, S., and AKI, K. (1977), *Fault Planes with Barriers: A Versatile Earthquake Model*, J. Geophys. Res. *82*, 5648–5670.

ELLSWORTH, W. L., and BEROZA, G. C. (1995), *Seismic Evidence for an Earthquake Nucleation Phase*, Science *268*, 851–855.

FEHLER, M., HOSHIBA, M., SATO, H., and OBARA, K. (1992), *Separation of Scattering and Intrinsic Attenuation for the Kanto-Tokai Region, Japan, Using Measurements of S-wave Energy versus Hypocentral Distance*, Geophys. J. Int. *108*, 787–800.

HANKS, T. C. (1982), f_{max}, Bull. Seismol. Soc. Am. *72*, 1867–1879.

HANKS, T. C., and MCGUIRE, R. K. (1981), *The Character of High Frequency Strong Ground Motion*, Bull. Seismol. Soc. Am. *71*, 2071–2095.

HOSHIBA, M., SATO, H., and FEHLER, M. (1991), *Numerical Basis of the Separation of Scattering and Intrinsic Absorption from full Seismogram Envelope—A Monte-Carlo Simulation of Multiple Isotropic Scattering*, Meteorol. Geophysics *42*, 65–91.

JAUMÉ, S. C., *Stress transfer, dynamic triggering, and stress correlations: How earthquake occurrence effects the timing and slip of subsequent earthquakes*. In *1-st ACES Workshop Proc.* (ed. Mora, P.) (APEC Cooperation for Earthquake Simulation, Brisbane, Australia 1999) pp. 273–279.

JIN, A., MAYEDA, K., ADAMS, D., and AKI, K. (1994), *Separation of Intrinsic and Scattering Attenuation in Southern California Using TERRAscope Data*, J. Geophys. Res. *99*, 17,835–17,848.

JOHNSON, A. M., FLEMING, R. W., and CRUIKSHANK, K. M. (1994), *Analysis of Structures Formed during the 28 June 1992 Landers-Big Bear, California, Earthquake Sequence*, Bull. Seismol. Soc. Am. *84*, 499–510.

KNOPOFF, L., and LEE, M. W., *The Self-organization of Aftershocks*. In *1-st ACES Workshop Proc.* (ed. Mora, P.) (APEC Cooperation Earthquake Simulation, Brisbane, Australia 1999) pp. 463–464.

LI, Y. G., LEARY, P. C., AKI, K., and MALIN, P. E. (1990), *Seismic Trapped Modes for in the San Andreas Fault Zones*, Science *249*, 763–766.

LI, Y. G., AKI, K., ADAMS, D., HASEMI, A., and LEE, W. H. K. (1994a), *Seismic Guided Waves Trapped in the Fault Zone of the Landers, California, Earthquake of 1992*, J. Geophys. Res. *99*, 11,705–11,722.

LI, Y. G., VIDAL, J. E., AKI, K., MARONEY, C., and LEE, W. H. K. (1994b), *Fine Structure of the Landers Fault Zone; Segmentation and the Rupture Process*, Science *256*, 367–370.

LI, Y. G., and VIDAL, J. E. (1996), *Low-velocity Fault Zone Guided Waves; Numerical Investigations of Trapping Efficiency*, Bull. Seismol. Soc. Am. *86*, 371–378.

LI, Y. G., ELLSWORTH, W. L., THURBER, C. H., MALIN, P. E., and AKI, K. (1997a), *Fault-zone Guided Waves from Explosions in the San Andreas Fault Strands near Anza, California*, Bull. Seismol. Soc. Am. *87*, 210–221.

LI, Y. G., VERNON, F. L., and AKI, K. (1997b), *San Jacinto Fault-zone Guided Waves; A Discrimination for Recently Active Fault Strands near Anza, California*, J. Geophys. Res. *102*, 11,689–11,701.

LI, Y. G., VIDAL, J. E., AKI, K., XU, F., and BURDETTE, T. (1998a), *Evidence of Shallow Fault Zone Strengthening after the 1992 M7.5 Landers, California Earthquake*, Science *279*, 217–219.

LI, Y. G., AKI, K., VIDAL, J. E., and ALVAREZ, M. G. (1998b), *A Delineation of the Nojima Fault Ruptured in the M7.2 Kobe, Japan, Earthquake of 1995 Using Fault Zone Trapped Waves*, J. Geophys. Res. *103*, 7247–7263.

MADARIAGA, R., OLSEN, K., and ARCHULETA, R. (1998), *Modeling Dynamic Rupture in a 3D Earthquake Fault Model*, Bull. Seismol. Soc. Am. *88*, 1182–1197.

MATSU'URA, M., *Physical modeling and simulation of the earthquake cycles*. In *1-st ACES Workshop Proc.* (ed. Mora, P.) (APEC Cooperation for Earthquake Simulation, Brisbane, Australia 1999), 159–160.

MITCHELL, B. J., and ROMANOWICZ, B. (1998), *Q of the Earth*, Pure appl. geophys. *153*, 235–713.

PAPAGEORGIOU, A. S., and AKI, K. (1983), *A Specific Barrier Model for the Quantitative Description of Inhomogeneous Faulting and Prediction of Strong Motion, Part 1. Description of the Model*, Bull. Seismol. Soc. Am. *73*, 693–722. *Part 2. Applications of the Model*, Bull. Seismol. Soc. Am. *73*, 953–978.

PHILLIPS, W. S., and AKI, K. (1986), *Site Amplification of Coda Waves from Local Earthquakes in Central California*, Bull. Seismol. Soc. Am. *76*, 627–648.

OHNAKA, M. (1996), *Nonuniformity of the Constitutive Law Parameters for Shear Rupture and Quasistatic Nucleation to Dynamic Rupture; a Physical Model of Earthquake Generation Processes*, Proc. Natl. Acad. Sci. U.S.A. *93*, 3795–3802.

OHNAKA, M., and SHEN, L. F. (1999), *Scaling of the Shear Rupture Process from Nucleation to Dynamic Propagation; Implications of Geometric Irregularity of the Rupturing Surfaces*, J. Geophys. Res. *104*, 817–844.

OHNAKA, M., and YAMASHITA, T. (1989), *A Cohesive Zone Model for Dynamic Shear Faulting Based on Experimentally Inferred Constitutive Relation and Strong Motion Source Parameters*, J. Geophys. Res. *94*, 4089–4104.

SATO, H., and FEHLER, M., *Seismic Wave Popagation and Scattering in the Heterogeneous Earth* (Springer-Verlag, New York 1998) 308 pp.

SIEH, K., JONES, L., HAUKSSON, E., HUDNUT, K., EBERHART-PHILLIPS, D., HEATON, T., HOUGH, S., HUTTON, K., KANAMORI, H., LILJE, A., LINDVALL, S., MCGILL, S., MORI, J., RUBIN, C., SPOTILA, J., STOCK, J., THIO, H. K., TREIMAN, J., WERNICKE, B., and ZACHARIASEN, J. (1993), *Near-field Investigations of the Landers Earthquake Sequence, April to July, 1992*, Science *260*, 171–176.

SINGH, S. K., and HERRMANN, R. B. (1983), *Regionalization of Crustal Coda Q in the Continental United States*, J. Geophys. Res. *88*, 527–538.

TSAI, Y. B., and AKI, K. (1969), *Simultaneous Determination of the Seismic Moment and Attenuation of Seismic Surface Waves*, Bull. Seismol. Soc. Am. *59*, 275–287.

WALD, D. J., and HEATON, T. H. (1995), *Spatial and Temporal Distribution of Slip for the 1992 Landers, California, Earthquake*, Bull. Seismol. Soc. Am. *84*, 668–691.

YOSHIMOTO, K., SATO, H., IIO, Y., ITO, H., OHMINATO, T., and OHTAKE, M. (1998), *Frequency-dependent Attenuation of High-frequency P and S Waves in the Upper Crust in Western Nagano, Japan*, Pure appl. geophys. *153*, 489–502.

(Received September 20, 1999, revised March 31, 2000, accepted April 6, 2000)

To access this journal online:
http://www.birkhauser.ch

Pure appl. geophys. 157 (2000) 2259–2282
0033–4553/00/122259–24 $ 1.50 + 0.20/0

❘ Pure and Applied Geophysics

A Physical Scaling Relation Between the Size of an Earthquake and its Nucleation Zone Size

Mitiyasu Ohnaka[1]

Abstract—A specific model of the earthquake nucleation that proceeds on a non-uniform fault is put forward to explain seismological data on the nucleation in terms of the underlying physics. The model is compatible with Gutenberg-Richter's similarity law for earthquake frequency-magnitude relation. A theoretical approach in the framework of fracture mechanics, based on a laboratory-based slip-dependent constitutive law, leads to the conclusion that the earthquake moment M_o scales with the third power of the critical slip displacement D_c and the critical size $2L_c$ (L_c, half-length) of the nucleation zone. This scaling relation quantitatively explains seismological data published, and it predicts that $2L_c$ is of the order of 10 km for earthquakes with $M_o = 10^{21}$ Nm, 1 km for earthquakes with $M_o = 10^{18}$ Nm, and 100 m for earthquakes with $M_o = 10^{15}$ Nm, under the assumption that the breakdown stress drop $\Delta\tau_b = 10$ MPa. However, L_c depends on not only D_c but also $\Delta\tau_b$, so that the scaling relation between L_c and D_c may be violated by $\Delta\tau_b$, because $\Delta\tau_b$ potentially takes any value in a wide range from 1 to 10^2 MPa, depending on the seismogenic environment. The good agreement between the theoretical relation and observed results suggests that a large earthquake may result from the failure of a large patch of high rupture growth resistance, whereas a small earthquake may result from the breakdown of a small patch of high rupture growth resistance. The present result encourages one to pursue the prediction capability for large earthquakes.

Key words: Earthquake nucleation, inhomogeneous fault, high rupture growth resistance, a slip-dependent constitutive law.

Introduction

Physical nature of the shear rupture nucleation on an inhomogeneous fault has been revealed by high-resolution laboratory experiments (Ohnaka and Kuwa-hara, 1990; Ohnaka and Shen, 1999). In particular, Ohnaka and Shen (1999) have conclusively demonstrated that the rupture nucleation consists of two phases: an initial, stable and quasi-static phase, and the subsequent, unstable and accelerating phase, and that the nucleation zone size and its duration are consistently scaled in terms of the slip-dependent constitutive law that governs the rupture process. One of the constitutive law parameters, which is referred to as the critical slip displacement, has been found to be a scaling parameter prescribed by the character-

[1] Earthquake Prediction Research Center, Earthquake Research Institute, The University of Tokyo, Yayoi 1-1-1, Bunkyo-ku, Tokyo 113-0032, Japan. E-mail: ohnaka-m@eri.u-tokyo.ac.jp

istic length representing geometric irregularity of the rupture surfaces. These provide a basis for modeling the earthquake nucleation in terms of the underlying physics, and allow one to discuss a scaling relation between the sizes of an earthquake and its nucleation zone specifically. The nucleation is defined here as the transition process from an initial phase of stable, quasi-static rupture up to the critical stage beyond which the rupture propagates at a high-speed close to sonic velocities.

As discussed in an earlier paper (OHNAKA and SHEN, 1999), there are two classes of physical quantities inherent in the shear rupture: scale-dependent quantities, and scale-independent quantities. The scale-dependent quantities include the nucleation zone size, its duration, the shear rupture energy, and the seismic moment. OHNAKA and SHEN (1999) have shown that scale dependence of scale-dependent physical quantities is commonly ascribed to the scale dependence of the critical slip displacement. The critical slip displacement D_c is defined as the slip displacement required for the local strength in the breakdown zone behind the rupture front to degrade to a residual friction stress level. The laboratory experiments have also demonstrated that D_c is greatly affected by the characteristic length representing geometric irregularity of the rupturing surfaces. On the other hand, theoretical analyses (RICE, 1980; OHNAKA and YAMASHITA, 1989) show that D_c is directly related to the breakdown zone size, which is relevant to the critical size of the nucleation zone. This suggests that the size of an earthquake should be related to the critical size of the nucleation zone, given that the larger the entire fault size, the larger characteristic length is in general included in the fault zone.

The objective of this paper is to show theoretically, on the basis of a specific model of the nucleation based on the laboratory experiments, how the size of a mainshock earthquake scales with the critical size of the nucleation zone. More specifically, a scaling relation that the earthquake moment is proportional to the third power of the critical size of the nucleation zone will be derived, based on a physical model of the nucleation. ELLSWORTH and BEROZA (1995) have empirically found, among other things, that the mainshock seismic moment scales with the seismic nucleation zone size, and hence the relation derived theoretically will be compared with seismological data on earthquake nucleation, particularly data analyzed by ELLSWORTH and BEROZA (1995). It will be shown that the theoretical relation agrees quantitatively with those data on earthquake nucleation.

A Constitutive Law for Earthquake Rupture

There are increasing amounts of evidence that the earthquake rupture that takes place in the brittle layer in the earth's crust is a mixed process between what is called frictional slip failure and fracture of intact rock mass, so that the constitutive law for the earthquake rupture should be formulated as a unifying law that governs both frictional slip failure and shear fracture of intact rock. In fact, shear fracture strength of intact material is the upper end member of the strength of frictional slip failure on the mating surfaces of the same material (OHNAKA, 1996; OHNAKA *et al.*, 1997). The slip-dependent constitutive law is a unifying law that governs both frictional slip failure and shear fracture of intact rock. We will derive a scaling relation that exists between the seismic moment of a mainshock earthquake and its nucleation zone size, by assuming a laboratory-based slip-dependent constitutive law as the governing law for the earthquake rupture in the framework of fracture mechanics.

The slip-dependent constitutive formulation presumes the slip displacement to be an independent and essential variable, and the rate- or time-dependence to be of secondary significance. That is, the shear traction is expressed as a function of the slip displacement in this formulation (Fig. 1), and the parameters prescribing the law, such as the peak shear strength τ_p, the breakdown stress drop $\Delta\tau_b$, and the critical slip displacement D_c, are assumed to be an implicit function of the slip velocity or time (OHNAKA *et al.*, 1997; OHNAKA, 1998). In Figure 1, τ_i denotes the initial strength on the verge of slip at the rupture front, τ_p the peak shear strength attained at the slip displacement D_a, τ_r the residual frictional stress, and D_c the critical slip displacement defined as the minimum amount of slip required for the shear strength to degrade to τ_r. The breakdown stress drop $\Delta\tau_b$ is defined as the shear stress difference between τ_p and τ_r.

The slip-dependent constitutive law is prescribed by a set of constitutive parameters (τ_i, τ_p, $\Delta\tau_b$, D_a, D_c), and the values of these parameters are affected by seismogenic environments. In particular, the rupturing surfaces are in mutual

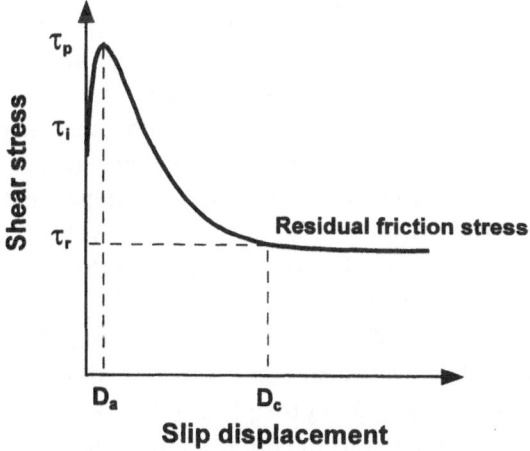

Figure 1

A slip-dependent constitutive relation for the shear rupture. τ_i denotes the initial shear stress on the verge of slip, τ_p denotes the peak shear strength, τ_r denotes the residual friction stress, D_a denotes the slip displacement at which the peak strength is attained, and D_c denotes the critical slip displacement.

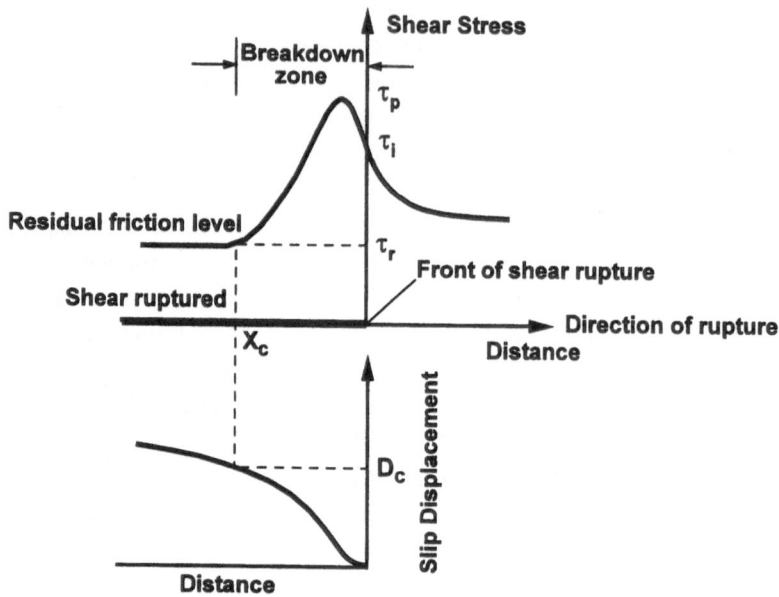

Figure 2

A physical model of the breakdown zone behind the rupture front, derived from a constitutive relation as shown in Figure 1. τ_i denotes the initial shear stress on the verge of slip at the rupture front, τ_p denotes the peak shear strength, τ_r denotes the residual friction stress, D_c denotes the critical slip displacement, and X_c denotes the breakdown zone size.

contact and are interacting throughout the breakdown process of the shear rupture, and hence D_a and D_c are severely influenced by geometric irregularity of the rupturing surfaces. As noted in the previous section, D_c is scale-dependent, and the constitutive law for shear rupture includes this scaling parameter. In this sense, the constitutive law is also scale-dependent. This scale-dependence is a very important property that plays an essential role in scaling scale-dependent physical quantities inherent in the shear rupture (OHNAKA and SHEN, 1999).

The breakdown zone is defined as the zone behind the rupture front over which the shear strength degrades transitionally to a residual friction stress level along the fault. Figure 2 shows a model of the breakdown zone behind a rupture front, which can be specified once a slip-dependent constitutive relation as shown in Figure 1 is given. The breakdown zone size X_c in the phase of dynamic rupture propagating at a high speed V_c is closely related to the critical slip displacement D_c by the relation (OHNAKA and YAMASHITA, 1989):

$$\frac{D_c}{X_c} = k\frac{\Delta\tau_b}{\mu} \tag{1}$$

where μ is the rigidity, and k is a well-defined, dimensionless quantity. The dimensionless quantity k has been calculated as

$$k = \frac{\Gamma}{\pi^2 \xi C(V_c)} \tag{2}$$

for a dynamic slip-weakening model (OHNAKA and YAMASHITA, 1989). Here, ξ in equation (2) represents a numerical parameter, $C(V_c)$ represents a known function of the rupture velocity V_c and Γ is a dimensionless parameter defined by (OHNAKA and YAMASHITA, 1989)

$$\Gamma = \int_0^1 \frac{\sigma(Y)}{\sqrt{Y}} \, dY \tag{3}$$

where $\sigma(Y)$ is the non-dimensional shear strength at a non-dimensional distance Y measured from the rupture front in the breakdown zone. $C(V_c)$ has a different functional form according to either in-plane shear (mode II) or anti-plane shear (mode III) (see OHNAKA and YAMASHITA, 1989). Relation (1) will be used later for deriving a scaling relation between the seismic moment of a mainshock earthquake and its nucleation zone size.

An Earthquake Nucleation Model

There is commanding evidence that earthquake faults in the seismogenic layer are inherently inhomogeneous. For instance, an "asperity" (KANAMORI and STEWART, 1978; KANAMORI, 1981) or "barrier" (AKI, 1979, 1984) on earthquake faults is a local patch of high rupture growth resistance on the fault. Such a local patch of high rupture growth resistance may be a mainfestation of geometric structure and/or irregularity of the fault (zone). An earthquake fault in general exhibits a geometrically irregular structure of various scales, and high resistance of the rupture growth can be attained at portions of fault bend or stepover, and at interlocking asperities on the fault surfaces with geometric irregularity.

Recent high-resolution laboratory experiments (OHNAKA and SHEN, 1999) have demonstrated that fault inhomogeneity plays an important role in scaling the rupture nucleation process; in other words, the characteristic length representing geometric irregularity of the rupturing surfaces is a key to scaling scale-dependent physical quantities inherent in the rupture, including the nucleation. It is therefore unrealistic to assume a uniform fault, and hence we assume an inhomogeneous fault model. For simplicity, however, the fault model assumed here comprises local, strong patches of high rupture growth resistance, and the remaining weak portion. There is a physical constraint to be imposed on the strength of such local patches; that is, the upper limit of the patch strength equals the shear strength of intact rock at lithospheric conditions. Such a local patch of high rupture growth resistance on the fault may be a physical manifestation of what has been called "barrier" (AKI, 1979, 1984) or "asperity" (KANAMORI and STEWART, 1978; KANAMORI, 1981). The size of a local patch of high rupture growth resistance may represent a characteristic distance on the fault.

We presume that the earthquake source at shallow depths is a shear rupture instability that takes place on such an inhomogeneous fault in the seismogenic layer. The laboratory experiments (OHNAKA and KUWAHARA, 1990; OHNAKA and SHEN, 1999) reveal that the rupture begins to nucleate at a position of the lowest resistance of rupture growth (or weakest toughness), and that the nucleation necessarily proceeds toward the remaining unbroken area of higher rupture growth resistance on the fault. One may argue that the rupture should nucleate at a position of high stress concentration. However, this is not the case for the rupture that nucleates on an inhomogeneous fault where the strength distribution is non-uniform, because the stress cannot concentrate beyond the strength. It will thus be reasonable to assume that the earthquake nucleation begins to occur, as a consequence of the tectonic loading, somewhere at a weakest portion on the fault, and that the nucleation is necessarily required to proceed toward the remaining unbroken area of higher rupture growth resistance. At this stage, the nucleation proceeds stably and quasi-statically, because no elastic strain energy stored in the medium surrounding the fault has been released. For the rupture to develop spontaneously, the elastic strain energy stored in the surrounding medium needs to be released.

As noted in the previous section, it has been demonstrated conclusively (OHNAKA and SHEN, 1999) that the nucleation process consists of two phases: an initial, stable and quasi-static phase (phase I), and the subsequent, unstable and accelerating phase (phase II). Phase I is a steady rupture growth controlled by the rate of an applied load, such as the tectonic loading. On the other hand, phase II is a spontaneous rupture extension driven by the release of the elastic strain energy stored in the surrounding medium (OHNAKA and SHEN, 1999). This physically means that the rupture cannot begin to propagate abruptly at a high speed close to sonic velocities immediately after the stored elastic strain energy is released. When the constitutive property of the fault is inhomogeneous, the rupture is required to grow at accelerating speeds from the quasi-static phase to the phase of high-speed rupture propagation. It can be inferred for major earthquakes that the time required for the rupture to grow from a slow phase of the rupture velocity being of the order of a few mm/s to the phase of high-speed rupture close to the shear wave velocity, is of the order of a few to a few tens of hours (OHNAKA and SHEN, 1999).

If the fault is tough enough, a large amount of the elastic strain energy can be stored in the medium surrounding the fault. However, if the fault is very weak everywhere on the entire fault, an adequate amount of the strain energy to bring about a large earthquake cannot be stored. We assume that patches of high rupture growth resistance on a fault are tough enough for an adequate amount of the elastic

A Rupture Nucleation Model

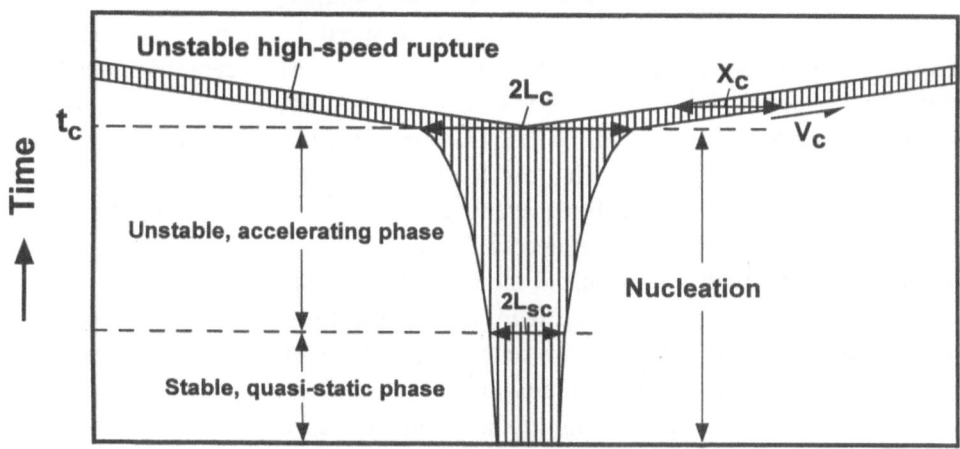

Figure 3

Schematic diagram of a rupture nucleation model. In this model, the rupture initially grows stably and quasi-statically at a steady speed to a critical length $2L_{sc}$, from which it extends spontaneously at accelerating speeds reaching another critical length $2L_c$ at a time $t = t_c$. Beyond the critical length $2L_c$, the rupture propagates bi-directionally at a constant, high-speed V_c close to the shear-wave velocity. The hatched portion represents the zone in which the breakdown (or slip-weakening) is proceeding with time. X_c denotes the breakdown zone size, and $2L_c$ denotes the critical size of the nucleation zone.

strain energy to be stored, but that the remaining portion of the fault is too weak to sustain an adequate amount of the elastic strain energy. In this case, the rupture may not be able to propagate spontaneously until the critical condition at which one of the patches of high rupture growth resistance is broken down by slow growth of rupture nucleation is met. If the patch size is geometrically large, the total amount of the elastic strain energy stored in the surrounding medium is large, so that the resulting earthquake will be large. On the other hand, a large amount of D_c is by definition required for the breakdown of a geometrically large patch, and the large amount of D_c necessarily leads to a large size of the nucleation zone. This qualitatively predicts that the size of an earthquake scales with the nucleation zone size.

With all the above mentioned in mind, we assume a specific rupture nucleation model shown in Figure 3. The hatched portion in Figure 3 shows the zone in which the breakdown (or slip-weakening) is proceeding with time. In this model, the nucleation initially proceeds stably and quasi-statically at a steady speed V_{st} (phase I) up to a critical length L_{sc}, beyond which the nucleation extends spontaneously at accelerating speeds (phase II) to another critical length L_c (at a time $t = t_c$ in Fig. 3). It has been found that the rupture growth rate at phase II increases with an increase in the rupture growth length according to a power law (OHNAKA and SHEN, 1999). After the rupture growth length has reached the critical length L_c at $t = t_c$, the rupture propagates at a constant, high-speed V_c close to the shear-wave velocity. This model is based on the fact revealed with recent high-resolution laboratory experiments (OHNAKA and SHEN, 1999). The behavior of rupture growth changes at the critical length $2L_{sc}$ (L_{sc}, half-length) from a quasi-static phase controlled by the rate of the tectonic loading to a self-driven, dynamic phase controlled by the inertia. Note, however, that the critical size of the nucleation zone is not defined here as $2L_{sc}$, but defined as $2L_c$, so as to make it possible to compare the present model with seismological data on the nucleation.

Although the model shown in Figure 3 is bi-directional, we can also assume a similar model in which the rupture extends uni-directionally. Since whether the rupture extends bi-directionally or uni-directionally is extraneous to the subject to be discussed here, we simply employ a bi-directional rupture model discussed previously (OHNAKA, 1996; OHNAKA and SHEN, 1999). In the present bi-directional rupture model, the critical size $2L_c$ (L_c, half-length) of the nucleation zone is related to the breakdown zone size X_c in the phase of dynamic, high-speed rupture as follows:

$$L < X_c \quad \text{for} \quad t < t_c \left. \right\}.$$
$$L_c = X_c \quad \text{at} \quad t = t_c$$

$$(4)$$

We thus have from (1) and (4)

$$L_c = \frac{1}{k} \frac{\mu}{\Delta \tau_b} D_c \qquad (5)$$

which shows that the critical size L_c of the nucleation zone can be expressed explicitly in terms of the slip-dependent constitutive law parameters D_c and $\Delta \tau_b$ (OHNAKA and SHEN, 1999). It may be worthwhile noting here that $\Delta \tau_b$ and D_c represent the breakdown stress drop and the critical slip displacement, respectively, around the critical stage $t = t_c$ beyond which the rupture propagates at a high-speed V_c close to sonic velocities.

Rupture Growth Resistance

The rupture growth resistance is a distinct physical quantity, in the framework of fracture mechanics, defined as the shear rupture energy required for the rupture front to further grow. The shear rupture energy G_c is defined by (PALMER and RICE, 1973)

$$G_c = \int_0^{D_c} [\tau(D) - \tau_r] \, dD \tag{6}$$

where $\tau(D)$ represents a slip-dependent constitutive relation, which governs the relation between the shear traction τ and the slip displacement D on the fault in the breakdown zone behind the rupture front. Once a slip-weakening constitutive relation $\tau(D)$ is specified, G_c can be calculated from (6), and expressed in terms of the constitutive law parameters $\Delta\tau_b$ and D_c as (OHNAKA and YAMASHITA, 1989)

$$G_c = \frac{1}{2} \Gamma \Delta\tau_b D_c \tag{7}$$

where Γ has been defined in equation (3).

Γ can be regarded as virtually constant. If a simplified, linear slip-weakening relation is assumed, it is derived from (6) that Γ is exactly unity. However, laboratory-based slip-dependent constitutive relations are found to be nonlinear. OHNAKA and YAMASHITA (1989) have estimated Γ to be about $1/2$ from (3) by numerical calculation, using nonlinear constitutive relations observed during dynamic frictional slip failure on a simulated fault in the laboratory. On the other hand, $\Gamma = 1$ has been evaluated from experimental data on nonlinear constitutive relations for the shear fracture of intact rock sample tested at lithospheric conditions (OHNAKA et al., 1997). It is thus confirmed that the assumption that Γ is of the order of unity is valid even for nonlinear constitutive relations, and hence, $\Gamma = 1$ will be used later in the present analysis.

We note here the energy required for fault rupture. The shear rupture energy $\overline{G_c}$ averaged over the entire fault area S is defined by:

$$\overline{G_c} = \frac{1}{S} \int_S G_c \, dS. \tag{8}$$

Let S_{A1} be the area of the geometrically largest patch of high rupture growth resistance (which may be called asperity) on the fault. Equation (8) may be rewritten as follows:

$$\overline{G_c} = \frac{1}{S} \left[\int_{S_{A1}} G_c \, dS + \int_{S - S_{A1}} G_c \, dS \right] \tag{9}$$

where the first term of the right-hand side of equation (9) denotes the integral over the area S_{A1} of the geometrically largest asperity on the fault, and the second term denotes the integral over the rest $S - S_{A1}$ of the fault area. The first term of the right-hand side of equation (9) is a fraction a (< 1) of $S\overline{G_c}$, so that we have

$$\overline{G_c} = \frac{1}{aS} \int_{S_{A1}} G_c \, dS = \left(\frac{S_{A1}}{aS} \right) \overline{G_c^{A1}} \tag{10}$$

where

$$\overline{G_c^{A1}} = \frac{1}{2} \Gamma \Delta \tau_b^{A1} D_c^{A1}. \tag{11}$$

Here, $\Delta \tau_b^{A1}$ and D_c^{A1} represent the breakdown stress drop and the critical slip displacement, respectively, required for the geometrically largest asperity on the fault to break down.

Equation (7) or (11) shows that the shear rupture energy is directly related to the critical slip displacement. This indicates that G_c is necessarily scale-dependent, because D_c is scale-dependent. The scale-dependence of D_c has been more fully discussed by OHNAKA (1998), and OHNAKA and SHEN (1999). The scale-dependence of G_c may be understood from the following consideration. Real rupture surfaces cannot be plane, but they have geometric irregularity (or roughness). The rougher the ruptured surfaces, the larger the real surface area becomes. However, the irregularity of real rupture surfaces is not taken into consideration to evaluate G_c, and hence G_c may be called the apparent rupture energy. It thus follows that the apparent rupture energy G_c becomes larger as the rupture surfaces become rougher (OHNAKA et al., 1997).

More specifically, the rupture surfaces in general have a band-limited fractal nature (see Discussion section). In this case, there is a characteristic length scale representing geometric irregularity of the rupture surfaces, and this characteristic length becomes longer on the rougher rupture surfaces. Since D_c scales with the characteristic length, G_c also scales with the characteristic length. This will be more fully discussed elsewhere, since in-depth discussion about this is beyond the scope of the present paper. A large-scale fault tends to include a large characteristic length scale. Thus, G_c defined by (6) or (7) is not only dependent on the fault material property, but also scale-dependent.

A Scaling Relation between Seismic Moment and Critical Size of Nucleation Zone

We first consider how the seismic moment M_o of a mainshock earthquake scales with the critical slip displacement D_c. Let $\bar{D}$ be the slip amount averaged over the entire fault area S, and $\overline{\Delta \tau}$ be the stress drop $\Delta \tau$ averaged over S. The seismic moment M_o is defined by (AKI, 1966)

$$M_o = \mu \bar{D} S. \tag{12}$$

The earthquake rupture finally arrests when the driving force has become lower than the rupture growth resistance. The condition of the rupture arrest may thus be written as $L(\Delta \tau)^2 \leq \kappa \mu G_c$ (κ being a constant), if the breakdown zone size X_c behind the rupture front is sufficiently small compared with the final fault length L. The sign of equality in this equation represents the condition at which the rupture grows quasi-statically. However, this may be regarded as the 'critical' condition below which the dynamic rupture must arrest. In the situation that the driving force (or $\Delta\tau$) gradually diminishes with distance toward the fault end, or that the resistance G_c gradually increases with outward distance from the fault end, we thus expect that the following relation:

$$(\overline{\Delta\tau})^2 = \kappa\mu\frac{\overline{G_c}}{L} \tag{13}$$

holds between $\overline{G_c}$, $\overline{\Delta\tau}$, and L.

If we further assume the following scaling relations:

$$S = c_1 L^2 \tag{14}$$

and

$$\bar{D} = c_2 L \tag{15}$$

where c_1 and c_2 are numerical constants, we have from (10)–(15)

$$M_o = c_1 c_2 \left(\frac{\kappa\Gamma}{2}\right)^3 \left(\frac{S_{A1}}{aS}\right)^3 \left(\frac{\mu}{\overline{\Delta\tau}}\right)^3 \left(\frac{\Delta\tau_b^{A1}}{\overline{\Delta\tau}}\right)^3 \mu(D_3^{A1})^3. \tag{16}$$

On the right-hand side of equation (16), the asperity area S_{A1}, the entire fault area S, and the critical slip displacement D_c^{A1} are scale-dependent, while the rest of the parameters are scale-independent. If it is assumed that the ratio $S_{A1}/(aS) = \overline{G_c}/\overline{G_c^{A1}}$ is scale-independent, we find that the critical slip displacement D_c^{A1} is the only scale-dependent parameter on the right-hand side of equation (16). If we assume that the stress drop $\overline{\Delta\tau}$ averaged over the fault area and the breakdown stress drop $\Delta\tau_b$ in the breakdown zone behind the rupture front are virtually constant (in a statistical sense), the equation (16) predicts theoretically that the mainshock seismic moment is proportional to the third power of the critical slip displacement at a geometrically largest patch of high rupture growth resistance on the fault.

The scaling relation $M_o \propto D_c^3$ at an asperity of the geometrically largest size on the fault, is well grounded, because D_c is by definition the critical slip displacement required for the breakdown of a local patch of the high rupture growth resistance, and because a large amount of the critical slip displacement is required for the breakdown of a patch of geometrically large size.

Similarly, from (5) and (16), we have the relation between the seismic moment M_o and the critical size $2L_c$ of the nucleation zone:

$$M_o = c_1 c_2 \left(\frac{k\kappa\Gamma}{4}\right)^3 \left(\frac{S_{A1}}{aS}\right)^3 \left(\frac{\Delta\tau_b^{A1}}{\overline{\Delta\tau}}\right)^6 \mu(2L_c)^3. \tag{17}$$

Equations (16) and (17) lead to the conclusion that the seismic moment is primarily prescribed by the property of the geometrically largest asperity on the fault, and that the seismic moment be proportional to the third power of the critical slip displacement and the critical size of the nucleation zone, if the assumptions made are physically reasonable.

The relations (16) and (17) have been derived theoretically by assuming an asperity model of earthquake fault, and scaling laws, in the framework of fracture mechanics based on a slip-dependent constitutive law. To what extent these theoretical relations are justified can be checked by comparing them with seismological data on the nucleation. This will be done in next section.

Comparison of Theoretical Relations with Seismological Data

Slip-dependent constitutive law parameters have been estimated for actual earthquakes by PAPAGEORGIOU and AKI (1983), ELLSWORTH and BEROZA (1995), and IDE and TAKEO (1997). PAPAGEORGIOU and AKI (1983) estimated the breakdown stress drop $\Delta\tau_b$, the critical slip displacement D_c, the breakdown zone size X_c, and the shear rupture energy G_c for earthquakes with moderate to large earthquakes, based on a specific barrier model for earthquake faulting. ELLSWORTH and BEROZA (1995) analyzed near-source recordings of the slow initial P wave for earthquakes with a wide magnitude range 2.6 to 8.1, and they evaluated the breakdown stress drop $\Delta\tau_b$ and the nucleation zone size L_c. IDE and TAKEO (1997) estimated constitutive relations for the 1995 Kobe earthquake from near-field seismic waves by waveform inversion and the solution of elastodynamic equations using a finite difference method. The basic parameters of the present concern, $\Delta\tau_b$, D_c, X_c, and L_c, together with the seismic moment M_o, are compiled for these earthquakes in Table 1.

Since D_c for earthquakes was not evaluated by ELLSWORTH and BEROZA (1995), we have evaluated this parameter for earthquakes listed in Table 1 in their paper. D_c can be evaluated from equation (5). In order to evaluate $\Delta\tau_b$ and L_c, Ellsworth and Beroza assumed that the longitudinal wave velocity $V_P = 6$ km/s, the shear wave velocity $V_S = V_P/1.73$, the rupture velocity $V = 0.8V_S$, and $\mu = 30,000$ MPa. Under these assumptions, and assuming that $\Gamma/\xi = 3.3$ (OHNAKA and YAMASHITA, 1989), we have $k = 2.9$ for an in-plane shear crack (mode II), and $k = 3.5$ for an anti-plane shear crack (mode III). With this in mind, we have assumed that $k = 3$ for a circular crack model employed by Ellsworth and Beroza, and evaluated D_c for those earthquakes from (5). The evaluated values for D_c are listed in Table 1.

For the estimate of L_c for the Kobe earthquake, we have presumed that at its hypocenter, whose depth was determined to be 16 km, the nucleation reached the critical stage beyond which the rupture propagated at a high speed V_c. From (5), we have estimated $L_c = 1700$ m using the values $\Delta\tau_b = 3$ MPa and $D_c = 0.5$ m evaluated by IDE and TAKEO (1997). They suggest, however, that 0.5 m is the upper limit of D_c at a deeper part of the fault where the nucleation must have reached the critical stage. In this case, 1700 m will be the upper limit of L_c for the Kobe earthquake. If $\Delta\tau_b = 3$ MPa and, for instance, $D_c = 0.3$ m are assumed, we have $L_c = 1000$ m. These suggest that L_c for the Kobe earthquake was of the order of 1×10^3 m.

Table 1

Constitutive parameters for earthquakes

Event	M_o (Nm)	$\Delta\tau_b$ (MPa)	D_c (m)	X_c (m)	L_c (m)	References
Fort Tejon 1857	$(5.3-9.0) \times 10^{20}$	50–70	3–4	1000–2000		PAPAGEORGIOU and AKI (1983)
Kern County 1952	2.0×10^{20}	68	3	1000		PAPAGEORGIOU and AKI (1983)
San Fernando 1971	1.2×10^{19}	48	1	500		PAPAGEORGIOU and AKI (1983)
Borrego Mountain 1968	6.3×10^{18}	30–40	0.4	600		PAPAGEORGIOU and AKI (1983)
Long Beach 1933	2.8×10^{18}	20	0.4	600		PAPAGEORGIOU and AKI (1983)
Parkfield 1966	1.4×10^{18}	20	0.4	500		PAPAGEORGIOU and AKI (1983)
19 Sep 1985	1.4×10^{21}	5.0	3.2		6300	ELLSWORTH and BEROZA (1995)
28 Jun 1992	9×10^{19}	4.1	1.4		3400	ELLSWORTH and BEROZA (1995)
25 Apr 1989	2.4×10^{19}	3.4	2.9×10^{-1}		850	ELLSWORTH and BEROZA (1995)
18 Oct 1989	2.8×10^{19}	1.9	4.2×10^{-1}		2200	ELLSWORTH and BEROZA (1995)
17 Jan 1994	1×10^{19}	40	2.4		600	ELLSWORTH and BEROZA (1995)
15 Oct 1979	6×10^{18}	14	1.4		1000	ELLSWORTH and BEROZA (1995)
9 Jun 1980	4.8×10^{18}	6.0	2.5×10^{-1}		410	ELLSWORTH and BEROZA (1995)
24 Oct 1993	5.8×10^{18}	5.5	2.9×10^{-1}		520	ELLSWORTH and BEROZA (1995)
28 Jun 1992	2×10^{18}	8.1	4.5×10^{-1}		560	ELLSWORTH and BEROZA (1995)
23 Apr 1992	1.4×10^{18}	17	2.6×10^{-1}		150	ELLSWORTH and BEROZA (1995)
31 May 1990	7.5×10^{17}	64	8.3×10^{-1}		130	ELLSWORTH and BEROZA (1995)
29 Jun 1992	4.8×10^{17}	2.9	1.2×10^{-1}		420	ELLSWORTH and BEROZA (1995)
28 Jun 1991	2.8×10^{17}	18	8.6×10^{-1}		480	ELLSWORTH and BEROZA (1995)
20 Mar 1994	8.9×10^{16}	42	9.7×10^{-1}		230	ELLSWORTH and BEROZA (1995)
3 Dec 1988	2.4×10^{16}	15	3.3×10^{-1}		220	ELLSWORTH and BEROZA (1995)
16 Jan 1993	2.4×10^{16}	8.8	2.6×10^{-1}		300	ELLSWORTH and BEROZA (1995)
14 Nov 1993	2.0×10^{16}	1.1	4.5×10^{-2}		410	ELLSWORTH and BEROZA (1995)
11 Aug 1993	1.3×10^{16}	8.3	2.2×10^{-1}		270	ELLSWORTH and BEROZA (1995)
3 Feb 1994	2×10^{15}	9.6	7.0×10^{-2}		73	ELLSWORTH and BEROZA (1995)
6 Feb 1994	1.4×10^{15}	8.0	3.7×10^{-2}		46	ELLSWORTH and BEROZA (1995)
2 Feb 1994	5×10^{14}	2.3	1.5×10^{-2}		66	ELLSWORTH and BEROZA (1995)
27 Oct 1991	3.5×10^{14}	6.5	3.4×10^{-2}		52	ELLSWORTH and BEROZA (1995)
6 Feb 1994	3.5×10^{14}	8.3	3.0×10^{-2}		36	ELLSWORTH and BEROZA (1995)
26 Oct 1992	2.5×10^{14}	100	2.8×10^{-2}		28	ELLSWORTH and BEROZA (1995)
1 Feb 1994	2.5×10^{14}	3.4	1.1×10^{-2}		32	ELLSWORTH and BEROZA (1995)
2 Jan 1990	1.8×10^{14}	8.2	3.2×10^{-2}		39	ELLSWORTH and BEROZA (1995)
4 Feb 1994	1.8×10^{14}	8.4	2.9×10^{-2}		35	ELLSWORTH and BEROZA (1995)
8 Mar 1994	1.3×10^{14}	0.4	4×10^{-3}		100	ELLSWORTH and BEROZA (1995)
30 Jan 1988	8.1×10^{13}	60	1.1×10^{-1}		18	ELLSWORTH and BEROZA (1995)
8 Nov 1992	7.9×10^{12}	5.4	5.9×10^{-3}		11	ELLSWORTH and BEROZA (1995)
Kobe 17 Jan 1995	1.9×10^{19}	1.5	1		1000	IDE and TAKEO (1997)
		3	<0.5		1700	

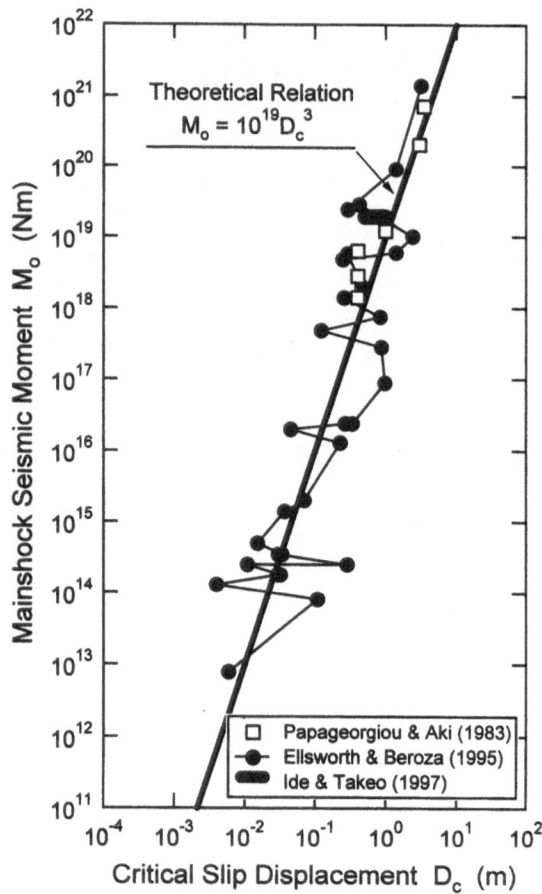

Figure 4

A plot of the seismic moment M_o against the critical slip displacement D_c for earthquakes listed in Table 1. The theoretical scaling relation, denoted by a thick line in the figure, is compared with seismological data.

SHIBAZAKI and MATSU'URA (1998) computed the far-field velocity waveform from a circular fault model with the slip-time function obtained by the numerical simulation of a slip failure event governed by a slip-dependent constitutive law, and they demonstrated that the seismic nucleation phase defined by ELLSWORTH and BEROZA (1995) on a seismogram corresponds to the critical nucleation phase at $t = t_c$ shown in Figure 3, at the earthquake source. This demonstrates that the critical size $2L_c$ of the nucleation zone defined here should equal twice the nucleation zone radius estimated by ELLSWORTH and BEROZA (1995), and concurrently, justifies the presumption that the critical size of the nucleation zone defined in this paper is compared directly with the seismic nucleation zone size estimated by Ellsworth and Beroza.

Figure 4 shows a plot of the logarithm of the mainshock seismic moment M_o against the logarithm of the critical slip displacement D_c for earthquake data compiled in Table 1. In Figure 4, white squares are data points taken from PAPAGEORGIOU and AKI (1983), black circles from ELLSWORTH and BEROZA (1995), and a black rectangle from IDE and TAKEO (1997). A thick straight line in the figure indicates the theoretical scaling relation:

$$M_o = 1 \times 10^{19}(D_c)^3 \tag{18}$$

where M_o and D_c are measured in Nm and m, respectively. It will be shown later that relation (18) is derived from the theoretical equation (16). It is found from Figure 4 that there is a trend for earthquake data that the seismic moment increases with an increase in the critical slip displacement, although there is a considerable fluctuation. This empirical trend is compared with the theoretical scaling relation (18), and we find that the trend agrees well with the theoretical scaling relation.

Figure 5 presents a plot of the logarithm of the seismic moment M_o against the logarithm of the breakdown stress drop $\Delta\tau_b$ for earthquake data compiled in Table 1. In contrast with Figure 4, it is found from Figure 5 that there is no such trend for the relation between M_o and $\Delta\tau_b$, indicating that the breakdown stress drop is independent of the seismic moment. This justifies the presumption that the breakdown stress drop is scale-independent. It is seen from Figure 5 that the average value for $\Delta\tau_b$ is roughly 10 MPa, though $\Delta\tau_b$ fluctuates in a range from 1 to 100 MPa. This average value will be used later to derive the scaling relation (18) from the theoretical relation (16).

In Figure 6, the logarithm of the mainshock seismic moment M_o is plotted against the logarithm of the critical size $2L_c$ of the nucleation zone for earthquakes compiled in Table 1. Figure 6 shows that the mainshock seismic moment scales with the critical size of the nucleation zone. This scaling relation has been found empirically by ELLSWORTH and BEROZA (1995). The nucleation zone size L_c has not been estimated for earthquakes analyzed by PAPAGEORGIOU and AKI (1983). Note, however, that the nucleation zone size L_c (half-length) equals the breakdown zone size X_c in the context of the present model. Contemplating that L_c is of the order of X_c, M_o for earthquakes analyzed by Papageorgiou and Aki has been overplotted in Figure 6 against $2L_c$ calculated from relation (4). White squares in Figure 6 are data points from PAPAGEORGIOU and AKI (1983), black circles from ELLSWORTH and BEROZA (1995), and a black rectangle from IDE and TAKEO (1997). A thick straight line in the figure represents the theoretical scaling relation:

$$M_o = 1 \times 10^9(2L_c)^3 \tag{19}$$

where M_o and L_c are measured in Nm and m, respectively. This relation is derived from equation (17), which will be shown below. Figure 6 shows good agreement between the theoretical prediction and seismological data.

We discuss here how relations (18) and (19) can be derived from the theoretical relations (16) and (17), respectively. To derive relations (18) and (19), we must assume appropriate values for the parameters: $c_1, c_2, S_{A1}/(aS), \mu, \overline{\Delta\tau}$, and $\Delta\tau_b^{A1}$, under the constraint that $\overline{\Delta\tau} < \Delta\tau_b^{A1}$. We assume here $c_1 = 0.5$ and $c_2 = 5 \times 10^{-5}$ in view of the values given by ABE (1975). We further assume $\mu = 30,000$ MPa, $\overline{\Delta\tau} = 3$ MPa, and $\Delta\tau_b^{A1} = 10$ MPa. Although we at present have no specific information as to what values both S_{A1}/S and a should take, it will not be unreasonable to assume $S_{A1}/S = 0.4$ and $a = 0.6$ (and hence $S_{A1}/(aS) = 2/3$) for the order-of-estimates. Under these assumptions, and given that $k = 3$, $\Gamma = 1$, and $\kappa = 2$, equation (16) is reduced to the relation: $M_o = 0.8 \times 10^{19}D_c^3 \approx 1 \times 10^{19}D_c^3$, and equation (17) is reduced to the relation: $M_o = 1.0 \times 10^9(2L_c)^3$. It has thus been demonstrated that equations (18) and (19) are derived from (16) and (17), respectively. The agreement

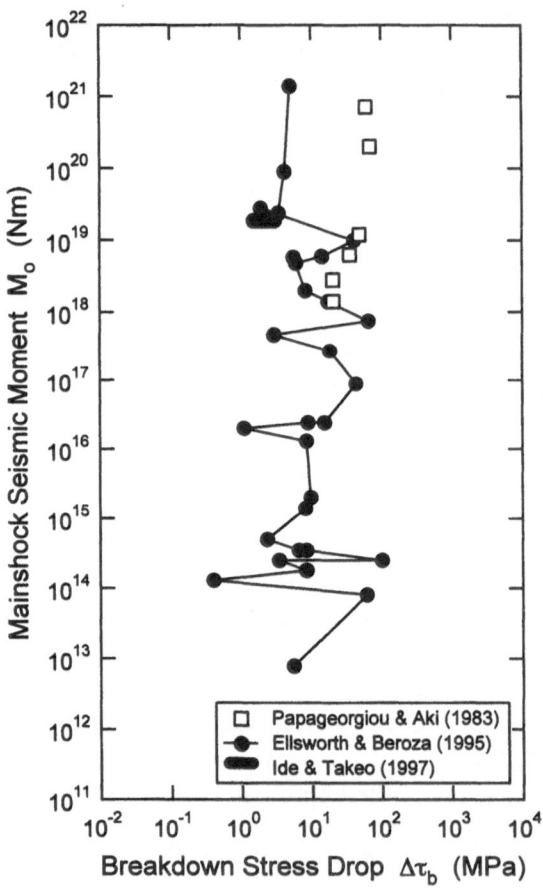

Figure 5
A plot of the seismic moment M_o against the breakdown stress drop $\Delta\tau_b$ for earthquakes listed in Table 1.

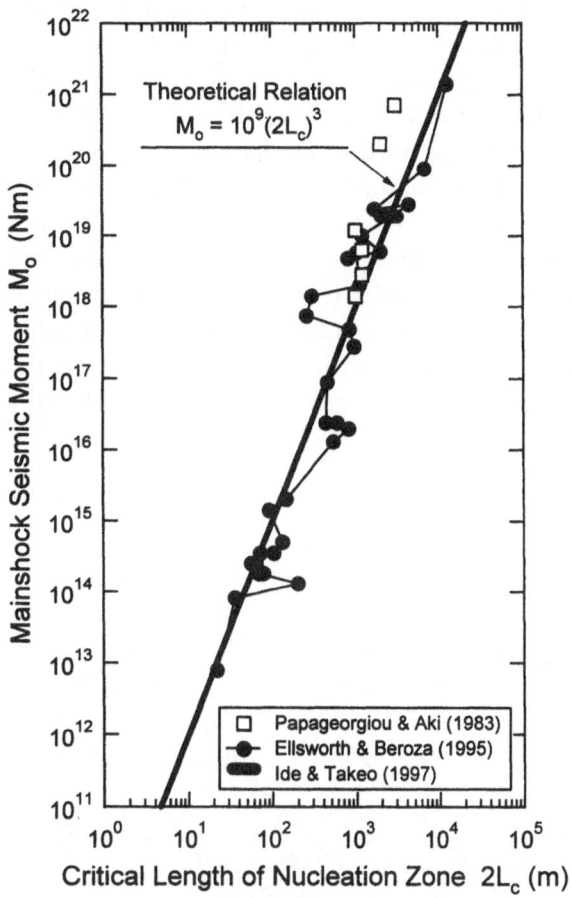

Figure 6
A plot of the seismic moment M_o against the critical size $2L_c$ of the nucleation zone for earthquakes listed in Table 1. The theoretical scaling relation denoted by a thick line in the figure is compared with seismological data.

of both equations (18) and (19) with seismological data suggests that the assumption of $S_{A1}/(aS) = 2/3$ is not unreasonable for earthquakes compiled in Table 1.

From (19) we have, for instance, $2L_c = 10$ km for earthquakes with $M_o = 10^{21}$ Nm, $2L_c = 1$ km for earthquakes with $M_o = 10^{18}$ Nm, and $2L_c = 100$ m for earthquakes with $M_o = 10^{15}$ Nm. These are estimates for the nucleation zone size for 'normal' earthquakes for which $\Delta\tau_b$ has been assumed to be 10 MPa. In reality, however, individual earthquakes have different values for $\Delta\tau_b$ (see Fig. 5), and equation (5) indeed shows that L_c depends on both D_c and $\Delta\tau_b$. Although $\Delta\tau_b$ is scale-independent, this does not mean that $\Delta\tau_b$ has a constant value for any earthquake. If a strong patch of high rupture growth resistance whose strength equals shear strength of intact rock mass is broken down, $\Delta\tau_b$ necessarily has a high value of the order of 10 to 100 MPa, depending on the individual, seismogenic environment (OHNAKA *et al.*, 1997). By contrast, if frictional slip failure occurs along a pre-existing fault of very low rupture growth resistance, $\Delta\tau_b$ may have a low value of the order of 1 MPa or less (OHNAKA and SHEN, 1999). It thus follows that $\Delta\tau_b$ can take any value in a wide range over 10^2 MPa, depending on the seismogenic environment. Hence, the proportional relationship between L_c and D_c may be severely violated by a fluctuation of $\Delta\tau_b$. Figure 7 illustrates how the proportional relation between L_c and D_c is violated by the fluctuation of $\Delta\tau_b$ for actual earthquakes. In Figure 7, white squares denote earthquake data from PAPAGEOR-GIOU and AKI (1983), black circles from ELLSWORTH and BEROZA (1995), and a black rectangle from IDE and TAKEO (1997). Four straight lines in Figure 7 represent theoretical scaling relations between L_c and D_c under the assumption that the parameter $\Delta\tau_b$ has constant values: 0.1, 1, 10, and 100 MPa, respectively.

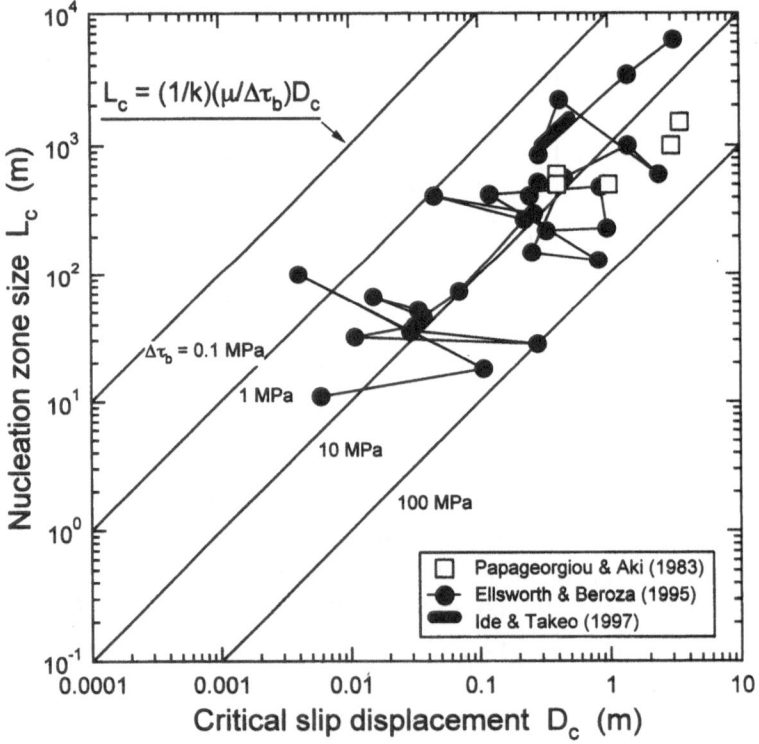

Figure 7
Scaling relation between L_c and D_c. Four straight lines in the figure represent theoretical relations between L_c and D_c with $\Delta\tau_b$ having constant values: 0.1, 1, 10, and 100 MPa, respectively. White squares, black circles, and a black rectangle denote earthquake data listed in Table 1. The scaling relation between L_c and D_c for actual earthquakes is violated by a fluctuation of $\Delta\tau_b$.

Discussion

We have proposed a fault model that the nucleation zone size and the final size of the resulting earthquake are both prescribed by a common patch (or asperity) of high rupture growth resistance on the fault. In this model, the nucleation includes the process during which the elastic strain energy sustained by a patch of high rupture growth resistance is released by failure of the patch. If the patch size is geometrically large, the resulting earthquake is large, because a geometrically larger patch of high rupture growth resistance can sustain a larger amount of the elastic strain energy, and because the failure of the geometrically larger patch results in the release of the larger amount of the elastic strain energy stored in the surrounding medium. On the other hand, a large amount of the critical slip displacement is required for the large patch of high rupture growth resistance to break down, resulting in the large size of the nucleation zone. Although the model may be overly simplified, it explains published data on the earthquake nucleation quantitatively, and allows one to provide a consistent comprehension for the earthquake nucleation in terms of the underlying physics. This suggests that an essential element of physical nature of the earthquake generation that occurs on a fault characterized by inhomogeneity has been incorporated into the model.

The present model implies that regions of a few kilometers in dimension on a seismogenic fault would slip, initially at a very slow and steady rate, and subsequently at accelerating rates, over distances ranging one meter during the nucleation of a large earthquake. Note that such slips at an initially slow and steady rate, and subsequently at accelerating rates over distances of the order of a meter can occur with or without producing micro-earthquakes, depending on the fault zone structure and ambient conditions (such as temperature). For simplicity, we take for instance a non-uniform fault which has a sizeable asperity region on which irregularities (or micro-asperities) of short wavelength components are superimposed. In this case, a slow slip failure in the asperity region necessarily brings about fracture of micro-asperities in the region, and hence it carries micro-earthquakes (immediate foreshocks). A model for this has been proposed in an earlier paper (OHNAKA, 1992), and immediate foreshock activities induced during the nucleation process have been discussed for certain earthquakes (OHNAKA, 1993; DODGE et al., 1995, 1996; SHIBAZAKI and MATSU'URA, 1995). It has been demonstrated in the laboratory that micro-seismic activities are indeed induced during the slip failure nucleation.

There is a pervasive idea that natural faults have a fractal nature at all scales, and that earthquake phenomena are explained by the model of self-organized criticality. If this is the case, earthquakes are unpredictable catastrophes. However, there is commanding counter-evidence against the above pervasive idea. Firstly, it is true that natural fault surfaces exhibit self-similarity over finite bandwidths (SCHOLZ and AVILES, 1986; AVILES et al., 1987; OKUBO and AKI, 1987), but they cannot be self-similar at all scales. The self-similarity is limited by the depth of seismogenic layer and fault segment size. When self-similarity of the fault surfaces is band-limited, a different fractal dimension can be calculated for each band bounded by upper- and lower-corner wavelengths (AVILES et al., 1987; OKUBO and AKI, 1987), and the corner wavelength that separates the neighboring two bands is a characteristic length representing geometric irregularity of the fault surfaces. In particular, pre-existing, matured faults, such as the San Andreas Fault, have a range of characteristic length scales departed from the self-similarity (AVILES et al., 1987; OKUBO and AKI, 1987; AKI, 1992, 1996), and not only the depth of seismogenic layer and fault segment size, but also barrier or asperity size on the fault, and the thickness of fault zone will be representative characteristic length scales (AKI, 1992, 1996). The earthquake generation process and its eventual size are necessarily prescribed and characterized by these characteristic length scales (SCHOLZ, 1982, 1994; ROMANOWICZ, 1992; AKI, 1992, 1996; MATSU'URA and SATO, 1997). Secondly, a large earthquake can take place on such a pre-existing, large-scale fault, only after a large amount of the elastic strain energy has been stored in the medium surrounding the fault. The elastic strain energy is accumulated with the tectonic stress buildup after the earthquake occurrence. However, a prolonged time period is needed for the strain energy to be again stored up to a critical level which has the potential to produce an ensuing large earthquake on the same fault. These indicate that the model of self-organized criticality is not applicable to large earthquakes (see also KNOPOFF, 1996).

The well-known Gutenberg-Richter power-law relation for earthquakes is a scale-independent relation, and this scale-independence implies that the underlying physics is to be found in scale-independent processes (KNOPOFF, 1996). Thus, the Gutenberg-Richter relation has been explained by the model of self-organized criticality (e.g., BAK and TANG, 1989). I wish to show that the Gutenberg-Richter relation:

$$N(M) \propto 10^{-bM} \tag{20}$$

where $N(M)$ is the cumulative number of earthquakes with magnitudes greater than M, and b is a numerical constant close to unity, can be derived from the present model, if a power-law relation holds for the asperity (or patch) size distribution (AKI, 1981):

$$N(R) \propto R^{-d} \tag{21}$$

where $N(R)$ is the cumulative number of asperities whose characteristic size is greater than R, and d is the fractal dimension. The power-law relation with d close to 2 has commonly been observed for crater size distribution, and fragment size distribution resulting from the crushing of a heterogeneous body. Since D_c is by definition the critical amount of the displacement required for fracture of an asperity with the characteristic size R in the present context, a proportional relationship holds between D_c and R; that is,

$$D_c \propto R \tag{22}$$

On the other hand, the seismic moment M_o is related to the magnitude M by (KANAMORI and ANDERSON, 1975)

$$M_o \propto 10^{1.5M}. \tag{23}$$

We thus have from (18), (21), (22), and (23)

$$N(M) \propto 10^{-(d/2)M} \tag{24}$$

and from (20) and (24)

$$b \equiv \frac{d}{2} \tag{25}$$

which agrees with the relation derived by AKI (1981). When $b \approx 1$ we have $d \approx 2$ for the asperity size distribution from (25). The present model thus leads to the conclusion that the b-value is a manifestation of the fractal dimension for the asperity size distribution arising from heterogeneities inherent in the fault zone and the seismogenic layer.

The present model indicates that "an earthquake knows from the very beginning how large it is going to be," and it has been shown that the model is indeed applicable to a class of actual earthquakes complied in Table 1. Theoretically, this implies that an earthquake is basically predictable. Practically, however, an earthquake may still not be predictable, unless it is known beforehand how non-uniformly the rupture growth resistance is distributed on the real fault (or fault network) in the seismogenic layer, and unless the ongoing nucleation can successfully be identified and monitored by any observational means. Nevertheless, the present result encourages us to pursue the prediction capability for large earthquakes.

Conclusions

Assuming a specific model of the earthquake nucleation on a non-uniform fault, a scaling relation between the seismic moment of a mainshock earthquake and the critical slip displacement, or the critical size of the nucleation zone has been derived theoretically in the framework of fracture mechanics based on a laboratory-based slip-dependent constitutive law. The present approach leads to the conclusion that the earthquake moment M_o should be related to the critical slip displacement D_c by equation (16), and the critical size $2L_c$ (L_c, half-length) of the nucleation zone by equation (17). Equations (16) and (17) are reduced to equations (18) and (19), respectively, for 'normal' earthquakes for which local stress drop $\Delta\tau_b$ in the breakdown zone behind the rupture front and the stress drop $\overline{\Delta\tau}$ averaged over the entire fault area S have been assumed to be 10 MPa and 3 MPa, respectively. The scaling relations derived well explain seismological data published. In particular, equation (19) predicts that $2L_c$ is of the order of 10 km for earthquakes with $M_o = 10^{21}$ Nm, 1 km for earthquakes with $M_o = 10^{18}$ Nm, and 100 m for earthquakes with $M_o = 10^{15}$ Nm. Since L_c depends on not only D_c but also $\Delta\tau_b$, the scaling relation between L_c and D_c may be violated by $\Delta\tau_b$. This results because $\Delta\tau_b$ potentially takes any value in a wide range from 1 to 10^2 MPa, depending on the seismogenic environment. The good agreement between the theoretical relations and seismological data suggests that a large earthquake may result from the failure of a large patch of high rupture growth resistance which is capable of sustaining an adequate amount of the elastic strain energy stored in the surrounding medium, wheareas a small earthquake may result from the breakdown of a small patch of high rupture growth resistance. The Gutenberg-Richter frequency-magnitude relation can be derived from the present model. The present result encourages one to pursue the prediction capability for large earthquakes.

Acknowledgements

I am grateful to Raul Madariaga and anonymous reviewers for reviewing the original manuscript, whose critical and constructive comments contributed to enhancing the value of the manuscript.

REFERENCES

ABE, K. (1975), *Reliable Estimation of the Seismic Moment of Large Earthquakes*, J. Phys. Earth *23*, 381–390.

AKI, K. (1966), *Generation and Propagation of G Waves from the Niigata Earthquake of June 16, 1964: Part 2. Estimation of Earthquake Moment, Released Energy and Stress Drop from the G Wave Spectra*, Bull. Earthq. Res. Inst., Univ. Tokyo *44*, 73–88.

AKI, K. (1979), *Characterization of Barriers on an Earthquake Fault*, J. Geophys. Res. *84*, 6140–6148.

AKI, K. *A probabilistic synthesis of precursor phenomena*. In *Earthquake Prediction: An International Review* (eds. Simpson, D. W., and Richards, P. G.) (American Geophysical Union, Washington, DC 1981) pp. 566–574.

AKI, K. (1984), *Asperities, Barriers, Characteristic Earthquakes and Strong Motion Prediction*, J. Geophys. Res. *89*, 5867–5872.

AKI, K. (1992), *Higher-order Interrelations between Seismogenic Structures and Earthquake Processes*, Tectonophysics *211*, 1–12.

AKI, K. (1996), *Scale-dependence in Earthquake Phenomena and its Relevance to Earthquake Prediction*, Proc. Natl. Acad. Sci. USA *93*, 3740–3747.

AVILES, C. A., SCHOLZ, C. H., and BOATWRIGHT, J. (1987), *Fractal Analysis Applied to Characteristic Segments of the San Andreas Fault*, J. Geophys. Res. *92*, 331–344.

BAK, P., and TANG, C. (1989), *Earthquakes as a Self-organized Critical Phenomenon*, J. Geophys. Res. *94*, 15,635–15,637.

DODGE, D. A., BEROZA, G. C., and ELLSWORTH, W. L. (1995), *Evolution of the 1992 Landers, California, Foreshock Sequence and its Implications for Earthquake Nucleation*, J. Geophys. Res. *100*, 9865–9880.

DODGE, D. A., BEROZA, G. C., and ELLSWORTH, W. L. (1996), *Detailed Observations of California Foreshock Sequences: Implications for the Earthquake Initiation Process*, J. Geophys. Res. *101*, 22,371–22,392.

ELLSWORTH, W. L., and BEROZA, G. C. (1995), *Seismic Evidence for an Earthquake Nucleation Phase*, Science *268*, 851–855.

IDE, S., and TAKEO, M. (1997), *Determination of Constitutive Relations of Fault Slip Based on Seismic Wave Analysis*, J Geophys. Res. *102*, 27,379–27,391.

KANAMORI, H., *The nature of seismic patterns before large earthquakes*. In *Earthquake Prediction: An International Review* (eds. Simpson, D. W., and Richards, P. G.) (American Geophysical Union, Washington, DC 1981) pp. 1–19.

KANAMORI, H., and ANDERSON, D. L. (1975), *Theoretical Basis of Some Empirical Relations in Seismology*, Bull. Seismol. Soc. Am. *65*, 1073–1095.

KANAMORI, H., and STEWART, G. S. (1978), *Seismological Aspects of the Guatemala Earthquake of February 4, 1976*, J. Geophys. Res. *83*, 3427–3434.

KNOPOFF, L. (1996), *A Selective Phenomenology of the Seismicity of Southern California*, Proc. Natl. Acad. Sci. USA *93*, 3756–3763.

MATSU'URA, M., and SATO, T. (1997), *Loading Mechanism and Scaling Relations of Large Interplate Earthquakes*, Tectonophysics *277*, 189–198.

OHNAKA, M. (1992), *Earthquake Source Nucleation: A Physical Model for Short-term Precursors*, Tectonophysics *211*, 149–178.

OHNAKA, M. (1993), *Critical Size of the Nucleation Zone of Earthquake Rupture Inferred from Immediate Foreshock Activity*, J. Phys. Earth *41*, 45–56.

OHNAKA, M. (1996), *Nonuniformity of the Constitutive Law Parameters for Shear Rupture and Quasistatic Nucleation to Dynamic Rupture: A Physical Model of Earthquake Generation Processes*, Proc. Natl. Acad. Sci. USA *93*, 3795–3802.

OHNAKA, M. (1998), *Earthquake generation processes and earthquake prediction: Implications of the underlying physical law and seismogenic environments*. In *Long-term Earthquake Forecasts* (eds. Ishibashi, K., Ikeda, Y., Satake, K., Hirata, N., and Matsu'ura, M.), Special Issue of J. Seismol. Soc. Japan, Ser. 2 *50*, 129–155.

OHNAKA, M., and KUWAHARA, Y. (1990), *Characteristic Features of Local Breakdown near a Crack-tip in the Transition Zone from Nucleation to Unstable Rupture during Stick-slip Shear Failure*, Tectonophysics *175*, 197–220.

OHNAKA, M., and SHEN, L.-F. (1999), *Scaling of the Shear Rupture Process from Nucleation to Dynamic Propagation: Implications of Geometric Irregularity of the Rupturing Surfaces*, J. Geophys. Res. *104*, 817–844.

OHNAKA, M., and YAMASHITA, T. (1989), *A Cohesive Zone Model for Dynamic Shear Faulting Based on Experimentally Inferred Constitutive Relation and Strong Motion Source Parameters*, J. Geophys. Res. *94*, 4089–4104.

OHNAKA, M., AKATSU, M., MOCHIZUKI, H., ODEDRA, A., TAGASHIRA, F., and YAMAMOTO, Y. (1997), *A Constitutive Law for the Shear Failure of Rock under Lithospheric Conditions*, Tectonophysics *277*, 1–27.

OKUBO, P. G., and AKI, K. (1987), *Fractal Geometry in the San Andreas Fault System*, J. Geophys. Res. *92*, 345–355.

PALMER, A. C., and RICE, J. R. (1973), *The Growth of Slip Surfaces in the Progressive Failure of Over-consolidated Clay*, Proc. Roy. Soc. Lond. *A332*, 527–548.

PAPAGEORGIOU, A. S., and AKI, K. (1983), *A Specific Barrier Model for the Quantitative Description of Inhomogeneous Faulting and the Prediction of Strong Ground Motion. Part II. Applications of the Model*, Bull. Seismol. Soc. Am. *73*, 953–978.

RICE, J. R., *The mechanics of earthquake rupture*. In *Physics of the Earth's Interior, Proc. Int. Sch. Phys. Enrico Fermi* (eds. Dziewonski, A. M., and Boschi, E.) (North-Holland, Amsterdam 1980) pp. 555–649.

ROMANOWICZ, B. (1992), *Strike-slip Earthquakes on Quasi-vertical Transcurrent Faults: Inferences for General Scaling Relations*, Geophys. Res. Lett. *19*, 481–484.

SCHOLZ, C. H. (1982), *Scaling Laws for Large Earthquakes: Consequences for Physical Models*, Bull Seismol. Soc. Am. *72*, 1–14.

SCHOLZ, C. H. (1994), *A Reappraisal of Large Earthquake Scaling*, Bull. Seismol. Soc. Am. *84*, 215–218.

SCHOLZ, C. H., and AVILES, C. A., *The fractal geometry of faults and faulting*. In *Earthquake Source Mechanics, Geophys. Monogr. Ser. 37* (eds. Das, S., Boatwright, J., and Scholz, C. H.) (AGU, Washington DC 1986) pp. 147–155.

SHIBAZAKI, B., and MATSU'URA, M. (1995), *Foreshocks and Pre-shocks Associated with the Nucleation of Large Earthquakes*, Geophys. Res. Lett. *22*, 1305–1308.

SHIBAZAKI, B., and MATSU'URA, M. (1998), *Transition Process from Nucleation to High-speed Rupture Propagation: Scaling from Stick-slip Experiments to Natural Earthquakes*, Geophys. J. Int. *132*, 14–30.

(Received September 30, 1999, revised March 25, 2000, accepted April 6, 2000)

To access this journal online:
http://www.birkhauser.ch

Pure appl. geophys. 157 (2000) 2283–2302
0033–4553/00/122283–20 $ 1.50 + 0.20/0

© Birkhäuser Verlag, Basel, 2000

❘Pure and Applied Geophysics

Regional Difference in Scaling Laws for Large Earthquakes and its Tectonic Implication

YOSHIHIRO FUJII[1] and MITSUHIRO MATSU'URA[2]

Abstract—We compiled 67 large earthquakes which occurred at and around plate boundaries for the last 140 yrs, and classified them into four groups; interplate strike-slip events, intraplate strike-slip events, underthrust events at island-arc subduction zones, and underthrust events at continental-margin subduction zones. For each group of earthquakes we examined relations between seismic moment M_0, fault length L, fault width W and average fault slip D, and found the following scaling laws. In the case of interplate strike-slip events, the well-known L-cubed dependence of seismic moment breaks down when L exceeds 30 km, because the extent of the seismogenic zone is limited in depth (≤ 12 km). For large events ($L \geq 30$ km), D and M_0 increase with L as $D = \overline{\Delta\tau}L/\mu(\alpha L + \beta)$ and $M_0 = \overline{\Delta\tau}\overline{W}L^2/(\alpha L + \beta)$, respectively, where the mean fault width $\overline{W}$ is 12 km and the mean stress drop $\overline{\Delta\tau}$ is 1.8 MPa. Here μ, α and β are structural parameters. For intraplate strike-slip events we obtained nearly the same relations, except for significantly higher stress drop (3.1 MPa). The difference in stress drop between interplate and intraplate events may be ascribed to the difference in stress accumulation rates and thus the recurrence time of earthquakes. In the case of underthrust events at island-arc subduction zones we also found the saturation of fault width ($\overline{W} = 120$ km) and the breakaway from the L-cubed dependence of M_0 for events larger than $L = 200$ km. If we consider the average dip-angle of plate boundaries at island-arc subduction zones to be 20–30°, this indicates that the extent of the seismogenic zone in depth is limited to 40–60 km. In the case of continental-margin subduction zones, on the other hand, we could not find the saturation of fault width nor the breakaway from the L-cubed dependence of M_0 from the analysis of the present data set ($W \leq 200$ km, $L \leq 1000$ km). For sufficiently large earthquakes, in general, the downward rupture growth is limited to a certain depth due to the existence of a ductile unstressed region which extends under the brittle seismogenic zone. Since the brittle-ductile transition occurs at 300–400°C, the difference in the lower limit of the seismogenic zones between tectonically different regions may be attributed to the difference in thermal state there.

Key words: Scaling law, large earthquake, plate boundary, seismic zone, thermal structure.

Introduction

In order to characterize a seismic source, we generally use seismic moment M_0, fault area S (or fault length L and fault width W), average fault slip D, and average

[1] Department of Earth and Planetary Science, University of Tokyo, 7-3-1, Hongo, Bunkyo-ku, Tokyo 113-0033, Japan. E-mail: gidai@camp.eps.s.u-tokyo.ac.jp
[2] Department of Earth and Planetary Science, University of Tokyo, 7-3-1 Hongo, Bunkyo-ku, Tokyo 113-0033, Japan. E-mail: matsuura@eps.s.u-tokyo.ac.jp

stress drop $\Delta\tau$, all of which are directly observable quantities except for $\Delta\tau$. To date various empirical relations between these parameters have been proposed by many investigators. Among them the most widely accepted relation is the $S^{3/2}$ dependence of seismic moment M_0 (Fig. 1). On the basis of a classical theory of circular cracks in a uniformly stressed elastic medium, AKI (1972) has demonstrated that the $S^{3/2}$ dependence of M_0 can be reasonably explained by assuming the average stress drop to be nearly constant over a broad range of source dimension. In the case of non-circular faults, as demonstrated by KANAMORI and ANDERSON (1975), the $S^{3/2}$ dependence of M_0 still holds, if the aspect ratio (W/L) of actual faults is nearly constant as indicated by ABE (1975). Assuming a constant aspect ratio of faults, we can read the $S^{3/2}$ dependence of M_0 as the L-cubed dependence of M_0. However, as pointed out first by SCHOLZ (1982) and later by SHIMAZAKI (1986) and ROMANOW-ICZ (1992), the general L-cubed dependence of M_0 breaks down for large strike-slip earthquakes on quasi-vertical faults. This breakdown in the moment-length relation results from the fact that at least the two basic assumptions, fault aspect ratio to be constant and initial stress field to be uniform, do not hold for large strike-slip earthquakes.

For small- and medium-sized earthquakes, the stress accumulation around the source region may be regarded as uniform, thus we can apply the classical theory

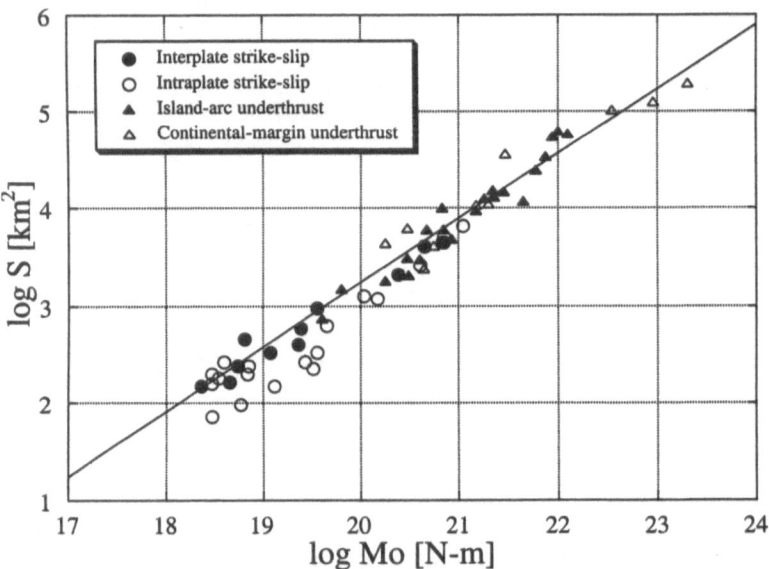

Figure 1

Plots of fault area S versus seismic moment M_0 for all earthquakes used in the present study. The solid and open circles indicate interplate and intraplate strike-slip events, respectively, and the solid and open triangles indicate island-arc and continental-margin underthrust events, respectively. The $S^{3/2}$ dependence of M_0 is denoted by the solid line.

of a shear crack in a uniformly stressed elastic medium. In this case, on the assumption of constant stress drop, the average fault slip D scales with the fault width W (W-model), and we obtain the L-cubed dependence of M_0. As an observational fact, however, we know that the fault width W is saturated for large earthquakes. If the downward rupture growth is forcibly stopped at a certain depth by the existence of a strong barrier, we can still apply the conventional W-model and obtain the linear L-dependence of M_0 as claimed by ROMANOWICZ (1992, 1994). In reality the downward rupture growth is limited to a certain depth because of the existence of a ductile unstressed region which extends under the brittle seismogenic zone (SIBSON, 1984; MARONE and SCHOLZ, 1988). In this case we can no longer apply the crack theory in a uniformly stressed elastic medium, and so the conventional W-model. Rather we should apply the theory of an in-plane shear crack in a uniformly stressed thin elastic layer, as pointed out by SCHOLZ (1982). According to the crack theory in a uniformly stressed elastic layer, the average fault slip D must scale with the fault length L (L-model), and consequently we obtain the L-squared dependence of M_0 as claimed by SCHOLZ (1982, 1994a,b).

Scaling Relations Expected from a Theoretical Model

From theoretical points of view, as discussed above, Romanowicz's W-model seems to be unreasonable, and Scholz's L-model to be reasonable. Observed data certainly indicate the L-squared dependence of M_0 for moderately large earthquakes, but not for very large earthquakes. The observed moment-length relation for very large earthquakes is linear as claimed by ROMANOWICZ (1992). Thus the data appear to support Romanowicz's W-model.

The key to this puzzle is in the mechanism of tectonic loading at plate boundaries, as demonstrated by MATSU'URA and SATO (1997). At plate boundaries, in general, stress accumulation on a fault plane is caused by viscous drag at the base of the lithosphere (base loading) and dislocation pile-ups at horizontal edges of the fault (edge loading). According to a theoretical model of tectonic loading by MATSU'URA and SATO (1997), the stress accumulation rate $\dot{\tau}$ after the occurrence of a large earthquake at transcurrent plate boundaries is given by

$$\dot{\tau} = V_{pl}\mu(\alpha + \beta/L), \tag{1}$$

where V_{pl} is a relative plate velocity, μ is the rigidity of the lithosphere, and α and β are structural parameters which depend on the thickness of the lithosphere and the viscosity of the asthenosphere. The values of α and β are numerically computed by the theoretical model of tectonic loading (MATSU'URA and SATO, 1997). In the above equation, the first term on the right-hand side corresponds to the effect of base loading, and the second term to the effect of edge loading. Then, given an average coseismic stress drop $\Delta\tau$, we can calculate the recurrence time, $T \equiv \Delta\tau/\dot{\tau}$ of earthquakes as

$$T = \Delta\tau L / V_{pl}\mu(\alpha L + \beta). \tag{2}$$

For large interplate earthquakes the average coseismic fault slip D must be given by the product of the recurrence time T and the relative plate velocity V_{pl},

$$D = \Delta\tau L / \mu(\alpha L + \beta). \tag{3}$$

Consequently the seismic moment, $M_0 \equiv \mu D W L$, can be written as

$$M_0 = \Delta\tau W L^2 / (\alpha L + \beta). \tag{4}$$

Therefore, if we assume the average stress drop $\Delta\tau$ and the fault width W to be constant, the theoretical model of tectonic loading gives the L-squared dependence of M_0 for moderately large earthquakes and the linear L-dependence of M_0 for very large earthquakes.

Data Sets

The purpose of the present study is to examine the validity of the theoretically expected scaling relations, Eqs. (3) and (4), through the detailed analysis of observed data. For this purpose, first we compile the source parameters of large earthquakes which occurred at and around plate boundaries. Next we classify those events into several groups according to the type of faulting and tectonic setting, because the average stress drop and the maximum fault width will be strongly affected by them.

The earthquakes used for the present analysis are sufficiently large, shallow events. From the earthquake catalogs compiled by PURUCARU and BERCKHEMER (1982), SATO (1989), and WELLS and COPPERSMITH (1994), first we selected 46 strike-slip events ($M_s \geq 6.0$) and 44 underthrust events ($M_s \geq 7.0$) which occurred at and around plate boundaries. For each of these earthquakes we examined the reliability and consistency of reported source parameters, seismic moment M_0, fault length L, fault width W and average fault slip D, by checking published results. The criterion used for checking reliability is whether or not the source parameter has been directly determined from observed (seismological, geodetical, geological and tsunami) data. The criterion used for checking consistency is whether or not the set of source parameters satisfies the relation $M_0 \equiv \mu D W L$. Even for large earthquakes there is a moderate number of cases in which all the source parameters have been independently determined from observed data. There are many cases in which the seismic moment M_0 and the fault area $S (= W \times L)$ have been directly determined and the average fault slip D has been estimated from these parameters by using the relation $M_0 \equiv \mu D W L$. We accepted these data sets for the present analysis, because

Table 1

The source parameters of the earthquakes used for the present analysis

No.	Date	Location	Lat.	Long.	M_s	M_0 (10^{20} N-m)	L (km)	W (km)	S (km²)	D (m)	Reference
A: Interplate strike-slip earthquakes											
1	1857 0109	San Andreas	35.0	−119.0	8.3	7.0	380	12	4560	4.6(a)	55
2	1906 0418	San Andreas	38.0	−123.0	7.7	6.7	450	10	4500	4.5(a)	55, 63
3	1939 1226	Anatolia	39.5	38.5	7.8	4.5	350	12	4200	2.4(a)	10, 13, 64
4	1943 1126	Anatolia	41.0	34.0	7.4	2.6	280					10
5	1944 0201	Anatolia	41.5	32.5	7.3	2.4	175	12	2100	2.4(a)	10, 13, 64
6	1967 0722	Anatolia	40.7	30.7	7.1	0.36	80	12	960	1.2(a)	10, 13, 64
7	1942 1220	Anatolia	40.5	36.5	7.1	0.25	50	12	600	0.9(a)	10, 13, 64
8	1940 0519	San Andreas	32.9	−115.5	7.2	0.23	45	9	405	1.8(b)	21
9	1992 0313	Anatolia	39.7	39.6	6.8	0.12	30	11	330	1.0(e)	14
10	1979 1015	San Andreas	32.6	−115.3	6.7	0.064	35	13	455	0.4(b)	11
11	1987 1124	San Andreas	33.0	−115.8	6.6	0.055	24	10	240	0.5(c)	7
12	1980 0609	San Andreas	32.2	−115.0	6.4	0.048	23			0.6(b)	57
13	1966 0628	San Andreas	35.9	−120.3	6.4	0.046	24	7	168	0.6(b)	38
14	1984 0424	San Andreas	37.2	−121.8	6.1	0.023	25	6	150	0.4(c)	12, 17
B: Intraplate strike-slip earthquakes											
1	1949 0822	Queen Charlotte	53.6	−133.3	8.1	11	440	15	6600	5.7(e)	47, 60
2	1958 0710	Queen Charlotte	58.3	−136.5	7.9	7.0	350	15	5250	4.8(e)	47
3	1972 0730	Queen Charlotte	56.8	−135.9	7.4	4.0	180	15	2700	5.0(e)	47
4	1891 1027	Japan	35.6	136.6	8.0	1.5	80	15	1200	3.0(a)	44
5	1992 0628	California	34.2	−116.4	7.6	1.1	85	15	1275	2.9(e)	56
6	1927 0307	Japan	35.6	135.1	7.4	0.46	33	19	627	3.7(c)	40
7	1943 0910	Japan	35.5	134.2	7.2	0.36	33	10	330	2.6(c)	42
8	1948 0628	Japan	36.1	136.2	7.1	0.33	23	10	230	2.5(c)	43
9	1930 1126	Japan	35.0	139.0	7.0	0.27	22	12	264	3.0(c)	6
10	1978 0114	Japan	34.8	139.3	6.6	0.13	15	10	150	2.5(c)	48
11	1980 0629	Japan	34.9	139.2	6.2	0.07	20	12	240	1.1(c)	61
12	1931 0921	Japan	36.2	139.2	6.7	0.068	20	10	200	1.0(b)	3
13	1974 0509	Japan	34.6	138.9	6.5	0.059	12	8	96	1.2(c)	6
14	1990 0220	Japan	34.7	139.3	6.4	0.04	19	14	266	0.5(e)	26
15	1969 0909	Japan	35.8	137.1	6.6	0.035	18	10	180	0.6(e)	45
16	1986 0721	California	37.5	−118.4	6.2	0.03	22	9	198	0.5(e)	18, 49
17	1963 0326	Japan	35.8	135.8	6.5	0.03	20	8	160	0.6(e)	4
18	1984 0914	Japan	35.8	137.6	6.1	0.03	12	6	72	1.4(e)	62

Table 1 (*continued*)

No.	Date	Location	Lat.	Long.	M_s	M_0 (10^{20} N-m)	L (km)	W (km)	S (km²)	D (m)	Reference
C: Underthrust events at island-arc subduction zones											
1	1965 0204	Central Aleutians	51.3	178.6	8.2	125	580	100	58000	4.8(e)	59, 65
2	1938 1110	Eastern Aleutians	55.5	−158.4	8.1	100	525	120	63000	3.2(e)	60
3	1957 0309	Central Aleutians	51.6	−175.4	8.1	88	550	100	55000	3.2(e)	16, 31
4	1963 1013	South Kuril	44.9	149.6	8.1	75	275	125	34375	4.3(e)	16, 37
5	1946 1221	Japan	33.1	135.8	8.0	60	250	100	25000	3.7(c)	9, 66
6	1958 1106	South Kuril	44.4	148.6	8.1	44	150	80	12000	5.1(e)	25
7	1968 0516	Japan	41.0	143.6	8.1	28	150	100	15000	4.1(e)	33
8	1952 0304	Japan	42.5	143.0	8.3	23	130	100	13000	3.5(e)	8
9	1969 0812	South Kuril	43.4	147.8	8.2	22	180	85	15300	2.9(e)	2
10	1971 0726	Solomon Islands	−4.9	153.2	7.7	18	180	70	12600	2.8(e)	52
11	1944 1207	Japan	33.8	136.0	7.8	15	120	80	9600	3.1(e)	34
12	1923 0901	Japan	35.4	139.2	8.0	8.4	93	53	4929	4.6(c)	41
13	1938 1105	Japan	37.0	141.7	7.5	7.0	100	60	6000	2.3(e)	5
14	1963 1020	South Kuril	44.9	150.3	7.2	7.0	100	60	6000	2.3(e)	24, 27
15	1973 0617	Japan	43.0	146.0	7.4	6.7	100	100	10000	1.0(d)	8
16	1938 1105	Japan	37.2	141.8	7.5	4.8	100	60	6000	1.6(e)	5
17	1938 0523	Japan	36.5	141.3	7.4	4.0	75	40	3000	2.7(e)	5
18	1978 0612	Japan	38.2	142.2	7.5	3.1	61	34	2074	2.2(e)	53
19	1965 0811	Vanuatu Islands	−15.5	166.9	7.3	3.0	63	50	3150	1.5(e)	22
20	1968 0401	Japan	32.5	132.3	7.6	1.8	56	32	1792	1.6(e)	54
21	1986 1114	Taiwan	24.0	121.8	7.8	1.7	68				28, 30
22	1985 0703	New Britain	−4.4	152.8	7.2	0.65	50	30	1500	0.9(e)	46
23	1970 0726	Japan	32.2	131.7	7.0	0.41	31	24	744	1.0(c)	54
D: Underthrust events at continental-margin subduction zones											
1	1960 0522	South Chile	−38.2	−73.5	8.5	2000	1000	200	200000	20(e)	36, 50
2	1964 0328	Central Alaska	61.1	−147.6	8.4	900	700	180	126000	14(e)	32, 50
3	1952 1104	Kamchatka	52.8	159.5	8.2	350	650	160	104000	6.7(e)	15, 35
4	1979 1212	Colombia-Ecuador	1.6	−79.4	7.6	29	280	130	36400		29
5	1966 1017	Northern Peru	−10.9	−78.8	7.8	20	80	140	11200	3.6(e)	1
6	1974 1003	Northern Peru	−12.4	−77.7	7.6	15	180	60	10800	2.8(e)	20
7	1971 1215	Kamchatka	56.0	163.2	7.5	6.7	50				67
8	1971 0708	Central Chile	−32.5	−71.2	7.7	5.6	70	60	4200	2.7(e)	23, 39
9	1966 1228	Central Chile	−25.5	−70.7	7.7	4.5	80	30	2400	2.4(e)	19
10	1973 0130	Mexico	18.4	−103.2	7.3	3.0	90	70	6300	1.4(e)	51
11	1978 1129	Mexico	15.8	−96.8	7.3	3.0	90	70	6300	1.0(e)	58
12	1979 0314	Mexico	17.3	−101.4	7.4	1.8	70	64	4480	0.8(e)	58

(a) Determined from surface offsets. (b) Determined from seismic wave data. (c) Determined from geodetic data. (d) Determined from tsunami data.
(e) Estimated from M_0 and S with $M_0 \equiv \mu DS$.

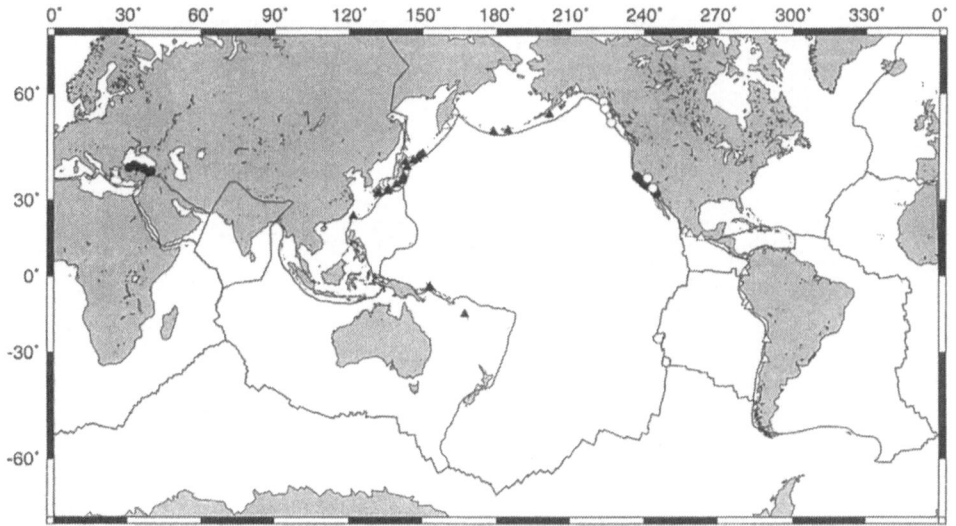

Figure 2

A map showing the epicenter distribution of all earthquakes used in the present study. The solid and open circles indicate epicenters of interplate and intraplate strike-slip events, respectively, and the solid and open triangles indicate epicenters of island-arc and continental-margin underthrust events, respectively.

the relation $M_0 \equiv \mu DWL$ is not the empirical relation to be examined but the definition of seismic moment.

After a critical examination of reliability and consistency, we finally selected 67 earthquakes and classified them into four groups, A) interplate strike-slip earthquakes, B) intraplate strike-slip earthquakes, C) underthrust earthquakes at island-arc subduction zones, and D) underthrust earthquakes at continental-margin subduction zones, according to the type of faulting and tectonic setting. The source parameters of these earthquakes used for the present analysis are listed in Table 1, in which the numbers in the last column correspond to those in the list of references in Appendix 1. For the average fault slip D we noted how it was determined or estimated. In Figure 2 the epicenters of the 67 earthquakes are plotted with plate boundaries. Here, the solid and open circles indicate the interplate and intraplate strike-slip events, respectively. The solid and open triangles indicate the underthrust events at island-arc and continental-margin subduction zones, respectively.

Analysis of Data

Initially we show the plots of seismic moment M_0 versus fault length L for all events in Figure 3, in which the solid and open circles indicate the interplate and

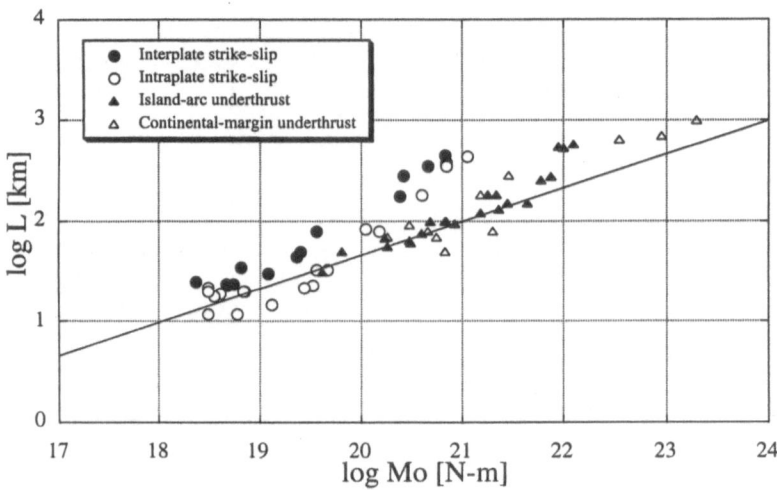

Figure 3

Plots of seismic moment M_0 versus fault length L for all earthquakes used in the present study. The solid and open circles indicate interplate and intraplate strike-slip events, respectively, and the solid and open triangles indicate island-arc and continental-margin underthrust events, respectively. The solid line denotes the L-cubed dependence of M_0.

intraplate strike-slip events, respectively, and the solid and open triangles indicate the underthrust events at island-arc and continental-margin subduction zones, respectively. From this diagram we can find the rough trend of L-cubed dependence of M_0 (solid line) as a whole, but on closer inspection we can recognize that each group of events has its own moment-length relation which deviates from the solid line. For this reason we classified the events into the four groups according to the type of faulting and tectonic setting. In the following part of this section we examine relations between seismic moment M_0, fault length L, fault width W and average fault slip D for each group of events, and make clear differences in scaling relations between the four groups of events.

Interplate Strike-slip Earthquakes

Figure 4 shows the plots of seismic moment M_0 versus fault length L for the interplate strike-slip events, which occurred along the San Andreas Fault and the Anatolia Fault. From this diagram we can see that the trend of moment-length relation abruptly changes around $M_0 = 10^{19}$ N-m. The abrupt change in trend is clearly caused by the saturation of fault width ($\overline{W} = 12$ km) for large events ($L \geq 30$ km), shown in Figure 5a. The fault-length dependence of seismic moment for large events gradually changes from L^2 to L, as predicted from the theoretical model of tectonic loading by MATSU'URA and SATO (1997).

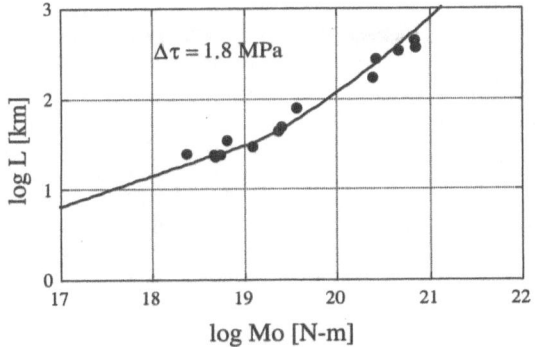

Figure 4

Plots of seismic moment M_0 versus fault length L for the interplate strike-slip events. The solid line denotes the theoretical M_0-L curve calculated from Eq. (4) with the mean stress drop $\overline{\Delta\tau} = 1.8$ MPa and the mean fault width $\overline{W} = 12$ km.

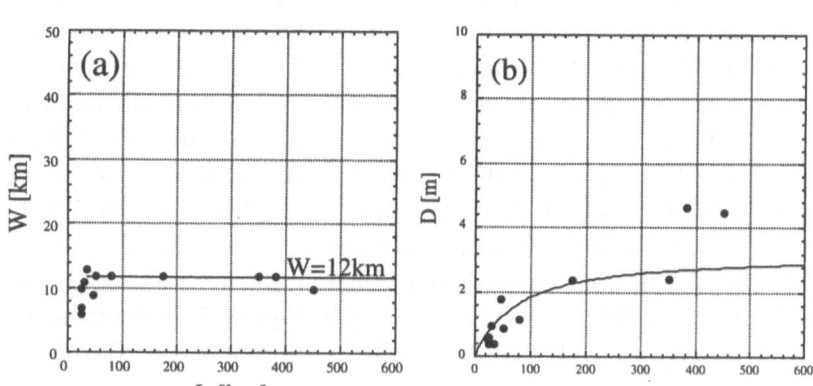

Figure 5

Plots of W vs. L (left) and D vs. L (right) for the interplate strike-slip events. (a) The solid line denotes the mean fault width for large events. (b) The solid line denotes the theoretical D-L curve calculated from Eq. (3) with the mean stress drop $\overline{\Delta\tau} = 1.8$ MPa.

Now we try to fit the theoretical moment-length curve in Eq. (4) to the observed data, using the typical values of structural parameters in Table 2, where the values of α and β are theoretically computed from the other three parameters, the thickness H of the lithosphere, the rigidity μ of the lithosphere and the viscosity η of the asthenosphere, in MATSU'URA and SATO (1997). In the present case, since seismic moment M_0, fault length L and fault width W are given for each event, we can calculate the average stress drop $\Delta\tau$ of each event by the equation

$$\Delta\tau = M_0(\alpha L + \beta)/WL^2. \tag{5}$$

Table 2

The values of structural parameters used for the computation of the theoretical
moment-length curves

Lithosphere's thickness	$H = 30$ km
Lithosphere's rigidity	$\mu = 40$ GPa
Asthenosphere's viscosity	$\eta = 10^{20}$ Pa-s
Derived parameter	$\alpha = 1.4 \times 10^{-2}$ km^{-1}
Derived parameter	$\beta = 1.0$

Subsequently, taking the arithmetic mean of the calculated stress drops for all large events ($L \geq 30$ km), we obtain the mean stress drop $\overline{\Delta\tau}$ of the interplate strike-slip earthquakes as 1.8 MPa. The solid curve in Figure 4 is the theoretical moment-length curve calculated from Eq. (4) with the mean stress drop $\overline{\Delta\tau} = 1.8$ MPa and the mean fault width $\overline{W} = 12$ km, that fits the observed data very well.

The theoretical model of tectonic loading expects another scaling relation between average fault slip D and fault length L (Eq. 3). Figure 5b shows the plots of D vs. L for the interplate strike-slip events. From this diagram we can see that the average fault slip D increases with L even after the saturation of W. Namely, for large events, the average fault slip D does not scale with W. In the same diagram we can also discover that the increase rate drops for very large events. Namely, the average fault slip D does not scale with L for very large events. Such a general tendency in the observed D-L relation can be well explained by the theoretical relation in Eq. (3). The solid curve in Figure 5b is the theoretical D-L curve calculated from Eq. (3) with the means stress drop $\overline{\Delta\tau} = 1.8$ MPa.

Intraplate Strike-slip Earthquakes

In Figure 6 we present the plots of M_0 vs. L for the intraplate strike-slip events, which occurred outside plate boundaries themselves. The observed moment-length relation for the intraplate events is very similar to that for the interplate events in Figure 4. The abrupt change in trend due to the saturation of fault width ($\overline{W} = 15$ km) occurs around $M_0 = 2 \times 10^{19}$ N-m. The systematic difference between them is only that the intraplate events have significantly large seismic moments in comparison with the interplate events. This systematic difference in M_0-L relation may be ascribed to the difference in average stress drop $\Delta\tau$ between the interplate events and the intraplate events, because the difference in fault width W is not so large between them (Figs. 5a and 7a).

In the same way as in the case of interplate earthquakes, we obtain the mean stress drop $\overline{\Delta\tau}$ of the intraplate strike-slip earthquakes as 3.1 MPa, which is about

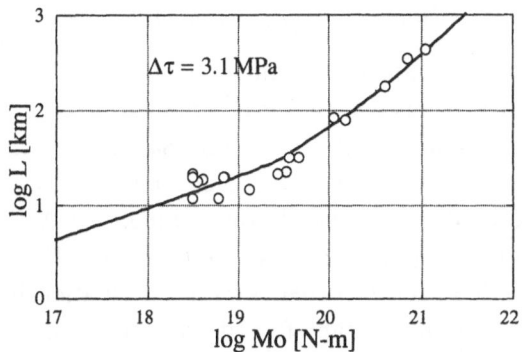

Figure 6

Plots of seismic moment M_0 versus fault length L for the intraplate strike-slip events. The solid line denotes the theoretical M_0-L curve calculated from Eq. (4) with the mean stress drop $\overline{\Delta\tau} = 3.1$ MPa and the mean fault width $\overline{W} = 15$ km.

twice as large as that of the interplate earthquakes. In Figure 6 we show the theoretical moment-length curve calculated from Eq. (4) with the mean stress drop $\overline{\Delta\tau} = 3.1$ MPa and the mean fault width $\overline{W} = 15$ km, which fits the observed data very well. In Figure 7b we show the plots of D vs. L for the intraplate strike-slip events together with the theoretical D-L curve calculated from Eq. (3) with the mean stress drop $\overline{\Delta\tau} = 3.1$ MPa. From this diagram we can see that the theoretical curve fits the observed data very well. Good agreement of the theoretical curves and the observed data suggests that the essential mechanism of tectonic loading for intraplate strike-slip earthquakes is similar to that for interplate strike-slip earthquakes.

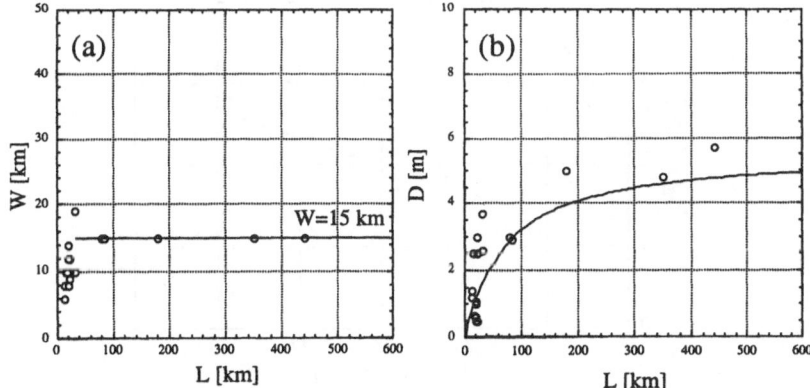

Figure 7

Plots of W vs. L (left) and D vs. L (right) for the intraplate strike-slip events. (a) The solid line denotes the mean fault width for large events. (b) The solid line denotes the theoretical D-L curve calculated from Eq. (3) with the mean stress drop $\overline{\Delta\tau} = 3.1$ MPa.

Underthrust Earthquakes at Island-arc and Continental-margin Subduction Zones

In Figure 8 we show the plots of seismic moment M_0 vs. fault length L for (a) the island-arc underthrust events and (b) the continental-margin underthrust events. In the case of island-arc underthrust events (Fig. 8a), we can find the breakaway from the L-cubed dependence of M_0 for $L \geq 200$ km, which corresponds to the saturation of fault width ($\overline{W} = 120$ km) shown in Figure 9a. We show the plots of D vs. L for the island-arc underthrust events in Figure 9b. From this diagram we can see that the average fault slip D increases with L even after the saturation of W, however its increase rate decreases rapidly and the saturation of average fault slip ($\overline{D} = 5$ m) occurs for very large earthquakes.

In the case of continental-margin underthrust events (Fig. 8b), we could not find the breakaway from the L-cubed dependence of M_0 from the analysis of the present data set ($W \leq 200$ km, $L \leq 1000$ km). In Figures 10a and b we show the plots of W

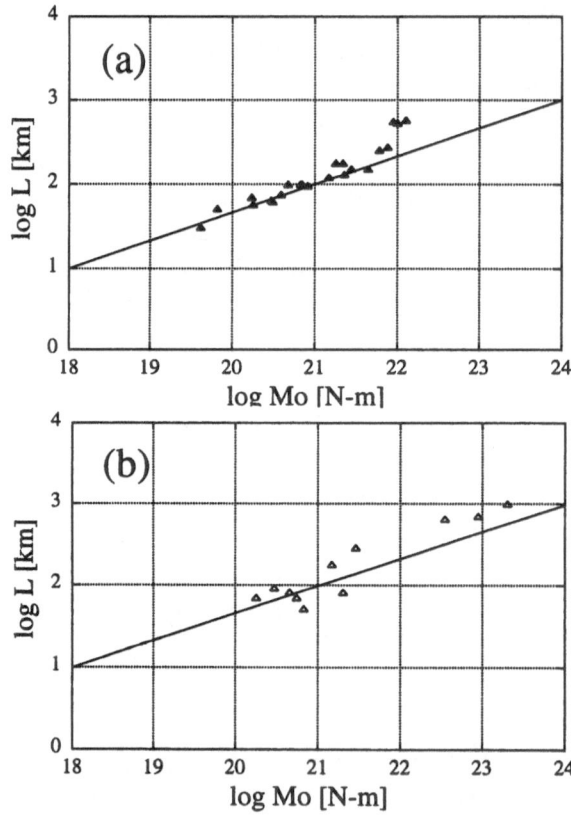

Figure 8

Plots of seismic moment M_0 versus fault length L for (a) the island-arc underthrust events and (b) the continental-margin underthrust events. In either case the solid line denotes the L-cubed dependence of M_0.

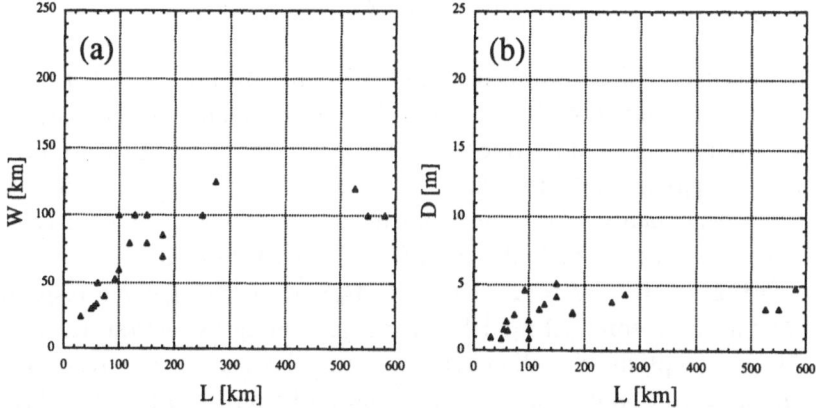

Figure 9
Plots of *W* vs. *L* (left) and *D* vs. *L* (right) for the island-arc underthrust events.

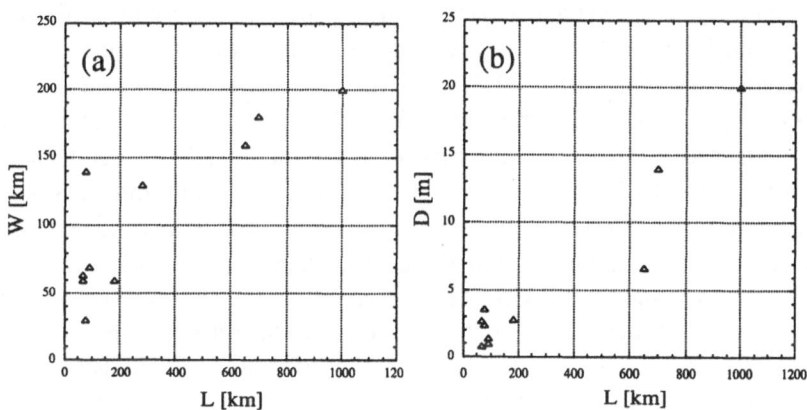

Figure 10
Plots of *W* vs. *L* (left) and *D* vs. *L* (right) for the continental-margin underthrust events.

vs. *L* and *D* vs. *L*, respectively, for the continental-margin underthrust events. From Figure 10a we can see that fault width *W* increases linearly with *L* over the entire range of $L \leq 1000$ km. From Figure 10b we can see that the average fault slip *D* also increases linearly with *L* over the entire range of $L \leq 1000$ km. Namely, for the continental-margin underthrust events, the saturation of fault width and fault slip does not occur in the range of $L \leq 1000$ km. This is the most essential difference of the continental-margin underthrust events from the island-arc underthrust events.

Discussion and Conclusions

We examined relations between seismic moment M_0, fault length L, fault width W and average fault slip D for four different earthquake groups; interplate strike-slip events, intraplate strike-slip events, island-arc underthrust events, and continental-margin underthrust events. In the case of interplate strike-slip events, the L-cubed dependence of seismic moment breaks down when L exceeds 30 km, because of the saturation of fault width ($\overline{W} = 12$ km). For large events, D and M_0 increase with L as $D = \overline{\Delta\tau}L/\mu(\alpha L + \beta)$ and $M_0 = \overline{\Delta\tau}WL^2/(\alpha L + \beta)$, respectively, as expected from the theoretical model of tectonic loading by MATSU'URA and SATO (1997). If we accept these scaling relations, the classical scaling relation between seismic moment and fault area $M_0 = cS^{3/2}$ should be replaced by $M_0 = \overline{\Delta\tau}S^2/(\alpha S + \beta\overline{W})$ for large earthquakes. The slope of this theoretical curve in the plot of log M_0 vs. log S gradually decreases in the range of 1–2.

The observed scaling relations for intraplate strike-slip events are very similar to those for interplate strike-slip events. This suggests the existence of similarity in the tectonic loading mechanism between them. Systematic differences in scaling relations between interplate events and intraplate events are consistently explained by the differences in stress drop, 1.8 MPa for interplate events and 3.1 MPa for intraplate events, which may be ascribed to the difference in stress accumulation rates and therefore the recurrence time of earthquakes (KANAMORI and ALLEN, 1986). In the case of island-arc underthrust events we can still find the saturation of fault width ($\overline{W} = 120$ km) and the breakaway from the L-cubed dependence of M_0 for $L \geq 200$ km. This indicates that the extent of the seismogenic zone in depth is limited to approximately 50 km. The observed D-L relation illustrates that the average fault slip D increases with L even after the saturation of W, nonetheless its increase rate soon declines and the saturation of average fault slip ($\overline{D} = 5$ m) occurs for very large earthquakes. In the case of continental-margin underthrust events, on the other hand, we could not find the breakaway from the L-cubed dependence of M_0 from the analysis of the present data set ($W \leq 200$ km, $L \leq 1000$ km). The observed W-L relation shows the linear increase of W with L, and the observed D-L relation also shows the linear increases of D with L, over the whole range of $L \leq 1000$ km. This indicates that the vertical extent of the seismogenic zone at continental-margin subduction zones is substantially deeper than that at island-arc subduction zones.

For sufficiently large earthquakes, in general, the downward rupture growth is limited to a certain depth because of the existence of a ductile unstressed region extending under the brittle seismogenic zone. The brittle-ductile transition occurs at 300–400°C (STESKY, 1975; TSE and RICE, 1986; BLANPIED *et al.*, 1991). This temperature corresponds to a depth of 12–15 km (MARONE and SCHOLZ, 1988) at and around transcurrent plate boundaries and 40–50 km at island-arc subduction boundaries (HYNDMAN and WANG, 1993; TICHELAAR and RUFF, 1993; HYND-

MAN *et al.*, 1995). Therefore, we may conclude that the observed difference in the lower limit of the seismogenic zone between tectonically different regions is attributed to the difference in thermal state there.

Acknowledgements

We thank anonymous reviewers for their useful suggestions for improving the manuscript.

Appendix A

1. ABE, K. (1972), *Mechanisms and Tectonic Implications of the 1966 and 1970 Peru Earthquake*, Phys. Earth Planet. Int. *5*, 3670–379.

2. ABE, K. (1973), *Tsunami and Mechanism of Great Earthquake*, Phys. Earth Planet, Int. *7*, 143–153.

3. ABE, K. (1974a), *Seismic Displacement and Ground Motion near a Fault. The Saitama Earthquake of September 21, 1931*, J. Geophys. Res. *79*, 4393–4399.

4. ABE, K. (1974b), *Fault Parameters Determined by Near- and Far-field Data. The Wakasa Bay Earthquake of March 26, 1963*, Bull. Seismol Soc. Am. *64*, 1369–1382.

5. ABE, K. (1977), *Tectonic Implications of the Large Shioya-Oki Earthquakes of 1938*, Tectonophysics *41*, 269–289.

6. ABE, K. (1978), *Dislocations, Source Dimensions and Stresses Associated with Earthquakes in the Izu Peninsula, Japan*, J. Phys. Earth *26*, 253–274.

7. AGNEW, D. C., and WYATT, F. K. (1989), *The 1987 Superstition Hills Earthquake Sequence, Strains and Tilts at Pinon Flat Observatory*, Bull. Seismol. Soc. Am. *79*, 2, 480–492.

8. AIDA, I. (1978), *Reliability of a Tsunami Source Model Derived from Fault Parameters*, J. Phys. Earth *26*, 57–73.

9. AIDA. I. (1981), *Numerical Simulation of the off-Nankaido Tsunami*, Bull. Earthq. Res. Inst. Tokyo Univ. *56*, 713–730 (in Japanese).

10. AMBRASEYS, N. N. (1988), *Engineering Seismology*, Earthq. Eng. Struct. Dvn. *17*, 1–105.

11. ARCHULETA, R. J. (1984), *A Faulting Model for the 1979 Imperial Valley Earthquake*, J. Geophys. Res. *89*, 4559–4585.

12. BAKUN, W. H., CLARK, M. M., COCKERHAM, R. S., ELLSWORTH, W. L., LINDTH, A. G., PRESCOTT, W. H., SHAKAL, A. F., and SPUDICH, P. (1984), *The Morgan Hill, California, Earthquake*, Science *225*, 288–291.

13. BARKA, A. (1996), *Slip Distribution along the North Anatolian Fault Associated with the Large Earthquakes of the Period 1939 to 1967*, Bull. Seismol. Soc. Am. *86*, 1238–1254.

14. BARKA, A., and EYDÖGAN, H. (1993), *The Erzincan Earthquake of 13 March 1992 in Eastern Turkey*, Terra Nova *5*, 190–194.

15. BEN-MENAHEM, A., and TOKSÖZ, M. N. (1963), *Source Mechanism from Spectrums of Long-period Surface Waves*, J. Geophys. Res. *68*, 5207–5222.

16. BEN-MENAHEM, A., and ROSENMAN, M. (1972), *Amplitude Patterns of Tsunami Waves from Submarine Earthquakes*, J. Geophys. Res. *77*, 3097–3128.

17. BEROZA, G. C., and SPUDICH, P. (1988), *Linearized Inversion for Fault Rupture Behavior: Application to the 1984 Morgan Hill, California, Earthquake*, J. Geophys. Res. *93*, 6275–6296.

18. COCKERHAM, R. S., and CORBETT, E. J. (1987), *The July 1986 Chalfant Valley, California, Earthquake Sequence Preliminary Results*, Bull. Seismol. Soc. Am. *77*, 280–289.

19. DESCHAMPS, A., LYON-CAEN, H., and MADARIAGA, R. (1980), *Etude du tremblement de Terre de Taltal (Chili 1976) à Partir des ondes sismiques de longue période*, Ann. Geophys. *36*(2), 179–190.

20. DEWEY, J. W., and SPENCE, W. (1979), *Seismic Gaps and Source Zones of Recent Large Earthquakes in Coastal Peru*, Pure appl. geophys. *117*, 1148–1171.

21. DOSER, D. I. (1990), *Source Characteristics of Earthquakes along the Southern San Jacinto and Imperial Fault Zones*, Bull. Seismol. Soc. Am. *80*, 1099–1177.

22. EBEL, J. E. (1980), *Source Processes of The 1965 New Hebrides Islands Earthquakes from Teleseismic Waveforms*, Geophys. J. R. Astron. Soc. *63*, 381–403.

23. EISENBERG, et al. (1972), *The July 8, 1971 Chilean Earthquake*, Bull. Seismol. Soc. Am. *62*, 423–430.

24. FUKAO, Y. (1979), *Tsunami Earthquakes and Subduction Processes near Deep-sea Trenches*, J. Geophys. Res. *84*, 2303–2314.

25. FUKAO, Y., and FURUMOTO, M. (1979), *Stress Drops, Wave Spectra and Recurrence Intervals of Great Earthquakes Implications of the Etorofu Earthquake of 1958 November 6*, Geophys. J. R. Astr. Soc. *57*, 23–40.

26. FUKUYAMA, E., and MIKUMO, T. (1993), *Dynamic Rupture Analysis: Inversion for the Source Process of the 1990 Izu-Oshima, Japan, Earthquake (M = 6.5)*, J. Geophys. Res. *98*, 6529–6542.

27. FURUMOTO, M. (1979), *Initial Phase Analysis of R Waves from Great Earthquakes*, J. Geophys. Res. *84*, 6867–6874.

28. GOLDSTEIN, P., and AREHLETA, R. J. (1991), *Deterministic Frequency-wave Number Methods and Direct Measurements of Rupture Propagation during Earthquakes Using a Dense Array: Data Analysis*, J. Geophys. Res. *96*, 6187–6198.

29. HERD, D. G., YOUD, T. L., MEYER, H., ARANGO, C. J. L., PERSON, W. J., and MENDOZA, C. (1981), *The Great Tumaco, Colombia Earthquake of 12 December 1979*, Science *221*, 441–445.

30. HWANG, L. J., and KANAMORI, H. (1989), *Teleseismic and Strong-motion Source Spectra from Two Earthquakes in Eastern Taiwan*, Bull. Seismol. Soc. Am. 79, 935–944.

31. JOHNSON, J. M., TANIOKA, Y., RUFF, L. J., SATAKE, K., KANAMORI, H., and SYKES, L. R. (1994), *The 1957 Great Aleutian Earthquake*, Pure appl. geophys. 142, 3–28.

32. KANAMORI, H. (1970), *The Alaskan Earthquake of 1964: Radiation of Long-period Surface Waves and Source Mechanisms*, J. Geophys. Res. 75, 5029–5040.

33. KANAMORI, H. (1971), *Focal Mechanism of the Tokachi-oki Earthquake of May 16, 1968: Contortion of the Lithosphere at a Junction of Two Trenches*, Tectonophysics 12, 1–13.

34. KANAMORI, H. (1972), *Tectonic Implications of The 1944 Tonankai and the Nankaido Earthquakes*, Phys. Earth Planet. Int. 5, 129–139.

35. KANAMORI, H. (1976), *Re-examination of The Earth's Free Oscillations Excited by The Kamchatka Earthquake of November 4, 1952*, Phys. Earth Planet. Int. 11, 216–226.

36. KANAMORI, H., and ANDERSON, D. L. (1975), *Theoretical Basis for Some Empirical Relations in Seismology*, Bull. Seismol. Soc. Am. 65, 1073–1095.

37. KIKUCHI, M., and FUKAO, Y. (1987), *Inversion of Long-period P Waves from Great Earthquakes along Subduction Zones*, Tectonophysics 144, 231–247.

38. LINDH, A. G., and BOORE, D. M. (1981), *Control of Rupture by Fault Geometry during the 1966 Parkfield Earthquake*, Bull. Seismol. Soc. Am. 71, 95–116.

39. MALGRANGE, M., DESCHAMPS, A., and MADARIAGA, R. (1981), *Thrust and Extensional Faulting under the Chilean Coast: 1965, 1971 Aconcagua Earthquakes*, Geophys. J. R. Astron. Soc. 66, 313–331.

40. MATSU'URA, M. (1977), *Inversion of Geodetic Data, II: Optimal Model of Conjugate Fault System for the 1927 Tango Earthquake*, J. Phys. Earth 25, 233–255.

41. MATSU'URA, M., and IWASAKI, T. (1983), *Study on Coseismic and Postseismic Crustal Movements Associated with the 1923 Kanto Earthquake*, Tectonophysics 97, 201–215.

42. MATSU'URA, M., and HASEGAWA, Y. (1987), *A Maximum Likelihood Approach to Nonlinear Inversion under Constraints*, Phys. Earth Planet. Int. 47, 179–187.

43. MATSU'URA, M., and HASEGAWA, Y. (1989), *Fault Model of the 1948 Fukui Earthquake Determined from Geodetic Data*, Chikyu 11, 42–46 (in Japanese).

44. MIKUMO, T., and ANDO, M. (1976), *A Search into the Faulting Mechanism of the 1891 Great Nobi Earthquake*, J. Phys. Earth 22, 87–108.

45. MIKUMO, T. (1973), *Faulting Mechanism of the Gifu Earthquake September 9, 1969, and Some Related Problems*, J. Phys. Earth 21, 191–212.

46. MORI, J. (1989), *The New Ireland Earthquake of July 3, 1985 and Associated Seismicity near The Pacific-Solomon Sea-Bismarck Sea Triple Junction*, Phys. Earth Planet. Int. 55, 144–153.

47. NISHENKO, S. P., and JAKOB, K. H. (1990), *Seismic Potential of the Queen Charlotte-Alaska-Aleutian Seismic Zone*, J. Geophys. Res. *95*, 2511–2532.

48. Okada, Y. (1978), *Fault Mechanism of the Izu-Oshima-kinkai Earthquake of 1978, as Inferred from the Crustal Movement Data*, Bull. Earthq. Res. Inst. *53*, 823–840.

49. PACHECO, J. F., and NABELEK, J. L. (1988), *Source Mechanisms of Three Moderate California Earthquakes of July 1986*, Bull. Seismol. Soc. Am. *78*, 1907–1929.

50. PLAFKER, G. (1972), *Alaskan Earthquake of 1964 and Chilean Earthquake of 1960: Implications for Arc Tectonics*, J. Geophys. Res. *77*, 901–925.

51. REYES, A., BRUNE, J. N., and CINNA LOMNITZ (1979), *Source Mechanism and Aftershock Study of the Colombia, Mexico Earthquake of January 10, 1973*, Bull. Seismol Soc. Am. *69*, 1819–1840.

52. SCHWARTZ, S. Y., LAY, T., and RUFF, L. J. (1989), *Source Process of Great 1971 Solomon Islands Doublet*, Phys. Earth Planet. Int. *56*, 294–310.

53. SENO, T. *et al.* (1979), *Rupture Process of The Miyagi-Oki, Japan, Earthquake of June 12, 1978*, Phys. Earth Planet. Int. *23*, 39–61.

54. SHIONO, K. *et al.* (1980), *Tectonics of the Kyushu-Ryukyu Arc as Evidenced from Seismicity and Focal Mechanism of Shallow to Intermediate-depth Earthquakes*, J. Phys. Earth *28*, 17–43.

55. SIEH, K. E. (1978), *Slip along the San Andreas Fault Associated with the Great 1857 Earthquake*, Bull. Seismol. Soc. Am. *68*, 1421–1448.

56. SIEH, K., JONES, L., HAUKSSON, E., HUDNUT, K., EBERHART-PHILLIPS, D., HEATON, T., HOUGH, S., HUTTON, K., KANAMORI, H., LILJE, A., LINDVALL, S., MCGILL, S. F., MORI, J., RUBIN, C., SPOTILA, A., STOCK, J., THIO, H. K., TREIMAN, J., WERNICKE, B., and ZACHARIASEN, J. (1993), *Near-field Investigations of the Landers Earthquake Sequence. April to July 1992*, Science *260*, 171–176.

57. SILVER, P., and MASUDA, T. (1985), *A Source Extent Analysis of the Imperial Valley Earthquake of October 15, 1979, and Victoria Earthquake of June 9, 1980*, J. Geophys. Res. *90*, 7639–7651.

58. SINGH, S. K., ASTIZ, L., and HAVSKOV, J. (1980), *Seismic Gaps and Recurrence Periods of Large Earthquakes along the Mexican Subduction Zone: A Reexamination*, Bull. Seismol. Soc. Am. *71*, 827–843.

59. SPENCE, W. (1977), *The Aleutian Arc: Tectonic Blocks, Episodic Subduction, Strain Diffusion, and Magma Generation*, J. Geophys. Res. *82*, 213–230.

60. SYKES, L. R. (1971), *Aftershock Zones of Great Earthquake, Seismic Gaps, and Earthquake Prediction for Alaska and the Aleutians*, J. Geophys. Res. *76*, 8021–8041.

61. TAKEO, M. (1988), *Rupture Process of the 1980 Izu-Hanto-Toho-Oki Earthquake*, Bull. Seismol. Soc. Am. *78*, 1074–1091.

62. TAKEO, M., and MIKAMI, N. (1987), *Inversion of Strong Motion Seismograms for the Source Process of the Nagonoken-seibu Earthquake of 1984*, Tectonophysics *144*, 271–285.

63. THATCHER, W. *et al.* (1997), *Resolution of Fault Slip along the 470-km-long Rupture of the Great 1906 San Francisco Earthquake and its Implications*, J. Geophys. Res. *102*, 5353–5367.

64. WESNOUSKY, S. (1986), *Earthquakes, Quaternary Faults and Seismic Hazard in California*, J. Geophys. Res. *91*, 12,587–12,631.

65. WU, F. T., and KANAMORI, H. (1973), *Source Mechanism of February 4, 1965, Rat Island Earthquake*, J. Geophys. Res. *78*, 6082–6092.

66. YABUKI, T., and MATSU'URA, M. (1992), *Geodetic Data Inversion Using a Bayesian Information Criterion for Spatial Distribution of Fault Slip*, Geophys. J. Int. *109*, 363–375.

67. ZOBIN, V. M., and SIMBRIEVA, I. G., (1977), *Focal Mechanism of Earthquakes in the Kamchatka-commander Region and Heterogeneities of the Active Seismic Zone*, Pure appl. geophys. *115*, 283–299.

REFERENCES

ABE, K. (1975), *Reliable Estimation of the Seismic Moment of Large Earthquake*, Phys. Earth Planet. Int. *23*, 381–390.

AKI, K. (1972), *Earthquake Mechanism*, Tectonophysics *13*, 423–446.

BLANPIED, M. L., LOCKNER, D. A., and BYERLEE, J. D. (1991), *Fault Stability Inferred from Granite Sliding Experiments at Hydrothermal Conditions*, Geophys. Res. Lett. *18*, 609–612.

KANAMORI, H., and ALLEN, C. R. (1986), *Earthquake repeat time and average stress drop*. In *Earthquake Source Mechanics* (eds. Das, S., Boatwrite, J., and Scholz, C. H.), AGU Geophys. Mono. *37*, 227–236.

KANAMORI, H., and ANDERSON, D. L. (1975), *Theoretical Basis for Some Empirical Relations in Seismology*, Bull. Seismol. Soc. Am. *65*, 1073–1095.

HYNDMAN, R. D., and WANG, K. (1993), *Thermal Constraints on the Zone of Major Thrust Earthquake Failure: The Cascadia Subduction Zone*, J. Geophys. Res. *98*, 2039–2060.

HYNDMAN, R. D., WANG, K., and YAMANO, M. (1995), *Thermal Constraints on the Seismogenic Portion of the Southwestern Japan Subduction Thrust*, J. Geophys. Res. *100*, 15,373–15,392.

MARONE, C., and SCHOLZ, C. H. (1988), *The Depth of Seismic Faulting and the Upper Transition from Stable to Unstable Slip Regimes*, Geophys. Res. Lett. *15*, 621–624.

MATSU'URA, M., and SATO, T. (1997), *Loading Mechanism and Scaling Relations of Large Interplate Earthquakes*, Tectonophysics *227*, 189–198.

PURUCARU, G., and BERCKHEMER, H. (1982), *Quantitative Relations of Seismic Source Parameters and a Classification of Earthquakes*, Tectonophysics *84*, 57–128.

ROMANOWICZ, B. (1992), *Strike-slip Earthquakes on Quasi-vertical Transcurrent Faults: Inferences for General Scaling Relations*, Geophys. Res. Lett. *19*, 481–484.

ROMANOWICZ, B. (1994), *Comment on "A Reappraisal of Large Earthquake Scaling by C. Scholz"*, Bull. Seismol. Soc. Am. *84*, 1675–1676.

SATO, R., *Handbook for Fault Parameters of Earthquakes in Japan* (Kashima Publisher, Tokyo 1989).

SCHOLZ, C. H. (1982), *Scaling Laws for Large Earthquakes and Consequences for Physical Models*, Bull. Seismol. Soc. Am. *72*, 1–14.

SCHOLZ, C. H. (1994a), *A Reappraisal of Large Earthquake Scaling*, Bull. Seismol. Soc. Am. *84*, 215–218.

SCHOLZ, C. H. (1994b), *Reply to comments on "A Reappraisal of Large Earthquake Scaling by C. Scholz"*, Bull. Seismol. Soc. Am. *84*, 1677–1678.

SIBSON, R. H. (1984), *Roughness at the Base of the Seismogenic Zone: Contributing Factors*, J. Geophys. Res. *89*, 5791–5799.

SHIMAZAKI, K. (1986), *Small and large earthquakes: The effects of the thickness of the seismogenic layer and the free surface*. In *Earthquake Source Mechanics* (eds. Das, S., Boatwrite, J., and Scholz, C. H.), AGU Geophys. Mono. *37*, 209–216.

STESKY, R. (1975), *The Mechanical Behavior of Fault Rocks at High Temperature and Pressure*, Ph.D. Thesis, Mass. Inst. of Technol., Cambridge.

TICHELAAR, B. W., and RUFF, L. J. (1993), *Depth of Seismic Coupling along Subduction Zones*, J. Geophys. Res. *98*, 2017–2037.

TSE, S. T., and RICE., J. R. (1986), *Crustal Earthquake Instability in Relation to the Depth Variation of Frictional Slip Properties*, J. Geophys. Res. *91*, 9452–9472.

WELLS, D. L., and COPPERSMITH, K. J. (1994), *New Empirical Relationships among Magnitude, Rupture Length, Rupture Width, and Surface Displacement*, Bull. Seismol. Soc. Am. *84*, 974–1002.

(Received July 30, 1999, revised March 21, 2000, accepted April 6, 2000)

To access this journal online:
http://www.birkhauser.ch

Pure appl. geophys. 157 (2000) 2303–2322
0033–4553/00/122303–20 $ 1.50 + 0.20/0

❘ Pure and Applied Geophysics

Continuous GPS Array and Present-day Crustal Deformation of Japan

TAKESHI SAGIYA,[1] SHIN'ICHI MIYAZAKI[1] and TAKASHI TADA[1]

Abstract—A GPS array with about 1,000 permanent stations is under operation in Japan. The GPS array revealed coseismic deformations associated with large earthquakes and ongoing secular deformation in the Japanese islands. Based on daily coordinate data of the GPS stations, strain rate distribution is estimated. Most regions with a large strain rate are related to plate boundaries and active volcanoes. In addition, the Niigata-Kobe Tectonic Zone (NKTZ) is recognized as a region of large strain rate along the Japan Sea coast and in the northern Chubu and Kinki districts. This newly found tectonic zone may be related to a hypothetical boundary between the Eurasian (or Amurian) and the Okhotsk (or North America) plates. Precise observation of crustal deformation provides important boundary conditions on numerical modeling of earthquakes and other crustal activities. Appropriate computation methods of continuous deformation field are directly applicable to data assimilation for such numerical simulations.

Key words: GPS, crustal deformation, the Japanese islands, data assimilation.

1. Introduction

Precise information concerning surface deformation of the earth is indispensable to numerical simulation of earthquakes and of other tectonic activity. Surface deformation can be interpreted in relation to an internal mechanical process of the earth, i.e., stress distribution or fault slips, using the elastic dislocation theory (e.g., OKADA, 1985). Assuming proper structures and boundary conditions, we can calculate a surface deformation pattern with numerical methods such as the finite element and the finite difference methods. Mainly because of poor quality of both structural and deformation data, however, realistic modeling of crustal activities has been difficult.

One of the disadvantages of crustal deformation data has been their temporal sparseness. Conventional geodetic measurements over a wide area require a large number of trained staff to conduct surveys and extended time to complete. In addition, limited accuracy of conventional geodetic measurements made it impossible to resolve significant interseismic deformation (order of 0.1 ppm/year) during a short time period such as 5–10 years.

---

[1] Crustal Dynamics Laboratory, Geographical Survey Institute, Kitosato-1, Ibaraki 305-0811, Japan. E-mail: sagiya@gsi-mc.go.jp

Space-based GPS (Global Positioning System) has been developed since the late 1970s. With its improving accuracy and development of portable receivers and antennas, GPS has become vastly popular in the study of geodesy and crustal deformation in the 1990s. GPS has realized not only precise relative positioning over a long distance, but also nearly continuous monitoring of crustal deformation. In addition, because of its relatively low cost, we can construct dense permanent GPS arrays to monitor crustal deformation in many locations. Among them, Japan (e.g., TADA *et al.*, 1997) and southern California (e.g., PRESCOTT, 1996; WATKINS *et al.*, 1997) are the places of two major regional GPS arrays in the world. Both regions contain considerable seismic risks and very large population. Monitoring and modeling of crustal deformation are very important for the estimation and mitigation of seismic hazard. These GPS arrays play a central role of providing precise information regarding ongoing crustal deformation for those purposes.

In this paper, we will briefly describe the continuous GPS array in Japan, and discuss present-day deformation of Japan based on the GPS data. The Japanese islands are situated at a complex plate boundary region with four tectonic plates (Fig. 1). Plate interaction causes frequent occurrence of large interplate as well as intraplate earthquakes. As a result of past earthquakes, there are numerous active fault traces throughout the Japanese islands (Fig. 1), implying style and rate of long-term deformation. Comparison of the present-day deformation with such a long-term deformation is important. In addition, comparison with conventional geodetic survey results is also meaningful. These comparisons imply that short-term deformation revealed by GPS is not temporal but closely relates to tectonic activities through the Quaternary period. Such information is essentially important for the long-term forecast of crustal activities, and should be utilized in various simulation studies as a boundary condition.

2. Continuous GPS Array of Japan

Geographical Survey Institute (GSI) of Japan started construction of its continuous GPS array in 1993. Operation of some 200 permanent stations began in the middle of 1994. GSI continued construction of additional stations, and the network numbered about 1,000 sites throughout the Japanese islands as of 1999 (Fig. 2). The GPS array was named GEONET (GPS Earth Observation NETwork). Details of the operation and analysis of the GPS array are described elsewhere (MIYAZAKI *et al.*, 1996, 1997, 1998). In addition, TADA *et al.* (1997) summarized initial results with GEONET. Here we briefly describe how the GPS array is operated.

The standard design of a GEONET station is a 5-meter tall stainless pillar equipped with a dual-frequency GPS antenna at the top, and a GPS receiver as well as a modem inside. The pillar is built on a foundation of concrete block buried in soil. All stations have power and telephone lines for continuous operation and

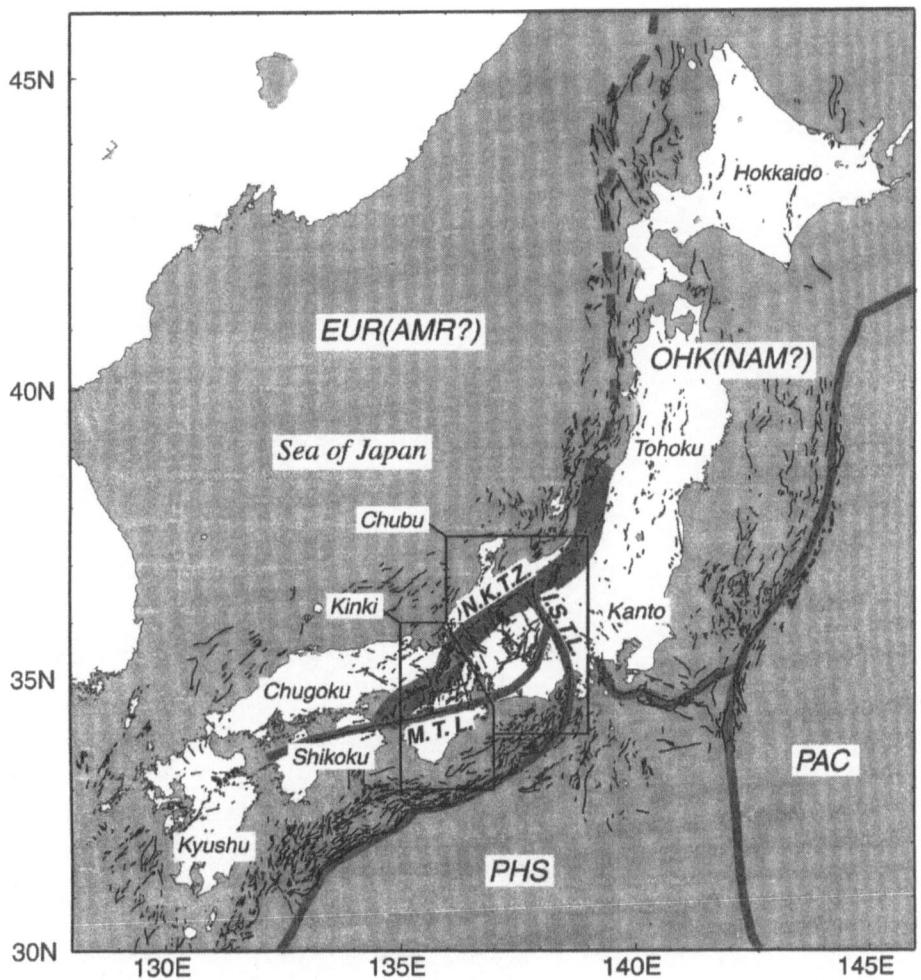

Figure 1

Tectonic map of Japan. Japanese district names are shown. Thin black lines denote surface traces of active faults (RESEARCH GROUP FOR ACTIVE FAULTS OF JAPAN, 1991). Gray lines denote plate boundaries and major tectonic lines. Dashed gray line denotes a possible plate boundary between the Eurasian (or Amurian) and the Okhotsk (or North American) plates. Thick gray line indicates location of the Niigata-Kobe Tectonic Zone (NKTZ) proposed in this paper. PAC: Pacific plate, PHS: Philippine Sea plate, EUR: Eurasian plate, AMR: Amurian plate, OHK: Okhotsk plate, NAM: North American plate, I.S.T.L.: Itoigawa-Shizuoka Tectonic Line, M.T.L.: Median Tectonic Line.

remote control. GPS phase data emitted from GPS satellites are captured by the GPS antenna and stored in the GPS receiver every 30 seconds. Stored phase data are transferred to the control system located in the central office of GSI once daily.

Routine process of GEONET yields precise daily coordinates of all 1,000 sites. Presently, the Bernese GPS software (ROTHACHER and MERVART, 1996) is used for

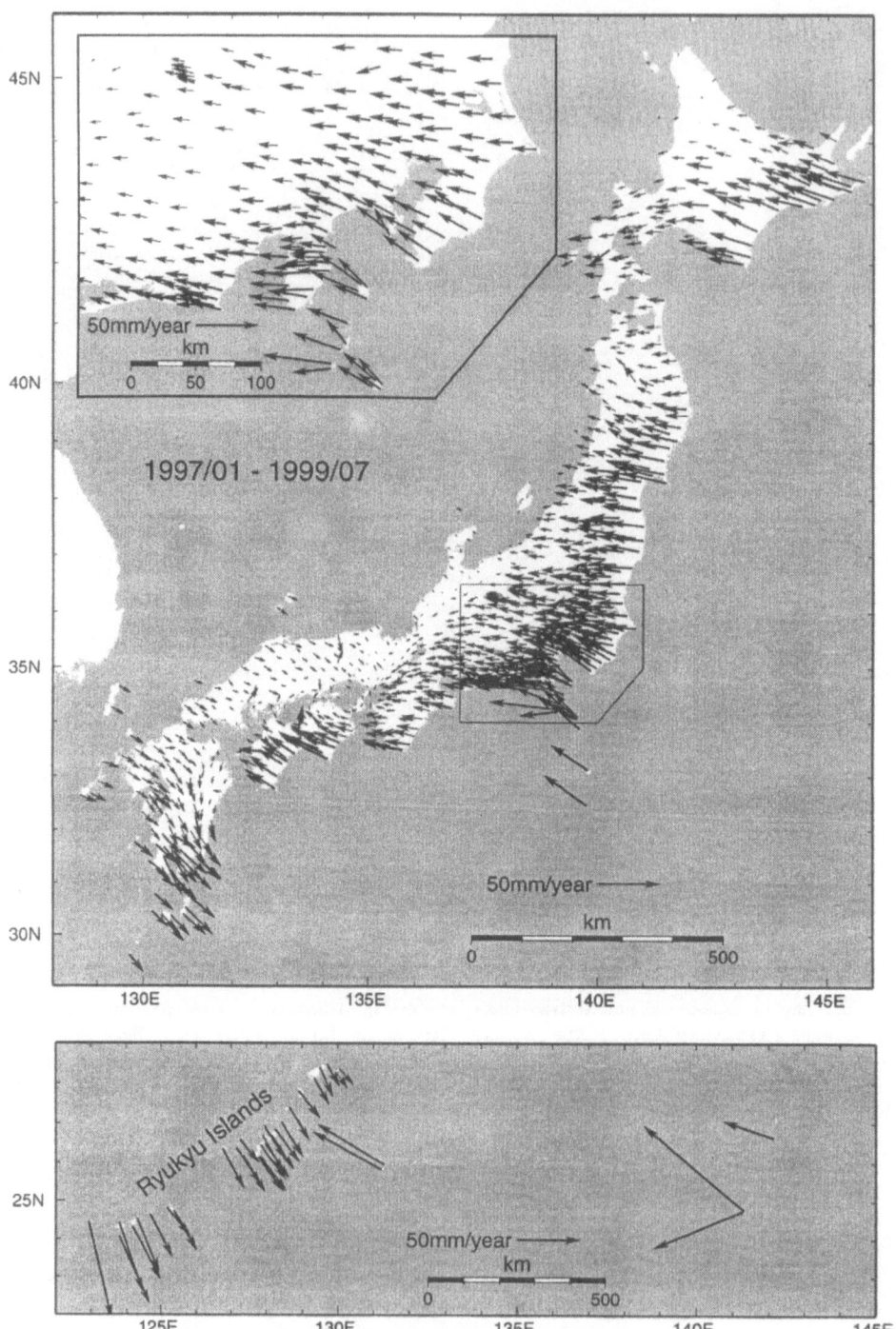

Figure 2
Horizontal displacement rate vectors of continuous GPS sites. All vectors are relative to the stable
Eurasian plate. Inset shows magnification of central Japan.

routine analysis. Since the final coordinate solution is calculated based on the IGS (International GPS Service for Geodynamics) precise orbit of GPS satellites, it takes about two weeks to derive the final solution after the observation. GSI also calculates a rapid solution of coordinates using IGSs rapid as well as predicted orbits. This rapid solution is obtained about 30 hours after the observation and is utilized for daily monitoring of crustal activities. Repeatability of the rapid solution is somewhat inferior to the final solution because of the limited accuracy of the orbital information. In the following sections, our discussion will be based solely on the final routine solutions.

3. Deformation of the Japanese Islands

3.1. Displacement Rate Distribution

In order to discuss tectonic deformation of the Japanese islands, we estimated the secular displacement rate of GPS stations based on daily coordinate solutions. We fitted a following function to daily time series data of the i-th coordinate component of the n-th station.

$$x_n^i(t) = a_n^i + b_n^i t + c_n^i \sin(2\pi t) + d_n^i \cos(2\pi t) + \sum_{k=1}^{m} e_{n,k}^i H(t - t_n^k). \tag{1}$$

Here, the first two terms on the right side correspond to a linear trend, the next two terms to a sinusoidal annual variation, and the last term to coseismic steps. Coseismic steps are described as a sum of Heaviside functions $\Sigma_{k=1}^{m} e_{n,k}^i H(t - t_n^k)$, where $e_{n,k}^i$ is a magnitude of the k-th step which occurred at t_n^k. The coefficient b_n^i is an estimate of displacement rate. We applied this expression to all three components separately, and estimated displacement rate components for all stations by solving least-squares problems.

Figure 2 shows horizontal displacement rate vectors for all stations. Referring to the kinematic reference frame by HEKI (1996), the vectors are calculated as those with respect to the stable part of the Eurasian plate. These velocity vectors are estimated using daily coordinates from January 1997 to July 1999. Approximately 600 stations have been operated since April 1996. However, we did not use coordinate data in 1996 because there were two large earthquakes ($M6.6$ on October 19 and $M6.6$ on December 3) southeast off Kyushu, accompanied by significant postseismic deformation (NISHIMURA et al., 1998). Coordinate data after January, 1997 may still contain postseismic effects of those earthquakes. In addition, there were seismic swarm activities near the Izu Peninsula (March, 1997 and April–May, 1998), and two large earthquakes in Kyushu ($M6.3$ on

March 26, 1998 and $M6.2$ on May 13, 1998, NISHIMURA *et al.*, 1998), for which we estimated coseismic steps as formulated in (1). After the removal of a fitted curve from original time series data, the standard deviation of residuals is 2–6 mm for horizontal components and 10–15 mm for vertical components. Since most of the coordinate time series are well approximated by equation (1), we may think that the velocity data shown in Figure 2 are representative of a present secular deformation of the Japanese islands.

Characteristic deformation pattern of the Japanese islands is explicitly shown in Figure 2. However, since the absolute value of deformation rate at each site depends on the reference frame we take, interpretation of the result needs special attention. The following are major characteristics of the velocity field in Figure 2. (1) Japan is undergoing east-west compressive deformation. (2) Western Japan, which has been assumed to be a part of the Eurasian plate (e.g., SENO *et al.*, 1996), has eastward velocity relative to the Eurasian plate. This implies that the western part of Japan belongs to another plate, possibly the Amurian plate (ZONENSHAIN and SAVOSTIN, 1981; WEI and SENO, 1998). (3) Pacific coast of northeastern Japan is strongly displaced to the east. This deformation can be attributed to the subduction of the Pacific plate and mechanical coupling at the plate interface (ITO *et al.*, 2000). (4) Southern coast of southwestern Japan, except for Kyushu, is displaced towards the northwest. This is attributed to the subduction of the Philippine Sea plate and interplate coupling (SAGIYA, 1995, 1999; ITO *et al.*, 1999). (5) On the other hand, southern Kyushu and the Ryukyu Islands are displaced trench-ward to the southeast. Compared with the areas along the Nankai Trough, this implies variability of plate interaction at subduction zones. This trench-ward motion may correspond to the backarc opening process at the Okinawa Trough.

3.2. Strain Rate Distribution

As will be noticed in Figure 2, velocity data at GPS sites are sometimes affected by the instability of GPS monument or other local effects. In addition, in order to utilize GPS data for simulation studies, we often need to speculate deformation at points where we have no data. For these purposes, estimation of continuous distribution of crustal deformation has essential importance.

We applied a method by SHEN *et al.* (1996) to estimate distribution of strain rate based on the displacement rate vectors shown in Figure 2. Strain rate tensor is independent of the reference frame, and suitable for the discussion of deformation. In this method, horizontal displacement rate components (u, v), strain rate components $(\dot{e}_{xx}, \dot{e}_{xy}, \dot{e}_{yy})$ and a rotation rate ω at a particular point $x_i = (x_i, y_i)$ are related with an observed displacement rate (U, V) at an observation point $X = (X, Y)$ as follows:

$$\begin{pmatrix} U \\ V \end{pmatrix} = \begin{pmatrix} 1 & 0 & \Delta x_i & \Delta y_i & 0 & \Delta y_i \\ 0 & 1 & 0 & \Delta x_i & \Delta y_i & -\Delta x_i \end{pmatrix} \begin{pmatrix} u \\ v \\ \dot{e}_{xx} \\ \dot{e}_{xy} \\ \dot{e}_{yy} \\ \omega \end{pmatrix} + \begin{pmatrix} \varepsilon_x^i \\ \varepsilon_y^i \end{pmatrix}. \tag{2}$$

Here, $\Delta x_i = X - x_i$, and $\Delta y_i = Y - y_i$, and $\dot{e}_{xy} = \frac{1}{2}(\partial u/\partial y) + (\partial v/\partial x)$. $(\varepsilon_x^i, \varepsilon_y^i)$ are the observational errors and they are weighted depending on the distance between the observation point X and the calculation point x_i in the following formula.

$$\varepsilon_{x,y}^i = \sigma_{x,y}^i \exp(\Delta R_i^2/2D^2). \tag{3}$$

In the above equation, $\sigma_{x(y)}^i$ is an original observational error of the x (or y) component of the displacement rate, $\Delta R_i = |X - x_i|$, and D is a parameter which controls weight among the observations and is called the Distance Decaying Constant (DDC). An appropriate value of DDC may vary on both tectonic deformation features and spacing of observational points. We tested several values for the DDC and applied 35 km as a standard value. Stations within a distance of $2D$ from the calculation point are used to estimate strain rate components. In order to eliminate outliers in observed data, we conducted the following screening process. We estimated the displacement rate vector at the identical coordinates of each GPS site without using the observed data at the site, and compared the estimated value with the observation. If the residual exceeds some threshold (here 50 times the nominal estimation error), then the observed data at the GPS site are eliminated and are not used for strain rate calculation. The number of eliminated sites is 103 in this case. Although the number of eliminated sites seems large, the estimated strain rate distribution changes insignificantly even if we include all the velocity data.

In Figures 3–5, we plotted estimated distribution of plotted principal strain rate axes, dilatation rate, and maximum shear strain rate. Dilatation rate (Δ) and shear strain rate (Σ) at each point are calculated as follows.

$$\begin{aligned} \Delta &= \dot{e}_{xx} + \dot{e}_{yy} \\ \Sigma &= \sqrt{\dot{e}_{xy}^2 + (\dot{e}_{xx} - \dot{e}_{yy})^2/4} \end{aligned}. \tag{4}$$

Figure 3 is a plot of estimated principal axes of strain rates. Large strain rate (> 0.1 ppm/y) is found in the southeastern Hokkaido, the central part of northeastern Honshu, the northern part of central Honshu, southern coast of Honshu, all of Shikoku, and southern Kyushu. Except for a few regions such as the Izu Peninsula and the Japan Sea coast in central Honshu, most of these regions are related to plate subduction. It should be noted that deformation pattern laterally varies even

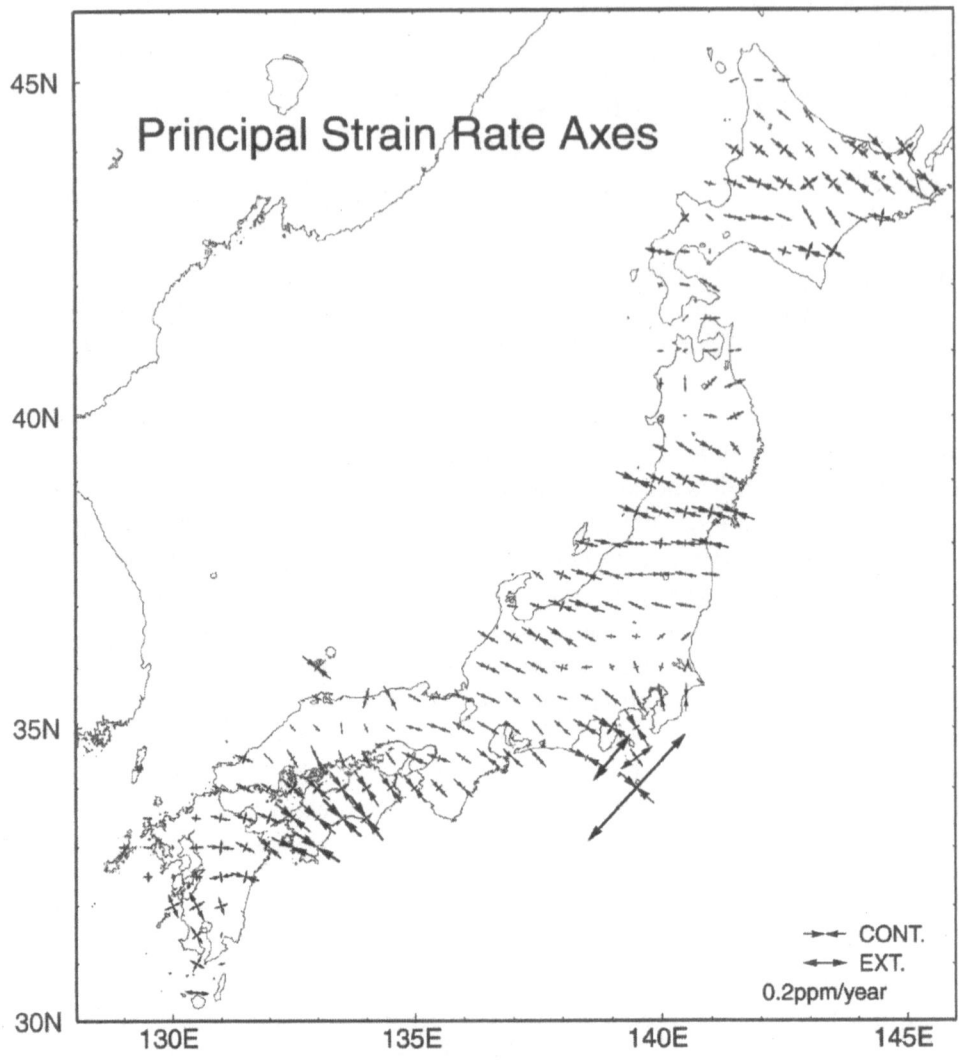

Figure 3
Estimated principal axes of strain rates.

along the same plate boundary. For example, in Tohoku, the central part (38N–39N) has the largest strain rate. Strain rate becomes smaller in both the south and the north. In northern Tohoku, significant postseismic deformation occurred after the 1994 Sanriku earthquake (HEKI *et al.*, 1997). Although coordinates data analyzed here are those from 1997 to 1999 and deformation itself is quite steady, small strain rates there may imply existence of postseismic effects. An obstacle to solve this problem is that we do not know the deformation pattern before the 1994 earthquake because the main shock occurred only 3 months after we started operating the GPS array. Continued effort of GPS monitoring will resolve temporal changes of postseismic effects. On the other hand, in southern Tohoku and northern Kanto, small strain rate has no reason but a weak coupling at the plate boundary. Intervention of the subducted Philippine Sea plate between the Pacific plate and the Japanese islands may be effective in decreasing the surface deformation in this region. Numerical modeling efforts will be necessary to validate such a speculation.

Figures 4 and 5 are estimated distributions of dilatation rate and maximum shear strain rate, respectively. Dilatation rate can be considered as representative of

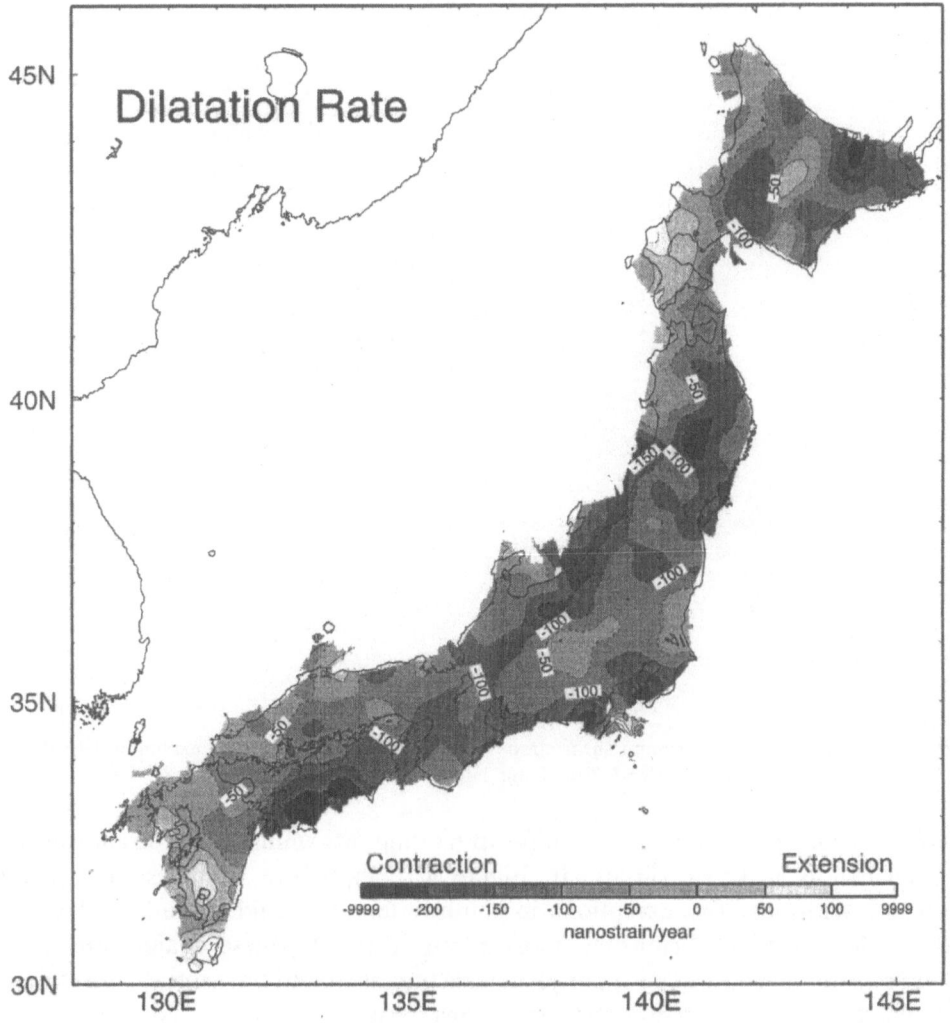

Figure 4
Estimated distribution of dilatational strain rate.

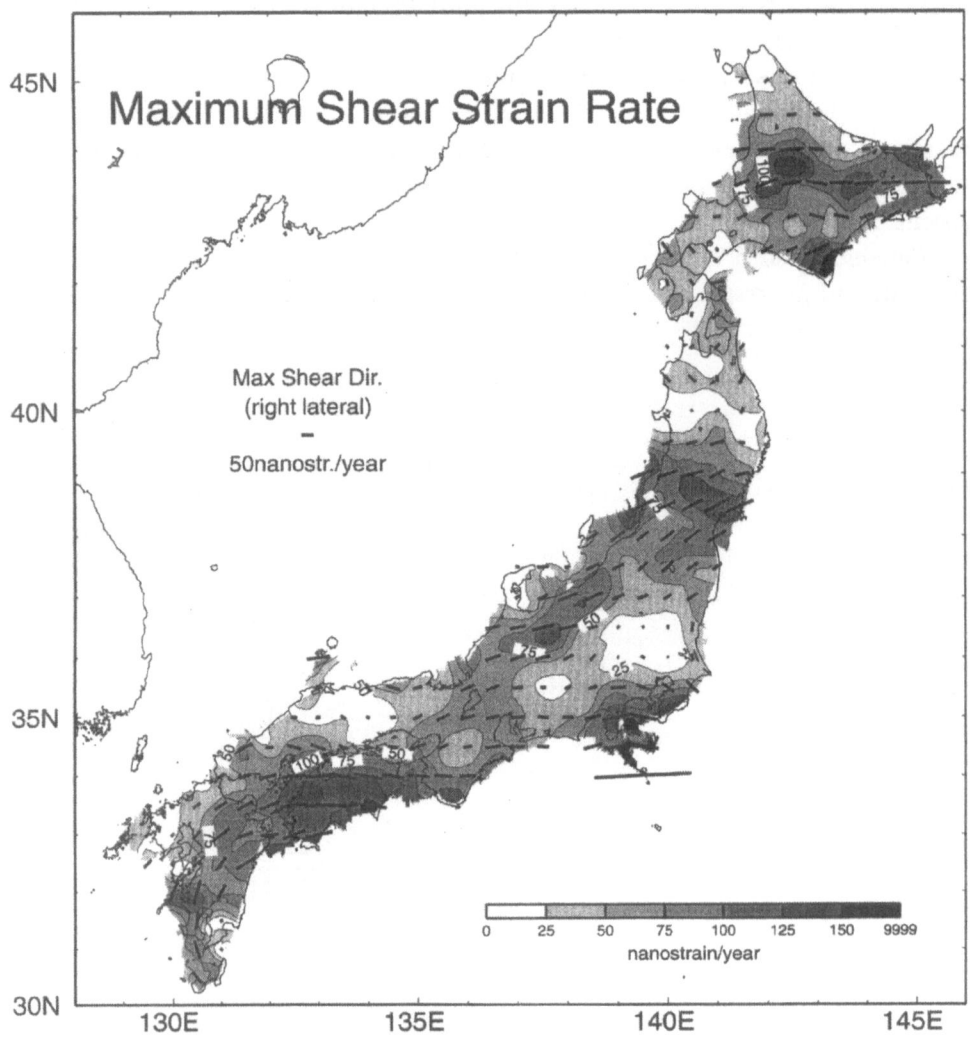

Figure 5

Estimated distribution of maximum shear strain rate. Bars denote direction of maximum right-lateral shear. Left-lateral shear is the largest in the conjugate direction.

horizontal deformation related to dip-slip faulting. Maximum shear strain rate can be related to strike-slip faults. In Figure 4, compressive tectonics of Japan is evident. With very few exceptions in southwestern Hokkaido, southern Kyushu, and around the Izu Peninsula, most parts of the Japanese islands are under compression. Large (> 0.1 ppm/y) contraction is found in central and eastern Hokkaido, central Tohoku, the Japan Sea coast of southern Tohoku and its extension to the southwest, and all along the southern coast of central and southwestern Japan. Maximum shear strain rate in Figure 5 displays a similar tendency, however the distribution is somehow different from that of the dilatation rate in Hokkaido and Tohoku. Also, the magnitude of maximum shear strain rate is basically smaller than that of the dilatation rate, implying compressive tectonics of the Japanese islands. These two scalars can show complementary distribution in principle as schematically illustrated in Figure 6.

Large strain rates along the Pacific coast of Hokkaido and Tohoku districts are caused by interaction between the subducting Pacific plate and the crust of northeastern Japan (ITO *et al.*, 2000). Another large strain rate belt is found along the southern Kanto, southern Chubu, and Shikoku districts, which can be interpreted as a result of the subduction of the Philippine Sea plate (SAGIYA, 1995, 1998, 1999; ITO *et al.*, 1999).

Concentrated deformation along the Japan Sea coast is an important new finding derived from the GPS array. There has been arguments relating to incipient subduction and the formation of a new plate boundary in this region (NAKAMURA, 1983; KOBAYASHI, 1983; SENO *et al.*, 1996). Present GPS results basically support these hypotheses, but are different from previous arguments in two aspects. First, the Itoigawa-Shizuoka Tectonic Line (ISTL) has been assumed as the possible plate boundary in central Japan while the GPS result implies the boundary traverses the northern part of Chubu and Kinki districts to the vicinity of Kobe. The other point to be noted is that this boundary is not a line, but a zone with considerable width. Such a spread deformation pattern is not similar to that of subduction zones, but similar to continental deformation. We name this deformation belt the Niigata-

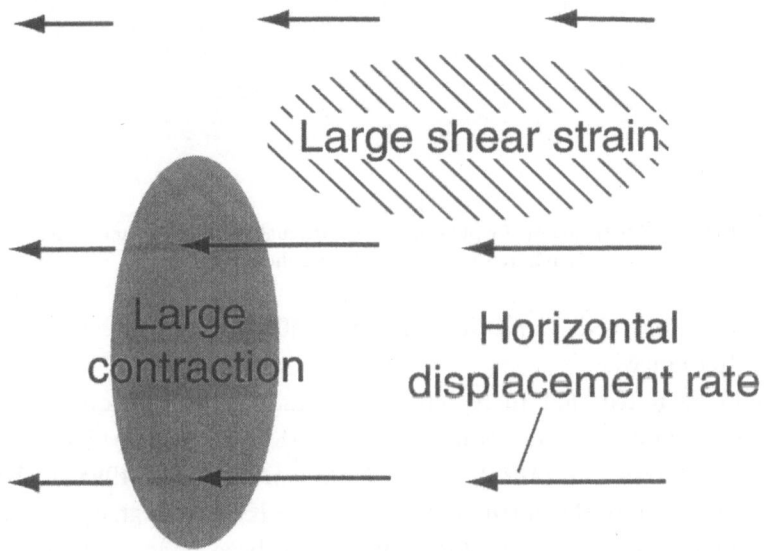

Figure 6
Schematic explanation for complementary distribution of dilatational and shear strain rates.

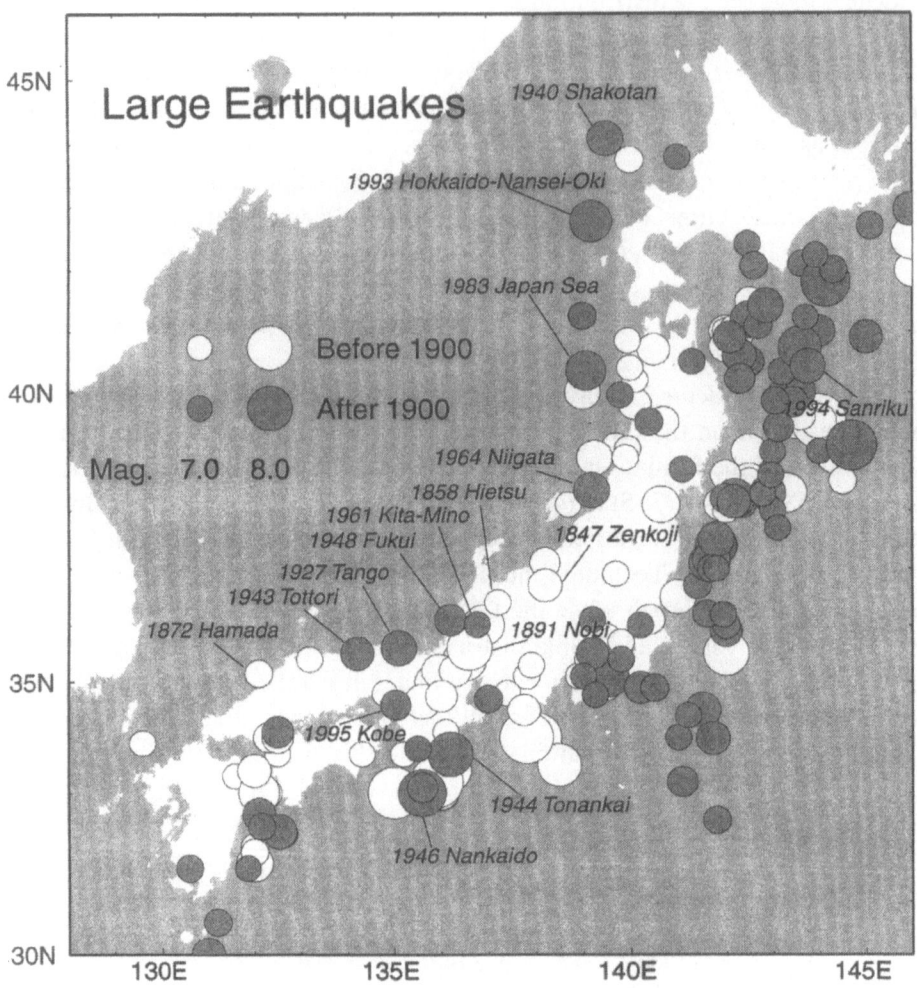

Figure 7
Distribution of large ($M > 7$) shallow (depth < 60 km) earthquakes in and around Japan. Earthquakes referred in this paper are annotated.

Kobe Tectonic Zone (NKTZ). NKTZ is about 500 km long in NE–SW direction and about 100 km wide.

It should be noted that there have been many large earthquakes along this tectonic zone such as the 1964 Niigata ($M7.5$), 1847 Zenkoji ($M7.4$), 1858 Hietsu ($M7.1$), 1961 Kita-Mino ($M7.0$), 1891 Nobi ($M8.0$), and 1995 Kobe ($M7.2$) earthquakes (Fig. 7). In the northern extension of NKTZ, though deformation there cannot be inferred by land-based GPS, there have been large earthquakes such as 1983 Japan Sea ($M7.7$), 1993 Hokkaido Nansei-Oki ($M7.8$), and 1940 Shakotan ($M7.5$) earthquakes, implying the existence of an active deformation zone.

Deformation style inferred from the strain rate distribution matches source mechanisms of these large earthquakes and the strike of active faults (Fig. 1) well. The northern part of NKTZ is dominated by a large east-west compression, and the 1964 Niigata earthquake had a thrust mechanism (SATAKE and ABE, 1983) with an east-west compression. Large earthquakes in the southwestern half of NKTZ mainly have strike-slip source mechanisms. Shear strain rates in this region are relatively larger than in surrounding regions. Though this region also has large compression rates, the absolute stress level as well as the vertical stress control an actual source mechanism. Therefore what we can affirm is that strain accumulation related to earthquake source mechanisms is faster in NKTZ than in other regions. Good correspondence with the historical seismicity implies that the present deformation field may represent tectonic deformation during at least several hundred years. HASHIMOTO (1990) first noticed this deformation zone based on his analysis of conventional survey data. Now, with precise data from the dense GPS array, we can explicitly identify the deformation zone and clarify characteristics of the ongoing deformation.

JACKSON et al. (1997) pointed out that regions of large shear strain in southern California correspond to recent large earthquakes. If that interpretation is right, the present deformation pattern may reflect postseismic deformation rather than long-term tectonic deformation. Although we cannot preclude both possibilities, the 1999 Hector Mine earthquake ($M7.1$) occurred where the strain rate was very high in JACKSON et al. (1997), implying correlation between short-term deformation and large earthquakes. The relation between the short-term deformation and a tectonic loading process of large earthquakes will be an important question for simulation studies to answer.

Existence of NKTZ is apparent from the deformation data. However, the underlying mechanism is not easy to understand. Unless we assume a heterogeneous crustal structure, it will be impossible to explain the sharp contrast in deformation rate. This is a very interesting problem which we will discuss at another occasion.

Other regions also have a sizable strain rate. In the southwestern part of Hokkaido, significant extensional strain rates are estimated. There was a $M7.8$ Hokkaido Nansei-Oki earthquake in July 1993 (e.g., TANIOKA et al., 1995). Assuming the source model by TANIOKA et al. (1995), SAGIYA et al. (1999) calculated a postseismic deformation pattern caused by viscoelastic relaxation. The result seems consistent with the present strain rate distribution, at least qualitatively. On the contrary, large compressive strain rates are found in eastern Hokkaido. Since there was a $M8.1$ Shikotan earthquake in October 1994 and the eastern part of Hokkaido was significantly displaced (TSUJI et al., 1995), we may examine the viscoelastic relaxation process here, too.

In southern Kyushu we found significantly large shear strain rates and extensional dilatation rates. This area recently experienced two large earthquakes in 1998

(*M*6.3 on March 26 and *M*6.2 on May 13, NISHIMURA *et al.*, 1998), with the maximum shear strain rate located around the epicenters of these earthquakes. Directivity of the strain rates (NE–SW compression, NW–SE expansion) is consistent with the source mechanism of the seismic events. Although we have modeled the coseismic deformation due to the two earthquakes, postseismic effects may still remain. Or the strain rate pattern may reflect the tectonic loading process responsible for the two events. There are also active volcanoes such as the Sakurajima volcano in southern Kyushu. Extensional strain in this region can also be interpreted in relation to magma intrusion of the volcanoes.

KATO *et al.* (1998) estimated a strain field of the Japanese islands based on similar GPS data such as we used. They estimated the empirical covariance function from the GPS velocity data in the Japanese islands, and applied it to estimate a spatially continuous velocity field. They calculated the strain rate field as a spatial derivative of the velocity field. Although their method is mathematically simple, their empirical covariance matrix has a much wider spatial correlation, and the resultant velocity field looks considerably smoother than those obtained in this study. Important tectonic features of deformation that we discussed above were not clear in their results.

4. Discussion

4.1. Comparison with Historical Survey Data

GEOGRAPHICAL SURVEY INSTITUTE (1996) compiled triangulation and trilateration data over the Japanese islands, and obtained strain distribution accumulated during the last 100 years. Figure 8 shows the distribution of strain rate estimated by dividing the total strain by the elapsed time. Before it is compared with GPS results, it must be noted that strain rate during 100 years may be significantly contaminated by effects of large earthquakes. In addition, the triangulation data may contain systematic scale errors since the survey result was mostly constrained by angular measurements. In spite of these problems, strain rate distributions for two different time scales (2 years and 100 years) have common features. Similar NW–SE compressive strain is dominant along NKTZ. Minor deformation in the Kanto area is another common feature. Along the Nankai Trough, strain rate in southern Chubu and southwestern Shikoku appears alike for both plots. This is because the plate boundary in these regions has not ruptured in the last 100 years (e.g., SAGIYA and THATCHER, 1999). On the other hand, strain rate patterns are different in southern Kinki and southeastern Shikoku, which were subjected to coseismic deformation of the 1944 Tonankai (*M*8.0) and 1946 Nankaido (*M*8.1) earthquakes. Strain rates in the Tohoku area are also very different for two periods. North-south oriented expansion is dominant for the longer period, while NW–SE contraction presently prevails in the same area. Since this region sustained many large earthquakes both on the plate boundary and inland (Fig. 7) during last 100 years, it may be difficult to define an average long-term deformation pattern. Differences are also found in eastern Chugoku and Kyushu areas. In Chugoku, predominant east-west contraction was found in the longer term, although not in the GPS result. In Kyushu, north-south extension was found to be concentrated in the central part in triangulation data, which is an important basis for a hypothesis concerning the opening of the Beppu-Shimabara graben (TADA, 1985). The GPS result also shows extensional strain, however, the deformation pattern looks different.

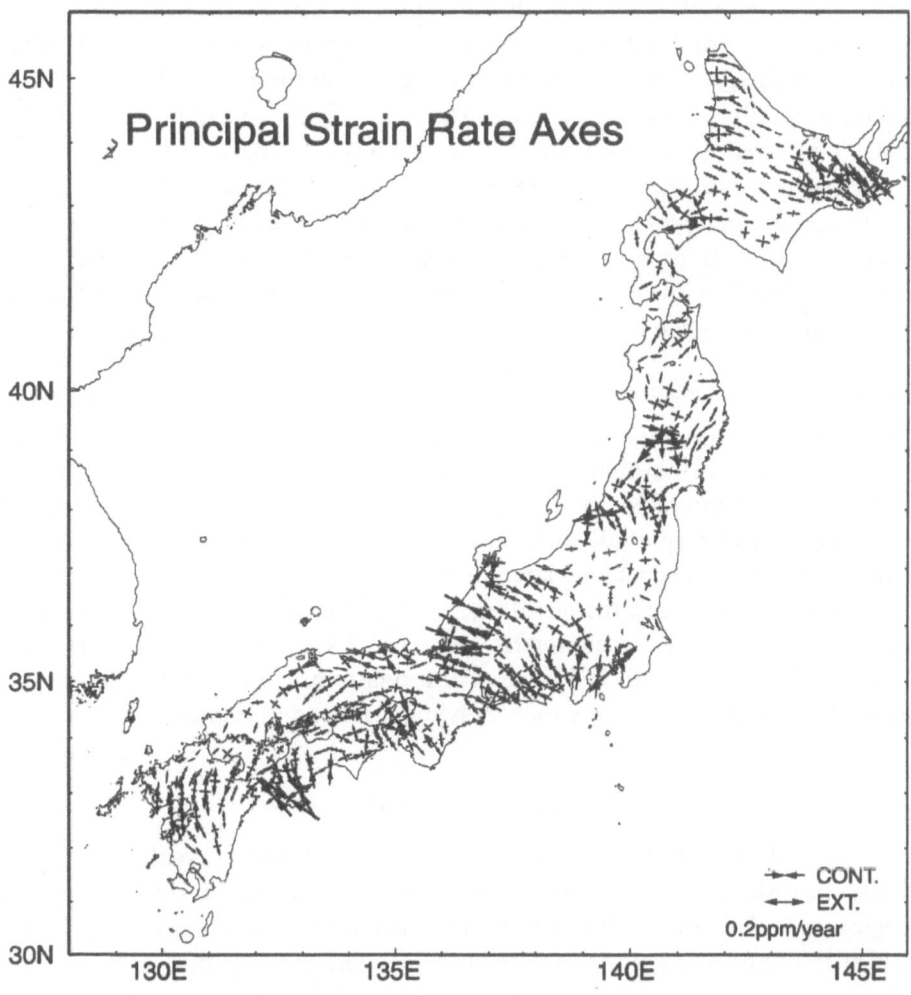

Figure 8
Distribution of principal strain rate axes obtained by comparison of conventional triangulation survey results (GEOGRAPHICAL SURVEY INSTITUTE, 1996). No correction was made for coseismic deformation due to large earthquakes between two surveys.

Both differences and similarities in two strain rate plots for different time periods strongly indicate that it is essentially important to understand the relationship between the instantaneous deformation pattern and the long-term deformation. Considering the time scale of deformation and the temporal sampling interval with continuous GPS array, we will surely progress significantly to solve this problem in the next few decades.

4.2. Strain Rate of Japan

Many authors have discussed the average strain rate of the Japanese islands. WESNOUSKY *et al.* (1982) analyzed historical earthquake records and slip rates of Quaternary faults, and estimated the average rate of seismic moment release. According to their analysis, strain rates of the Japanese islands for the last 400 years and those during the late Quaternary are mutually consistent and are about 0.02–0.03 ppm/year. ISHIKAWA and HASHIMOTO (1999) analyzed the triangulation data for the last 100 years and concluded that the geodetically determined strain rates are 0.1–0.3 ppm/year, nearly one order of magnitude larger than the geological value. SHEN-TU *et al.* (1995) also conducted the strain rate analysis and compared estimates with different data sources. Their conclusion was that geodetically estimated strain rates are 3 to 5 times larger than estimates based on seismic and geological data.

We found from GPS data, which represents virtually instantaneous strain rates, that strain rate distribution is significantly heterogeneous. Regions with large strain rates, such as NKTZ, have strain rates larger than 0.1 ppm/year, while we found very little deformation in southern Kanto, southern Chubu and Chugoku districts. There is an order of magnitude difference in strain rates among different regions. It is a matter of course that major plate boundary regions are affected by interseismic strain accumulation toward the next large earthquake, and that the nominally large strain rates there will not accumulate long-term. Except for such plate boundary regions, average strain rate is less than 0.1 ppm/year, and will be closer to the seismic and geological estimates than previous geodetic estimates.

4.3. Inconsistency with the Long-term and Seismicity Data

We have discussed that the present deformation field revealed by GPS is basically consistent with the long-term deformation deduced from seismic and geological data. However, there are examples demonstrating inconsistency between the short-term and the long-term deformations. The first example is found in the middle part of the Itoigawa-Shizuoka Tectonic Line. Average slip rate of the active faults there is estimated as large as 8–10 mm/year, based on trench excavation survey (OKUMURA *et al.*, 1994) and geomorphologic studies (IKEDA and YONEKURA, 1986). Present strain rate around the fault segments, based on the GPS data, is of the order of 0.01 ppm/year and the two crustal blocks across the faults have no significant relative motion. If both observations are correct, strain rates must have a significant temporal variation, or we must invent a special loading mechanism for the fault system. Similar discussion is applicable to other active faults located in the central Chubu district.

Another example of inconsistency is an occurrence of large earthquakes along the Japan Sea coast of the Chugoku district. There have been a series of large earthquakes such as 1872 Hamada ($M7.1$), 1927 Tango ($M7.3$) and 1943 Tottori ($M7.2$) earthquakes along the coast whereas present strain rates there are considerably smaller than in surrounding regions. Recurrence history of these earthquakes in a geological time scale is not well known. Thus those earthquakes might occur once every 10,000 years. Since the present strain rates cannot represent absolute stress or strain but changes in strain or stress only, comparison with seismicity data needs special care.

From these examples, we recognize that extrapolation of the present deformation pattern to a longer time period is not straightforward. Historical earthquake records, geological estimates of strain rates, and conventional geodetic survey data still hold useful information, which GPS data cannot provide easily. Such an inconsistency between different time scales will be an important constraint for numerical models of crustal activities.

4.4. Application to Simulation: Assimilation

One of the great advantages of continuous GPS observation is that we can make observations far more frequently than before. This enables us to feed observation data into numerical calculations and compare them with those expected from the calculation. This process is generally called assimilation, and it is an important aspect in numerical modeling of weather forecasts, oceanic flows, and so on. Crustal deformation is the most reliable and useful observation for data assimilation to date.

However, continuous GPS observation for earthquake simulation is not the same as meteorological observation for weather forecasting. For example, hourly observation of temperature or barometric pressure will show significant variation even in one day. Conversely, daily coordinates of GPS stations generally do not change considerably day by day. This means that even if we feed new data into simulation, information gain by the new data is not always significant. In this sense, assimilation of GPS data need not to be real-time.

In numerical simulations such as the finite element method and the finite difference method, the model region is divided into numerous elements, or it consists of a large number of grid points. The data assimilation process in such numerical simulations requests the provision of information, such as the displacement rate or strain rate, to each grid point. In most cases, the number of observation points is expected to be much smaller than the number of grid points. We need to interpolate available observation data to estimate the most plausible value at each grid point. The strain analysis method we employed in this paper will facilitate this purpose.

Finally, one of the great advantages of GPS observations is that they are three-dimensional measurements. Namely, we can make not only horizontal but also vertical observations. Although the present accuracy of vertical measurement by GPS is moderate, use of vertical information from GPS observations is expected to be very useful and it will greatly benefit simulation studies.

5. Conclusion

Daily coordinate data obtained by the Japanese continuous GPS array are analyzed to estimate the secular deformation field of the Japanese islands. Significant strain accumulation is found in plate boundary regions as well as inland areas. Among them, the Niigata-Kobe Tectonic Zone can be identified as a zone with high strain rates. This zone is about 100 km wide and 500 km long. Since large historical earthquakes and active faults are concentrated in this zone, it is considered to have played a central role in the deformation of the Japanese inland. GPS-based short-term deformation is basically consistent with the long-term deformation data based on conventional geodetic surveys, historical earthquakes, and Quaternary active faults. Uniformity of tectonic deformation over a geological time scale is implied. However, there are exceptions where short-term deformation is incompatible with long-term deformation. Simulation studies of the Japanese islands will be necessary to test the possibility of temporal changes in deformation pattern. Possessing continuous GPS data, simulation studies of tectonic deformation and earthquake occurrence will be improved through the assimilation procedure. Now is the time to proceed toward a realistic earthquake simulation that is capable of forecasting future seismic activity.

Acknowledgements

One of the authors (T. S.) is grateful to ACES for the travel support enabling attendance at the inaugural workshop of ACES held at Brisbane and Noosa Heads, Australia. We thank the GPS group at GSI for their efforts in constructing continuous GPS sites and managing daily data processing. Norihiko Ishikawa of GSI is acknowledged for the use of his analysis result of the triangulation data. Careful review by Bernard Minster enhanced this paper. Finally, we thank Peter Mora, editor of this special issue, for his kind encouragement to submit this work.

REFERENCES

GEOGRAPHICAL SURVEY INSTITUTE (1996), *Horizontal Crustal Deformations in the Japanese Islands*, Rep. Coord. Comm. Earthquake Predict. *55*, 658–665 (in Japanese).

HASHIMOTO, M. (1990), *Horizontal Strain Rates in the Japanese Islands during Interseismic Period Deduced from Geodetic Surveys (Part 1): Honshu, Shikoku and Kyushu*, Zisin *43*, 13–26 (in Japanese with English abstract).

HEKI, K. (1996), *Horizontal and Vertical Crustal Movements from Three-dimensional Very Long Baseline Interferometry Kinematic Reference Frame: Implication for the Reversal Timescale Revision*, J. Geophys. Res. *101*, 3187–3198.

HEKI, K., MIYAZAKI, S., and TSUJI, H. (1997), *Silent Fault Slip Following an Interplate Thrust Earthquake at the Japan Trench*, Nature *386*, 595–598.

IKEDA, Y., and YONEKURA, N. (1986), *Determination of Late Quaternary Rates Net Slip on Two Major Fault Zones in Central Japan*, Bull. Dept. Geogr. Univ. Tokyo *18*, 49–63.

ISHIKAWA, N., and HASHIMOTO, M. (1999), *Average Horizontal Crustal Strain Rates in Japan during Interseismic Period Deduced from Geodetic Surveys (Part 2)*, Zisin *52*, 299–315.

ITO, T., YOSHIOKA, S., and MIYAZAKI, S. (1999), *Interplate Coupling in Southwest Japan Deduced from Inversion Analysis of GPS Data*, Phys. of the Earth and Planet. Inter. *115*, 17–34.

ITO, T., YOSHIOKA, S., and MIYAZAKI, S. (2000), *Interplate Coupling in Northeast Japan Deduced from Inversion Analysis of GPS Data*, Earth and Planet. Sci. Lett. *176*, 117–130.

JACKSON, D. D., SHEN, Z., POTTER, D., GE, X., and SUNG, L. (1997), *Southern California Deformation*, Science *277*, 1621–1622.

KATO, T., EL-FIKY, G. S., OWARE, E. N., and MIYAZAKI, S. (1998), *Crustal Strains in the Japanese Islands as Deduced from Dense GPS Array*, Geophys. Res. Lett. *25*, 3445–3448.

KOBAYASHI, Y. (1983), *Beginning of Plate Subduction*, Earth Monthly *3*, 510–518 (in Japanese).

MIYAZAKI, S., TSUJI, H., HATANAKA, Y., ABE, K., YOSHIMURA, A., KAMADA, K., KOBAYASHI, K., MORISHITA, H., and IIMURA, Y. (1996), *Establishment of the Nationwide GPS Array (GRAPES) and its Initial Results on the Crustal Deformation of Japan*, Bull. Geogr. Surv. Inst. *42*, 27–41.

MIYAZAKI, S., SAITO, T., SASAKI, M., HATANAKA, Y., and IIMURA, Y. (1997), *Expansion of GSI's Nationwide GPS Array*, Bull. Geogr. Surv. Inst. *43*, 23–34.

MIYAZAKI, S., HATANAKA, Y., SAGIYA, T., and TADA, T. (1998), *The Nationwide GPS Array as an Earth Observation System*, Bull. Geogr. Surv. Inst. *44*, 11–22.

NAKAMURA, K. (1983), *Possible Nascent Trench along the Eastern Japan Sea as the Convergent Boundary between Eurasian and North American Plates*, Bull. Earthq. Res. Inst. *58*, 711–722 (in Japanese with English abstract).

NISHIMURA, S., ANDO, M., and MIYAZAKI, S. (1998), *Interplate Coupling along the Nankai Trough and Southeastward Motion along Southern Part of Kyushu*, Zisin *51*, 443–456 (in Japanese with English abstract).

OKADA, Y. (1985), *Surface Deformation due to Shear and Tensile Faults in a Half-space*, Bull. Seismol. Soc. Am. *75*, 1135–1154.

OKUMURA, K., SHIMOKAWA, K., YAMAZAKI, H., and TSUKUDA, E. (1994), *Recent Surface Faulting Events along the Middle Section of the Itoigawa-Shizuoka Tectonic Line—Trenching Survey of the Gofukuji Fault near Matsumoto, Central Japan*, Zisin *46*, 425–438 (in Japanese with English abstract).

PRESCOTT, W. H. (1996), *Satellites and Earthquakes: A New Continuous GPS Array for Los Angeles, Yes, It will Radically Improve Seismic Risk Assessment for Los Angeles*, EOS Trans. AGU *77* (43), 417.

RESEARCH GROUP FOR ACTIVE FAULTS OF JAPAN, *Active Faults in Japan: Sheet Maps and Inventories* (revised edition) (Univ. Tokyo Press, Tokyo 1991).

ROTHACHER, M., and MERVART, L. (ed.), *Bernese GPS Software Version 4.0* (Astronomical Institute, University of Berne 1996).

SAGIYA, T. (1995), *Crustal Deformation Cycle and Interplate Coupling in Shikoku, Southwest Japan*, Ph.D. Thesis, Univ. of Tokyo, 164 pp.

SAGIYA, T. (1998), *Interplate Coupling and Plate Tectonics at the Northern End of the Philippine Sea Plate Deduced from Continuous GPS Data*, Bull. Earthq. Res. Inst. *73*, 275–290 (in Japanese with English abstract).

SAGIYA, T. (1999), *Interplate Coupling in the Tokai District, Central Japan Deduced from Continuous GPS Data*, Geophys. Res. Lett. *26*, 2315–2318.

SAGIYA, T., and THATCHER, W. (1999), *Coseismic Slip Resolution along a Plate Boundary Megathrust: The Nankai Trough, Southwest Japan*, J. Geophys. Res. *104*, 1111–1129.

SAGIYA, T., MIYAZAKI, S., and TADA, T. (1999), *Crustal Strain Rate of the Japanese Islands deduced from GPS Observation*, Abstracts of the 1999 Japan Earth and Planetary Sciences Joint Meeting, Sk-006.

SATAKE, K., and ABE, K. (1983), *A Fault Model for the Niigata, Japan, Earthquake of June 16, 1964*, J. Phys. Earth *31*, 217–223.

SENO, T., SAKURAI, T., and STEIN, S. (1996), *Can the Okhotsk Plate be Discriminated from the North American Plate?*, J. Geophys. Res. *101*, 11,305–11,315.

SHEN, Z., JACKSON, D. D., and GE, B. X. (1996), *Crustal Deformation Across and Beyond the Los Angeles Basin from Geodetic Measurements*, J. Geophys. Res. *101*, 27,957–27,980.

SHEN-TU, B., HOLT, W. E., and HAINES, A. J. (1995), *Intraplate Deformation in the Japanese Islands: A Kinematic Study of Intraplate Deformation at a Convergent Plate Margin*, J. Geophys. Res. *100*, 24,275–24,293.

TADA, T. (1985), *Spreading of the Okinawa Trough and its Relation to the Crustal Deformation in the Kyushu (2)*, Zisin *38*, 1–12.

TADA, T., SAGIYA, T., and MIYAZAKI, S. (1997), *The Deforming Japanese Islands as Viewed with GPS*, Kagaku *67*, 917–927 (in Japanese).

TANIOKA, Y., SATAKE, K., and RUFF, L. J. (1995), *Total Analysis of the 1993 Hokkaido Nansei-oki Earthquake Using Seismic Wave, Tsunami, and Geodetic Data*, Geophys. Res. Lett. *22*, 9–12.

TSUJI, H., HATANAKA, Y., SAGIYA, T., and HASHIMOTO, M. (1995), *Coseismic Crustal Deformation from the 1994 Hokkaido-Toho-Oki Earthquake Monitored by a Nationwide Continuous GPS Array in Japan*, Geophys. Res. Lett. *22*, 1669–1672.

WATKINS, M. M., BOCK, Y., HUDNUT, K. W., and PRESCOTT, W. H. (1997), *The Southern California Integrated GPS Network: Status Report*, EOS Trans. AGU, 1997 AGU Spring Meeting, S105.

WEI, D., and SENO, T. (1998), *Determination of the Amurian plate motion*. In *Mantle Dynamics and Plate Interactions in East Asia*, Geodynam. Series *27* (eds. Flower, M., Chung, S., Lo, C., and Lee, T.) pp. 337–346.

WESNOUSKY, S. G, SCHOLZ, C. H., and SHIMAZAKI, K. (1982), *Deformation of an Island Arc: Rates of Moment Release and Crustal Shortening in Intraplate Japan Determined from Seismicity and Quaternary Fault Data*, J. Geophys. Res. *87*, 6829–6852.

ZONENSHAIN, L. P., and SAVOSTIN, L. A. (1981), *Geodynamics of the Bikal Rift Zone and Plate Tectonics of Asia*, Tectonophys. *76*, 1–45.

(Received November 11, 1999, revised March 21, 2000, accepted April 7, 2000)

To access this journal online:
http://www.birkhauser.ch

Pure appl. geophys. 157 (2000) 2323–2349
0033–4553/00/122323–27 $ 1.50 + 0.20/0

Implications of a Statistical Physics Approach for Earthquake Hazard Assessment and Forecasting

V. G. Kossobokov,[1] V. I. Keilis-Borok,[1] D. L. Turcotte[2] and
B. D. Malamud[2]

Abstract—There is accumulating evidence that distributed seismicity is a problem in statistical physics. Seismicity is taken to be a type example of self-organized criticality. This association has important implications regarding earthquake hazard assessment and forecasting. A characteristic of a thermodynamic system is that it exhibits a background noise that is self-organized. In the case of a dilute gas, this self-organization is the Maxwell–Boltzmann distribution of molecular velocities. In seismicity, it is the Gutenberg–Richter frequency-magnitude scaling; this scaling is fractal. Observations favor the hypothesis that smaller earthquakes in moderate-sized regions occur at rates that are only weakly dependent on time. Thus, the rate of occurrence of smaller earthquakes can be extrapolated to assess the hazard of larger earthquakes in a region. We obtain the rate of occurrence of earthquakes with $m > 4$ in $1° \times 1°$ areas from the NEIC catalog. Using only this data we produce global maps of the seismic hazard. Observations also favor the hypothesis that the stress level at which an earthquake occurs is a second-order critical point. As a critical point is approached, correlations extend over increasingly larger distances. In terms of seismicity, the approach to a critical point is associated with an increase in the rate of occurrence of intermediate-sized earthquakes prior to a large earthquake. This precursory activation has been shown to exhibit power-law scaling and to occur over a region about ten times larger than the rupture length of the large earthquake. Analyses of the spinoidal behavior associated with second-order critical points predict the power-law increase in seismic activity prior to a characteristic earthquake. This precursory activation provides the basis for intermediate-range earthquake forecasting.

Key words: Earthquake hazard assessment, self-organized critical behavior, inverse-cascade model, spinoidal behavior.

1. Introduction

Earthquakes constitute a major hazard in many of the earth's regions. The plate-tectonics hypothesis explains the concentration of earthquakes at plate boundaries, but significant numbers of earthquakes occur within plate interiors. The hypothesis of stick-slip behavior and elastic rebound explains the physics of

[1] International Institute of Earthquake Prediction Theory and Mathematical Geophysics, Russian Academy of Sciences, Warshavskoe sh. 79, kor. 2, Moscow 113556, Russia Federation. E-mails: volodya@mitp.ru and vkborok@mitp.ru
[2] Department of Geological Sciences, Cornell University, Ithaca, NY 14853-1504 U.S.A. E-mails: turcotte@geology.cornell.edu and bruce@malamud.com

earthquakes in a general way. Earthquakes occur on pre-existing faults and stick-slip events occur if the friction on the faults is velocity weakening. When the slip-event (earthquake) occurs, the elastic strains in the adjacent rock are relieved, generating seismic waves and heat. The laws of elasticity can be solved in a self-consistent manner to obtain periodic earthquake cycles.

However, earthquakes do not occur in periodic cycles. The earth's crust is extremely complex, and faults and earthquakes in a region occur on widely ranging scales. Considerable evidence exists that faults and earthquakes interact on a range of scales, from thousands of kilometers to millimeters or less. Evidence in support of this hypothesis stems from the universal validity of scaling relations. The most famous of these is the Gutenberg–Richter frequency-magnitude relation (GUTEN-BERG and RICHTER, 1954)

$$\log \dot{N}_{CE} = -bm + \log \dot{a} \tag{1}$$

where $\dot{N}_{CE}$ is the (cumulative) number of earthquakes with a magnitude greater than m occurring in a specified area and time, and b and $\dot{a}$ are constants. This relation is valid for earthquakes both regionally and globally. The constant b or "b-value" varies from region to region, but is generally in the range $0.8 < b < 1.2$ (FROHLICH and DAVIS, 1993). The constant $\dot{a}$ is a measure of the regional level of seismicity. There are a variety of measures for the magnitude, including local, body-wave, surface-wave, and moment magnitude (LAY and WALLACE, 1995). In general, for small-intensity earthquakes ($m < 5.5$), these different magnitude measures give approximately equivalent results.

Although the Gutenberg–Richter frequency-magnitude relation was originally developed as an empirical relation, we now recognize that it belongs to a broad range of natural phenomena that exhibit fractal scaling (TURCOTTE, 1989, 1997). For earthquakes, fractal scaling implies the validity of the relation

$$\dot{N}_{CE} = C A_E^{-\alpha} \tag{2}$$

where $\dot{N}_{CE}$ is the (cumulative) number of earthquakes with rupture area greater than A_E occurring in a specified area and time; C and α are constants with $D = 2\alpha$ the fractal dimension. AKI (1981) showed that (1) and (2) are entirely equivalent with

$$\alpha = b = D/2. \tag{3}$$

Thus, the universal applicability of the Gutenberg–Richter relation implies universal fractal behavior of earthquakes. Other natural phenomena that satisfy fractal scaling relations include landslides (PELLETIER *et al.*, 1997) and forest fires (MALA-MUD *et al.*, 1998).

The fractal behavior of seismicity can be associated with chaotic behavior and self-organized criticality (TURCOTTE, 1999a). This association is illustrated by the behavior of slider-block models. These models are considered analogs for the behavior of faults in the earth's crust. The simplest example is a single slider block of mass m pulled over a surface by a spring attached to a constant-velocity driver plate. The interaction of the block with the surface is controlled by friction. Many friction laws have been proposed; the simplest is the static-dynamic friction law. If the block is stationary, the static frictional force is F_s, if the block is slipping the dynamical frictional force is F_d. If $F_s > F_d$ stick-slip behavior is obtained, the motion of the block comprises periodic slip events.

The behavior of a pair of slider blocks pulled over a surface and connected by a connector spring was studied in detail by HUANG and TURCOTTE (1990). The equations of motion for the two blocks were solved simultaneously. Solutions were governed by two parameters: the stiffness of the system, k_c/k_p (k_c spring constant of the connector spring and k_p the spring constant of the puller springs) and the ratio of static to dynamic friction, F_s/F_d. For some values of these parameters, deterministic chaos was found. Chaotic behavior requires some asymmetry in the problem, i.e., $F_{s1} \neq F_{s2}$. The period doubling route to chaos was observed with positive values of the Lyapunov exponent in the chaotic regions. The behavior of the pair of slider blocks is very similar to the behavior of the logistic map (MAY, 1976).

The chaotic behavior of the low-dimensional Lorenz equations (LORENZ, 1963) is now accepted as evidence that the behavior of the atmosphere and oceans is chaotic. Similarly, the chaotic behavior of a pair of slider blocks is evidence that earthquakes exhibit chaotic behavior. It is instructive to make comparisons between the behavior of the earth's atmosphere and the behavior of the earth's crust. Massive numerical simulations are routinely used to forecast the weather. In many cases, they are quite accurate on time scales of 24 to 48 hours, however on the scale of weeks, they are of little value. The motions of the storm systems are relatively stable, considering the complexity involved. In many cases, the paths of major storms such as hurricanes can be predicted with considerable accuracy, but in other cases there are major uncertainties.

What about the earth's crust? Forecasting or predicting an earthquake is quite different than forecasting the path and intensity of a hurricane (TURCOTTE, 1991). The hurricane exists but the earthquake does not exist until it happens. GELLER et al. (1997) have argued, based on the chaotic behavior of the earth's crust, that "earthquakes cannot be predicted." This is certainly true in the sense that the exact time of occurrence of an earthquake cannot be predicted. However, this is also true of hurricanes, the exact path of a hurricane cannot be predicted. Nevertheless, a probabilistic forecast of hurricane paths with a most probable path are routinely made, and their use is of great value in terms of requiring evacuations and in other preparations. An essential question concerning earthquakes is whether similar useful probabilistic forecasts can be made. In fact, this is already being done in terms of hazard assessments. The rate of occurrence of small earthquakes is extrapolated to estimate the rate of occurrence of larger earthquakes.

The applicability of power-law (fractal) scaling to earthquakes can be justified in terms of scale-invariance; the power-law distribution is the only distribution that does not require a characteristic length scale. However, there is accumulating evidence that a more fundamental reason exists. In the past ten years, a variety of numerical models have been found to exhibit a universal behavior that has been called self-organized criticality. In self-organized criticality the "input" to a complex system is slow and steady; whereas the output is a series of events or "avalanches" that follow power-law (fractal) frequency-size statistics. Regional seismicity is often taken as a naturally occurring example of self-organized criticality. The input is the slow and steady motion of the tectonic plates and the output is the earthquakes.

The concept of self-organized criticality (TURCOTTE, 1999b) evolved from the "sandpile" model proposed by BAK *et al.* (1988). In this model there is a square grid of boxes and at each time step a particle is dropped into a randomly selected box. When a box accumulates four particles, they are redistributed to the four adjacent boxes, or in the case of edge boxes, they are lost from the grid. Since only nearest-neighbor boxes are involved in a redistribution, this is a cellular-automata model. Redistributions can lead to further instabilities and avalanches of particles in which many particles may be lost from the edges of the grid. The input is the steady-state addition of particles. A measure of the state of the system is the average number of particles in the boxes. This "density" fluctuates about a quasi-equilibrium value. Each of the multiple redistributions during a time step contributes to the size of the model "avalanche." One measure of the size of a model avalanche is given by the number of particles lost from the grid during each sequence of redistributions; an alternative measure is given by the number of boxes that participate in the redistributions.

This model is called a "sandpile" model because of the resemblance to an actual sandpile on a table. The randomly dropped particles in the model are analogous to the addition of particles to an actual sandpile; the model avalanches are analogous to sand avalanches down the sides of the sandpile. In some cases, the sand avalanches lead to the loss of particles off the table. Extensive numerical studies of the "sandpile" model were carried out by KADANOFF *et al.* (1989). They found that the noncumulative frequency-size distribution of avalanches satisfies (2) with $\alpha \approx$ 1.0.

A second example of "self-organized criticality" is the behavior of large arrays of slider blocks. The slider block model with a pair of slider blocks considered above can be extended to include large numbers of slider blocks. Multiple slider-block simulations were first considered by BURRIDGE and KNOPOFF (1967). OTSUKA (1972) considered a two-dimensional array of slider blocks and obtained power-law distributions for the size of slip events.

CARLSON and LANGER (1989) considered long linear arrays of slider blocks with each block connected by springs to the two neighboring blocks and to a constant-velocity driver. They used a velocity-weakening friction law and considered up to 400 blocks. Slip events involving numerous blocks were observed, the motions of all blocks involved in a slip event were coupled, and the applicable equations of motion had to be solved simultaneously. Because of the strong similarities, these are often known as molecular-dynamic simulations. Although the system is completely deterministic, the behavior was apparently chaotic. Frequency-size statistics were obtained for slip events. The events fell into two groups. In the first group, smaller events obeyed a power-law (fractal) relationship with a slope near unity. In the second group, there was an anomalously large number of large events that included all the slider blocks. The observed behavior was characteristic of self-organized criticality. The motion of the driver plate is the steady input. The slip events are the avalanches with a fractal distribution.

NAKANISHI (1991) studied multiple slider-block models using the cellular-automata approach. A linear array of slider blocks was considered but only one block was allowed to move in a slip event. The slip of one block could lead to the instability of either or both of the adjacent blocks, which would then be allowed to slip in a subsequent step or steps, until all blocks were again stable. BROWN et al. (1991) proposed a modification of this model involving a two-dimensional array of blocks. The use of the cellular-automata approach greatly reduces the complexity of the calculations and the results using the two approaches are generally very similar. A wide variety of slider-block models have been proposed and studied; these have been reviewed by CARLSON et al. (1994) and TURCOTTE (1997).

The standard multiple slider-block model consists of a square array of slider blocks as illustrated in Figure 1. Each block with mass m is attached to the driver plate with a driver spring, spring constant k_p. Adjacent blocks are attached to each other with connector springs, spring constant k_c. A block remains stationary as long as the net force on the block is less than the static resisting force, F_s. If the static frictional force is greater than the dynamic frictional force, $F_s > F_d$, stick-slip behavior is observed. In order to carry out a simulation it is necessary to specify the stiffness, k_c/k_p, the friction, F_s/F_d, and the area of the square array.

HUANG et al. (1992) carried out numerous simulations on a square array of blocks using stick-slip friction and a cellular-automata approach. Their noncumulative frequency-area statistics for model slip events are given in Figure 2. The number of slip events per time step with area A_e, N_e/N_0, is given as a function of A_e. Results are given for a stiffness $k_c/k_p = 30$, friction $F_s/F_d = 1.5$, and four grid sizes, 20×20, 30×30, 40×40, and 50×50. There is good agreement with the power-law relation (2) taking $\alpha \approx 1$. For stiff systems, k_c/k_p large, the entire grid of slider blocks is strongly correlated and large slip events including all blocks occur regularly. These are the peaks for $A_e = 400$, 900, and 1600, illustrated in Figure 2. For soft systems, k_c/k_p relatively small, no large events occur.

There are strong similarities between the behavior of the sandpile model and the slider-block model. In both cases, smaller slip events have a noncumulative power-

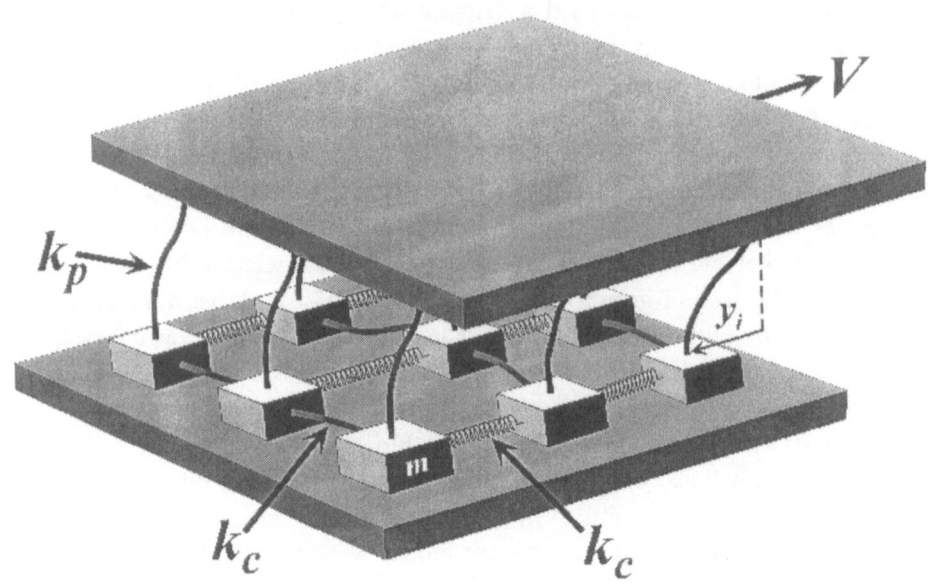

Figure 1

Illustration of the two-dimensional slider-block model. An array of blocks, each with mass m, is pulled across a surface by a driver plate at a constant velocity, V. Each block is coupled to the adjacent blocks with either leaf or coil springs (spring constant k_c), and to the driver plate with leaf springs (spring constant k_p).

law frequency-area distribution with a slope near unity. Whereas the sandpile model is stochastic in the selection of boxes, the slider-block model is fully deterministic. The slider-block model provides a bridge between chaotic behavior (two slider blocks) and self-organized critical behavior (large numbers of slider blocks). Adjacent solutions for the chaotic behavior of a pair of slider blocks have an exponential divergence. Adjacent solutions for large numbers of slider blocks have a power-law divergence.

The power-law distribution of avalanches in the sandpile and slider-block models can be explained in terms of an inverse cascade (TURCOTTE *et al.*, 1999). A metastable cluster is a group of boxes in the sandpile model or a group of slider blocks in the slider-block model over which an avalanche will spread once it is initiated. The inverse cascade is the coalescence of smaller metastable regions to form larger metastable regions. This coalescence gives a power-law distribution of metastable cluster sizes. Avalanches sample this distribution and thus also have a power-law distribution of sizes. However, significant numbers of metastable clusters are lost only from the largest cluster sizes, and these losses terminate the cascade and the region of power-law scaling.

Since the concept of self-organized criticality was first introduced, earthquakes have been identified as an example of this phenomenon in nature (BAK and TANG,

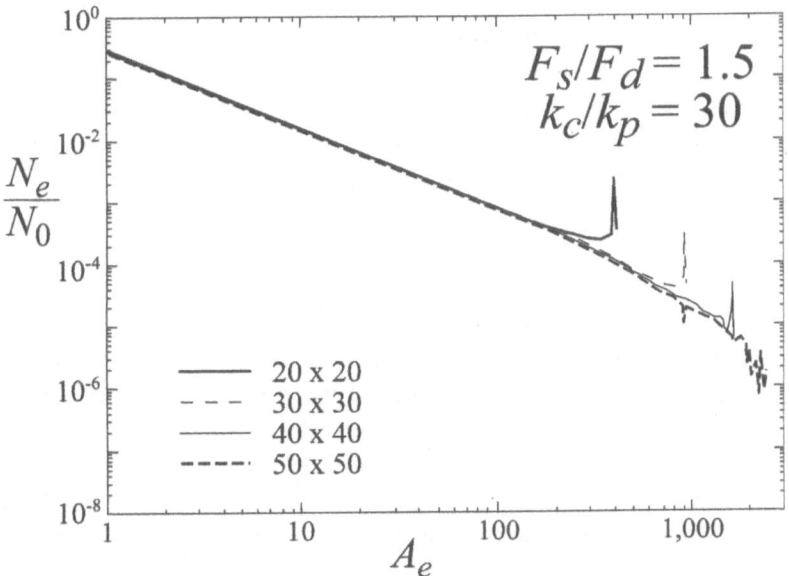

Figure 2

Results for a two-dimensional slider-block model with multiple blocks (HUANG et al., 1992). The ratio of the number of slip events, N_e, with area A_e, to the total number of slip events N_0, is plotted against A_e the number of blocks involved in an event (HUANG et al., 1992). Results are given for systems with stiffness $k_c/k_p = 30$, friction $F_s/F_d = 1.5$, and grid sizes 20×20, 30×30, 40×40, and 50×50. The peaks at $A_e = 400$, 900, and 1600, correspond to catastrophic slip events involving the entire system.

1989). These authors also pointed out the similarity in slopes between self-organized critical behavior and earthquakes. However, since the model data are noncumulative, the agreement with the cumulative earthquake data must be considered fortuitous. The power-law exponents are significantly higher for the application than for the model.

There is also accumulating observational evidence that the occurrence of an earthquake is equivalent to a critical point. The evidence for precursory seismic activation prior to an earthquake will be discussed in Section 3 of this paper. This activation is the primary basis for the intermediate-range earthquake prediction algorithms developed by Keilis-Borok and his associates (KEILIS-BOROK, 1990, 1996; KEILIS-BOROK and KOSSOBOKOV, 1990; KEILIS-BOROK and ROTWAIN, 1996; KNOPOFF et al., 1996; KOSSOBOKOV et al., 1999). This activation is associated with the approach of the "tuning parameter," the regional stress, to the critical value that will result in an earthquake. As the critical point is approached, the correlation length increases. For an earthquake, the correlation length is the size of the region over which precursory seismic activation occurs. BOWMAN et al. (1998) give observational evidence that the correlation length is about ten times the length of the rupture zone. It appears reasonable to hypothesize that seismicity in the earth's crust is a thermodynamic system. Small earthquakes represent a background noise that is independent of time and does correlate with the "seismic cycle." The rate of occurrence of small earthquakes is proportional to the rate of increase of the regional stress (the tuning parameter), and this is proportional to the rate of occurrence of the larger earthquakes. This hypothesis is the basis of our approach to probabilistic seismic hazard assessment given in the next section.

2. Hazard Assessment

The validity of the Gutenberg–Richter frequency-magnitude relation (1) for regional seismicity has been recognized for approximately 50 years. The use of this relation to extrapolate the rate of occurrence of small earthquakes to larger earthquakes (CORNELL, 1968) has been routinely incorporated into many regional seismic hazard assessments in a number of developed countries. Probably the best-documented probabilistic seismic hazard assessment was developed for the eastern United States by the Lawrence Livermore National Laboratory under a contract from the U.S. Nuclear Regulatory Commission. The documentation runs to eight volumes with some 3000 pages (BERNREUTER *et al.*, 1989). The objective was to assess the seismic hazard for nuclear power plant operation in the eastern United States.

In their study, the use of the Gutenberg–Richter frequency-magnitude relation (1) was an essential feature of the hazard assessment. However, a standard approach to its use could not be agreed upon. As a first step, the eastern United States was divided into 35 zones based on geological considerations. "Expert opinion" was then used to obtain values for the parameters $\dot{a}$ and b in (1). For each of the 35 zones, eleven "experts" were asked to provide their "preferred" values of these parameters based on their own estimates. The different experts used different selections of earthquakes, different corrections to magnitudes (for instrumental and historic events) and different treatments of aftershocks. As an example, for Zone 1 (New England), in the equation $\log N_{CE} = -bm + a$, values of a ranged from 2.68 to 4.39 and the values of b ranged from 0.60 to 1.15. Note that a in their equation is for the entire period whereas our $\log \dot{a}$ in (1) is per year. The report concluded that the frequency of occurrence predictions in a region could have errors as large as factors of 40–100.

The United States Geological Survey prepares seismic hazard maps for the United States. The most recent version (FRANKEL *et al.*, 1996) systematically utilized the Gutenberg–Richter relation (1). Details of the approach for the eastern United States have been given by FRANKEL (1995). First, the region was divided into 11 km by 11 km cells. Then the number of earthquakes N_i in each cell i with magnitudes greater than a given m was determined. Thereafter the grid of N_i values was spatially smoothed using a Gaussian function with a correlation distance c. A map of smoothed 10^a values was prepared (FRANKEL, 1995, Fig. 4) taking $m = 3$ (1924–1991) and $c = 50$ km. The optimal c-value of about 50 km was found through trial and error.

In this paper, we propose a systematic approach to the use of the Gutenberg–Richter relation (1) to estimate the seismic hazard globally, using a method very similar to the one just discussed (FRANKEL, 1995). In particular, we must choose an earthquake catalog and a lower-magnitude cutoff that gives reasonable completeness for a specified period of time. Then we must also choose an area over which to average the observed seismicity. Finally, we directly use the Gutenberg–Richter relation (1) to extrapolate the occurrence of small earthquakes in order to assess the hazard of larger earthquakes.

Before discussing the global approach we consider two specific examples of the application of the Gutenberg–Richter relation to earthquake hazard assessment. We consider the regional seismicity in two localities, Southern California and the New Madrid seismic zone.

The cumulative frequency-magnitude distribution of seismicity in Southern California is given in Figure 3. The data, from the Southern California Seismographic Network (SCSN CATALOG, 1995) are for the period 1932–1994. The

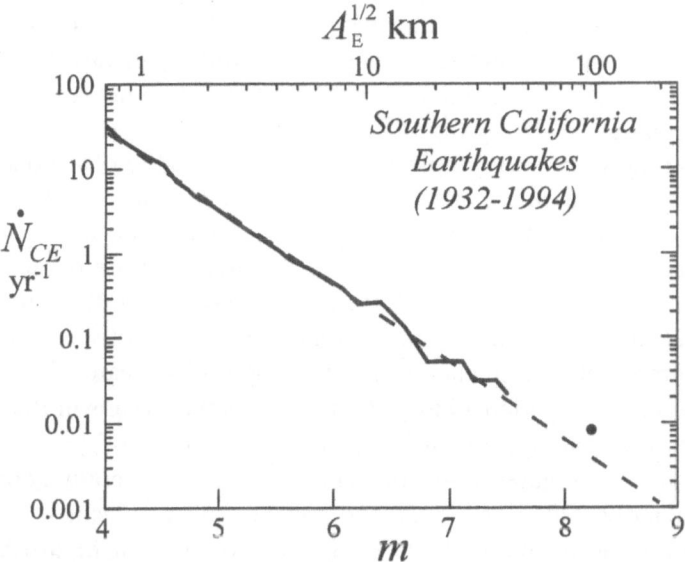

Figure 3

Cumulative number of earthquakes per year, $\dot{N}_{CE}$, occurring in Southern California with magnitudes greater than m as a function of m. Also given is the square-root of the equivalent rupture area. The solid line is for the time period 1932–1994, with data obtained from the SCSN CATALOG (1995). The straight dashed line is the Gutenberg–Richter relation (1) with $b = 0.92$ and $\dot{a} = 1.4 \times 10^5$ yr^{-1}. The solid circle is the observed occurrence rate of great earthquakes in Southern California (SIEH, 1978; SIEH et al., 1989).

catalog contains a variety of magnitude measures. The cumulative number of earthquakes per year, $\dot{N}_{CE}$, with magnitudes greater than m, are given as a function of m. Over the range $4.0 < m < 7.5$, the data in Figure 3 are in excellent agreement with the straight dashed line, obtained by taking the Gutenberg–Richter relation (1) with $b = 0.92$ and $\dot{a} = 1.4 \times 10^5$ yr^{-1}.

Also included in Figure 3 is a solid circle representing the value associated with great earthquakes on the southern section of the San Andreas Fault. Dates for ten large earthquakes on this fault section have been obtained from radiocarbon dating of liquefaction features (SIEH, 1978). The mean repeat time is 132 years, giving $\dot{N}_{CE} = 0.0076$ yr^{-1}. SIEH *et al.* (1989) estimated that the last great earthquake, in 1857, had a magnitude $m = 8.25$. Taking these values for $\dot{N}_{CE}$ and m, we obtain the solid circle in Figure 3. An extrapolation of the Gutenberg–Richter statistics (the dashed line in Fig. 3) appears to make a reasonable prediction of great earthquakes on this section of the San Andreas Fault. It must be noted that the extrapolation of the levels of regional seismicity to regional "characteristic" earthquakes is controversial (DAVISON and SCHOLZ, 1985; PACHECO *et al.*, 1992; SCHOLZ, 1997). But, even without this extrapolation to "characteristic" earthquakes, this assessment provides valuable information on the regional seismic hazard for moderate-sized earthquakes.

An essential question concerning the extrapolation of background seismicity to estimate the seismic hazard, is the time dependence of the background seismicity. If the background seismicity increased systematically during an earthquake cycle, this increase could be used for earthquake prediction. There is no evidence for such a systematic increase.

The second question is the statistical temporal variability of regional seismicity. To address this question, we consider the time dependence of the background seismicity in Southern California. The frequency-magnitude distributions of the regional seismicity in Southern California on a yearly basis are plotted in Figure 4 using data obtained from the SCSN CATALOG (1995). Again, this catalog contains a variety of magnitude measures. For each individual year between 1980–1994, the cumulative number of earthquakes $\dot{N}_{CE}$ with magnitudes greater than m is plotted as a function of m. The period 1980–1994 taken together results in the Gutenberg–Richter power-law relation (1) with $b = 1.05$ and $\dot{a} = 2.06 \times 10^5$ yr^{-1}, shown as the solid straight lines in Figures 4a–c. In Figure 4, there is generally good agreement between each individual year's data and the Gutenberg–Richter relation (solid straight line) for the period 1980–1994. The exceptions can be attributed to the aftershock sequences of the 1987 Whittier, 1992 Landers, and 1994 Northridge earthquakes.

With aftershocks removed, the background seismicity in Southern California illustrated in Figure 4 is nearly uniform from year to year, and is not a function of time. Small earthquakes behave like a thermal background noise. This is observational evidence that the earth's crust is continuously on the brink of failure

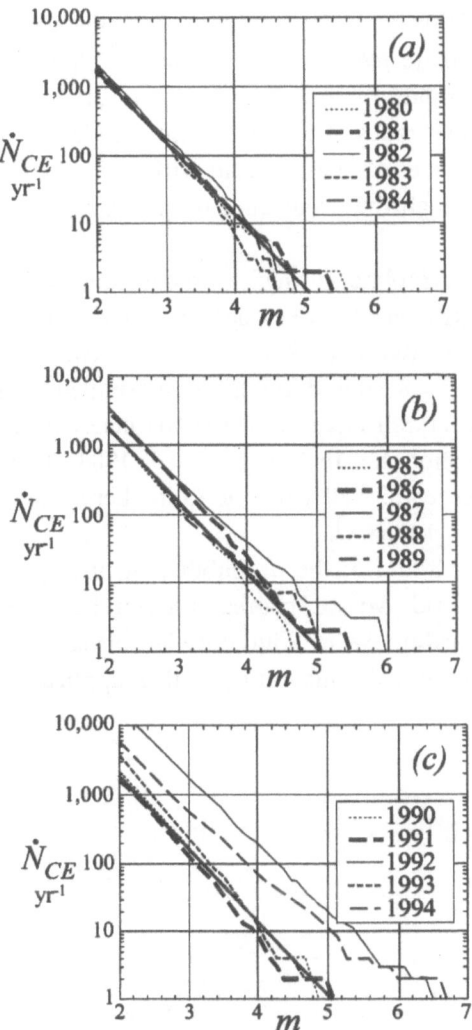

Figure 4

Cumulative number of earthquakes per year, $\dot{N}_{CE}$, occurring in Southern California with magnitudes greater than m as a function of m. Fifteen individual years are considered (SCSN CATALOG, 1995): (a) 1980–1984, (b) 1985–1989, (c) 1990–1994. The solid straight-line in (a) to (c) is the Gutenberg–Richter relation (1) with $b = 1.05$ and $\dot{a} = 2.06 \times 10^5$ yr^{-1}, the best-fit to all data during 1980–1994. The larger number of earthquakes in 1987, 1992, and 1994 can be attributed to the aftershocks of the Whittier, Landers, and Northridge earthquakes, respectively. If aftershocks are excluded, the background seismicity in Southern California is nearly uniform in time.

(SCHOLZ, 1991). Further evidence for this comes from induced seismicity. Whenever the crust is loaded, whether in a tectonically active area or not, earthquakes are induced. Examples of loading include the filling of a reservoir behind a newly completed dam or the high-pressure injection of fluids in a deep well.

Since the eastern United States is a plate interior, the concept of rigid plates would preclude seismicity in the region. However, the plates act as stress guides. The forces that drive plate tectonics are applied at plate boundaries. Because the plates are essentially rigid, these forces are transmitted through their interiors. However, the plates have zones of weakness that will deform under these forces and earthquakes result. Thus, earthquakes occur within the interior of the surface plates of plate tectonics, although the occurrence frequencies are considerably lower than at plate boundaries.

As an example of intraplate seismicity we consider the New Madrid, Missouri, seismic zone. Three great earthquakes occurred here during the winter of 1811–1812. JOHNSTON and SCHWEIG (1996) suggest that two of these earthquakes had moment magnitudes $m = 8$. Based on both instrumental and historical records, JOHNSTON and NAVA (1985) have given the frequency-magnitude distribution for earthquakes in this region for the period 1816–1983. Their results are given in Figure 5. The data correlate well with the Gutenberg–Richter relation (1) taking $b = 0.90$ and $\dot{a} = 2.24 \times 10^3 \ \text{yr}^{-1}$.

Comparing Figures 3 and 5, the probability of having a moderate-sized earthquake in the New Madrid, Missourri area is about 1/60 the probability of having a similar magnitude earthquake in Southern California. Since the b-values for the two regions are about equal, this ratio is also applicable to other earthquake

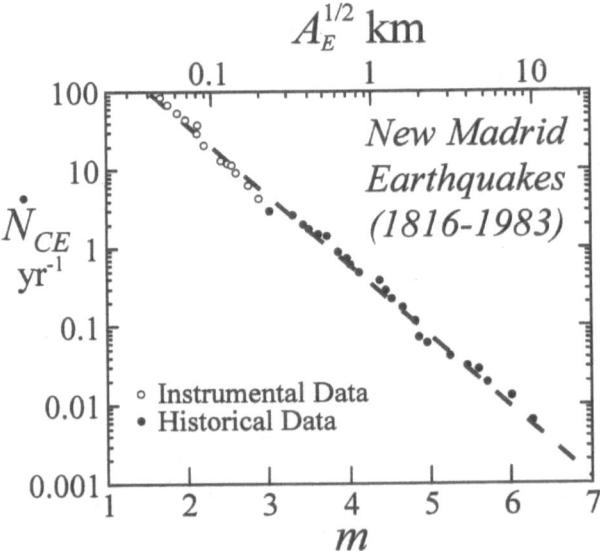

Figure 5

The cumulative number of earthquakes per year, $\dot{N}_{CE}$, occurring in the New Madrid, Missouri seismic zone (1816–1983) with magnitudes greater than m as a function of m. The open circles represent instrumental data and the solid circles historical data (JOHNSTON and NAVA, 1985). The straight dashed line is the Gutenberg–Richter relation (1) with $b = 0.90$ and $\dot{a} = 2.24 \times 10^3 \ \text{yr}^{-1}$.

magnitudes. This extrapolation is the basis of our global seismic hazards assessment. The results given in Figure 5 also provide further support for the hypothesis that the background seismicity in a region is independent of time. The instrumental data given in Figure 5 span a period of about 30 years and are in excellent agreement with the historical data that span 120 years.

The results given above support the use of the Gutenberg–Richter relation (1) to assess the global seismic hazard. We base our systematic global assessment of the seismic hazard on the epicenters of earthquakes with magnitudes $m \geq 4$, where the data were obtained from the NEIC Global Hypocenter Data Base (GHDB, 1989) for the period 1964–1995. We considered both a magnitude 5 cutoff and a magnitude 4 cutoff. The advantage of the magnitude 5 cutoff is that we could be certain that the database is globally complete for earthquakes $m \geq 5$ during this period (ENGDAHL et al., 1998). The advantage of the magnitude 4 cutoff is that there are many more earthquakes on which to base our statistics. Since our primary goal is to provide the best seismic hazard assessment in populous regions, and since these regions generally have seismic networks that provide a complete catalog for $m \geq 4$ during this period, we have chosen this value for our cutoff. However, it should be emphasized that in some highly populated areas (for instance, some localities in the third world), the catalog is not complete down to magnitude 4 and the seismic hazard will be underestimated.

We include earthquakes from all depths in our analyses. On the downdip side of subduction zones this will lead to an overestimate of the seismic hazard. However, some relatively deep earthquakes can result in substantial damage and casualties.

In order to provide a map of the seismic hazard on a global basis, the surface of the earth was divided into $1° \times 1°$ regions and the number of earthquakes per year with magnitudes $m \geq 4$ in each region was determined. We defined the seismic intensity factor, I_4, as the number of magnitude $m \geq 4$ earthquakes that occurred in a given $1° \times 1°$ region per year, normalized by the cosine of the latitude. Normalization was due to areas changing with respect to changing latitude. For instance, take two $1° \times 1°$ regions, one at the equator and the other at 60° latitude. If the equator region has twice as many earthquakes as the one at 60°, then the density of earthquakes per area is equal.

The choice of $1° \times 1°$ regions may seem arbitrary. However, this is the minimum area over which consistent maps of free-air gravity can be made by averaging available measurements (TALWANI, 1970). Averaging of gravity observations is necessary because local free-air gravity measurements are very "noisy" due to uncompensated topography. When averages of measurements are made over $1° \times 1°$ areas this topographic noise is removed. In addition, our studies using different-sized areas for seismic averaging indicate that a $1° \times 1°$ area is optimal in terms of reducing noise and maximizing spatial resolution. There appears to be a tectonic variability on scales less than about 100 km that is reflected both in the uncompensated topography and the spatial variability of seismicity. FRANKEL (1995) found that a Gaussian correlation length of 50 km was optimal for smoothing seismicity, similar to our choice.

A global map of the seismic intensity factor is given in Figure 6a. The boundaries of the tectonic plates are clearly defined. Seismicity is particularly intense in subduction zones (i.e., the ring of fire around the Pacific) as expected. A broad band of seismicity extends from Southern Europe to Southeast Asia, which is associated with continent–continent collision zones between the Eurasian plate and the African, Arabian, and Indian plates. The minimum value of the seismic intensity factor considered is $I_4 = 1/32$ yr^{-1}, one magnitude $m \geq 4$ earthquake in the 32-year period considered. The maximum value is about $I_4 = 40$ yr^{-1}, forty magnitude $m \geq 4$ earthquakes per year. The range of I_4 is over three orders of magnitude.

In Figure 7, two regional maps of the seismic intensity factor are given, the United States in Figure 7a and Europe in Figure 7b. For the United States, the intense seismicity on the boundary between the Pacific and North American plates (including the San Andreas Fault) is clearly illustrated. The distributed seismicity in the western United States is also shown. The most intense seismicity in the eastern United States is the New Madrid seismic zone (31 N, 90 W) as discussed above. For Europe, the Aegean region has particularly intense seismicity, however high levels of seismicity extend throughout the Mediterranean region.

The basis for using the seismic intensity factor, I_4, is illustrated in Figure 8. For five different values of I_4, the cumulative number of earthquakes per year in a $1° \times 1°$ area, $\dot{N}_{CE}$, with magnitudes greater than m is plotted as a function of m. The different lines are derived by using the Gutenberg–Richter relation (1) with $b = 0.9$, and calculating the constant $\dot{a}$ from $\log \dot{a} = 3.6 + \log I_4$ (TURCOTTE, 1999a). In Figure 8, given an area with $I_4 = 1$ yr^{-1}, an earthquake with $m \geq 6$ has a return period of 63 years ($\dot{N}_{CE} = 0.016$) and one with $m \geq 8$ has a return period of 4000 yrs ($\dot{N}_{CE} = 0.00025$).

It can be argued that the use of earthquakes with $m \geq 4$ for such a short period (37 years) would not represent the long-term seismic hazard. In order to address this concern, we used the NEIC Global Hypocenter Data Base (GHDB, 1989) to determine the largest earthquake occurring in each $1° \times 1°$ area for the period 1900–1997. These data are given in Figure 6b. Comparing Figures 6a and 6b, it is seen that there is a strong correlation between the value of the earthquake intensity factor, I_4, calculated using 32 years of data, and the largest earthquake to have occurred in each $1° \times 1°$ area, using 98 years of data.

Figure 6

(a) Global map of the seismic intensity factor I_4, the average annual number of earthquakes during the period 1964–1995 with magnitude $m \geq 4$ in each $1° \times 1°$ cell. (b) The maximum magnitude earthquake that occurred in each $1° \times 1°$ cell during the period 1900–1997. Data for both (a) and (b) are from the NEIC Global Hypocenter Data Base (GHDB, 1989).

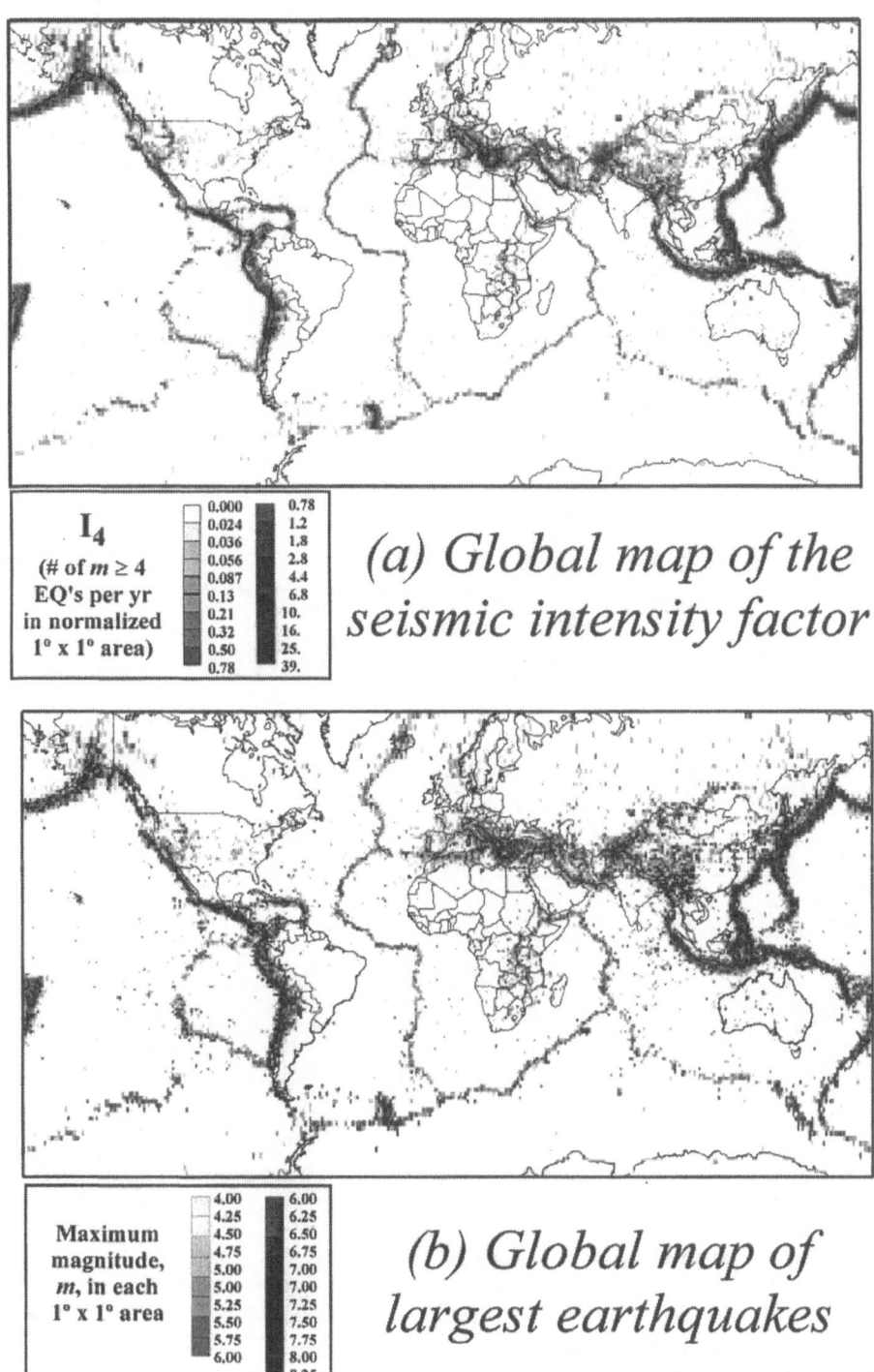

I_4 (# of $m \geq 4$ EQ's per yr in normalized $1° \times 1°$ area)	0.000 0.024 0.036 0.056 0.087 0.13 0.21 0.32 0.50 0.78	0.78 1.2 1.8 2.8 4.4 6.8 10. 16. 25. 39.

(a) Global map of the seismic intensity factor

Maximum magnitude, m, in each $1° \times 1°$ area	4.00 4.25 4.50 4.75 5.00 5.00 5.25 5.50 5.75 6.00	6.00 6.25 6.50 6.75 7.00 7.00 7.25 7.50 7.75 8.00 8.25

(b) Global map of largest earthquakes

Fig. 6.

A number of developed countries have prepared seismic hazard maps. In many cases, as discussed above, the local Gutenberg–Richter relation has been one of the inputs into the assessments. For the United States, the U.S. Geological Survey's

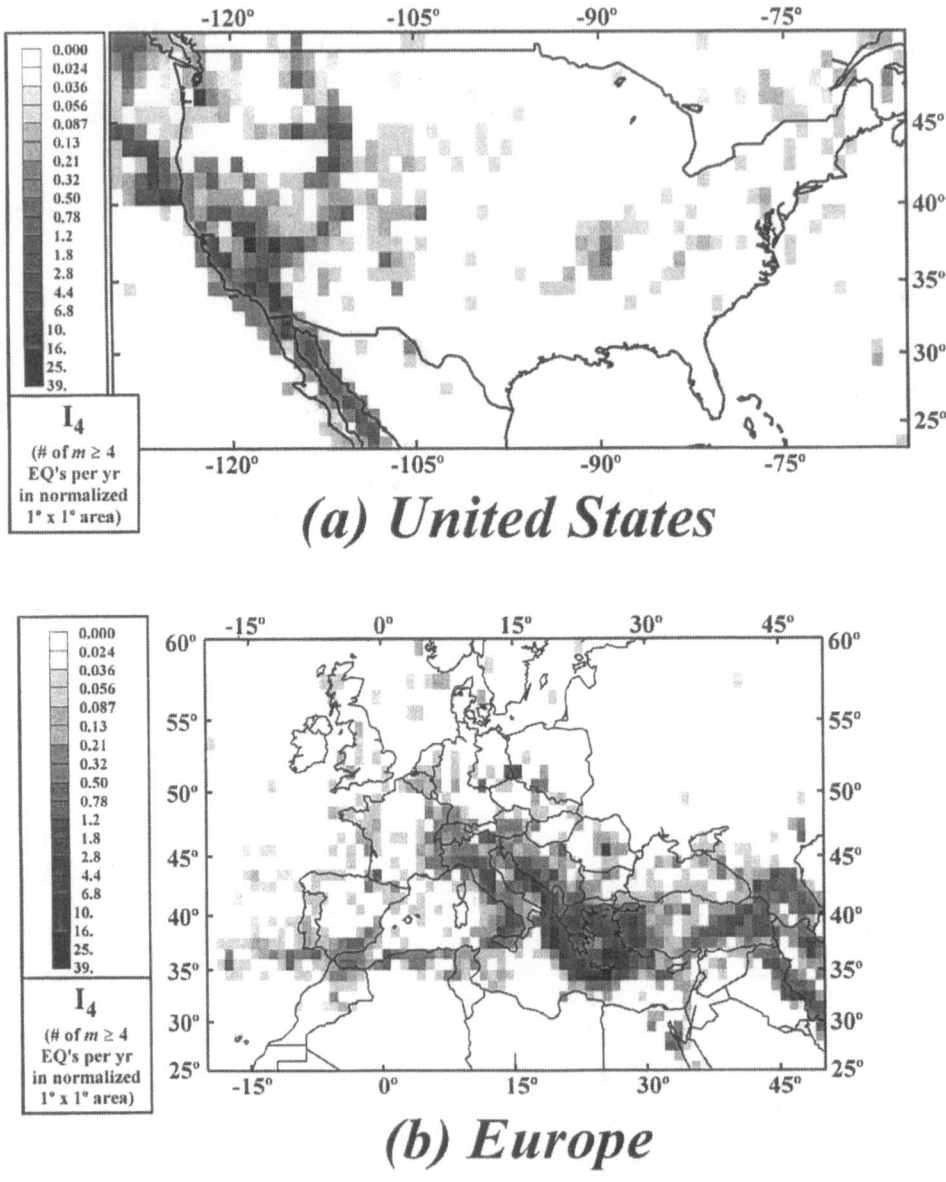

Figure 7
Regional maps of the seismic intensity factor I_4, the average annual number of earthquakes during 1964–1995 with magnitudes $m \geq 4$ in each normalized $1° \times 1°$ cell for (a) United States and (b) Europe. Data for both (a) and (b) are from the NEIC Global Hypocenter Data Base (GHDB, 1989).

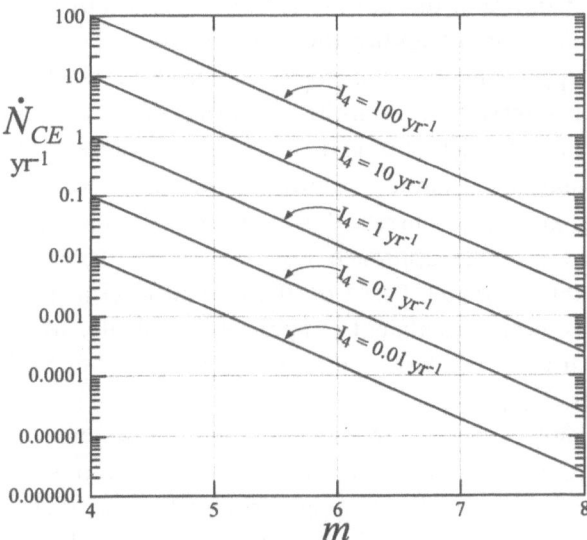

Figure 8
Similarity basis for extending the seismic intensity factor I_4 to higher earthquake magnitudes. See text for discussion.

June 1996 maps (FRANKEL *et al.*, 1996) are available on the internet. Seismic hazard maps generally give the probability of exceeding a specified peak ground acceleration during a specified time interval. The data used to formulate the maps include the historic seismicity discussed in this paper, maps of active faults, and levels of seismic attenuation. Taking 0.1 g as the reference acceleration, the return times for an earthquake that exceeds this value are the following (FRANKEL *et al.*, 1996): 23 years in Los Angeles, 27 years in San Francisco, 56 years in Seattle, 140 years in Salt Lake City, 150 years in Portland, 360 years in Memphis, 770 years in New York, 2900 years in Chicago, and 24,000 years in St. Paul.

We compare our approach for seismic hazard estimation to that used by the USGS by comparing probabilities in two regions: Southern California and the New Madrid seismic zone. We assume that the 23-year return time for Los Angeles is appropriate for Southern California and from Figure 3 find that the recurrence interval of 23 years ($\dot{N}_{CE} = 0.043$ yr^{-1}) corresponds to a magnitude $m \approx 7$ earthquake. From Figure 5, the recurrence time for a magnitude $m = 7$ earthquake in the New Madrid area is about 900 years ($\dot{N}_{CE} \approx 0.0011$ yr^{-1}); the USGS hazard assessment gives a return time, for 0.1 g acceleration, of 360 years for this region. The hazard assessment maps are reasonably consistent with the seismic intensity maps. The hazard assessment maps give a higher relative risk in aseismic zones because they include seismic attenuation that is less in aseismic zones. Comparisons for other regions can be made, based on the seismic intensity factors given in Figure 6a.

The primary advantage of our approach is that it is totally based on a generally accepted data set. There are no ambiguities with regard to the technique. This is not the case for other approaches to assessing seismic hazard, which combine geophysical and geological observations in arbitrary ways. Different studies give different weights to historical and paleo-seismic data, and to the presence of "active" faults. Considering the many uncertainties regarding fault depth, seismic attenuation, the available database, and the occurrence of an earthquake on a particular fault, we believe that the very simple approach we propose for assessing seismic hazard provides a reasonable approach based on the present level of knowledge. Another advantage of our approach is that it is applicable globally. Many countries with severe seismic hazards do not have the resources to carry out probabilistic seismic hazard assessments.

There are certainly objections to our approach. These include:

(1) The time interval used (in our case 32 yrs) is too short to establish a long-term rate of seismic activity. Strong earthquakes and their associated aftershocks will create local anomalies in the maps. We partially argue against this disadvantage, noting that with aftershocks removed, the background seismicity in Southern California (Fig. 4) on a year-to-year basis is essentially constant. In addition, there is close correspondence between the historical and recent instrumental data for New Madrid (Fig. 5). Observations favor the hypothesis that the rate of occurrence of smaller earthquakes in a region is only weakly dependent on time if the area is sufficiently large. If the smaller earthquakes correlated with the earthquake cycle, then they could be used for the temporal prediction of earthquakes. This is clearly not the case.

(2) The large "characteristic" earthquakes do not fall on an extrapolated Gutenberg–Richter curve. We have shown that this extrapolation is reasonably good for Southern California (Fig. 3). In addition, even if large earthquakes fall off the extrapolation, the hazard-assessment approach given above will be valuable for moderate-sized earthquakes.

(3) The extrapolation does not specify an upper limit to the expected earthquakes in a region. This is certainly true, but is a universal problem in seismic hazard assessment.

As discussed above, we are not the first to employ this technique. FRANKEL (1995) used a similar technique to contour values of 10^a ($\log N_{CE} = -bm + a$) for the eastern United States. His values of 10^a represent the number of earthquakes that occur in each 11 km × 11 km square grid in 60 years with $m > 0$. Note that in our nomenclature, $10^a = \tau \dot{a}$ where τ is the number of years considered, and $\dot{a}$ is the same as used in (1). His map (FRANKEL, 1995, Fig. 4) gives $10^a = 32$ for the New Madrid area, 8 for the Charleston area, 16 for eastern Tennessee, and 8 for New York City. Since 10^a is proportional to our I_4, these results can be compared to the results given in Figure 7a. Our results tend to give far greater variability between these regions.

3. Seismic Activation

We have discussed how small earthquakes can be used to quantify the hazard associated with large earthquakes. An important question is whether small earthquakes can be used to forecast the temporal occurrence of large earthquakes. As shown in Figure 4, the occurrence of the smallest earthquakes seems to have very little temporal dependence. However, there is accumulating evidence that there may be an activation of intermediate-sized earthquakes prior to a great earthquake. The occurrence of a relatively large number of intermediate-sized earthquakes in northern California prior to the 1906 San Francisco earthquake has been noted (SYKES and JAUMÉ, 1990). It has also been proposed that there is a power-law increase in seismicity prior to a major earthquake. This was first proposed by BUFE and VARNES (1993). They considered the cumulative amount of Benioff strain (square-root of seismic energy) in a specified region. They show that an accurate retrospective prediction of the Loma Prieta earthquake could be made, assuming a power-law temporal increase in Benioff strain prior to the earthquake.

Systematic increases in intermediate level seismicity before a large earthquake have been proposed by several authors (VARNES, 1989; BUFE et al., 1994; KNOPOFF et al., 1996; VARNES and BUFE, 1996; BREHM and BRAILE, 1998, 1999; JAUMÉ and SYKES, 1999). A systematic study of the optimal spatial region and magnitude range to obtain the power-law seismic activation has been carried out by BOWMAN et al. (1998). Four examples of their results are given in Figure 9. Clear increases in seismic activity prior to the Kern County, Loma Prieta, Landers, and Coalinga earthquakes are illustrated. In each case data for the cumulative Benioff strain ε is compared with the empirical relation (solid line)

$$\varepsilon[t] = \varepsilon_0 - B(t_0 - t)^s \tag{4}$$

where t is the time measured forward from the previous characteristic earthquake, t_0 is the time interval between characteristic earthquakes, ε_0 is the cumulative Benioff strain when the characteristic earthquake occurs, and B and s are positive constants used to fit the data. The comparison with the Kern County (July 21, 1952) data is made with $s = 0.30$, with Loma Prieta (October 18, 1989) $s = 0.28$, with Landers (June 28, 1992) $s = 0.18$, and with Coalinga (May 2, 1983) $s = 0.18$.

BOWMAN et al. (1998) also found ξ the optimal radius (the correlation length) for the precursory activation. This optimal radius is given as a function of earthquake magnitude in Figure 10. The dependence on the square-root of rupture area is also shown. The radius over which activation occurs is about ten times the length of rupture, $\xi \approx 10A_E^{1/2}$. DOBROVOLSKY et al. (1979) and KEILIS-BOROK and KOSSOBOKOV (1990) reported earlier a similar scaling for the maximum distance between an earthquake and its precursors, using pattern recognition techniques.

The observations of seismic activation given above are consistent with results obtained from studies of cellular-automata slider-block models using a mean-field

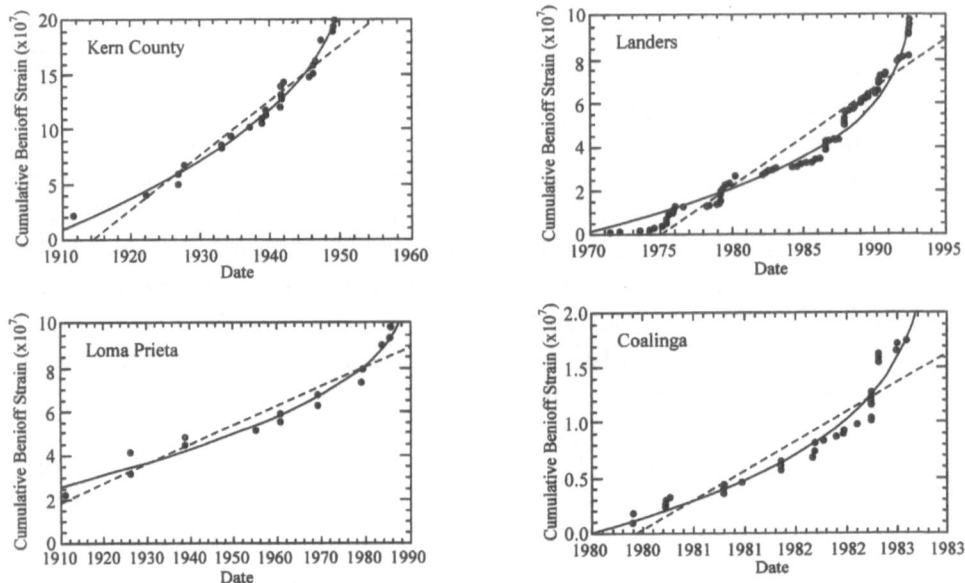

Figure 9

Power-law increases in the cumulative Benioff strains prior to four major earthquakes in California
(BOWMAN *et al.*, 1998). Data points are cumulative Benioff strains $\varepsilon[t]$ prior to each earthquake. Clear
increases in seismic activity prior to the 1952 Kern County, 1989 Loma Prieta, 1992 Landers, and 1983
Coalinga earthquakes are illustrated. In each of the four examples, the data have been correlated (solid
lines) with the power-law relation given in (4). Dashed straight lines represent a best-fit constant rate of
seismicity.

approach (RUNDLE *et al.*, 1996, 1997a,b, 1999; FISHER *et al.*, 1997). This approach
involves concepts applied to equilibrium thermodynamics and the approach to a
second-order phase transition through a spinoidal.

In the scaling associated with the approach to the critical point, the correlation
length scales with the "rupture" length. This scaling follows directly from the
spinoidal equation (RUNDLE *et al.*, 1997, Eq. 54). The details of this approach are
given in RUNDLE *et al.* (2000). This is the scaling associated with the optimal
activation length given in Figure 10. In addition, the rate of occurrence of events as
the critical point is approached can be obtained (RUNDLE *et al.*, 2000, Eq. 16). This
result can explain the rate of seismic activation given in (4).

4. Forecasting

Based on pattern recognition algorithms a number of intermediate-range earth-
quake prediction algorithms have been developed at the International Institute of
Earthquake Prediction Theory and Mathematical Geophysics in Moscow (KEILIS-
BOROK, 1990; KEILIS-BOROK and ROTWAIN, 1990; KEILIS-BOROK and KOS-

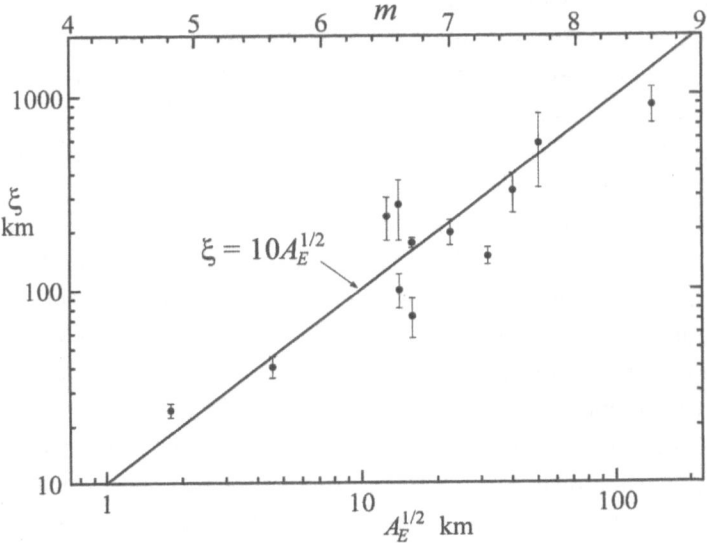

Figure 10

The optimum radius (correlation length) ξ for precursory seismic activation is given as a function of the square-root of the rupture area, $A_E^{1/2}$, and the magnitude, m, for twelve major earthquakes (Bowman *et al.*, 1998).

SOBOKOV, 1990). The pattern recognition includes seismic activation, quiescence (SCHREIDER, 1990), increases in the clustering of events, and changes in aftershock statistics (MOLCHAN *et al.*, 1990). The first algorithm, $M8$, was developed to make intermediate-term predictions of the largest earthquake ($m > 8$). This method utilizes overlapping circles of seismicity with diameters of 384, 560, 854, and 1333 km for earthquakes, with magnitudes 6.5, 7.0, 7.5, and 8.0, respectively. Within each circle four quantities are determined. The first three are measures of intermediate levels of seismicity and the fourth is a measure of aftershock activity.

The first quantity that must be specified is the lower magnitude cutoff m_{min} for earthquakes to be considered in the circle. Two magnitude cutoffs are considered for each circular region. The long-term number of earthquakes per year, $\dot{N}$, in the circle, with magnitudes greater than m_{min} is determined; $m_{min\ 10}$ corresponds to $\dot{N} = 10$, and $m_{min\ 20}$ corresponds to $\dot{N} = 20$. The first quantity $\dot{N}_1(t)$ is the number of earthquakes per year in the circle with magnitudes greater than $m_{min\ 10}$ and $\dot{N}_2(t)$ is the number of earthquakes per year with magnitudes greater than $m_{min\ 20}$ (both estimates over the preceding six-year window). The second quantity is the trend in activity $L_1 = d\dot{N}_1/dt$ and $L_2 = d\dot{N}_2/dt$ for running six-year windows. Clearly $\dot{N}(t)$ and $L(t)$ are strongly correlated. The third quantity, $Z_1(t)$ or $Z_2(t)$, is the ratio of the average linear dimension of rupture to the average separation between earthquakes for a running six-year window in a circle. The final measure is the number of aftershocks in a specified magnitude range and time window following a main shock. This quantity $B(t)$ is a measure of aftershock activation.

Since $\dot{N}$, L, and Z are determined both for $\dot{N} = 10$ and 20 earthquakes per year there are seven time series to be considered. An earthquake alarm or time of increased probability (*TIP*) is issued if 6 of the 7 quantities, including B, exceed their average values by a specified value of 75% for B and 90% for the others. In order to issue an alarm, these conditions must be satisfied for two successive time periods and the alarm lasts for five years. Details of this algorithm have been given by KEILIS-BOROK (1996). Two examples of the application of this algorithm are given in Figure 11 (KEILIS-BOROK, 1996; KOSSOBOKOV *et al.*, 1999) for the 1989 Loma Prieta and the 1992 Landers earthquakes.

In order to refine the area in which a *TIP* is declared, the *MSc* (Mendocino scenario) algorithm was developed from studies of the seismicity prior to the Eureka (Cape Mendocino) earthquake (1980, $m = 7.2$). Using earthquakes with smaller magnitudes than in the $M8$ algorithm and utilizing transitions from excess seismicity to quiescence, this algorithm reduces the area over which a *TIP* is declared.

The algorithms described above were developed utilizing pattern recognition approaches. Although they have had demonstrated predictive successes (KOS-SOBOKOV *et al.*, 1999; ROTWAIN and NOVIKOVA, 1999), their use remains quite controversial (ENEVA and BEN-ZION, 1997). The main difficulty is that, although success to failure ratios of predictions is quite high, the time and spatial windows of alarms are also quite high.

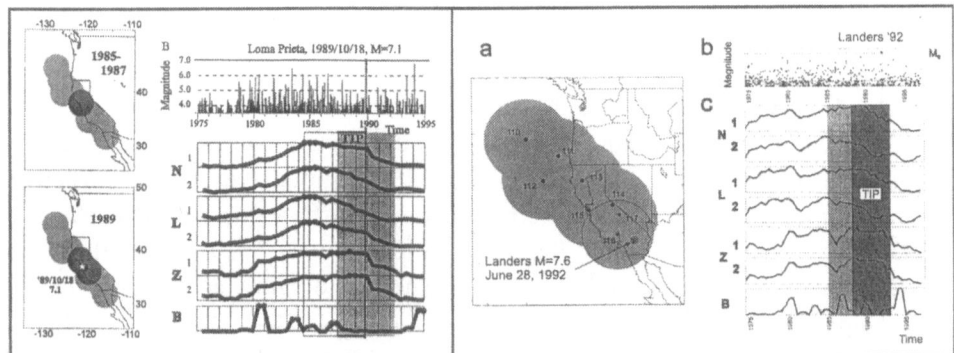

Figure 11

Applications of the $M8$ algorithm to (1) the $m = 7.1$ Loma Prieta earthquake (KEILIS-BOROK, 1996) and (2) the $m = 7.6$ Landers earthquake (KOSSOBOKOV *et al.*, 1999). For the Landers earthquake, part (a) of the right-half of the figure, the circular regions where "Times of Increased Probability" (*TIP*s) have been declared, are shown. In (b) the seismicity in the relevant region is shown. In (c) the values of the seven parameters $\dot{N}_1$, $\dot{N}_2$, L, L_1, L_2, Z_1, Z_2, and B are shown along with the time interval of the *TIP*.

There are clearly very strong similarities between the $M8$ algorithm and the seismic activation hypothesis. Consider the Loma Prieta earthquake, the intermediate-sized events that led to the $M8$ *TIP* illustrated in Figure 11 were the same events that generated the increase in Benioff strain illustrated in Figure 9. Both of these approaches are based on the concept that correlation lengths increase before major earthquakes (HARRIS, 1998).

5. Conclusions

We conclude that distributed seismicity is an example of both self-organized critical behavior and second-order critical-point behavior. Evidence for self-organizing critical behavior is the applicability under a wide variety of conditions of the Gutenberg–Richter frequency-magnitude relation. This is a power-law (fractal) scaling between the number of earthquakes and their rupture size. The number of small earthquakes in a region appears to be approximately constant in time. If this number varied with the temporal cycle of the characteristic earthquake in the region, then the number could be used for earthquake forecasting; this is clearly not the case. However, using Gutenberg–Richter scaling, the number of smaller earthquakes in a region can be extrapolated to assess the hazard of larger earthquakes in the region. Whether this can be done for the characteristic earthquake in the regional is controversial, however it certainly can be done for intermediate-sized earthquakes. Based on the extrapolation of the frequency of small earthquakes to larger earthquakes, we have given global maps of the seismic hazard. These utilize 32 years of instrumental seismic data, and averages are carried out over $1° \times 1°$ regions. These regions are selected because they remove the tectonic "grain."

It is widely accepted that the number of intermediate-sized earthquakes in a region increases prior to the characteristic earthquake for the region. This precursory activation has been shown to exhibit power-law scaling and to occur over a region about ten times larger than the rupture size of the characteristic earthquake. This precursory activation is the primary basis for the use of the $M8$ algorithm that has been developed empirically for earthquake forecasting.

Both the Gutenberg–Richter scaling and the precursory activation are consistent with the application of condensed-matter theory to regional seismicity. The scaling of small earthquakes in a region is equivalent to the thermal fluctuations in solids, liquids, and gases. This behavior falls under the general class of phenomena that exhibit self-organized critical behavior. Examples include the sandpile, slider-block, and forest-fire models. These models are characterized by a fractal. frequency-area distribution of avalanches; model sand slides, slip events, and forest fires. These models can be explained in terms of an inverse-cascade involving the coalescence of metastable clusters.

The characteristic earthquake appears to behave as a second-order phase change. As the phase change is approached the scale of the region over which correlated activity occurs (the correlation length) increases. In terms of earthquakes, this correlated region is the region of precursory activation. Analyses of the spinoidal behavior associated with second-order phase transitions predict the power-law increase in seismic activity prior to a characteristic earthquake.

Acknowledgements

The authors wish to acknowledge many useful discussions with Gleb Morein, John Rundle and Charlie Sammis. This manuscript benefited significantly from the detailed and thorough comments of two anonymous reviewers. The authors received support from NSF grant EAR 9804859.

REFERENCES

AKI, K., *A probabilistic synthesis of precursory phenomena.* In *Earthquake Prediction* (eds. Simpson, D. W., and Richards, P. G.) (American Geophysical Union, Washington, D.C. 1981) pp. 566–574.

BAK, P., TANG, C., and WIESENFELD, K. (1988), *Self-organized Criticality,* Phys. Rev. A*38*, 364–374.

BAK, P., and TANG, C. J. (1989), *Earthquakes as a Self-organized Critical Phenomenon,* J. Geophys. Res. *94*, 15,635–15,637.

BERNREUTER, D. L., SAVY, J. B., MENSING, R. W., and CHEN, J. C., *Seismic Hazard Characterization of 69 Nuclear Plant Sites East of the Rocky Mountains* (U.S. Nuclear Regulatory Commission NUREG/CR-5250 UCID-21517, 8 volumes, Washington, D.C. 1989).

BOWMAN, D. D., OUILLON, G., SAMMIS, C. G., SORNETTE, A., and SORNETTE, D. (1998), *An Observational Test of the Critical Earthquake Concept,* J. Geophys. Res. *103*, 24,359–24,372.

BREHM, D. J., and BRAILE, L. W. (1998), *Intermediate-term Earthquake Prediction Using Precursory Events in the New Madrid Seismic Zone,* Bull. Seismol. Soc. Am. *88*, 564–580.

BREHM, D. J., and BRAILE, L. W. (1999), *Intermediate-term Earthquake Prediction Using the Modified Time-to-failure Method in Southern California,* Bull. Seismol. Soc. Am. *89*, 275–293.

BROWN, S. R., SCHOLZ, C. H., and RUNDLE, J. B. (1991), *A Simplified Spring-block Model of Earthquakes,* Geophys. Res. Lett. *18*, 215–218.

BUFE, C. G., and VARNES, D. J. (1993), *Predictive Modeling of the Seismic Cycle of the Greater San Francisco Bay Region,* J. Geophys. Res. *98*, 9871–9883.

BUFE, C. G., NISHENKO, S. P., and VARNES, D. J. (1994), *Seismicity Trends and Potential for Large Earthquakes in the Alaska-Aleutian Region,* Pure appl. geophys. *142*, 83–99.

BURRIDGE, R., and KNOPOFF, L. (1967), *Model and Theoretical Seismicity,* Bull. Seismol. Soc. Am. *57*, 341–371.

CARLSON, J. M., and LANGER, J. S. (1989), *Mechanical Model of an Earthquake Fault,* Phys. Rev. A*40*, 6470–6484.

CARLSON, J. M., LANGER, J. S., and SHAW, B. E. (1994), *Dynamics of Earthquake Faults,* Rev. Mod. Phys. *66*, 657–670.

CORNELL, A. C. (1968), *Engineering Seismic Risk Analysis,* Bull. Seismol. Soc. Am. *58*, 1583–1606.

DAVISON, F., and SCHOLZ, C. H. (1985), *Frequency-moment Distribution of Earthquakes in the Aleutian Arc: A Test of the Characteristic Earthquake Model,* Bull. Seismol. Soc. Am. *75*, 1349–1362.

DOBROVOLSKY, I. R., ZUBKOV, S. I., and MIACHKIN, V. I. (1979), *Estimation of the Size of Earthquake Preparation Zones,* Pure appl. geophys. *117*, 1025–1044.

ENEVA, M., and BEN-ZION, Y. (1997), *Techniques and Parameters to Analyze Seismicity Patterns Associated with Large Earthquakes*, J. Geophys. Res. *102*, 17,785–17,795.

ENGDAHL, E. R., VAN DER HILST, R., and BULAND, R. (1998), *Global Teleseismic Earthquake Relocation with Improved Travel Times and Procedures for Depth Determination*, Bull. Seismol. Soc. Am. *88*, 722–743.

FISHER, D. S., DAHMEN, K., RAMANATHAN, S., and BEN-ZION, Y. (1997), *Statistics of Earthquakes in Simple Models of Heterogeneous Faults*, Phys. Rev. Lett. *78*, 4885–4888.

FRANKEL, A. F. (1995), *Mapping Seismic Hazard in the Central and Eastern United States*, Seis. Res. Lett. *60* (4), 8–21.

FRANKEL, A. F., MUELLER, C., BARNHARD, T., PERKINS, D., LEYENDECKER, E. V., DICKMAN, N., HANSON, S., and HOPPER, M., *National Seismic Hazard Maps* (USGS Open-File Report 1996) pp. 96–532.

FROHLICH, C., and DAVIS, S.D. (1993), *Teleseismic b Values; or, Much Ado About 1.0*, J. Geophys. Res. *98*, 631–644.

GELLER, R. J., JACKSON, D. D., KAGAN, Y. Y., and MULARGIA, F. (1997), *Earthquakes Cannot Be Predicted*, Science *275*, 1616–1617.

GHDB (1989), *Global Hypocenter Data Base (GHDB)*, CD ROM (NEIC/USGS, Denver, Colorado, 1989) and its updates through 1997.

GUTENBERG, B., and RICHTER, C. F., *Seismicity of the Earth and Associated Phenomenon, 2nd ed* (Princeton University Press, Princeton 1954).

HARRIS, R. A. (1998), *Forecasts of the 1989 Loma Prieta, California, Earthquake*, Bull. Seismol. Soc. Am. *88*, 898–916.

HUANG, J., and TURCOTTE, D. L. (1990), *Are Earthquakes an Example of Deterministic Chaos?*, Geophys. Res. Lett. *17*, 223–226.

HUANG, J., NARKOUNSKAIA, G., and TURCOTTE, D. L. (1992), *A Cellular-automata, Slider-block Model for Earthquakes II. Demonstration of Self-organized Criticality for a 2-D System*, Geophys. J. Int. *111*, 259–269.

JAUMÉ, S. C., and SYKES, L. R. (1999), *Evolving Towards a Critical Point: A Review of Accelerating Seismic Moment/Energy Release Prior to Large and Great Earthquakes*, Pure appl. geophys. *155*, 279–305.

JOHNSTON, A. C., and NAVA, S. J. (1985), *Recurrence Rates and Probability Estimates for the New Madrid Seismic Zone*, J. Geophys. Res. *90*, 6737–6753.

JOHNSTON, A. C., and SCHWEIG, E. S. (1996), *The Enigma of the New Madrid Earthquakes of 1811–1812*, An. Rev. Earth Planet. Sci. *24*, 339–384.

KADANOFF, L. P., NAGEL, S. R., WU, L., and ZHOU, S. M. (1989), *Scaling and Universality in Avalanches*, Phys. Rev. A*39*, 6524–6533.

KEILIS-BOROK, V. I. (1990), *The Lithosphere of the Earth as a Nonlinear System with Implications for Earthquake Prediction*, Rev. Geophys. *28*, 19–34.

KEILIS-BOROK, V. I. (1996), *Intermediate-term Earthquake Prediction*, Proc. Natl. Acad. Sci. *93*, 3748–3755.

KEILIS-BOROK, V. I., and KOSSOBOKOV, V. G. (1990), *Premonitory Activation of Earthquake Flow. Algorithm M8*, Phys. Earth Planet. Int. *61*, 73–83.

KEILIS-BOROK, V. I., and ROTWAIN, I. M. (1990), *Diagnosis of Time of Increased Probability of Strong Earthquakes in Different Regions of the World: Algorithm CN*, Phys. Earth Planet. Int. *61*, 57–72.

KNOPOFF, L., LEVSHINA, T., KEILIS-BOROK, V. I., and MATTONI, C. (1996), *Increased Long-range Intermediate-magnitude Earthquake Activity Prior to Strong Earthquakes in California*, J. Geophys. Res. *101*, 5779–5796.

KOSSOBOKOV, V. G., ROMASHKOVA, L. L., KEILIS-BOROK, V. I., and HEALY, J. H. (1999), *Testing Earthquake Prediction Algorithms: Statistically Significant Advance Prediction of the Largest Earthquakes in the Circum-Pacific, 1992–1997*, Phys. Earth Planet. Int. *111*, 187–196.

LAY, T., and WALLACE, T. C., *Modern Global Seismology* (Academic Press, San Diego 1995).

LORENZ, E. N. (1963), *Deterministic Nonperiodic Flow*, J. Atmos. Sci. *20*, 130–141.

MALAMUD, B. D., MOREIN, G., and TURCOTTE, D. L. (1998), *Forest Fires: An Example of Self-organized Critical Behavior*, Science *281*, 1840–1842.

MAY, R. M. (1976), *Simple Mathematical Models with Very Complicated Dynamics*, Nature *261*, 459–467.

MOLCHAN, G. M., DMITRIEVA, O. E., ROTWAIN, I. M., and DEWEY, J. (1990), *Statistical Analysis of the Results of Earthquake Prediction, Based on Bursts of Aftershocks*, Phys. Earth Planet. Int. *61*, 128–139.

NAKANISHI, H. (1991), *Statistical Properties of the Cellular Automata Model for Earthquakes*, Phys. Rev. *A43*, 6613–6621.

OTSUKA, M. (1972), *A Simulation of Earthquake Occurrence*, Phys. Earth Planet. Int. *6*, 311–315.

PACHECO, J., SCHOLZ, C. H., and SYKES, L. R. (1992), *Changes in Frequency-size Relationship from Small to Large Earthquakes*, Nature *355*, 71–73.

PELLETIER, J. D., MALAMUD, B. D., BLODGETT, T., and TURCOTTE, D. L. (1997), *Scale-invariance of Soil Moisture Variability and its Implications for the Frequency-size Distribution of Landslides*, Eng. Geol. *48*, 255–268.

ROTWAIN, I., and NOVIKOVA, O. (1999), *Performance of the Earthquake Prediction Algorithm CN in 22 Regions of the World*, Phys. Earth Planet. Int. *111*, 207–213.

RUNDLE, J. B., GROSS, S., KLEIN, W., FERGUSON, C., and TURCOTTE, D. L. (1997a), *The Statistical Mechanics of Earthquakes*, Tectonophys. *277*, 147–164.

RUNDLE, J. B., KLEIN, W., and GROSS, S. (1996), *Dynamics of a Traveling Density Wave Model for Earthquakes*, Phys. Rev. Lett. *76*, 4285–4288.

RUNDLE, J. B., KLEIN, W., and GROSS, S. (1999), *Physical Basis for Statistical Patterns in Complex Earthquake Populations: Models, Predictions, and Tests*, Pure appl. geophys. *155*, 575–607.

RUNDLE, J. B., KLEIN, W., GROSS, S., and FERGUSON, C. D. (1997b), *The Traveling Density Wave Model for Earthquakes and Driven Threshold Systems*, Phys. Rev. *E56*, 293–307.

RUNDLE, J. B., KLEIN, W., TURCOTTE, D. L., and MALAMUD, B. D. (2000), *Precursory Seismic Activation and Critical Point Phenomena*, Pure appl. geophys. *157*, 2165–2182.

SCHOLZ, C. H., in *Spontaneous Formation of Space Time Structure and Criticality* (eds. Riste, T., and Sherrington, D.) (Kluwer, Amsterdam 1991) pp. 41–56.

SCHOLZ, C. H. (1997), *Size Distributions for Large and Small Earthquakes*, Bull. Seismol. Soc. Am. *87*, 1074–1077.

SCHREIDER, S. (1990), *Formal Definition of Premonitory Seismic Quiescence*, Phys. Earth Planet. Int. *61*, 113–127.

SCSN CATALOG. *Southern California Seismographic Network Catalog in Electronic Format at the Southern California Earthquake Center (SCEC) Data Center* (California Institute of Technology, Pasadena, California 1995).

SIEH, K. E. (1978), *Slip Along the San Andreas Fault Associated with the Great 1857 Earthquake*, Bull. Seismol. Soc. Am. *68*, 1421–1448.

SIEH, K. E., STUIVER, M., and BRILLINGER, D. (1989), *A More Precise Chronology of Earthquakes Produced by the San Andreas Fault in Southern California*, J. Geophys. Res. *94*, 603–623.

SYKES, L. R., and JAUMÉ, S. C. (1990), *Seismic Activity on Neighboring Faults as a Long-term Precursor to Large Earthquakes in the San Francisco Bay Area*, Nature *348*, 595–599.

TALWANI, M., *Gravity. In The Sea* (ed. Maxwell, A. E.), vol. 4, part I (Wiley-Interscience, New York 1970) pp. 251–297.

TURCOTTE, D. L. (1989), *A Fractal Approach to Probabilistic Seismic Hazard Assessment*, Tectonophys. *167*, 171–177.

TURCOTTE, D. L. (1991), *Earthquake Prediction*, An. Rev. Earth Planet. Sci. *19*, 263–281.

TURCOTTE, D. L., *Fractals and Chaos in Geology and Geophysics*, 2nd ed (Cambridge University Press, Cambridge 1997).

TURCOTTE, D. L. (1999a), *Seismicity and Self-organized Criticality*, Phys. Earth Planet. Int. *111*, 275–293.

TURCOTTE, D. L. (1999b), *Self-organized Criticality*, Rep. Prog. Phys. *62*, 1377–1429.

TURCOTTE, D. L., MALAMUD, B. D., MOREIN, G., and NEWMAN, W. I. (1999), *An Inverse-cascade Model for Self-organized Critical Behavior*, Physica *A268*, 629–643.

VARNES, D. J. (1989), *Predicting Earthquakes by Analyzing Accelerating Precursory Seismic Activity*, Pure appl. geophys. *130*, 661–686.

VARNES, D. J., and BUFE, C. G. (1996), *The Cyclic and Fractal Seismic Series Preceding an m_b 4.8 Earthquake on 1980 February 14 near the Virgin Islands*, Geophys. J. Int. *124*, 149–158.

(Received August 12, 1999, revised May 3, 2000, accepted May 5, 2000)

To access this journal online:
http://www.birkhauser.ch

Pure appl. geophys. 157 (2000) 2351–2364
0033–4553/00/122351–14 $ 1.50 + 0.20/0

Pure and Applied Geophysics

Application of Linked Stress Release Model to Historical Earthquake Data: Comparison between Two Kinds of Tectonic Seismicity

CHUNSHENG LU[1,2] and DAVID VERE-JONES[1]

Abstract—The linked stress release model, incorporating a slow buildup of stress within a seismic region, its stochastic release through earthquakes and transfer between seismic regions, is applied to fit historical data from two typical kinds of seismicity: earthquakes occurring in intraplate (North China) and plate boundary (New Zealand) regions. The best model among different modifications of the basic model, which may reflect on a possible geophysical mechanism for earthquake occurrences, is obtained in terms of Akaike information criterion. For both tectonic regions studied, the linked stress release model fits the New Zealand data better than a collection of independent simple models, but is nearly indistinguishable from the simple stress release model in the case of North China. The seismicity in a plate boundary region due to subduction is more active and complex than that in an intraplate region due to collision between tectonic plates. The results highlight the major differences in tectonic seismicity, especially the heterogeneities of tectonic stress fields, and dynamic triggering mechanism with evidence that the crust may lie in a near-critical state.

Key words: Historical seismicity, tectonic stress fields, stress release, stress transfer, spatial interaction, triggering mechanism.

1. Introduction

Based on field observation and crustal deformation measurement of the 1906 San Francisco earthquake, REID (1911) proposed an elastic rebound theory of earthquake origins. According to the theory, stress in a seismically active region accumulates due to relative movement of faults. When the stress exceeds a certain threshold, for example the strength of rock media, an earthquake occurs and the accumulated strain energy is released in the form of seismic waves. Although this model and its modifications (so-called time- and slip-predictable models by fixing only strength or residual stress) have been widely used in long-term prediction (SHIMAZAKI and NAKATA, 1980), real sequences of large earthquakes are fundamentally more complicated. We cannot exactly determine, at least at present, the

---

[1] School of Mathematical and Computing Sciences, Victoria University of Wellington, Wellington 600, New Zealand. E-mails: cslu@mcs.vuw.ac.nz, dvj@mcs.vuw.ac.nz
[2] LNM, Institute of Mechanics, Academia Sinica, Beijing 100080, China.

strength of rock media in a seismic region. In other words, what we may know is that the risk of an earthquake occurrence will increase as the stress accumulates. Through a development of the Markov model suggested by KNOPOFF (1971), the stress release model, a stochastic version of elastic rebound theory, has been developed. It incorporates a deterministic buildup of stress within a region and its stochastic release through earthquakes, and has been applied to the statistical analysis of historical earthquake data in China, Japan and Iran (VERE-JONES, 1978; VERE-JONES and DENG, 1988; ZHENG and VERE-JONES, 1991, 1994).

All the models based on elastic rebound theory suggest that a large earthquake should be followed by a period of quiescence, whereas in reality a strong earthquake can be followed by a period of activation and sometimes other earthquakes of comparable magnitude (GABRIELOV and NEWMAN, 1994). This seems most plausibly consistent with a view of the events taking place in the earth's crust as forming part of a tightly linked, near-critical process, exhibiting the self-similarity, long-range correlation and power-law distributions which are characteristic of a physical process in near-critical state (VERE-JONES, 1976; TAKAYASU and MATSUZAKI, 1988; TURCOTTE, 1992; MAIN, 1996; JAUMÉ and SYKES, 1999). Here, it is indeed observed that large events in one part of a region are noticeably often followed by large events in other quite distant parts of the region. One of the possible mechanisms is the competition between local strengthening and weakening through transfer and interaction of tectonic stress, these effects combining to cause a triggering mechanism of earthquake occurrences.

Although the stress field within a seismic region can be extracted from different kinds of information, the temporal variations of seismicity or historical earthquake data may most directly reflect the nature of earthquake-generating stress (ZHAO et al., 1990). In this paper we will use the linked stress release model, a natural extension of the simple stress release model incorporating the stress transfer among subregions, to interpret the historical earthquake catalogues from China and New Zealand. The results obtained underline the major differences in tectonic structures and earthquake mechanisms between the two regions.

2. Linked Stress Release Model

In the univariate stress release model, the regional stress level $X(t)$ increases deterministically between two earthquakes and releases stochastically as a scalar Markov process. The evolution of stress versus time is assumed to follow the equation

$$X(t) = X(0) + \rho t - S(t) \tag{1}$$

where $X(0)$ is the initial stress level, ρ is the constant loading rate from external tectonic forces, and $S(t) = \Sigma_{t_i < t} S_i$, where t_i, S_i are the origin time and stress release associated with the i-th event (VERE-JONES and DENG, 1988; ZHENG and VERE-JONES, 1991, 1994).

Obviously, issues of stress readjustment and transfer cannot be considered in this simple framework. In order to take into account interactions among different subregions due to stress transfer, the evolution of stress $X_i(t)$ in the i-th subregion versus time is rewritten as

$$X(t, i) = X(0, i) + \rho_i t - \sum_j \theta_{ij} S(t, j) \qquad (2)$$

where $S(t, j)$ is the accumulated stress release from events within the subregion j over the period $(0, t)$, and the coefficient θ_{ij} measures the fixed proportion of stress drop from events, initiated in the subregion j, which is transferred to the subregion i. Here, θ_{ij} may be positive or negative, resulting in damping or excitation respectively. It is convenient, if ignoring aftershocks, to set $\theta_{ii} = 1$ for all i. We call this new version a linked (or coupled) stress release model (SHI et al., 1998; LIU et al. 1998; LU et al., 1999a,b,c). If $\theta_{ij} = 0$ for all $i \neq j$ in (2), the model is reduced to an independent combination of simple forms as in (1).

The value of stress release during an earthquake can be estimated from its magnitude in terms of the relation $M = \frac{2}{3} \log_{10} E + \text{const}$ (KANAMORI and ANDERSON, 1975), where E is the released energy during an earthquake. For simplicity, the stress drop S during an earthquake is presumed to be proportional to the square root of the released energy, i.e., $S \propto E^{1/2}$. Then, we have the formula

$$S = 10^{0.75(M - M_0)} \qquad (3)$$

where M_0 is the threshold magnitude.

The probability intensity of an earthquake occurrence is controlled by a risk function $\Psi(x)$. Generally speaking, the risk function should be increasing nonlinearly with the stress level x. If the solid media had an exact critical strength, $\Psi(x)$ would have to be zero until x reached the critical strength, and infinite beyond it. By contrast, a finite constant value of $\Psi(x)$ corresponds to a pure random process in which the occurrence of events is independent of the stress level. Thus, a simple choice of $\Psi(x)$ is the exponential function $\Psi(x) = \exp(\mu + \nu x)$, where μ and ν are constants and indicate the background and sensitivity to risk, respectively. This is a convenient compromise between time-predictable and purely random (Poisson) processes.

It is further assumed that the probability distribution of earthquake sizes is independent of stress level (and as a default governed by the standard Gutenberg-Richter law). The key for statistical analysis is that the data in historical earthquake catalogues can be treated as a marked point process in time-stress space with the conditional intensity function

$$\lambda(t, i) = \exp\left\{a_i + v_i\left[\rho_i t - \sum_j \theta_{ij} S(t, j)\right]\right\} \tag{4}$$

where $\alpha_i = \mu_i + v_i X_i(0)$, v_i, ρ_i and θ_{ij} are the parameters to be fitted. We choose to parameterise the intensity in this form because it is more amenable to physical intuition (i.e., the ρ_i are input rates and the v_i are sensitivities to increase in stress). A simpler parameterisation can be recovered by setting $b_i = v_i \rho_i$, $c_{ij} = \theta_{ij}/\rho_i$. Estimates of the parameters are found by maximising the log-likelihood

$$\log L = \sum_i\left[\sum_j \log \lambda(t_j, i) - \int_{T_1}^{T_2} \lambda(t, i)\, dt\right] \tag{5}$$

where the observation interval (T_1, T_2) contains events at times $T_1 < t_j < T_2$ $(j = 1, 2, \ldots, N)$, and N is the number of events. This can be done numerically by using routines in the statistical seismology library (SSLib) developed recently (HARTE, 1998).

With discretion in the degree of interaction and subregions, we have a number of possible models. The choice among these models will be based on the Akaike information criterion (AIC), which is defined as

$$\text{AIC} = -2 \log \hat{L} + 2k \tag{6}$$

where $\log \hat{L}$ is the maximum log-likelihood for a given model and k is the number of parameters to be fitted in the model (AKAIKE, 1977). This represents a rough way of compensating for the effect of adding parameters, and is a useful heuristic measure of the relative effectiveness of different models, in avoiding overfitting. For example, the simple stress release model ($\theta_{ij} = 0$ for $i \neq j$ in (4)) with three parameters as against the Poisson model with only one (α_i) or the Poisson with exponential trend with two (α_i, $\beta_i = \rho_i v_i$), must demonstrate a significantly better fit to justify the additional parameters. In typical cases, model differences which would be significant at around the 5% confidence level correspond to differences in AIC values of around 1.5 or 2. The best model is that for which AIC has the smallest value. However, we should caution that the AIC values obtained here should be used as rough guides only, since the amount of historical earthquake data is not too large and the distribution of the log-likelihood is nonstandard (DALEY and VERE-JONES, 1988).

Using different combinations of the parameters in (4), we can examine different stress interaction mechanisms. If we assume all the parameters $\theta_{ij} \neq 0$, long-range interaction is allowed; otherwise, if we only let $\theta_{ij} \neq 0$ for neighbouring regions, short-range interaction will predominate. On the other hand, special combinations of these parameters, such as the risk level v_i, the loading rate ρ_i etc., can be used to identify tectonic features and difference between different tectonic seismicity.

3. Applications to Historical Catalogues

Several versions of the linked stress release model have been implemented to both the historical earthquake catalogues and synthetic ones generated by geophysical models (SHI *et al.*, 1998; LIU *et al.*, 1998; IMOTO *et al.*, 1999; LU *et al.*, 1999a,b,c). When we analyse the real data, considerable caution should be taken to the completeness of historical catalogues used in the modelling. Here, the data require expert scrutiny and advice in aspects such as the analysis and selection of individual events, choice of time and magnitude thresholds, interpretation of magnitudes and epicentres, etc. We also need to stress that in all the data sets studied, aftershocks should be deliberately excluded from the catalogues since the model framework presented above is only suitable to the analysis of mainshocks (UTSU *et al.*, 1995). Fortunately, since the data in the two cases to be studied are restricted to large events, the numbers of aftershocks exceeding the threshold are usually very small, consequently their exclusion or inclusion should not have a major effect on the analysis.

3.1 Chinese Data (North China)

In general, the Chinese historical catalogue is one of the oldest and most extensive in existence. The region selected here represents a substantial part of northern China, and comprises the northeastern coastal region, excluding the Yellow Sea but including the Bohai Sea, and the main part of the Ordos Plateau, including in particular its western and southern boundaries. It is essentially bounded by latitudes 32°N and 42°N, longitudes 104°E and 120°E, and covers about the 500-year period 1480-present. The catalogue for this region is believed to be complete for events with $M \geq 6.0$ (GU, 1983; ZHENG and VERE-JONES, 1991). Here, $M_0 = 6.0$ is used as the lower threshold (see Figs. 1 and 2).

Geophysical considerations suggest that the region should be treated as containing at least two major components: the coastal region and the Ordos Plateau, roughly separated by the Taihung Mountains. To the east of this region the principle stress is compressive, oriented in the E–W direction, and corresponds to pressure from the Pacific Plate. To its west, the stress is mainly in a N–S direction and driven by pressure from the Indian Plate and Tibetan Plateau (LI and LIU, 1986; MA *et al.*, 1990).

As illustrated in Figures 1 and 2, the subregions and data set used are the same as in ZHENG and VERE-JONES (1991, 1994), except that several current events have been added to the catalogue. The total number of events is 66, 33 in the western part and 33 in the eastern part.

The three kinds of basic models (i.e., the Poisson model, the simple stress release model and linked stress release model), are applied to the analysis of each subregion. The results are set out in Table 1. In terms of the difference between

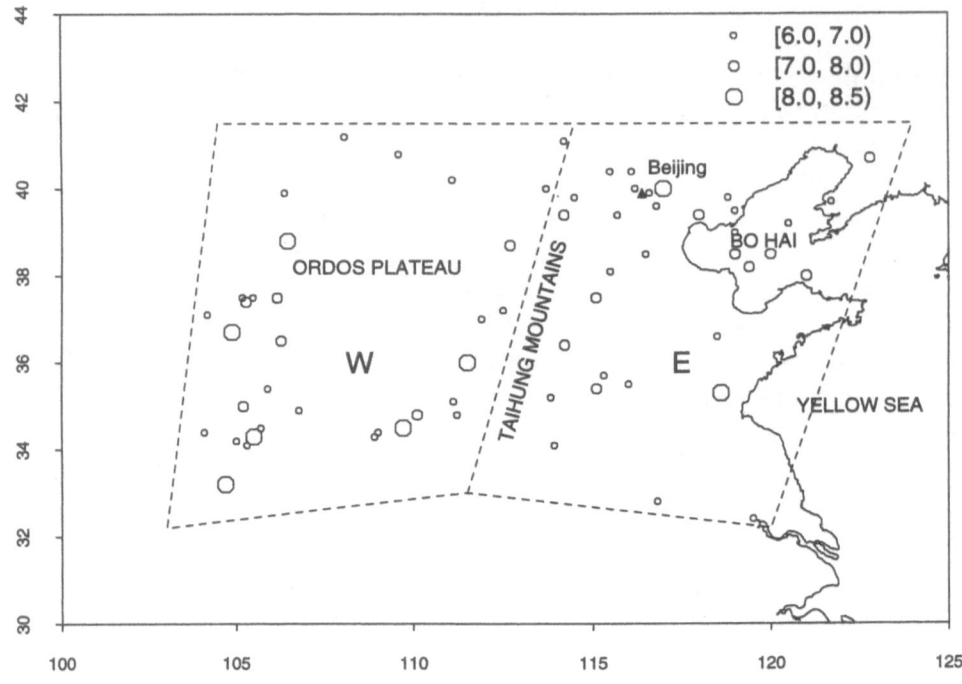

Figure 1

The epicentre distribution of major earthquakes with magnitude $M \geq 6.0$ and the definition of subregions of the North China historical earthquakes. The region is divided into two subregions: the western part (W) and the eastern part (E), respectively. The total number of events is 66, 33 in the western part and 33 in the eastern part. Please note that the subregion boundaries (dashed line) are used only as an eye's guide.

AIC values, it is obvious that the simple stress release model fits the data better than the Poisson model. As discussed above, we can choose different combinations of the parameters in (4) to fit the data. Here, the best fitting model, having the smallest AIC value, is obtained when we let all the initial risk levels (α_i), the risk sensitivities (ν_i) and the loading rates (ρ_i) be equal for both subregions. In fact, this is a natural choice if we compare the parameters fitted by simple stress release models. Although the difference between the AIC values is too small to clearly indicate a preferred model, the similarity in the parameter values for the two regions, and the fact that fewer parameters ($k = 5$) are needed in the linked stress release model, suggest that the latter model provides the more effective description of the data. Thus it seems that northern China can be viewed as a single region with similar seismicity and modest interactions between two subregions.

The risk functions calculated by using the fitted parameters are shown in Figure 3. The negative value of θ_{12} implies that events occurring in subregion E tend to trigger events in subregion W. Therefore, the main compressive stress may be in the

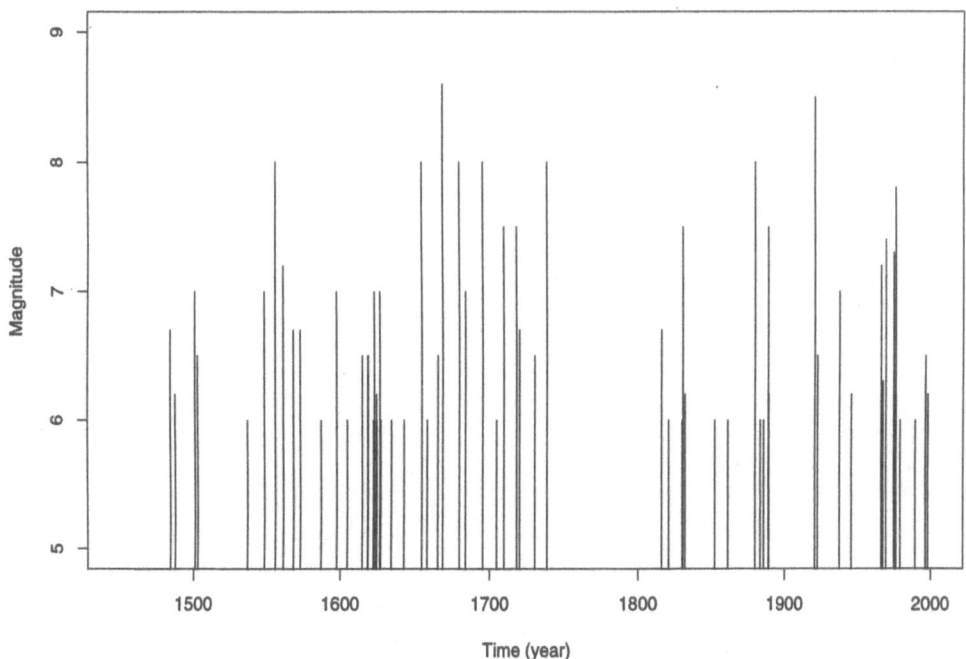

Figure 2

The magnitude versus time of the North China historical earthquakes with magnitude $M \geq 6.0$ during the period from 1480 to 1998.

Table 1

The fitted parameters and AIC values obtained by using the Poisson model, the simple stress release model (SRM) and the linked stress release model (LSRM) to the North China historical earthquake data

Model	Subregion i	α_i	v_i	ρ_i	θ_{i1}	θ_{i2}	k	AIC
Poisson	1 (W)	−2.755					2	499.72
	2 (E)	−2.755						
SRM	1 (W)	−3.192	0.0042	3.599			6	490.24
	2 (E)	−3.457	0.0043	3.031				
LSRM	1 (W)	−3.273	0.0042	3.033	1.000	−0.231	5	488.96
	2 (E)	−3.273	0.0042	3.033	0.025	1.000		

E–W direction, in agreement with the conclusion from the viewpoint of plate tectonics (LI and LIU, 1986).

3.2 New Zealand Data

Compared to the Chinese historical catalogue, the New Zealand catalogue is relatively incomplete, and has a short record history. Although the main upgrade of

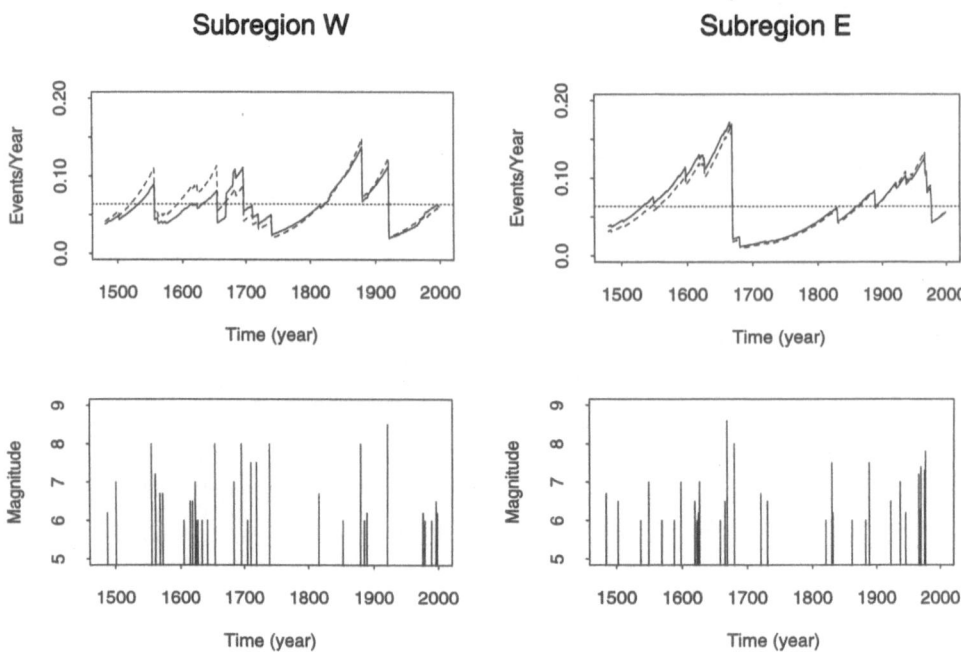

Figure 3

The risk function (events/year) versus time (year) for each subregion calculated by the linked stress release model (solid line), the simple stress release model (dashed line), and the Poisson model (dotted line). For comparison the earthquake versus time for each subregion is also plotted.

seismological recording in New Zealand occurred in the late 1980s (ANDERSON and WEBB, 1994), the catalogue with the magnitude $M \geq 6.0$ should be reasonably complete for shallow events since 1840. To decrease the chance of bias from the historical records, especially from the pre-instrumental earthquake catalogues, a lower threshold of $M_0 = 6.2$ has been used, as shown in Figures 4 and 5.

As an active island arc structure, New Zealand seems unlikely to consist of a few relatively well-defined seismic components, but rather of a closely interacting, highly fragmented ensemble within which a few larger units are embedded. The seismicity of the region studied here (see Fig. 4) is dominated by the boundaries of subducting zone of the Pacific plate beneath the Australian plate. The definition of appropriate subregions, which must satisfy both geophysical and statistical requirements, is in no case trivial. One way that might be used to define subregions is by the application of some clustering algorithm, with boundaries drawn equidistant between neighbouring clusters. An additional consideration is that subregions must include sufficient observations to allow the numerical parameter fitting procedure to converge. The choice of the two subregions (N and S) provided in the present study, as illustrated in Figure 4, is not claimed to be optimal, or more exactly

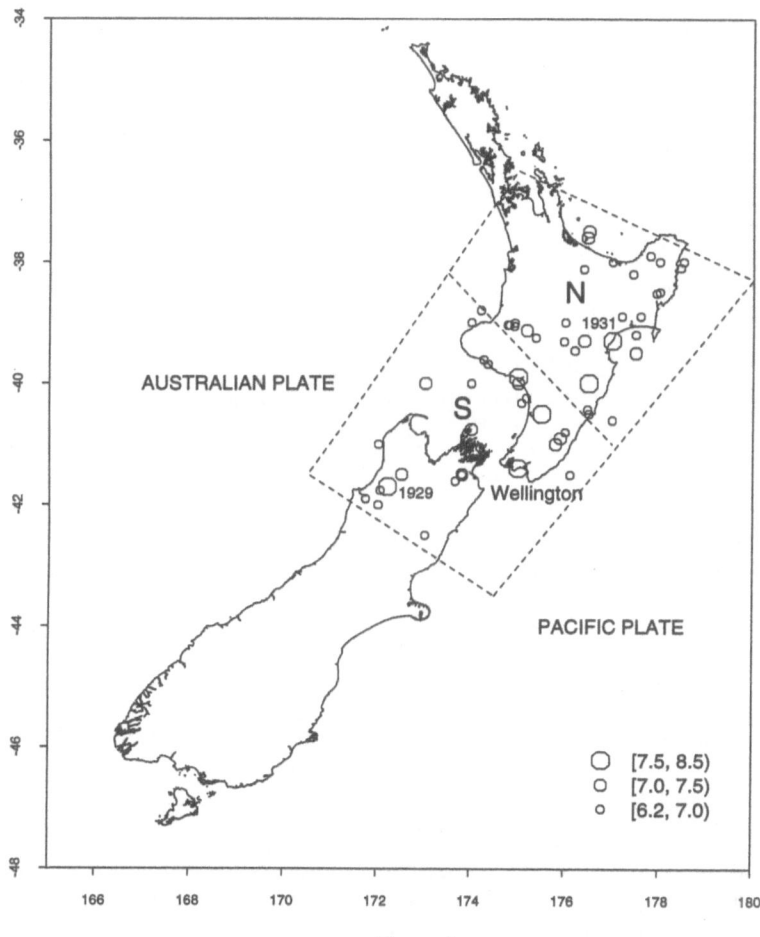

Figure 4

The epicentre distribution of major earthquakes with magnitude $M \geq 6.2$ and the definition of subregions of the New Zealand historical earthquakes. The region is divided into two subregions: the northern part (N) and the southern part (S), respectively. The total number of events is 65, 31 in the northern part and 34 in the southern part. Please note that the subregion boundaries (dashed line) are used only as an eye's guide

speaking, is just a preliminary attempt. Here, the total number of events is 65, 31 in the northern part and 34 in the southern part.

As with the analysis of the North China data, three basic models are applied to each subregion. The results are shown in Table 2. The simple stress release model for each subregion is an improvement on the Poisson model, and further substantial improvement is obtained by the linked stress release model in which the difference of the AIC values is about 8.80 or 9.2, although more parameters ($k = 8$ or $k = 7$) are needed in the latter.

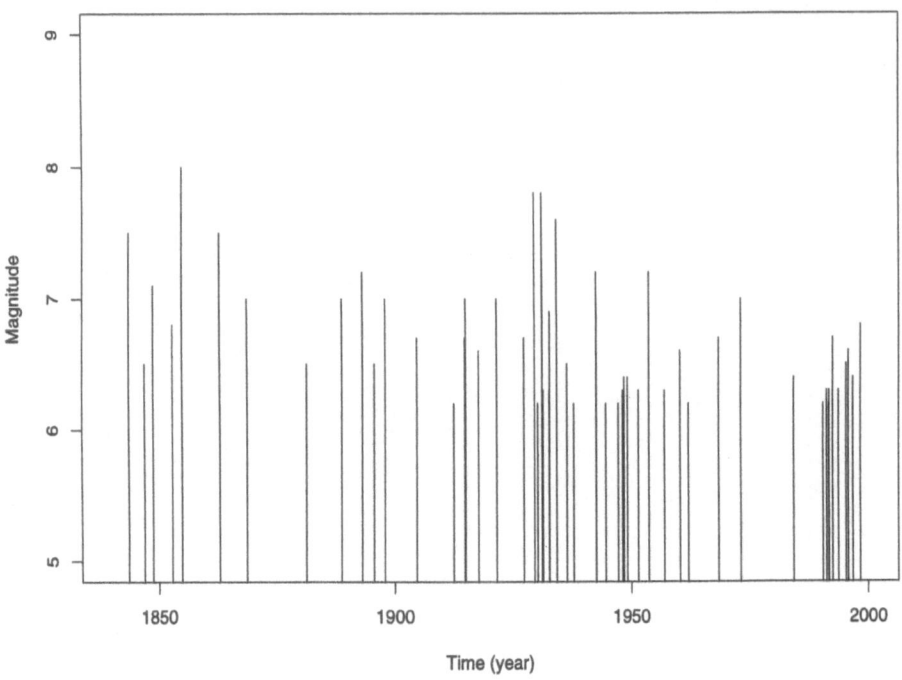

Figure 5
The magnitude versus time of the New Zealand historical earthquakes with magnitude $M \geq 6.2$ during
the period spanning 1840 to 1998.

Table 2

*The fitted parameters and AIC values obtained by using the Poisson model, the simple stress release
model (SRM) and the linked stress model (LSRM) to the New Zealand historical earthquake data,
where CRLSRM represents the common loading rates introduced in the linked stress release model*

Model	Subregion i	α_i	ν_i	ρ_i	θ_{i1}	θ_{i2}	k	AIC
Poisson	1 (S)	−1.629					2	339.44
	2 (N)	−1.536						
SRM	1 (S)	−1.132	0.0051	5.968			6	330.46
	2 (N)	−3.130	0.0026	10.391				
LSRM	1 (S)	−0.896	0.0056	4.578	1.000	−0.261	8	321.66
	2 (N)	−7.893	0.0127	0.772	−1.179	1.000		
CRLSRM	1 (S)	−0.851	0.0035	1.182	1.000	−0.866	7	321.26
	2 (N)	−7.908	0.0129	1.182	−1.110	1.000		

The risk functions calculated by using the fitted parameters are shown in Figure
6. The seismicity of the two subregions is very different. The risk in subregion N is
higher than that in subregion S. This is well consistent with the recent results

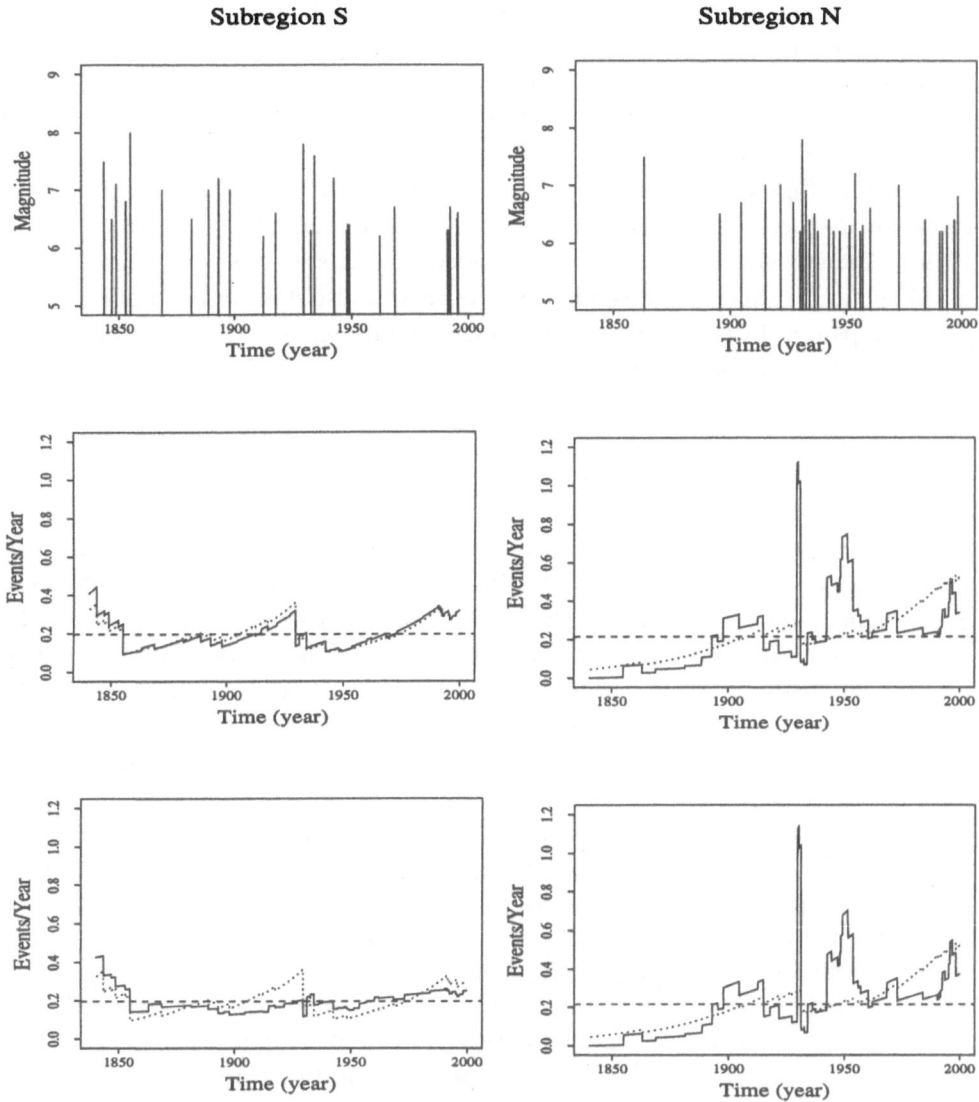

Figure 6

The risk function (events/year) versus time (year) for each subregion calculated by the linked stress release model (solid line) without (middle) or with (bottom) the loading rates ρ_i being constrained to be equal, the simple stress release model (dashed line), and the Poisson model (dotted line). For comparison, the earthquake versus time for each subregion is also plotted.

discovered by the $M8$ and MSc (i.e., the Mendocino Scenario) algorithms (KOS-SOBOKOV *et al.*, 1999). Carefully checking the parameters in Table 2, we find both the interaction coefficients (θ_{21} and θ_{12}) are negative, which indicates that events occurring in one subregion will increase the risk and trigger events in the other subregion. In particular the very large value of θ_{21}, coupled with the low value of

ρ_2, suggests that a large part of the risk in the lower North Island is due to the triggering effect of events in the South Island. This feature is retained even in the third model considered, where (as in the study of Japanese data by LU et al. (1999a)) the loading rates ρ_i are constrained to be equal. This results in a model with one less parameter and marginally better AIC than the unconstrained model. As shown in Figure 6, the Hawke's Bay earthquake ($M = 7.8$, 2 Feb., 1931) in the subregion N seems to be triggered by the Buller earthquake ($M = 7.8$, 16 June, 1929) in the subregion S (Fig. 4). Such triggering effects may be a common feature of subduction seismicity since similar behaviour was also discovered in our analysis of the Japanese historical data (Tokyo-Kamakura region) (ZHENG and VERE-JONES, 1994; LU et al., 1999a).

4. Discussion and Conclusion

In this paper, the linked stress release model, incorporating stress transfer and spatial interaction, is proposed and applied to the historical earthquake data from North China and New Zealand. The results show that in both cases studied, the linked stress release model fits the New Zealand data better than a collection of independent simple models, but is nearly indistinguishable from the simple stress release model in the case of North China. Moreover, the differences in the results for the two regions appear to reflect the differences in the tectonic settings of the regions. As we have seen, the seismicity in the plate boundary (subduction) region is more active and complex than that in the intraplate (plate collision) region. Thus more parameters (different loading rates, risk levels, etc.) in the models are needed for the description of the seismicity at plate boundary. As MATSU'URA and SATO (1997) pointed out that, for example, the assumption of uniform loading may not be appropriate for large interplate earthquakes.

In the linked stress release model proposed in this paper, the stress field is treated as a scalar which is representative of the whole region being studied. This is only likely to be a reasonable approximation if the real stress field has a dominant principal component which is not greatly variable, either in strength or direction, across the region. Theoretically speaking, it seems not to be difficult to introduce the stress field as a tensor in the framework of the model presented here, but the real data that can be used are not large enough for such a statistical analysis.

We should also note that the length scales in the two regions are substantially different. The dimension of one subregion in North China is approximately corresponding to the entire New Zealand region. In order to consider the possible scale-dependent effect, we further divided the North China region into four subregions: two in the western part and two in the eastern part (ZHENG and VERE-JONES, 1994). This gives us an opportunity to check whether there is a long-range stress interaction among subregions. The results are similar to those obtained above, i.e., no clearly preferred model emerges.

In summary, the results furnish a possible hint that the crust may lie in a near-critical state, so that an earthquake occurring in one region could trigger another earthquake distant from it within the region. The model itself provides a simple paradigm whereby spatio-temporal complexity of seismicity can be related to both the dynamics and heterogeneities in a seismic region (KAGAN, 1994; BEN-ZION, 1996). Despite the crudity of the model in physical terms, it has the advantage of fitting simple physical ideas into a stochastic framework, thereby allowing it to be objectively fitted and tested on real data. On the other hand, we can also apply this model to the synthetic catalogues, especially to those generated by considering special tectonic regions (ROBINSON and BENITES, 1996) or geophysical mechanism (MORA and PLACE, 1998). Using the fitted parameters, we can forecast the long-term risk in a seismic region (see OGATA, 1981) and improve simulation models for earthquake process.

Acknowledgements

This work is supported by the Marsden Fund administered by the Royal Society of New Zealand. We would like to thank Y. Shi, X. Zheng, L. Ma, M. Bebbington, H. Takayasu, and especially D. Harte for many valuable discussions and assistance. The paper has benefited from the comments of two anonymous referees.

REFERENCES

ANDERSON, H., and WEBB, T. (1994), *New Zealand Seismicity: Patterns Revealed by the Upgraded National Seismograph Network*, New Zeal. J. Geol. Geophys. *37*, 477–493.

AKAIKE, H., *On entropy maximisation principle*. In *Applications of Statistics* (ed. Krishnaiah, P. R.) (North Holland, Amsterdam 1977) pp. 27–41.

BEN-ZION, Y. (1996), *Stress, Slip, and Earthquakes in Models of Complex Single-fault Systems Incorporating Brittle and Creep Deformations*, J. Geophys. Res. *101*, 5677–5706.

DALEY, D., and VERE-JONES, D., *An Introduction to the Theory of Point Processes* (Springer, Berlin 1988) 702 pp.

GABRIELOV, A., and NEWMAN, W. I., *Seismicity modelling and earthquake prediction: A review*. In *Nonlinear Dynamics and Predictability of Geophysical Phenomena* (eds. Newman, W. I., Gabrielov, A., and Turcotte, D. L.) (Am. Geophys. Union, Washington, D.C. 1994) pp. 7–13.

GU, G. X., *Chinese Earthquake Catalogue* (in Chinese) (Science Press, Beijing 1983) 895 pp.

HARTE, D. (1998), *Documentation for the Statistical Seismology Library*, Rep. Sch. Math. Comput. Sci., Victoria Univ. Wellington *10*, 1–101.

IMOTO, M., MAEDA, K., and YOSHIDA, A. (1999), *Use of Statistical Models to Analyse Periodic Seismicity Observed for Clusters in the Kanto Region, Central Japan*, Pure appl. geophys. *155*, 609–624.

JAUMÉ, S. C., and SYKES, L. R. (1999), *Evolving towards a Critical Point: A Review of Accelerating Seismic Moment/Energy Release Prior to Large and Great Earthquakes*, Pure appl. geophys. *155*, 279–306.

KAGAN, Y. Y. (1994), *Observational Evidence for Earthquakes as a Nonlinear Dynamic Process*, Physica D *77*, 160–192.

KANAMORI, H., and ANDERSON, D. L. (1975), *Theoretical Basis of Some Empirical Relations in Seismology*, Bull. Seismol. Soc. Amer. *65*, 1073–1095.

KNOPOFF, L. (1971), *A Stochastic Model for the Occurrence of Main-sequence Earthquakes*, Rev. Geophys. Space Phys. *9*, 175–188.

KOSSOBOKOV, V. G., ROMASHKOVA, L. L., KEILIS-BOROK, V. I., and HEALY, J. H. (1999), *Testing Earthquake Prediction Algorithms: Statistically Significant Advance Prediction of the Largest Earthquakes in the Circum-Pacific, 1992–1997*, Phys. Earth Planet. Int. *111*, 187–196.

LI, F. G., and LIU, G. X. (1986), *Stress State in the Upper Crust of the China Mainland*, J. Phys. Earth (Suppl.) *34*, S71–S80.

LIU, J., VERE-JONES, D., MA, L., SHI, Y., and ZHUANG, J. (1998), *The Principle of Coupled Stress Release Model and Its Application*, Acta Seismol. Sinica *11*, 273–281.

LU, C., HARTE, D., and BEBBINGTON, M. (1999a), *A Linked Stress Release Model for Historical Japanese Earthquakes: Coupling Among Major Seismic Regions*, Earth Planets Space *51*, 907–916.

LU, C., VERE-JONES, D., and TAKAYASU, H. (1999b), *Avalanche Behaviour and Statistical Properties in a Microcrack Coalescence Process*, Phys. Rev. Lett. *82*, 347–350.

LU, C., VERE-JONES, D., TAKAYASU, H., TRETYAKOV, A. YU., and TAKAYASU, M. (1999c), *Spatio-temporal Seismicity in an Elastic Block Lattice Model*, Fractals *7*, 301–311.

MA, Z., FU, Z., ZHANG, Y., WANG, C., ZHANG, G., and LIU, D. *Earthquake Prediction: Nine Major Earthquakes in China (1966–1976)* (Seismological Press, Beijing 1990) 332 pp.

MAIN, I. (1996), *Statistical Physics, Seismogenesis, and Seismic Hazard*, Rev. Geophys. *34*, 433–462.

MATSU'URA, M., and SATO, T. (1997), *Loading Mechanism and Scaling Relations of Large Interplate Earthquakes*, Tectonophysics *277*, 189–198.

MORA, P., and PLACE, D. (1998), *Numerical Simulation of Earthquake Faults with Gauge: Towards a Comprehensive Explanation for the Heat Flow Paradox*, J. Geophys. Res. *103*, 21,067–21,089.

OGATA, Y. (1981), *On Lewis's Simulation Method for Point Processes*, IEEE Trans. Inf. Theory *27*, 23–31.

REID, H. F. (1911), *The Elastic-rebound Theory of Earthquakes*, Univ. Calif. Publ. Geol. Sci. *6*, 413–444.

ROBINSON, R., and BENITES, R. (1996), *Synthetic Seismicity Models for the Wellington Region, New Zealand: Implications for the Temporal Distribution of Large Events*, J. Geophys. Res. *101*, 27,833–27,844.

SHI, Y., LIU, J., VERE-JONES, D., ZHUANG, J., and MA, L. (1998), *Application of Mechanical and Statistical Models to Study of Seismicity of Synthetic Earthquakes and the Prediction of Natural Ones*, Acta Seismol. Sinica *11*, 421–430.

SHIMAZAKI, K., and NAKATA, T. (1980), *Time Predictable Recurrence Model for Large Earthquakes*, Geophys. Res. Lett. *12*, 717–719.

TAKAYASU, H., and MATSUZAKI, M. (1988), *Dynamical Phase Transition in Threshold Elements*, Phys. Lett. A *131*, 244–247.

TURCOTTE, D. L. *Fractals and Chaos in Geology and Geophysics* (Cambridge University Press, Cambridge 1992) 221 pp.

UTSU, T., OGATA, Y., and MATSU'URA, R. S. (1995), *The Centenary of the Omori Formula for a Decay Law of Aftershock Activity*, J. Phys. Earth *43*, 1–33.

VERE-JONES, D. (1976), *A Branching Model for Crack Propagation*, Pure appl. geophys. *114*, 711–726.

VERE-JONES, D. (1978), *Earthquake Prediction: A Statistician's View*, J. Phys. Earth *26*, 129–146.

VERE-JONES, D, and DENG, Y. L. (1988), *A Point Process Analysis of Historical Earthquakes from North China*, Earthquake Res. China *2*, 165–181.

ZHAO, Z., OIKE, K., MATSUMURA, K., and ISHIKAWA, Y. (1990), *Stress Field in the Continental Part of China Derived from Temporal Variations of Seismic Activity*, Tectonophysics *178*, 357–372.

ZHENG, X., and VERE-JONES, D. (1991), *Application of Stress Release Models to Historical Earthquakes from North China*, Pure appl. geophys. *135*, 559–576.

ZHENG, X., and VERE-JONES, D. (1994), *Further Applications of the Stochastic Stress Release Model to Historical Earthquake Data*, Tectonophysics *229*, 101–121.

(Received August 7, 1999, revised March 13, 2000, accepted April 6, 2000)

Pure appl. geophys. 157 (2000) 2365–2383
0033–4553/00/122365–19 $ 1.50 + 0.20/0

⎮Pure and Applied Geophysics

Development of a New Approach to Earthquake Prediction: Load/Unload Response Ratio (LURR) Theory

XIANG-CHU YIN,[1,2] YU-CANG WANG,[1] KE-YIN PENG,[1,2] YI-LONG BAI,[1] HAI-TAO WANG[1] and XUN-FEI YIN[1,3]

Abstract—The seismogenic process is nonlinear and irreversible so that the response to loading is different from unloading. This difference reflects the damage of a loaded material. Based on this insight, a new parameter-load/unload response ratio (LURR) was proposed to measure quantitatively the proximity to rock failure and earthquake more than ten years ago. In the present paper, we review the fundamental concept of LURR, the validation of LURR with experimental and numerical simulation, the retrospective examination of LURR with new cases in different tectonic settings (California, USA, and Kanto region, Japan), the statistics of earthquake prediction in terms of LURR theory and the random distribution of LURR under Poisson's model. Finally we discuss LURR as a parameter to judge the closeness degree to SOC state of the system and the measurement of tidal triggering earthquake.

The Load/Unload Response Ratio (LURR) theory was first proposed in 1984 (YIN, 1987). Subsequently, a series of advances were made (YIN and YIN, 1991; YIN, 1993; YIN et al., 1994a,b, 1995; MARUYAMA, 1995). In this paper, the new results after 1995 are summarized (YIN et al., 1996; WANG et al., 1998a, 1999; ZHUANG and YIN, 1999).

Key words: Load/Unload Response Ratio (LURR), earthquake prediction, random distribution of LURR, intermittent SOC, tidal triggering earthquake.

1. Introduction

It is recognized by many scientists that the physical essence of an earthquake is precisely the failure or instability of the focal media accompanied by a rapid release of energy. Therefore the preparation process of an earthquake is exactly the deformation and damage process of the focal media.

---

[1] LNM (Laboratory for Non-linear Mechanics), Institute of Mechanics, CAS, Beijing 100036, China.

[2] CAP (Center for Analysis and Prediction), CSB (China Seismological Bureau), Beijing 100080, China.

[3] Water Resources and Hydropower Planning and Design General Institute, MWR (Ministry of Water Resources), Beijing 100011, China. E-mail: XYC: yinxc@btamail.net.cn; WYC: yin@lnm.imech.ac.cn; PKY: xcyin@public.bta.net.cn; BYL: baiyl@lnm.imech.ac.cn; WHT: yinxc@btamail.net.cn; YXF: xfyin@263.net

From the microscopic viewpoint, the damage process for geo-material (rock) has incredible richness in complexity (MEAKIN, 1991; BAI *et al.*, 1994; KRAJCIN-NOVIC, 1996). In any rock there must be a large number of disordered defects (cracks) with different size, shape and orientation. The damage process includes nucleation, growth, interaction, coalition and cascade of cracks. It is an irreversible, nonequilibrium and nonlinear one, which has been intensively studied for decades but a lot of fundamental questions remain still unsolved. The problem of damage and failure in solid mechanics is as difficult as the problem of turbulence in fluid mechanics, and this is the inherent difficulty of earthquake prediction.

Since the problem of damage and failure for solids is one of scientific and technological importance so that a suite of effective phenomenological methods have been developed, to which the key is the constitutive relationship or the constitutive curve of materials (JAEGER and COOK, 1976). From the macroscopic viewpoint the constitutive curve is a comprehensive description of the mechanical property of the materials. A typical constitutive curve for focal media (rock) is shown in Figure 1. For generality, in Figure 1 the ordinate denotes general load P instead of stress σ and the abscissa is the response R to P instead of strain ε. If the load acting on the material increases monotonously, the material will experience the regimes of elastic, damage and failure or destabilization. The most essential characteristic of the elastic regime is its reversibility; i.e., the positive process and the contrary process are reversible. In other words, the loading modulus and the unloading one are equal to each other. Contrary to the elastic regime, the damage one is irreversible and the loading response is different from the unloading one or the loading modulus should be different from the unloading one. This difference indicates the deterioration of materials due to damage.

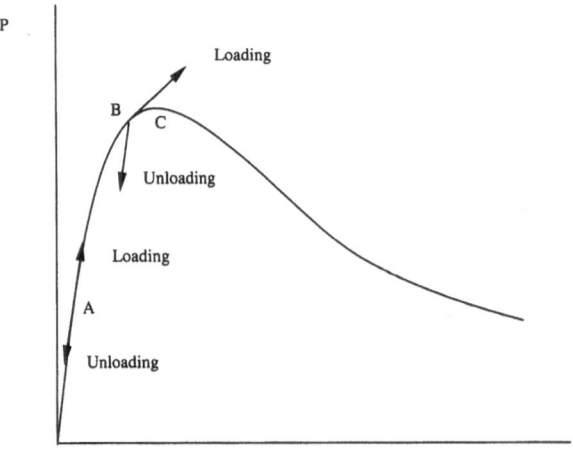

Figure 1
The constitutive curve of focal zone.

In order to measure quantitatively the difference, two parameters are defined as follows. The first one is the response rate X defined as

$$X = \lim_{\Delta P \to 0} \frac{\Delta R}{\Delta P}, \tag{1}$$

where ΔP and ΔR denote the increments of load P and response R, respectively. The second one is the Load/Unload Response Ratio (LURR) Y

$$Y = \frac{X_+}{X_-}, \tag{2}$$

where X_+ and X_- refer to response rate under loading and unloading condition, respectively.

It is clear that $Y = 1$ for the elastic regime since $X_+ = X_-$ and $Y > 1$ for the damage regime due to $X_+ > X_-$. The more seriously damaged the material, the larger the Y value will become. As the media approach failure the Y value becomes increasingly larger so that the Y value (LURR) could measure the proximity to failure and also acts as a precursor for earthquake prediction.

In continuum damage mechanics, the damage degree of material is measured by damage variable or damage parameter D. There are many ways to define D—from scalar to high order tensor (KRAJCINNOVIC, 1996). A direct way is to define D as the relative variation of the effective stiffness tensor. However for a simple condition, it can be simplified. According to LEMAITRE's definition (LAMAITRE, 1987),

$$D = E_0 - E/E_0, \tag{3}$$

where modulus E_0 denotes the Young's modulus for the original material and E means the Young's modulus in the damaged state. It is easy to derive the relation between D and Y in some simplest condition (say uniaxial tension or compression). If we take the stress as P and strain as R in (1) and (2), then

$$Y = E_-/E_+. \tag{4}$$

Assuming the unloading modulus is equal to E_0 and the loading modulus of damaged one E_0 is E, then

$$Y = E_0/E.$$

Therefore we can derive a very simple relation between Y and D as:

$$D = 1 - 1/Y. \tag{5}$$

If we adopt other more complicated definitions of damage, the relation between Y and D is no longer as simple as (5), although there is still a functional relation between Y and D. In other words, Y is actually another damage variable.

In fact, the reduction of modulus is due to the existence of microcracks in the material. Many scientists have studied the relation between modulus and the contained cracks. According to ODA (1983), the increase of the compliance tensor (M_{ijkl} the reciprocal of stiffness—$\varepsilon_{ij} = M_{ijkl}\sigma_{kl}$) could be expressed as

$$M_{ijkl} - M^0_{ijkl} = (\Delta/4)(\delta_{il}F_{jk} + \delta_{jl}F_{ik} + \delta_{jk}F_{il} + \delta_{ik}F_{jl}), \tag{6}$$

where M^0_{ijkl} denotes the compliance tensor of the undamaged medium, M_{ijkj} denotes the compliance tensor of the damaged medium, δ_{ij} is Kronecker delta, F_{ij} are the components of the fabric tensor which are defined as

$$F_{ij} = \frac{\pi N}{V} \int_0^\infty \int_\Omega a^3 D(n_k, a) n_i n_j \, d\Omega \, da, \tag{7}$$

where a is the size of crack, D is a crack distribution density function, n denotes the unit normal vector and $d\Omega$ is the spherical surface element.

It is obvious that not only the calculation of the fabric tensor is very complicated, but also it is much too difficult to obtain enough data for calculating F_{ij} at present. It is known that a crack or seismic fault with size a corresponds to an earthquake with specified magnitude and energy (KANAMORI and ANDERSON, 1975). Therefore it would be better to define the Y value directly by the seismic energy as follows:

$$Y = \frac{\left(\sum_{i=1}^{N^+} E_i^m\right)_+}{\left(\sum_{i=1}^{N^-} E_i^m\right)_-} \tag{8}$$

where E denotes seismic energy which can be calculated according to the Gutenberg-Richter formula (KANAMORI and ANDERSON, 1975; BULLEN and BOLT, 1985), the sign " + " means loading and " − " means unloading, $m = 0$ or $1/3$ or $1/2$ or $2/3$ or 1. When $m = 1$, E^m is exactly the energy itself; $m = 1/2$, E^m denotes the Benioff strain; $m = 1/3, 2/3$, E^m represents the linear scale and area scale of the focal zone, respectively; $m = 0$, Y is equal to N^+/N^-, and N^+ and N^- denote the number of earthquake which occurred during the loading and unloading duration, respectively.

In order to predict earthquakes in terms of parameter Y (LURR), a solution should be found as to load and unload the crustal blocks hundreds of kilometers in size and select proper parameters as the R (response) to calculate Y. These problems have been elucidated in our previous papers (YIN and YIN, 1991; YIN, 1993; YIN *et al.*, 1994a, 1995; MARUYAMA, 1995).

Here we just add two notes. The first one is the explanation of symbol Y. If the seismic energy is selected as R, then the LURR is denoted by Y_m instead of Y, Y_m is

$$Y_m = \frac{\left(\sum\limits_{i=1}^{N+} E_i^m\right)_+}{\left(\sum\limits_{i=1}^{N-} E_i^m\right)_-}. \tag{9}$$

When the ground water lever is selected as R, the LURR is denoted by Y_w, for Coda Q by Y_{cq}, wave velocity ratio by $Y_{v\text{-}r}$, crust deformation by Y_d, geomagnetism by Y_{gm}, etc.

The second one is the criteria to judge loading and unloading. We adopt the Coulomb failure hypothesis to judge loading or unloading according to the sign of the increment of Coulomb failure stress which is denoted by CFS in recent literature (e.g., HARRIS, 1998; REASENBERG and SIMPSON, 1992).

$$CFS = \tau_n + f\sigma_n, \tag{10}$$

where f, τ_n and σ_n stand for inner frictional coefficient, shear stress and normal stress (positive in tension) respectively, n is the normal of the fault plane on which the CFS reaches its maximum. This plane is parallel to the second principal stress (middle principle stress) σ_2 and the angle θ between the fault plane and the minimum principle stress (maximum compressive stress) σ_3 satisfies the following relation:

$$\tan 2\theta = 1/f. \tag{11}$$

ΔCFS is the increment of CFS. If the increment of Coulomb failure stress $\Delta CFS > 0$, it is referred to as loading; otherwise $\Delta CFS < 0$ is referred to as unloading.

It is well known that the resultant stress σ_{ij} in the crust consists of tectonic stress σ_{ij}^T and the tide induced stress σ_{ij}^t. Since the level of σ_{ij}^T (in the order of 10^6–10^8 Pa) is considerably higher than the level of σ_{ij}^t (10^3–10^4 Pa) so the directions of the principle stress of the crust resultant stress and then the direction of n can be determined by the tectonic stress only. However, the change rate of tidal induced stress is much larger than the change rate of the tectonic stress (VIDALI et al., 1998) thus ΔCFS is mainly due to tidal-induced stress which could be calculated precisely. The calculation of elastic deformation of the earth can be formulated as a system of six differential equations of first order. Following and improving Molodensksy-Takeuchi's work, we calculate the tide-induced stress components of any section in the crust in terms of the Runge-Kutta numerical method (MELCHIOR, 1978; YIN and YIN, 1991). The shear and normal stress on the fault plane with normal n can be obtained by stress tensor transform after which the ΔCFS can be calculated easily according (10).

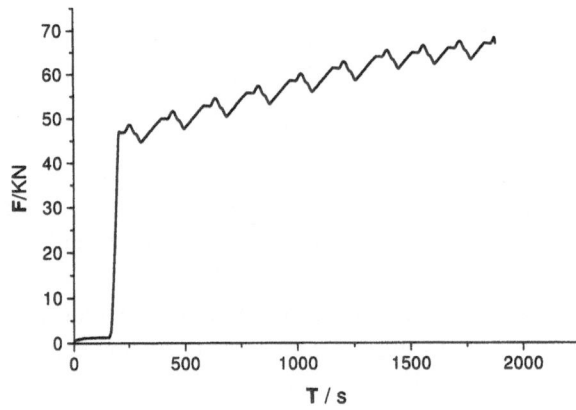

Figure 2
The *F-t* (load-time) curve.

2. *Validation of LURR Theory*

A. *Laboratory Simulation*

In order to test the validity of LURR theory, a series of rock fracture experiments have been conducted (SHI *et al.*, 1994; WANG *et al.*, 1998b). The specimens are rock (marble or sandstone) columns under uniaxial compression. The experiments were conducted in LNM (Laboratory for Non-linear Mechanics of Continuous Media), Institute of Mechanics, CAS and the Center Lab of USTC with test machine MTS-50. The load increases with constant rate superposing a harmonic force which simulates the tidal force (Fig. 2).

The force, displacement and the travel time are measured during the entire experiment and the variation of Young's modulus or wave velocity is taken as *R*. The experimental results are shown in Figure 3. It indicates that when the stress is at low level, the *Y* value is always close to 1 and as soon as the specimen is damaged the *Y* value rises. Finally the specimen fractures and *Y* reaches a high value (considerably larger than 1). These results suggest that LURR is available as a precursor for the fracture of a brittle solid and also for an earthquake.

It is worthwhile to note that under uniaxial compression the rock material shows brittle behavior and the damage regime is very narrow so that the *Y* value increases very steeply (refer to Fig. 5).

B. *Numerical Simulation*

Based on the lattice model of MORA and PLACE (1993), DEM (CUNDALL and STRACK, 1979) and MD (Molecular Dynamics approach), we developed a discrete model to simulate the damage and fracture of brittle solid (especially under compression) and in the interim measure the variation of LURR. In our model

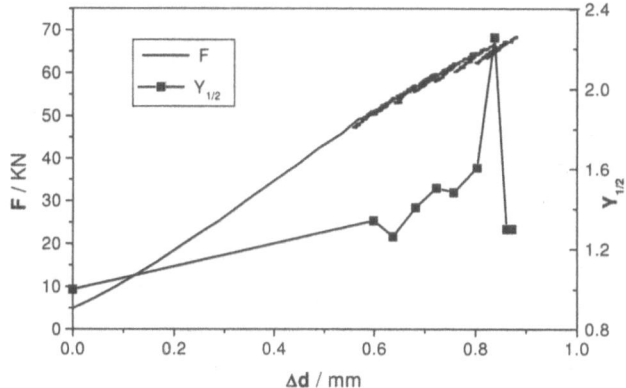

Figure 3
The variation of LURR with compressive displacement (Y-Δd curve) during the process of damage and fracture for a marble specimen.

(WANG et al., 1999, 2000, this issue) the medium is discretized into a number of round particles which are arranged into a triangular lattice. Many particles form a mesoscopic unit with different shape and size. Any two neighboring particles interact by radial force, tangential force and bending moment. Any particle must obey the conversation laws of momentum and angular momentum. Subsequently the equations of conversation law of momentum and angular momentum can be calculated step by step with a suitable time step and the results describe the whole process underlying the damage process or earthquake process. According to equation (8), the variation of Y (LURR) during the entire process can be calculated. Figure 4 is one of the simulation results. They substantiate once again that the parameter Y (LURR) is actually a quantitative indicator which mirrors the

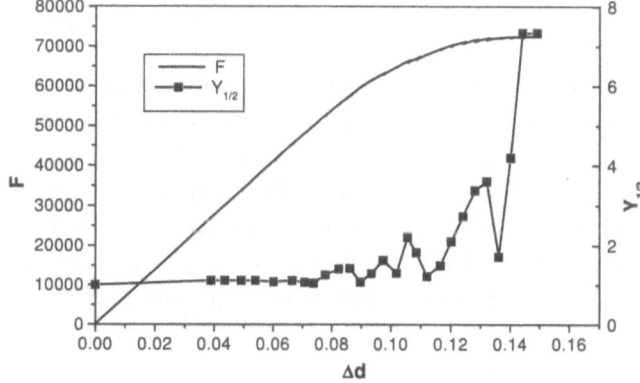

Figure 4
The numerical simulation results of variation in Y during the process of damage and fracture for rock.

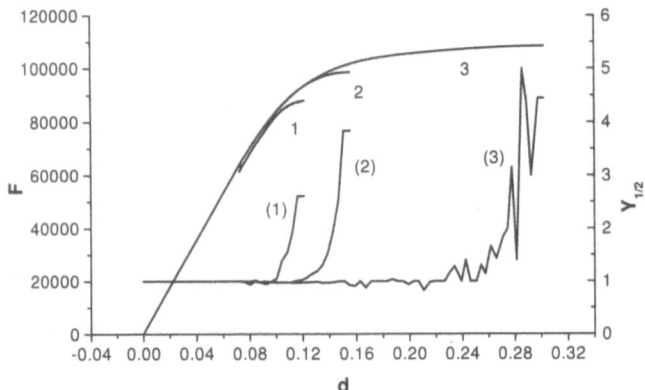

Figure 5
The influence of material heterogeneity on the variation Y during the damage process.

closeness degree to instability so that it could be a precursor for earthquake prediction.

Figure 5 shows us the influence of brittleness of the medium on the behavior including the variation of Y. The more brittle the medium is, the more sharply the Y value increases with the stress. The position of Figure 3 relates to the curve 1 or 2 in Figure 5. The influence of material heterogeneity and other factors on the damage process and the variation of Y during the process (WANG *et al.*, 1999, 2000, this issue), are also studied.

In addition, the chains network model (LIANG *et al.*, 1996) has been adopted to simulate the damage-fracture process for rock and the evolution of LURR. The simulation results, in terms of two models, coincide beautifully.

3. *Retrospective Examination*

Although the results of laboratory experiment and numerical simulation support the validity of LURR theory very well, the most convincing way to validate the LURR theory is the retrospective examination of LURR theory with real seismic data. Hundreds of cases have been studied with the data in China (YIN *et al.*, 1995).

In our study, usually $m = 1/2$, spatial windows of $1° \times 1°$ to $2° \times 2°$ and temporal windows of several months to years are chosen (similarly hereinafter). Since the size of the seismogenic region scales with the size of the ensuing main shock, the linear scale L of the spatial window seems to be selected according to the formula below (WANG, 1999)

$$\log L(\text{km}) = 0.5 M_s - 0.8 \tag{12}$$

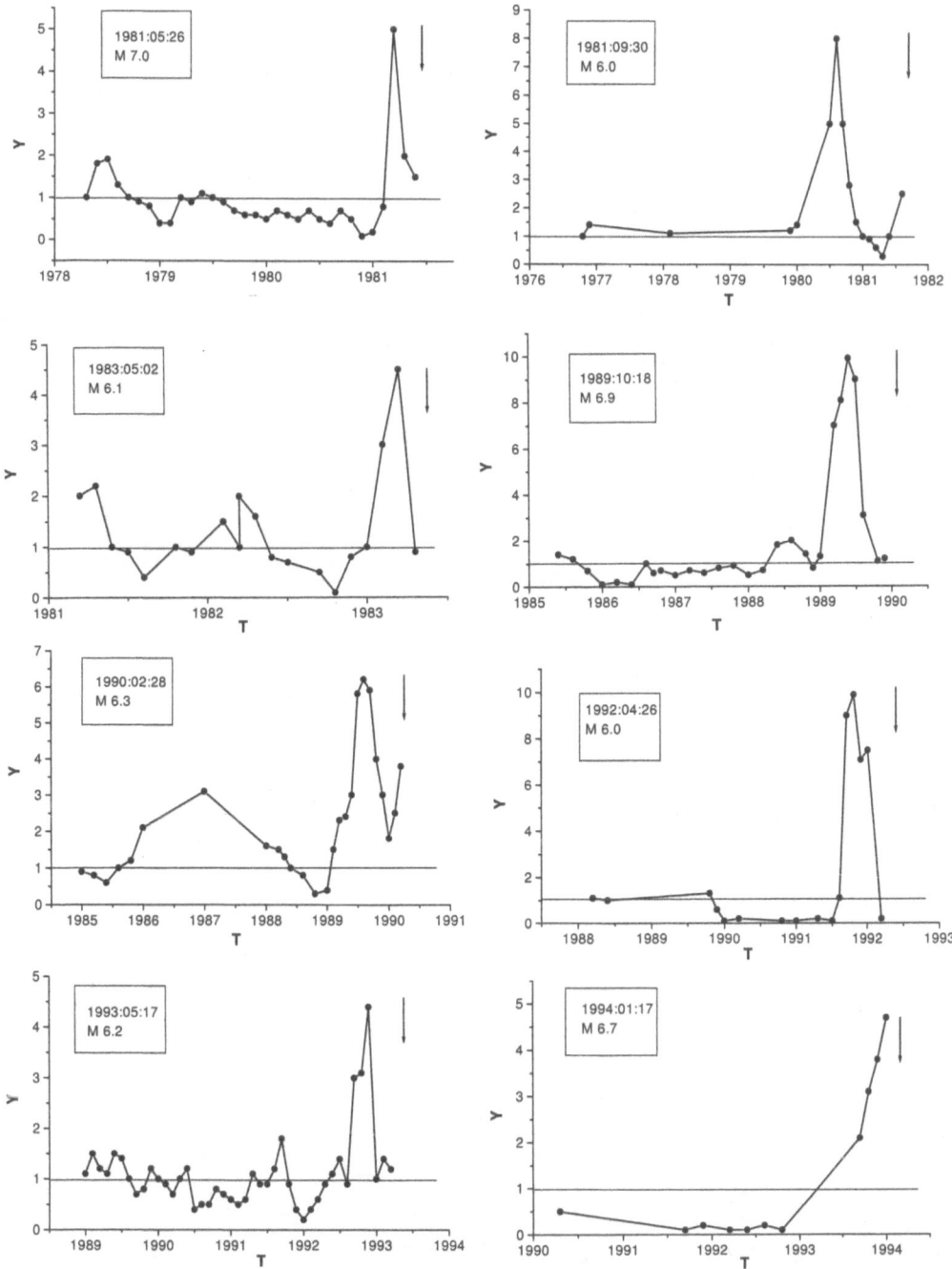

Figure 6

Variation in $Y_{1/2}$ with time before all strong earthquakes that occurred in southern California from 1980 to 1994 (the figure in box denotes year, month, day of occurrence and the magnitude, respectively).

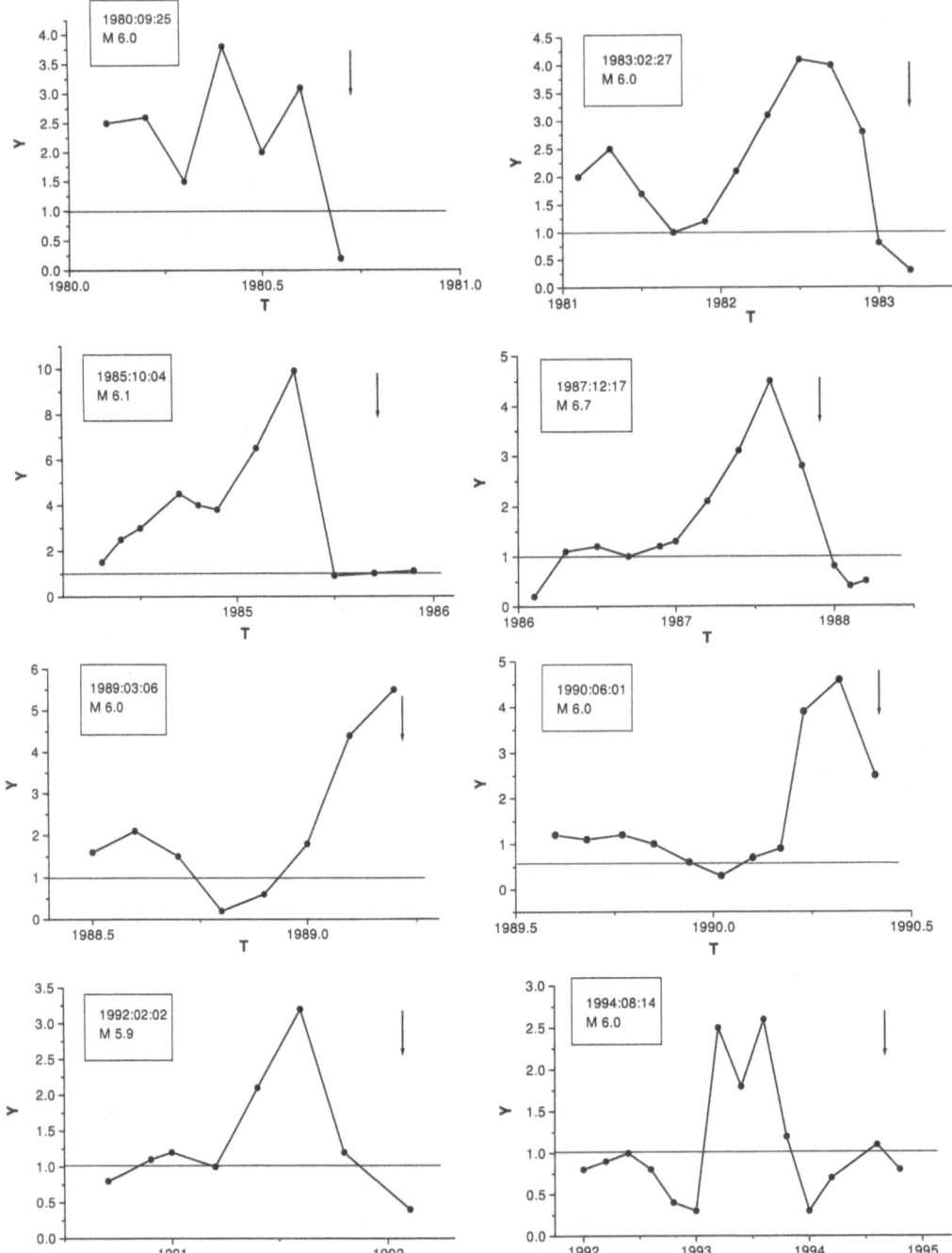

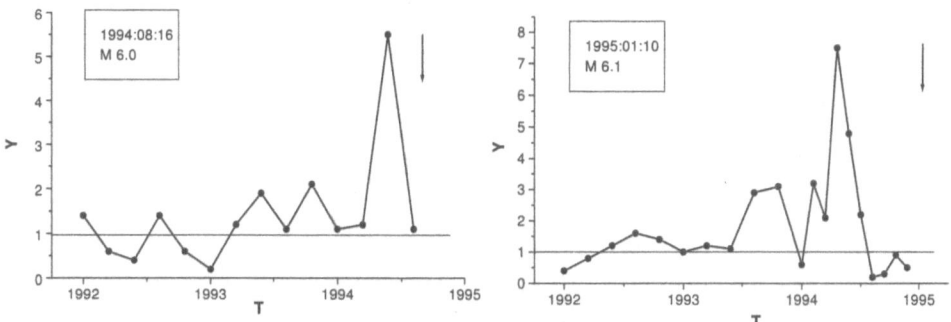

Figure 7

Variation in $Y_{1/2}$ with time prior to all moderate earthquakes that occurred in the Kanto region (Japan) from 1980 to 1995 (the figure in the box denotes year, month, day of occurrence and the magnitude, respectively).

where M_s is the magnitude of the predicted main shock. The temporal window is also related to the magnitude of the predicted main shock. Another consideration is that the size of the sample, the number of earthquakes in the selected spatial and temporal windows, should be large enough. Otherwise the calculated value of LURR might be without enough confidence (refer to Fig. 10).

The results of retrospective examination indicate that the highest proportion (more than 80%) of main shocks is preceded by a period during which the Y values increase markedly and remain high values ($Y_m > 1$, more precisely $Y_m > Y_m^c$, refer to part 5 of this paper). In contrast, we selected seven "stable" regions with low seismicity (no earthquake with $M \geq 4$ occurred in the examined period) on the Chinese mainland and analyzed the variation in Y for more than two decades (from 1970 to 1992). For all seven regions, the Y value always fluctuates slightly around 1 during the entire duration (YIN *et al.*, 1995).

Subsequently we have conducted further retrospective examinations for a series of strong earthquakes occurring in southern California (USA) and Kanto region (Japan). The results are shown in Figures 6 and 7, respectively. For most of the cases, the Y values are significantly larger than 1.

In addition, we have examined the variation of Y in the Parkfield section of the San Andreas Fault from 1970 to 1999. It can be seen from Figure 8 that through the entire period, the Y value is near unity, except for the Y value in 1982–1983 which reaches to Y_m^c and is a precursor for the swarm (the strongest earthquake with $M = 6.3$ and many earthquakes with $M > 5$ involved) which occurred in the region which is located to the northeast, and the distance between the region and Parkfield is less than 50 kilometers. The result from Figure 8 provides an explanation of why the expected earthquake around 1987 (BAKUN and LINDH, 1985; ROELOFFS and LANGBEIN, 1994) has not occurred yet.

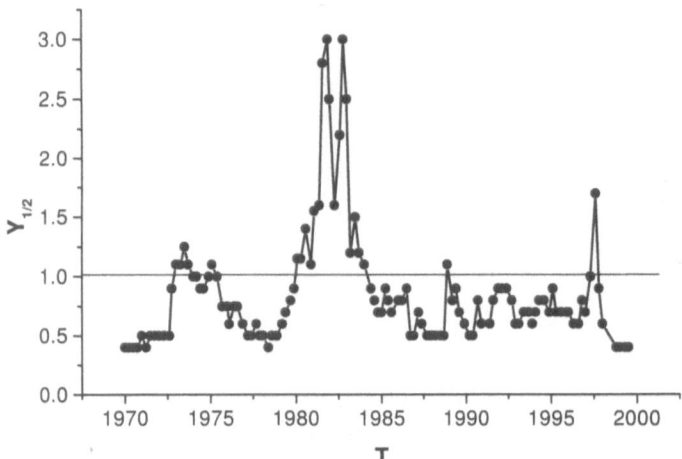

Figure 8
Variation in $Y_{1/2}$ in Parkfield region from 1970 to present.

It is well known that the tectonic regimes of southern California, Kanto and the Chinese mainland are quite different. The San Andreas Fault is a typical transform fault, Japan is located in a complex tectonic setting—interaction of four plates (Pacific, North American, Eurasian and Philippine Sea Plates) and most of the earthquakes occurring on the Chinese mainland are intraplate ones. The entire results of retrospective examination in different tectonic settings suggest that the Y value indeed indicates the proximity to instability of a crust block for a specific region, and the LURR theory could pave the way to earthquake prediction.

4. Tentative Practice of Earthquake Prediction

To date, all earthquake predictions in terms of LURR theory performed by X. C. Yin and his group (not including the predictions by other people outside our group) with formal documentation, are listed in Table 1. The predicted cases are not enough for statistical testing, but Table 1 indicates that we have indeed conducted successfully some earthquake predictions (intermediate term prediction—the predicted time scale ranges from 1 month to about 1 year), using LURR with a relatively high accuracy rate.

As an example, it is worth describing the prediction of the Kanto earthquake (1996.09.11, M_s 6.6, 35.5°N, 140.9°E). In early 1996, a scientist from JMA (Japanese Meteorological Agency) asked the first author of this paper to assist them in predicting the seismic tendency for some regions in Japan (Wakayama, Kanto, etc.) and offered us the data. After calculation and analysis of the variation of LURR—$Y(t)$ for these regions, we predicted that "a strong earthquake with a

Table 1

Statistics of EQ prediction using LURR

Prediction cases	Accuracy rating
1 Northridge EQ (1994.01.17, M_s 6.6) Predicted on 1993.10.28 that a "a medium EQ (which means M_s 6–6.5) would occur in SA6 region within about 1 year from now on" (A letter to Dr. Eric Bergman in ISOP, USGS).	C
2 Kanto EQ (1996.09.11, M_s 6.6, 35.5°N, 140.9°E) Predicted in April and May of 1996 that "a strong EQ with magnitude about M6 could occur in Kanto region (35°–36°N; 139°–141°E) within 1 year." (A Fax to Dr. Hosono Kohji, JMA and a paper published in Earthquake Research and China, both Chinese and English versions)	C
3 Predicted on 1994.09.20 that a felt EQ (M_L 4–5) would occur within 1 month in BJC Region (38.5°–41°N, 115.5°–117.5°E) The predicted EQ did not occur in the expected period. Beijing EQ (1994.12.23, M_L 4.3.40.6°N, 115.6°E)	F
4 Predicted on 1994.10.31 that a felt EQ (M_L 4–5) would occur in 2 months in BJC Region (38.5°–41°N, 115.5°–117.5°E) Tangshan EQ (1998.04.14, M_L 5.0, 39.7°N, 118.5°E)	C
5 Predicted on 1998.04.06 that a felt EQ about M_L 5 will occur within 2 months in BJE Region (38.5°–41°N, 117°–120°E)	C
6 Predicted on 1998.04.20 that an EQ with M 5–6 will occur within 2 months or slightly longer in BJC region. The predicted EQ has not occurred in the expected period.	F
7 Predicted on 1995.11.06 the seismic tendency of main China in 1996 (exactly 1995.11.07–1997.01.31).	C:3, F:0, M:0
8 Predicted on 1996.11.06 the seismic tendency of mainland China in 1997 (exactly 1996.11.07–1998.01.31.	C:5, F:2, M:0

Notes: EQ: earthquake. C: Correct prediction. F: False, the predicted EQ has not occurred. M: Missing, strong EQ occurred, but no prediction.

magnitude nearing 6 could occur in the Kanto region (35°–36°N; 139°–141°E) within one year." We faxed it to Dr. HOSONO (JMA) and wrote a paper (both in Chinese and English) based on the above research, and submitted it to ERC (Earthquake Research in China) in April 1996 (Chinese version) and May 1996 (English version). The Chinese version of this paper was published in ERC (No. 3, 1, Sep., 1996) and its English version was published in ERC (No. 4, 1996).

According to our experience, we have made preliminary conclusions as follows:

A. If the Y value is low for a region, we are fully confident that no strong earthquakes will occur in the near future (say several months) in this region.

B. If the Y value is high enough for a region, there are several possibilities:

1. In the majority of cases, a strong earthquake or earthquakes occur in the predicted window (time window: about 1 year, space window: about 100 km, and its magnitude relates to the areas of high Y value).

2. Sometimes the strong earthquake or earthquakes do not occur in the predicted window, but in the neighborhood of the window (not far from the window)

3. In rare cases, no strong event occurs for a prolonged time (say after 1 year or longer).

5. The Random Distribution of LURR

The preparation and occurrence process of earthquakes is controlled not only by deterministic dynamical law but also is affected by stochastic or disorder factors. Therefore we must study the influence of random factors on LURR in order to judge whether the height the Y value reaches can be considered as a precursor under the specified confidence (e.g., 0.95) (ZHUANG and YIN, 1999).

The Poisson model and the binomial model are used to describe the occurrence times of earthquakes. In order to save space, only the results of the Poisson model are mentioned below. We assume that earthquakes in a region obey the following basic assumption:

1. The earthquakes occur consistent with a Poisson process with a constant rate λ. The number of earthquakes occurring in the time interval $[0, T]$ has a Poisson distribution with expectation λT, i.e.,

$$\Pr\{N = n\} = \frac{(\lambda T)^n}{n!} e^{-\lambda t}. \tag{13}$$

2. The distribution of the magnitudes obeys the Gutenberg-Richter law, i.e., an exponential distribution with the probability density function.

3. The probabilities of an earthquake falling in a loading period and unloading period are equal, both $1/2$.

Based on the above assumptions, a simulation algorithm for computing the distribution and the confidence bands of the LURR is outlined below:

1. For each time interval (assumed as unit time interval), simulate two random variables P, Q belonging to a Poisson distribution of rate $\lambda/2$, where λ is the occurrence rate of earthquakes. P, Q can be regarded as the number of loading earthquakes and unloading earthquakes occurring in the time interval, respectively.

2. According to the given b-value, simulate P magnitudes for the loading earthquake and Q magnitudes for the unloading earthquakes.

3. Calculate the Y-value.

4. Repeat steps 1 to 3 for one million times, and draw the histogram of Y-values, which could be regarded as the probability density function (p.d.f) of LURR. Figure 9 displays an example for $b = 1$, $\lambda T = 40$ and $m = 1/2$.

5. Find the 0.90, 0.95 and 0.99 confidence bands from the p.d.f of LURR, respectively. Figure 10 is an example of such kind of simulated results. For example, for the condition: Occurrence rate $= 50$, $b = 1$, $m = 1/2$, Confidence 95%, $Y_{1/2}$ value should be equal or greater than 2.4 which denoted Y_m^c.

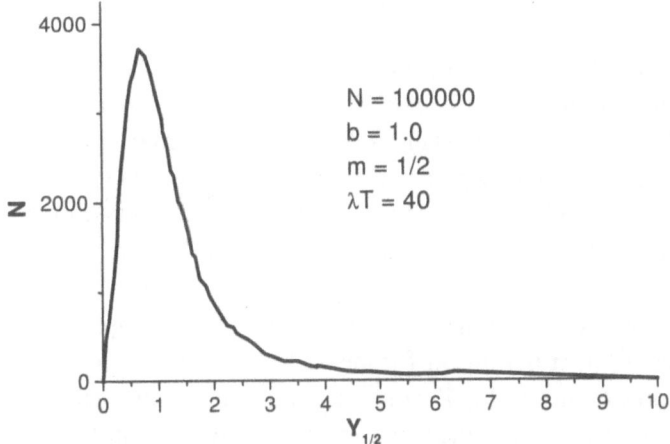

Figure 9
The random distribution of $Y_{1/2}$ under the specified condition.

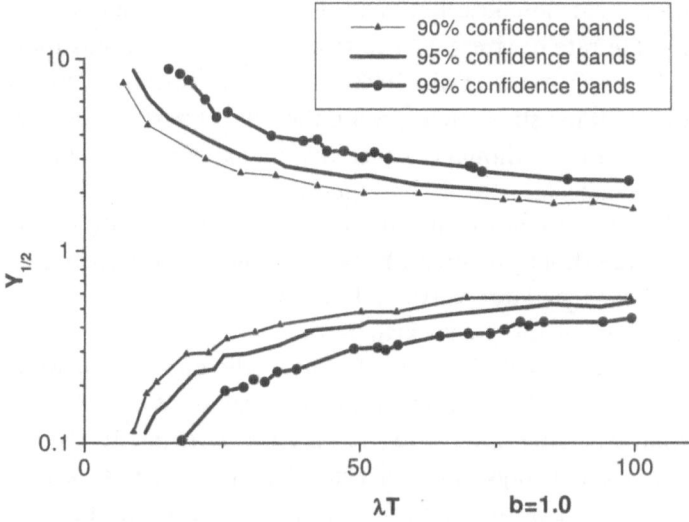

Figure 10
Variation of $Y_{1/2}$ with an occurrence rate under the Poisson model for specified confidence.

The results establish that the variation of LURR is controlled by the occurrence rate for the Poisson model and the parameter S ($S = b/m$, where b is the constant in the Gutenberg-Richter law and m is the power in (9)). The larger the occurrence rate and the S-value, the more stable or more concentrated around 1 the Y value is.

6. Discussion and Conclusion

The question of tidal triggering of earthquakes has been a controversial topic for which quite different even opposite opinions appeared in the literature (COT-TON, 1922; KNOPOFF, 1964; EMTER, 1998). There is remarkable new interest in it, based on new data and viewpoints (HARRIS, 1998; VIDALI *et al.*, 1998). In our opinion, whether the tidal stress triggers earthquakes depends on many factors such as tectonic setting, time and so on. Hence it is inappropriate to seek a stereotyped conclusion in all cases.

It is worthwhile to mention that the Y_m value (for example, $Y_0 = N^+/N^-$) is a suitable and convenient parameter for measuring tidal triggering of earthquakes. As shown in Figures 6 and 7, the Y values are different in different periods (every Y_m value is calculated for a specified period ranging from months to years). For the duration in which the $Y_m < Y_m^c$, the tidal stress does not trigger earthquakes, only in case of $Y_m > Y_m^c$, does tidal triggering of earthquakes take place.

The explanation of the phenomena mentioned above is simple. The tidal stress is very small: several orders smaller than the crust tectonic stress. Consequently it only could trigger earthquakes, but cannot by itself engender earthquakes in spite of the fact that the rate of stress change from earth tides exceeds that from tectonic stress accumulation (VIDALI et al., 1998). For the condition of low tectonic stress, the total stress (tectonic stress superposing the tidal stress) is still low, tidal stress is difficult to trigger any earthquake, hence the Y_m value must be low ($Y_m < Y_m^c$). On the contrary, when tectonic stress is high enough, it is very sensitive to any tiny extrinsic disturbance. Consequently the tidal stress may easily trigger earthquakes so that the Y value should be high ($Y_m > Y_m^c$) which generally appears prior to the occurrence of strong or large earthquakes and persists months to years and even decades (in the case of the great earthquake).

The question mentioned above is relevant to the question of whether the whole crust is in the SOC (self-organized criticality) state. We prefer the intermittent criticality hypothesis (BOWMAN et al., 1998; HUANG et al., 1998) to the hypothesis that the entire crust always stays in critical state. When a large or even great earthquake occurs just now in a region, a large portion of the cumulated energy and stress will be released so that the tectonic stress becomes low. In this case, the crust of this region is not in the critical state, its Y value is low ($Y_m < Y_m^c$) and tidal stress is difficult to trigger any earthquake. Thereafter the tectonic stress is accumulated gradually in this region which approaches the critical state (of course this process is not monotonic and linear). Finally it reaches the critical state in such a situation, the tectonic stress is high and the system must be sensitive to any tiny extrinsic disturbance, then the tidal stress easily triggers earthquakes so that the Y value should be high ($Y_m > Y_m^c$). Therefore the Y value is also a parameter to represent the approaching degree to critical state.

To sum up, the LURR theory was proposed on the basis of the constitutive relation (macroscopic phenomenological methodology). Later research brought insight to the mesoscopic viewpoint, though it still leaves many questions for future research to answer.

The LURR theory is still young therefore it has a broad room to develop. Besides seismic energy (8), many other geophysical parameters concerning the seismogenic process such as Coda Q, ratio of velocity, level of groundwater, cubic strain tilt and strain in crust, geomagnetism parameters etc. could also be the R (response) to define LURR (CHEN et al., 1994; YANG et al., 1994; CHEN and YIN, 1995; WANG, 1995; ZHEN, 1996; ZHANG, 1995). All of them have yielded interesting results and other new parameters being the new R (responses), are constantly emerging.

On the other hand, LURR could be applied not only to natural earthquake prediction but also to forecasting of other geological disasters such as RIS (CHEN and YIN, 1995), MIS landslide (HUANG and QU, 1997).

As discussed above, LURR could also be an indicator measuring the proximity to the SOC state and the tectonic stress level for a specified region.

Acknowledgements

This research is supported by the Natural Science Foundation of China (Grant No. 19732060), the key project of the ninth five-year-plan of MOST (Ministry of Science and Technology) and CSB (China Seismological Bureau). We are grateful to PETER MORA for his precious suggestions. The authors are also grateful to Bernard Minster and Steven Jaumé for their enhancement of the present paper.

REFERENCES

BAI, Y. L., LU, C. S., KE, F. J., and XIA, M. F. (1994), *Evolution Induced Catastrophe*, Phys. Lett. *A185*, 196–201.

BOWMAN, D. D., OUILLEN, G., SAMMIS, C. G., SORNETTE, A., and SORNETTE, D. (1998), *An Observational Test of the Critical Earthquake Concept*, J. Geophys. Res. *103*, 24,359–24,372.

BAKUN, W. H., and LINDH, A. G. (1985), *The Parkfield, California, Earthquake Prediction Experiment*, Science *229*, 619–624.

BULLEN, K. E., and BOLT, B. A., *An Introduction to the Theory of Seismology* (Cambridge University Press, Cambridge 1985).

CHEN, J. M., ZHANG, S. D., and YANG, L. Z. (1994), *Seismic Abnormalities in LURR of Groundwater Lever*, Earthquake *1*, 73–78.

CHEN, X. Z., and YIN, X. C. (1995), *Applications of LURR Theory to the Earthquake Prediction for Reservoir-induced Earthquakes*, Earthquake Research in China *11*, 361–367.

COTTON, L. A. (1922), *Earthquake Frequency, with Special Reference to Tidal Stresses in the Lithosphere*, Bull. Seismol. Soc. Am. *12*, 47–198.

CUNDALL, P. A., and STRACK, O. D. L. (1979), *A Discrete Element Model for Granular Assemblies*, Geotechnique *29*, 47–65.

EMTER, D., *Tidal triggering of earthquakes and volcanic events.* In *Tidal Phenomena* (ed. Wilhelm, A.) (Elsevier, Amsterdam 1998).

HARRIS, R. A. (1998), *Introduction to Special Section: Stress Triggers, Stress Shadows, and Implication for Seismic Hazard*, J. Geophys. Res. *103*, 24,347–24,358.

HUANG, Y., SALEUR, H., SAMMIS, C. G., and SORNETTE, D. (1998), *Precursor, Aftershocks, Criticality and Self-organization Criticality*, Europhys. Lett. *41*, 43–48.

HUANG, R. Q., and QU, O., *Scientific Analysis Principle and its Applications to Generalized System in Engineering Geology* (Geological Press, Beijing 1997) (in Chinese).

JAEGER, J. C., and COOK, N. G. W., *Fundamentals of Rock Mechanics* (Chapman and Hall, London 1976).

KANAMORI, H., and ANDERSON, D. L. (1975), *Theoretical Basis of Some Empirical Relation in Seismology*, Bull. Seismol. Soc. Am. *65*, 1073–1096.

KNOPOFF, L. (1964), *Earth Tide as a Triggering Mechanism for Earthquakes*, Bull. Seismol. Soc. Am. *54*, 1865–1870.

KRAJCINNOVIC, D., *Damage Mechanics* (Elsevier, Amsterdam 1996).

LAMAITRE, J., *Formulation and identification of damage kinetic constitutive equations.* In *Continuum Damage Mechanics* (ed. Krajcinnovic, D.) (Springer-Verlag, Wien, New York 1987).

LIANG, N., LIU, Q., LI, J., and SONG, H., *A Chains network model simulating meso-mechanics behavior and micro damage evolution of in situ reinforced ceramics.* In *Advanced in Engineering Plasticity and Its Applications* (eds. Abe, T., and Tsuta, T.) (Pergamon, Amsterdam 1996) pp. 141–146.

MARUYAMA, T. (1995), *Earthquake Prediction in China*, Zisin *19* (5), 68–76.

MEAKIN, P. (1991), *Model for Material Failure and Deformation*, Science *252*, 226–234.

MELCHIOR, P., *The Tide of the Planet Earth* (Pergamon Press, New York 1978).

MORA, P., and PLACE, D. (1993), *A Lattice Solid Model for the Nonlinear Dynamics of Earthquakes*, Int. J. Mod. Phys. *C4*, 1059–1074.

ODA, M. (1983), *A Method for Evaluating the Effect of Crack Geometry on the Mechanical Behavior of Cracked Rock Mass*, Mech. Mater. *2*, 163–171.

REASENBERG, P. A., and SIMPSON, R. W. (1992), *Response of Regional Seismicity to the Static Stress Change Produced by the Loma Preita Earthquake*, Science *255*, 1687–1690.

ROELOFFS, E. A., and LANGBEIN, J. (1994), *The Earthquake Prediction Experiment at Parkfield, California*, Rev. Geophys. *32*, 315–336.

SHI, X., XU, H., WAN, Y., LU, Z., and CHEN, X. (1994), *The Rock Fracture under Simulated Tide Force—Laboratory Study on the Loading and Unloading Response Ratio (LURR) Theory*, Acta Geophys. Sin. *37*, 631–636.

VIDALI, J. E., AGNEW, D. C., JOHNSTON, M. J. S., and OPPENHEIMER, D. H. (1998), *Absence of Earthquake Correlation with Earth Tides: An Indication of High Preseismic Fault Stress Rate*, J. Geophys. Res. *103*, 24,567–24,572.

WANG, T. W. (1995), *The Application of Load/Upload Response Ratio in Earthquake Prediction by Magnetic Method*, Observation and Research of Seismic Geomagnetism *16*, 26–29 (in Chinese).

WANG, H. T. (1999), *The Synthetic Application of LURR Theory to Earthquake Prediction*, Doctoral thesis of Institute of Geophysics, Chinese Seismological Bureau.

WANG, H. T., PENG, K. Y., ZHANG, Y. X., WANG, Y. C., and YIN, X. C. (1998a), *Characters of Variation of LURR During the Earthquake Sequence of Xinjiang*, Chinese Sci. Bull. *43*, 1752–1755.

WANG, Y. C., YIN, X. C., and WANG, H. T. (1998b), *The Simulation of Rock Experiment on Load/Upload Response for Earthquake Prediction*, Earthquake Research in China *14* (2), 126–130.

WANG, Y. C., YIN, X. C., and WANG, H. T. (1999), *Numerical Simulation on Load/Upload Response Ratio (LURR) Theory*, ACTA Geophys. Sin. (or Chinese Journal of Geophysics) *43*, 511–522.

WANG, Y. C., YIN, X. C., KE, F. J., XIA, M. F., and PENG, K. Y., (2000), *Numerical Simulation of Rock Failure and Earthquake Process on Mesoscopic Scale*, Pure appl. geophys. *157*, 1905–1928.

YANG, L. Z., HE, S. H., and XI, Q. W. (1994), *Study on the Variation of the Property of Rock Elasticity by Load/Unload Response Ratio of Tidal Volume Strain*, Earthquake Research in China *10*, 90–94.

YIN, X. C. (1987), *A New Approach to Earthquake Prediction*, Earthquake Research in China *3*, 1–7 (in Chinese with English abstract).

YIN, X. C. (1993), *A New Approach to Earthquake Prediction*, Russia's Nature *1*, 21–27 (in Russian).

YIN, X. C., and YIN, C. (1991), *The Precursor of Instability for Nonlinear System and Its Application to Earthquake Prediction*, Science in China *34*, 977–986.

YIN, X. C., YIN, C., and CHEN, X. Z. (1994a), *The precursor of instability for nonlinear system and its application to earthquake prediction — The load-unload response ratio theory*. In *Nonlinear Dynamics and Predictability of Geophysical Phenomena* (eds. Newman, W. I., Gabrelov, A., and Turcotte, D. L.), Geophysical Monograph, *83*, IUGG Volume *18*, 55–60, 1994.

YIN, X. C., CHEN, X. Z., SONG, Z. P., and YIN, C. (1994b), *The Load-unload Response Ratio Theory and its Application to Earthquake Prediction*, J. Earthq. Predict. Res. *3*, 325–333.

YIN, X. C., CHEN, X. Z., SONG, Z. P., and YIN, C. (1995), *A New Approach to Earthquake Prediction: The Load/Unload Response Ratio (LURR) Theory*, Pure appl. geophys. *145*, 701–715.

YIN, X. C., SONG, Z. P., and WANG, Y. C. (1996), *The Temporal Variation of LURR in Kanto and Other Regions in Japan and Its Application to Earthquake Prediction*, Earthquake Research in China *10*, 381–385.

ZHANG, J. H. (1995), *The Analysis of the Geomagnetic Abnormalities in Load/Unload Response Ratio Method*, Observation and Research of Seismic Geomagnetism *16*, 61–63 (in Chinese).

ZHEN, X. P. (1996), *Earthquake and Load/Unload Response Ratio of Terrestrial Magnetic Field to Solar Wind*, Observation and Research of Seismic Geomagnetism *17*, 49–53 (in Chinese).

ZHUANG, J. C., and YIN, X. C. (1999), *Random Distribution of the Load/Unload Response Ratio (LURR) Under Assumptions of Poisson Model*, Earthquake Research in China *15*, 128–138.

(Received August 5, 1999, revised February 26, 2000, accepted April 28, 2000)

 To access this journal online:
http://www.birkhauser.ch

Notes to Authors

PAGEOPH welcomes original contributions in English (and occasionally in French and German) on all aspects of geophysics. All manuscripts should be submitted to the Regular Issues Editor-in-Chief, in triplicate, formatted with double spacing and wide margins. For further details see the following paragraphs.

Format of Manuscripts

Length and Page Charges: A paper should not exceed 16 printed pages including tables and figures. For articles exceeding 16 printed pages the authors will be charged sFr. 80.00 for each additional page. No page charges, except those for color prints, are required for contributors to special issues.

Title Page: This should include the the complete title, full names and addresses of all authors. In addition, corresponding authors should provide their fax number and e-mail address if they are available.

Abbreviated Title: It is necessary to indicate an abbreviated title, which will be used as a running head (no more than 50 characters including spaces).

Abstract: The abstract should be in English, and in the language of the text, if different. It has to be of no more than 10 sentences and should be concise and self-contained.

Keywords: Up to 6 keywords should be listed, suitable for incorporation into information-retrieval systems.

Text: The text must include a citation for each item listed under References; the approximate position of each figure should be indicated in the text. The metric system should be used throughout the text, figures, and tables.

Tables: Tables are to be presented on separate pages, with a brief title for each.

Figures: Figure captions and legends are to be typed on a separate page or pages as the last element in the manuscript. Make sure that line thickness and lettering allow an adequate size reduction. Heliographic or photocopies are not suitable for reproduction. Highquality, glossy, photographic prints must be submitted. Color prints are permitted but authors will be charged for them.

References: They are to be listed in alphabetical order in the following style:
Journal article: Haurwitz, B., and Cowley, A.D. (1973), The Diurnal and Semidurnal Barometric Oscillations, Global Distribution and Annual Variation, Pure Appl. Geophys. 102, 193-222.

Whole book: Bath, M., Introduction to Seismology (Birkhäuser, Basel 1973).

Article in a book: Haurwitz, B., and Cowley, A.D., Barometric oscillations, In Introduction to Seismology (ed. Bath. M.) (Birkhäuser, Basel 1973) pp. 193-222.

Submission of Manuscripts

Manuscripts must be submitted in triplicate, formatted with double line spacing and wide margins. Copies of the figure should be attached at the end of the manuscript. Original, high-quality, glossy figures may be submitted later. All manuscripts pages, including references, tables, and captions, should be numbered consecutively, starting with the title page as page one.

All manuscripts should be submitted to Regular Issues
Editor-in-Chief
Brian Mitchell
Department of Earth & Atmospheric Sciences
Saint Louis University
3507 Laclede Avenue
St. Louis, MO 63103, USA
e-mail: mitchell@eas.slu.edu

Delivering manuscripts in diskette form may substantially facilitate the publication process provided certain points are taken into consideration:

Texts on diskette should be delivered in either DOS or Macintosh format. They should be saved and delivered in two separate versions:
• with standard text format as offered by your word processing program, and
• in addition in Rich Text Format (RTF) or, as a last resort, as an ASCII file.
Numerous word processing programs offer these options when saving the text.
The final hard copy of the manuscript should be submitted together with the diskette.
The electronic and printed version must be absolutely identical. All pictorial and graphic illustrations should be delivered as hard copy originals and must be 200% of the final printed size. Digital drawings and graphs should be submitted in Encapsulated PostScript (EPS) or Tag Image File Format (TIFF) form. Do not fail to include a hard copy for ready viewing. Back-up copies of the diskettes must be kept. Diskettes must be adequately protected for transport.

Galley Proofs
Unless indicated otherwise, galley proofs will be sent to the first-named author directly from Birkhäuser Verlag AG and should be returned with the least possible delay. Textual alterations made in the galley proof stage will be charged to the author. One copy of the corrected proof is to be returned immediately to
Editorial Office
Attn. Mrs. Renate D'Arcangelo
Harvard University
233 Pierce Hall, 29 Oxford Street
Cambridge, MA 02138, USA

The editorial office assumes no responsibility for delayed proofs, errors in the original manuscript, or major alterations in proofs for any reason.

Reprints
The authors will receive 50 reprints of each article without charge. Additional reprints may be ordered in lots of 50 when the final corrected page proofs are returned. Orders submitted thereafter are subject to considerably higher rates.